# Lecture Notes in Computer Science 13258

More information about this series at https://link.springer.com/bookseries/558

José Manuel Ferrández Vicente ·
José Ramón Álvarez-Sánchez ·
Félix de la Paz López · Hojjat Adeli (Eds.)

# Artificial Intelligence in Neuroscience

## Affective Analysis and Health Applications

9th International Work-Conference on the Interplay
Between Natural and Artificial Computation, IWINAC 2022
Puerto de la Cruz, Tenerife, Spain, May 31 – June 3, 2022
Proceedings, Part I

*Editors*
José Manuel Ferrández Vicente
Universidad Politécnica de Cartagena
Cartagena, Spain

Félix de la Paz López
Universidad Nacional de Educación
a Distancia
Madrid, Spain

José Ramón Álvarez-Sánchez
Universidad Nacional de Educación
a Distancia
Madrid, Spain

Hojjat Adeli
Ohio State University
Columbus, OH, USA

ISSN 0302-9743 ISSN 1611-3349 (electronic)
Lecture Notes in Computer Science
ISBN 978-3-031-06241-4 ISBN 978-3-031-06242-1 (eBook)
https://doi.org/10.1007/978-3-031-06242-1



This Springer imprint is published by the registered company Springer Nature Switzerland AG
The registered company address is: Gewerbestrasse 11, 6330 Cham, Switzerland

# Preface

The main topic of these IWINAC 2022 books is the study of intelligent systems inspired by the natural world, in particular biology. Several algorithms and methods and their applications are discussed, including evolutionary algorithms. Bio-inspired intelligent systems have thousands of useful applications in fields as diverse as machine learning, biomedicine, control theory, telecommunications, and, why not, music and art. These books covers both the theory and practice of bio-inspired artificial intelligence, along with providing a bit of the basis and inspiration for the different approaches. This is a discipline that strives to develop new computing techniques through observing how naturally occurring phenomena behave to solve complex problems in various environmental situations. Brain-inspired computation is one of these techniques that covers multiple applications in very different fields. Through the International Work-Conference on the Interplay between Natural and Artificial Computation (IWINAC) we provide a forum in which research in different fields can converge to create new computational paradigms that are on the frontier between neural and biomedical sciences and information technologies.

As a multidisciplinary forum, IWINAC is open to any established institutions and research laboratories actively working in the field of natural or neural technologies. But beyond achieving cooperation between different research realms, we wish to actively encourage cooperation with the private sector, particularly SMEs, as a way of bridging the gap between frontier science and societal impact.

In this edition, four main themes outline the conference topics: neuroscience, affective computing, robotics, and deep learning.

1) Machine learning holds great promise in the development of new models and theories in the field of neuroscience, in conjunction with traditional statistical hypothesis testing. Machine learning algorithms have the potential to reveal interactions, hidden patterns of abnormal activity, brain structure and connectivity, and physiological mechanisms of the brain and behavior. In addition, several approaches for testing the significance of the machine learning outcomes have been successfully proposed to avoid "the dangers of spurious findings or explanations void of mechanism" by means of proper replication, validation, and hypothesis-driven confirmation. Therefore, machine learning can effectively provide relevant information to take great strides toward understanding how the brain works. The main goal of this field is to build a bridge between two scientific communities, the machine learning community, including lead scientists in deep learning and related areas within pattern recognition and artificial intelligence, and the neuroscience community. Artificial intelligence has become the ultimate scale to test the limits of technological advances in dealing with life science challenges and needs. In this sense, the interplay between natural and artificial computation is expected to play a most relevant role in the diagnosis, monitoring, and treatment of neurodegenerative diseases, using the advanced computational solutions provided by machine learning and data science. This requires

interchanging new ideas, launching projects and contests, and, eventually, creating an inclusive knowledge-oriented network with the aim of empowering researchers, practitioners and users of technological solutions for daily life experience in the domains of neuromotor and linguistic competence functional evaluation, clinical explainability, and rehabilitation by interaction with humans, robots, and gaming avatars, not being strictly limited to only these but also open to other related fields. The use of machine learning-based precision medicine in monitoring daily life activity and providing well-being conditions to especially sensitive social sectors is one of the most relevant objectives. Case study descriptions involving neurodegenerative diseases (Alzheimer's disease, fronto-temporal dementia, cerebrovascular damage and stroke, autism, Parkinson's disease, amyotrophic lateral sclerosis, multiple sclerosis, Huntington's chorea, etc.) are included in the proceedings. Mild cognitive impairment (MCI) is considered the stage between the mental changes that are seen between normal ageing and early-stages of dementia. Indeed, MCI is one of the main indicators of incipient Alzheimer's disease (AD) among other neuropsychological diseases. The growth of these diseases is generating a great interest in the development of new effective methods for the early detection of MCI because, although no treatments are known to cure MCI, this early diagnosis would allow early intervention to delay the effects of the disease and accelerate progress towards effective treatment in its early stages. Although there have been many years of research, the early identification of cognitive impairment, as well as the differential diagnosis (to distinguish significant causes or typologies for its treatment), are problems that have been addressed from different angles but are still far from being solved. Diverse types of tests have already been developed, such as biological markers, magnetic resonance imaging, and neuropsychological tests. While effective, biological markers and magnetic resonance imaging are economically expensive, invasive, and require time to get a result, making them unsuitable as a population screening method. On the other hand, neuropsychological tests have a reliability comparable to biomarker tests, and are cheaper and quicker to interpret.

2) Emotions are essential in human-human communication, cognition, learning, and rational decision-making processes. However, human-machine interfaces (HMIs) are still not able to understand human sentiments and react accordingly. With the aim of endowing HMIs with the emotional intelligence they lack, the science of affective computing focuses on the development of artificial intelligence by means of the analysis of affects and emotions, such that systems and devices could be able to recognize, interpret, process, and simulate human sentiments.

   Nowadays, the evaluation of electrophysiological signals plays a key role in the advancement towards that purpose since they are an objective representation of the emotional state of an individual. Hence, the interest in physiological variables like electroencephalograms, electrocardiograms, or electrodermal activity, among many others, has notably grown in the field of affective states detection. Furthermore, emotions have also been widely identified by means of the assessment of speech characteristics and facial gestures of people under different sentimental conditions. It is also worth noting that the development of algorithms for the classification of affective states in social media has experienced a notable increase in recent years. In this sense, the language of posts included in social networks, such as Facebook or Twitter, is evaluated with the aim of detecting the sentiments of the users of those media tools. For this edition, the theme of affective computing and sentiment analysis was intended to be a meeting point for researchers who are interested in any of the areas of expertise related to sentiment analysis, including those seeking to initiate their studies and those currently working on these topics. Hence, papers introducing new proposals based on the analysis of physiological measures, facial recognition, speech recognition, or natural language processing in social media are included as examples of affective computing and sentiment analysis.

3) Over the last decade there has been an increasing interest in using machine learning, and in the last few years deep learning methods, combined with other vision techniques to create autonomous systems that solve vision problems in different fields. Therefore, a special session was organized to serve researchers and developers publishing original, innovative, and state-of-the art algorithms and architectures for real-time applications in the areas of computer vision, image processing, biometrics, virtual and augmented reality, neural networks, intelligent interfaces, and biomimetic object-vision recognition.

   The aim was to provide a platform for academics, developers, and industry-related researchers belonging to the vast communities of the neural network, computational intelligence, machine Learning, deep learning, biometrics, vision systems, and robotics fields to, discuss, share, experience, and explore traditional and new areas of computer vision, machine learning, and deep learning which can be combined to solve a range of problems. The objective of the session was to integrate the growing international community of researchers working on the application of machine learning and deep learning methods in vision and robotics to facilitate a fruitful discussion on the evolution of these technologies and the benefits to society.

4) Finally, deep learning has meant a breakthrough in the artificial intelligence community. The best performances attained so far in many fields, such as computer vision or natural language processing, have been overtaken by these novel paradigms to a point that only ten years ago was just science fiction. In addition, this technology has been open sourced by the main IA companies; hence, making it quite straightforward to design, train, and integrate deep-learning based systems. Moreover, the amount of data available every day is not only enormous but also growing at an exponential rate. In recent years there has been an increasing interest in using machine learning methods to analyse and visualize massive data generated from very different sources and with many different features: social networks, surveillance systems, smart cities, medical diagnosis, business, cyberphysical systems, or media digital data. This topic was selected to serve researchers and developers publishing original, innovative, and state-of-the art machine learning algorithms and architectures to analyse and visualize large amounts of data.

The wider view of the computational paradigm gives us more elbow room to accommodate the results of the interplay between nature and computation. The IWINAC forum thus becomes a methodological approximation (a set of intentions, questions, experiments, models, algorithms, mechanisms, explanation procedures, and engineering and computational methods) to the natural and artificial perspectives of the mind embodiment problem, both in humans and in artifacts. This is the philosophy that continues in

IWINAC meetings, the interplay between the natural and the artificial, and we face this same problem every two years, although last year we had to postpone the conference due to the COVID-19 pandemic. This synergistic approach will permit us not only to build new computational systems based on the natural measurable phenomena but also to understand many of the observable behaviors inherent to natural systems.

The difficulty of building bridges between natural and artificial computation is one of the main motivations for the organization of IWINAC events. The IWINAC 2022 proceedings contain the works selected by the Scientific Committee from more than 200 submissions, after the refereeing process. The first volume, entitled Artificial Intelligence in Neuroscience: Affective Analysis and Health Applications, includes all the contributions mainly related to new tools for analyzing neural data, or detecting emotional states, or interfacing with physical systems. The second volume, entitled Bioinspired Systems and Applications: from Robotics to Ambient Intelligence, contains the papers related to bioinspired programming strategies and all the contributions oriented to the computational solutions to engineering problems in different application domains, such as biomedical systems or big data solutions.

An event of the nature of IWINAC 2022 cannot be organized without the collaboration of a group of institutions and people who we would like to thank now, starting with UNED and the Universidad Politécnica de Cartagena. The collaboration of the Universidad de La Laguna (ULL) was crucial, as was the efficient work of the Local Organizing Committee, chaired by Josefa Dorta Luis and Pedro Gómez Vilda with the close collaboration of Carmen Victoria Marrero Aguilar, (UNED), Carolina Jorge Trujillo (ULL), and Chaxiraxi Díaz Cabrera (ULL). In addition to our universities, we received financial support from the Universidad de La Laguna, Ayuntamiento Puerto de la Cruz, Programa de Grupos de Excelencia de la Fundación Séneca, and Apliquem Microones 21 s.l.

We want to express our gratitude to our invited speakers Hojjat Adeli, from Ohio State University (USA), Rafael Rebolo, from the Instituto de Astrofísica de Canarias (Spain), Manuel de Vega, from the Universidad de La Laguna (Spain), Athanasios Tsanas, from the University of Edinburgh (UK), and Luis M. Sarro, from the Universidad Nacional de Educacion a Distancia (Spain) for accepting our invitation and for their magnificent plenary talks. We would also like to thank the authors for their interest in our call and the effort in preparing the papers, conditio sine qua non for these proceedings. We thank the Scientific and Organizing Committees, in particular the members of these committees who acted as effective and efficient referees and as promoters and managers of preorganized sessions on autonomous and relevant topics under the IWINAC global scope. Our sincere gratitude goes also to Springer, for the continuous receptivity, help, and collaboration in all our joint editorial ventures on the interplay between neuroscience and computation.

Finally, we want to express our special thanks to BCD Travel (formerly Viajes Hispania), our technical secretariat, and to Chari García, Ana María García, and Juani Blasco, for making this meeting possible, and for arranging all the details that comprise the organization of this kind of event.

We would like to dedicate these two volumes of the IWINAC proceedings in memory of Professor Mira.

June 2022

José Manuel Ferrández Vicente
José Ramón Álvarez-Sánchez
Félix de la Paz López
Hojjat Adeli

# Organization

## General Chair

| | |
|---|---|
| José Manuel Ferrández Vicente | Universidad Politécnica de Cartagena, Spain |

## Organizing Committee

| | |
|---|---|
| José Ramón Álvarez-Sánchez | Universidad Nacional de Educación a Distancia, Spain |
| Félix de la Paz López | Universidad Nacional de Educación a Distancia, Spain |

## Honorary Chairs

| | |
|---|---|
| Hojjat Adeli | Ohio State University, USA |
| Rodolfo Llinás | New York University, USA |
| Zhou Changjiu | Singapore Polytechnic, Singapore |

## Local Organizing Committee

| | |
|---|---|
| Josefa Dorta Luis | Universidad de La Laguna, Spain |
| Pedro Gómez-Vilda | Universidad Politécnica de Madrid, Spain |
| Victoria Marrero Aguiar | Universidad Nacional de Educación a Distancia, Spain |
| Carolina Jorge Trujillo | Universidad de La Laguna, Spain |
| Chaxiraxi Díaz Cabrera | Universidad de La Laguna, Spain |

## Invited Speakers

| | |
|---|---|
| Hojjat Adeli | Ohio State University, USA |
| Rafael Rebolo | Instituto de Astrofísica de Canarias, Spain |
| Athanasios Tsanas | University of Edinburgh, UK |
| Manuel de Vega | Universidad de La Laguna, Spain |
| Luis M. Sarro | Universidad Nacional de Educación a Distancia, Spain |

## Field Editors



## International Scientific Committee

Margarita Bachiller Mayoral, Spain
Francisco Bellas, Spain
Emilia I. Barakova, The Netherlands
Guido Bologna, Switzerland
Paula Bonomini, Argentina
Enrique J. Carmona Suárez, Spain
José Carlos Castillo, Spain
Germán, Castellanos-Dominguez, Colombia
Sung-Bae Cho, South Korea
Ricardo Contreras, Chile
Jose Manuel Cuadra Troncoso, Spain
Félix de la Paz López, Spain
Javier de Lope, Spain
Enrique Domínguez, Spain
Eduardo Fernández, Spain
Richard J. Duro, Spain
Antonio Fernández-Caballero, Spain
José Manuel Ferrandez, Spain
Victor Fresno, Spain
Jose Garcia-Rodriguez, Spain
Pedro Gómez-Vilda, Spain
Pascual González, Spain
Juan M. Gorriz, Spain
Manuel Graña, Spain
César Hervás Martínez, Spain
Tom Heskes, The Netherlands
Joost N. Kok, The Netherlands
Krzysztof Kutt, Poland
Markus Lappe, Germany
Emilio Letón Molina, Spain
Maria Teresa López Bonal, Spain
Tino Lourens, The Netherlands
Angeles Manjarrés, Spain
Jose Manuel Molina Lopez, Spain
Oscar Matínez Mozos, Japan
Rafael Martínez Tomás, Spain
Juan Morales Sánchez, Spain
Grzegorz J. Nalepa, Poland
Elena Navarro, Spain
Andrés Ortiz García, Spain
José Palma, Spain
Francisco Peláez, Brazil
Maria Pinninghoff, Chile
Javier Ramírez, Spain
Andoni Razvan, USA

Mariano Rincón Zamorano, Spain
Victoria Rodellar, Spain
Camino Rodríguez Vela, Spain
Pedro Salcedo Lagos, Chile
Ángel Sanchez, Spain
Eduardo Sánchez Vila, Spain
José Luis Sancho-Gómez, Spain
José Santos Reyes, Spain
Antonio Sanz, Spain
Luis M. Sarro, Spain
Andreas Schierwagen, Germany
Guido Sciavicco, Italy
Fermín Segovia, Spain
Antonio J. Tallón-Ballesteros, Spain
Javier Toledo-Moreo, Spain
Jan Treur, The Netherlands
Ramiro Varela Arias, Spain
Hujun Yin, UK
Yu-Dong Zhang, UK

# Contents – Part I

## Machine Learning in Neuroscience

**Neuromotor and Cognitive Disorders**

**Affective Analysis**

**Health Applications**

# Contents – Part II

**Deep Learning**

## Artificial Intelligence Applications

# Machine Learning in Neuroscience

# ConvNet-CA: A Lightweight Attention-Based CNN for Brain Disease Detection

Hengde Zhu[1], Jian Wang[1], Shui-Hua Wang[1], Yu-Dong Zhang[1(✉)], and Juan M. Górriz[2(✉)]

[1] School of Computing and Mathematical Sciences, University of Leicester, Leicester LE1 7RH, UK
yudongzhang@ieee.org

[2] Department of Signal Theory, Networking and Communications, University of Granada, 52005 Granada, Spain
gorriz@ugr.es

**Abstract.** Attention-based convolutional networks have attracted great interest in recent years and achieved great success in improving representation capability of networks. However, most attention mechanisms are complicated and implemented by introducing a large number of extra parameters. In this study, we proposed a lightweight attention-based convolutional network (ConvNet-CA) that has a low computation complexity yet a high performance for brain disease detection. ConvNet-CA weights the importance of different channels in features maps and pays more attention to important channels by introducing an efficient channel attention mechanism. We evaluated ConvNet-CA on a publicly accessible benchmark dataset: Whole Brain Atlas. The brain diseases involved in this study are stroke, neoplastic disease, degenerative disease, and infectious disease. The experimental results showed that ConvNet-CA achieved highly competitive performance over state-of-the-art methods on distinguishing different types of brain diseases, with an overall multi-class classification accuracy of 94.88 ± 3.64%.

**Keywords:** Deep learning · Attention mechanism · Medical image

## 1 Introduction

Brain disease is one of the most dangerous diseases which threaten human's health. It can cause headaches, coma, visual impairment, and movement disorder. There are varieties of brain diseases, and usually, doctors cannot distinguish them on the surface. Magnetic resonance imaging (MRI) is acknowledged as an ideal method to scan the brain's inner structure. Through MRI, clinicians can observe and assess the inner condition of the brain. In this way, clinicians decide whether the patient has a brain disease and which one the patient is suffering [1].

Although MRI scans enable clinicians to examine the brain's inner structure and condition, it is still not easy to detect pathological brains. The difference


J. M. Ferrández Vicente et al. (Eds.): IWINAC 2022, LNCS 13258, pp. 3–12, 2022.
https://doi.org/10.1007/978-3-031-06242-1_1

between healthy brains and pathological brains can be very subtle [12]. Even the most experienced clinicians cannot avoid mistakes in brain examination. Therefore, using a computer-aided diagnosis system to help brain disease detection is becoming increasingly necessary.

There exist some methods that apply artificial intelligence to detect pathological brains automatically. Wang, et al. [15] proposed a pathological brain detection method based on stationary wavelet entropy. Nayak, et al. [9] designed a brain disease detection approach using improved particle swarm optimization and evolutionary extreme learning machine. These methods were effective in the model training and have achieved relatively high accuracy, but to extract more representative features from brain medical images, we opt to utilise deep learning to process medical images. Deep learning has been widely applied in image classification tasks in recent years. Due to its flexibility, it can be adapted to special problems [3]. As one of the most commonly used models, convolutional neural networks (CNNs) are often adopted as the backbone model [13]. Although they have shown superiority over traditional machine learning-based methods, the training is usually time-consuming, especially when a CNN has lots of layers. Our goal is to propose a lightweight CNN that is easy to train, while achieving better results than popular CNNs.

In this study, we introduced an efficient channel attention module that can be easily integrated into CNN architectures and proposed a novel lightweight attention-based CNN for brain disease detection. Our proposed model, named ConvNet-CA, has a concise structure and performs better than many large-scale CNNs [2,5,7,11]. The main contribution of this study is we designed an efficient and lightweight model for pathological brain detection with high accuracy.

## 2 Data

In this study, we include a total number of 197 axial-oriented MRI T2-weighted images of healthy brain and pathological brains. Each image has a consistent dimension of $256 \times 256$ pixels. These images are acquired from a publicly accessible dataset: the Whole Brain Atlas [8]. There are five categories of brain images in our study, including the normal brain, stroke, neoplastic disease, degenerative disease, and infectious disease. Statistic information of the dataset is listed in Table 1. Some samples from the dataset are shown in Fig. 1. The detection of brain disease is considered as a 5-class classification task.

## 3 Methodology

This study applies advanced deep learning technology to diagnose brain disease based on MRI scans. Our model is based on a CNN architecture and the Efficient Channel Attention (ECA) module [14]. The overall structure of our proposed model ConvNet-CA is shown in Fig. 3. It consists of three convolutional layers for feature extraction, with each layer followed by a ECA module to refine feature maps. The ECA modules only introduce a few parameters while allowing the

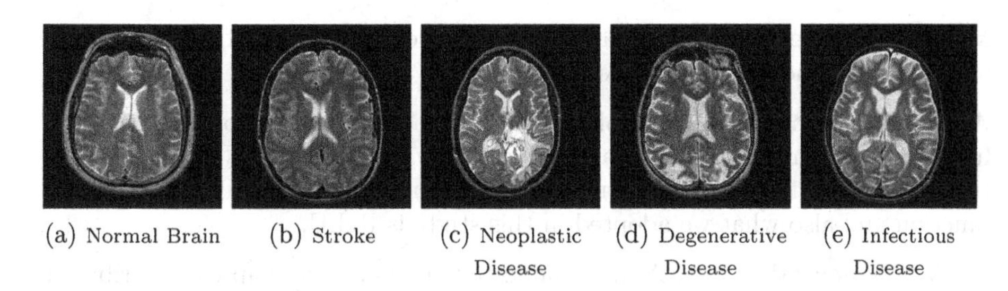

(a) Normal Brain (b) Stroke (c) Neoplastic Disease (d) Degenerative Disease (e) Infectious Disease

**Fig. 1.** Brain MRI samples from the dataset

**Table 1.** Statistic information of the dataset.

| Categories | Samples |
|---|---|
| Normal brain | 20 |
| Stroke | 72 |
| Neoplastic disease | 31 |
| Degenerative disease | 41 |
| Infectious disease | 33 |

model to focus on more important channels of feature maps. Pooling layers are added to summarise re-fined feature maps and reduce the size of feature maps. In this section, Conv-CA is explained in details.

### 3.1 Convolutional Neural Network

Researchers applied CNNs and developed hundreds of variants intending to solve almost every vision task. Although there existed some competitors in recent years such as graph neural networks and transformers, CNNs remain the mostly-adopted backbone of network architecture in the computer vision field. A CNN architecture typically has four elements, convolutional layer, pooling, activation function, and fully connected layer. Besides, dropout is one of the most applied techniques used in CNN training.

**Convolutional Layer.** Convolutional layers are the most important parts of CNNs. They are used for feature extraction. In most cases, a CNN consists of many convolutional layers. These convolutional layers form a hierarchy that enables a CNN to extract deeper and deeper features of input data. In a convolutional layer, kernels move upon every region of the input data in a fixed order to produce feature maps. These feature maps are then fed to the next convolutional layer as new input.

**Pooling.** As a down-sampling method, pooling is used for removing redundant information in feature maps and preserving the most valuable information. Max pooling and average pooling are two of the most commonly pooling methods.

Max pooling selects the maximal value as the representation of a local region, while average pooling utilises the mean value of a local region.

**Activation Function.** A convolutional layer is often followed by an activation function that introduces nonlinearity. In this way, it enables CNNs to better portray the data distribution of the real world. The most commonly used activation function and also what we adopted in this study is ReLU.

**Fully Connected Layer.** A fully connected layer aims to map the distributed feature representation to sample annotation space. A fully connected layer is usually placed at the end of CNNs. It acts as a classifier of the model. In this research, we applied a fully connected layer at the end of the model structure.

**Dropout.** When training CNNs, we randomly select parts of neurons in hidden layers and set them to zero [10]. This operation is called dropout. Dropout is widely used in training deep neural networks for relieve overfitting. In this study, the dropout method is applied in the fully connected layer.

ConvNet-CA adopts a lightweight CNN as the backbone which consists of three convolutional layers with ReLU, three max pooling layers, and one fully connected layer with a dropout strategy.

### 3.2 Channel Attention Mechanism

Attention mechanism has been one of the most prominent progress researchers made in the deep learning over the past few years. It enable neural networks to focus on the most valuable information. In image analysis, spatial attention and channel attention are two of the most widely used attention mechanisms. Since Hu, et al. [6] proposed SENet where a channel attention module proved its potential, researchers have been looking for more advanced methods to make the best use of channel attention in feature extraction.

The channel attention module of SENet, named SE module, is divided into two steps, squeeze and excitation (see Fig. 2a). In the squeeze step, global average pooling summarises each feature map and compresses its dimension to $1 \times 1$. Supposing that the input block $\omega$ has a dimension of $H \times W \times C$, where $H$ denotes height, $W$ denotes width, and $C$ denotes the number of channels, the squeeze step can be described as

$$G_{squeeze}(\omega_c) = \frac{1}{H \times W} \sum_{i=1}^{H} \sum_{j=1}^{W} \omega_c(i, j) \tag{1}$$

The squeeze step produces a global representation vector $z \in \mathcal{R}^{1 \times 1 \times C}$. This vector does not directly weight the importance of channels. Instead, two fully connected layers are employed to learn importance of different channels. The first fully connected layer is applied to the global representation vector, which reduces its dimension to $1 \times 1 \times \frac{C}{r}$ with a reduction ratio $r$. Then a ReLU is adopted for nonlinear activation, and a second fully connected layer is applied to regain its dimension to $1 \times 1 \times C$. In the end, a *sigmoid* function is employed to activate the vector, which is later multiplied by the original feature maps to acquire weighted feature maps. Supposing that the weights of the first and second fully connected layer are denoted as $V_1 \in \mathcal{R}^{\frac{C}{r} \times C}$ and $V_2 \in \mathcal{R}^{C \times \frac{C}{r}}$, the ReLU function is denoted as $\delta$ and the *sigmoid* function is denoted as $\sigma$, the excitation step can be described as

$$G_{excitation}(z, V) = \sigma(V_2 \delta(V_1 z)). \tag{2}$$

As a SE module contains two fully connected layers that contain a massive number of parameters, it can result in slowing down the model training. In addition, the SE module endures a dimension reduction and a dimension increment, which might cause loss of information from original feature maps [14]. To limit model complexity and avoid dimension variation, we find this ECA module [14] that applies one convolution operation to generate the channel attention vector. The ECA module is more efficient as it has much fewer parameters. It can also preserve information better than the SE module as it avoids dimension reduction.

Similar to the SE module, the ECA module first applies a global average pooling layer to compress feature maps to a global representation vector $z \in \mathcal{R}^{1 \times 1 \times C}$. A convolution layer with a kernel size of $k$ and zero-padding of $\lfloor \frac{k}{2} \rfloor$ is then applied to the compressed vector to produce a channel attention vector with the size of the global representation vector. This operation captures local interaction across channels with a coverage of $k$. It is efficient that it only introduces $k$ number of extra parameters. Given the channel dimension $C$, to adaptively determine the size of coverage $k$, the authors of the ECA module proposed a nonlinear mapping [14] as below

$$k = |\frac{\log_2(C)}{\gamma} + \frac{b}{\gamma}|_{odd}. \tag{3}$$

where $\gamma$ and $b$ are two constants.

The process of the ECA module is illustrated in a more understandable way in Fig. 2b. The input block is down-sampled by a global average pooling layer. A convolution operation is performed on the compressed vector and generates the channel attention vector without dimension reduction. Then, a *sigmoid* function is applied to introduce nonlinearity. Finally, the channel attention vector is multiplied by original feature maps to obtain the weighted feature maps. It is observed that the ECA module has fewer steps and does not have the process of dimension reduction. Therefore, ConvNet-CA adopts ECA modules to perform channel attention.

### 3.3 ConvNet-CA

As it is shown in Fig. 3, ConvNet-CA has a very simple and elegant architecture. It consists of three blocks and one fully connected output layer. In the first block, we employ a 3 × 3 convolution followed by ReLU. Then we apply the first channel attention module. At the end of the first block, there exists a max-pooling layer. Convolution, channel attention module, and max-pooling layer constitute the first block. The second and the third blocks repeat the first block's structure.

The main difference is the number of kernels in convolutional layers. After these three main blocks, we adopt a fully connected layer with dropout to output classification probabilities.

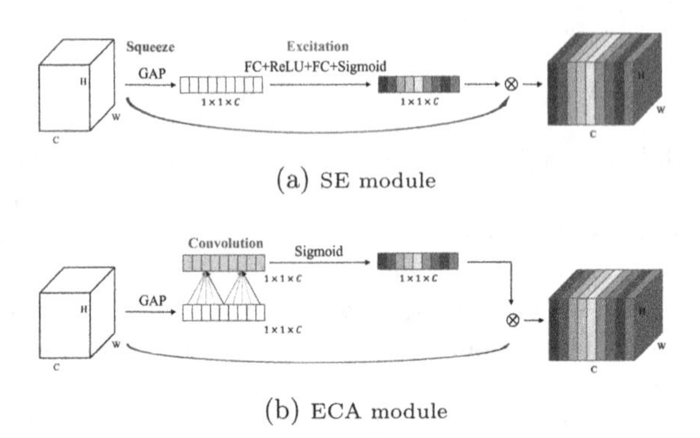


(a) SE module

(b) ECA module

**Fig. 2.** Comparison of ECA module with SE module

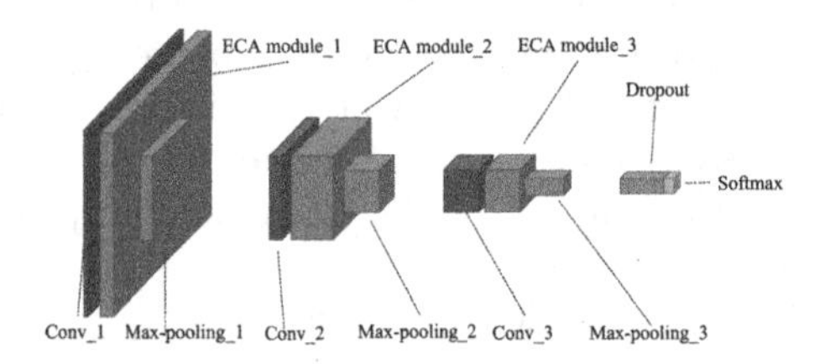


**Fig. 3.** Overall structure of ConvNet-CA

### 3.4 Evaluation Metrics

Due to the small size of the dataset, we adopt the 5-fold cross-validation scheme to evaluate models' performance and generalization ability. The dataset is partitioned into five subsets. Each subset is called a fold. We perform five iterations of training and testing. In each iteration, one fold of data is used in testing, while the remaining folds are used in training. The averaged model performance on five iterations is calculated as the 5-fold cross-validated performance.

Four common evaluation metrics, including accuracy (ACC), precision (PRE), sensitivity (SEN), and f1-score (F1s), are used to compare classification performance among different methods. True Positive (TP), False Positive (FP), True Negative (TN), and False Negative (FN) of predicted results are introduced to define above evaluation metrics. The definitions are as follows

$$ACC = \frac{TP + TN}{TP + FP + TN + FN} \tag{4}$$

$$PRE = \frac{TP}{TP + FP} \tag{5}$$

$$SEN = \frac{TP}{TP + FN} \tag{6}$$

$$F1s = 2 \times \frac{(PRE * SEN)}{(PRE + SEN)} \tag{7}$$

As this is a multi-class classification task, PRE, SEN, and F1s are first calculated for each class separately, and then the unweighted means of them, termed as $\text{PRE}_{macro}$, $\text{SEN}_{macro}$, and $\text{F1s}_{macro}$, are used to give an overall performance.

## 4 Experiments and Results

### 4.1 Experiment Set-Up

The study conducted experiments on a NVIDIA TESLA P100 GPU with 16 GB RAM provided by Kaggle. The hyperparameter settings of ConvNet-CA are given in Table 2. It is worth noting that other methods of comparison used in this study are trained under the same settings of optimizer, learning rate, batch size and training epochs (see Table 3).

**Table 2.** Hyperparameter settings of ConvNet-CA.

| Hyperparameter | Value | Hyperparameter | Value |
|---|---|---|---|
| Optimizer | Adam | Training epochs | 100 |
| Learning rate | 0.0001 | Activation function | ReLU |
| Dropout | 0.2 | Kernel size | 3 |
| Batch size | 1 | Number of Conv layers | 3 |

### 4.2 Performance on Multi-class Classification

We compared ConvNet-CA with four popular CNNs with strong representation capability. These networks are derived from Xception [2], Inception [11], ResNet50 [5], DenseNet121 [7]. As the dataset used in this study is a small dataset, directly training deep networks on it can lead to the overfitting problem. Thus, these networks were pre-trained on a large dataset, ImageNet, to learn how to capture representative features from images and then re-trained on our medical dataset. To adapt our dataset, the original fully connected layers at the end of these networks are replaced with a global average pooling layer and two new fully connected layers to perform a 5-class classification task. Except for the new fully connected layers, parameters of other layers of networks are frozen. These networks are re-trained to perform a domain-specific task.

The performance of different networks is summarised in Table 3. ConvNet-CA achieved the best performance over all metrics, with the classification accuracy of $94.88 \pm 3.64\%$, the macro precision of $96.50 \pm 2.74\%$, the macro sensitivity of $93.34 \pm 5.01\%$, and the macro f1-score of $94.21 \pm 4.45\%$. The deviation of accuracy and precision of ConvNet-CA are the smallest, indicating that its performance was more stable than other methods. In contrast, most deep networks achieved over 80.00% of all metrics, apart from ResNet50. The total number of parameters and FLOPs of different networks are shown in Table 4. It shows that ConvNet-CA has much fewer parameters and requires less computational resources. Its performance indicates that it can capture useful pattern more effectively.

**Table 3.** Classification performance. Metrics displayed in 'mean±standard deviation' format.

| Method | ACC(%) | $PRE_{macro}$(%) | $SEN_{macro}$(%) | $F1s_{macro}$(%) |
|---|---|---|---|---|
| Xception | $82.24 \pm 5.31$ | $80.73 \pm 5.68$ | $81.90 \pm 4.82$ | $78.54 \pm 4.82$ |
| InceptionV3 | $83.73 \pm 2.20$ | $86.98 \pm 3.18$ | $82.92 \pm 3.80$ | $82.67 \pm 3.52$ |
| ResNet50 | $60.96 \pm 8.09$ | $56.24 \pm 4.03$ | $57.66 \pm 4.64$ | $49.09 \pm 4.03$ |
| DenseNet121 | $81.28 \pm 7.76$ | $80.99 \pm 7.48$ | $81.86 \pm 5.61$ | $80.52 \pm 7.49$ |
| ConvNet-CA | **94.88±3.64** | **96.50±2.74** | **93.34±5.01** | **94.21±4.45** |

**Table 4.** Total number of parameters and FLOPs.

| Method | Params(M) | FLOPs(G) |
|---|---|---|
| Xception | 22.96 | 11.95 |
| InceptionV3 | 23.91 | 7.73 |
| ResNet50 | 25.69 | 10.13 |
| DenseNet121 | 8.09 | 7.45 |
| ConvNet-CA | **0.80** | **1.75** |

### 4.3 The Effectiveness of Channel Attention Mechanism

We designed experiments to study the effectiveness of channel attention used in ConvNet-CA. We denote ConvNet as the backbone of the ConvNet-CA, where channel attention is removed. The comparison of performance between ConvNet and ConvNet-CA is shown in Fig. 4. Experiment results show that the introduction of channel attention led to an overall performance increase of around 3.00%. The channel attention allows the network to focus on more useful information. It is worth noting that channel attention only introduces a few parameters. The total number of parameters for ConvNet and ConvNet-CA are 798,085 and 798,098, respectively. The slight difference in the number of parameters and the significant performance improvement demonstrated the effectiveness of channel attention.

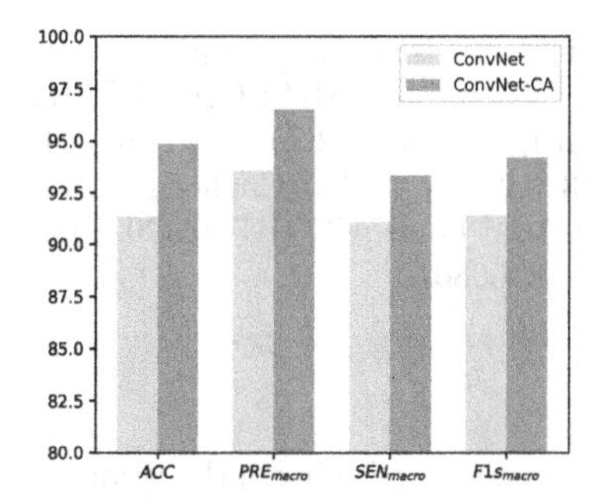


**Fig. 4.** Comparison between ConvNet-CA and ConvNet

### 4.4 Comparison with State-of-the-Art Methods

The overall performance of ConvNet-CA is compared with several state-of-the-art methods using the same dataset: VMD+SNPE+ANOVA [4] and FCEntF-II + K-ELM [9]. All mentioned methods above classify the brain MRI scans into five categories. To avoid the overfitting problem, both of our study and the study of Nayak, et al. [9] adopted a 5-fold cross validation scheme while a 10-fold cross validation scheme is adopted in the study of Gudigar, et al. [4]. From the Table 5, it can be observed that our method ConvNet-CA obtained the highest classification accuracy of 94.88% among the three methods.

**Table 5.** Comparison with state-of-the-art methods.

| Study | Method | ACC(%) |
|---|---|---|
| Gudigar et al. [4] | VMD+SNPE+ANOVA | 90.68 |
| Nayak et al. [9] | FCEntF-II + K-ELM | 93.00 |
| Our approach | ConvNet-CA | **94.88** |

## 5 Conclusion

This study proposed a lightweight attention-based CNN for brain disease detection. Compared with other deep networks with a large amount of parameters, our lightweight model can capture features more efficiently and effectively from a specific dataset, with higher performance over four evaluation metrics in the study. This network integrates an efficient attention mechanism to assign different importance to different channels of feature maps. It enables the model to pay more attention to most important channels. Experimental results demonstrated the effectiveness of the channel attention mechanism used in this study which led to significant performance enhancements. In the future, we shall apply our proposed model to more medical datasets and improve the model's generalization ability. The visualisation of the model is another research direction that would help us understand how the network works.

**Acknowledgements.** This paper is partially supported by Medical Research Council Confidence in Concept Award, UK (MC_PC_17171), Royal Society International Exchanges Cost Share Award, UK (RP202G0230), British Heart Foundation Accelerator Award, UK (AA/18/3/34220), Global Challenges Research Fund (GCRF), UK (P202PF11), Sino-UK Industrial Fund, UK (RP202G0289), and Hope Foundation for Cancer Research, UK (RM60G0680).

## References

1. Atlas, S.W.: Magnetic Resonance Imaging of the Brain and Spine, vol. 1. Lippincott Williams & Wilkins, Philadelphia (2009)
2. Chollet, F.: Xception: deep learning with depthwise separable convolutions. In: Proceedings of the IEEE Conference on Computer Vision and Pattern Recognition, pp. 1251–1258 (2017)
3. Górriz, J.M., et al.: Artificial intelligence within the interplay between natural and artificial computation: advances in data science, trends and applications. Neurocomputing **410**, 237–270 (2020)
4. Gudigar, A., Raghavendra, U., Ciaccio, E.J., Arunkumar, N., Abdulhay, E., Acharya, U.R.: Automated categorization of multi-class brain abnormalities using decomposition techniques with MRI images: a comparative study. IEEE Access **7**, 28498–28509 (2019)
5. He, K., Zhang, X., Ren, S., Sun, J.: Deep residual learning for image recognition. In: Proceedings of the IEEE Conference on Computer Vision and Pattern Recognition, pp. 770–778 (2016)
6. Hu, J., Shen, L., Sun, G.: Squeeze-and-excitation networks. In: Proceedings of the IEEE Conference on Computer Vision and Pattern Recognition, pp. 7132–7141 (2018)
7. Huang, G., Liu, Z., Van Der Maaten, L., Weinberger, K.Q.: Densely connected convolutional networks. In: Proceedings of the IEEE Conference on Computer Vision and Pattern Recognition, pp. 4700–4708 (2017)
8. Johnson, K.A., et al.: The whole brain atlas (2001)
9. Nayak, D.R., Dash, R., Majhi, B.: Discrete ripplet-II transform and modified PSO based improved evolutionary extreme learning machine for pathological brain detection. Neurocomputing **282**, 232–247 (2018)
10. Srivastava, N., Hinton, G., Krizhevsky, A., Sutskever, I., Salakhutdinov, R.: Dropout: a simple way to prevent neural networks from overfitting. J. Mach. Learn. Res. **15**(1), 1929–1958 (2014)
11. Szegedy, C., Vanhoucke, V., Ioffe, S., Shlens, J., Wojna, Z.: Rethinking the inception architecture for computer vision. In: Proceedings of the IEEE Conference on Computer Vision and Pattern Recognition, pp. 2818–2826 (2016)
12. Tofts, P.: Quantitative MRI of the Brain: Measuring Changes Caused by Disease. Wiley, USA (2005)
13. Wang, J., Zhu, H., Wang, S.H., Zhang, Y.D.: A review of deep learning on medical image analysis. Mob. Netw. Appl. **26**(1), 351–380 (2021)
14. Wang, Q., Wu, B., Zhu, P., Li, P., Zuo, W., Hu, Q.: ECA-net: efficient channel attention for deep convolutional neural networks. In: 2020 IEEE CVF Conference on Computer Vision and Pattern Recognition (CVPR). IEEE (2020)
15. Wang, S., Du, S., Atangana, A., Liu, A., Lu, Z.: Application of stationary wavelet entropy in pathological brain detection. Multimed. Tools Appl. **77**(3), 3701–3714 (2016). https://doi.org/10.1007/s11042-016-3401-7

# Temporal Phase Synchrony Disruption in Dyslexia: Anomaly Patterns in Auditory Processing

Marco A. Formoso[1,5(✉)], Andrés Ortiz[1,5], Francisco J. Martínez-Murcia[2,5], Diego Aquino Brítez[3,5], Juan José Escobar[3], and Juan Luis Luque[4]

[1] Communications Engineering Department, University of Málaga, 29004 Málaga, Spain
marco.a.formoso@ic.uma.es
[2] Department of Signal Theory, Communications and Networking, University of Granada, 18060 Granada, Spain
[3] Department of Computer Architecture and Technology, University of Granada, 18014 Granada, Spain
[4] Department of Developmental and Educational Psychology, University of Málaga, 29071 Málaga, Spain
[5] Andalusian Data Science and Computational Intelligence Institute (DaSCI), Granada, Spain

**Abstract.** The search for a dyslexia diagnosis based on exclusively objective methods is currently a challenging task. Usually, this disorder is analyzed by means of behavioral tests prone to errors due to their subjective nature; e.g. the subject's mood while doing the test can affect the results. Understanding the brain processes involved is key to proportionate a correct analysis and avoid these types of problems. It is in this task, biomarkers like electroencephalograms can help to obtain an objective measurement of the brain behavior that can be used to perform several analyses and ultimately making a diagnosis, keeping the human interaction at minimum. In this work, we used recorded electroencephalograms of children with and without dyslexia while a sound stimulus is played. We aim to detect whether there are significant differences in adaptation when the same stimulus is applied at different times. Our results show that following this process, a machine learning pipeline can be built with AUC values up to 0.73.

**Keywords:** EEG · Hilber transform · Dyslexia · Neural adaptation

## 1 Introduction

Developmental Dyslexia (DD) is a learning disorder with an estimate prevalence of 7% [17]. It is characterized by slow and inaccurate word recognition and by poor spelling and decoding abilities, despite individuals having adequate intelligence and sensory abilities. Diagnosis is usually made by means of behavioral


J. M. Ferrández Vicente et al. (Eds.): IWINAC 2022, LNCS 13258, pp. 13–22, 2022.
https://doi.org/10.1007/978-3-031-06242-1_2

tests. These tests, although performed by specialists, are not free from human error. For example, children's mood can affect results and the test analysis is subjective to some degree. In addition, this type of testing can only be done by children who already have writing and reading skills, thus limiting the impact of a possible intervention. This is why approaches using biomarkers are studied; they are an objective way to obtain information of the processes underlying the disorder and at the same time provide a method to diagnose those children in pre-read age, giving the opportunity to make an early intervention. Several biomarkers such as those obtained from electroencephalograms (EEG) and magnetoencephalograms (MEG), among others, have been used in the literature to further study the disorder and its causes. Additionally, it is common to derive connectivity metrics from these markers to determine how brain areas collaborate. A review of some of these metrics can be found in [10]. In this work, we use spectral phase lag index (PLI) that it is described in more detail in the methods section.

As mentioned above, the use of biomarkers such as EEG and MEG is widely used in the search for answers in neurological disorders and brain behavior. Proof of that are the works [1,2,8,9,13] where the signal brains are used to explore different diseases such as Alzheimer's disease, Parkinsonian syndromes or epileptic disorders. In a closer approach to this work, where auditory stimuli are used to trigger a reaction in the brain in the study of DD, Molinaro et al. [14] found that atypical neural entrainment at different rates may arise in affected subjects. Other works try to found the origin of DD is caused by the atypical synchronization in the right hemisphere [4,6]. Typically, these works are based on composing a set of features, from the entire EEG and comparing the differences between groups, without taking into account the possible brain adaptation of the brain over time.

In this work, we hypothesize that a stimulus adaptation must occur and be different in the control group and the dyslexic one, thus providing a way to differentiate both groups. This has to be reflected by a change in the brain behavior that can be analyzed through EEG signals. The metrics used here are typically used to measure the degree of synchronization between different EEG channels and bands, whereas in this work we take a different approach. Instead, we split the EEG in several segments and compute the same metrics this time between different segments representing different time slots.

The rest of the paper is organized as follows: In the next section, the materials and methods used in this work are explained, as well as the data acquisition. In addition, the synchrony metric is presented and the classification pipeline is detailed. The next section shows the results obtained applying the so-called methodology, and in the last section are discussed.

# 2 Methods

## 2.1 Data

The EEG data was recorded[1] by the *Leeduca* group at the University of Málaga. The signals were recorded by a Brainvision actiCHamp Plus with a 32 channels amplifier with a sampling rate 500 Hz. The montage 10–20 standardized system can be consulted in Fig. 1. EEG were obtained while several auditory stimuli were presented to the subject. These stimuli consisted of white noise, amplitude modulated at 4.8 Hz, 16 Hz and 20 Hz. They were presented to the subject in 2.5-min sessions each in the next order: 4.8 Hz–16 Hz–20 Hz–20 Hz–16 Hz–4.8 Hz for a total of 15 min. Stimuli were determined by expert linguistic psychologists studying the main frequency components present in voice, corresponding to syllables and phonemes.

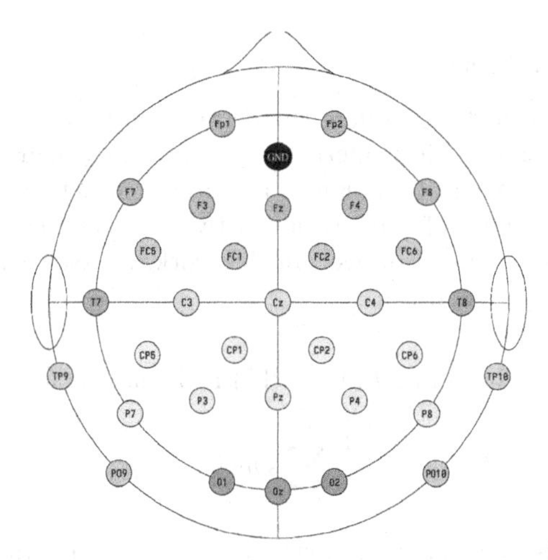


**Fig. 1.** Electrode montage in the extended 10–20 system used in the experiments. All 32 channels plus GND. Cz is used as reference.

In the next table we show the subjects recorded in this experiment. They were extracted from a cohort (N = 700) of children from different primary schools of Andalucía (Spain). Comorbidities with other neurodevelopmental disorders such as Language Impairment (LI), Speech Sound Disorder (SSD), Attention Deficit Hyperactivity Disorder (ADHD), Autism, and other auditory or visual sensory deficit disorders were taken into account in the screening process, along with information about other relevant conditions which can affect reading achievement, as immigration or bilingualism [3] (Table 1).

[1] The study was carried out with the understanding and written consent of each child's legal guardian and in the presence thereof, and was approved by the Medical Ethical Committee of the Malaga University (ref. 16-2020-H) and according to the dispositions of the World Medical Association Declaration of Helsinki.

**Table 1.** Database. Age range: 88–100 ($t(1) = -1.4, p > 0.05$).

| Group | Male/Female | Mean age (months) |
|---|---|---|
| Control | 17/15 | $94.1 \pm 3.3$ |
| Dyslexia | 7/9 | $95.6 \pm 2.9$ |

The raw EEGs were preprocessed in order to remove eye-blinking artifacts related as well as impedance variations due to movements. Also, ocular artifacts were removed by source separation using Independent Component Analysis (ICA) [11]. Finally, all channels were reference to the Cz channel and band-pass filtered by means of a finite response filter (FIR) to extract 5 different brain waves: Delta, 1.5–4 Hz; Theta, 4–8 Hz, Alpha, 8–13 Hz; Beta, 13–30 Hz; and Gamma, 30–80 Hz.

### 2.2 Connectivity Metric

As stated in the introduction section, there are several metrics that can be used to assess the connectivity. These metrics represent the synchronization strength between two signals. When both signals are from distinct brain areas, this can be seen as a measurement of cooperation between the areas, allowing a way of studying the brain behavior. However, in this work, we use them to compare the same signal at different times.

**Phase Lag Index.** Phase Lag Index (PLI) is defined as follows:

$$PLI_{xy} = \left| \frac{1}{N} \sum_{t=1}^{N} sgn(imag(S_{txy}) \right| \tag{1}$$

where $imag(S_{txy})$ is the imaginary part of the cross-spectral density at time $t$ and $sgn$ is the sign function ($-1$ for negative values, $+1$ for positive values and 0 for zero values). The cross-spectral density $S_{txy}$ can be computed as $S_{txy} = abs(x)abs(y)e^{i}\theta_x - \theta_y$. This is a metric that mitigates the effects of volume conduction. That is, spurious connectivity caused by the recording of the same source by two different electrodes [18]. These connections will have phase lags of zero or $\pi$.

**Hilbert Transform.** Usually, the connectivity metrics use the instantaneous amplitude or phase, that can be computed from the analytic version of the raw (time) signal. This is a complex version of the signal obtained by means of the Hilbert Transform (HT). Once computed, it is possible to obtain the instantaneous amplitude and phase. Instantaneous frequency can be also obtained by differentiating the instantaneous phase. Hilbert Transform is defined as follows:

$$\mathcal{H}[x(t)] = \frac{1}{\pi} \int_{-\infty}^{+\infty} \frac{x(t)}{t - \tau} d\tau \tag{2}$$

and the analytic signal $z_i(t)$ for a signal $x(t)$ can be obtained as

$$z_i(t) = x_i(t) + j\mathcal{H}\{x_i(t)\} = a(t)e^{j\phi(t)} \quad (3)$$

From $z_i(t)$, it is straightforward to compute the instantaneous amplitude as

$$a(t) = \sqrt{re(z_i(t))^2 + im(z_i(t))^2} \quad (4)$$

and the instantaneous, unwrapped phase is

$$\phi(t) = tan^{-1}\frac{im(z_i(t))}{re(z_i(t))} \quad (5)$$

### 2.3 Classification Pipeline

After processing the EEGs as shown in Sect. 2.1. The next steps in the pipeline are summarized in Fig. 2 and are as follows:

- First, we split and compute the metrics for each subject. We split the signal into 10 segments. That is, for each subject, we have 2 sets 40 Hz EEG records. We call 40 Hz UP (obtained by applying the auditory stimuly in an ascending frequency way, up 40 Hz) 40 Hz DOWN (obtained during the application of auditory stimuli in a descending frequency way, down to 4.8 Hz). Each one composed of 32 EEG channels, filtered to obtain 5 different bands. Following this method, we compute the metrics described in Sect. 2.2 between the first segment and the rest of them by pairs, thus obtaining 9 metric values for every channel and band.
- The following steps are done inside a cross-validation loop:
  - In this step, we performed the selection of the best channels and bands separately for the dyslexic and control groups. We do this by an anomaly detection approach using a one class support vector machine (OCSVM). The goal is to detect which channels present significant differences between the applications of the different stimuli, 40 Hz [UP—DOWN]. For each channel and band, we train a OCSVM with the UP data and test it with the DOWN data. If there are differences, the DOWN data should be detected almost entirely as outliers. To ensure that those anomalies are not occurring by chance, a permutation test is performed. The criterion is based on the following equation:

$$\frac{PERMS + 1}{N + 1} \quad (6)$$

    where $N$ is the number of permutation test run (1000) and $PERMS$ are the number of those permutations whose classification score is equal or higher than the ground classification score. The threshold at which we consider a channel significant is 0.05.

- The last stage is the classification with two different classifiers: KNeighborsClassifier and Support Vector Machine (SVM). We take the most significant channels from the previous step and use them to create a mask that is applied to the dataset. The significant channels are multiplied by 1 and non-significant by 0. In order to avoid biasing the data, the mask for dyslexia and the control group is applied to each subject regardless of the group to which he or she belongs, although data are duplicated.

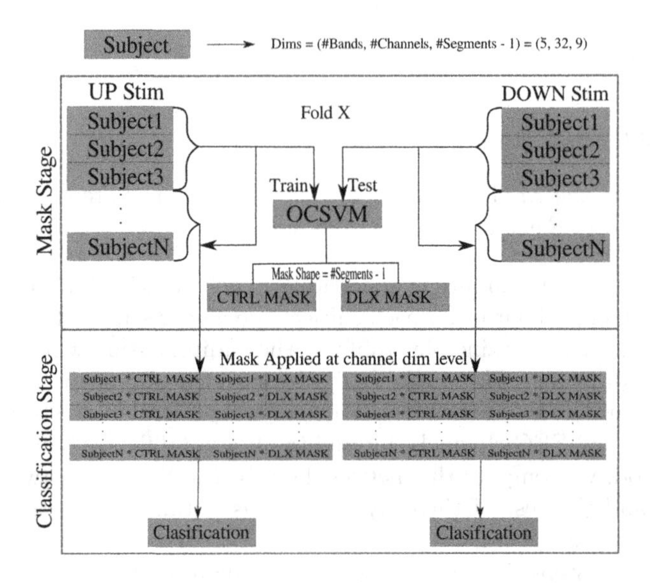


**Fig. 2.** Pipeline overview. The classification task is done per band and for every stimulus (40 HzUP, 40 HzDOWN).

## 3 Results

In this section, the results of applying the methods of the last section are presented. We carried out experiments exploring all the possible combinations of bands and classifiers for 40 Hz stimulus. The classification strategy follows a cross validation pattern with 5 folds. This is true for both, mask and classification stage (Fig. 2).

In Table 2 the results of these combinations are shown. The best overall values are obtained in the Alpha band. Although both of the classifiers used show similar AUC results, KNN discriminates better between both groups, thus yielding better sensitivity values. This is better seen in the Fig. 3.

**Table 2.** Results

| Band | Classifier | AUC | Sensitivity | Specificity | Accuracy |
|---|---|---|---|---|---|
| **Alpha** | **KNN** | **0.76 ± 0.15** | **0.82 ± 0.15** | **0.76 ± 0.18** | **0.73 ± 0.11** |
| | SVM | 0.71 ± 0.15 | 0.07 ± 0.13 | 0.97 ± 0.07 | 0.71 ± 0.07 |
| Beta | KNN | 0.48 ± 0.06 | 0.20 ± 0.25 | 0.78 ± 0.25 | 0.60 ± 0.12 |
| | SVM | 0.54 ± 0.13 | 0.00 ± 0.00 | 1.00 ± 0.00 | 0.71 ± 0.05 |
| Delta | KNN | 0.50 ± 0.13 | 0.20 ± 0.27 | 0.89 ± 0.17 | 0.69 ± 0.11 |
| | SVM | 0.59 ± 0.11 | 0.00 ± 0.00 | 1.00 ± 0.00 | 0.71 ± 0.05 |
| Gamma | KNN | 0.66 ± 0.13 | 0.47 ± 0.45 | 0.79 ± 0.18 | 0.67 ± 0.05 |
| | SVM | 0.37 ± 0.14 | 0.20 ± 0.4 | 0.80 ± 0.4 | 0.60 ± 0.20 |
| Theta | KNN | 0.46 ± 0.09 | 0.00 ± 0.0 | 0.91 ± 0.17 | 0.65 ± 0.11 |
| | SVM | 0.54 ± 0.09 | 0.00 ± 0.00 | 1.00 ± 0.00 | 0.71 ± 0.05 |

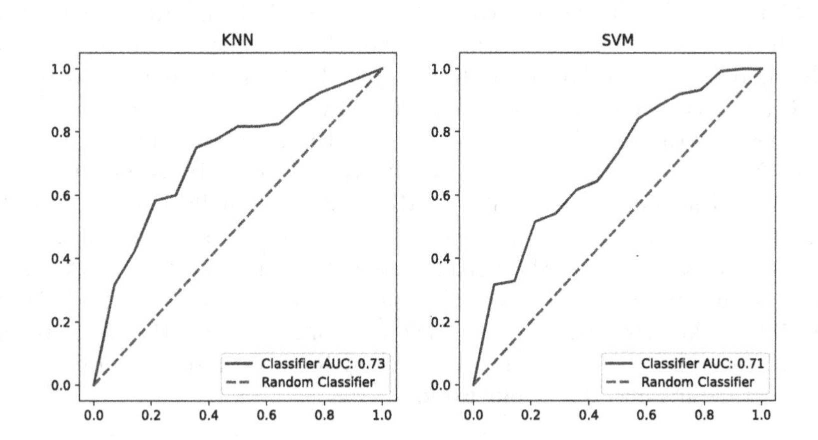


**Fig. 3.** ROC comparative between KNN and SVM for the Alpha band

## 4 Discussion

The proposed method seeks to find differences in adaptation through time of the brain when exposed to certain stimuli. To this end, a metric usually used to determine the synchronization of different brain regions is applied but between different time instants instead. First, we get the most discriminative channels and bands through an anomaly detection approach. Then, a mask is applied to the entire dataset to highlight these channels and bands and proceed with the classification stage. The results show the best values for the Alpha band and using a KNN classifier.

**Table 3.** Comparative table

| Method | Channels | Accuracy | Sensitivity | Specificity | AUC |
|---|---|---|---|---|---|
| MRI + SVC [19] | T1-MRI | 0.8 ± * | 0.82 ± * | 0.78 ± * | * |
| MEG + SVC + GC [5] | 253 | 0.63 ± 0.04 | 0.64 ± 0.04 | 0.65 ± 0.04 | * |
| MEG + SVC + GE [5] | 253 | 0.94 ± 0.02 | 0.93 ± 0.01 | 0.93 ± 0.02 | * |
| MEG + SVC + CI [5] | 253 | 0.80 ± 0.01 | 0.80 ± 0.02 | 0.79 ± 0.02 | * |
| MEG + SVC + wIFCG [5] | 253 | 0.97 ± 0.01 | 0.96 ± 0.02 | 0.95 ± 0.02 | * |
| EEG + SVC [16] (Writing) | 32 | 0.59 ± * | 0.64 ± * | 0.53 ± * | * |
| EEG + SVC [16] (Typing) | 32 | 0.78 ± * | 0.88 ± * | 0.66 ± * | * |
| EEG + OCSVC [15] | 32 | 0.71 ± * | 0.53 ± * | 0.78 ± * | 0.79 ± * |
| EEG + DAE [12] | 32 | 0.56 ± * | 0.76 ± * | 0.66 ± * | 0.74 ± * |
| **Proposed (EEG + KNN)** | **32** | **0.73 ± 0.15** | **0.82 ± 0.15** | **0.76 ± 0.18** | **0.73 ± 0.11** |

Although the majority of works focused in dyslexia are based on an exploratory analysis, it is worth noting the efforts made in the search of an automatic diagnosis method using biomarkers. In Table 3 our method is compared with previous ones in this context. Works varying in the use of biomarkers from using structural imaging [19], MEG [5] to EEG [7,12,15,16]. The best values are often found when the features are extracted from MEGs, but it requires a 252 channel acquisition system, and MEG data is usually harder to obtain. Works using EEG like in [16] use interactive task like writing and typing, limiting the diagnosis age like the behavioural tests. The use of auditory stimuli overcomes this limitation and avoid possible bias introduced by the task. Works in which these kinds of stimuli are applied are found in works [12,15]. Although all of them search for the differences/synchrony in between several brain regions at a same time instant, unlike our work.

## 5 Conclusions

In this work, we present a method to detect differences in how dyslexic children vs non-dyslexic adapt to auditory stimuli at different frequencies (4.8 Hz, 16 Hz, 20 Hz) while recording an EEG. The same frequency stimulus were presented to the children twice. Then, we measure the synchrony through time at both stimulus application and seek for differences in adaptation. To this end, PLI is used as connectivity metric and a two steps pipeline is built to automate the classification process. The first step is an anomaly detection approach step to detect which channels and bands present the most significant differences, and the last step being a classification step. We found that there are differences in the Alpha band that allow a classifier to distinguish between control and dyslexic group with sensitivity values up to 0.82, specificity up to 0.76 and AUC of 0.76. These differences imply that dyslexic children adapt different when certain stimulus is applied to them and afterwards.

As a future work, a more intensive exploratory analysis of the rest of the stimuli is planned, as well as try other metrics others than PLI.

**Acknowledgments.** This work was supported by projects PGC2018-098813-B-C32, PGC2018-098813-B-C31 (Spanish "Ministerio de Ciencia, Innovación y Universidades"), UMA20-FEDERJA-086 and P18-RT-1624 (Consejería de economía y conocimiento, Junta de Andalucía) and by European Regional Development Funds (ERDF) as well as the BioSiP (TIC-251) research group. M. A. Formoso Grant PRE2019-087350 funded by MCIN/AEI/ 10.13039/501100011033 by "ESF Investing in your future". Work by F.J.M.M. was supported by the MICINN "Juan de la Cierva - Incorporación" Fellowship. We also thank the *Leeduca* research group and Junta de Andalucía for the data supplied and the support. Funding for open access charge: Spanish "Ministerio de Ciencia, Innovación y Universidades", and European Regional Development Funds (ERDF).

## References

1. Adeli, H., Zhou, Z., Dadmehr, N.: Analysis of EEG records in an epileptic patient using wavelet transform. J. Neurosci. Methods **123**(1), 69–87 (2003)
2. Bell, M.A., Cuevas, K.: Using EEG to study cognitive development: issues and practices. J. Cogn. Dev. **13**(3), 281–294 (2012)
3. De Vos, A., Vanvooren, S., Vanderauwera, J., Ghesquière, P., Wouters, J.: A longitudinal study investigating neural processing of speech envelope modulation rates in children with (a family risk for) dyslexia. Cortex **93**, 206–219 (2017)
4. Di Liberto, G., Peter, V., Kalashnikova, M., Goswami, U., Burnham, D., Lalor, E.: Atypical cortical entrainment to speech in the right hemisphere underpins phonemic deficits in dyslexia. NeuroImage **175**, 70–79 (2018)
5. Dimitriadis, S.I., Simos, P.G., Fletcher, J.M., Papanicolaou, A.C.: Aberrant resting-state functional brain networks in dyslexia: symbolic mutual information analysis of neuromagnetic signals. Int. J. Psychophysiol. **126**, 20–29 (2018)
6. Flanagan, S., Goswami, U.: The role of phase synchronisation between low frequency amplitude modulations in child phonology and morphology speech tasks. J. Acoust. Soc. Am. **143**, 1366–1375 (2018). https://doi.org/10.1121/1.5026239
7. Formoso, M.A., Ortiz, A., Martinez-Murcia, F.J., Gallego, N., Luque, J.L.: Detecting phase-synchrony connectivity anomalies in EEG signals. Application to dyslexia diagnosis. Sensors **21**(21), 7061 (2021)
8. Gálvez, G., Recuero, M., Canuet, L., Del-Pozo, F.: Short-term effects of binaural beats on eeg power, functional connectivity, cognition, gait and anxiety in Parkinson's disease. Int. J. Neural Syst. **28**(05), 1750055 (2018). pMID: 29297265
9. Górriz, J.M., et al.: Artificial intelligence within the interplay between natural and artificial computation: advances in data science, trends and applications. Neurocomputing **410**, 237–270 (2020)
10. Hülsemann, M.J., Naumann, E., Rasch, B.: Quantification of phase-amplitude coupling in neuronal oscillations: comparison of phase-locking value, mean vector length, modulation index, and generalized-linear-modeling-cross-frequency-coupling. Front. Neurosci. **13**, 573 (2019)
11. Li, R., Principe, J.C.: Blinking artifact removal in cognitive EEG data using ICA. In: 2006 International Conference of the IEEE Engineering in Medicine and Biology Society, pp. 5273–5276 (2006)
12. Martinez-Murcia, F.J., et al.: EEG connectivity analysis using denoising autoencoders for the detection of dyslexia. Int. J. Neural Syst. **30**(07), 2050037 (2020)
13. Mirzaei, G., Adeli, A., Adeli, H.: Imaging and machine learning techniques for diagnosis of Alzheimer's disease. Rev. Neurosci. **27**(8), 857–870 (2016)

14. Molinaro, N., Lizarazu, M., Lallier, M., Bourguignon, M., Carreiras, M.: Out-of-synchrony speech entrainment in developmental dyslexia. Hum. Brain Mapp. **37**, 2767–2783 (2016)
15. Ortiz, A., López, P.J., Luque, J.L., Martínez-Murcia, F.J., Aquino-Britez, D.A., Ortega, J.: An anomaly detection approach for dyslexia diagnosis using EEG signals. In: Ferrández Vicente, J.M., Álvarez-Sánchez, J.R., de la Paz López, F., Toledo Moreo, J., Adeli, H. (eds.) IWINAC 2019, Part I. LNCS, vol. 11486, pp. 369–378. Springer, Cham (2019). https://doi.org/10.1007/978-3-030-19591-5_38
16. Perera, H., Shiratuddin, M.F., Wong, K.W., Fullarton, K.: EEG signal analysis of writing and typing between adults with dyslexia and normal controls. Int. J. Interact. Multimed. Artif. Intell. **5**(1), 62 (2018)
17. Peterson, R., Pennington, B.: Developmental dyslexia. Lancet **379**, 1997–2007 (2012)
18. Stam, C.J., Nolte, G., Daffertshofer, A.: Phase lag index: assessment of functional connectivity from multi channel EEG and MEG with diminished bias from common sources. Hum. Brain Mapp. **28**(11), 1178–1193 (2007)
19. Tamboer, P., Vorst, H., Ghebreab, S., Scholte, H.: Machine learning and dyslexia: classification of individual structural neuro-imaging scans of students with and without dyslexia. NeuroImage: Clin. **11**, 508–514 (2016)

# CAD System for Parkinson's Disease with Penalization of Non-significant or High-Variability Input Data Sources

Diego Castillo-Barnes[1](✉), J. Merino-Chica[1], R. Garcia-Diaz[1], C. Jimenez-Mesa[1], Juan E. Arco[2], J. Ramírez[1], and J. M. Górriz[1]

[1] Department of Signal Theory, Telematics and Communications, University of Granada, Periodista Daniel Saucedo Aranda, 18071 Granada, Spain
diegoc@ugr.es

[2] Department of Communications Engineering, University of Malaga, Blvr. Louis Pasteur 35, 29004 Malaga, Spain
http://sipba.ugr.es/

**Abstract.** In the last decade, the progressive development of new machine learning schemas in combination with novel biomarkers have led us to more accurate models to diagnose and predict the evolution of neurological disorders like Parkinson's Disease (PD). Though some of these previous work have attempted to combine multiple input data sources, many studies are critical of their lack of robustness when combining several input sources that with a high variability and/or not statistically significant. In order to minimize this problem, we have develop a Computer-Aided-Diagnosis (CAD) system for PD able to combine multiple input data sources underestimating those data types with poor classification rates and high variability. This model has been evaluated using FP-CIT SPECT and MRI images from healthy control subjects and patients with Parkinson's Disease. As shown by our results, the cross-validation model proposed here does not only preserves the performance of our CAD system (93.8% of balanced accuracy) but also minimizes its variability even despite the input data sources poorly statistically significant.

**Keywords:** Machine learning · Ensemble learning · Neuroimaging · Parkinson's disease · Multimodal analysis · Computer-aided-diagnosis

## 1 Introduction

### 1.1 Parkinson's Disease

Parkinson's disease (PD) is the second most prevalent neurodegenerative disorder among middle-aged and elderly population and it is expected that in 2030 will be an incidence rate of 1 cases per 120 people over the age of 65 years [18]. Although its pathogenesis is still unclear, this disease is characterized by a progressive loss of dopaminergic neurons in the nigrostriatal pathway. As these cells


J. M. Ferrández Vicente et al. (Eds.): IWINAC 2022, LNCS 13258, pp. 23–33, 2022.
https://doi.org/10.1007/978-3-031-06242-1_3

are in charge of producing dopamine, whose role is fundamental in the coordination and generation of muscular movements, patients with PD experiment motor symptoms such as tremor, ridigity, bradykinesia or postural inestability; and non-motor like anxiety, depression or lack of emotion expressiveness [9].

One problem related to the effective management of PD patients is our capability to diagnose them in the early stages of the disease. It is known that about 65–80% of the dopaminergic neurons are already lost when first symptoms appear. Therefore, most of the tests used for the medical diagnosis of PD patients are based on the identification of intensity patterns on FP-CIT SPECT scans. As FP-CIT SPECT scans make use of a radioligand that presents a high binding affinity for dopamine transporters in the brain, these kind of tests give us a quantitative distribution of neurotransmitters in a patient's brain [23]. An example of two FP-CIT SPECT scans has been depicted in Fig. 1. As shown, whereas the Healthy Control (HC) subject presents a striatum (caudate and putamen) area c-shaped, highlighted and symmetrical; the same region from the patient with PD seems to be round and poorly illuminated.

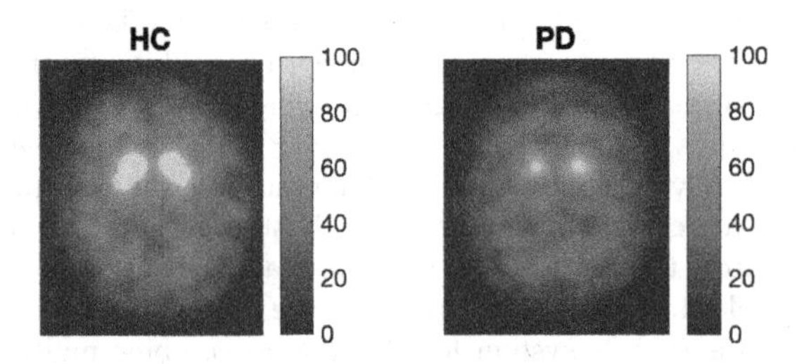


**Fig. 1.** Axial representation of two FP-CIT SPECT scans from a HC subject (left) and a patient with PD (right).

### 1.2 CAD Systems for Parkinson's Diagnosis

Despite some scans might not present evidences of dopaminergic deficit[1], intensity distributions near *striatum* region are frequently used as input data for Computer-Aided-Diagnosis (CAD) systems based on Machine Learning techniques [26]. This provides us automated solutions for PD diagnosis, and reduces the risk of misdiagnosis [8].

### 1.3 Ensemble Learning for Multimodal Data Analysis

Although there are many CAD systems to classify brain images for PD diagnosis, most of them have not attempted to combine two or more imaging modalities [1,12,17]. If we focus on multimodal CAD systems for PD, we can observe two

[1] Patients with parkinsonism labeled as SWEDD (*Scan Without Evidence of Dopaminergic Deficit*).

main strategies: on the one hand, works that have tried to extract imaging features (e.g. volume descriptors from regions of interest) [11]; and, on the other hand, ensemble learning models able to combine the output of different classifiers trained with different input data sources [2,13,20,24]. Regarding this second group, most of these proposals have been based on majority voting [14]. Nevertheless, when the number of input data sources is high and the reliability of many of them is poor, these solutions tend to underfitting. Thus, although there are works that solve this problem by a weighted majority voting schema, in some cases even introducing a windowing technique to weight non-linearly heavier the input data sources with higher classification rates [2], these solutions do not take into account the variability of classifications in a Cross-Validation (CV) loop. In this context, for this work we have proposed a nested CV schema, similar to the one presented in [2], which also penalizes the influence of input sources with high variability.

## 2 Materials and Methods

### 2.1 Parkinson's Progression Markers Initiative

**About PPMI.** Data used in the preparation of this article were obtained from the Parkinson's Progression Markers Initiative (PPMI) database[2]. For up-to-date information on the study, visit ppmi-info.org. PPMI -a public-private partnership- is funded by the Michael J. Fox Foundation for Parkinson's Research and funding partners listed on ppmi-info.org/about-ppmi/who-we-are/study-sponsors.

Informed consents to clinical testing and neuroimaging prior to participation of the PPMI cohort were obtained and approved by the institutional review boards (IRB) of all participating institutions. The PPMI obtained written informed consent from all study participants before enrolled in the Initiative. More information are available in ppmi-info.org/participants.

**Demographics.** For this work we have used a total of 232 MRI and FP-CIT SPECT images from HC and PD subjects whose demographics are included in Table 1. Note that though the PPMI includes more than 800 subjects, in order to combine all the information from both modalities we have discarded all input subjects whose differences between MRI and SPECT acquisition dates were longer than 15 days.

[2] Available at: ppmi-info.org/access-dataspecimens/download-data.

**Table 1.** Demographics. Age is given in terms of mean ± standard deviation.

| | Male | | Female | | Both | |
|---|---|---|---|---|---|---|
| | % | Age | % | Age | % | Age |
| HC | 70.0 | $62.6 \pm 11.6$ | 30.0 | $61.2 \pm 7.5$ | 17.2 | $62.2 \pm 10.5$ |
| PD | 64.6 | $62.8 \pm 9.4$ | 35.4 | $61.6 \pm 8.2$ | 69.1 | $62.4 \pm 9.0$ |
| SWEDD | 59.4 | $63.2 \pm 10.8$ | 40.6 | $58.5 \pm 10.5$ | 13.7 | $61.3 \pm 11.0$ |

### 2.2 Image Preprocessing

**Spatial Normalization.** All brain scans have been spatially registered using the software tool SPM12[3]. Following the recommendations given in [3] we have compared both affine and non-linear spatial registrations to a reference template in Montreal Neurological Institute (MNI) standard space. For that, we have made use of the MNI152 template included in SPM12 for MRI images and a functional template generated *ad-hoc* as explained in [15] for FP-CIT SPECT scans. After being co-registered and averaged, all input scans presented the same size ($121 \times 145 \times 121$ voxels) and a voxel-size of $1.5 \times 1.5 \times 1.5$ mm.

**Intensity Normalization.** Once spatially normalized, FP-CIT SPECT scans have been intensity normalized by means of an $\alpha$-stable normalization procedure described in [16]. This method consists of the linear transformation given in (1) where $\gamma^*$ and $\mu^*$ represent the mean of $\gamma$ (dispersion) and $\mu$ (location) parameters obtained from all the input scans.

$$Y = \frac{\gamma^*}{\gamma} X + \left(\mu^* - \frac{\gamma^*}{\gamma} \mu\right) \tag{1}$$

Figure 2 includes a montage showing differences between 8 FP-CIT SPECT scans randomly selected before/after intensity normalization using $\alpha$-stable distributions.

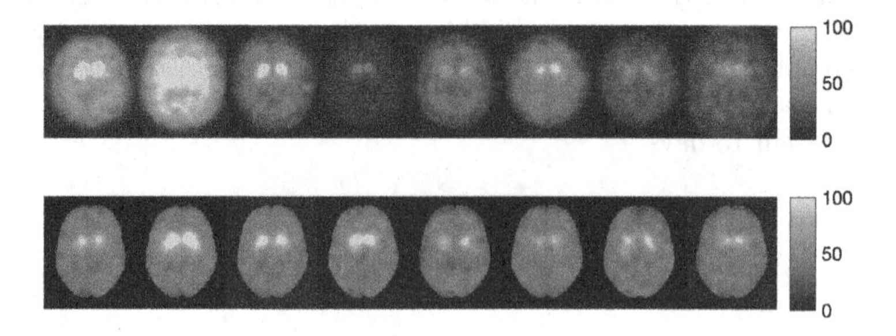


**Fig. 2.** Set of 8 FP-CIT SPECT scans randomly selected (up) and their respective versions once spatially and intensity normalized (down).

[3] Available at www.fil.ion.ucl.ac.uk/spm/software/spm12/.

### 2.3 Feature Selection and Dimensionality Reduction Algorithms

In order to avoid potential overfitting, it is recommended to reduce the number of input features given to classifiers [7]. For that, we have made use of a two-levels procedure which includes a feature selection step, using ANOVA [21]; and a dimensionality reduction procedure by means of Partial Least Squares (PLS) [25]. In any case, *p*-value limit has been set to 0.001 and the number of principal components for PLS will never be greater than the number of input samples given to each classifier.

### 2.4 Classification Schema

The main proposal of this work is the classification schema depicted in Fig. 3. This proposal combines several input data sources discarding those ones less representative for PD diagnosis.

Once preprocessed, the MRI and FP-CIT SPECT images from HC and PD subjects in the PPMI dataset are standardized. Then, we calculate a Region Of Interest (ROI) centered at striatum area but also including a margin to analyze surrounding intensity levels following the recommendations given in [4]. Note as MRI and FP-CIT SPECT images are co-registered to the reference MNI template, this ROI is common to all input scans. From here on we will consider each input data type separately. Within the external loop, each subset pass through a feature selection, a dimensionality reduction and a classification stage. Whereas in the inner loop, each subset is again separated into two new training/test subsets. Then, it proceeds with the same selection, reduction and classification methods. This proposal generates two different metrics: on the one hand, the classification parameters of the inner loop, $w_{\text{int}}$, and on the other hand the predicted labels in the outer loop, $l_{\text{ext}}$. Note that all individual classifications have been carried out using Support-Vector-Machine (SVM) classifiers [19]. Specifically, we have made use of Linear Kernel functions, to make the hyperplane decision boundary between input classes.

For this work, we have implemented a total of 4 ensemble learning models to be compared between them. This includes the classic Majority Voting (MV) proposal; a Weighted Majority Voting (WMV) method using the balanced accuracy[4] calculated in the internal loop to ponderate the influence of each imaging modality to the final decision; and a WMV schema using a non-linear function, as those depicted in Fig. 4, to overweight the influence of highly relevant input data sources [2]. In addition to this, and to avoid input data sources with a high variance, in all these scenarios we have divided the balanced accuracies calculated in the inner loop by their standard deviation. This increases the robustness of our solution as the ensemble learning model focuses on input features more common from each class.

[4] Arithmetic mean between sensitivity and specificity.

## 3 Results

For this work, we have made use of MRI and FP-CIT SPECT scans from subjects in the PPMI dataset to develop a multimodal CAD system for PD based on ensemble learning. Once the input scans have been preprocessed, they are used as inputs for the classification schema depicted in Fig. 3. Following the recommendations given in [10], both external and internal CV loops are based on a 10-fold CV schema. Regarding the feature selection step using ANOVA, we have selected the 40 best input features with $p_{\text{value}} < 0.001$. As the number of input features matching this condition were larger than the number of input samples, we have also made use of PLS. In this case, in order not to lose the predictive power of our original model, we have selected $n_{\text{Comp}} = 10$ components.

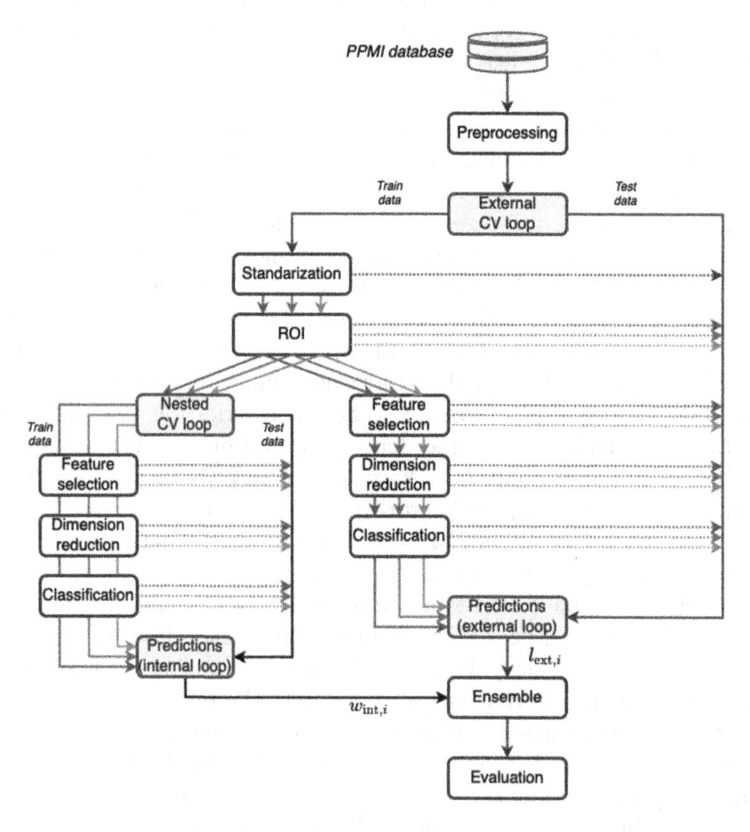


**Fig. 3.** Classification schema proposed when combining the three types of input data: FP-CIT SPECT (red), MRI GM (blue) and MRI WM (green) images (Color figure online).

Final classification results obtained for the GM/WM MRI and FP-CIT SPECT modalities resulted as summarized in Table 2, while the final performance of the ensemble model remainded as follows in Table 3. Furthermore, a representation of the Receiver Operating Characteristic (ROC) has been included in Fig. 5.

**Table 2.** Classification results obtained in the external CV loop.

| | Accuracy | Sensitivity | Specificity |
|---|---|---|---|
| $\text{MRI}_{\text{GM}}$ | $51.3\% \pm 18.1\%$ | $47.5\% \pm 24.8\%$ | $55.0\% \pm 22.9\%$ |
| $\text{MRI}_{\text{WM}}$ | $58.8\% \pm 18.6\%$ | $55.0\% \pm 30.7\%$ | $62.5\% \pm 24.2\%$ |
| $\text{SPECT}_{\text{FP-CIT}}$ | $93.8\% \pm 8.8\%$ | $92.5\% \pm 12.0\%$ | $95.0\% \pm 10.5\%$ |

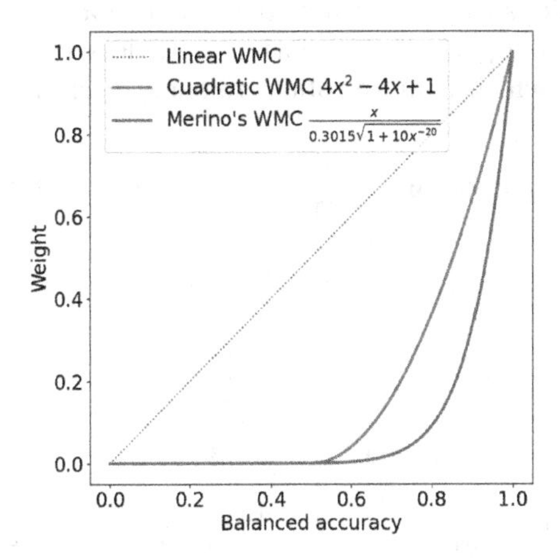


**Fig. 4.** Weighted majority voting with using different non-linear windowing functions.

**Table 3.** Classification results obtained using different ensemble proposals.

| | Accuracy | Sensitivity | Specificity |
|---|---|---|---|
| Classic MV | $73.7\% \pm 16.0\%$ | $67.5\% \pm 26.4\%$ | $80.0\% \pm 19.7\%$ |
| Linear WMV | $86.2\% \pm 12.4\%$ | $85.0\% \pm 12.9\%$ | $87.5\% \pm 17.6\%$ |
| Merino's WMV | $93.8\% \pm 8.8\%$ | $92.5\% \pm 12.0\%$ | $95.0\% \pm 10.5\%$ |

## 4 Discussion

Multimodal analysis of brain images for PD is challenging because of factors such as a limited spatial/temporal resolution, the lack of quantification, and imaging distortions [6,26]. Nevertheless, we can find many multimodal works that have tried to combine different biomarkers by ensemble learning trying to improve the classification results obtained when analyzing subjects with PD only using FP-CIT SPECT scans [26]. Though most of them are based on Majority Voting [14], these kind of solutions tend to underfitting when combining several non-related input sources [2]. As expected, when using the MV and the linear WMV models for ensemble learning, this behaviour is also confirmed. Indeed, owing to results in Tables 2 and 3, we can observe that voting from MRI-T1 GM and MRI-T1 WM modalities increase the failure rate of the predictive model by 20.1% when considering the MV schema and by 7,6% for the WMV model. Fortunately, the application of non-linear windowing functions like those represented in Fig. 4, suppress the effect of non-relevant input data sources (93.8% of balanced accuracy) and preserves its variability (8.8%). Though results obtained when using a simple cuadratic form are equivalent to Merino's function, we have decided to include both expressions to propose a smoothed version of our windowed WMV ensemble model (keeping some information from less relevant input sources), and a restrictive one which is more robust to slight variations in our dataset.

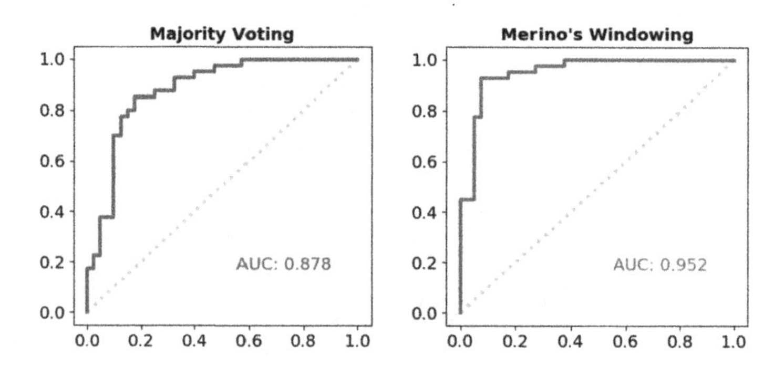


**Fig. 5.** ROC curves obtained for MV and WMV with Merino's windowing experiments.

In comparison to other previous works such as [2], as we are estimating the balanced accuracy divided by its standard deviation, we are penalizing any input source whose variability is high. This strategy give us a robust version of our ensemble learning model as predictions become more sensitive to those input features more frequently found in HC and PD classes. Unfortunately, the lack of reliable input biomarkers sources has led to a model which only focuses on FP-CIT SPECT data. Indeed, despite there are two previous works claiming that MRI images could be used for PD prediction using the Voxel-Based-Morphometry (VBM) approach, [5,22], as our proposal is focused on the ensemble learning methodology, we have decided not to include this method. Besides, owing to the reduced number of input samples available, we have avoided the use this kind of solutions and Neural Networks to avoid the potential overfitting of these models and to preserve the physical interpretability of our results.

## 5 Conclusions

In this work, we have proposed an improved ensemble learning methodology based on weighted majority voting to assist in the diagnosis of PD. This approach is composed by a double cross-validation loop that computes the performance of each input data source and weights its influence using the balanced accuracy parameter using a non-linear windowing technique. In this context, the performance of each input data source is divided by its variability to increase the ensemble learning model robustness.

The approach has been evaluated using three different functional and structural brain imaging modalities. Nevertheless, it is completely flexible as it allows working with any number of input sources, and because it can be adapted to integrate any other feature selection/extraction, classification and/or regression algorithms.

**Acknowledgments.** This work was supported by the MCIN/ AEI/10.13039/5011000 11033/ and FEDER "Una manera de hacer Europa" under the RTI2018-098913-B100 project; by the Consejería de Economía, Innovación, Ciencia y Empleo (Junta de Andalucía) and FEDER under CV20-45250, A-TIC-080-UGR18, B-TIC-586-UGR20 and P20-00525 projects; and by the Ministerio de Universidades under the FPU18/04902 grant given to C. Jimenez-Mesa and the Margarita-Salas grant to J.E. Arco.

## References

1. Augimeri, A., et al.: CADA—computer-aided DaTSCAN analysis. EJNMMI Phys. **3**(1), 1–13 (2016). https://doi.org/10.1186/s40658-016-0140-9
2. Castillo-Barnes, D., et al.: Robust ensemble classification methodology for I123-Ioflupane SPECT images and multiple heterogeneous biomarkers in the diagnosis of Parkinson's disease. Front. Neuroinform. **12**, 53 (2018). https://doi.org/10.3389/fninf.2018.00053
3. Castillo-Barnes, D., et al.: Comparison between affine and non-affine transformations applied to $I^{[123]}$-FP-CIT SPECT images used for Parkinson's disease diagnosis. In: Ferrández Vicente, J.M., Álvarez-Sánchez, J.R., de la Paz López, F., Toledo Moreo, J., Adeli, H. (eds.) IWINAC 2019, Part I. LNCS, vol. 11486, pp. 379–388. Springer, Cham (2019). https://doi.org/10.1007/978-3-030-19591-5_39
4. Castillo-Barnes, D., et al.: Morphological characterization of functional brain imaging by isosurface analysis in Parkinson's disease. Int. J. Neural Syst. **30**(09), 2050044 (2020). https://doi.org/10.1142/s0129065720500446
5. Cigdem, O., Beheshti, I., Demirel, H.: Effects of different covariates and contrasts on classification of Parkinson's disease using structural MRI. Comput. Biol. Med. **99**, 173–181 (2018). https://doi.org/10.1016/j.compbiomed.2018.05.006
6. Górriz, J.M., et al.: Artificial intelligence within the interplay between natural and artificial computation: advances in data science, trends and applications. Neurocomputing **410**, 237–270 (2020). https://doi.org/10.1016/j.neucom.2020.05.078
7. Hawkins, D.M.: The problem of overfitting. J. Chem. Inf. Comput. Sci. **44**(1), 1–12 (2003). https://doi.org/10.1021/ci0342472

8. Hustad, E., Skogholt, A.H., Hveem, K., Aasly, J.O.: The accuracy of the clinical diagnosis of Parkinson disease. The HUNT study. J. Neurol. **265**(9), 2120–2124 (2018). https://doi.org/10.1007/s00415-018-8969-6
9. Iarkov, A., Barreto, G.E., Grizzell, J.A., Echeverria, V.: Strategies for the treatment of Parkinson's disease: beyond dopamine. Front. Aging Neurosci. **12**, 4 (2020). https://doi.org/10.3389/fnagi.2020.00004
10. Kohavi, R.: A study of cross-validation and bootstrap for accuracy estimation and model selection. In: Proceedings of the 14th International Joint Conference on Artificial Intelligence, IJCAI 1995, Quebec, Canada, vol. 2, pp. 1137–1145 (August 1995)
11. Martins, R., et al.: Automatic classification of idiopathic Parkinson's disease and atypical parkinsonian syndromes combining [11c]raclopride PET uptake and MRI grey matter morphometry. J. Neural Eng. **18**(4), 046037 (2021). https://doi.org/10.1088/1741-2552/abf772
12. Nicastro, N., et al.: Classification of degenerative parkinsonism subtypes by support-vector-machine analysis and striatal $^{123}$I-FP-CIT indices. J. Neurol. **266**(7), 1771–1781 (2019). https://doi.org/10.1007/s00415-019-09330-z
13. Ramírez, J., et al.: Ensemble of random forests one vs. rest classifiers for MCI and AD prediction using ANOVA cortical and subcortical feature selection and partial least squares. J. Neurosci. Methods **302**, 47–57 (2018). https://doi.org/10.1016/j.jneumeth.2017.12.005
14. Rokach, L.: Pattern Classification Using Ensemble Methods. World Scientific Publishing Company, Singapore (2009)
15. Salas-Gonzalez, D., et al.: Building a FP-CIT SPECT brain template using a posterization approach. Neuroinformatics **13**(4), 391–402 (2015). https://doi.org/10.1007/s12021-015-9262-9
16. Salas-Gonzalez, D., et al.: Linear intensity normalization of FP-CIT SPECT brain images using the $\alpha$-stable distribution. NeuroImage **65**, 449–455 (2013). https://doi.org/10.1016/j.neuroimage.2012.10.005
17. Salvatore, C., et al.: Machine learning on brain MRI data for differential diagnosis of Parkinson's disease and progressive supranuclear palsy. J. Neurosci. Methods **222**, 230–237 (2014). https://doi.org/10.1016/j.jneumeth.2013.11.016
18. Savica, R., et al.: Time trends in the incidence of Parkinson disease. JAMA Neurol. **73**(8), 981 (2016). https://doi.org/10.1001/jamaneurol.2016.0947
19. Schoölkopf, B.: Learning with Kernels - Support Vector Machines, Regularization, Optimization, and Beyond. MIT Press, Cambridge (2002)
20. Segovia, F., et al.: Multivariate analysis of 18f-DMFP PET data to assist the diagnosis of parkinsonism. Front. Neuroinform. **11**, 23 (2017). https://doi.org/10.3389/fninf.2017.00023
21. Shaw, R.G., Mitchell-Olds, T.: Anova for unbalanced data: an overview. Ecology **74**(6), 1638–1645 (1993). https://doi.org/10.2307/1939922
22. Solana-Lavalle, G., Rosas-Romero, R.: Classification of PPMI MRI scans with voxel-based morphometry and machine learning to assist in the diagnosis of Parkinson's disease. Comput. Methods Programs Biomed. **198**, 105793 (2021). https://doi.org/10.1016/j.cmpb.2020.105793
23. Vlaar, A.M., et al.: Diagnostic value of 123i-ioflupane and 123i-iodobenzamide SPECT scans in 248 patients with parkinsonian syndromes. Eur. Neurol. **59**(5), 258–266 (2008). https://doi.org/10.1159/000115640
24. Wingate, J., Kollia, I., Bidaut, L., Kollias, S.: Unified deep learning approach for prediction of Parkinson's disease. IET Image Process. **14**(10), 1980–1989 (2020). https://doi.org/10.1049/iet-ipr.2019.1526

25. Wold, S., Ruhe, A., Wold, H., W. J. Dunn, I.: The collinearity problem in linear regression. The partial least squares (PLS) approach to generalized inverses. SIAM J. Sci. Stat. Comput. **5**(3), 735–743 (1984). https://doi.org/10.1137/0905052
26. Zhang, Y.D., et al.: Advances in multimodal data fusion in neuroimaging: overview, challenges, and novel orientation. Inf. Fusion **64**, 149–187 (2020). https://doi.org/10.1016/j.inffus.2020.07.006

# Automatic Classification System for Diagnosis of Cognitive Impairment Based on the Clock-Drawing Test

C. Jiménez-Mesa[1,2(✉)], Juan E. Arco[3], M. Valentí-Soler[4], B. Frades-Payo[4], M. A. Zea-Sevilla[4], A. Ortiz[1,3], M. Ávila-Villanueva[5], Diego Castillo-Barnes[1,2], J. Ramírez[1,2], T. del Ser-Quijano[4], C. Carnero-Pardo[5], and J. M. Górriz[1]

[1] Data Science and Computational Intelligence (DASCI) Institute, Granada, Spain
[2] Signal Theory, Telematics and Communications Department, University of Granada, Granada, Spain
carmenj@ugr.es
[3] Communications Engineering Department, University of Málaga, Málaga, Spain
[4] Fundacion CIEN, Madrid, Spain
[5] FIDYAN Neurocenter, Granada, Spain
http://sipba.ugr.es/

**Abstract.** The prevalence of dementia is currently increasing worldwide. This syndrome produces a deterioration in cognitive function that can not be reverted. However, an early diagnosis can be crucial for slowing its progress. The Clock Drawing Test (CDT) is a widely used paper-and-pencil test for cognitive assessment in which an individual has to manually draw a clock on a paper during a certain time. Nevertheless, there are a lot of scoring systems for this test and most of them depend on the subjective assessment of the expert. This study proposes a computer-aided diagnosis (CAD) system based on deep learning in order to automate the diagnosis of cognitive impairment (CI) from the result of the CDT. This is addressed by employing a preprocessing pipeline in which the clock is detected and centered, as well as binarized for decreasing the computational burden. Then, the resulting image is fed into a Convolutional Neural Network (CNN), which is used to identify the informative patterns within the CDT drawings that are relevant for the assessment of the patient's cognitive status. Performance is evaluated in a real context where differentiating between CI patients and controls. The proposed method provides an accuracy of 68.62% in this classification task, with an AUC of 74.53%. A validation method using resubstitution with upper bound correction is also discussed.

**Keywords:** Alzheimer's disease · Clock Drawing Test · Cognitive impairment · Deep learning · Image processing · Machine learning

## 1 Introduction

Dementia, which includes Alzheimer's disease (AD), is one of the most common neurological syndromes in the world [11]. One of the first phases of diagnosis is a


J. M. Ferrández Vicente et al. (Eds.): IWINAC 2022, LNCS 13258, pp. 34–42, 2022.
https://doi.org/10.1007/978-3-031-06242-1_4

cognitive assessment. Among the various tests performed for clinical diagnosis, the Clock Drawing Test (CDT) outstands for its simplicity [17]. This test is commonly performed in conjunction with other screening methods, such as Mini-Mental State Examination (MMSE) [14] with the aim of improving the reliability of the diagnosis.

### 1.1 Clock Drawing Test

The CDT consists on a paper-and-pencil cognitive screening tool which allows to detect several cognitive changes related to spacial vision, frontal lobe execution or memory, among others [6]. In the test, the patient has to draw a clock including the numbers 1 to 12 and a specific position of the clock hands, ten past eleven. Once this is done, a physician has to evaluate the drawing and establish a score. This score is correlated with the patient's cognitive status, reflecting the progress of cognitive impairment (CI). The scoring decision by the physicist is a subjective decision. Therefore, with the machine learning (ML) techniques developed in recent years, more and more studies are proposing an automatic scoring system instead of a manual system [4,12].

### 1.2 Artificial Intelligence

The use of ML and deep learning (DL) models for classification tasks [7] is increasing. In particular, in the field of neuroimaging, the development of computer aided diagnosis (CAD) systems applying such algorithms is very common. Support Vector Machines (SVM) are one of the most widely used ML approaches [10]. Meanwhile, within deep learning, the base model in image classification is the convolutional neural network (CNN) [13]. On the basis of this, a large number of more complex architectures have been proposed, such as DenseNet-121 [9] or Mobilenetv2 [15].

### 1.3 Automatic CDT Scoring Systems

Previous works related to CDT classification are divided in two categories. The ones based on the original CDT and on digital Clock Drawing Test (dCDT), which applies a digital ballpoint pen instead of paper and pencil [12]. According to Chan et al. [3], the pooled sensitivity and specificity obtained screening mild cognitive impairment (MCI) in previous studies using dCDT is higher than CDT. Specifically, the former is associated with a sensitivity and specificity of 0.86% and 0.92%, and the latter with 0.63% and 0.77%, respectively. Nevertheless, it is important to note that not all medical centres have the financial means to digitalise this type of test. Therefore, it would be best to move forward in the study of the cheaper and simpler version of the CDT, the paper-and-pencil based one.

Previous works include both ML [5] and DL [4] approaches. A growing number of works are implementing deep learning-based architectures in recent years [1,2].

## 2 Materials and Methods

### 2.1 CDT Database

In the collected clock drawings, participants were given an A4 size paper and a pencil and asked to draw a clock with the clock hands pointing to ten past eleven. Once the clock drawings are finished by the subjects, a score is given from 0 to 7. A score of 7 would mean a perfect draw of the clock, so it is related to a healthy subject who does not suffer from dementia or other similar syndromes. By contrast, a score of 0 indicates that the subject is completely unable to draw a clock.

The database applied in this work consists of the CDT draws from patients in the Department of Neurology of FIDYAN Neurocenter (Granada and Malaga, Spain). The database consists of 1520 patients, with an average age of 70.85 years and a ratio between females and males of 1.15. Nevertheless, since 459 samples are controls and 1061 are CI patients, the database was balanced to 918 subjects. Their demographical information is summarized in Table 1.

**Table 1.** Demographics of the subjects contained in the database. CI means cognitive impairment subject, S denotes superior education, NS means non-superior education, M represents male and F means female.

| | CI | Controls | Total |
|---|---|---|---|
| Number of participants | 459 | 459 | 918 |
| Age | 73.71 ± 9.27 | 64.63 ± 14.63 | 69.17 ± 13.06 |
| Education (S/NS) | 220/239 | 255/204 | 475/443 |
| Sex (M/F) | 221/238 | 223/236 | 444/474 |

### 2.2 Image Preprocessing

The paper-and-pencil draws of patients were scanned in order to digitise the images. This scanned image consists of the sheet where the patient makes the drawing. This image can contain non-relevant information such as previous drawing attempts or comments from the clinician, in addition to some identifiers related to the subject. For this reason, we applied a preprocessing pipeline in order to isolate the region of interest (the clock drawing) for further analysis.

Figure 1 summarizes the different preprocessing steps. No patient information is reported in the Figure. First, the original RGB images are converted to grayscale ones. After that, a binarisation and an edge filling steps are employed to detect the objects contained in the image and to identify if they belong to the region of interest. Our algorithm properly recognizes elements even when they are drawn outside the clock face, which is not unusual for numbers 12, 3, 6 and 9. Finally, the image is cropped and resized to 224 × 224, one of the most widely used sizes in image classification. The use of binary images implies that the intensity of the pixels are either 0 or 1.

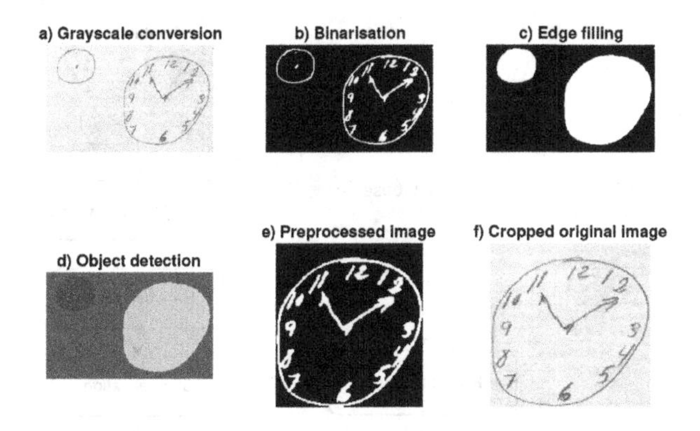


**Fig. 1.** Preprocessing diagram of the clock test drawings.

## 2.3 Classification Model

**Deep Learning Approach.** The main approach proposed is based on a convolutional neural network (CNN). It was designed for differentiating between cognitively impaired and healthy subjects from the resulting CDT. A schematic of the network is depicted in Fig. 2, which includes four 2D convolutional blocks (convolutional layer, batch normalization, rectified linear (ReLU) activation function and a maxpooling layer) and three fully connected layers. Dropout [18] is applied in combination with the linear layers to prevent overfitting, whereas a final softmax layer was added to the model to predict the probability of each sample belonging to the two classes under analysis.

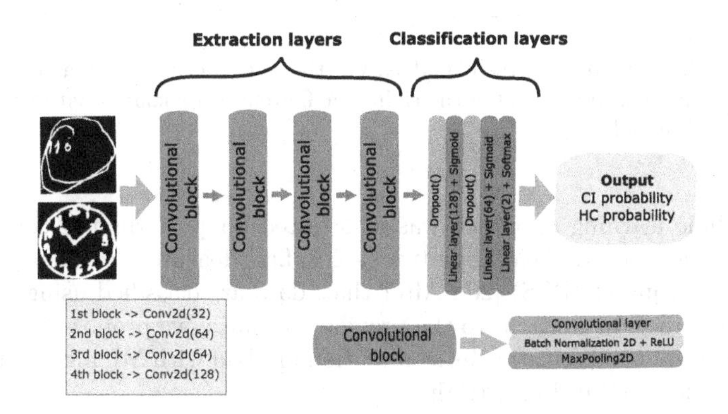


**Fig. 2.** Diagram of the classification process.

The implemented classification scheme applies k-Fold cross-validation for the estimation of metrics of interest. The fitting of the model is performed with the training data, while the results obtained are analysed with the test data. This flowchart is shown in Fig. 3 (left).

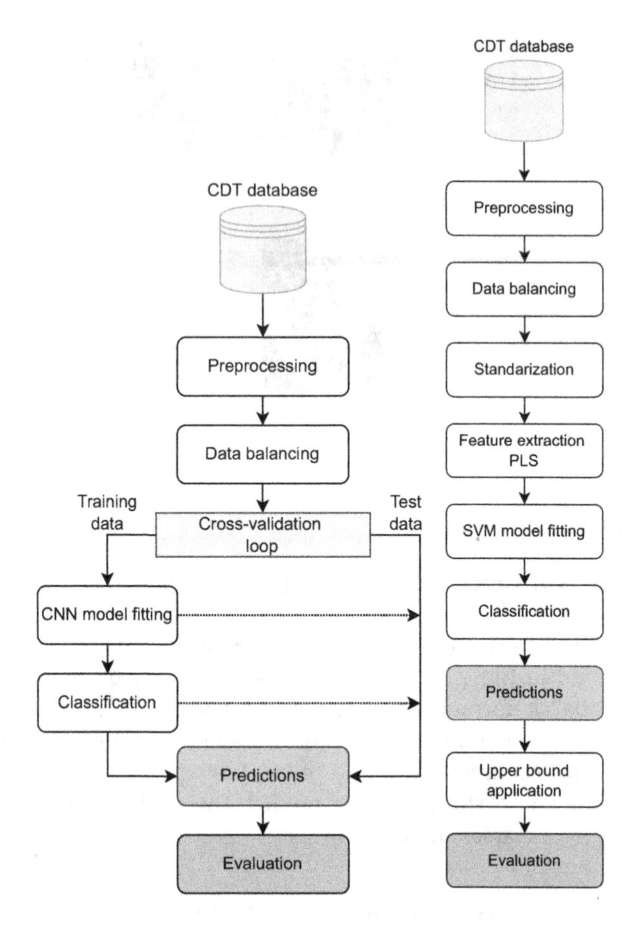


**Fig. 3.** Flowchart of the analysed models. On the left, the flowchart associated with the deep learning-based model. On the right, the flowchart associated with the machine learning-based model.

A machine learning approach was also tested for performance comparative purposes. This method is based on two stages: first, features are extracted using Partial Least Squares (PLS) [20]. After that, data are classified using SVM [16] with linear kernel. In contrast to the previous scenario, we propose to use upper bound-corrected resubstitution as a validation method with ML algorithms. This flowchart is depicted in Fig. 3 (right).

The upper bound can be seen as the difference between empirical and actual errors, $\mu \geq |E_{act}(f(x)) - E_{emp}(f(x))|$. Thus, the actual accuracy obtained using the whole dataset as training and test set could be limited by the upper bound proposed by Vapnik [19], which is defined as:

$$\mu_{VC} \leq \sqrt{\frac{h\left(\ln\left(\frac{2n}{h}\right)+1\right)-\ln\left(\frac{\eta}{4}\right)}{n}} \tag{1}$$

where $\eta$ is the significance level and $n$ is the size of the training set. The VC dimension is represent by $h$ and is equal to $d+1$ for linear functions, as in this case, being $d$ the features dimension.

This upper bound could be seen as a theoretical classification limit, in this case for linear classifiers, which allows the use of all accessible data to establish the metrics of interest. Besides accuracy, sensitivity and specificity can be limited with this limit considering that its associated errors are partial errors of the classification error.

## 3 Results

To estimate the classification performance of the proposed method in the CDT database, two strategies are applied. A 5-fold cross validation strategy in conjunction with the CNN model, and the resubstitution estimation of the actual error on the training set in the ML approach, as shown in Fig. 3.

The hyperparameters associated with the CNN were optimized as follows. The dropout selected value was 0.5. Adam optimization algorithm with a learning rate of 0.001 was applied. The loss function employed was Binary Cross-Entropy (BCE), whereas the system was trained during 70 epochs, and a batch size of 1.

In the ML approach, the number of features extracted by PLS was 2 and the significance level selected 0.05. The upper bound related to this values together with a sample size of 918 is 0.2207 per unit.

Final results obtained by each approach are shown in Table 2. In the CNN scenario, other metrics are estimated. The positive predictive value (PPV) and negative predictive value (NPV) were $69.47 \pm 1.72$ and $68.11 \pm 2.74$, respectively. The ROC curve obtained is depicted in Fig. 4.

## 4 Discussion

In this work, a classification framework for the automatic classification of CDT is proposed. This approach is based on a preprocessing pipeline where images are cropped, centered and binarized. Once these images are standardized, they are entered into both a ML and a DL approaches that identify the most relevant features of each individual class. The performance of this alternative was evaluated in a scenario where differentiating between patients who suffered from cognitive impairment and controls.

**Table 2.** Classification results obtained using CNN and SVM with their different validation methods. The symbol "–" represents that no information is available for that metric. **Experiment where the same 734 subsets of samples than in the training set of the 5-Fold experiment are evaluated. Its upper bound is 0.2431.

| Experiment | Validation method | Acc (%) | Spec (%) | Sens (%) | AUC (%) |
|---|---|---|---|---|---|
| CNN | 5-Fold | 68.62 ± 1.57 | 70.60 ± 3.48 | 66.67 ± 5.21 | 74.53 ± 1.87 |
| PLS+SVM | Resubstitution | 97.71 | 98.25 | 97.16 | 99.61 |
| | Res. with upper bounding | 75.64 | 76.18 | 75.09 | – |
| PLS+SVM** | Res. with upper bounding | 74.25 ± 0.53 | 74.33 ± 0.46 | 74.17 ± 0.70 | – |

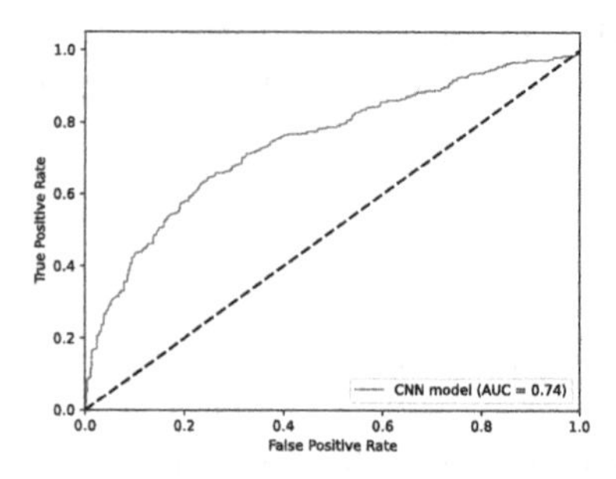


**Fig. 4.** ROC curve obtained by the CNN approach.

The results obtained indicate that the methodology proposed is close to the expected performance according to Chan et al. [3], who establish a mean sensitivity and specificity of 0.63% and 0.77% when the paper-and-pencil CDT is analysed. Moreover, the database used in this study is relatively small. A larger database, which we are currently processing, will allow us to increase the performance obtained and to study in greater depth the differences between the drawings made by subjects with cognitive impairment and controls. Previous studies have developed systems for the automatic diagnosis of cognitive impairment from the clock-drawing test. Most of these works rely on the use of a digital version of the CDT. This allows a greater variety of features to be employed, resulting in a considerable increase in performance. Nevertheless, these devices are not common in clinical centers, and even unaffordable for hospitals of some regions. Therefore, it is also necessary to continue the development of automatic classification systems for the original CDT.

The different validation methods applied suggest that the performance associated with cross-validation and resubstitution with upper bound correction is similar. The latter offers the advantage of being able to use the complete database for the evaluation of the results. The upper bound correction implies not to consider the empirical error obtained but the actual one, setting an upper limit on the accuracy. The upper bound used in this work is highly conservative. There are less conservative ones [8] but the assumptions on which they are based need to be analysed in more detail to assess its validity in this type of database.

## 5 Conclusion

In this work, we propose a method for the automatic diagnosis of cognitive impairment based on the clock-drawing test. This is addressed by employing a preprocessing in which the clock is detected and centered, in addition to be binarized in order to reduce the computational cost of the subsequent mathematical operations. Then, a CNN is used to find the relevant patterns of information that characterize CI patients and controls. Moreover a linear machine learning approach is tested to optimise the sample size of the database. The performance achieved is in line with what is expected to be obtained using an analogical version of the CDT. Our results manifest the suitability of our method in hospitals and medical clinics worldwide, especially in those regions with low resources.

**Acknowledgments.** This work was supported by the MCIN/ AEI/10.13039/ 501100011033/ and FEDER "Una manera de hacer Europa" under the RTI2018-098913-B100 project, by the Consejeria de Economia, Innovacion, Ciencia y Empleo (Junta de Andalucia) and FEDER under CV20-45250, A-TIC-080-UGR18, B-TIC-586-UGR20 and P20-00525 projects, and by the Ministerio de Universidades under the FPU18/04902 grant given to C. Jimenez-Mesa and the Margarita-Salas grant to J.E. Arco

## References

1. Amini, S., et al.: An artificial intelligence-assisted method for dementia detection using images from the clock drawing test. J. Alzheimer's Dis. **83**(2), 581–589 (2021). https://doi.org/10.3233/JAD-210299
2. Binaco, R., et al.: Machine learning analysis of digital clock drawing test performance for differential classification of mild cognitive impairment subtypes versus Alzheimer's disease. J. Int. Neuropsychol. Soc. **26**(7), 690–700 (2020). https://doi.org/10.1017/s1355617720000144
3. Chan, J.Y.C., et al.: Evaluation of digital drawing tests and paper-and-pencil drawing tests for the screening of mild cognitive impairment and dementia: a systematic review and meta-analysis of diagnostic studies. Neuropsychol. Rev. 1–11 (2021). https://doi.org/10.1007/s11065-021-09523-2
4. Chen, S., Stromer, D., Alabdalrahim, H.A., Schwab, S., Weih, M., Maier, A.: Automatic dementia screening and scoring by applying deep learning on clock-drawing tests. Sci. Rep. **10**(1), 1–11 (2020). https://doi.org/10.1038/s41598-020-74710-9
5. Davoudi, A., et al.: Classifying non-dementia and Alzheimer's disease/vascular dementia patients using kinematic, time-based, and visuospatial parameters: the digital clock drawing test. J. Alzheimer's Dis. **82**(1), 47–57 (2021). https://doi.org/10.3233/JAD-201129
6. Freedman, M., Leach, L., Kaplan, E., Winocur, G., Shulman, K., Delis, D.C.: Clock Drawing: A Neuropsychological Analysis. Oxford University Press, USA (1994)
7. Górriz, J.M., et al.: Artificial intelligence within the interplay between natural and artificial computation: advances in data science, trends and applications. Neurocomputing **410**, 237–270 (2020)
8. Górriz, J.M., Ramirez, J., Suckling, J.: On the computation of distribution-free performance bounds: application to small sample sizes in neuroimaging. Pattern Recognit. **93**, 1–13 (2019). https://doi.org/10.1016/j.patcog.2019.03.032

9. Huang, G., Liu, Z., Van Der Maaten, L., Weinberger, K.Q.: Densely connected convolutional networks. In: Proceedings of the IEEE Conference on Computer Vision and Pattern Recognition, pp. 4700–4708 (2017)
10. Jimenez-Mesa, C., et al.: Optimized one vs one approach in multiclass classification for early Alzheimer's disease and mild cognitive impairment diagnosis. IEEE Access **8**, 96981–96993 (2020). https://doi.org/10.1109/access.2020.2997736
11. Knopman, D.S., et al.: Alzheimer disease. Nat. Rev. Dis. Primers **7**(1), 1–21 (2021). https://doi.org/10.1038/s41572-021-00269-y
12. Müller, S., Preische, O., Heymann, P., Elbing, U., Laske, C.: Increased diagnostic accuracy of digital vs. conventional clock drawing test for discrimination of patients in the early course of Alzheimer's disease from cognitively healthy individuals. Front. Aging Neurosci. **9**, 101 (2017). https://doi.org/10.3389/fnagi.2017.00101
13. Ortiz, A., Munilla, J., Górriz, J.M., Ramírez, J.: Ensembles of deep learning architectures for the early diagnosis of the Alzheimer's disease. Int. J. Neural Syst. **26**(07), 1650025 (2016). https://doi.org/10.1142/s0129065716500258
14. Palsetia, D., Rao, G.P., Tiwari, S.C., Lodha, P., De Sousa, A.: The clock drawing test versus mini-mental status examination as a screening tool for dementia: a clinical comparison. Ind. J. Psychol. Med. **40**(1), 1–10 (2018)
15. Sandler, M., Howard, A., Zhu, M., Zhmoginov, A., Chen, L.C.: Mobilenetv2: inverted residuals and linear bottlenecks. In: Proceedings of the IEEE Conference on Computer Vision and Pattern Recognition, pp. 4510–4520 (2018)
16. Schölkopf, B., Smola, A.J., Bach, F., et al.: Learning with Kernels: Support Vector Machines, Regularization, Optimization, and Beyond. MIT Press, Cambridge (2002)
17. Shulman, K.I.: Clock-drawing: is it the ideal cognitive screening test? Int. J. Geriatr. Psychiatry **15**(6), 548–561 (2000). https://doi.org/10.1002/1099-1166(200006)15:6⟨548::aid-gps242⟩3.0.co;2-u
18. Srivastava, N., Hinton, G., Krizhevsky, A., Sutskever, I., Salakhutdinov, R.: Dropout: a simple way to prevent neural networks from overfitting. J. Mach. Learn. Res. **15**(1), 1929–1958 (2014)
19. Vapnik, V., Levin, E., Cun, Y.L.: Measuring the VC-dimension of a learning machine. Neural Comput. **6**(5), 851–876 (1994)
20. Wold, S., Ruhe, A., Wold, H., W. J. Dunn, I.: The collinearity problem in linear regression. The partial least squares (PLS) approach to generalized inverses. SIAM J. Sci. Stat. Comput. **5**(3), 735–743 (1984). https://doi.org/10.1137/0905052

# Unraveling Dyslexia-Related Connectivity Patterns in EEG Signals by Holo-Hilbert Spectral Analysis

Nicolás J. Gallego-Molina[1,3(✉)], Andrés Ortiz[1,3], Francisco J. Martínez-Murcia[2,3], and Ignacio Rodríguez-Rodríguez[1,3]

[1] Communications Engineering Department, University of Málaga, 29004 Málaga, Spain
njgm@ic.uma.es

[2] Department of Signal Theory, Communications and Networking, University of Granada, 18060 Granada, Spain

[3] Andalusian Data Science and Computational Intelligence Institute (DaSCI), Granada, Spain

**Abstract.** Neuronal oscillations provide relevant information that helps to understand the neural mechanisms underlying cognitive processes and neural disorders. EEG and MEG methods record these brain oscillations and offer an invaluable insight into healthy and pathological brain function. These signals are helpful to study and achieve an objective and early diagnosis of neural disorders as Developmental Dyslexia (DD). DD early diagnosis is a challenging task that makes possible the application of individualized intervention tasks to dyslexic children in the early stages. In this work, we use EEG signals to explore the neural basis of DD and progress towards an early differential diagnosis. This is achieved by studying Cross-Frequency Coupling (CFC) mechanisms, such as the Phase-Amplitude Coupling (PAC). We apply a recent approach to infer CFC dynamics, the Holo-Hilbert Spectral Analysis (HHSA). This is a further step to overcome the limitations of current PAC methods. We pursue the HHSA over an EEG dataset from the Leeduca project. Then, Holo-Hilbert spectrums are used to explore the changes and patterns of PAC in DD. Finally, the discriminatory capability of Holo-Hilbert spectrums is validated by machine learning techniques.

**Keywords:** HHSA · PAC · EEG · Machine learning · Dyslexia diagnosis

## 1 Introduction

The study of neural oscillations in the human brain contributes to the understanding of the neural mechanisms behind brain pathologies and conditions. These brain oscillations are captured by EEG and MEG signals, providing a novel understanding of the importance of the neurophysiological relevance of


J. M. Ferrández Vicente et al. (Eds.): IWINAC 2022, LNCS 13258, pp. 43–52, 2022.
https://doi.org/10.1007/978-3-031-06242-1_5

functional connectivity in the brain [8,21]. The consequences and manifestation of many neurological disorders and diseases are usually the first discovered. However, the neural basis has yet to be confirmed in many cases. Furthermore, it would help to achieve an objective and early diagnosis, which is currently a challenging task. This is the situation of Developmental Dyslexia (DD). It is a learning disability [2,30] characterized by phonemic awareness and phonological processing difficulties, such as poor word decoding, spelling and reading fluency. It is not related to mental age or inadequate schooling and has a crucial social impact. DD is the most prevalent learning disability affecting between 5% and 12% of the population [28]. Most researchers consider that these difficulties have their origin in cognitive deficits in the phonological component of language [30]. Current research findings on the neural basis in DD are related to the entrainment phenomenon to the speech amplitude envelope in the auditory cortex [10,12,23]. The most prevalent hypothesis of this 'phonological deficit' is the Temporal Sampling Framework (TSF) by Goswami [9]. TSF joins difficulties in processing rise time of the amplitude envelope with impaired temporal sampling by low-frequency oscillatory mechanisms. Thus, accurate speech encoding is achieved by neural oscillations in the brain linked to amplitude patterns at different temporal rates [11].

One way to study these neural oscillations on the phonological processing in DD is through the PAC. It is a type of CFC, that describes interactions between different frequency bands. In particular, PAC characterizes the statistical dependence between the phase of a low-frequency brain oscillation and the amplitude of the high-frequency component across different spatial and temporal scales [4]. This is a good measure to reveal the functional connectivity in brain cognitive processes [4,17]. Various methods have been used to calculate the PAC; the most used are: Phase-Locking Value (PLV) [24], Modulation Index (MI) [31] and Mean Vector Length (MVL) [3]. However, there is no consensus to measure PAC [17]. These methods have limitations, such as the filter frequency bandwidth to extract the bands, and time window size used in the estimation process [7]. Furthermore, these techniques may lead to spurious or apparent CFC results related to non-stationarities and non-linearities in the neural oscillations that cannot be linked to underlying physiological interactions [1,5,18].

We can use decomposition methods that do not assume a linear and stationary nature for neural oscillations to deal with these limitations. This is the case of Empirical Modal Decomposition (EMD) [15], which is the base of the HHSA proposed by Huang et al. [14]. HHSA is a non-linear analysis tool that resolves the identification of intrinsic amplitude modulations [19]. HHSA has been applied in recent works for the study of neural oscillations in EEG signals: to analyze the modulation components from sensorimotor Mu rhythm induced by slow and fast-rate repetitive movements [13], in the steady-state visually evoked potential for study the human visual system's response to the envelope of amplitude-modulated signals [25], to characterize the CFC in the visual working memory [20], and a comparative of HHSA with other PAC measure methods in simulated and measured steady-state visual evoked potentials [19].

We apply HHSA on EEG signals from dyslexic subjects and normal readers. These signals were acquired during a passive task, where children listened to amplitude-modulated white noise. These auditory stimuli resemble the sampling processes performed in the brain for language processing. Then, we obtain Holo-Hilbert Spectrums (HHS) from these signals, where we directly quantify PAC in brain oscillations. Finally, we use these HHS to train a Support Vector Machine (SVM) algorithm and perform classification to find discriminant patterns in HHS, i.e., having different PAC functional connectivity for each group. The rest of the paper is as follows. In Sect. 2, the database and methods used are presented. Section 3 lists the principal results and Sect. 4, describes the conclusions.

## 2 Materials and Methods

The HHS can represent the interactions between neural oscillations, despite the non-linear and non-stationary characteristics of the EEG signals. The Leeduca research group provided the present experiment EEG data at the University of Málaga. The EEG dataset is obtained from 48 participants among dyslexic and control subjects. Participants were right-handed, native Spanish speakers. They had a normal or corrected-to-normal vision and no hearing impairment. All dyslexic children in this study had received a formal diagnosis of dyslexia at school. None of the skilled readers reported having reading or spelling difficulties or receiving a proper prior diagnosis of dyslexia. These subjects were matched in age. The control group (32 skilled readers) had a mean age of $94.1 \pm 3.3$ months. The dyslexic group (16 dyslexic readers) mean age was $95.6 \pm 2.9$ months. The experiment was conducted in the presence of each child's legal guardians and with their understanding and written consent.

EEG signals were acquired using 32 active electrodes (Brain-Products actiCAP). It was set at a sampling rate 500 Hz. These electrodes were located in the 10–20 standardized system. Participants underwent 15-minute sessions in which they were presented white noise auditory stimuli modulated at 4.8, 16, 40 Hz sequentially in ascending and descending order, for 2.5 min each. EEG signals were preprocessed using EEGLAB to remove artifacts related to eye blinking and impedance variations due to movement. Artifacts corresponding to eye blinking were removed by blind source separation using Independent Component Analysis (ICA). The signal from each channel was independently normalized to zero mean and unit variance. Then, it was referenced to the Cz electrode. Baseline correction was also applied. After the preprocessing, we have used software developed in *Python* to process the EEG signals, and the HHSA is based on the *EMD* package [29]. The general process to conduct the HHSA is represented in Fig. 1.

### 2.1 Holo-Hilbert Spectral Analysis

The PAC dynamics between the EEG frequency bands are revealed using Holo-Hilbert spectral analysis. The HHSA considers the non-linearities and non-

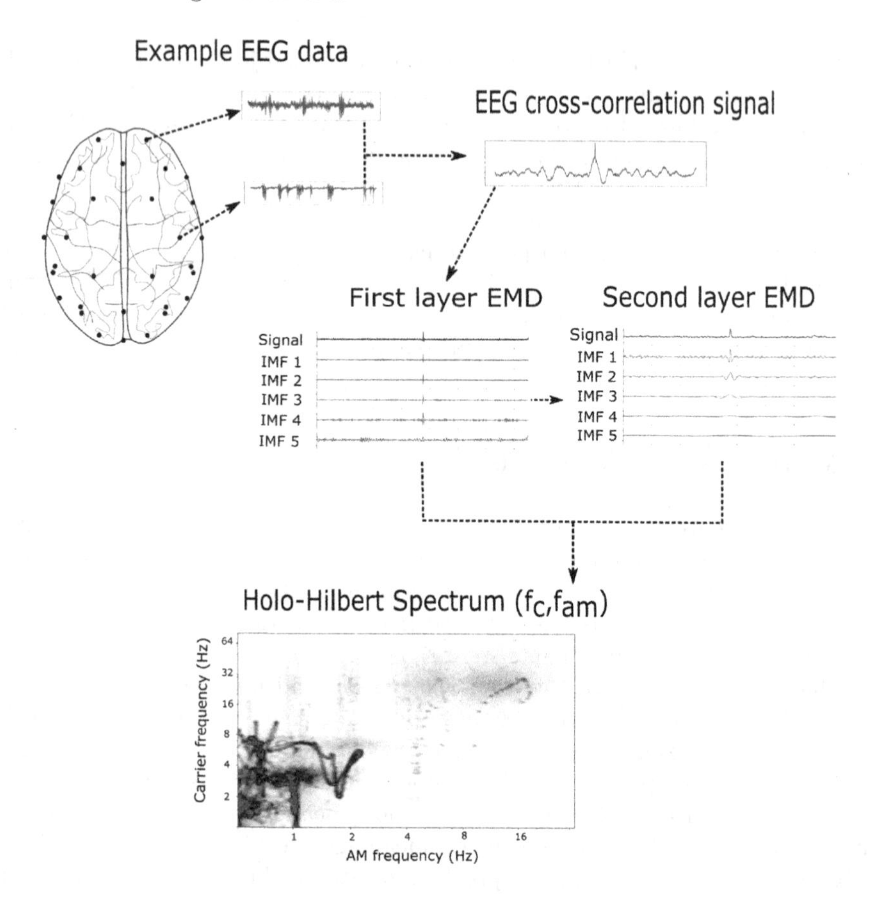


**Fig. 1.** Holo-Hilbert Spectral Analysis. First, the cross-correlation signal is calculated from two EEG signals. This is the signal in which the HHSA is applied. Then, the first layer IMFs are obtained by masking EMD. The second layer masking EMD is applied to the amplitude function of each first layer IMF. Finally, the information acquired by the HHSA is integrated into a spectral representation named the Holo-Hilbert spectrum.

stationarities of the data. It provides a complete information about all oscillatory processes in the EEG signals [14]: additive and multiplicative, intra-mode and inter-mode, stationary and non-stationary, linear and non-linear interactions. HHSA consists of a multiple-layer nested application of EMD [14,15]. We have used a two-nested layer EMD as in other recent works [19,20,25]. We have employed masking EMD [6,32], an enhanced algorithm to resolve the mode-mixing problem in EMD. The general process of HHSA is summarized as follows:

1. The first layer masking EMD is applied to EEG signals. Decomposing the original signal into a layer of several Intrinsic Mode Functions (IMF) [15]. Now,

the signal can be expressed as:

$$x(t) = \sum_{j=1}^{n} C_j(t) + r_n = \sum_{j=1}^{n} a_j(t)cos\theta_j(t) + r_n \quad (1)$$

where $x(t)$ is the original signal, $C_j(t)$ is the $j_{th}$ IMF, and $r_n$ is the trend free of oscillatory characteristics. Each IMF can be expressed as $a_j(t)cos\theta_j(t)$, where $a_j(t)$ is the amplitude function and $\theta_j(t)$ is the phase function. $\theta_j(t)$ is determined by using Normalized Hilbert Transform (NHT) [16].

2. The second layer EMD is implemented by applying masking EMD to the amplitude function $a_j(t)$ of each first layer IMF. The amplitude functions are decomposed into the second layer IMFs as

$$a_j(t) = \sum_{k=1}^{m} C_{jk}(t) + R_jm = \sum_{k=1}^{m} a_{jk}(t)cos\Theta_{jk}(t) + R_{jm} \quad (2)$$

where $a_{jk}(t)$ are the amplitude functions of the second layer IMFs, $C_{jk}(t)$. $\Theta_{jk}(t)$ is the second layer phase function, and $R_{jm}$ is the second layer trend. It is usual to refer to the instantaneous frequency obtained from the first layer EMD as the "carrier frequency" ($f_c$) and the second layer instantaneous frequency as the "amplitude modulated (AM) frequency" ($f_am$) [19,20].

The information acquired by the HHSA is integrated into a spectral representation named the Holo-Hilbert spectrum. This is a three-dimensional $(f_{am}, f_c, time)$ spectrum projected from the instantaneous frequencies and amplitudes of the two-layer IMFs. A common practice is to marginally sum the HHS over the time space [19]. This aids interpretability by obtaining two-dimensional HHS, where each axis corresponds to the carrier and AM frequencies.

### 2.2 Functional Connectivity and Classification

The HHSA describes the cross-frequency dynamics of neural oscillations [19]. In this work we have used the HHSA to overcome the limitations of existing methods to measure PAC. We have explored the functional connectivity between EEG channels by obtaining HHSs from the cross-correlation signal between the $j_{th}$ and the $i_{th}$ EEG signals. Where $j$ is the EEG channel of interest and $i$ is the EEG channel from other brain regions [13]. This way, we have $N \times N - 1$ HHSs for each subject, with $N = 31$ effective channels.

A feature pre-selection stage is applied over the resulting HHSs by means of Mann-Whitney U test, selecting the most significant parts in the HHSs with $p{-}value < 0.05$. These features are further used to train an SVM classifier. Thus, we converted all HHS obtained for a subject to a one-dimensional vector, kept the zones with differential information and use them to classify between controls and dyslexic subjects. Cross-validation is applied to evaluate the performance of the classification strategy, using a stratified k-fold cross-validation scheme with $k = 5$.

## 3 Results and Interpretation

In this section, we show the results of the HHSA for the exploratory analysis of the PAC functional connectivity in the 4.8 Hz stimulus. First, we present the classification results to demonstrate the discriminative capabilities of the HHS. Since HHSA is applied to the correlation signals of two EEG channels, we have N-1 HHS for each channel. These sets of HHS are used in an SVM to classify between controls and dyslexics. The performance is assessed with a cross validation scheme. Table 1 shows the metrics obtained, and Fig. 2 illustrates the ROC curve.

**Table 1.** Dyslexia classification results obtained in other works.

| Author | Sample size | Accuracy | Sensitivity | Specificity | AUC |
|---|---|---|---|---|---|
| Ortiz et al. [26] | 48 | 0.78 | 0.66 | 0.81 | 0.83 |
| Martinez-Murcia et al. [22] | 48 | 0.74 | 0.596 | 0.79 | 0.762 |
| Perera et al. [27] | 32 | 0.72 | 0.76 | 0.67 | 0.86 |
| Proposed | 48 | 0.74 | 0.75 | 0.73 | 0.84 |

Furthermore, in Table 1 we make a comparison with other dyslexia studies. These have a similar sample size and achieve accuracy and AUC in the same range as our proposed method. In [26] the accuracy obtained is higher, and [27] presents the best AUC. However, our method achieve a more balanced trade-off between sensitivity and specificity.

Finally, we outline the information contained in the Holo Hilbert spectra. Each HHS depicts the PAC between a channel of interest and another EEG channel. In Fig. 3 we present the results for the T8 electrode. It is located in the scalp part of the temporal lobe, very close to the superior temporal gyrus, which contains Brodmann areas 41, 42 and 22. The first two are parts of the primary auditory cortex and the latter is involved in speech processing. HHSs in Fig. 3 offer the average HHS for dyslexic and control groups, aiding the visual exploration. The PAC represented by HHSs shows greater activation in Theta and Delta frequency bands for each cross-correlation signal. We can observe that the patterns in HHS fluctuate in this range of frequency bands for dyslexic and control subjects. In some cases the activation is greater and in others it covers a larger area of the HHS. These fluctuations expose potentially differential information about PAC connectivity between each group.

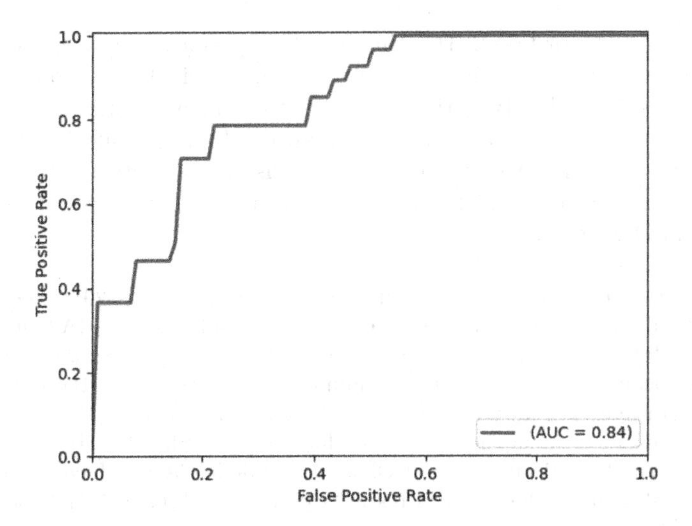


**Fig. 2.** ROC curve after stratified k-fold cross-validation with $k = 5$.

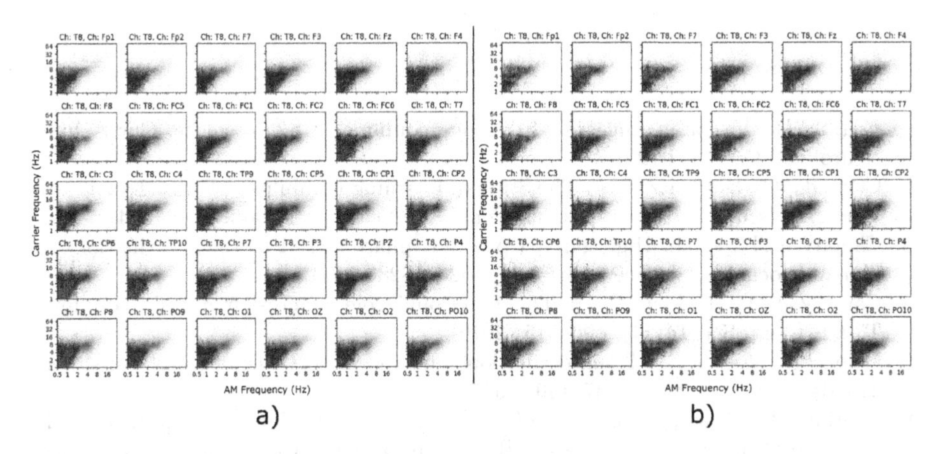


**Fig. 3.** Average Holo-Hilbert spectrums for the cross-correlation signals of EEG channel T8 with each other EEG channel. a) Controls. b) Dyslexic subjects.

## 4 Conclusions and Future Work

In this work we have performed a classification based on a novel method, HHSA, and SVM. The HHSA overcomes the limitations of current PAC methods. Thus, revealing information about the non-linear and non-sinusoidal features of oscillations in brain EEG signals. It provides a way to study cross-frequency interactions and the functional role of neural oscillations. The results of the HHSA are represented in Holo-Hilbert spectra, that show PAC dynamics free of spurious results in dyslexic and control subjects. The differential patterns in PAC are assessed with a classification performed using an SVM algorithm. Reaching an AUC up to 0.84 for the 4.8 Hz stimulus. This performance demonstrates the effectivity of our work in the differential diagnosis of DD. We plan to apply this method to the other two stimuli EEG data as future works. Improving the classification stage to increase the performance achieved would be worthwhile. We also aim to continue the exploratory analysis of PAC with the HHSA. Thus, extracting information about the neural implications of PAC from the HHSs in controls and DD subjects.

**Acknowledgments.** This work was supported by projects PGC2018-098813-B-C32 (Spanish "Ministerio de Ciencia, Innovación y Universidades"), UMA20-FEDERJA-086 and P18-RT-1624 (Consejería de economía y conocimiento, Junta de Andalucía), and by European Regional Development Funds (ERDF) as well as the BioSiP (TIC-251) research group. We gratefully acknowledge the support of NVIDIA Corporation with the donation of one of the GPUs used for this research. Work by F.J.M.M. was supported by the IJC2019-038835-I MICINN "Juan de la Cierva - Incorporación" Fellowship. We also thank the Leeduca research group and Junta de Andalucía for the data supplied and support.

## References

1. Aru, J., et al.: Untangling cross-frequency coupling in neuroscience. Curr. Opin. Neurobiol. **31**, 51–61 (2015)
2. Cortiella, C., Horowitz, S.H.: The State of Learning Disabilities: Facts, Trends and Emerging Issues, 3rd edn. National Center for Learning Disabilities (2014)
3. Canolty, R.T., et al.: High gamma power is phase-locked to theta oscillations in human neocortex. Science **313**(5793), 1626–1628 (2006)
4. Canolty, R.T., Knight, R.T.: The functional role of cross-frequency coupling. Trends Cogn. Sci. **14**(11), 506–515 (2010)
5. Cole, S.R., Voytek, B.: Brain oscillations and the importance of waveform shape. Trends Cogn. Sci. **21**(2), 137–149 (2017)
6. Deering, R., Kaiser, J.: The use of a masking signal to improve empirical mode decomposition. In: Proceedings of IEEE International Conference on Acoustics, Speech, and Signal Processing, ICASSP 2005, vol. 4, pp. iv/485–iv/488 (March 2005)
7. Dvorak, D., Fenton, A.A.: Toward a proper estimation of phase–amplitude coupling in neural oscillations. J. Neurosci. Methods **225**, 42–56 (2014)
8. Górriz, J.M., Ramírez, J., Ortíz, A., Martínez-Murcia, F.J., Segovia, F., Suckling, J., et al.: Artificial intelligence within the interplay between natural and artificial computation: advances in data science, trends and applications. Neurocomputing **410**, 237–270 (2020)
9. Goswami, U.: A temporal sampling framework for developmental dyslexia. Trends Cogn. Sci. **15**(1), 3–10 (2011)
10. Goswami, U.: A neural basis for phonological awareness? An oscillatory temporal-sampling perspective. Curr. Dir. Psychol. Sci. **27**(1), 56–63 (2018)

11. Goswami, U.: A neural oscillations perspective on phonological development and phonological processing in developmental dyslexia. Lang. Linguist. Compass **13**(5), e12328 (2019)
12. Goswami, U., Huss, M., Mead, N., Fosker, T.: Auditory sensory processing and phonological development in high IQ and exceptional readers, typically developing readers, and children with dyslexia: a longitudinal study. Child Dev. **92**, 1083–1098 (2020)
13. Hsu, H.T., Lee, W.K., Shyu, K.K., Yeh, T.K., Chang, C.Y., Lee, P.L.: Analyses of EEG oscillatory activities during slow and fast repetitive movements using Holo-Hilbert spectral analysis. IEEE Trans. Neural Syst. Rehabil. Eng. **26**(9), 1659–1668 (2018)
14. Huang, N.E., et al.: On Holo-Hilbert spectral analysis: a full informational spectral representation for nonlinear and non-stationary data. Philos. Trans. R. Soc. A: Math. Phys. Eng. Sci. **374**(2065), 20150206 (2016)
15. Huang, N.E., et al.: The empirical mode decomposition and the Hilbert spectrum for nonlinear and non-stationary time series analysis. Proc. R. Soc. Lond. Ser. A: Math. Phys. Eng. Sci. **454**, 903–995 (1998)
16. Huang, N.E., Wu, Z., Long, S.R., Arnold, K.C., Chen, X., Blank, K.: On instantaneous frequency. Adv. Adapt. Data Anal. **01**(02), 177–229 (2009)
17. Hülsemann, M.J., Naumann, E., Rasch, B.: Quantification of phase-amplitude coupling in neuronal oscillations: comparison of phase-locking value, mean vector length, modulation index, and generalized-linear-modeling-cross-frequency-coupling. Front. Neurosci. **13**, 573 (2019)
18. Hyafil, A.: Misidentifications of specific forms of cross-frequency coupling: three warnings. Front. Neurosci. **9**, 370 (2015)
19. Juan, C.H., et al.: Revealing the dynamic nature of amplitude modulated neural entrainment with Holo-Hilbert spectral analysis. Front. Neurosci. **15**, 977 (2021)
20. Liang, W.K., Tseng, P., Yeh, J.R., Huang, N.E., Juan, C.H.: Frontoparietal beta amplitude modulation and its interareal cross-frequency coupling in visual working memory. Neuroscience **460**, 69–87 (2021)
21. Lopes da Silva, F.: EEG and MEG: relevance to neuroscience. Neuron **80**(5), 1112–1128 (2013)
22. Martinez-Murcia, F.J., et al.: EEG connectivity analysis using denoising autoencoders for the detection of dyslexia. Int. J. Neural Syst. **30**(07), 2050037 (2020)
23. Molinaro, N., Lizarazu, M., Lallier, M., Bourguignon, M., Carreiras, M.: Out-of-synchrony speech entrainment in developmental dyslexia. Hum. Brain Mapp. **37**(8), 2767–2783 (2016)
24. Mormann, F., et al.: Phase/amplitude reset and theta–gamma interaction in the human medial temporal lobe during a continuous word recognition memory task. Hippocampus **15**(7), 890–900 (2005)
25. Nguyen, K.T., et al.: Unraveling nonlinear electrophysiologic processes in the human visual system with full dimension spectral analysis. Sci. Rep. **9**(1), 16919 (2019)
26. Ortiz, A., López, P.J., Luque, J.L., Martínez-Murcia, F.J., Aquino-Britez, D.A., Ortega, J.: An anomaly detection approach for dyslexia diagnosis using EEG signals. In: Ferrández Vicente, J.M., Álvarez-Sánchez, J.R., de la Paz López, F., Toledo Moreo, J., Adeli, H. (eds.) IWINAC 2019, Part I. LNCS, vol. 11486, pp. 369–378. Springer, Cham (2019). https://doi.org/10.1007/978-3-030-19591-5_38
27. Perera, H., Shiratuddin, M.F., Wong, K.W., Fullarton, K.: EEG signal analysis of writing and typing between adults with dyslexia and normal controls. Int. J. Interact. Multimed. Artif. Intell. **5**(1), 62 (2018)

28. Peterson, R.L., Pennington, B.F.: Developmental dyslexia. The Lancet **379**(9830), 1997–2007 (2012)
29. Quinn, A.J., Lopes-dos-Santos, V., Dupret, D., Nobre, A.C., Woolrich, M.W.: EMD: empirical mode decomposition and Hilbert-Huang spectral analyses in Python. J. Open Source Softw. **6**(59), 2977 (2021)
30. Thambirajah, M.S.: Developmental dyslexia: an overview. Adv. Psychiatr. Treat. **16**(4), 299–307 (2010)
31. Tort, A.B.L., et al.: Dynamic cross-frequency couplings of local field potential oscillations in rat striatum and hippocampus during performance of a T-maze task. Proc. Natl. Acad. Sci. **105**(51), 20517–20522 (2008)
32. Tsai, F.F., Fan, S.Z., Lin, Y.S., Huang, N.E., Yeh, J.R.: Investigating power density and the degree of nonlinearity in intrinsic components of anesthesia EEG by the Hilbert-Huang transform: an example using ketamine and alfentanil. PLOS ONE **11**(12), e0168108 (2016)

# Inter-channel Granger Causality for Estimating EEG Phase Connectivity Patterns in Dyslexia

Ignacio Rodríguez-Rodríguez[1,3(✉)], A. Ortiz[1,3], Marco A. Formoso[1,3], Nicolás J. Gallego-Molina[1,3], and J. L. Luque[2]

[1] Communications Engineering Department, University of Málaga, 29004 Málaga, Spain
ignacio.rodriguez@ic.uma.es

[2] Department of Basic Psychology, University of Málaga, 29019 Málaga, Spain

[3] Andalusian Data Science and Computational Intelligence Institute (DaSCI), Granada, Spain

**Abstract.** Methods like Electroencephalography (EEG) and magnetoencephalogram (MEG) record brain oscillations and provide an invaluable insight into healthy and pathological brain function. These signals are helpful to study and achieve an objective and early diagnosis of neural disorders as Developmental Dyslexia (DD). An atypical oscillatory sampling could cause the characteristic phonological difficulties of dyslexia at one or more temporal rates; in this sense, measuring the EEG signal can help to make an early diagnosis of DD. The LEEDUCA study conducted a series of EEG experiments on children listening to amplitude modulated (AM) noise with slow-rhythmic prosodic (0.5–1 Hz) to detect differences in perception of oscillatory sampling that could be associated with dyslexia. The evolution of each EEG channel has been studied in the frequency domain, obtaining the analytical phase using the Hilbert transform. Subsequently, the cause-effect relationships between channels in each subject have been reflected thanks to Granger causality, obtaining matrices that reflect the interaction between the different parts of the brain. Hence, each subject was classified as belonging or not to the control group or the experimental group. For this purpose, two ensemble classification algorithms were compared, showing that both can reach acceptable classifying performance in delta band with an accuracy up to 0.77, recall of 0.91 and AUC of 0.97 using Gradient Boosting classifier.

**Keywords:** EEG · Dyslexia · Granger causality · Hilbert · Ada Boost · Gradient Boosting

## 1 Introduction

Developmental dyslexia (DD) refers to a learning difficulty that hampers learners' reading skills acquisition irrespective of their mental age or level of schooling. It generally results in mild to severe reading difficulties and can cause unintelligible hand-writing, letter migration within or between words, and frequent misspelling of words. Depending on the type of test battery employed, 5% to 12% of any given population can be


J. M. Ferrández Vicente et al. (Eds.): IWINAC 2022, LNCS 13258, pp. 53–62, 2022.
https://doi.org/10.1007/978-3-031-06242-1_6

affected by DD [17]. Recent advances in the field of artificial intelligence have been widely taken up in the field of neuroscience [11].

While diagnosing DD generally relies on behavioral tests that assess learners' reading and writing skills, these tests may be subject to exogenous influences, such as the learner's attitude or state of mind, causing flawed diagnoses on a fundamental level [22]. Accordingly, there is an urgent need for the development of additional objective metrics that can provide more precise diagnoses at an earlier stage. Numerous techniques for the collection of functional brain data have demonstrated their usefulness in the field of neuroscience, including functional magnetic resonance imaging (fMRI) [3], magnetoencephalography (MEG) and the more recent functional near-infrared spectroscopy (fNIRS). However, electroencephalography (EEG) remains the most common and cost-effective detector of cortical brain activity while also offering enhanced temporal resolution.

Recent research has highlighted a potential biological mechanism underlying DD, with newly developed models pointing to atypical dominant neural entrainment situated in the brain's right hemisphere [4] for three main rhythm categories, namely at the slow rhythmic prosodic (0.5–1 Hz), syllabic (4–8 Hz) and phoneme (12–40 Hz) levels [6]. According to this stream of research, learners with DD perform atypical oscillatory sampling using at least one temporal rate, introducing phonological difficulties in the comprehension of certain linguistic units, e.g. phonemes or syllables.

As not all frequency bands (i.e. Delta, Theta, Alpha, Beta, and Gamma) equally experience this phenomenon, a separate investigation of these bands via EEG in this context may represent a valuable contribution to the DD research field. Hence, prior studies have explored the fundamental underpinnings of DD using either EEG or MEG and employing speech-based stimuli, assuming that DD grounds in a diminished awareness of the individual units of speech [14]. For example, [18] revealed that the differences between the dyslexic group and the control group are located in the preferred entrainment phase of the Delta band following the visual and auditory stimulus.

In light of this, the field of neuroscience is breaking new ground with the advent of connectivity analysis [21], referring to the analysis of measures linking two signals that have been acquired through separate channels, e.g. correlation, covariance and causality. Using these parameters in the context of brain signals emanating from different regions provides an inkling of the underlying neural network, thereby upholding the hyper-connected model of the brain. In the context of DD, EEG is frequently drawn upon to examine the brain's functional network connectivity and organization. Functional connectivity here refers to when activity across different brain areas is coordinated when the participant is focused on performing a task or remains at rest. Previous studies have developed a variety of EEG-based techniques to measure functional connectivity, especially for the identification of the patterns of neurological conditions, e.g. Parkinson's disease [2]. Meanwhile, in the field of cognitive neuroscience, brain connectivity has been utilized to pinpoint which areas of the brain are active during learning and language acquisition [20]. Measures of spectral connectivity can be deduced from variations in the frequency, phase, and power spectrum. In line with this, techniques have been developed to determine how electrodes lying on the same frequency band relate from a statistical point of view [25].

Estimating two channels' connectivity involves the analysis of each of these channels' phases separately. When a signal is extracted from an electrode, its phase changes over time, and thus the phase, $\phi(t)$, must be computed for each channel i, which is the instantaneous phase. Hereby, the instantaneous phase is captured via a Hilbert transform, calculated based on the band-pass filtered signals.

This method, as outlined above, enable us to determine the phase value for each time point, facilitating the estimation of the inter-channel correlation and causality. For example, [15] employed it to identify changes in epileptic patients' phase synchronization, evidencing that characteristic synchronization changes often precede epileptic seizures. Subsequently, the inter-channel connectivity is determined through the estimation of the cause-effect relationship. The statistical hypothesis test known as the Granger causality test reveals whether a time series is a factor, thereby helping to predict the characteristics of further time series. Although it was first introduced in the field of economics in the 1980s, Granger causality has shown to be highly useful in numerous other areas, including neuroscience [5]. In relation to EEG it has, for example, contributed to research on emotion recognition [9], the brain's response to the stimulation of the vagus nerve [24], and the perception of pain [23].

Based on the prior characterization, this study draws on machine learning classification algorithms to estimate whether DD can be identified based on a subject's Granger causality matrices. Machine learning (ML) techniques are eminently suited to this type of investigation due to the nature and characteristics of EEG signal classification and the complexity of the problem posed here [12].

In summary, this work aims to show how low-level auditory processing initiates different patterns of connectivity in the involved brain networks. It examines this connectivity by applying Granger causality to the phase synchronization observed between the EEG channels.

## 2 Material and Methods

### 2.1 Data Acquisition

This study utilizes EEG data from the University of Málaga's Leeduca Study Group [16]. These data were taken from 97 participants, comprising 67 skilled readers and 30 dyslexic readers, who were age-matched ($t(1) = -1.4, p > 0.05$, age range: 88–100 months). The study obtained the understanding and written consent of the participants' legal guardians, who were also present at all times during the experiment. The participants were each presented with an auditory stimulus during a 15-min session. Specific stimuli comprising band-limited white noise, with amplitude modulation at different frequencies, were used to explore the synchronicity patterns induced by low-level auditory processing. A Brainvision actiCHamp Plus with 32 active electrodes (actiCAP, Brain Products GmbH, Germany) was used to capture the EEG signals at a sampling rate 500 Hz. The participants were presented with the stimuli modulated 4 Hz (prosodic frequency) for 2.5 min at a time. The choice of stimulus was made in accordance with expert linguistic psychologists knowledgeable of the main frequency components present in the human voice that correspond to words. The participants were all righthanded native Spanish speakers without hearing impairment and with normal or corrected-to-normal vision. All children in the dyslexic group had received a formal diagnosis of dyslexia during their schooling. None of the children in the skilled reader group reported difficulties with reading or spelling, nor had any previously received a formal diagnosis of dyslexia. We used the 10–20 standardized system when choosing the locations of the 32 electrodes.

### 2.2 Preprocessing

First, the EEG signals were preprocessed to eliminate all artefacts due to eye-blinking as well as all movement-related impedance variation. Independent component analysis (ICA) [13] based on eye movements recorded through the EOG channel was used to remove ocular artefacts, while movement or noise-related artefacts were removed by eliminating the relevant EEG segments. Subsequently, the channels were referenced to the Cz channel. Next, the EEG channels were put through a band-pass filter, aiming to extract in-formation in line with the five EEG frequency bands (Delta, 1.5–4 Hz; Theta, 4–8 Hz, Alpha, 8–13 Hz; Beta, 13–30 Hz; and Gamma, 30–80 Hz). Notably, as the signals' phase is of interest to this research, it was not possible to use filtering processes, e.g. Infinite Impulse Response (IIR) filters, that would trigger phase distortion. Thus, Finite Impulse Response (FIR) filters were used as these produce a constant phase lag, which can be later subjected to correction. In concrete terms, the two-way zero-phase lag band-pass FIR Least-Squares filter used here passes the signal forward and backward through the filter, achieving a zero-lag phase in the overall filtering process and thereby compensating for the produced phase lag [19]. We used low-pass filtering with a threshold 80 Hz, adding 50 Hz notch filter in the preprocessing stage to remove this frequency component.

### 2.3 Hilbert Transform

Through a Hilbert transform, a real signal can be transformed into an analytic signal, referring to a complex-valued time series with no negative frequency components. This facilitates the computation of the time-varying amplitude, phase and frequency from the analytic signal, respectively termed the instantaneous amplitude, phase and frequency.

Hilbert Transform (HT) is defined for a signal $x(t)$:

$$\mathcal{H}\left[x(t)\right] = \frac{1}{\pi} \int_{-\infty}^{+\infty} \frac{x(t)}{t-\tau} d\tau \tag{1}$$

and the analytic signal $z_i(t)$ for a signal $x(t)$ can be obtained as

$$z_i\left(t\right) = x_i\left(t\right) + j\mathcal{H}x_i\left(t\right) = a(t)e^(j\phi(t)) \tag{2}$$

From $z_i(t)$, it is straightforward to compute the instantaneous amplitude as

$$a(t) = \sqrt{re(z_i(t))^2 + im(z_i(t))^2} \tag{3}$$

and the instantaneous, unwrapped phase is

$$\phi\left(t\right) = tan^{-1}\frac{im\left(z_i\left(t\right)\right)}{re\left(z_i\left(t\right)\right)} \tag{4}$$

Using the method described above delivers the phase value for each time point, facilitating the estimation of the inter-channel synchronization based on the phase variation.

### 2.4 Granger Causality Test

First developed by econometrician Clive Granger, the concept of Granger causality [10] is frequently used in research to describe the causal interactions that appear among continuous-valued time series. Grounding in a statistical hypothesis test, it fundamentally asserts that "the past and present may cause the future, but the future cannot cause the past". It further states that knowledge of a cause will inform the prediction of the cause's future effects to an extent beyond that of auto-regression. In more concrete terms, the variable $x$ will Granger-cause $y$ if the auto-regression for $y$ based on past $x$ and $y$ values shows significantly greater accuracy than one using solely past $y$ values. As an example, if $x_t$ and $y_t$ are two stationary time-series sequences, then $x_{t-k}$ and $y_{t-k}$ are, respectively, the past $k$ values of $x_t$ and $y_t$. We then perform Granger causality using the following two regressions:

$$\widehat{y}_{t1} = \sum_{k=1}^{l} a_k y_{t-k} + \varepsilon_t \tag{5}$$

$$\widehat{y}_{t2} = \sum_{k=1}^{l} a_k y_{t-k} + \sum_{k=1}^{w} b_k x_{t-k} + \eta_t \tag{6}$$

where $\widehat{y}_{t1}$ and $\widehat{y}_{t2}$ represent the fitting values by the first and second regression, respectively; l and w are, respectively, the maximum numbers of the lagged observations of $x_t$ and $y_t$, , $a_k$; $b_k \in \mathbb{R}$ are regression coefficient vectors estimated by least squares; and $\varepsilon_t$, $\eta_t$ are white noises (prediction errors). Notably, $w$ can be infinite, however, due to the finite nature of the data given, $w$ is here considered finite and given a length significantly below that of the time series, as estimated using model selection, e.g. the Akaike information criterion (AIC) [1]. Subsequently, we apply the F-test to receive a p-value showing whether Eq. (5) produces an improved regression model compared to Eq. (6) with statistical significance. If this is the case, we may state that $x$ Granger-causes $y$.

### 2.5 Machine Learning Classification

Ensemble methods produce multiple models, thereafter integrating them for better results. Weighted majority voting (or sum) is used to combine their predictions to derive a final prediction. Each boosting iteration involves data modifications encompassing the application of $w_1, w_2, \ldots, w_n$ to each training sample. At the onset, the weights are set to $w_i = 1/N$, meaning the first step essentially uses the original data to train a weak learner. The sample weights are modified separately, with the learning algorithm being reapplied to the reweighted data, at each successive iteration. For each step, the incorrectly predicted training examples as per the boosted model in the previous step receive increased weights, while those that were correctly predicted receive decreased weights. Thus, over multiple iterations, difficult-to-predict examples increasingly gain influence, forcing the following weak learners to emphasize the examples that the prior ones missed. In general, more accurate solutions are delivered by ensemble methods compared to single models, and they excel at enhancing binary prediction for small data sets. This study utilizes the sklearn library in Python to compare an Ada Boost classifier with a Gradient Boosting classifier. The Ada Boost [7] classifier - a meta-estimator - first fits a classifier to the data and subsequently fits further copies of the classifier to the same data, whereby the weights of the incorrectly classified instances are modified to make the following classifiers concentrate on problematic ones. Meanwhile, the Gradient Boosting classifier [8] produces an additive model using a forward stage-wise construction, permitting arbitrary differentiable loss function optimization. For each stage, n regression trees are fit to the multinomial or deviance binomial loss function's negative gradient. In the special case of binary classification, a single regression tree is used. In order to find the better set of parameters, we used the GridSearchCV library was employed, using a cross-validation method with 20 folds and a parameters grid. These grids are exposed in Table 1.

**Table 1.** Parameters grid of machine learning classifiers.

| Algorithm | Parameter | Range |
|---|---|---|
| Ada Boost | n_estimators | 1 to 25 |
| | Learning rate | 1 to 3.5 |
| | Boosting algorithm | SAMME, SAMME.R |
| Gradient Boosting | N_estimators | 1 to 10 |
| | Loss | deviance, exponential |
| | Learning rate | 0.05 to 0.5 |
| | Criterion | friedman_mse, squared_error, mse, mae |
| | Min_samples_split | 0.1 to 3 |
| | Min_samples_leaf | 0.1 to 3 |
| | Max_depth | 1 to 4 |

## 3 Results

The methodological process described above first obtains the signal of each EEG channel transformed to its analytical phase. With this, the Granger causality matrices are calculated and can be the input of the classification algorithms (Fig. 1). Two algorithms have been compared, obtaining the best combination of parameters for each frequency band. The chosen parameter set is shown in Tables 2 and 3.

The results (Table 4) show that the best results are obtained in the beta band, in general, reaching an accuracy of 0.79 with Gradient Boost. With this algorithm the precision and recall results are also good, with an AUC of 0.97. In this band Ada Boost

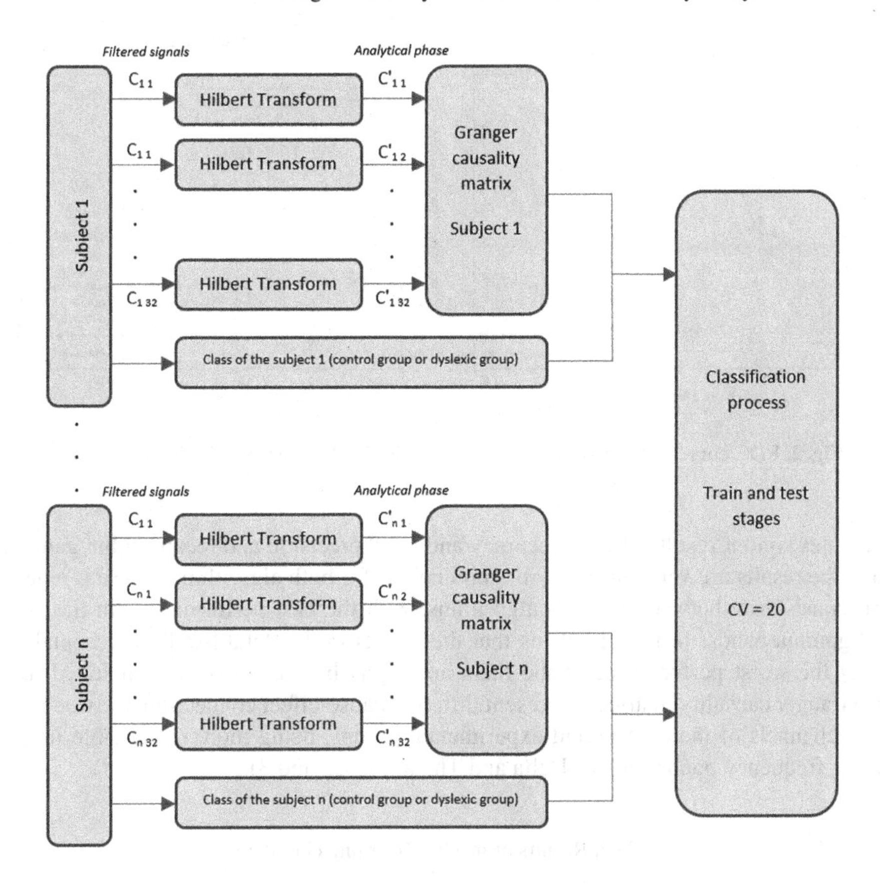


**Fig. 1.** Methodology schema

**Table 2.** Best parameter set for Ada Boosting

| Band | Algorithm | Learning rate | Estimators |
|---|---|---|---|
| Delta | Samme | 1 | 18 |
| Theta | Samme R | 1.75 | 4 |
| Alpha | Samme R | 1.75 | 16 |
| Beta | Samme R | 2.5 | 3 |
| Gamma | Samme | 1 | 13 |

**Table 3.** Best parameter set for Gradient Boosting

| Band | Criterion | Learning rate | Loss | Max depth | Max features | Min samples leaf | Min samples split | Estimators |
|---|---|---|---|---|---|---|---|---|
| Delta | friedman_mse | 0.5 | deviance | 1 | auto | 0.1 | 0.1 | 1 |
| Theta | friedman_mse | 0.5 | exponential | 2 | log2 | 0.1 | 0.1 | 3 |
| Alpha | friedman_mse | 0.5 | exponential | 4 | sqrt | 0.1 | 0.1 | 9 |
| Beta | mae | 0.1 | deviance | 2 | auto | 0.1 | 0.1 | 9 |
| Gamma | friedman_mse | 0.5 | deviance | 1 | auto | 0.1 | 0.1 | 9 |

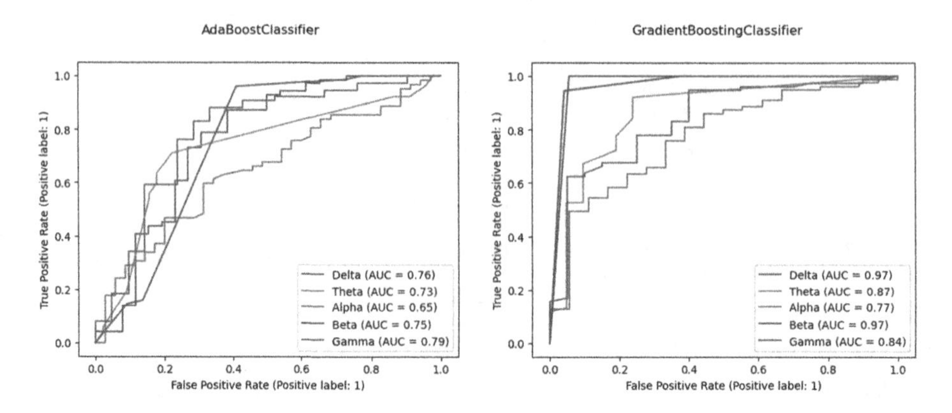


**Fig. 2.** ROC curves (Ada Boost)

**Fig. 3.** ROC curves (Gradient Boosting)

generates similar results of 0.77 accuracy and good precision and recall. In the gamma band the results are very similar, with good results for both algorithms. There is a general consistency between the two algorithms, with the best performances in the beta and gamma bands. In both it seems that the results in the delta band are acceptable, being the worst performance in the theta and alpha bands. It can be concluded that the Granger causality matrices represent different cause-effect connections between the EEG channels of the control and experimental groups, being more remarkable in the higher frequency bands such as Delta and Theta (Figs. 2 and 3).

**Table 4.** Results of machine learning classifiers

| Algorithm | Band | Accuracy | Precision | Recall | AUC |
|---|---|---|---|---|---|
| Ada Boost | Delta | $0.75 \pm 0.16$ | $0.86 \pm 0.15$ | $0.85 \pm 0.16$ | 0.76 |
| | Theta | $0.61 \pm 0.26$ | $0.83 \pm 0.12$ | $0.62 \pm 0.29$ | 0.73 |
| | Alpha | $0.61 \pm 0.17$ | $0.83 \pm 0.14$ | $0.68 \pm 0.21$ | 0.65 |
| | Beta | $0.77 \pm 0.15$ | $0.84 \pm 0.14$ | $0.89 \pm 0.18$ | 0.75 |
| | Gamma | $0.78 \pm 0.18$ | $0.91 \pm 0.09$ | $0.90 \pm 0.17$ | 0.79 |
| Gradient Boost | Delta | $0.77 \pm 0.19$ | $0.80 \pm 0.17$ | $0.91 \pm 0.14$ | 0.97 |
| | Theta | $0.70 \pm 0.22$ | $0.80 \pm 0.15$ | $0.87 \pm 0.22$ | 0.87 |
| | Alpha | $0.73 \pm 0.16$ | $0.85 \pm 0.13$ | $0.89 \pm 0.14$ | 0.77 |
| | Beta | $0.79 \pm 0.17$ | $0.84 \pm 0.14$ | $0.89 \pm 0.19$ | 0.97 |
| | Gamma | $0.74 \pm 0.17$ | $0.90 \pm 0.11$ | $0.89 \pm 0.17$ | 0.84 |

## 4 Conclusions and Future Work

With the methodology described above, it can be concluded that the causality matrices between channels allow the classification of subjects with and without DD. The results indicate that there is a distinction in the causality matrices of some subjects and others: this implies that the interconnection of the different zones is not the same in subjects with DD.

This methodology is not only relevant for diagnostic purposes but also for the study of the differences in the brain processes developed during auditory processing in controls and dyslexic children, paving the way to go deeper into the biological basis of dyslexia.

The delta band seems to be where these differences are most accentuated, and allows a clearer classification. Thus, a precision of 0.77 and AUC of 0.97 are achieved using the Gradient Boosting classifier. However, the performance ratios are also acceptable in the beta and gamma bands, which indicates that at these frequencies a different interconnection of the brain areas is also taking place. It is noteworthy that the results are consistent with the two classifiers used, which further supports the idea that the most influential bands in DD are those identified.

We intend to test the validity of the proposed method with two other stimuli, corresponding to syllabic (4–8 Hz) and phoneme (12–40 Hz) levels. This will give us an idea of the different brain interconnections at different frequencies.

**Acknowledgments.** This work was supported by projects PGC2018-098813-B-C32 & RTI2018-098913-B100 (Spanish "Ministerio de Ciencia, Innovaci6n y Universidades"), P18-RT-1624, UMA20-FEDERJA-086, CV20-45250, A-TIC-080-UGR18 and P20 00525 (Consejería de econnomía y conocimiento, Junta de Andalucía) and by European Regional Development Funds (ERDF). M.A. Formoso work was supported by Grant PRE2019-087350 funded by MCIN/AEI/10.13039/501100011033 by "ESF Investing in your future". Work of J.E. Arco was supported by Ministerio de Universidades, Gobierno de España through grant "Margarita Salas".

## References


1. Akaike, H.: A new look at the statistical model identification. IEEE Trans. Autom. Control **19**(6), 716–723 (1974)
2. Chaturvedi, M., et al.: Phase lag index and spectral power as QEEG features for identification of patients with mild cognitive impairment in Parkinson's disease. Clin. Neurophysiol. **130**(10), 1937–1944 (2019)
3. Clark, K.A., et al.: Neuroanatomical precursors of dyslexia identified from pre-reading through to age 11. Brain **137**(12), 3136–3141 (2014)
4. Di Liberto, G.M., Peter, V., Kalashnikova, M., Goswami, U., Burnham, D., Lalor, E.C.: Atypical cortical entrainment to speech in the right hemisphere underpins phonemic deficits in dyslexia. Neuroimage **175**, 70–79 (2018)
5. Ding, M., Chen, Y., Bressler, S.L.: 17 Granger causality basic theory and application to neuroscience. In: Handbook of Time Series Analysis Recent Theoretical Developments and Applications, vol. 437 (2006)
6. Flanagan, S., Goswami, U.: The role of phase synchronisation between low frequency amplitude modulations in child phonology and morphology speech tasks. J. Acoust. Soc. Am. **143**(3), 1366–1375 (2018)
7. Freund, Y., Schapire, R.E.: A decision-theoretic generalization of on-line learning and an application to boosting. J. Comput. Syst. Sci. **55**(1), 119–139 (1997)
8. Friedman, J.H.: Stochastic gradient boosting. Comput. Stat. Data Anal. **38**(4), 367–378 (2002)

9. Gao, Y., Wang, X., Potter, T., Zhang, J., Zhang, Y.: Single-trial EEG emotion recognition using Granger Causality Transfer entropy analysis. J. Neurosci. Methods **346**, 108904 (2020)
10. Granger, C.W.: Investigating causal relations by econometric models and cross-spectral methods. Econometrica J. Econome Soc. **37**(3), 424–438 (1969)
11. Górriz, J.M., Ramírez, J., Ortíz, A., Martínez-Murcia, F.J., et al.: Artificial intelligence within the interplay between natural and artificial computation: advances in data science, trends and applications. Neurocomputing **410**, 237–270 (2020). https://doi.org/10.1016/j.neucom.2020.05.078. https://www.sciencedirect.com/science/article/pii/S0925231220309292
12. LeCun, Y., Bengio, Y., Hinton, G.: Deep learning. Nature **521**(7553), 436–444 (2015)
13. Li, R., Principe, J.C.: Blinking artifact removal in cognitive EEG data using ICA. In: 2006 International Conference of the IEEE Engineering in Medicine and Biology Society, pp. 5273–5276. IEEE (2006)
14. Molinaro, N., Lizarazu, M., Lallier, M., Bourguignon, M., Carreiras, M.: Out-of-synchrony speech entrainment in developmental dyslexia. Hum. Brain Mapp. **37**(8), 2767–2783 (2016)
15. Mormann, F., Lehnertz, K., David, P., Elger, C.E.: Mean phase coherence as a measure for phase synchronization and its application to the EEG of epilepsy patients. Physica D Nonlinear Phenomena **144**(3–4), 358–369 (2000)
16. Ortiz, A., Martinez-Murcia, F.J., Luque, J.L., Gim'enez, A., Morales-Ortega, R., Ortega, J.: Dyslexia diagnosis by EEG temporal and spectral descriptors an anomaly detection approach. Int. J. Neural Syst. **30**(07), 2050029 (2020)
17. Peterson, R.L., Pennington, B.F.: Developmental dyslexia. Lancet **379**(9830), 1997–2007 (2012)
18. Power, A.J., Mead, N., Barnes, L., Goswami, U.: Neural entrainment to rhythmic speech in children with developmental dyslexia. Front. Hum. Neurosci. **7**, 777 (2013)
19. Robertson, D.G.E., Dowling, J.J.: Design and responses of Butterworth and critically damped digital filters. J. Electromyogr. Kinesiol. **13**(6), 569–573 (2003)
20. Romeo, R.R., et al.: Language exposure relates to structural neural connectivity in childhood. J. Neurosci. **38**(36), 7870–7877 (2018)
21. Schmidt, C., Piper, D., Pester, B., Mierau, A., Witte, H.: Tracking the reorganization of module structure in time-varying weighted brain functional connectivity networks. Int. J. Neural Syst. **28**(04), 1750051 (2018)
22. Thompson, P.A., Hulme, C., Nash, H.M., Gooch, D., Hayiou-Thomas, E., Snowling, M.J.: Developmental dyslexia predicting individual risk. J. Child Psychol. Psychiatry **56**(9), 976–987 (2015)
23. Tripanpitak, K., He, S., Sonmezicsik, I., Morant, T., Huang, S.Y., Yu, W.: Granger causality-based pain classification using EEG evoked by electrical stimulation targeting nociceptive Aδ and C fibers. IEEE Access **9**, 10089–10106 (2021)
24. Uchida, T., Fujiwara, K., Inoue, T., Maruta, Y., Kano, M., Suzuki, M.: Analysis of VNS effect on EEG connectivity with granger causality and graph theory. In: 2018 Asia-Pacific Signal and Information Processing Association Annual Summit and Conference (APSIPA ASC), pp. 861–864. IEEE (2018)
25. Unde, S.A., Shriram, R.: Coherence analysis of EEG signal using power spectral density. In: 2014 Fourth International Conference on Communication Systems and Network Technologies, pp. 871–874. IEEE (2014)

# Automatic Diagnosis of Schizophrenia in EEG Signals Using Functional Connectivity Features and CNN-LSTM Model

Afshin Shoeibi[1(✉)], Mitra Rezaei[2], Navid Ghassemi[3], Zahra Namadchian[4], Assef Zare[4], and Juan M. Gorriz[5]

[1] Faculty of Electrical Engineering, FPGA Lab, K. N. Toosi University of Technology, Tehran, Iran
afshin.shoeibi@gmail.com

[2] Electrical and Computer Engineering Department, Tarbiat Modaters University, Tehran, Iran

[3] Computer Engineering Department, Ferdowsi University of Mashhad, Mashhad, Iran

[4] Faculty of Electrical Engineering, Gonabad Branch, Islamic Azad University, Gonabad, Iran

[5] Department of Signal Theory, Networking and Communications, Universidad de Granada, Granada, Spain

**Abstract.** Schizophrenia (SZ) is a mental disorder that threatens the health of many people around the world. People with schizophrenia always suffer from symptoms that include hallucinations and loss of coordination between thoughts and feelings. Using deep learning and connectivity characteristics, we present a method to detect SZ from electroencephalography (EEG) signals. In this study, the data set of the Institute of Psychiatry and Neurology in Warsaw, Poland has been selected and used for experiments. First, the EEG signals are divided into 25-second time frames during the preprocessing step. Then, in the feature extraction step, deep learning (DL) and functional connectivity features (FCF) are used simultaneously. The DL model includes a CNN-LSTM network, and the functional connectivity techniques include the synchronization likelihood (SL), Fuzzy SL (FSL), and simplified interval type-2 FSL (SIT2FLS) methods. In this step, the DL features and the each functional connectivity features are combined using a concatenate layer and finally classified by the sigmoid activation layer. To better evaluate the results, K-Fold with $K = 5$ was used in the classification step. The results show that the proposed method has been able to achieve 99.43% accuracy.

**Keywords:** Schizophrenia · Diagnosis · EEG · Functional connectivity features · Simplified interval type-2 FSL · CNN-LSTM


J. M. Ferrández Vicente et al. (Eds.): IWINAC 2022, LNCS 13258, pp. 63–73, 2022.
https://doi.org/10.1007/978-3-031-06242-1_7

# 1 Introduction

Today, many people around the world deal with SZ, which significantly affects their daily lives. This mental disorder with destructive effects on the brain causes problems in expressing emotions, understanding facts, concentration, and speech of patients [14,18]. The World Health Organization (WHO) estimates that approximately 21 million people worldwide live with SZ [14,18].

Numerous clinical and imaging modalities for the diagnosis of SZ have been proposed [18]. Among the proposed methods, neuroimaging techniques are of great importance for physicians. Neuroimaging modalities are divided into two categories: structural and functional techniques [18]. Structural imaging modalities include structural magnetic resonance imaging (sMRI) and diffusion tensor imaging (DTI) [14]. Also, the most important functional imaging modalities include EEG [16], functional near-infrared spectroscopy (fNIRS), magnetoencephalography (MEG), and functional MRI (fMRI) [14].

EEG recording is one of the most popular functional neuroimaging techniques that provide crucial information to physicians about brain function at the time of SZ [14,18]. Specialist physicians benefit from EEG recordings, which are inexpensive and provide important information from different parts of the brain [14,18]. In addition to all the benefits of EEG, smoothness poses challenges for physicians. EEG signals have multiple channels, which complicates the data [18]. Also, EEG signals are sensitive to noise, which leads to incorrect SZ detection.

There is a lot of research currently being conducted on artificial intelligence (AI) techniques for addressing the challenges of SZ detection [5–8]. Shalbaf et al. [15] used the ResNet-18 model along with an support vector machine (SVM) classifier to SZ detection. To extract the specificity of the EEG signals have been investigated. Similarly, references [10,11] have used different types of CNN models in SZ detection. Some studies have also used CNN-RNN models to SZ detection from EEG signals, yielding satisfactory results [18,19].

In this paper, a new method for SZ detection from EEG signals is introduced. First, the EEG signals are decomposed into 25-second time frames in the preprocessing step. Then, DL and connectivity function features are extracted from each EEG signal frame simultaneously. As mentioned earlier, the proposed DL model includes the CNN-LSTM network, and the functional connectivity features also include the SL [4], FSL [4], and SIT2FSL. After that, the DL and functional connectivity features are merged using a concatenate layer. Finally, the sigmoid method is used to classify the features.

In the following, other parts of the article are stated. A proposed method for detecting SZ is discussed in Sect. 2. In the Sect. 3, we discuss the results of the experiments. Finally, the Conclusion, Discussion, and future works are presented in Sect. 4.

## 2 Material and Methods

The proposed method of detecting SZ from EEG signals is presented in this section. The block diagram of the proposed method is shown in Fig. 1. According to Fig. 1, EEG signals are first applied to the input of the CAD system. The EEG signals are then preprocessed. The third step involves feature extraction using ML and DL techniques. In the fourth step, the concatenate layer is used to combine the features. In the fifth step, feature classification is performed using fully connected layer with the sigmoid activation function.

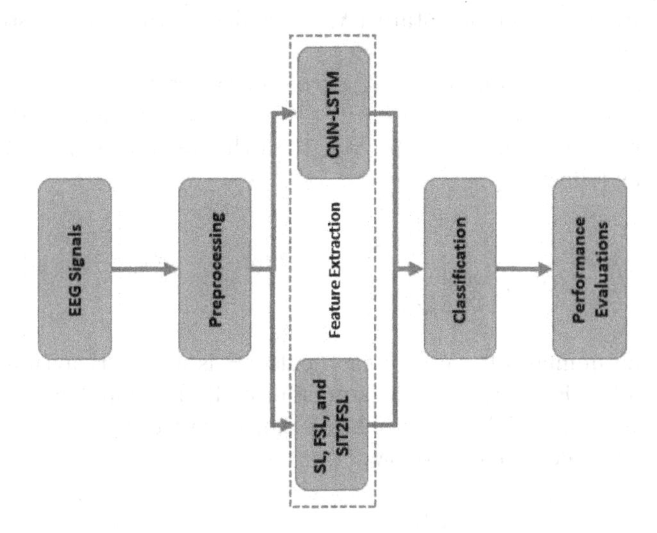


**Fig. 1.** The block diagram of proposed method.

### 2.1 Dataset

In this paper, we used the Institute of Psychiatry and Neurology in Warsaw, Poland dataset [12]. This dataset includes EEG signals from 14 patients ranging in age from 27.9 to 28.3 years. Additionally, EEG signals were recorded from 14 normal individuals of the same age and gender. EEG signals were recorded using standard 10–20. Also, since the sampling frequency of each EEG signal 250 Hz, for each frame of the signal, a frame with dimensions of $6250 * 19$ is created [12]. The number of signal recording electrodes after initial preprocessing is 19. In this dataset, all EEG signals are segmented into 25-s.

### 2.2 Feature Extraction

In this section, the proposed feature extraction method is presented. First, handcrafted features include SL [4], FSL [4], and SIT2FSL features are extracted from EEG signals. In the following, a CNN-LSTM model with proposed layers is used. Finally, handcrafted and DL features are combined to achieve high performance.

**Synchronization Likelihood.** The Takens theorem is used to display paths in the SL calculation [4]. Reconstruction of paths based on chaos theory requires the delay time determination ($T$) and an embedding dimension ($d$) to create state space. In such a state space, the nth state of $X_{k,n}$ from the kth path or $X_k$ is shown as follows [4]:

$$X(k,n) = (x_{k,n}, x_{k,n+T}, \dots, x_{k,n+(d-1)T}) \tag{1}$$

where $X_{k,n}$ is the $n^{th}$ state of the $k$ time series ($k = 1, 2, \dots, M$, when $M$ systems are paired) [4]. For each time series, the similarity of the states is calculated by comparing with the reference state ($X_{k,n}$), which is the central state of the window $W_{w_1}^{w_2}(k,n)$ [17].

Window $W_{w_1}^{w_2}(k,n)$ around state $X_{k,n}$ contains all states $X_{k,m}$ in which index $m$ satisfies the condition $w_1 < |n-m| < w_2$ [4]. The probability that a state $X_{k,m}$ is closer to the reference state $X_{k,n}$ than the distance $\varepsilon_{k,n}$ in the window is calculated as follows [4]:

$$P_{k,n}^{\varepsilon_{k,n}} = \frac{1}{2(w_2 - w_1)} \sum_{\substack{m=1 \\ W_1<|n-m|<W_2}}^{N} \Theta(\varepsilon_{k,n} - |X_{k,n} - X_{k,m}|) \tag{2}$$

where $N$ is the number of states of each path, $\Theta$ is a step Heaviside function, and the symbol $|.|$ Expresses Euclidean distance [4]. The number of time series when the states $X_{k,n}$ and $X_{k,m}$ in the Euclidean sense are closer to the Euclidean distance $\varepsilon_{k,n}$ are calculated as follows [4]:

$$H_{n,m} = \sum_{k=1}^{M} \Theta(\varepsilon_{k,n} - |X_{k,n} - X_{k,m}|) \tag{3}$$

The similarity of the $k^{th}$ time series ($k = 1, 2, \dots, M$) to be similar to the other $M-1$ time series in relation to the state pair ($X_{k,n}, X_{k,m}$) is calculated as follows [4]:

$$S_{k,n,m} = \frac{H_{n,m} - 1}{M - 1} \Theta(\varepsilon_{k,n} - |X_{k,n} - X_{k,m}|) \tag{4}$$

So the similarity of the kth time series synchronization in the window is $S_{k,n}$ [4]:

$$S_{k,n} = \frac{\sum_{\substack{m=1 \\ W_1<|n-m|<W_2}}^{N} S_{k,n,m}}{\sum_{m=1}^{N} \Theta(\varepsilon_{k,n} - |X_{k,n} - X_{k,m}|)} \tag{5}$$

The expression $S_{k,n}$ can be changed from $p_{ref}$ to 1 and determines how intense the kth time series is in synchronizing with the other time series in the $W_{w_1}^{w_2}$ window [4].

**Fuzzy Synchronization Likelihood.** The main limitation with much of the literature regarding SL is that the degree of similarities is not quantified and considered in the algorithm. To address this problem, an improved measure of generalized synchronization known as Fuzzy SL (FSL) has been proposed based on the theory of fuzzy logic and Gaussian membership function. This method could effectively detect similar patterns and the degree of their similarities, and in comparison, with conventional SL, it employs fuzzy membership functions (MF) instead of Heaviside step function and is able to measure the interdependencies more accurately and reliably [2] and [4]. In this section, FSL approach is described.

In the window $W_{w_1}^{w_2}$, a Type-1 Fuzzy (T1 F) Gaussian MF with center at the reference state and a standard deviation of $\varepsilon_{k,n}$ is employed:

$$\mu_{k,n}(X_{k,m}) = \exp\left(-\frac{|X_{k,m} - X_{k,n}|^2}{\varepsilon_{k,n}}\right) \tag{6}$$

where $\mu_{k,n}(X_{k,m})$ is the membership value of the state $X_{k,m}$ in the $d$ dimensional Gaussian function of the time series $k$ in the window and $|X_{k,m} - X_{k,n}|$ indicates Euclidean distance between in the states $X_{k,m}$ and $X_{k,n}$. Figure 2 shows this T1F Gaussian MF.

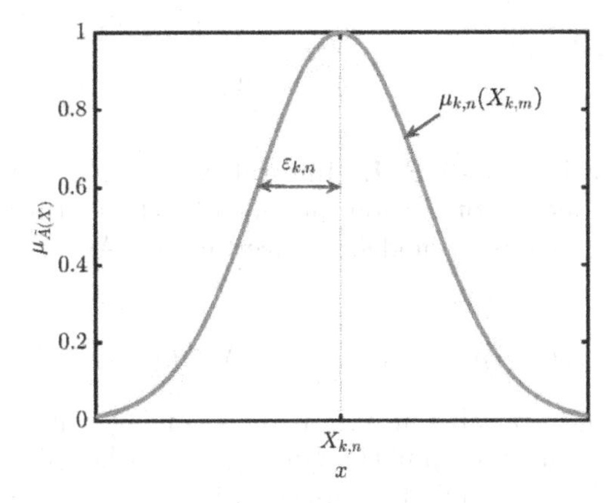


**Fig. 2.** T1F Gaussian MF $\mu_{k,n}(X_{k,m})$.

Then by replacing Heaviside function in Eq. (7) with MF in (11), the probability that a state $X_{k,m}$ is closer to the reference state $X_{k,n}$ than a distance $\varepsilon_{k,n}$ in the window is computed as follows [3]:

$$P_{k,n}^{\varepsilon_{k,n}} = \frac{1}{2(w_2 - w_1)} \sum_{\substack{m=1 \\ W_1 < |n-m| < W_2}}^{N} \mu_{k,n}(X_{k,m}) \tag{7}$$

In the FSL approach where closeness is measured with the MF $\mu$ the number of time series whose embedded states $X_{k,n}$ and $X_{k,m}$ are closer than the threshold $\varepsilon_{k,n}$ is

$$H_{n,m} = \sum_{k=1}^{N} \mu_{k,n}(X_{k,m}) \tag{8}$$

In contrast to $H$ computed in the conventional SL method Eq. 8, the crisp hard limiter Heaviside step function ($\Theta$) is exchanged with the soft Gaussian membership function ($\mu$). Therefore, substitution of $\Theta$ from Eq. (13) into Eqs. (9) and (10) results in synchronization likelihood of the $k^{th}$ time series in the window $W_{w_1}^{w_2}(k, n)$ as follows [1]:

$$S_{k,n} = \frac{\sum_{\substack{m=1 \\ W_1<|n-m|<W_2}}^{N} \left[\frac{H_{n,m}-\mu_{k,n}(X_{k,m})}{M-1} . \mu_{k,n}(X_{k,m})\right]}{\sum_{\substack{m=1 \\ W_1<|n-m|<W_2}}^{N} \mu_{k,n}(X_{k,m})} \tag{9}$$

Sliding the window over all the time series (shifting the window with a step), computing the FSL in each shifted window and average the FSLs of all shifted windows lead to an overall fuzzy set [3] and [1]. In the case of two coupled systems ($M = 2$), Eqs. (13) and (14) will simplify to [1]:

$$S_{1,n} = S_{2,n} = \frac{1}{2P_{ref}(w_2 - w_1)} \sum_{\substack{m=1 \\ W_1<|n-m|<W_2}}^{N} [\mu_{1,m}(X_{1,m}), \mu_{2,n}(X_{2,m})] \tag{10}$$

**Simplified Interval Type-2 FSL.** Interval type-2 fuzzy sets (IT2 FSs) due to their additional degrees of freedom provided by the footprint of uncertainty (FOU) outperform T1 FSs in modeling uncertainties. An IT2 FS ($\hat{A}$) is represented as [9]

$$\tilde{A} = \{(x, u), \mu_{\tilde{A}}(x, u) = 1 | \forall x \in X, \forall u \in J_x \subseteq [0, 1]\} \tag{11}$$

where $x$ is the primary variable in $X$, $u$ is the secondary variable which has the domain $J_x \subseteq [0, 1]$, and the amplitude of $\mu_{\tilde{A}}(x, u)$ is known as the secondary grade. Figure 3 indicates an IT2 F Gaussian MF.

Taking advantage of this feature, the Simplified interval type-2 FSL (SIT2FSL) method is proposed in this section. In the window $W_{w_1}^{w_2}(k, n)$, a Gaussian MF of IT2 FS with center at the reference state and a standard deviation of $\varepsilon_{k,n}$ which is defined on the interval $[\underline{\varepsilon}_{k,n}, \overline{\varepsilon}_{k,n}]$($\varepsilon_{k,n} \in [\underline{\varepsilon}_{k,n}, \overline{\varepsilon}_{k,n}]$) is applied [9]:

$$\mu_{k,n}(X_{k,m}) = \exp\left(-\frac{|X_{k,m} - X_{k,n}|^2}{\varepsilon_{k,n}}\right) \in [\underline{\mu}_{k,n}, \overline{\mu}_{k,n}] \tag{12}$$

where

$$\underline{\mu}_{k,n}(X_{k,m}) = \exp\left(-\frac{|X_{k,m} - X_{k,n}|^2}{\underline{\varepsilon}_{k,n}}\right) \tag{13}$$

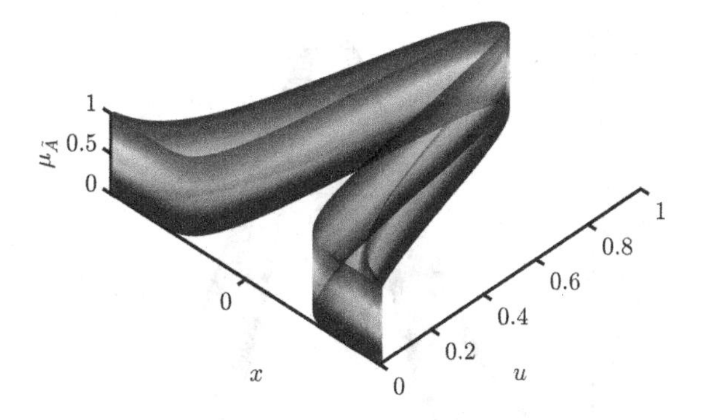


**Fig. 3.** IT2F Gaussian MF.

$$\overline{\mu}_{k,n}(X_{k,m}) = \exp\left(-\frac{|X_{k,m} - X_{k,n}|^2}{\overline{\varepsilon}_{k,n}}\right) \tag{14}$$

In this approach the IT2 MF is implemented by two embedded T1 MFs. The SIT2 Gaussian MF is depicted in Fig. 4. Since the MF value changes over an interval, the final MF could be calculated based on their weighted average as follows:

$$\mu_{k,n}(X_{k,m}) = \gamma_{k,n}.\underline{\mu}_{k,n}(X_{k,m}) + (1-\gamma_{k,n}).\overline{\mu}_{k,n}(X_{k,m}), \qquad \gamma_{k,n} \in (0,1) \tag{15}$$

$\mu_{k,n}(X_{k,m})$ is the membership value of the state $X_{k,m}$, the time series $k$ in the window $|X_{k,m} - X_{k,n}|$ demonstrates Euclidean distance between in the states $X_{k,m}$ and $X_{k,n}$. The probability values $P_{k,n}^{\varepsilon_{k,n}} \in [\underline{P}_{k,n}^{\varepsilon_{k,n}}, \overline{P}_{k,n}^{\varepsilon_{k,n}}]$ are obtained as follows:

$$\begin{aligned} P_{k,n}^{\varepsilon_{k,n}} &= \frac{1}{2(w_2 - w_1)} \sum_{\substack{m=1 \\ W_1<|n-m|<W_2}}^{N} \mu_{k,n}(X_{k,m}) \\ &= \frac{1}{2(w_2 - w_1)} \sum_{\substack{m=1 \\ W_1<|n-m|<W_2}}^{N} \left[\gamma_{k,n}.\underline{\mu}_{k,n}(X_{k,m}) + (1-\gamma_{k,n}).\overline{\mu}_{k,n}(X_{k,m})\right] \\ &= \gamma_{k,n}\underline{P}_{k,n}^{\varepsilon_{k,n}} + (1-\gamma_{k,n})\overline{P}_{k,n}^{\varepsilon_{k,n}} \end{aligned} \tag{16}$$

where

$$\underline{P}_{k,n}^{\varepsilon_{k,n}} = \frac{1}{2(w_2 - w_1)} \sum_{\substack{m=1 \\ W_1<|n-m|<W_2}}^{N} \exp\left(-\frac{|X_{k,m} - X_{k,n}|^2}{\underline{\varepsilon}_{k,n}}\right) \tag{17}$$

$$\overline{P}_{k,n}^{\varepsilon_{k,n}} = \frac{1}{2(w_2 - w_1)} \sum_{\substack{m=1 \\ W_1<|n-m|<W_2}}^{N} \exp\left(-\frac{|X_{k,m} - X_{k,n}|^2}{\overline{\varepsilon}_{k,n}}\right) \tag{18}$$

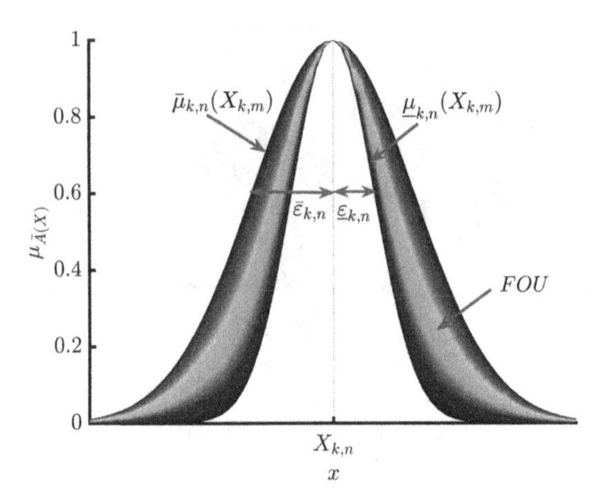


**Fig. 4.** SIT2F Gaussian MF.

Similarly, in the SIT2 FSL method where closeness is measured with the MF $\mu$ the number of time series whose embedded states $X_{k,n}$ and $X_{k,m}$ are closer than the threshold $\varepsilon_{k,n}$ is as

$$H_{n,m} = \sum_{k=1}^{N} \mu_{k,n}(X_{k,m}) = \sum_{k=1}^{N} \left[\gamma_{k,n} \cdot \underline{\mu}_{k,n}(X_{k,m}) + (1 - \gamma_{k,n}) \cdot \overline{\mu}_{k,n}(X_{k,m})\right] \quad (19)$$

$$= \gamma_{k,n} \underline{H}_{n,m} + (1 - \gamma_{k,n}) \overline{H}_{n,m} \quad (20)$$

where

$$\underline{H}_{n,m} = \sum_{k=1}^{N} \underline{\mu}_{k,n}(X_{k,m}) \quad (21)$$

and

$$\overline{H}_{n,m} = \sum_{k=1}^{N} \overline{\mu}_{k,n}(X_{k,m}) \quad (22)$$

Using the same procedure as for FSL, IT2 fuzzy synchronization likelihood of the $k^{th}$ time series in the window $w_1 < |n - m| < w_2$ is described by:

$$s_{k,n} = \frac{\sum_{\substack{m=1 \\ W_1 < |n-m| < W_2}}^{N} \left[\frac{H_{n,m} - (\gamma_{k,n} \cdot \underline{\mu}_{k,n}(X_{k,m}) + (1-\gamma_{k,n}) \cdot \overline{\mu}_{k,n}(X_{k,m}))}{M-1} \left(\gamma_{k,n} \cdot \underline{\mu}_{k,n}(X_{k,m}) + (1 - \gamma_{k,n}) \cdot \overline{\mu}_{k,n}(X_{k,m})\right)\right]}{\sum_{\substack{m=1 \\ W_1 < |n-m| < W_2}}^{N} \left(\gamma_{k,n} \cdot \underline{\mu}_{k,n}(X_{k,m}) + (1 - \gamma_{k,n}) \cdot \overline{\mu}_{k,n}(X_{k,m})\right)} \quad (23)$$

Sliding the window over all the time series with the same time step, computing the IT2 FSL in each shifted window and average the IT2 FSLs of all shifted windows also results in an overall IT2 FSL.

### 2.3 CNN-LSTM Model

Before feeding data to LSTM part of the network, each of signals (with length of 625) is divided into segments of length 25. In doing this, each segment has an overlap of 25 with the next one, thus 24 segments are extracted; then a 1D-CNN is applied on these segments, and their output is used for input of LSTM. According to Table 1, the connectivity and DL features are first merged by the Concatenate layer. Next, two dense layers with ReLU activation function and a sigmoid layer are placed, in which the sigmoid function is for classifying two classes.

**Table 1.** The proposed CNN-LSTM model

| Layers | Filters | Kernel size | Stride | Padding | Activation | Input shape | Output shape |
|---|---|---|---|---|---|---|---|
| Input Data | – | – | – | – | – | (50,19) | (50,19) |
| ID Conv | 16 | 5 | 1 | Same | ReLU | (50,19) | (50,16) |
| BatchNorm | – | – | beta = 0.99 | – | – | (50,16) | (50,16) |
| ID Conv | 32 | 5 | 1 | Same | ReLU | (50,16) | (50,16) |
| Dropout | – | – | Rate = 0.5 | – | – | (50,16) | (50,32) |
| MaxPooling | – | 2 | 1 | – | – | (50,32) | (25,32) |
| 1D Conv | 64 | 3 | 1 | Same | ReLu | (25,64) | (25,64) |
| Flatten | – | – | – | – | – | (25,64) | (:,1600) |
| LSTM | 1 | 100 | – | – | – | (24,1600) | (:,100) |
| Input 2 | – | – | – | – | – | (,19) | (:,19) |
| Concatenate | | | LSTM and Input 2 | | | (:,100) & (:,19) | (:,119) |
| Dense | 256 | – | – | – | ReLU | (:,119) | (:,256) |
| Dropout | – | – | Rate = 0.5 | – | – | (:,256) | (:,256) |
| Dense | 128 | – | – | – | ReLU | (:,256) | (:,128) |
| Dropout | – | – | Rate = 0.5 | – | – | (:,128) | (:,128) |
| Dense | 1 | – | – | – | Sigmoid | (:,128) | (:,1) |

## 3 Experiment Result

In this paper, hardware with Nvidia 1070 GPU specifications, 16 GB RAM, and Core i7 CPU are used to implement CADS SZ detection. Also, Sikit-learn toolbox [13], and TensorFlow-2 [20] have been used to implement the desired algorithms. Each of the signals in the dataset is divided into 25-s time intervals to implement the proposed method. Next, functional and connectivity-based SL, FSL, and SIT2FSL features along with DL features are extracted from EEG signals. These features are then combined and finally classified by the sigmoid activation function. Table 2 shows the results of combining SL, FSL, and SIT2FSL methods alongside the CNN-LSTM model. As shown in Table 2, it can be seen that the SIT2FSL Feature with CNN-LSTM model was able to achieve the highest accuracy.

**Table 2.** Results of the proposed method

| Method | Results (%) | | | |
|---|---|---|---|---|
| | Acc | Sens | Spec | F1-score |
| CNN-LSTM | 99.15 ± 0.20 | 99.12 ± 0.18 | 99.18 ± 0.21 | 99.14 ± 0.31 |
| SL + CNN-LSTM | 99.21 ± 0.27 | 99.20 ± 0.30 | 99.22 ± 0.29 | 99.21 ± 0.41 |
| FSL + CNN-LSTM | 99.30 ± 0.15 | 99.32 ± 0.22 | 99.28 ± 0.17 | 99.29 ± 0.28 |
| SIT2FSL + CNN-LSTM | 99.43 ± 0.13 | 99.46 ± 0.12 | 99.40 ± 0.15 | 99.42 ± 0.18 |

## 4 Discussion, Conclusion, and Future Works

In this paper a new method for SZ detection from EEG signals is presented utilizing functional connectivity features and CNN-LSTM model. The proposed method consists of four parts: dataset, preprocessing, feature extraction, and classification. As a first step, the dataset of the Institute of Psychiatry and Neurology in Warsaw, Poland was selected for the experiments. In the second step, preprocessing of each of the EEG signals of this dataset is performed. In the third step, feature extraction is performed simultaneously by SL, FSL, SIT2FSL, and CNN-LSTM model, and then these features are combined. Finally, the sigmoid activation function in the CNN-LSTM model is used for classification. The results show that combining SIT2FSL techniques with CNN-LSTM model has been able to achieve 99.43% accuracy. The method proposed in this paper can be used in a hospital facility as an ancillary software in the future work. The proposed method can be implemented on software and hardware platforms, and this can help physicians diagnose SZ quickly. Other DL models including 2D-CNN, RNNs, AEs, CNN-RNN, and CNN-AE can be used for future work [16,17].

## References


1. Ahmadlou, M., Adeli, H.: Fuzzy synchronization likelihood with application to attention-deficit/hyperactivity disorder. Clin. EEG Neurosci. **42**(1), 6–13 (2011)
2. Ahmadlou, M., Adeli, H.: Visibility graph similarity: a new measure of generalized synchronization in coupled dynamic systems. Physica D Nonlinear Phenomena **241**(4), 326–332 (2012)
3. Ahmadlou, M., Adeli, H.: Complexity of weighted graph: a new technique to investigate structural complexity of brain activities with applications to aging and autism. Neurosci. Lett. **650**, 103–108 (2017)
4. Ahmadlou, M., Adeli, H., Adeli, A.: Fuzzy synchronization likelihood-wavelet methodology for diagnosis of autism spectrum disorder. J. Neurosci. Methods **211**(2), 203–209 (2012)
5. Cortes-Briones, J.A., Tapia-Rivas, N.I., D'Souza, D.C., Estevez, P.A.: Going deep into schizophrenia with artificial intelligence. Schizophrenia Res. (2021)
6. de Filippis, R., et al.: Machine learning techniques in a structural and functional MRI diagnostic approach in schizophrenia: a systematic review. Neuropsychiatr. Dis. Treat. **15**, 1605 (2019)

7. Górriz, J.M., et al.: Artificial intelligence within the interplay between natural and artificial computation: advances in data science, trends and applications. Neurocomputing **410**, 237–270 (2020)
8. Lanillos, P., Oliva, D., Philippsen, A., Yamashita, Y., Nagai, Y., Cheng, G.: A review on neural network models of schizophrenia and autism spectrum disorder. Neural Netw. **122**, 338–363 (2020)
9. Liu, F., Mendel, J.M.: Encoding words into interval type-2 fuzzy sets using an interval approach. IEEE Trans. Fuzzy Syst. **16**(6), 1503–1521 (2008)
10. Naira, T., Alberto, C.: Classification of people who suffer schizophrenia and healthy people by EEG signals using deep learning (2020)
11. Oh, S.L., Vicnesh, J., Ciaccio, E.J., Yuvaraj, R., Acharya, U.R.: Deep convolutional neural network model for automated diagnosis of schizophrenia using EEG signals. Appl. Sci. **9**(14), 2870 (2019)
12. Olejarczyk, E., Jernajczyk, W.: Graph-based analysis of brain connectivity in schizophrenia. PLoS ONE **12**(11), e0188629 (2017)
13. Pedregosa, F., et al.: Scikit-learn: machine learning in Python. J. Mach. Learn. Res. **12**, 2825–2830 (2011)
14. Sadeghi, D., et al.: An overview on artificial intelligence techniques for diagnosis of schizophrenia based on magnetic resonance imaging modalities: methods, challenges, and future works. arXiv preprint arXiv:2103.03081 (2021)
15. Shalbaf, A., Bagherzadeh, S., Maghsoudi, A.: Transfer learning with deep convolutional neural network for automated detection of schizophrenia from EEG signals. Phys. Eng. Sci. Med. **43**(4), 1229–1239 (2020)
16. Shoeibi, A., et al.: Detection of epileptic seizures on EEG signals using ANFIS classifier, autoencoders and fuzzy entropies. Biomed. Signal Process. Control **73**, 103417 (2022)
17. Shoeibi, A., et al.: Applications of deep learning techniques for automated multiple sclerosis detection using magnetic resonance imaging: a review. Comput. Biol. Med. **136**, 104697 (2021)
18. Shoeibi, A., et al.: Automatic diagnosis of schizophrenia in EEG signals using CNN-LSTM models. Front. Neuroinform. **15** (2021)
19. Singh, K., Singh, S., Malhotra, J.: Spectral features based convolutional neural network for accurate and prompt identification of schizophrenic patients. Proc. Inst. Mech. Eng. Part H J. Eng. Med. **235**(2), 167–184 (2021)
20. Singh, P., Manure, A.: Introduction to TensorFlow 2.0. In: Learn TensorFlow 2.0, pp. 1–24. Apress, Berkeley, CA (2020). https://doi.org/10.1007/978-1-4842-5558-2_1

# Sleep Apnea Diagnosis Using Complexity Features of EEG Signals

Behnam Gholami[1], Mohammad Hossein Behboudi[2], Ali Khadem[3], Afshin Shoeibi[4](✉), and Juan M. Gorriz[5]

[1] Faculty of Science, Department of Physical Chemistry, K. N. Toosi University of Technology, Tehran, Iran
[2] School of Behavioral and Brain Science, University of Texas, Texas, USA
[3] Faculty of Electrical Engineering, Department of Biomedical Engineering, K. N. Toosi University of Technology, Tehran, Iran
[4] Faculty of Electrical Engineering, FPGA Lab, K. N. Toosi University of Technology, Tehran, Iran
afshin.shoeibi@gmail.com
[5] Department of Signal Theory, Networking and Communications, Universidad de Granada, Granada, Spain

**Abstract.** Sleep apnea syndrome is one the most prevalent sleep disorders. The accurate diagnosis and treatment of apnea by physicians can help to avoid its destructive effects in the long term. Electroencephalogram (EEG) records activity of the brain from different areas of scalp and can be an appropriate method to diagnose sleep apnea. In this work, we proposed a Computer Aided Diagnosis System (CADS) for sleep apnea based on complexity features of EEG. At first, EEG time series of 20 participants were decomposed into six frequency bands (delta, theta, alpha, sigma, beta, and gamma) by using bandpass Finite Impulse Response (FIR) filters. Then, complexity features such as fractals, Lempel-Ziv Complexity (LZC), entropies, and generalized Hurst exponent that was used for the first time to detect sleep apnea from EEG signals, were extracted from each frequency band. The minimum-redundancy maximum-relevance (mRMR) algorithm was applied to sort 120 features of three EEG channels. Finally, two popular classifiers, Support Vector Machine (SVM) and K-Nearest Neighbors (KNN), were used to detect sleep apnea. 99.33% accuracy was obtained using the SVM classifier and generalized hurst exponent had an effective contribution to detect apnea.

**Keywords:** Sleep apnea · Generalized Hurst exponent · Complexity features · EEG · SVM

## 1 Introduction

Sleep apnea is one of the most common sleep disorders. Based on the American Academy of Sleep Medicine (AASM), between 4% and 19% of the elderly

Supported by organization x.


J. M. Ferrández Vicente et al. (Eds.): IWINAC 2022, LNCS 13258, pp. 74–83, 2022.
https://doi.org/10.1007/978-3-031-06242-1_8

population diagnosed with sleep apnea disorder and it has been predicted that this percentage increases to about 49% for adult population. There are three different types of sleep apnea: Obstructive Sleep Apnea (OSA), Central Sleep Apnea (CSA), and Mixed Sleep Apnea (MSA). In sleep apnea syndrome, the airflow decreases nearly 10% in regular periods each takes long about 10 s [22]. When these short respiratory pauses occur, the brain makes a proper reaction by a partial awakeness and then goes back to sleep after the respiratory circulation went back to normal. People need sleep to recover their body and mind but apnea does not allow body to complete this cycle and the person feels exhausted after sleep. Furthermore, apnea may cause some problems such as depression, a rise in blood pressure, headache in the morning, and decline of concentration [1]. Early and accurate diagnosis of apnea can be useful for physicians to prevent more serious problems and remedy apnea. Polysomnography (PSG) is the perfect way to record the main physiological events that to sleep. Despite beneficial information of PSG, it is prone to human error because data segments are labeled by a human expert which is a time-consuming and expensive process [9]. Therefore, an automatic diagnostic system can be a proper alternative way. Some of these systems can be found in the literature, for instance, in [2], Electromyogram (EMG) and EEG signals, in [6],Electrocardiogram (ECG) signal and in [24], ECG, EMG and Electrooculography (EOG) signals were applied to automatically diagnose sleep apnea syndrome. The brain activity can be measured by the electric field over the scalp which is represented in EEG signals that have appropriate information from brain cortical activity [25]. Since EEG measures the brain activity with a high temporal resolution it is a proper method for sleep studies and its outstanding performance in detection of sleep disorders gained a lot of attention from sleep researchers. EEG is one of the biomedical signals of the PSG recording and can be used to continuously monitor events of the brain, thus may be a helpful method to detect apnea.

Brain is the most complex organ in the human body. This complexity also appears in EEG signals. Hence a complexity analysis of EEG signals seems to be helpful to study the brain function during sleep apnea events. Up to now, some studies have been conducted to diagnose sleep apnea from complexity of EEG signals. For example, in [35], variance and sample entropy were extracted from each frequency band, The neighbor composition analysis (NCA) was used as a feature selection method, then, Random Forest (RF), SVM and KNN classifications were used and average accuracy (88.99%) was acquired. In [27], an EEG channel was chosen and its signal was decomposed into frequency subbands. LZC feature was extracted from each sub-band to detect apnea. Finally, the highest accuracy (96%) was obtained using the KNN classification. In [21], the EEG signals were decomposed into five frequency bands. Then, the entropy features of beta and delta bands were used to get the accuracy of 87.64% using the KNN classifier. In [36], detrended fluctuation analysis (DFA) and the SVM classifier resulted in accuracy of 95.1% in diagnosis of sleep apnea.

In this study, different complexity features of EEG signals such as Hurst exponent, generalized Hurst exponent, Higuchi and Katz fractal, LZC and Shannon entropy, Renyi entropy, Tssalis entropy, threshold entropy, permutation entropy, spectral entropy, wavelet entropy, SURE entropy, norm entropy, log energy entropy, fuzzy entropy, multiscale entropy and sample entropy were extracted from each sub-band of EEG channels CZ-A1, Fp-A1, and O1-A2. It is noteworthy that, to the best of our knowledge, generalized hurst exponent was applied for the first time to detect sleep apnea from EEG. Conventional machine learning algorithms such as SVM and KNN were applied to classify the feature vectors. To reduce the complexity of the model and speed up the classification process, the mRMR feature selection algorithm was applied to select the best feature set for the classification. This paper is structured as follows. In Sect. 2, experimental EEG data and the proposed approach are described in detail. In Sect. 3, the results are illustrated, discussed and compared with related works. Finally, Sect. 4 will conclude the paper and suggest some future works.

## 2 Materials and Methods

### 2.1 Experimental Data

In this work, based on the new standard of the AASM 20 healthy subjects were recorded and 12 apneic people who suffered from different type of apnea (central, obstructive and mixed apnea and hypopnea). The age of normal subjects was between 20 and 65 years and for apneic people was between 41 and 70 years [5]. The sample frequency 200 Hz. Because of our aim, three different EEG channels (CZ-A1, FP1-A1 and O1-A1) were selected to study different regions of the brain (Fig. 1).

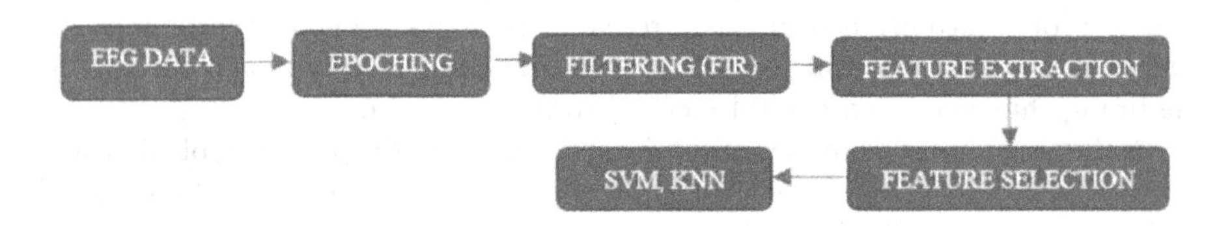


**Fig. 1.** Steps of sleep apnea detection.

### 2.2 Proposed Method

Firstly, by using a bandpass FIR filter with Hamming window each 5-second epoch of EEG signals were decomposed to 6 different frequency bands including delta ($\delta$, 0–4 Hz), theta ($\theta$, 4–8 Hz), alpha ($\alpha$, 8–12 Hz), sigma ($\sigma$, 12–15 Hz), beta ($\beta$, 15–25 Hz) and gamma ($\gamma$, 25–45 Hz). Then, complexity features such as generalized Hurst exponent, Higuchi and Katz fractal, LZC and entropies such as Shannon, Renyi, Tssalis, threshold, permutation, spectral, wavelet, SURE, norm, log energy, fuzzy, multiscale and sample were extracted from frequency sub-bands. To enhance the efficiency of classification, the mRMR algorithm was used to choose the most effective features. Finally, the SVM classifier with RBF function and the KNN were applied to discriminate apneic subjects from normal ones.

### 2.3 Feature Extraction

In Table 1, complexity features are explained.

**Table 1.** Complexity features that were used in details.

| Complexity feature | Formula | Description |
|---|---|---|
| LZC [27] | $C(n) = \frac{c(n)}{b(n)}$ | $b(n)$ gives the asymptotic behavior of $c(n)$ for a random string |
| Higuchi fractal [28] | $FD = \ln(L(k))/\ln(1/k)$ | $L(k)$ computes an average length for time series that have a similar time delay, $k_{\max} = 6$ |
| Katz fractal [19] | $D = \frac{\log(\frac{L}{a})}{\log(\frac{d}{a})}$ | $L$ is the length of the curve, $a$ is the average distance between successive points |
| Hurst exponent [19] | $H = \frac{\log(\frac{R}{S})}{\log T}$ | $T$ is the duration of the sample of data and $R/S$ the corresponding value of a rescaled range |
| Generalized Hurst exponent [11] | $K_q(\tau) \sim (\frac{\tau}{v})^{qH(q)}$ | $\tau$ is the time interval, denotes the time between observations $\tau$ can vary between $\nu$ and $\tau_{\max}$ |
| Shannon [20] | $En\ _{Shan} = -\Sigma_i(x_i^2)\log(x_i^2)$ | $x_i$ is the coefficient of $x$ |
| Log- energy [20] | $En\ _{thr} = \Sigma_i(x_i^2)$ | $x_i$ is the coefficient of $x$ |
| Threshold [20] | $En\ _{thr} = 1, if\vert x_i\vert > p$ $En\ _{thr} = 0, elswhere$ | $p = 0.2$ |
| Sure [20] | $En\ _{Su} = N - \#\{i\ such\ that\vert x_i\vert \leq p\} + \Sigma_i \min(x_i^2, p^2)$ | $p = 5.5$ and $x_i$ is the $i^{th}$ sample of the signal |
| Norm [20] | $En\ _{no} = \vert x_i\vert^p$ | $p = 1$ |
| Renyi [25] | $En\ _{Re} = \frac{1}{1-\alpha}\log(\Sigma_{i=1}^n x_i^a)$ | $\alpha = 0.5$ |
| Tssalis [19] | $En\ _{Ts} = (\Sigma_{i=1}^n x_i x_i^q)/(q-1)$ | $q = 1.5$ |
| Permutation [14] | $En\ _{per} = -\Sigma_{n=1}^{n!} x(\pi)\ln x(\pi)$ | $n$ is the embedding dimension and there will be $n!$ possible order patterns $\pi$ |
| Multiscale [26] | $En\ _{multi} = \frac{1}{\tau}\Sigma_{i=(j-1)\tau+1}^{j\tau} x$ | $\tau$ is the scale factor, $1 \leq j \leq N/\tau$, $N$ is data points |
| Spectral [12] | $En\ _{spec} = \frac{\Sigma P_K \log P_K}{\log n}$ | $n$ is the number of frequencies and $P_K$ are spectral amplitudes of $k$ frequencies region |
| Wavelet [10] | $E\ _{total} = \sum_{i=1}^{l+1} E_i = \vert cA_1\vert^2 + \sum_{j=1}^{1} \vert cD_j\vert^2$ $En\ _{wa} = -\Sigma_{i=1}^k P_i \ln P_i$ | $cA_1$, wavelet coefficient of each layer is $cD_{j,}$, $j = 1, 2, \ldots .l$ |

### 2.4 Feature Selection

It is notable that for removing irrelevant and inappropriate features we need a feature selection algorithm which is a very important step in machine learning and pattern recognition. After feature extraction and based on k-fold cross validation with k = 10, we considered 90% of data for test and 10% for train. Before the data is inputted to the classifier, mRMR algorithm was used to rank features. mRMR sorts features based on having the most relevance with labels and the least redundancy with the selected features. In this step, features that are ranked by mRMR will be fed to the classifier one after another. After adding each feature, the accuracy of the classification will be calculated. This process will be repeated until all the features enter the classifier. The optimum number of features is the first point in which the highest accuracy is obtained.

### 2.5 Classification

SVM is a powerful tool in machine learning to distinguish two different groups of data. SVM is able to separate nonlinear data by using the best vector with the most margin between groups. Data are located on high-dimensional space and an appropriate hyper-plane can distinguish data. KNN is a non-parametric approach that is applied to data analysis, machine learning, and pattern detection. The value of K is an important parameter to achieve superior detection. Similarity of labeled data to a test data sample is the basic of the performance of KNN [21]. In this work the value of K is 1.

## 3 Results and Discussion

Complexity features in EEG signal have been proposed to detect apnea in this work. Finally, we had 120 features that were extracted from 6 frequency bands. Accuracy of classification (SVM) is shown in Table 2:

**Table 2.** Accuracy using the SVM classifier with whole feature and the SVM classifier by mRMR.

| Channel | SVM | | mRMR + SVM | |
|---|---|---|---|---|
| | Feature | Accuracy % | Features | Accuracy % |
| Cz-A1 | 120 | 93.5 | 37 | 93.33 |
| Fp-A1 | 120 | 92 | 39 | 91.69 |
| O1-A1 | 120 | 92.5 | 24 | 89.49 |

According to Table 2, using CZ-A1, FP-A1 and O1-A1 channels, 93.5%, 92% and 92.5% accuracy were acquired by the SVM classifier respectively but using these channels, accuracy of 91.5%, 89.9% and 90% were acquired by the KNN classifier. Although the number of optimum features that were selected by the mRMR algorithm were nearly similar in two classifications, the highest accuracy (93.33%) was resulted in the CZ-A1 channel just with 37 features in SVM classifier. 39 and 24 features in FP-A1 and O1-A1 channels showed accuracy of 91.69% and 89.49% by SVM respectively. Using the KNN classifier, 36 features were selected in CZ-A1 with 90.21% accuracy and in FP-A1 and O1-A1, 39 and 19 features were chosen that resulted 87.92% and 87.92% accuracy, respectively.

To better study the, the efficient features and frequency bands were illustrated in below figures:

According to Figs. 2 and 3, for the SVM classification, the most effective features and the number of them were exactly similar those selected for KNN in CZ-A1.

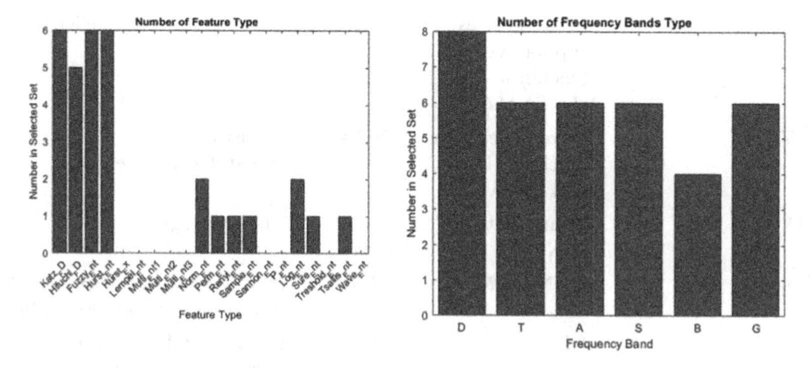


**Fig. 2.** Selected features and frequency bands by the mRMR + KNN classifier in CZ-A1 (Delta: D, Theta: T, Alpha: A, Sigma: S, Beta: B, Gamma: G).

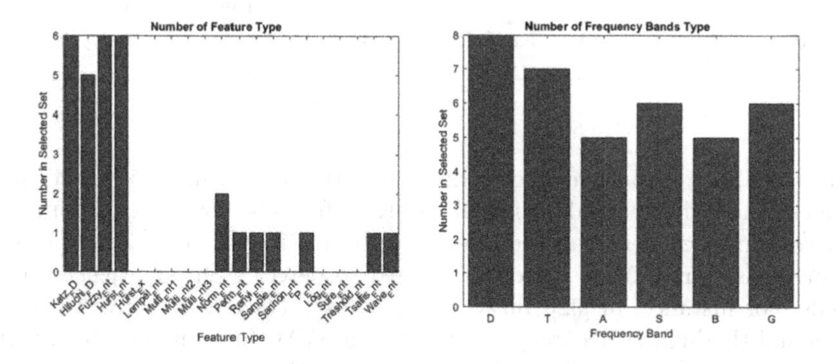


**Fig. 3.** Selected features and frequency bands by the mRMR + SVM classifier in CZ-A1.

**Table 3.** Some related works for detection of apnea by EEG signal.

| Reference | No. subjects | Classifier | The highest accuracy % | Features |
|---|---|---|---|---|
| [35] | 30 | RF, SVM, KNN | 88.99 | Sample entropy and variance |
| [27] | 16 | Discriminant analysis (DA), Decision tree (DT), Ensemble learning | 96 | LZC |
| [21] | 5 | KNN | 87.64 | Entropy in beta and delta band |
| [36] | 12 | SVM | 95.1 | Multifractal-Detrended fluctuation analysis (DFA) |
| [29] | 1 | Least squares support vector machines (LS-SVMs) | 96.25 | Autoregressive coefficients |
| [26] | 18 | extreme learning machine (ELM) and LS-SVM | 99.53 | Statistical and non-statistical measures |
| [15] | 15 | Neural network (NN) | 91 | Theta energy ratios |
| [23] | 14 | KNN, SVM | 84.83 | Mean and variance extracted from Delta band power ratio |
| [3] | 21 | KNN | 98.02 | Entropy and log-variance extracted from beta and delta band |
| [8] | 159 | SVM, RF | 90 | Entropy and Energy Heart rate Synchronization |
| [4] | 5 | SVM, KNN | 82.69 | Time domain, wavelets, and frequency domain |
| [31] | 30 | SVM, KNN, RF | 94.33 | Approximate entropy |
| [32] | 25 | SVM, KNN, RF | 93.19 | Fuzzy entropy |

In the Table 3, some related works that extracted features to detect apnea are mentioned. In this study, Higuchi and Katz fractals, fuzzy entropy and generalized hurst exponent had the most contribution to diagnose sleep apnea. In the previous works, fuzzy entropy was used to detect sleep apnea using EEG signals. For instance, in [32], fuzzy entropy was extracted from five frequency bands and the highest accuracy was 93.19% in SVM. In sigma and beta bands the fuzzy entropy in apneic subjects was larger than normal but there is an opposite trend in other bands. In [13], fuzzy entropy and wavelet transformation were extracted from EEG and 89% accuracy was obtained. In [16], the result of fuzzy entropy showed a clear interhemispheric asymmetry in apneic people. In [33], fuzzy entropy was extracted from EEG and ECG beside other features to diagnose apnea. The highest accuracy (96.24%) was acquired by one EEG and ECG channel. Also, different fractals were applied in EEG time series to detect sleep apnea. In [30], analysis of multifractal detrended fluctuation was used to extract features from EEG of 120 subjects and 95.83% accuracy was obtained by RF classifier. It was proved that EEG time series for apnea presented multifractality by [7]. However, generalized hurst exponent has been used to discriminate disorders by EEG time series [17,18], as far as we know it is the first time it has been used to diagnose apnea in this work and had a fruitful effect.

## 4 Conclusion

We investigated previous complexity features that were applied to detect apnea with a new one (generalized hurst exponent). The results showed that CZ-A1 channel that is located in the center of scalp presented higher accuracy than FP-A1 and O1-A1. Also, SVM distinguished better normal and abnormal subjects in comparison with KNN. It was proved that nonlinear features especially from delta band were the most effective features to detect sleep apnea. It can be useful if we investigate other concepts of chaos theory such as graph or connectivity between channels to study sleep apnea in the future works.

## References


1. Alqassim, S., et al.: Sleep apnea monitoring using mobile phones. In: International Conference on e-Health Networking, Applications and Services (Healthcom). IEEE (2012)
2. Azim, Md.R., et al.: Analysis of EEG and EMG signals for detection of sleep disordered breathing events. In: International Conference on Electrical and Computer Engineering (2010)
3. Bhattacharjee, A., et al.: Sleep apnea detection based on Rician modeling of feature variation in multiband EEG signal. IEEE J. Biomed. Health Inform. **23**(3), 1066–81074 (2018)
4. Bello, S.A., Alqasemi, U.: Computer Aided Detection of Obstructive Sleep Apnea from EEG Signals. SSRN 3890660 (2021)
5. Devuyst, S., Dutoit, T., Kerkhofs, M.: The DREAMS databases and assessment algorithm. Zenodo, Genève (2005)
6. Gutta, S., et al.: Cardiorespiratory model-based data-driven approach for sleep apnea detection. IEEE J. Biomed. Health Inform. **22**(4), 1036–1045 (2017)
7. Gaurav, G., Anand, R.S., Kumar, V.: EEG based cognitive task classification using multifractal detrended fluctuation analysis. Cogn. Neurodyn. **15**(6), 999–1013 (2021)
8. Jayaraj, R., Mohan, J.: Classification of sleep apnea based on sub-band decomposition of EEG signals. Diagnostics 11(9) (2021)
9. Kandel, E.R., et al.: Principles of Neural Science, vol. 3. McGraw-Hill, New York (2000)
10. Kang, J., et al.: EEG entropy analysis in autistic children. J. Clin. Neurosci. **62**, 199–206 (2019)

11. Karegar, F.P., Fallah, A., Rashidi, S.: ECG based human authentication with using Generalized Hurst Exponent. In: Iranian Conference on Electrical Engineering (ICEE) (2017)
12. Kannathal, N., et al.: Entropies for detection of epilepsy in EEG. Comput. Methods Programs Biomed. **80**(3), 187–194 (2005)
13. Lin, S.-Y., et al.: EEG signal analysis of patients with obstructive sleep apnea syndrome (OSAS) using power spectrum and fuzzy entropy. In: International Conference on Natural Computation, Fuzzy Systems and Knowledge Discovery (ICNC-FSKD) (2017)
14. Li, X., Ouyang, G., Richards, D.A.: Predictability analysis of absence seizures with permutation entropy. Epilepsy Res. **77**(1), 70–74 (2007)
15. Liu, D., Pang, Z., Lloyd, S.R.: A neural network method for detection of obstructive sleep apnea and narcolepsy based on pupil size and EEG. IEEE Trans. Neural Netw. **19**(2), 308–318 (2008)
16. Li, Y., et al.: Interhemispheric brain switching correlates with severity of sleep-disordered breathing for obstructive sleep apnea patients. Appl. Sci. **9**(8), 1568 (2019)
17. Lahmiri, S.: Generalized Hurst exponent estimates differentiate EEG signals of healthy and epileptic patients. Physica A Stat. Mech. Appl. **490**, 378–385 (2018)
18. Gorriz, J.M., et al.: Artificial intelligence within the interplay between natural and artificial computation: advances in data science, trends and applications. Neurocomputing **410**, 237–270 (2020)
19. Subha, D.P., et al.: EEG signal analysis: a survey. J. Med. Syst. **34**(2), 195–212 (2010)
20. Sharma, A., Amarnath, M., Kankar, P.K.: Feature extraction and fault severity classification in ball bearings. J. Vibr. Control **22**(1), 176–192 (2016)
21. Saha, S., et al.: An approach for automatic sleep apnea detection based on entropy of multi-band EEG signal. In: IEEE Region 10 Conference (TENCON) (2016)
22. Senaratna, C.V., et al.: Prevalence of obstructive sleep apnea in the general population: a systematic review. Sleep Med. Rev. **34**, 70–81 (2017)
23. Shahnaz, C., Minhaz, A.T., Ahamed, S.T.: Sub-frame based apnea detection exploiting delta band power ratio extracted from EEG signals. In: IEEE Region 10 Conference (TENCON) (2016)
24. Schluter, T., Conrad, S.: An approach for automatic sleep stage scoring and apnea-hypopnea detection. Front. Comput. Sci. **6**(2), 230–241 (2012)
25. Tibdewal, M.N., et al.: Multiple entropies performance measure for detection and localization of multi-channel epileptic EEG. Biomed. Signal Process. Control **38**, 158–167 (2017)
26. Taran, S., Bajaj, V.: Sleep apnea detection using artificial bee colony optimize Hermite basis functions for EEG signals. IEEE Trans. Instrum. Meas. **69**(2), 608–616 (2019)
27. Taran, S., et al.: Detection of sleep apnea events using electroencephalogram signals. Appl. Acoust. **181**, 108–137 (2021)
28. Uthayakumar, R.: Fractal dimension in Epileptic EEG signal analysis. In: Banerjee, S., Rondoni, L. (eds.) Applications of Chaos and Nonlinear Dynamics in Science and Engineering-Vol. 3, pp. 103–157. Springer, Heidelberg (2013). https://doi.org/10.1007/978-3-642-34017-8_4
29. Ubeyli, E.D., et al.: Analysis of sleep EEG activity during hypopnoea episodes by least squares support vector machine employing AR coefficients. Expert Syst. Appl. **37**(6), 4463–4467 (2010)

30. Wang, H., Guo, Z., Du, W.: Diagnosis of rolling element bearing based on multifractal detrended fluctuation analyses and continuous hidden Markov model. J. Mech. Sci. Technol. **35**(8), 3313–3322 (2021). https://doi.org/10.1007/s12206-021-0705-y
31. Wang, Y., et al.: An efficient method to detect sleep hypopnea-apnea events based on EEG signals. IEEE Access **9**, 641–650 (2020)
32. Wang, Y., et al.: A Robust Sleep Apnea-hypopnea Syndrome Automated Detection Method Based on Fuzzy Entropy Using Single Lead-EEG Signals (2021)
33. Xin, X., Yaru, Z., Sanli, Y., et al.: A New Method for Detecting Sleep Apnea. Research Square (2022)
34. Zadeh, L.A.: Fuzzy sets as a basis for a theory of possibility. Fuzzy Sets Syst. **100**, 9–34 (1999)
35. Zhao, X., et al.: Classification of sleep apnea based on EEG sub-band signal characteristics. Sci. Rep. **11**(1), 1–11 (2021)
36. Zhou, J., Wu, X., Zeng, W.: Automatic detection of sleep apnea based on EEG detrended fluctuation analysis and support vector machine. J. Clin. Monit. Comput. **29**(6), 767–772 (2015). https://doi.org/10.1007/s10877-015-9664-0

# Representational Similarity Analysis: A Preliminary Step to fMRI-EEG Data Fusion in MVPAlab

David López-García[1(✉)], J. M. González-Peñalver[1], J. M. Górriz[2], and María Ruz[1]

[1] Mind, Brain and Behavior Research Center (CIMCYC), University of Granada, Granada, Spain
{dlopez,mruz}@ugr.es

[2] Signal Theory, Telematics and Communications Department (TSTC), University of Granada, Granada, Spain
gorriz@ugr.es

**Abstract.** The study of brain cognitive function has recently expanded from classical univariate to multivariate analyses. In combination with different non-invasive neuroimaging modalities, these techniques unveil how cognitive processes are coded in space or in time. Moreover, recent trends allow fusion methods to combine signals of different nature and offer both spatial and temporal coherent information. This work reviews and implements in the MVPAlab Toolbox the Representational Similarity Analysis (RSA) for electroencephalography signals, which is a preliminary step to EEG-fMRI data fusion. To evaluate this methodology we have built a demo dataset from a pre-recorded EEG experiment designed to study differences in preparation between perceptual expectation and selective attention. We discuss the strengths and the versatility of this multivariate technique and its potential applications on multimodal data fusion. The complete source code is fully-integrated in the MVPAlab Toolbox, which increases the broad number of already implemented analyses and the versatility of the tool.

**Keywords:** Representational Similarity Analysis · RSA · Multimodal data fusion · MVPAlab · Electroencephalography

## 1 Introduction

The vast progress of science and technology occurred in the past few decades has witnessed the use of Machine Learning-based techniques in a wide range of scientific disciplines, including neuroscience [1]. Nowadays, the study of brain functioning employing magneto/electroencephalography (M/EEG) signals relies mostly on multivariate techniques, leaving behind classic univariate approaches such as Event-Related Potentials (ERPs). Multivariate approaches outperform univariate ones in terms of sensitivity detecting subtle changes in activations


J. M. Ferrández Vicente et al. (Eds.): IWINAC 2022, LNCS 13258, pp. 84–94, 2022.
https://doi.org/10.1007/978-3-031-06242-1_9

associated with specific information content in brain patterns. As a result, several Multivariate Pattern Analysis (MVPA) toolboxes, specifically designed for M/EEG signals, have recently emerged (e.g. The Amsterdam Decoding and Modeling Toolbox [2], MVPA-light [3], The Decision Decoding Toolbox [4], or The MVPAlab Toolbox [5], among others [6–8]).

However, due to the nature of the signal of current non-invasive neuroimaging techniques, these tools help neuroscientists to characterize cognitive processes either in a time or space-resolved way. While M/EEG signals provide exceptional temporal resolution but lack spatial resolution, functional Magnetic Resonance Imaging (fMRI), localizes brain activity changes at millimetric levels but with poor temporal resolution. Overcoming this dichotomy is one of the major current challenges in the field, with some trends betting on multimodal data fusion as a promising solution [9]. Fusing methods combine information of non-concurrent recordings from different neuroimaging modalities, preserving their individual strengths while overcoming their weaknesses.

Nevertheless, relating activity patterns between different modalities of brain-activity measurement is not a straight forward exercise, since those techniques measure signals of different nature (e.g. hemodynamic response vs. electrical activity), that have no direct correspondence. Therefore, a methodology that abstracts from the activity patterns *per se* is required to solve this correspondence problem. The main goal of this work is to implement Representational Similarity Analysis [10] (RSA) on electroencephalography signals, a methodological approach that abstracts from the signal space to a common similarity space, generating commensurable dissimilarity matrices for different experimental conditions and thus allowing fusion of different signals. These analyses have been fully-integrated in MVPAlab, an easy-to-use decoding toolbox for multidimensional M/EEG data. The inclusion of these analyses increases the MVPAlab Toolbox versatility and lays the foundation for the multimodal fusion methods, which represents one of the most important lines of development on the MVPAlab roadmap.

## 2 Materials and Methods

### 2.1 Materials

High-density electroencephalography signals were recorded from forty-eight participants to study differences in preparation between perceptual expectation and selective attention [data currently in preparation for submission]. To evaluate the performance of RSA analyses we compiled a sample EEG dataset including five different EEG data files corresponding to five participants randomly selected from the original sample. This demo dataset and the implemented code for the RSA analysis is publicly available in the MVPAlab's GitHub repository. As shown in Table 1, eight different experimental conditions were extracted for each participant. The experimental details of this study are specified in the following paragraphs.

***Experimental Design.*** The main task consisted on a cue-target sequence as shown in Fig. 1, where participants were required to discriminate the sex (male or female) of the upcoming target stimuli (a *face* or a *name*). Each participant repeated this sequence 640 times in total, divided into 32 blocks of 20 trials each. Two different types of blocks were defined: *attention* and *expectation*. In expectation blocks, the shape of the cue predicted the category of the upcoming stimulus (75% of *validity*). Thus, after a cue associated with faces, a face appeared with a 75% of probability. In attention blocks, the shape of the cue indicated the category of the upcoming stimulus to respond to (50% of *validity*).

**Table 1.** Extracted experimental conditions: Target-locked trials were grouped into eight different experimental conditions based on their category (name vs. face), block type (attention vs. expectation) and validity (valid vs. invalid).

| Valid trials | | Invalid trials | |
|---|---|---|---|
| Attention | Expectation | Attention | Expectation |
| `val_name_att` | `val_name_exp` | `inv_name_att` | `inv_name_exp` |
| `val_face_att` | `val_face_exp` | `inv_face_att` | `inv_face_exp` |

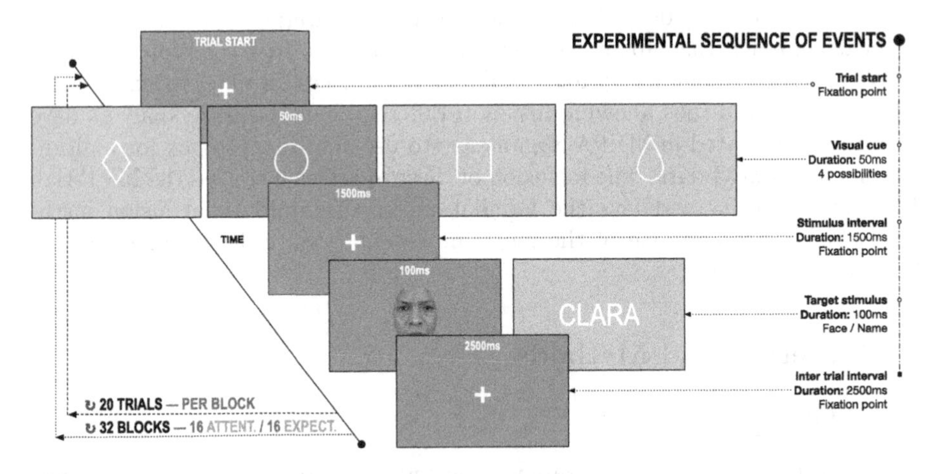


**Fig. 1. Experimental sequence of events.** The complete sequence is repeated 640 times, grouped into 32 blocks. After the stimulus presentation, participants had to respond *"yes"* or *"no"* (by pressing the key `A` or `L`) to the sex discrimination task. Block order, cue shape, stimulus category and response keys are fully counterbalanced across participants to avoid potential confounds. Each sequence was categorized based on its validity (valid or invalid trials), block type (attention or expectation trials) and its category (names or faces). Face images were extracted from *The Chicago face database.* [11]

***EEG Data Acquisition and Preprocessing.*** High-density electroencephalography was recorded from 64 electrodes mounted on an elastic cap (actiCap slim, Brain Products) in a magnetically shielded room at the Mind, Brain and Behavior Research Center (CIMCYC) of the University of Granada. Impedances were kept below 10k. EEG activity was referenced to the FCz electrode during the recording session and signals were digitalized at a sampling rate of 1 KHz.

Following previous literature [12,13], electroencephalography recordings were downsampled 256 Hz and digitally filtered between 0.1 and 120 Hz using a high and low-pass FIR filters, preserving the phase information. Power line interference and its harmonics [50 and 100 Hz] were removed using a notch filter. All channels were visually inspected and, on average, 1.85% of them were excluded due to excessive noise. EEG recordings were epoched [−1000, 2000 ms locked at the target presentation] extracting data only from correct trials. To remove blinks and eye movements from the remaining data, Independent Component Analysis (ICA) was computed using the *runica* algorithm from EEGLAB [14]. Artifactual components were rejected by visual inspection of the raw activity for each component, scalp maps and power spectrum. Then, an automatic trial rejection process was performed, pruning the data from no stereotypical artifacts. The trial rejection procedure was based on (1) abnormal spectra: the spectrum should not deviate from baseline by ±50 dB in the 0–2 Hz frequency window (which is optimal for localizing any remaining eye movements) and should not deviate by −100 dB or +25 dB in 20–40 Hz (useful for detecting muscle activity); (2) improbable data: the probability of occurrence of each trial was computed by determining the probability distribution of values across trials, with a rejection threshold set at ± (3) extreme values: all trials with amplitudes in any electrode out of ±150 µV range were automatically rejected. On average, a total of 8% of the trials were automatically removed in the trial rejection stage. Finally, previously excluded channels were reconstructed by spherical interpolation and the entire dataset was average re-referenced and baseline corrected [−200, 0 ms].

### 2.2 Methods

***Representational Similarity Analysis*** is a multivariate computational technique employed to reveal fundamental insights about how information is represented in the brain. Note that the term *representation* is usually interpreted as the activity pattern induced in the brain by certain experimental conditions or stimuli [10]. First, brain activity is recorded employing a neuroimaging technique while the participant is performing a task. During the task, the participants are exposed to several experimental conditions or stimuli that evoke different spatiotemporal activity patterns across the brain. The main goal of RSA analysis is to understand these spatiotemporal representations by abstracting from the signal space to a high-order and common representational space [15]. In order to do that, activity patterns associated with each pair of experimental conditions are related and visualized constructing 2-dimensional *Representational Dissimi-*

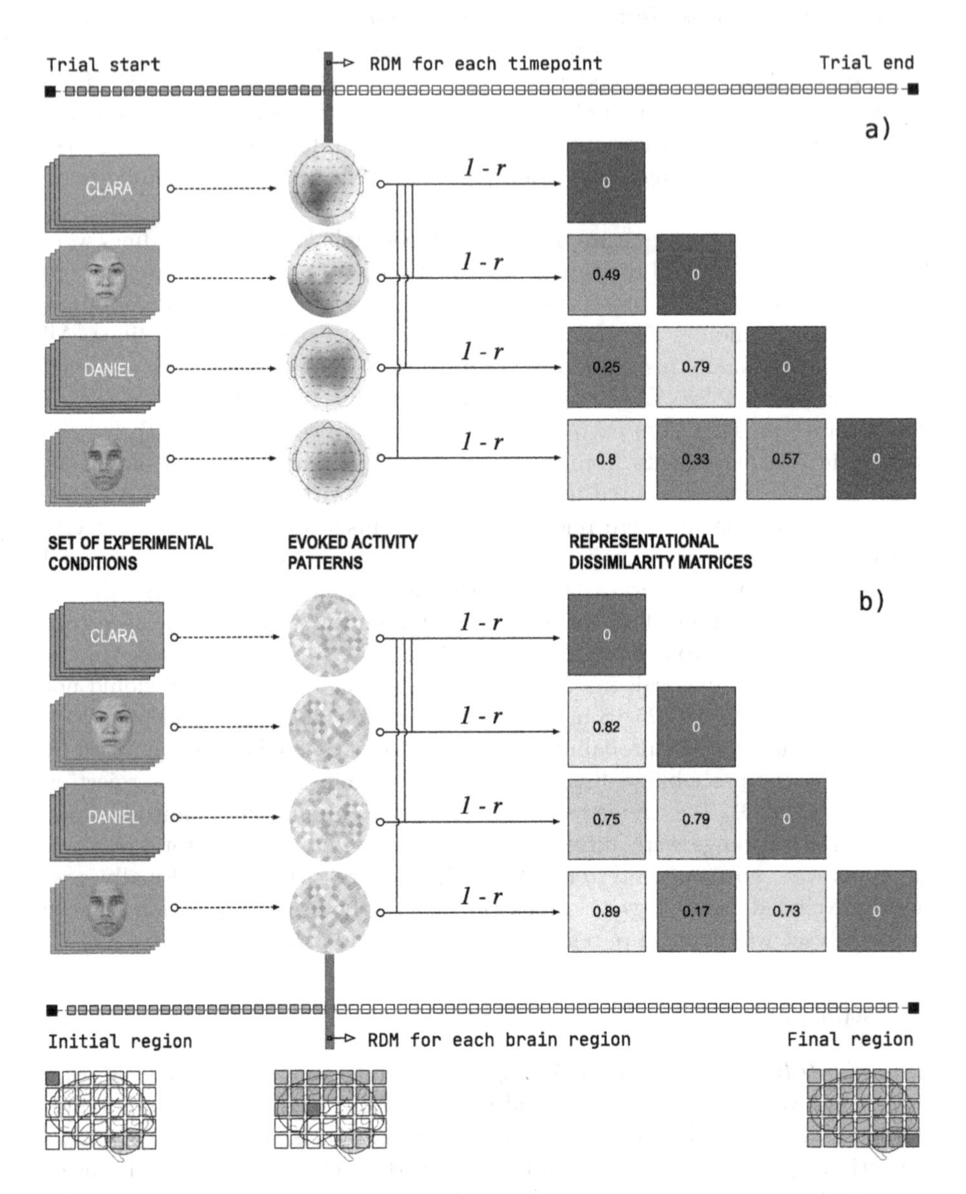


**Fig. 2. Representational Similarity Analysis scheme for different neuroimaging modalities.** During the experiment, participants are exposed to different stimuli or experimental conditions that elicit different condition-specific activity patterns in the brain. These patterns are then pairwise compared by computing a dissimilarity measure (e.g. 1− Pearson's $r$) and arranged in the so-called Representational Dissimilarity Matrix (RDM). Depending on the neuroimaging modality employed these RDMs can be generated in **(a)** a time or **(b)** space-resolved manner.

*larity Matrices* (RDMs). One of the most remarkable characteristic of DRMs is that they are comparable no matter the nature of the original data.

Therefore, representations across brain regions, temporal points, individuals, neuroimaging modalities or even animal species are now easily comparable in this high-order representational space by computing the correlation between different DRMs [16]. As stated in the Introduction section, this is one of the basic principles of multimodal data fusion: employing a common analysis framework that solves the intrinsic correspondence problem between different neuroimaging techniques. This approach is very flexible and versatile, allowing the comparison of DRMs in different contexts depending on the researcher's question and hypotheses [17].

***Representational Dissimilarity Matrices*** are the cornerstone of RSA technique. These matrices present pairwise dissimilarities between experimental conditions revealing how distinguishable are the brain-activity patterns associated with them. Thus, a DRM describes the geometry of the arrangement of these patterns in the representational space [15]. By definition, RDM are square matrices horizontally and vertically indexed by the set of stimuli (or experimental conditions), leading to a symmetric matrix along its diagonal (see Fig. 3). Zero values at the diagonal represent the dissimilarity obtained by comparing each experimental condition to itself while values outside the diagonal result from pairwise comparisons between all possible combinations of two different experimental conditions. As an example, Fig. 2 depicts how DRMs could be constructed from different data sources (M/EEG and fMRI). In our case, for temporal-accurate neuroimaging modalities such as EEG or MEG a neural RDM is generated for each time point and participant. Alternatively, in fMRI studies individual DRMs are usually constructed for each brain region and participant.

Dissimilarity between activity patterns can be assessed employing several dissimilarity measures. Selecting an adequate one is one of the most important decisions in RSA since it usually depends on the inferential aim of the analysis and the original data [18]. In the MVPAlab Toolbox we implemented three of the most popular measures to estimate neural dissimilarity matrices: the correlation distance (1− Pearson's $r$), the Euclidean distance and the Mahalanobis distance. First, in order to reduce the signal-to-noise ratio and to improve computational efficiency, trials belonging to same condition were randomly averaged in threes. Then, for each time-point and participant neural RDMs were estimated employing the previous dissimilarity measures.

***Theoretical RDM Models.*** Representational Dissimilarity Matrices can be generated not only from neural or behavioral sources but also from theoretical predictions. These are not empirical but conceptual RDMs and they are generated based on expected theoretical dissimilarities between experimental conditions. Hence, to test the RSA technique we built tree different conceptual RDMs based on (1) stimulus type: names vs. faces (2) block type: attention vs. expectation and (3) trial validity: valid vs. invalid. As shown in Fig. 3 a binary

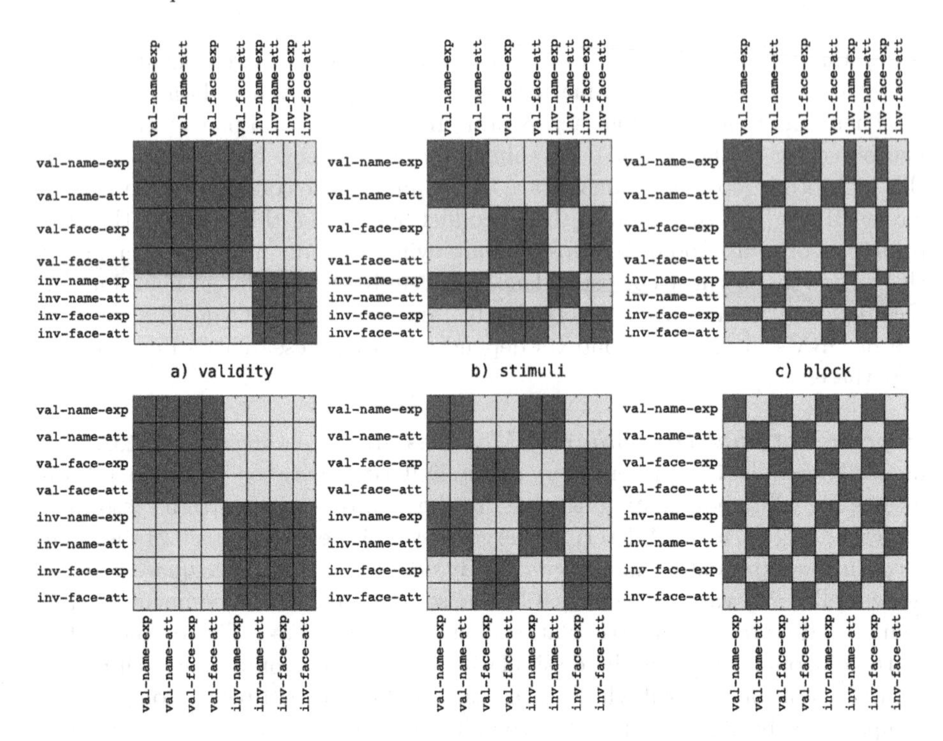


**Fig. 3. Theoretical RDMs models.** Three different conceptual RDMs built based on the three main variables in our data: trial validity (valid or invalid), stimulus category (names or faces), block type (attention or expectation). Note that each model can be constructed assuming identical number of trial per condition (bottom row of the figure), leading to symmetrical cell sizes, or adjusting the size of the cell according to the actual number of trials per condition (top row of the figure). (Color figure online)

value of expected dissimilarity [0—1] was assigned to each pair of experimental conditions, represented in dark and light green in the figure. For example, focusing on the stimulus category, we can assume that activity patterns elicited by faces are more similar between them than those elicited by names (Fig. 3b). Similarly, attending to the type of the block, we can hypothesise that activity patterns elicited by stimulus in attentional blocks differ from those elicited in expectation blocks (Fig. 3c), and the same occurs for validity (Fig. 3a).

***Second Order Dissimilarity Analysis.*** Once the neural and theoretical RDMs were estimated, we performed the second order dissimilarity analysis. Since an RDM is square and symmetrical matrix along its diagonal, both the diagonal and the upper triangle were removed for computational efficiency. Then, for each time-point, participant and theoretical model, the resulting matrices were vectorized and compared employing the Spearman correlation. As a result, we obtained a time series of Fisher's Z transformed correlation coefficients for each theoretical model and participant (see Fig. 4a). Finally, we computed a two-tailed cluster-based permutation analysis to test if the obtained coefficients were significantly higher or lower than zero ($10^5$ permuted iterations and $\alpha = 0.001$ for both group-average and cluster-size levels)

## 3 Results and Discussion

The estimated neural dissimilarity matrices time series for a specific participant is shown in the Fig. 4a. Three different time-points are depicted in the figure, $t = -80$ ms, $t = 220$ ms and $t = 380$ ms. As shown, before the stimulus presentation ($t = -80$ ms) the neural RDM do not visually present a clear pattern, an indication of low or zero correlation with any of the theoretical models as it will be outlined below. However, 220 ms–3800 ms after the stimulus presentation the RDMs depicts a clear dissimilarity pattern which resembles one of our theoretical models, which leads to positive correlation with that particular model. The euclidean distance was the selected dissimilarity measure, but the analysis was repeated using other measures such as 1− Pearson's $r$ obtaining equivalent results. Note that the cell sizes are not symmetrical due to the different number of trial per class, which means that the analyses were done in a trial-by-trial manner. Thus, the size of the matrices (and the experimental conditions distributions) is different for each participant, which implies that theoretical RDMs should also be specifically generated and adapted for them. This also means that a direct comparison of RDMs between participants (or neuroimaging modalities) is not allowed, weakening the actual potential of this analytic framework and its applications. This inconvenience can be easily solved by collapsing trial information and computing the analysis in a condition-by-condition manner. This way, if the same set of experimental conditions is employed this approach could combine information from different participants, neuroimaging modalities or even animal species.

The results of the second order Representational Similarity Analysis are depicted in the Fig. 4b and c. As shown, the Spearman correlation was computed between the neural RDMs and the three theoretical models for each participant in a time-resolved way. Statistically significant results ($\alpha = 0.001$) were obtained for our three theoretical models. Positive correlations were found practically the entire time windows for the stimulus category model, yielding a correlation coefficient of Fisher($\rho$)$\sim$ 0.8 in $t = 220$ ms after the stimulus onset, which indicates how differently the brain represents faces vs. non-faces information.

As stated before, the main goal of this work is to integrate the Representational Similarity Analysis in a user-friendly software tool such as the MVPAlab toolbox and discuss its potential applications, not just in M/EEG but also in combination with other neuroimaging techniques such as fMRI, preserving their individual strengths while overcoming their weaknesses. Therefore, this work increments the versatility of the MVPAlab Toolbox and more importantly, it serves as a stepping stone for developing forefront techniques such as multimodal fusion models.

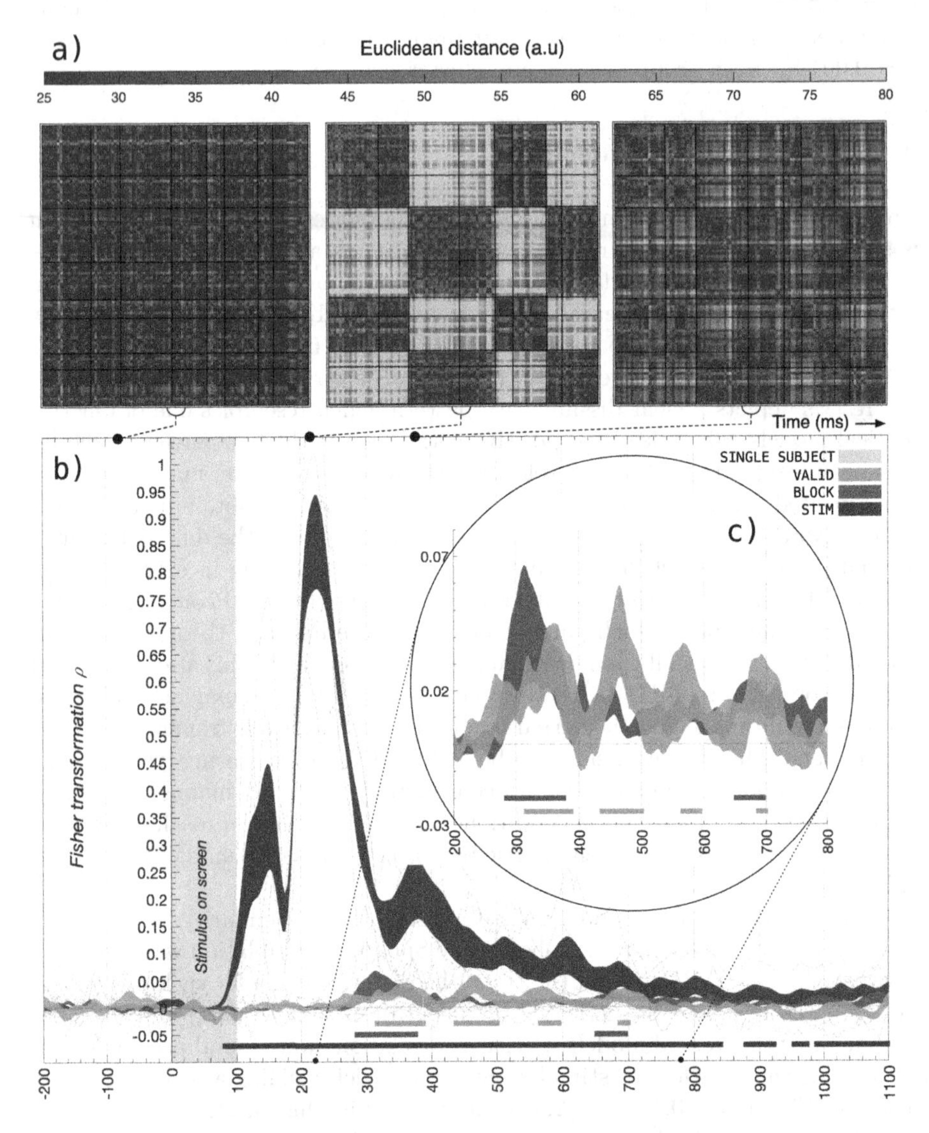


**Fig. 4. Representational similarity analyses results.** **(a)** Neural RDMs estimated through the euclidean distance for a specific participant at three different time-points: $t = -80\,\text{ms}$, $t = 220\,\text{ms}$ and $t = 380\,\text{ms}$. **(b)** Second order analysis: Time series of the group-averaged Fisher's Z transformation of the $\rho$ correlation coefficient. Shaded areas represent the Standard Error of the Mean for each theoretical model. Dashed grey lines represent the single participant results. Statistically significant time windows are highlighted using horizontal bars. **(c)** Enlarged version of the previous diagram to better appreciate the positive correlations between neural RDMs and theoretical block type and validity models.

## 4 Conclusions

The current work is a preliminary approximation to multimodal data fusion techniques based on Representational Similarity Analysis. We developed and tested this analytic framework for electroencephalography signals employing a sample dataset specifically designed to study differences in preparation mechanisms. One of the many potential applications of RSA analysis is to compute correlations between neural dissimilarity matrices and theoretical models based on our predictions to study how information is represented in the brain. Three theoretical models were designed attending to the three main variables in our experiment: stimulus category, block type and trial validity. Statistically significant positive correlations were found between the empirical and theoretical models, yielding a Fisher($\rho$)$\sim$0.8 for the stimulus category model 220 ms after the stimulus onset. This methodology has been fully-integrated in MVPAlab, an easy-to-use decoding toolbox for multidimensional data, increasing its versatility and laying the foundation for the multimodal fusion methods.

**Acknowledgments.** This research was supported by the Spanish Ministry of Science and Innovation under the PID2019-111187GB-I00 grant, by the MCIN/AEI/10.13039/50110 0 011033/ and FEDER "Una manera de hacer Europa" under the RTI2018-098913-B100 project, by the Consejería de Economía, Innovación, Ciencia y Empleo and FEDER under CV20-45250, A-TIC-080-UGR18, B- TIC-586-UGR20 and P20-00525 projects. The first author of this work is supported by a scholarship from the Spanish Ministry of Science and Innovation (BES-2017-079769).

## References

1. Górriz, J.M., et al.: Artificial intelligence within the interplay between natural and artificial computation: advances in data science, trends and applications. Neurocomputing **410**, 237–270 (2020)
2. Fahrenfort, J.J., van Driel, J., van Gaal, S., Olivers, C.N.L.: From ERPs to MVPA using the Amsterdam Decoding and Modeling toolbox (ADAM). Front. Neurosci. **12**, 368 (2018)
3. Treder, M.S.: MVPA-light: a classification and regression toolbox for multidimensional data. Front. Neurosci. **14**(June), 1–19 (2020)
4. Bode, S., Feuerriegel, D., Bennett, D., Alday, P.M.: The Decision Decoding ToolBOX (DDTBOX) - a multivariate pattern analysis toolbox for event-related potentials. Neuroinformatics **17**(1), 27–42 (2019)
5. López-García, D., Peñalver, J.M.G., Górriz, J.M., Ruz, M.: MVPAlab: a machine learning decoding toolbox for multidimensional electroencephalography data. Comput. Methods Programs Biomed. **214**, 106549 (2022)
6. Oostenveld, R., Fries, P., Maris, E., Schoffelen, J.M.: FieldTrip: open source software for advanced analysis of MEG, EEG, and invasive electrophysiological data. Comput. Intell. Neurosci. **2011**, 156869 (2011)
7. Hanke, M., et al.: PyMVPA: a unifying approach to the analysis of neuroscientific data. Front. Neuroinform. **3**, 1–13 (2009)
8. Gramfort, A., et al.: MEG and EEG data analysis with MNE-Python. Front. Neurosci. **7**, 1–13 (2013)

9. Cichy, R.M., Oliva, A.: A M/EEG-fMRI fusion primer: resolving human brain responses in space and time. Neuron **107**(5), 772–781 (2020)
10. Kriegeskorte, N., Mur, M., Bandettini, P.: Representational similarity analysis - connecting the branches of systems neuroscience. Front. Syst. Neurosci. **2**, 1–28 (2008)
11. Ma, D.S., Correll, J., Wittenbrink, B.: The Chicago face database: a free stimulus set of faces and norming data. Behav. Res. Methods **47**(4), 1122–1135 (2015). https://doi.org/10.3758/s13428-014-0532-5
12. López-García, D., Sobrado, A., González-Peñalver, J.M., Górriz, J.M., Ruz, M.: Multivariate pattern analysis of electroencephalography data in a demand-selection task. In: Ferrández Vicente, J.M., Álvarez-Sánchez, J.R., de la Paz López, F., Toledo Moreo, J., Adeli, H. (eds.) IWINAC 2019. LNCS, vol. 11486, pp. 403–411. Springer, Cham (2019). https://doi.org/10.1007/978-3-030-19591-5_41
13. López-García, D., Sobrado, A., Peñalver, J.M.G., Górriz, J.M., Ruz, M: Multivariate pattern analysis techniques for electroencephalography data to study flanker interference effects. Int. J. Neural Syst. **30**(7), 2050024 (2020)
14. Delorme, A., Makeig, S.: EEGLAB: an open source toolbox for analysis of single-trial EEG dynamics including independent component analysis. J. Neurosci. Methods **134**(1), 9–21 (2004)
15. Nili, H., Wingfield, C., Walther, A., Su, L., Marslen-Wilson, W., Kriegeskorte, N.: A toolbox for representational similarity analysis. PLoS Comput. Biol. **10**(4), e1003553 (2014)
16. Kriegeskorte, N., Kievit, R.A.: Representational geometry: integrating cognition, computation, and the brain. Trends Cogn. Sci. **17**(8), 401–412 (2013)
17. Popal, H., Wang, Y., Olson, I.R.: A guide to representational similarity analysis for social neuroscience. Soc. Cogn. Affect. Neurosci. **14**(11), 1243–1253 (2019)
18. Walther, A., Nili, H., Ejaz, N., Alink, A., Kriegeskorte, N., Diedrichsen, J.: Reliability of dissimilarity measures for multi-voxel pattern analysis. Neuroimage **137**, 188–200 (2016)

# Towards Mixed Mode Biomarkers: Combining Structural and Functional Information by Deep Learning

A. Ortiz[1,3(✉)], Juan E. Arco[1,3], Marco A. Formoso[1,3], Nicolás J. Gallego-Molina[1,3], Ignacio Rodríguez-Rodríguez[1], J. Martínez-Murcia[2,3], Juan M. Górriz[2,3], and Javier Ramírez[2,3]

[1] Communications Engineering Department, University of Málaga, 29004 Málaga, Spain
aortiz@ic.uma.es

[2] Department of Signal Theory, Communications and Networking, University of Granada, 18060 Granada, Spain

[3] Andalusian Data Science and Computational Intelligence Institute (DaSCI), Granada, Spain

**Abstract.** The transference of information between entities is one of the most popular application of deep learning. It has been used to generate a stylized version of an image by combining a source image to another that determines the *style* of the final result. In the field of neuroimaging, different modalities are frequently available, providing structural or functional information. Those modalities are usually analyzed separately, although it is possible to jointly use features extracted from structural and functional neuroimage to improve the classification performance in Computer Aided Diagnosis (CAD) tools. In this paper we propose a method based on the principles of neural style transfer to combine information from Magnetic resonance Imaging (MRI) and Positron Emission Tomography (PET), generating a new image that contains structural and functional information. The usefulness of this method has been assessed with images from the Alzheimer Disease Neuroimaging Initiative, demonstrating that using the new mixed mode image outperforms the classification accuracy obtained by individual MRI or PET images.

**Keywords:** Deep learning · Style transfer · Modality combination · PET

## 1 Introduction

Current neuroimaging techniques provide very useful information regarding the structure or functional state of the brain. Thus, features extracted from those modalities are usually exploited in Computer Aided Diagnosis (CAD) tools, since they figure out discriminative information which is highly valuable to diagnose different neurological and neurodegenerative diseases. Both image modalities provide different but complementary information: while structural MRI informs about the distribution of the different tissues in the brain (mainly gray matter and white matter), functional imaging such as PET informs about changes in cerebral blood flow and changes in cerebral glucose metabolism. This is an indicator of neuronal activity that can be figured out by means


J. M. Ferrández Vicente et al. (Eds.): IWINAC 2022, LNCS 13258, pp. 95–103, 2022.
https://doi.org/10.1007/978-3-031-06242-1_10

of different radiotracers. For instance, 18F-FDG-PET imaging has been extensively used for the diagnosis of neurodegenerative diseases such as the Alzheimer disease [1,4,6,8,10,13]. In a similar way, changes in different brain structures related to the progression of a neurodegenerative proccess can be revealed by structural MRI, and further used in differential diagnosis tasks [3,5,11–14]. These works use GM or WM images obtained by segmentation of MRI to classify controls and AD patients [3,5] or to compute Regions of Interest (ROI), searching for common patterns in Controls (CN) and AD subjects. Moreover, the construction of neurodegeneration models to study the progression of the disease usually requires the use of both, functional and structural information. The construction of multimodal models is generally addressed by combining features computed from functional and structural images [13,14], which in turn, consists on constructing two different models. This method has demonstrated its effectiveness in differential diagnosis tasks, but its use in exploratory analysis is limited since it is difficult to link structural and functional features. On the other hand, information fusion techniques aim to combine information from different sources to generate new knowledge, not explicitly present in each independent source. This could aim to jointly explot features of different nature with discriminative or exploratory purposes. In this way, different deep learning architectures have been developed to combine information, or to transfer information from one entity to another, especially for image or video data. This is the case of Neural Style Transfer [9], in which an input image is *stylized* by the so-called *style* image. This process consists on extracting features from both images and modify the input one according to a specific objective (loss) function that minimizes the similarity between features extracted at differnt abstraction levels.

In this paper we propose a method to combine structural information from MRI imaging with functional information from PET imaging, generating a new multi-modal image that condensates these two image modalities. The proposal is based on unsupervised learning, thus the process is performed for each image individually, without taking into account label information. The proposed methodology has been evaluated using structural and functional images from the Alzheimer Disease Neuroimaging Initiative (ADNI) [2]

The rest of the paper is organized as follows. After this introduction, Sect. 2 shows the methods used in this work, focusing on the deep learning architectures used. Section 3 presents the results in two ways. Firstly, some examples of generated mixed mode images are shown to visually assess the differential information figured out by the proposed method betwen different labelled images. Moreover, the discriminative power of the generated images are evaluated by a classification test. Finally, Sect. 4 shows the conclusions and future work.

## 2 Material and Methods

The method proposed in this work aims to combine functional and structural information figured out by imaging methods. Specifically, structural innformation is extracted from MRI imaging, while the functional status of the brain is figured out from PET imaging. These two modalities have been obtained from the Alzheimer's Disease Neuroimaging Initiative (ADNI), and preprocessed as explained in the following. The database used in this work contains multimodal PET/MRI image data from 138 subjects, comprising 68 Controls (CN), 70 AD from the ADNI database [2]. This repository was created to study the advance of the Alzheimer's disease, collecting structural MRI and functional imaging from Positron Emission Tomography (PET) as well as different biomarkers. Subject's demographics are shown in Table 1.

**Table 1.** Patient demographics

| Evaluation | Sex (M/F) | Mean Age $\pm$ Std | Mean MMSE $\pm$ Std |
|---|---|---|---|
| NC | 43/25 | $75.81 \pm 4.93$ | $29.06 \pm 1.08$ |
| AD | 46/24 | $75.33 \pm 7.17$ | $22.84 \pm 2.91$ |

MRI images from the ADNI database have been spatially normalized according to the VBM-T1 template and segmented into White Matter (WM) and Grey Matter (GM) tissues using the VBM toolbox for SPM [17].

This ensures each image voxel to correspond to the same anatomical position. After image registration, all the images from ADNI database were resized to $121 \times 145 \times 121$ voxels with voxel-sizes of 1.5 mm (Sagittal) $\times$ 1.5 mm (coronal) $\times$ 1.5 mm (axial). A nonlinear deformation field is estimated that best overlays the tissue probability maps on the individual sujects' image. The tissue probability maps provided by the International Consortium for Brain Mapping (ICBM) are derived from 452 T1-weighted scans, which were aligned with an atlas space, corrected for scan inhomogeneities, and classified into grey matter, white matter and cerebro-spinal fluid. Segmentation through SPM/VBM provides values in the range $[0, 1]$ indicating the membership probability to a specific tissue.

In the case of structural MRI, most part of the information is contanined in the distribution of the tissues in the brain. Gray matter (GM) makes up the outer most layer of the brain, consisting of neuronal cell bodies and then, it is directly related to the concentration of neurons in a specific area of the brain. This way, the segmented GM is used here to construct a GM density map, which informs of the neuronal concentration at each brain area. Brain areas were considered according to the AAL neuroanatomical atlas [15] which delimitates 116 regions. The atlas has been registered along with the MRI and PET images so that it is possible to use it to mask the brain for extracting information at different areas.

## 2.1 GM Density Map

GM density at each brain region has been computed volume in the $i$-region by using the following equation

$$D_i = \frac{\#voxels_i > thr}{\#Voxels\_region\_i} \tag{1}$$

The threshold $thr$ in Eq. 1, has been selected to 0.5, indicating that those volxels with a probability higher than 0.5 are considered as GM.

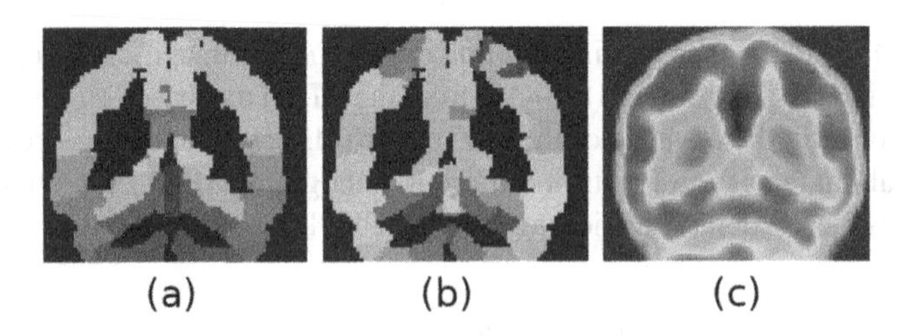


**Fig. 1.** AAL116 atlas (a), Gray Matter density map (b) and PET functional image (c) for a control subject

Figure 1(a) shows the AAL116 atlas co-registered along with PET images, Fig. 1(b) the GM density map computed using the method previously explained and Fig. 1(c) the PET functional image for a control subject.

In the following, we will explain the method used to fuse information from the GM density map (Fig. 1(b)) and PET functional image Fig. (1(c))

### 2.2 Feature Extraction Using Deep Learning

Image fusion can be seen as the process of transferring information from one image to another. Different methods based on deep learning architectures that use generative models have been developed to this end. One of the first references for that is the neural style transfer method [9]. This is based on the extraction of features at different abstraction levels for a reference image and the so-called *style* image, in such a way that features from the *style* image are transferred to the one used as reference. In that method, a pre-trained network, usually a VGG net [16] is used to extract features at different layers. Then, a specific loss function is computed regarding the similarity between the *stilyzed*, the *style* image and the so-called *content* image used as a reference. It is worth noting that VGG network is trained for image classification, and then, the training process is supervised.

In our case, we do not have a pre-trained network with 3D, PET images or MRI images. This way, we propose a method to compute 3D features by means unsupervised learning. This method is based on a 3D convolutional stacked autoencoder. The specific architecture used here is shown in Fig. 2.

Stacked autoencoders are usually based on using the bottleneck layer as input to another autoencoder and so on. This is performed in such a way that the hidden layers of the final network is composed of the bootleneck layers of the autoencoders previously trained. This final network is fine-tuned by supervised learning. In our implementation shown in Fig. 2, we use a different approach that only uses unsupervised learning to extract features at different abstraction levels and scales. Each of the autoencoders used here consist of four convolutional layers, two for the encoder the decoder subnetworks. Thus, once the autoencoder is trained, the feature map provided by the last convolutional layer in the autoencoder subnetwork is used as input for the next autoencoder. This layer will be further used during the inference to extract features.

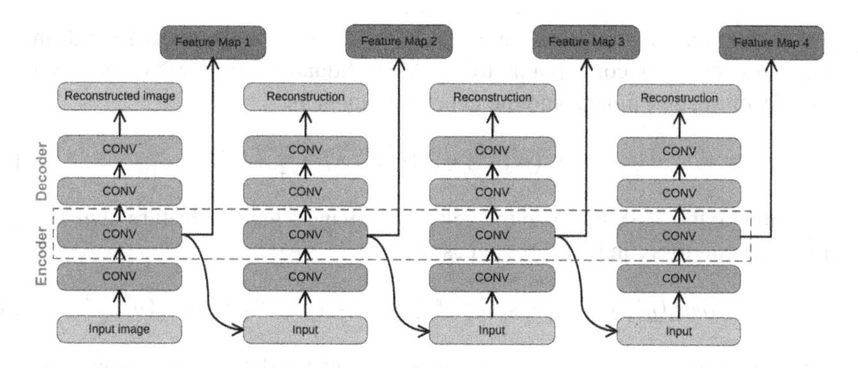


**Fig. 2.** Stacked autoencoder architecture used to extract features from 3D images

## 2.3 Mixed Mode Images: Image Fusion Procedure

Image fusion is addressed with a similar method that the one used in original neural style transfer [9] work. It is based on iteratively modify an image according to the gradients provided by a loss function. In Fig. 3, it can be seen that the second layer of each autoencoder is used to compose a new network. This network is only used to extract features in the same way that VGGnet is used in [9]. However, while VGGnet is based on supervised training, our method based on stacked autoencoeders, does not use label information in any part of the process. The stacked autoencoder is trained using the PET images in our database. Then, three loss functions are used to compute the loss between the input images and the feature maps.

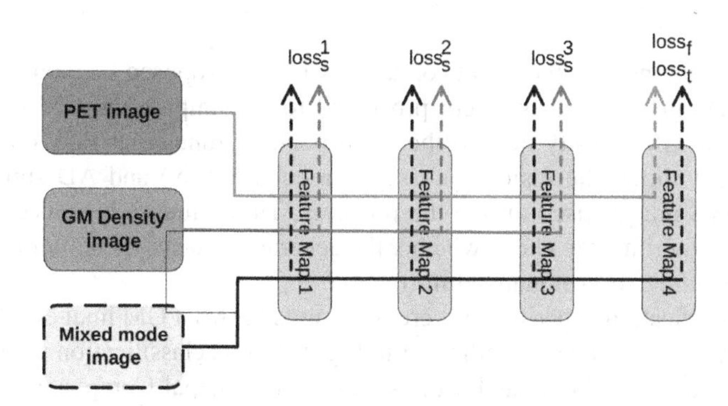


**Fig. 3.** Network architecture to generate the mixed mode image

These loss functions for the layer $l$ can be expressed as:

$$loss_s(g^l, f^l) = \|\mathcal{G}(g^l) - \mathcal{G}(f^l)\|_2^2 \tag{2}$$

where $\mathcal{G}$ is the Gram matrix, defined as $\mathcal{G}(A) = A \cdot A^T$ for the usual Euclidean dot product. Thus, this loss corresponds to the Mean Squared Error (MSE) between the gram matrix of image $f$ image and the generated image $g$.

$$loss_c(g^l, f^l) = \|(g^l) - (f^l)\|_2^2 \tag{3}$$

corresponding to the MSE between the generated image $g$ and the feature map $f$.

Thus, the total loss can be expressed as:

$$loss = loss_s(g^1, f^1) + loss_s(g^2, f^2) + loss_s(g^3, f^3) + loss_c(g^4, f^4) \tag{4}$$

Moreover, a total variation loss term that account for the variation among neighbouring voxels in the generated image is added to the final loss function, as

$$loss_{tv}(I) = \sum_{ijk} |I_{i+1,j,k} - I_{i,j,k}| + |I_{i,j+1,k} - I_{i,j,k}| + |I_{i,j,k+1} - I_{i,j,k}| \tag{5}$$

This aims to smooth the final result, facilitating the interpretability. In addition, each term of the final loss is weighted by a coefficient that determines the influence of that term during the optimization process. Consequently, the final loss function used is:

$$Loss = \alpha \cdot loss_s + \beta \cdot loss_c + \gamma \cdot loss_{tv} \tag{6}$$

The values of the weights $\alpha$, $\beta$ and $\gamma$ have been determined by experimentation ($\alpha = 100, \beta = 0.1, \gamma = 10$).

## 3 Results

In this section, we present the results obtained with the proposed method. The images for all subjects in the database were processed to 1) compute the GM density map and 2) fuse the GM density map to the corresponding functional PET image. As an example, Fig. 4 shows the fusion process for a control (CN) and AD subject, along with the source images used during the proccess. These images have been used in a classification experiment to check whether the generated images (hereafter mixed mode images) retained the discriminative information.

These classification experiments were performed for MRI GM images, PET images and the *mixed mode images*, as indicated in Fig. 5. In this classification pipeline, input images are treated in a Voxel as Features way, and Principal Component Analysis is used to reduce the dimensionality of the feature space [7] (which is considerably high in the VAF approach). Using this method, the input images are projected onto the space spanned by a number of prinicipal components (npc), generating a new *npc*-dimensional feature space.

All the results in this work were assessed by k-fold cross-validation (k=10) to ensure the independence between the training and testing subsets.

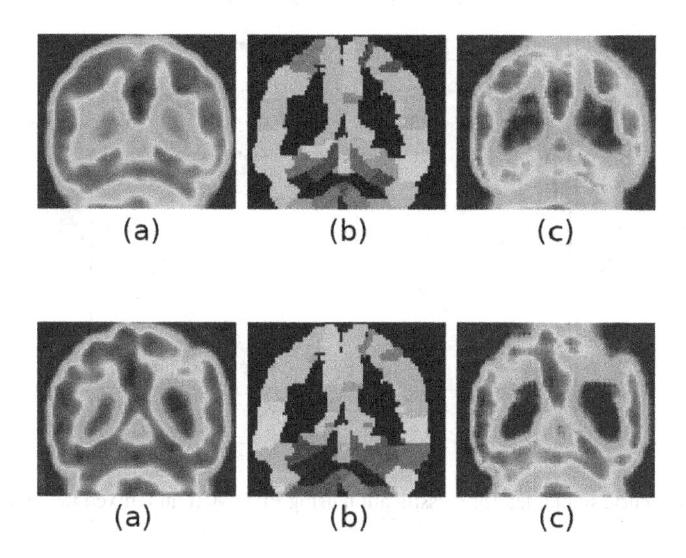


**Fig. 4.** Generated images for CN (first row) and AD subjects (second row). PET image, GM map are shown in (a) and (b), respectively. Generated image (mixed mode) using the proposed mehot is shown in (c).

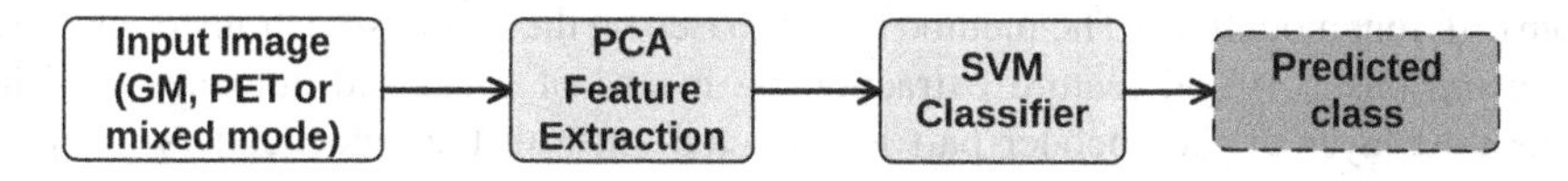


**Fig. 5.** Classification method. PCA is used to extract features from the input images, and SVM is used as classifier

**Table 2.** Classification results

| Image modality | Accuracy | Sensitivity | Specificity | AUC |
|---|---|---|---|---|
| PET | $0.84 \pm 0.08$ | $0.86 \pm 0.13$ | $0.82 \pm 0.10$ | $0.92 \pm 0.06$ |
| MRI GM | $0.81 \pm 0.08$ | $0.82 \pm 0.15$ | $0.80 \pm 0.11$ | $0.90 \pm 0.10$ |
| **Mixed Mode** | $\mathbf{0.87 \pm 0.10}$ | $\mathbf{0.85 \pm 0.10}$ | $\mathbf{0.88 \pm 0.12}$ | $\mathbf{0.93 \pm 0.08}$ |

As shown in Table 2 and Fig. 6 mixed mode images not only retain the discriminative information included in GM and PET images but also provide a higher performance that using single-modality images. This desmonstrates the viability of this method to fuse information from different sources and validates the discriminative capabilities of mixed mode images.

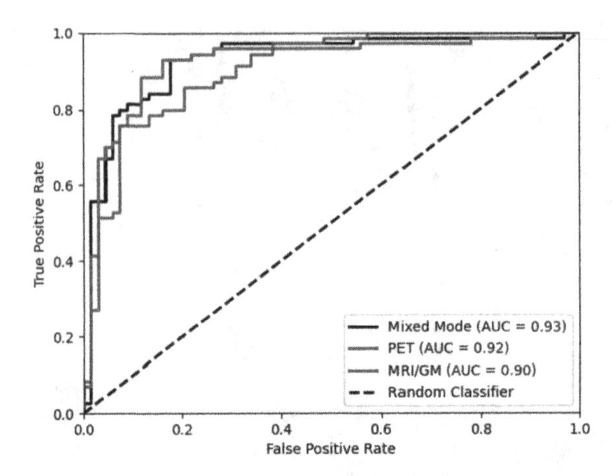


**Fig. 6.** Roc curves obtained for classification using PET, MRI and mixed mode images

## 4 Conclusions and Future Work

In this work, we propose a method based on deep learning to combine medical images from different modalities. The method used is based on the neural style transfer method, but using unsupervised feature extraction by means of a stacked autoencoder. The uppermost layer of the encoder part is extracted once all the autoencoders composing the stack are trained. These layers are used to compose a new network, in which act as feature extractors. Moreover, an image is generated using an optimization process which starts from the PET image, and modifying the values of the voxels aimimg to minimize a loss function that simultaneously improves the similarity between the generated image and the two modalities being combined.

The results obtained show the viability of the proposed method and provides the arena to continue exploring this line, not only with image data but also with other biomedical signals (for instance, time series data). Alternatively, the inclusion of an $\ell_1$ penalty term in the loss functions could help in the selection of the most representative voxels, as well as the implementation of supervised feature selection methods within the training subset such as supervised PCA instead of the unsupervised procedure used in this work.

**Acknowledgments.** This work was supported by projects PGC2018-098813-B-C32 and RTI2018-098913-B100 (Spanish "Ministerio de Ciencia, Innovación y Universidades"), UMA20-FEDERJA-086, CV20-45250, A-TIC-080-UGR18 and P20 00525 (Consejería de economía y conocimiento, Junta de Andalucía) and by European Regional Development Funds (ERDF). Work of J.E. Arco was supported by Ministerio de Universidades, Gobierno de España through grant "Margarita Salas".

## References


1. Alsop, D.C., Casement, M., de Bazelaire, C., Fong, T., Press, D.Z.: Hippocampal hyperperfusion in Alzheimer's disease. Neuroimage **42**(4), 1267–1274 (2008). http://www.sciencedirect.com/science/article/B6WNP-4SSG4W3-2/2/e86def44fdf4c58eb9cb2f58f16fcdc1
2. Alzheimer's Disease Neuroimaging Initiative (2021). http://adni.loni.ucla.edu/. Accessed 5 Nov 2021
3. Chyzhyk, D., Graña, M., Savio, A., Maiora, J.: Hybrid dendritic computing with kernel-LICA applied to Alzheimer's disease detection in MRI. Neurocomputing **75**(1), 72–77 (2012). https://doi.org/10.1016/j.neucom.2011.02.024
4. Chételat, G., et al.: Amyloid-PET and 18F-FDG-PET in the diagnostic investigation of Alzheimer's disease and other dementias. Lancet Neurol. **19**(11), 951–962 (2020)
5. Cuingnet, R., et al.: Alzheimer's Disease neuroimaging initiative: automatic classification of patients with Alzheimer's disease from structural MRI: a comparison of ten methods using the ADNI database. Neuroimage **56**(2), 766–781 (2010)
6. Górriz, J.M., et al.: Artificial intelligence within the interplay between natural and artificial computation: advances in data science, trends and applications. Neurocomputing **410**, 237–270 (2020)
7. Álvarez, I., et al.: Alzheimer's diagnosis using eigenbrains and support vector machines. In: Cabestany, J., Sandoval, F., Prieto, A., Corchado, J.M. (eds.) IWANN 2009. LNCS, vol. 5517, pp. 973–980. Springer, Heidelberg (2009). https://doi.org/10.1007/978-3-642-02478-8_122
8. Martinez-Murcia, F.J., Górriz, J.M., Ramírez, J., Ortiz, A., Disease Neuroimaging Initiative, et al.: A spherical brain mapping of MR images for the detection of Alzheimer's disease. Curr. Alzheimer Res. **13**(5), 575–588 (2016)
9. Jing, Y., Yang, Y., Feng, Z., Ye, J., Yu, Y., Song, M.: Neural style transfer: a review. IEEE Trans. Visual. Comput. Graph. **26**(11), 3365–3385 (2020)
10. Martinez-Murcia, F.J., Ortiz, A., Gorriz, J.M., Ramirez, J., Castillo-Barnes, D.: Studying the manifold structure of Alzheimer's disease: a deep learning approach using convolutional autoencoders. IEEE J. Biomed. Health Inform. **24**(1), 17–26 (2020)
11. Ortiz, A., Górriz, J.M., Ramírez, J., Martinez-Murcia, F.J., Alzheimer's Disease Neuroimaging Initiative et al.: Automatic ROI selection in structural brain MRI using SOM 3D projection. PLoS ONE **9**(4), e93851 (2014)
12. Ortiz, A., Górriz, J.M., Ramírez, J., Martínez-Murcia, F.J.: LVQ-SVM based CAD tool applied to structural MRI for the diagnosis of the Alzheimer's disease. Pattern Recogn. Lett. **34**(14), 1725–1733 (2013). https://doi.org/10.1016/j.patrec.2013.04.014
13. Ortiz, A., Munilla, J., Górriz, J.M., Ramírez, J.: Ensembles of deep learning architectures for the early diagnosis of the Alzheimer's disease. Int. J. Neural Syst. **26**(07), 1650025 (2016)
14. Ortiz, A., Munilla, J., Álvarez Illán, I., Górriz, J.M., Ramírez, J., Alzheimer's Disease Neuroimaging Initiative: Exploratory graphical models of functional and structural connectivity patterns for Alzheimer's disease diagnosis. Front. Comput. Neurosci. **9**, 132 (2015)
15. Rolls, E.T., Huang, C.C., Lin, C.P., Feng, J., Joliot, M.: Automated anatomical labelling atlas 3. Neuroimage **206**, 116189 (2020)
16. Simonyan, K., Zisserman, A.: Very deep convolutional networks for large-scale image recognition. arXiv preprint arXiv:1409.1556 (2014)
17. Structural Brain Mapping Group. Department of Psychiatry (2014). http://dbm.neuro.uni-jena.de/vbm8/VBM8-Manual.pdf. Accessed 10 Mar 2014

# Modelling the Progression of the Symptoms of Parkinsons Disease Using a Nonlinear Decomposition of 123I FP-CIT SPECT Images

Jose Antonio Simón-Rodríguez, Francisco Jesús Martinez-Murcia(✉), Javier Ramírez, Diego Castillo-Barnes, and Juan Manuel Gorriz

Department of Signal Theory, Networking and Communications, Andalusian Institute on Data Science and Computational Intelligence, University of Granada, Granada, Spain
josesimon@correo.ugr.es, fjesusmartinez@ugr.es

**Abstract.** Parkinson's Disease (PD) is one of the most relevant neurodegenerative disorder. It is mainly caused by a loss of dopamine neurons leading to a reduction in the neurotransmitter dopamine, which is essential in the control of movement. While the diagnosis of PD is mainly clinical, new markers are being used with high accuracy in the later stages of the disease, where symptoms are clear. However, the early stages of the disease, when symptoms start to evolve and treatments could potentially be more effective, are yet to be explored. In this work we explore the low-dimensional latent space of the Parkinson's Progression Markers Initiative (PPMI) DaTSCAN imaging dataset, with a twofold objective: to perform an early diagnosis of PD, and to link the low-dimensional representation of the images to symptomatology. Different unsupervised methods have been used to extract the features (ISOMAP and PCA), and the resulting space is evaluated by means of binary or multiclass classification, and linear regression, using Support Vector Machines (SVM). We obtained a diagnosis of PD with an Area Under the ROC Curve (AUC) above 0.94 for three different variables, and a relevant link between the Unified Parkinson's Disease Rating Scale (UPDRS) and the imaging composite features with $R^2 > 0.2$ even for a simple linear model. These results pave the way to explore latent representations in PD and study the progression of the disease and its symptomatology.

**Keywords:** Parkinson Disease (PD) · ISOMAP · Principal Component Analysis (PCA) · Support Vector Machine (SVM) · ROC Curve

## 1 Introduction

Parkinson's disease (PD) affects more than 6 million people worlwide [15], being the second most relevant neurodegenerative disorder after Alzheimer's Disease


J. M. Ferrández Vicente et al. (Eds.): IWINAC 2022, LNCS 13258, pp. 104–113, 2022.
https://doi.org/10.1007/978-3-031-06242-1_11

(AD). Parkinson's disease is a severe, progressive and chronic disease of unknown origin which is mainly caused by the progressive loss of dopaminergic neurons of the nigrostriatal pathway [11]. The loss of dopamine neurons leads to a reduction in the neurotransmitter dopamine, which is essential in the control of movement. Nowadays, PD is incurable, and when symptoms appear it means that neuronal destruction is excessive, and any possible treatments that may slow or stop the progression are mainly ineffective. Therefore the greatest challenge in this neurodegenerative disorder is an early diagnostic, even before the apparition of symptoms.

At present, the typical diagnosis of PD is clinical, and based on a subjective view of a patient's symptoms. There are however two main reasons that complicate the diagnosis of PD. First and foremost, because there are many similar movement disorders that conform what is frequently called parkinsoninsm, of different etiology. Secondly, the severity of these symptoms changes over time. For these reasons, approximately 25% of PD diagnoses are incorrect when compared to autopsy findings [10].

Diagnosis has improved with time, with the increasing use of biomarkers such as tau protein, present in Cerebro-Spinal Fluid (CSF). There are also many studies that aim at diagnosing PD using movement or tremor recorded in wearable sensors [5,7] or the smartphone data [16]. Or even studies that relate PD with speech faculties: [12].

Neuroimaging modalities yields early diagnosis of PD by providing noninvasive biomarkers. Among them, Single Photon Emission Computed Tomography (SPECT) uses highly specific radiopharmaceuticals –e.g., DaTSCAN– that bind to dopamine transporters in the striatum, making it possible to observe dopaminergic deficits in the brain. However, these images provide high dimensional features that could be better exploited by means of computers than by simple visual inspection. As a consequence, Machine Learning (ML) has thrived in the latest years, mainly to perform a differential diagnosis and enabling Computer Aided Diagnosis (CAD) systems [6,11,17]. However, there are a lesser amount of studies that try to make sense of PD progression using these images and other variables.

This study aims to develop an algorithm that can model the progression of PD, offering support for its early diagnosis. The proposed system is based on a non-linear decomposition of the SPECT images using unsupervised machine learning methods and the modelling of the composite variables via support vector machines (SVMs). The system is tested in two differential tasks: differential diagnosis (classification) and disease progression analysis (regression) by means of a longitudinal dataset, both using SVMs. The system is evaluated by means of stratified $k$-fold cross-validation using a different performance metrics in order to ensure the validity of the decomposition techniques for studying and diagnosing PD.

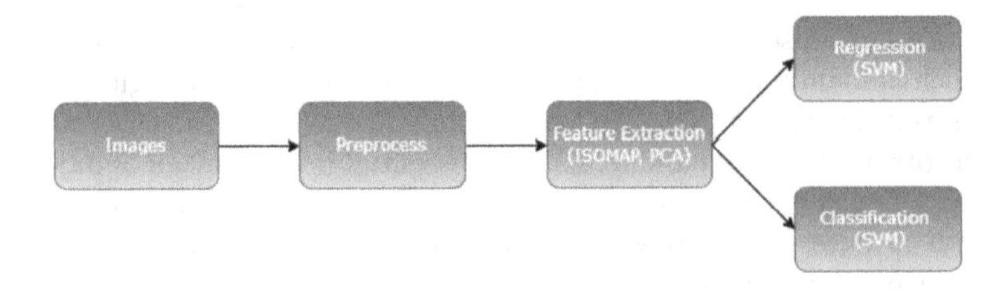


**Fig. 1.** Flowchart of the study

## 2 Methodology

### 2.1 Dataset Description and Image Preprocessing

Data used in the preparation of this article were obtained from the Parkinson's Progression Markers Initiative (PPMI) database (https://www.ppmi-info.org/accessdata-specimens/download-data). For up-to-date information on the study, visit www.ppmi-info.org.

The cohort used for this study is composed by those subjects initially diagnosed either as controls (CTL), with no evidence of neurodegenerative deficits, and Parkinson's Disease (PD) affected subjects, with different levels of severity. The CTL group consists of 101 males and 53 females, with 2 of them showing mild symptomatology (HY_on=1). The PD group consists of 284 males and 159 females. Table 1 shows the demographic analysis of the dataset. PD subjects and some subjects of the CTL group were clinically followed for up to 5 years, providing data for 1399 sessions, which will be evaluated afterwards to study the progression of imaging biomarkers and their relationship to PD-specific progression indicators.

**Table 1.** Demographics of the PPMI subjects included in the study.

| Prim. Diag. | Sex | HY | N | Age | UPDRS3 on | Cognitive state |
|---|---|---|---|---|---|---|
| PD | M | 1.0 | 124 | 60.42 [9.96] | 14.411 [5.98] | 1.091 [0.29] |
| | | 2.0 | 159 | 64.06 [8.79] | 25.346 [8.75] | 1.242 [0.44] |
| | | 3.0 | 1 | 67.74 [0.00] | 23.000 [0.00] | 1.000 [0.00] |
| | F | 1.0 | 76 | 57.00 [9.39] | 14.553 [6.51] | 1.000 [0.00] |
| | | 2.0 | 81 | 63.45 [9.26] | 24.617 [8.38] | 1.045 [0.21] |
| | | 3.0 | 2 | 63.11 [11.10] | 30.000 [8.49] | 1.000 [0.00] |
| CTL | M | 0.0 | 99 | 61.25 [11.03] | 1.133 [2.04] | 1.071 [0.26] |
| | | 1.0 | 2 | 61.07 [9.02] | 5.500 [4.95] | 1.000 [0.00] |
| | F | 0.0 | 53 | 59.21 [12.67] | 1.283 [2.33] | 1.000 [0.00] |

DaTSCAN images were preprocessed using affine registration to the MNI space as in [8]. Intensity normalization was based on a thresholding technique based on an alpha-stable modelling of the image intensity distribution using the algorithm proposed in [2]. Additionally, three categorical variables will be used for classification: approximate diagnosis at each visit (APPRDX), diagnosis at the first visit (PRIMDIAG) and the value for Hoehn and Yahr (HY) scale for PD, that ranges from 0 (no symptoms) to 3 (severe disability). APPRDX distinguishes between healthy patients, PD's and SWEDD patients, whereas PRIMDIAG just between PD and CTL. The Unified Parkinson's Disease Rating Scale (UPDRS) degree for PD, that measures the degree of symptomatology, is also used as a continuous variable in regression.

### 2.2 Manifold Learning

Manifold Learning builds on the assumption that real, highly dimensional data such as images lies on a lower-dimensional nonlinear manifold. The group of algorithms aimed at approximating this manifold and how real data is projected to it are known as Manifold Learning algorihtms [8].

This work follows the outline at Fig. 1, in which feature extraction via different dimensionality reduction techniques is achieved. After the image preprocessing described at Sect. 2, we apply two algorithms: Principal Component Analysis (PCA) or ISOMAP.

Principal Component Analysis (PCA) is a linear dimensionality reduction technique which allows to identify new linear subspaces. The principal components (PC) are each one of the spatial directions which maximizes the variance of the data while being orthogonal to each other, therefore they are uncorrelated variables [9].

In contrast to the linear nature of PCA, we have applied an isometric feature mapping, or Isomap, an algorithm that performs nonlinear modelling of a manifold extending metric multidimensional scaling via geodesic distances. The aim is to find a linear, lower dimensional subspace in which to embed the data in a high dimensional space, preserving the geodesic distance between the data [3].

We used the algorithm for computing Isomap as defined in [14], that is composed of three steps:

1. First, K-nearest neighbors are applied to construct a neighborhood graph.
2. Second, the shortest path between all pairs of point is calculated by estimating geodesic distances, usually using the Dijktra or Floyd-Warshall algorithms.
3. Finally, a d-dimensional Euclidean embedding is constructed by a partial eigenvalue decomposition (i.e., taking the d largest eigenvalues of the kernel.

### 2.3 Classification and Regression Experiments

Last step in our analysis is the application of machine learning modelling for predicting scores. We have used an algorithm that has proven robust in many applications: Support Vector Machines (SVM), which have been widely used in

Alzheimer's [13] or cancer [1]. We have used the implementation of LIBLINEAR for SVM classifiers (SVC) and SVM regression (SVR) [4].

Linear SVC is a particular case of SVR, as both try to predict a target variable $Y$ from a set of data $X$, which in our case are the coordinates of the projections of the images in the manifold. SVMs in general try to model the curve inherent to the trend of data, creating a linear hyperplane that in the case of classification, separates (theoretically) the data. In the case of SVR, it becomes a predictor of certain variables.

5-fold stratified cross-validation (CV) is used to obtain performance measures both for regression and classification. For the classification approach, the average and standard deviation of accuracy, sensitivity, specificity and balanced accuracy are provided, along with the ROC curve in each CV loop. In the case of regression, Mean Absolute Error (MAE), Mean Squared Error (MSE), Root Mean Squared Error (RMSE) and $R^2$ are used as performance measures.

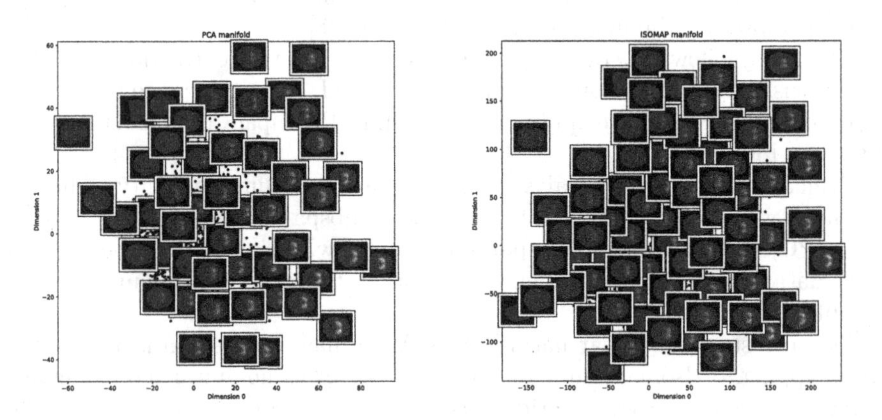


**Fig. 2.** DaTSCAN central slice placed in the corresponding coordinates of the first 2 dimensions of PCA (left) and Isomap (right).

## 3 Results and Discussion

First of all, we observe the differences between the linear decomposition with PCA and non-linear decomposition with Isomap in Fig. 2. First, two dimensions are shown for each methodology. Both approaches show similar trends in the first dimension (dimension 0), spanning from negative to positive values that are related to the intensity of the striatal region. However, the Isomap decomposes in a more uniformly distributed coordinate space, thanks to its nonlinear modelling, whereas PCA is more influenced by extreme values and outliers. Dimension 1, however, differs. In Isomap it is related to the intensity of the tails of the striatal region (putamen), whereas in PCA seems to measure roughly the asymmetry of the image.

With the aim of predicting the symptomatology of a patient just from image composite values, we trained a SVR with the projections of both Isomap and ICA. The performance of the SVR in predicting the variable UPDRS is shown at Table 2. There, the Isomap is shown a more accurate decomposition than the PCA, with no big differences between the two and three component decomposition, and higher R2 (above 0.2) than PCA.

**Table 2.** Performance of the SVR in predicting UPDRS from the 2 and 3-component PCA and Isomap projections.

| | ISOMAP 2 | ISOMAP 3 | PCA 2 | PCA 3 |
|---|---|---|---|---|
| MAE | 12.317 [0.749] | 12.318 [0.746] | 12.692 [0.730] | 12.692 [0,733] |
| R2 | 0.207 [0.063] | 0.206 [0.064] | 0.190 [0.0557] | 0.191 [0.056] |
| MSE | 261.594 [29.299] | 261.468 [29.310] | 266.494 [27.564] | 266.374 [27.516] |
| RMSE | 16.148 [0.917] | 16.144 [0.918] | 16.302 [0.849] | 16.299 [0.847] |

The resulting SVR model for Isomap and PCA is shown at Fig. 3. There, the black lines represent a perfect linear reconstruction. We can observe that the predictions using Isomap decomposition are more linear with respect to the real UPDRS score than those of the PCA. The error encountered is due mainly to larger UPDRS values, which involves a more severe symptomatology.

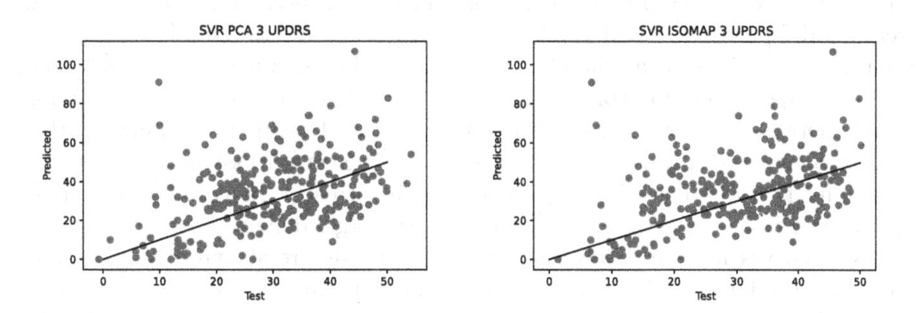


**Fig. 3.** Comparison of the predictions obtained by a SVR using the PCA and the Isomap decomposition with two components.

The composite imaging features obtained from PCA and Isomap are also used to predict three different targets: PRIMDIAG, APPRDX, and HY scale (see Sect. 2 for a description of these variables). To do so, a SVC is trained, and measures of Accuracy, Sensitivity, Specificity, and Balanced Accuracy (average of Sensitivity and Specificity) are reported in Table 3.

**Table 3.** Performance of the SVC when using the two- and three-dimensional projections of Isomap and PCA.

| Variable | Performance | ISOMAP 3 | PCA 3 | PCA 2 | ISOMAP 2 |
|---|---|---|---|---|---|
| APPRDX | Accuracy | 0.879 [0.028] | 0.856 [0.043] | 0.855 [0.041] | 0.871 [0.034] |
| | Bal. Accuracy | 0.882 [0.043] | 0.847 [0.052] | 0.847 [0.054] | 0.877 [0.048] |
| | Sensitivity | 0.886 [0.088] | 0.835 [0.091] | 0.835 [0.091] | 0.886 [0.097] |
| | Specificity | 0.878 [0.032] | 0.859 [0.046] | 0.858 [0.043] | 0.869 [0.040] |
| PRIMDIAG | Accuracy | 0.880 [0.031] | 0.857 [0.046] | 0.856 [0.044] | 0.873 [0.036] |
| | Bal. Accuracy | 0.751 [0.046] | 0.724 [0.054] | 0.722 [0.054] | 0.745 [0.048] |
| | Specificity | 0.520 [0.086] | 0.474 [0.099] | 0.470 [0.099] | 0.507 [0.091] |
| | Sensitivity | 0.982 [0.014] | 0.973 [0.015] | 0.973 [0.015] | 0.982 [0.015] |
| HY | Accuracy | 0.611 [0.038] | 0.594 [0.020] | 0.593 [0.022] | 0.606 [0.021] |
| | Bal. Accuracy | 0.466 [0.021] | 0.441 [0.019] | 0.438 [0.013] | 0.457 [0.009] |
| | Class 0 | 0.897 [0.066] | 0.817 [0.108] | 0.829 [0.086] | 0.903 [0.066] |
| | Class 1 | 0.143 [0.060] | 0.126 [0.039] | 0.086 [0.038] | 0.072 [0.040] |
| | Class 2 | 0.824 [0.065] | 0.821 [0.061] | 0.838 [0.059] | 0.852 [0.057] |
| | Class 3 | 0.000 [0.000] | 0.000 [0.000] | 0.000 [0.000] | 0.000 [0.000] |

It can be observed that the best performing decomposition is again Isomap with three dimensions. In this cases, it is able to approximate each subject's and session diagnosis with high accuracy. It is also capable of predicting primary diagnosis with high sensitivity (>0.98). For its part, the symptomatology as measured by HY, was predicted with a Multiclass one-vs-all SVC. The results were good for classes 0 and 2, but not for classes 1 and 3. This is relevant for interpreting this result, as class 0 are controls and class 2 has severe symptoms. Class 3 subjects were interpreted as class 2, probably because they show similar dopaminergic deficit, in contrast to controls. Class 1 is so close to class 0 that some subjects with primary diagnosis as CTL have class 1 assigned. However, it is very likely that some of these subjects are incorrectly interpreted as controls and other as class 2 because they show dopaminergic deficit.

ROC curves for the differential binary diagnosis are shown at Fig. 4. Individual curves for each CV fold are shown, and the mean ROC curve is shown in blue. There it can be seen that the AUC of our methodology is very high for providing a good differential diagnosis of PD based solely on imaging markers.

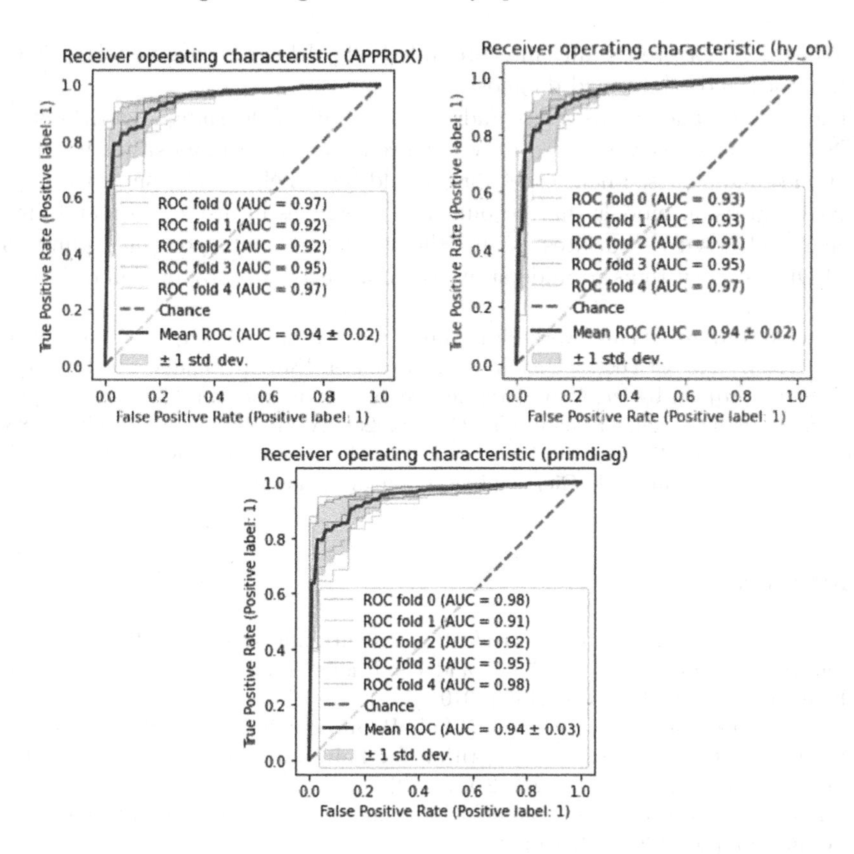


**Fig. 4.** ROC curves obtained with SVC applied to ISOMAP 3 output with the apporximate diagnosis, HY scale and primary diagnosis.

# 4 Conclusions

Due to the critical importance and the difficulty of diagnosing Parkinson's disease, it is necessary to find reliable and accurate methods of early diagnosis. The aim of this work is to explore the usefulness of nonlinear manifold learning methodologies to inform a new space in which imaging features relate to symptomatology of PD, allowing for the creation of longitudinal PD progression models. The proposed methodology, using Isomap as a manifold learning method, and Support Vector Machines (SVM) as machine learning models for classification and regression, yielded high performance in regression, binary classification, and multi-class classification, as in the case of the HY score.

Each of the dimensions of the nonlinear subspace found by Isomap can be related to relevant changes in the brain such as the concentration of dopamine transporters (DaT) in the striatum or DaT concentration at the putamen (as it is the case of dimension 1). This builds a machine learning model that, unlike many in the literature, is fully interpretable by healthcare professionals, paving the way for a more informed diagnosis.

The latent space of Isomap and other manifold learning algorithms in DaTSCAN images is yet to be fully explored. Other decomposition methods based on self-supervised neural networks could be of help here, along with more complex regression models that account for differences in distribution and hierarchical models with covariates, paving the way for new interpretable Computer Aided Diagnosis systems to understand the diagnosis and progression of PD.

**Acknowledgements.** This work was supported by the MCIN/ AEI/10.13039/ 501100011033/ and FEDER "Una manera de hacer Europa" under the RTI2018-098913-B100 project, by the Consejería de Economía, Innovacióon, Ciencia y Empleo (Junta de Andalucía) and FEDER under CV20-45250, A-TIC-080-UGR18, B-TIC-586-UGR20 and P20-00525 projects. Work by F.J.M.M. is supported by the MCIN AEI IJC2019-038835-I 'Juan de la Cierva - Incorporacion' fellowship.

## References

1. Artan, Y., et al.: Prostate cancer localization with multispectral MRI using cost-sensitive support vector machines and conditional random fields. IEEE Trans. Image Process. **19**(9), 2444–2455 (2010)
2. Castillo-Barnes, D., Martínez-Murcia, F.J., Ramírez, J., Górriz, J., Salas-Gonzalez, D.: Expectation-maximization algorithm for finite mixture of $\alpha$-stable distributions. Neurocomputing **413**, 210–216 (2020)
3. Cayton, L.: Algorithms for manifold learning. Univ. of California at San Diego Technical report 12(1–17), 1 (2005)
4. Fan, R.E., Chang, K.W., Hsieh, C.J., Wang, X.R., Lin, C.J.: Liblinear: a library for large linear classification. J. Mach. Learn. Res. **9**, 1871–1874 (2008)
5. Górriz, J.M., et al.: Artificial intelligence within the interplay between natural and artificial computation: advances in data science, trends and applications. Neurocomputing **410**, 237–270 (2020). https://doi.org/10.1016/j.neucom.2020.05.078
6. Illán, I., Górriz, J., Ramírez, J., Segovia, F., Jiménez-Hoyuela, J., Ortega Lozano, S.: Automatic assistance to Parkinson's disease diagnosis in datscan spect imaging. Med. Phys. **39**(10), 5971–5980 (2012)
7. Kubota, K.J., Chen, J.A., Little, M.A.: Machine learning for large-scale wearable sensor data in Parkinson's disease: Concepts, promises, pitfalls, and futures. Mov. Disord. **31**(9), 1314–1326 (2016)
8. Martinez-Murcia, F.J., Ortiz, A., Gorriz, J.M., Ramirez, J., Castillo-Barnes, D.: Studying the manifold structure of Alzheimer's disease: a deep learning approach using convolutional autoencoders. IEEE J. Biomed. Health Inform. **24**(1), 17–26 (2019)
9. Murcia, F.J.M.: Statistical neuroimage modeling, processing and synthesis based on texture and component analysis: tackling the small sample size problem. Ph.D. thesis, Universidad de Granada (2017)
10. Pahwa, R., Lyons, K.E.: Early diagnosis of Parkinson's disease: recommendations from diagnostic clinical guidelines. Am. J. Manag. Care **16**(4), 94–99 (2010)

11. Rojas, A., et al.: Application of empirical mode decomposition (EMD) on datscan spect images to explore Parkinson disease. Expert Syst. Appl. **40**(7), 2756–2766 (2013). https://doi.org/10.1016/j.eswa.2012.11.017, https://www.sciencedirect.com/science/article/pii/S0957417412012274
12. Sakar, B.E., et al.: Collection and analysis of a Parkinson speech dataset with multiple types of sound recordings. IEEE J. Biomed. Health Inform. **17**(4), 828–834 (2013)
13. Stoeckel, J., Fung, G.: Svm feature selection for classification of spect images of alzheimer's disease using spatial information. In: Fifth IEEE International Conference on Data Mining (ICDM 2005). IEEE (2005). 8-pp
14. Tenenbaum, J.B., De Silva, V., Langford, J.C.: A global geometric framework for nonlinear dimensionality reduction. Science **290**(5500), 2319–2323 (2000)
15. Vos, T., et al.: Global, regional, and national incidence, prevalence, and years lived with disability for 310 diseases and injuries, 1990–2015: a systematic analysis for the global burden of disease study 2015. The lancet **388**(10053), 1545–1602 (2016)
16. Zhan, A., et al.: Using smartphones and machine learning to quantify Parkinson disease severity: the mobile Parkinson disease score. JAMA Neurol. **75**(7), 876–880 (2018)
17. Zubal, I.G., Early, M., Yuan, O., Jennings, D., Marek, K., Seibyl, J.P.: Optimized, automated striatal uptake analysis applied to spect brain scans of Parkinson's disease patients. J. Nucl. Med. **48**(6), 857–864 (2007)

# Capacity Estimation from Environmental Audio Signals Using Deep Learning

C. Reyes-Daneri[1](✉), F. J. Martínez-Murcia[2,3], and A. Ortiz[1,3]

[1] Communications Engineering Department University of Málaga, 29004 Málaga, Spain
crisreyda@hotmail.com

[2] Department of Signal Theory, Communications and Networking University of Granada, 18060 Granada, Spain

[3] Andalusian Data Science and Computational Intelligence Institute (DaSCI), Jaén, Spain

**Abstract.** Estimating the capacity of a room or venue is essential to avoid overcrowding that could compromise people's safety. Having enough free space to guarantee a minimal safety distance between people is also essential for health reasons, as in the current COVID-19 pandemic. Already existing systems for automatic crowd counting are mostly based on image or video data, and some of them, using deep learning architectures. In this paper, we study the viability of already existing Deep Learning Crowd Counting systems and propose new alternatives based on new network architectures containing convolutional layers, exclusively based on the use of environmental audio signals. The proposed architecture is able to infer the actual capacity with a higher accuracy in comparison to previous proposals. Consequently, conclusions from the accuracy obtained with out approach are drawn and the possible scope of deep learning based crowd counting systems is discussed.

**Keywords:** Automated Crowd Counting · Capacity control · Convolutional Neural Networks · Regression

## 1 Introduction

Controlling people capacity at places has always been an important task to make for many companies and public organizations. Automated Crowd Counting has recently gained more prominence than ever since the COVID-19 world pandemic started, in order to stop the spread of this disease. Besides this, Crowd Counting systems have been used before for many other applications such as urban planning, public safety or event management and evaluation [7].

This automation has traditionally been made using images or video through Machine Learning, dealing with the problems of expensive camera systems and installations, privacy issues, and so on. The first approaches to this problem focused mainly on estimating the number of people through handcrafted features,


J. M. Ferrández Vicente et al. (Eds.): IWINAC 2022, LNCS 13258, pp. 114–124, 2022.
https://doi.org/10.1007/978-3-031-06242-1_12

such as histograms of oriented gradients [10] or real-time face detection [11]. The most up-to-date methods based on CNN (Convolutional Neural Networks) [5–7] have achieved remarkable improvements. These architectures proved to be very efficient approximating the number of individuals in large crowds. For example, the method proposed by Zhang et al. [12], made possible to detect accurately the number of people by scanning their appearance and their movement characteristics between frames of video. The problem with this type of models is that they cannot work on a static image, but require a series of dynamic frames.

The problem with all these models mentioned before is that they assume that crowds are composed of individual entities. This can cause a lot of trouble, mostly when the crowd is significantly dense. In these cases, occlusion usually occurs, making person-to-person counting a very difficult task. To address the occlusion problem, some researchers propose multi-view methods [13]. However, this is not the best alternative, since it is not always easy to simultaneously transport several cameras with specific calibration and parameters. In another of the analyzed works [2], one-second segments of audio are used to add complementary information to the image. This way, it is possible to estimate the number of people in a place in extreme conditions: occlusion, low illumination or low resolution. It was shown that these audio inclusions could generally improve AI accuracy for these scenarios.

Until now, however, the inclusion of audio to solve this problem has been limited to complement the visual information, but the exclusive use of audio for Crowd Counting has never been considered until now. In the present work, an exclusive audio-based Crowd Counting system is developed, so that it is not necessary a single image of the scene. A first approach to this was estimating the number of people with the sound signal volume, but given the high variability of possible scenarios, this could fail. The system here presented extracts audio components that help to perform Crowd Counting through CNN.

After this introduction section, Sect. 2 presents the materials needed and the methods used for this project. In Sect. 3, the neural network architectures used are listed and described. Section 4 presents the results obtained with every tested configuration. Finally, in Sect. 5 we discuss the results obtained and conclusions are drawn in Sect. 6.

## 2 Materials and Methods

### 2.1 The DISCO Dataset

The main objective of this project was to test Crowd Counting systems that included (as a complement or exclusively) audio. However, there exist very few datasets containing this type of data. At first, the idea of extracting the data from a surveillance camera was considered. This idea was dismissed due to lack of resources.

After considering various options, the dataset finally used for this work was the one named auDIoviSual Crowd cOunting (DISCO) dataset, firstly presented at [1]. This choice was made because this dataset includes audio as well as images (video frames) and manually estimated density maps for each of them. This dataset contains 1935 video frames and 1-s length audios, so each image in this dataset has it's associated audio corresponding to half a second before (t - 0.5 s) and half a second after (t + 0.5 s) the image was taken.

## 2.2 Data Processing

Before the data is fed to the neural networks, different preprocessing pipelines were established for each data modality:

### Image Processing

1. **Rescaling** of both the image and its respective density map. This way, images and maps of lower resolution and size than the original ones are loaded in exchange for faster computation. This process manages to reduce the size of the density maps, which take up approximately 16.6 MB each.
2. **Normalization** of images. Image data is processed better in the subsequent neural networks if its values are close to those of a normal or uniform distribution. This leads to a faster convergence and better precision in most machine learning approaches. The procedure is applied pixel by pixel on each image in the dataset, as the next expression shows:

$$in_j = (io_j - \bar{io})/\sigma \quad (1)$$

   where $\sigma$ and $\bar{io}$ are respectively the standard deviation and the mean of the image pixel values, $in_j$ is the value of each pixel of the normalized images and $io_j$ the value of each pixel of the original images of the set.
3. **Random flipping** of the images and their respective density maps. This is used during training for data augmentation, in order for the model to learn from more diverse images and therefore to become more robust. It is important to ensure that this flip is done jointly, so that the positions of the heads on the density maps match up with those on the pictures.

### Audio Processing

In order to test audio architectures, the audio data has been processed in different ways to find an optimal way for the system to extract as much relevant information as possible:

- **Audio vector**: The audio file is loaded, creating this way a data vector from the original digital audio file. Should the amplitude values of this vector be in an interval larger than $[-1, 1]$, a normalization might be applied. Otherwise, it is left as is.
- **Spectrogram**: this transform is commonly used for speech detection and processing. It consists of the application of a Fourier transform to overlapping segments of the audio vector, weighted by a window. The Hanning window has been used in this work [9]. This way, the frequency components of the

audio are shown with full detail in each subsegment. The result of this transformation is a two-dimensional matrix where its horizontal axis is time and its vertical axis is the modulus of the audio power at each frequency value in each segment.

- **Mel Spectrogram**: based on the previous transform, a non-linear transformation is applied to the frequency on the spectrogram [14]. This transformation aims to imitate the human ear behaviour, which does not perceive frequencies on a linear scale but has more difficulties differentiating high-frequency sounds than low-frequency sounds.
  This non-linear transformation is obtained from the following expression:

$$m = 2595 \log_{10}\left(1 + \frac{f}{700}\right) \quad (2)$$

  where $f$ is the frequency(Hz) and $m$ is the result of the conversion to Mel Scale (Mel).

# 3 Neural Network Architectures

In this section, the models based on neural networks used for image and audio scenarios will be listed and described:

## 3.1 Architectures for Image Data

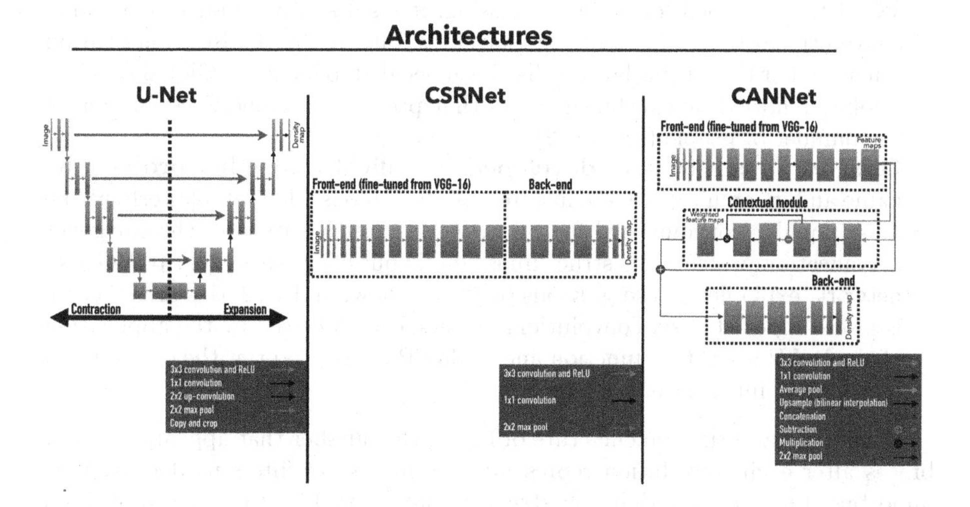


Fig. 1. Image architectures

- **U-Net** [4]: it is a convolutional neural network architecture that has proven to be very effective for tasks where the output is of a similar size to the input, such as image segmentation, image fusion, etc. This architecture owes its name to its 'U' shaped structure. As it is shown in the first architecture structure in Fig. 1, five levels are observed for U-Net, conforming two fundamental opposite parts: contraction and expansion of the data.
- **CSRNet** [5]: similar to UNet, CSRNet is also conformed by two parts (see Fig. 1):
  1. **A convolutional neural network**. This first part is used for two-dimensional feature extraction. In this case, this part corresponds to the first ten layers of a pre-trained VGG-16 network, using exclusively $3 \times 3$ kernel sizes and three pooling layers.
  2. **A dilated convolutional neural network**. It uses dilated $3 \times 3$ convolutional kernels in order to extract information of greater relevance and maintain the output resolution.
- **CANNet** [6]: as shown in Fig. 1, this architecture starts with the same front-end and ends with the same back-end configuration as CSRNet. The main difference between both of this architectures is that CANNet owns an intermediate module. This contextual module creates weighted feature maps from the features extracted in the front-end part. Then, these feature maps are fused together for its later insertion in the back-end part of the network.

### 3.2 Architectures for Audio Signals

- **VGGish** [2]: consists of a network based on Google's VGG-16 network, with the difference that the last three fully-connected layers are discarded. Since VGGish is intended for audio processing, it results of a six one-dimensional convolutional layers. Each of these layers is followed by the ReLU activation function. For the output layer, it has been decided to include a Global Average Pooling followed by two linear layers that predict the numeric estimation of the number of people (see Fig. 2).
- **CrisNet**: It is a network **developed specifically for this project**. Its name and its architecture are inspired by the famous AlexNet [8], perhaps the most popular convolutional neural network, intended to solve the Imagenet classification problem. Its structure is simple but effective, similar to VGGish network, explained in the previous point. As shown in Fig. 2, this architecture is also composed of six convolutional layers, each followed by the application of the ReLU activation function and a Max Pooling. To print the only output digit, a last linear layer is implemented.

To understand the nomenclature of Fig. 2, the number that appears between braces after each convolution represents the number of filters used in it. It is important to point out that the structure shown in Fig. 2 can be applied to any shape in which the audio input is processed. This way, the sizes of the convolutional kernels are not fixed, but will vary depending on the shape of the audio information that is provided to the network. In the case of the spectrogram, for example, two-dimensional kernels are used.

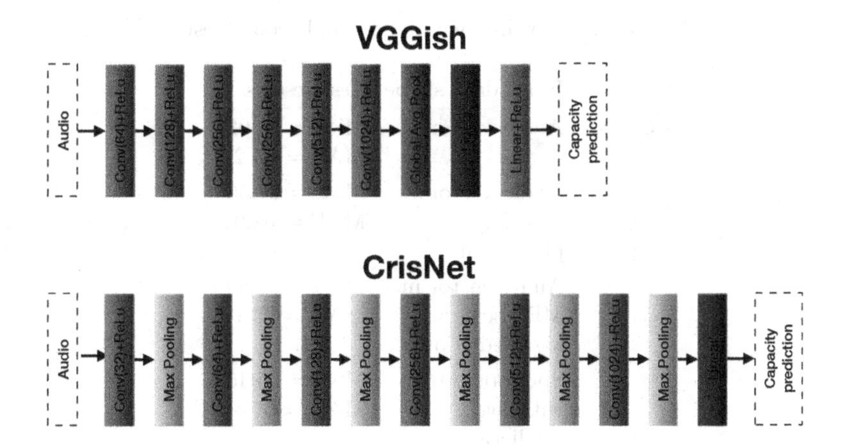


**Fig. 2.** Layers contained inside each audio architecture

## 4 Results

This section presents the results obtained for every neural network. Accuracy of the networks was measured via Mean Absolute Error (MAE) and Mean Squared Error (MSE). Whenever different input data formats were used, it is specified in the following tables. Table 1 and Table 2 show the results of image and audio tests respectively.

**Table 1.** Error for each image people count test

| Model | Test results |
|---|---|
| UNet | MSE = 43.97<br>MAE = 37.93 |
| CANNet | MSE = 41.43<br>MAE = 41.04 |
| CSRNet | MSE = 25.77<br>MAE = 20.19 |

Regarding the mean absolute error results of each of the image networks, the network that provides the best results is the CSRNet architecture, published by Yuhong Li et al. in 2018. This network [5] was specifically made for Crowd Counting in highly congested scenes, so there is no doubt this is the best option so far. Meanwhile, UNet and CANNet remain on a similar slightly worse line.

Due to the greater difficulty of audio Crowd Counting, more tests have been carried out in this scenario than in the previous image one. As the MAE and MSE results show for each of the audio networks, the best network configurations are the ones that use two-dimensional spectrograms instead of one-dimensional

**Table 2.** Error for each audio people count test

| Model | Input data shape | Test results |
|---|---|---|
| **VGGish** | Audio vector | MSE = 79.02<br>MAE = 65.22 |
| | Audio vector with log prediction | MSE = 87.56<br>MAE = 62.62 |
| | Audio vector in MEL scale with log prediction | MSE = 97.59<br>MAE = 73.12 |
| | Spectrogram with log prediction | MSE = 95.45<br>MAE = 67.56 |
| | MEL Spectrogram | MSE = 84.99<br>MAE = 70.56 |
| | MEL Spectrogram with log prediction | MSE = 106.9<br>MAE = 90.73 |
| **CrisNet** | Audio vector | MSE = 80.78<br>MAE = 68.07 |
| | Audio vector in MEL scale with log prediction | MSE = 79.84<br>MAE = 61.28 |
| | Spectrogram | MSE = 73.01<br>MAE = 56.21 |
| | Spectrogram with log prediction | MSE = 79.26<br>MAE = 59.49 |
| | MEL Spectrogram | MSE = 77.7<br>MAE = 61.18 |
| | MEL Spectrogram with log prediction | MSE = 183.8<br>MAE = 81.06 |

audio vectors, specially for CrisNet. Another important point to note is that MEL scale does not seem to help much for this Crowd Counting task.

## 5 Discussion

Our results support that there is a fundamental difference between image and audio prediction models. In all cases, image predictions are more accurate than audio predictions. Comparing Tables 1 and 2, all image tests have lower MSE and MAE values than any sound test. There are some factors that impact accuracy in image models, starting from the fact that the exact number is computed as the sum of all heads found in the images. Therefore, derivative measures depend on this -manual- estimation.

Second, predicting an image-like map from an image is a well studied problem. That is the case of predicting density maps from images, and not trying to predict a single number from the images, which could be a more difficult case.

The linear regression graphs belonging to each image model tested are represented in Fig. 3. For any of the three image networks tested, a high correlation is observed and, therefore, a high accuracy of all models. The CSRNet model, the best one tested so far, reaches a Pearson coefficient of R = 0.983.

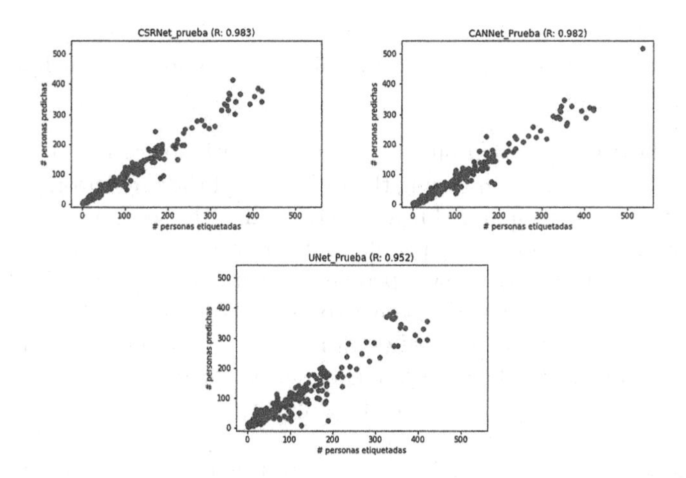


**Fig. 3.** Comparison between real and prediction values for image architectures

As for the audio models, the problem is far more complex that the image-to-image translation. This approach treats the problem as a regression problem, in which a bare number has to be predicted from different audio inputs (namely audio vector or image-like spectrograms). As aforementioned, predicting a bare number instead of a density map is also subject to more inaccuracies, especially with large ranges like our problem (crowds range from tens to hundreds of subjects).

In Fig. 4, the linear regression graphs belonging to some of the audio models tested are represented. Comparing this with Fig. 3, noticeably worse results are observed.

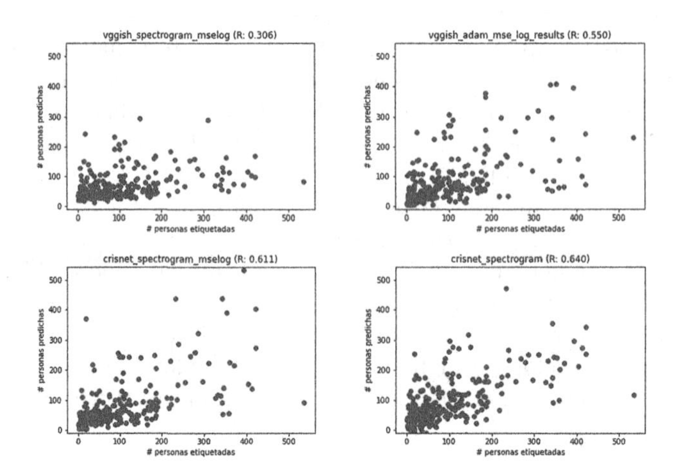


**Fig. 4.** Comparison between real and prediction values for audio architectures

The two best models turn out to be CrisNet with spectrogram (R = 0.640) and CrisNet with spectrogram using the logarithm of the prediction (R = 0.611). The first option presents a more linear behavior than the second one for scenarios that contain a large number of people and, in turn, the second one seems to work better in scenarios with fewer people. To increase the precision of CrisNet, a distinction could be made between two systems for two different scenarios: not using the logarithm of the prediction for scenarios where large crowds tend to accumulate and using it however for scenarios where large accumulations of people are not common. This way, the most accurate tool would be used for each case.

With these results obtained, it is still interesting considering the study of the use of audio for this type of problem, either by itself or as a refinement to the visual estimation, like it is done at [2].

## 6 Conclusions

The fact that the best results for the proposed audio counting systems have not been obtained should not be a reason for discouragement. Since possibilities are endless, there is always a chance of finding a better configuration. However, it is true that the proposed CrisNet models reach higher correlation values than the rest ($R > 0.6$) and that in their two modalities (linear and logarithmic prediction) have greater versatility in scenarios of high and low density of people respectively. The combination of these and other systems through an ensemble mechanism has not been tested yet, but it could be an interesting future line to develop.

The possible scope of application for these systems is interesting. In places where a microphone may be more discreet or more useful, the use of audio Crowd Counting systems could be chosen, taking into account that these systems are much more effective in settings with large crowds of people. Another application for this system could be for scenarios that by their nature do not require that much accuracy for the counting of people. For example, an audio Crowd Counting system could be installed in a not very busy urban place, in order to detect crowds that can imply a biological COVID-19 threat when the threshold of people capacity established as dangerous is exceeded. Also, if other complementary data (temperature, humidity) was available, it could help to improve the accuracy of these systems.

One of the main motivations for this project was the development of a system that controls the capacity of a health center. This idea arises due to the frequent saturation of many of the Spanish health centers as a result of the emergence of the global COVID-19 pandemic. Many of the patients that came to these centers took the risk of ending up being infected in such crowds. The idea is that they should be able to receive a remote notice to come only at times of less influx. The development of this type of applications would be an interesting possibility to consider in a future work.

**Acknowledgements.** This work was supported by projects PGC2018-098813-B-C32 (Spanish "Ministerio de Ciencia, Innovación y Universidades"), UMA20-FEDERJA-086 (Consejería de econnomía y conocimiento, Junta de Andalucía) and by European Regional Development Funds (ERDF), as well as the BioSiP (TIC-251) research group. Work by F.J.M.M. was supported by the MICINN "Juan de la Cierva - Incorporación" IJC2019-038835-I Fellowship.

## References

1. Wang, Q., et al.: Audiovisual crowd counting dataset (2020). https://doi.org/10.5281/zenodo.3828468
2. Wang, Q., et al.: Ambient sound helps: audiovisual crowd counting in extreme conditions (2020). https://arxiv.org/pdf/2005.07097.pdf
3. Hershey, S., et al.: CNN architectures for large-scale audio classification (2017). https://arxiv.org/pdf/1609.09430.pdf
4. Thomas, C.: U-Nets with ResNet Encoders and cross connections. Journal (2019). https://towardsdatascience.com/u-nets-with-resnet-encoders-and-cross-connections-d8ba94125a2c
5. Li, Y., Zhang, X., Chen, D.: CSRNet: dilated convolutional neural networks for understanding the highly congested scenes (2018). https://arxiv.org/pdf/1802.10062.pdf
6. Liu, W., Salzmann, M., Fua, P.: Context-aware crowd counting (2019). https://arxiv.org/pdf/1811.10452.pdf
7. Gorriz, J.M., et al.: Artificial intelligence within the interplay between natural and artificial computation: advances in data science, trends and applications. Neurocomputing **410**, 237–270 (2020). https://doi.org/10.1016/j.neucom.2020.05.078
8. Krizhevsky, A., Sutskever, I., Hinton, G.E.: ImageNet classification with deep convolutional neural networks (2012). https://proceedings.neurips.cc/paper/2012/file/c399862d3b9d6b76c8436e924a68c45b-Paper.pdf
9. Wen, H., et al.: Hanning self-convolution window and its application to harmonic analysis (2009). https://doi.org/10.1007/s11431-008-0356-6

10. Dalal, N., Triggs, B.: Histograms of oriented gradients for human detection (2005). https://hal.inria.fr/inria-00548512/document
11. Viola, P., Jones, M.J.: Robust real-time face detection (2004). https://www.face-rec.org/algorithms/boosting-ensemble/16981346.pdf
12. Zhang, Y., Zhou, D., Chen, S., Gao, S., Ma, Y.: Single-image crowd counting via multi-column convolutional neural network (2016). http://people.eecs.berkeley.edu/~yima/psfile/Single-Image-Crowd-Counting.pdf
13. Zhang, Q., Chan, A.B.: Wide-area crowd counting via ground-plane density maps and multi-view fusion CNNs (2019). http://visal.cs.cityu.edu.hk/static/pubs/conf/cvpr19-wacc.pdf
14. Zhang, B., Leitner, J., Thornton, S.: Audio recognition using MEL spectrograms and convolution neural networks. http://noiselab.ucsd.edu/ECE228_2019/Reports/Report38.pdf

# Covid-19 Detection by Wavelet Entropy and Self-adaptive PSO

Wei Wang[1], Shui-Hua Wang[1], Juan Manuel Górriz[2](✉), and Yu-Dong Zhang[1](✉)

[1] School of Computing and Mathematical Sciences, University of Leicester, University Road, Leicester LE1 7RH, UK
yudongzhang@ieee.org
[2] Department of Signal Theory, Networking and Communications, University of Granada, 52005 Granada, Spain
gorriz@ugr.es

**Abstract.** The rapid global spread of COVID-19 disease poses a huge threat to human health and the global economy. The rapid increase in the number of patients diagnosed has strained already scarce healthcare resources to track and treat Covid-19 patients in a timely and effective manner. The search for a fast and accurate way to diagnose Covid-19 has attracted the attention of many researchers. In our study, a deep learning framework for the Covid-19 diagnosis task was constructed using wavelet entropy as a feature extraction method and a feedforward neural network classifier, which was trained using an adaptive particle swarm algorithm. The model achieved an average sensitivity of 85.14% ± 2.74%, specificity of 86.76% ± 1.75%, precision of 86.57% ± 1.36%, accuracy of 85.95% ± 1.14%, and F1 score of 85.82% ± 1.30%, Matthews correlation coefficient of 71.95 ± 2.26%, and Fowlkes-Mallows Index of 85.83% ± 1.30%. Our experiments validate the usability of wavelet entropy-based feature extraction methods in the medical image domain and show the non-negligible impact of different optimisation algorithms on the models by comparing them with other models.

**Keywords:** COVID-19 · Wavelet entropy · Self-adaptive particle swarm optimization

## 1 Introduction

Since December 2019, the global economy and human health have been threatened by an epidemic that has been difficult to stop despite several national-level embargoes imposed by various countries [8]. The epidemic is caused by a virus called SARS-CoV-2 [23], named Coronavirus Disease-19 (Covid-19), which has non-specific symptoms of respiratory illness, including a persistent cough, fever and loss of taste, and can be life-threatening in severe cases. There are many


J. M. Ferrández Vicente et al. (Eds.): IWINAC 2022, LNCS 13258, pp. 125–135, 2022.
https://doi.org/10.1007/978-3-031-06242-1_13

types of Covid-19 vaccine that have been available for the past two years. However, the global epidemic has not improved significantly due to the "hesitation" to vaccinate and the rapid mutation of the SARS-CoV-2 [14].

The difficulty in preventing and controlling Covid-19 is mainly due to the high survival rate and infectivity of SARS-CoV-2. SARS-CoV-2 is transmitted primarily to droplet transmission and can survive for hours or even days depending on the physical characteristics of the surface and can therefore enter the respiratory tract through contact with human limbs and the mouth and nose, infecting the recipient [17]. Although effective preventive measures and quarantine policies can effectively stop the spread of Covid-19, the complexity of human social interaction makes it difficult to eliminate the virus at all. In addition, Covid-19 has an average incubation period of 5.2 days, which makes it extremely difficult to diagnose and track down infected individuals and does not allow isolation policies to control transmission effectively [11]. Furthermore, the availability of nucleic acid test kits (the primary tool used for Covid-19 diagnosis) is increasingly strained by many untargeted tests. Still, although medical imaging of the lungs can be an essential tool for Covid-19 detection, the relative shortage of specialist radiologists with large numbers of patients inevitably results in a high level of false positives, making the diagnosis of Covid-19 a significant problem for many hospitals in the control and treatment of Covid-19 [19]. Thus, discovering a fast, convenient and accurate method of Covid-19 diagnosis is considered by many researchers to be an essential research issue to assist in the fight against Covid-19.

The rapid development of computer vision technology has evolved from a simple image classification tool to one of the essential tools for image analysis in areas such as medical, robotics, and so on [6]. As a result, there are many Covid-19 diagnostic models and frameworks with good performance based on computer vision techniques. A radial basis function (RBF) based model proposed by Lu [13] have to ability to diagnose COVID-19 CT images. Chen [3] proposed a model with a feature extraction module based on the greyscale co-occurrence matrix (GLCM) and support vector machine (SVM) classifier. This model has got a promising performance in the COVID-19 chest CT images classification task. Yao and Han [21] has used wavelet entropy (WE) for feature extraction and combined biogeography-based optimisation (BBO) to deal with the COVID-19 chest CT images classification task. Their model (WE-BBO) has achieved an excellent performance. Also with a WE feature extraction module, Wang [20] used the Jaya algorithm as the training algorithm (WE-Jaya) and has got a considerable performance improvement in the Covid-19 diagnosis task. In this research, we build a WE based model with the Particle Swarm Optimisation algorithm to do the Covid-19 diagnosis task on this literature and have achieved performance improvements in all aspects compared with the above methods.

In the rest of the paper, Sect. 2 presents the dataset of our experiment. Section 3 introduces methods for our experiment. Finally, Sect. 4 presents and interpreted the result and compares it with other start-of-the-art methods.

## 2 Dataset

A chest CT slice dataset [20] containing 296 data samples was used in the experimental part of the research. Each sample in the dataset includes one chest CT image slice as a data feature and one nucleic acid assay result as a label. These CT images and nucleic acid test results were obtained from 132 subjects, 77 males and 55 females. The dataset contains two categories, 148 positive images (CT images from Covid-19 infected subjects) and 148 health control images (CT images from healthy people). The dataset is balanced, with each category containing 66 data samples. Figure 1 shows one sample from each of the two categories in the dataset.

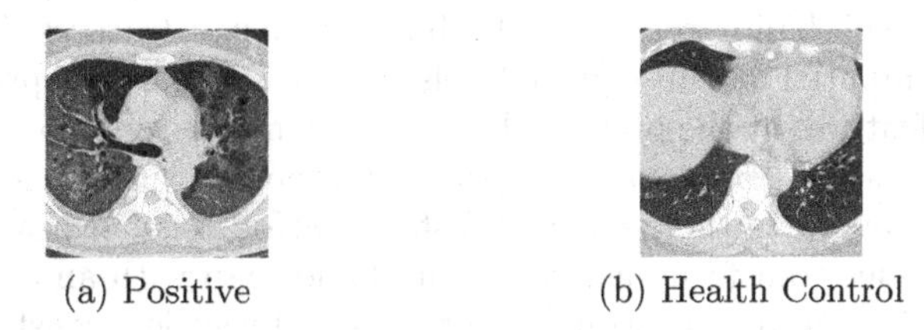

(a) Positive (b) Health Control

**Fig. 1.** Dataset samples of Health Control (a) and Positive (b) (COVID Patient)

## 3 Methodology

### 3.1 Wavelet Entropy

In image analysis, the spatial frequency information of an image signal can be extracted by using the Fourier transform to convert the image signal in the time domain to the frequency domain. The formula for the Fourier transform is shown in Eq. 1 [1].

$$\hat{f}(\omega) = \int_{\infty}^{\infty} f(t) e^{-i\omega t} \mathrm{d}t, \tag{1}$$

where $\omega$ represents the frequency, $t$ represents time.

However, the Fourier transform is flawed in dealing with real-world smooth signals, as it cannot capture the moment when the frequency occurs [9]. One solution is to replace the infinitely long triangular basis in the Fourier transform with a finite decay wavelet basis to obtain the time point at which a frequency occurs, thus extracting both frequency and time information. This replacement transform is called the wavelet transform. With the data $x(t)$ and the mother wavelet $\psi(t)$, the wavelet transform is defined as Eq. 2 [4].

$$W_{\psi} x(a, \tau) = \frac{1}{\sqrt{a}} \int_{-\infty}^{+\infty} x(t) \psi\left(\frac{t-\tau}{a}\right) \mathrm{d}t, \tag{2}$$

where a represents scale, $\tau$ represents shift parameter, $\psi$ refers to mother wavelet function, which can be dilated or translated by modulating a and $\tau$ following Eq. 3.

$$\psi_{a,\tau}(t) = \psi\left(\frac{t-\tau}{a}\right). \tag{3}$$

In this study, the three-level discrete wavelet transform (DWT) is used in the feature extraction stage to generate a series of wavelet basis functions by translating and expanding the mother wavelet. Based on this, the DWT transform coefficients are obtained using the inner product of the image signal and the wavelet function. The wavelet transform can be performed at multiple levels. One level of the wavelet transform will be calculated for the input map, LL1, LH1, HL1 and HH1, where LL1 represents the low-frequency features, and the other three sub-bands represent the high-frequency features. The wavelet transform is computed at subsequent levels for the sub-bands representing the low-frequency features at the previous level to obtain three sub-bands representing the high-frequency features and one sub-band representing the low-frequency features, respectively. As the wavelet transform introduces downsampling, each level of the wavelet transform acquires a sub-band map with an edge length of half the input [5]. Figure 2 illustrates the three-level wavelet transform process.

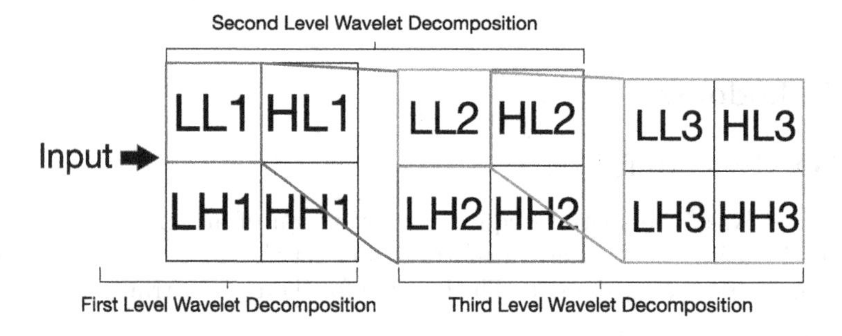


**Fig. 2.** Illustration of three-level decomposition of discrete wavelet transforms.

Despite the many advantages of the wavelet transform, the feature information it extracts is too redundant and contains a large amount of useless information. To prevent this problem from incurring unnecessary space and time costs, we use entropy to reduce the dimensionality of the features before feeding the information into the classifier. First introduced by Shannon [18], entropy values increase as the level of confusion rises and represent the average uncertainty of the information. In images, entropy represents the different probabilities of individual pixel grey levels, and specific texture patterns possess a particular entropy when they are repeated approximately. This method is widely used for the quantitative analysis of image details, and entropy values can effectively compare and classify image details. Equation 4 defines the calculation of the entropy [22].

$$S(\alpha) = -\sum_{i}^{n} \mathrm{P}(\alpha_i) \log_b \mathrm{P}(\alpha_i), \tag{4}$$

where $\alpha$ refers to grey levels and $\alpha_i$ are the coefficients of $\alpha$ on an orthonormal basis, P represents the probabilities of grey levels.

### 3.2 Feedforward Neural Network

Feedforward neural networks is a simple neural network structure and are one of the most widely used and rapidly developing artificial neural networks. The neurons in a feedforward neural network are arranged in layers, which are usually an input layer, several hidden layers and an output layer [7]. The neurons in each network layer are connected to the neurons in the previous layer only, accepting the previous layer's output as input and transmitting the output of the current layer to the next layer as input. The feedforward neural network maps the input $x$ to $y = f(x; \theta)$ and trains the value of the parameter $\theta$ to obtain the function with the lowest loss value, thus obtaining the output closest to the target value. The structure of the feedforward neural network is shown in Fig. 3.

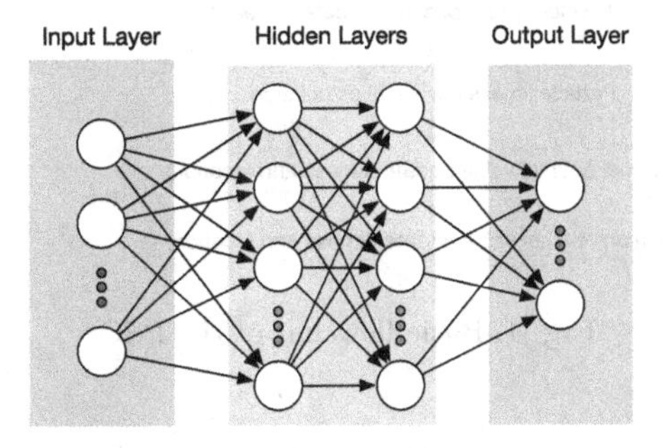


**Fig. 3.** Baisc structure of feedforward neural network

Inputting features into a feedforward neural network classifier for classification recognition is one of the most common classification methods and is used in a wide range of classification tasks in several domains. In this study, a one-hidden-layer feedforward neural network with 40 hidden neurons is involved as the classifier. Its input features are extracted from CT images by the wavelet entropy method, while the output is a diagnosis of Covid-19, positive or healthy.

### 3.3 Self-adaptive Particle Swarm Optimisation

The particle swarm optimisation (PSO) algorithm was first proposed by Kennedy and Eberhart [10]. It is known as one of the fastest-growing intelligent algorithms due to its ease of understanding, ease of implementation and global solid search capability [15]. Figure 4 shows the basic flow diagram of PSO, which starts with the initialisation of the particle population size, particle dimension, number of iterations, inertia weight, learning factor and iteration step size range, and outputs the optimal historical position and adaptation values for the population and individual particles for the first time when randomly initialising the velocity and position of each particle.

The core idea of PSO lies in the continuous motion of all particles, updating positions until convergence to the optimal position that can be found. The updating of particle positions in PSO follows Eq. 5.

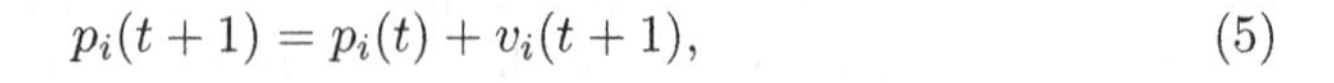

$$p_i(t+1) = p_i(t) + v_i(t+1), \quad (5)$$

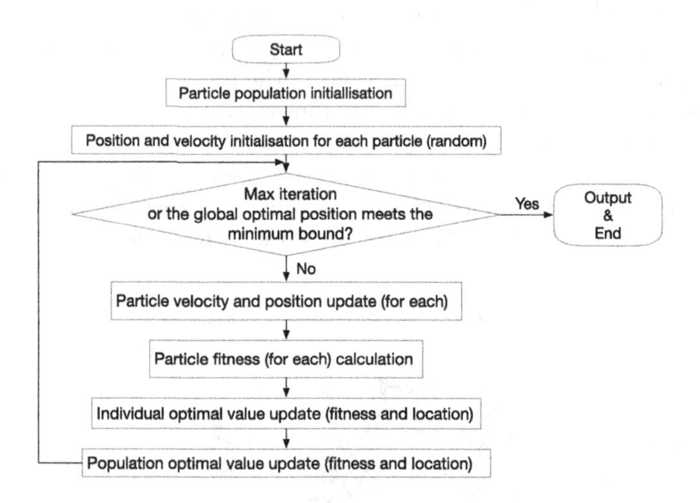


**Fig. 4.** Basic flowchar of the PSO

where t refers to time, p is a position vector that refers to the position of the particle i and v represents velocity which is calculated by Eq. 6

$$v_i(t+1) = wv_i(t) + c_1r_1(p_{i\text{best}} - p_i(t)) + c_2r_2(p_{g\text{best}}(t) - p_i(t)), \quad (6)$$

where $w$ refers to the inertia weight of a particle, $c_1$ and $c_2$ are accelerating constants, $r_1$ and $r_2$ are uniformly distributed random variables in the range of $[0, 1]$, $p_{i\text{best}}$ is position vector that refers to the best candidate solution of the particle $i$, $p_{g\text{best}}$ refers to the best position of all particles.

However, there are significant limitations in the movement paths of the particles in PSO, and the model can easily fall into a local optimum when the global optimum does not exist over the paths of the initial particle positions and the local optimum. Also, the behaviour of all particles moving towards the discovered optimal solution simultaneously is a waste of computational resources [2].

The Self-adaptive Particle Swarm Optimisation Algorithm (SaPSO) [12] used in this experiment solves these problems by giving the particles the ability to explore directions other than the path between the initial position and the optimal solution that can be found. In SaPSO, there are two states for each particle, the exploitative state and the explorative state. When a particle is in the exploitative state, it follows the traditional PSO, moving towards the current population optimal solution and the individual optimal solution. When a particle is in the explorative state, it moves away from the current individual optimal solution and the worst solution and searches for other directions. The velocity for updating particle positions in the explorative state is calculated following Eq. 7.

$$v_i(t+1) = w_i(t)v_i(t) - c_1r_1(p_{i\text{best}} - p_i(t)) - c_2r_2(p_{i\text{worst}} - p_i(t), \qquad (7)$$

where $p_{i\text{worst}}$ refers to the worst solution found by the particle $i$. All other variables represent the same to Eq. 6.

### 3.4 K-fold Cross-validation

Partitioning datasets according to data usage is an important method to avoid overfitting and improve the robustness of models in deep learning. However, it is often difficult to obtain large datasets in many domains. Therefore, the partitioning of the dataset can lead to too few training samples, which in turn leads to a higher chance of overfitting and lower robustness of the model. Cross-validation is one of the practical solutions to such problems, with its ability to improve data usage by repeating the disruption and partitioning steps. In this study, a 10-fold cross-validation [16] is used. The dataset is divided into ten copies, and one of the data sample sets is used as training data and the others as test data in each iteration without repetition to calculate the performance metric ($PM_i$, $i$ = 0, 1, ..., 9). The basic flow of 10-fold cross-validation is shown in Fig. 5.

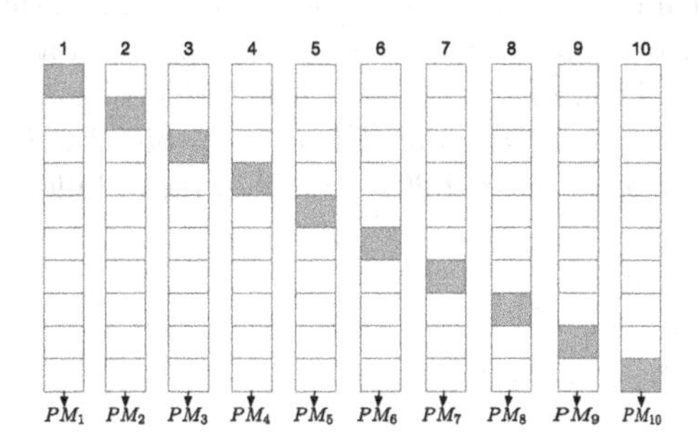


**Fig. 5.** Illustration of 10-fold cross-validation.

After obtaining the performance metrics $(PM_1, PM_2, \ldots, PM_{10})$ for ten iterations, the final performance metrics are calculated as $PM_{final} = \frac{1}{10}\sum_{i=1}^{10} PM_i$.

# 4 Experiment Results and Discussions

## 4.1 WE Results

The Fig. 6 shows a sample of the results of the three-stage wavelet transform of this experiment, where Fig. 6(a) is the input image, transformed by the one-stage wavelet transform into the image shown in Fig. 6(b). The upper left corner of its four grids is the high-frequency component, and the other three grids are all low-frequency components. After the second wavelet transform of the high-frequency components, the image shown in Fig. 6(c) is output, and similarly, the image shown in Fig. 6(d) is the final output of the three-stage wavelet transform, which is also the input of the classifier.

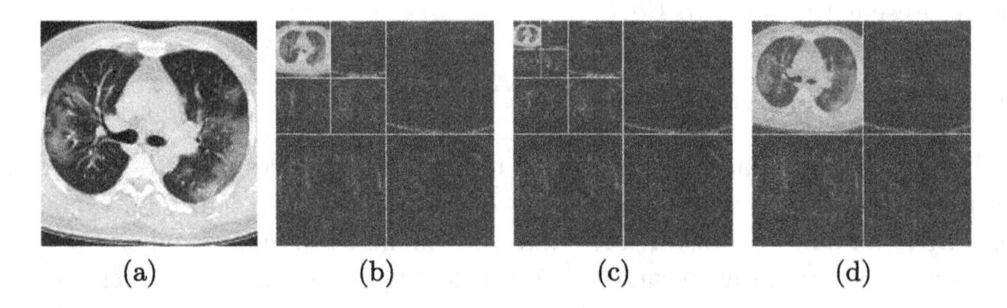


**Fig. 6.** Three-level Wavelet decomposition result sample

## 4.2 Statistical Results

The proposed model is built with wavelet entropy as a feature extraction module and the classifier is a single hidden layer feedforward neural network with forty hidden neurons. Model is trained with the self-adaptive particle swarm optimisation algorithm and used 10-fold cross-validation as the dataset splitting mechanism. The performance is shown in Table 1, with an average sensitivity of 85.14% ± 2.74%, specificity of 86.76% ± 1.75%, the precision of 86.57% ± 1.36%, the accuracy of 85.95% ± 1.14%, and F1 score of 85.82% ± 1.30%, Matthews correlation coefficient of 71.95 ± 2.26%, and Fowlkes-Mallows Index of 85.83% ± 1.30%.

**Table 1.** 10 runs of 10-fold cross-validation

| Run | Sen | Spc | Prc | Acc | F1 | MCC | FMI |
|---|---|---|---|---|---|---|---|
| 1 | 85.14 | 86.49 | 86.30 | 85.81 | 85.71 | 71.63 | 85.72 |
| 2 | 82.43 | 89.86 | 89.05 | 86.15 | 85.61 | 72.50 | 85.68 |
| 3 | 86.49 | 83.78 | 84.21 | 85.14 | 85.33 | 70.30 | 85.34 |
| 4 | 82.43 | 87.84 | 87.14 | 85.14 | 84.72 | 70.37 | 84.75 |
| 5 | 84.46 | 87.16 | 86.81 | 85.81 | 85.62 | 71.65 | 85.62 |
| 6 | 83.11 | 86.49 | 86.01 | 84.80 | 84.54 | 69.63 | 84.55 |
| 7 | 88.51 | 85.14 | 85.62 | 86.82 | 87.04 | 73.69 | 87.06 |
| 8 | 87.84 | 85.14 | 85.53 | 86.49 | 86.67 | 73.00 | 86.67 |
| 9 | 89.19 | 87.84 | 88.00 | 88.51 | 88.59 | 77.03 | 88.59 |
| 10 | 81.76 | 87.84 | 87.05 | 84.80 | 84.32 | 69.72 | 84.36 |
| Mean | 85.14±2.74 | 86.76±1.75 | 86.57±1.36 | 85.95±1.14 | 85.82±1.30 | 71.95±2.26 | 85.83±1.30 |

(Sen = Sensitivity; Spc = Specificity; Prc = Precision; Acc = Accuracy; F1 = F1 Score; MCC = Matthews correlation coefficient; FMI = Fowlkes-Mallows Index)

### 4.3 Comparison to State-of-the-Art Approaches

The proposed model has shown more promising performance compared to other state-of-art deep learning-based models in all aspects. The performance comparison is shown in Table 2, which demonstrates the great potential of the models resulting from the combination of the adaptive particle swarm optimisation algorithm and the wavelet entropy feature extraction module for medical image diagnosis tasks.

**Table 2.** Performance comparison to other methods

| Method | Sen | Spc | Prc | Acc | F1 | MCC | FMI |
|---|---|---|---|---|---|---|---|
| RBFNN | 66.89 ± 2.43 | 75.47 ± 2.53 | 73.23 ± 1.48 | 71.18 ± 0.80 | 69.88 ± 1.08 | 42.56 ± 1.61 | 69.97 ± 1.04 |
| WE-BBO | 72.97 ± 2.96 | 74.93 ± 2.39 | 74.48 ± 1.34 | 73.95 ± 0.98 | 73.66 ± 0.98 | 47.99 ± 2.00 | 73.66 ± 1.33 |
| GLCM-SVM | 72.03 ± 2.94 | 78.04 ± 1.72 | 76.66 ± 1.07 | 75.03 ± 1.12 | 74.24 ± 1.57 | 50.20 ± 2.17 | 74.29 ± 1.53 |
| WE-Jaya | 73.31 ± 2.26 | 78.11 ± 1.92 | 77.03 ± 1.35 | 75.71 ± 1.04 | 75.10 ± 1.23 | 51.51 ± 2.07 | 75.14 ± 1.22 |
| WE-SaPSO | 85.14 ± 2.74 | 86.76 ± 1.75 | 86.57 ± 1.36 | 85.95 ± 1.14 | 85.82 ± 1.30 | 71.95 ± 2.26 | 85.83 ± 1.30 |

## 5 Conclusions

With the help of artificial intelligence technology, medical images can be used to diagnose diseases more intelligently and efficiently. The proposed model has achieved a promising performance in chest CT image-based COVID-19 diagnosis, and it could be adapted to other disease diagnosis tasks based on medical radiological images after retraining. The ability to accurately diagnose new diseases in a very short period of time will be an important weapon for mankind in dealing with unknown diseases that may arise in the future.

## References


1. Allen, J.B., Rabiner, L.R.: A unified approach to short-time fourier analysis and synthesis. Proc. IEEE **65**(11), 1558–1564 (1977)
2. Chen, M.R., Li, X., Zhang, X., Lu, Y.Z.: A novel particle swarm optimizer hybridized with extremal optimization. Appl. Soft Comput. **10**(2), 367–373 (2010)
3. Chen, Y.: Covid-19 classification based on gray-level co-occurrence matrix and support vector machine. In: Santosh, K.C., Joshi, A. (eds.) COVID-19: Prediction, Decision-Making, and its Impacts. LNDECT, vol. 60, pp. 47–55. Springer, Singapore (2021). https://doi.org/10.1007/978-981-15-9682-7_6
4. Daqrouq, K., Sweidan, H., Balamesh, A., Ajour, M.N.: Off-line handwritten signature recognition by wavelet entropy and neural network. Entropy **19**(6), 252 (2017)
5. Ghazali, K.H., Mansor, M.F., Mustafa, M.M., Hussain, A.: Feature extraction technique using discrete wavelet transform for image classification. In: 2007 5th Student Conference on Research and Development, pp. 1–4 (2007). https://doi.org/10.1109/SCORED.2007.4451366
6. Górriz, J.M., et al.: Artificial intelligence within the interplay between natural and artificial computation: Advances in data science, trends and applications. Neurocomputing **410**, 237–270 (2020). https://doi.org/10.1016/j.neucom.2020.05.078, https://www.sciencedirect.com/science/article/pii/S0925231220309292
7. Han, F., Ling, Q.H., Huang, D.S.: An improved approximation approach incorporating particle swarm optimization and a priori information into neural networks. Neural Comput. Appl. **19**(2), 255–261 (2010)
8. Hotez, P.J., Fenwick, A., Molyneux, D.: The new covid-19 poor and the neglected tropical diseases resurgence. Infectious Diseases of Poverty **10**(1), 3 (2021). https://doi.org/10.1186/s40249-020-00784-2, `<Go to ISI>://WOS:000613237100001`
9. Karthiga, R., Narasimhan, K.: Automated diagnosis of breast cancer using wavelet based entropy features. In: 2018 Second International Conference on Electronics, Communication and Aerospace Technology (ICECA), pp. 274–279. IEEE (2018)
10. Kennedy, J., Eberhart, R.: Particle swarm optimization. In: Proceedings of ICNN 1995 - International Conference on Neural Networks, vol. 4, pp. 1942–1948 (1995). https://doi.org/10.1109/ICNN.1995.488968
11. Li, X.: Risk factors for severity and mortality in adult COVID-19 inpatients in Wuhan. J. Allergy Clin. Immunol. **146**(1), 110–118 (2020). https://doi.org/10.1016/j.jaci.2020.04.006
12. Li, X., Fu, H., Zhang, C.: A self-adaptive particle swarm optimization algorithm. In: 2008 International Conference on Computer Science and Software Engineering, vol. 5, pp. 186–189. IEEE (2008)
13. Lu, Z.: A pathological brain detection system based on radial basis function neural network. J. Med. Imaging Health Inform. **6**(5), 1218–1222 (2016)
14. Machingaidze, S., Wiysonge, C.S.: Understanding covid-19 vaccine hesitancy. Nat. Med. **27**(8), 1338–1339 (2021)
15. Poli, R.: Analysis of the publications on the applications of particle swarm optimisation. J. Artif. Evol. Appl. **2008**, 1–10 (2008)
16. Rajasekaran, S., Rajwade, A.: Analyzing cross-validation in compressed sensing with poisson noise. Signal Processing **182**, 9 (2021). https://doi.org/10.1016/j.sigpro.2020.107947, `<Go to ISI>://WOS:000618541700009`
17. SanJuan-Reyes, S., Gómez-Oliván, L.M., Islas-Flores, H.: Covid-19 in the environment. Chemosphere **263**, 127973 (2021)

18. Shannon, C.E.: A mathematical theory of communication. Bell Syst. Tech. J. **27**(4), 623–656 (1948). https://doi.org/10.1002/j.1538-7305.1948.tb00917.x
19. Song, Y., et al.: Deep learning enables accurate diagnosis of novel coronavirus (covid-19) with CT images. IEEE/ACM Trans. Comput. Biol. Bioinf. **18**(6), 2775–2780 (2021). https://doi.org/10.1109/TCBB.2021.3065361
20. Wang, W.: Covid-19 detection by wavelet entropy and jaya. Lect. Notes Comput. Sci. **12836**, 499–508 (2021)
21. Yao, X., Han, J.: COVID-19 detection via wavelet entropy and biogeography-based optimization. In: Santosh, K.C., Joshi, A. (eds.) COVID-19: Prediction, Decision-Making, and its Impacts. LNDECT, vol. 60, pp. 69–76. Springer, Singapore (2021). https://doi.org/10.1007/978-981-15-9682-7_8
22. Yildiz, A., Akin, M., Poyraz, M., Kirbas, G.: Application of adaptive neuro-fuzzy inference system for vigilance level estimation by using wavelet-entropy feature extraction. Expert Syst. Appl. **36**(4), 7390–7399 (2009)
23. Yuki, K., Fujiogi, M., Koutsogiannaki, S.: Covid-19 pathophysiology: a review. Clin. Immunol. **215**, 108427 (2020)

# RDNet: ResNet-18 with Dropout for Blood Cell Classification

Ziquan Zhu[1], Zeyu Ren[1], Shui-Hua Wang[1(✉)], Juan M. Górriz[2(✉)], and Yu-Dong Zhang[2(✉)]

[1] School of Computing and Mathematical Sciences, University of Leicester, Leicester LE1 7RH, UK
Shuihuawang@ieee.org

[2] Department of Signal Theory, Networking and Communications, University of Granada, 52005 Granada, Spain
gorriz@ugr.es, yudongzhang@ieee.org

**Abstract.** **(Aims)** Blood cells are hematopoietic pluripotent stem cells derived from bone marrow. Blood diseases occur primarily in the hematopoietic system and can affect the hematopoietic system with abnormal blood changes, characterized by anemia, bleeding, and fever. It is helpful for doctors to diagnose blood diseases by classifying blood cells. However, doctors take a lot of time and energy to classify blood cells. The classification process is easily disturbed by external factors, such as doctors' lack of rest, fatigue, etc. Many researchers use CNN to classify and detect red blood cells or white blood cells. However, using CNN has some problems in the classification or detection process. First, most researchers only classify blood cells into two categories, but there are many different types of blood cells. In addition, some studies are multi-classification of cells, but the results are often not ideal. **(Methods)** We propose a new model (RDNet) for the automatic classification of four types of blood cells to deal with these problems. The proposed RDNet selects the pre-trained ResNet-18 as the backbone. We transfer the pre-trained ResNet-18 because of the difference between the blood cell data set with the ImageNet data set. We add dropout to improve the classification performance. **(Results)** The accuracy of the proposed RDNet is 86.53%. The proposed RDNet obtains better accuracy than the transferred ResNet-18 because we add dropout in RDNet. Based on the accuracy, the proposed model is an effective tool to classify blood cells.

**Keywords:** Blood cells · Dropout · ResNet-18 · Transfer learning · Convolutional neural network

## 1 Introduction

Blood cells are hematopoietic pluripotent stem cells derived from bone marrow. In addition to having the ability to proliferate, stem cells can migrate


J. M. Ferrández Vicente et al. (Eds.): IWINAC 2022, LNCS 13258, pp. 136–144, 2022.
https://doi.org/10.1007/978-3-031-06242-1_14

out of bone marrow hematopoietic tissue under certain circumstances and form hematopoietic cell nodules with blood flow to extramedullary tissue, which is called colony-forming unit.

Blood diseases occur primarily in the hematopoietic system and can affect the hematopoietic system with abnormal blood changes, characterized by anemia, bleeding, and fever. The factors leading to blood diseases may be: 1 Malnutrition can lead to malnutrition anemia, 2 There may also be some external toxic effects, such as chemicals or radiation. It is helpful for doctors to diagnose blood diseases by classifying blood cells. However, it takes a lot of time and energy for doctors to classify blood cells. The classification process is easily disturbed by external factors, such as doctors' lack of rest, fatigue, etc.

More and more researchers use computer technology to classify blood cells. [16] developed a deep learning model, which was based on convolution neural networks (CNN). The CNN-based learning model was to classify the blood cell images. [10] proposed a novel model (BloosCaps) to classify blood cells. The proposed model achieved an accuracy of 99.3%. [14] presented a method- Canonical Correlation Analysis (CCA) to deal with the problems of multiple cells overlap. The results showed that the proposed model improved accuracy than other blood cell classification models. [5] provided the comparison of deep learning models and traditional approaches in the classification of white blood cells. The traditional approach got a 99.8% accuracy. The deep learning models obtained a 99% accuracy. [8] used Regional Convolutional Neural Networks (R-CNN) to identify white blood cells. The R-CNN obtained high accuracy in various types of white blood cells. [1] proposed a novel WBC nucleus segmentation model based on the k-means algorithm and color space conversion. The proposed model got a 98.61% accuracy. [9] proposed a model to classify blood cells. Support vector machine model was used in the proposed model. [13] proposed a model to classify dog red blood cell morphology. The proposed model was based on CNN. [2] presented a model to automatically count blood cells. The model combined CNN, instance segmentation, R-CNN, and transfer learning. [6] proposed RBCNet to detect and count blood images. The proposed RBCNet used U-Net and R-CNN. The study obtained an accuracy that was higher than 97%.

From the above analysis of the research, it can be concluded that many researchers use CNN to classify and detect red blood cells or white blood cells. However, there are some problems in the classification or detection process. First, most researchers only classify blood cells into two categories, but there are many different types of blood cells. In addition, some studies are multi-classification of cells, but the results are often not ideal. To deal with these problems, we propose a new model (RDNet) for the automatic classification of four types of blood cells. The proposed RDNet selects the pre-trained ResNet-18 as the backbone. We transfer the pre-trained ResNet-18 because of the difference between the blood cell data set with the ImageNet data set. The transferred pre-trained ResNet-18 is abbreviated as TRNet. We add dropout to improve the classification performance.

The organization of the rest paper is as follows. Section 2 discusses the blood cell data set used in this paper. The methodology is shown in Sect. 3. Section 4 is about the results. The conclusion is given in Sect. 5.

## 2 Materials

The data set used in this paper can be available on the Kaggle website. There are four types of blood cells, which are Eosinophil, Lymphocyte, Monocyte, and Neutrophil, respectively. Some figures of four types of blood cells are shown in Fig. 1.

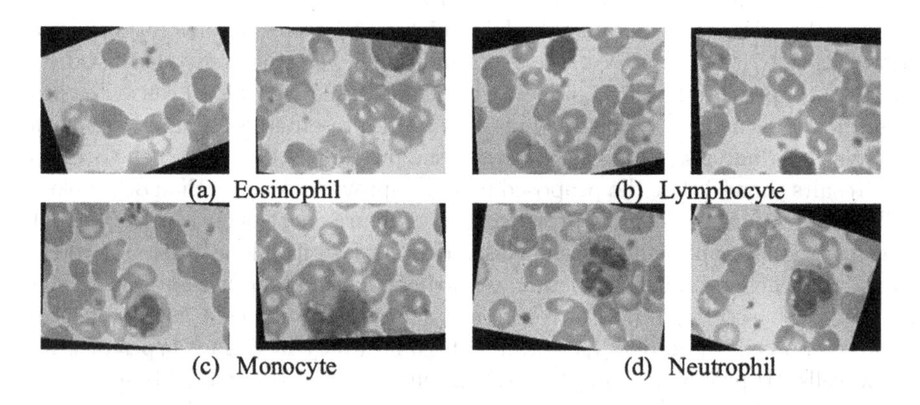


**Fig. 1.** Four types of blood cells in data set

## 3 Methodology

One of the most important steps in image classification [12] is feature extraction. But it is very difficult for researchers to extract useful information from images because images contain too much information. Previously, researchers generally manually extracted the features of images. However, manually extracting features is time-consuming and usually cannot get good results [11]. With the rapid development of artificial intelligence [3], researchers pay more attention to the feature extraction of images using computer vision technology. More and more excellent artificial intelligence models have been proposed, such as VGG [15], AlexNet [7], etc.

### 3.1 Proposed RDNet

A novel method (RDNet) is proposed for the automatic classification of blood cells. The pipeline of the proposed RDNet is shown in Fig. 2. The pseudocode of our model is given in Table 1. The ResNet-18 is pre-trained on the ImageNet data set. The pre-trained ResNet-18 is chosen as the backbone of the proposed RDNet.

We transfer the pre-trained ResNet-18. The transferred pre-trained ResNet-18 is abbreviated as TRNet. We remove the softmax and classification layer from the pre-trained ResNet-18. What's more, we add FC64, FC4, softmax, and classification layers to form the TRNet. We add dropout between FC64 and FC4 to improve the classification performance. The whole network is the proposed RDNet.

**Table 1.** The pseudocode of our model

**Step 1** *Load the pre-trained ResNet-18.*
**Step 2** *Transfer the pre-trained ResNet-18.*
  **Step 2.1** *Remove, softmax, and classification layer from the pre-trained ResNet-18.*
  **Step 2.2** *Add FC64, FC4, softmax, and classification layers.*
  **Step 2.3** *The whole network is named TRNet.RDNet*
**Step 3** *Add dropout between FC64 and FC4*
  **Step 3.1** *The whole network is named RDNet.*
**Step 4** *Train the proposed RDNet*
  **Step 4.1** *Input is the training set.*
  **Step 4.2** *Target is the labels of the training set.*
**Step 5** *Test the trained RDNet on the test set.*
**Step 6** *Report the classification performance of the trained RDNet.*

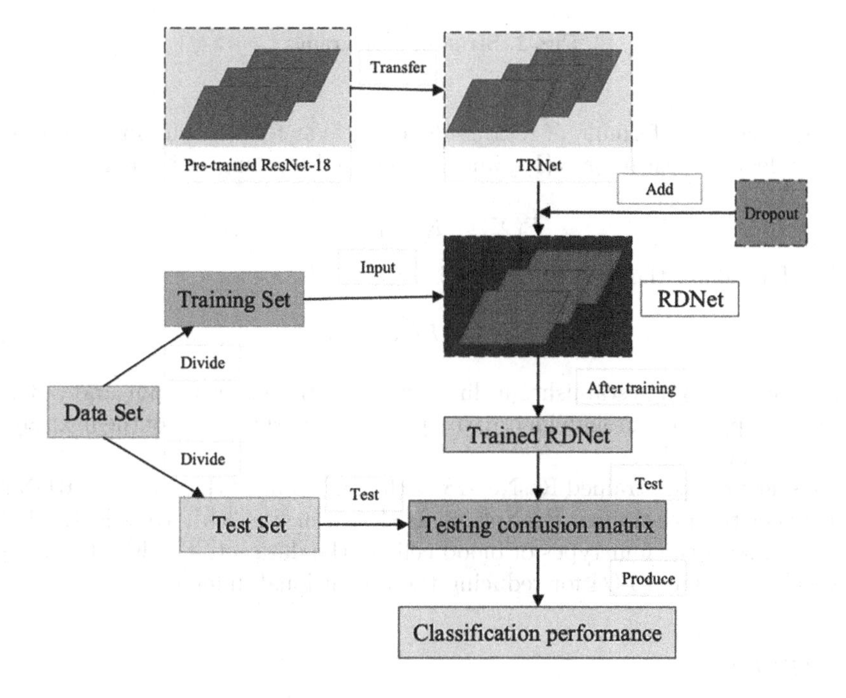


**Fig. 2.** The pipeline of the proposed RDNet

### 3.2 The Backbone of the Proposed RDNet

More information can usually be obtained with the deeper network. However, with the deepening of the network, the optimization effect worsens, and the accuracy of test data and training data is reduced. This is because the deepening of the network will cause the problems of gradient explosion and gradient disappearance. [4] proposed a new model (residual learning) to reduce these problems. The structure of residual learning is shown in Fig. 3(a).

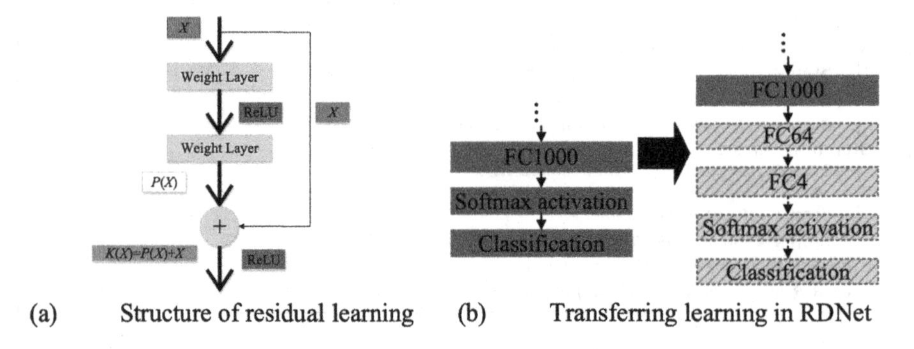


**Fig. 3.** Structural diagrams

The calculation formula of residual learning is as follows. Suppose the input is X, the learned feature is K(X), and the residual learning result is P(X):

$$P(X) = K(X) - X \tag{1}$$

The formula of the learned feature is:

$$K(X) = P(X) + X \tag{2}$$

Residual learning establishes a direct connection between input and output (identity mapping). Through identity mapping, the performance of the next layer will not decline at least.

We select the pre-trained ResNet-18 as the backbone of the proposed RDNet. We transfer the pre-trained ResNet-18, as shown in Fig. 3(b). We add the FC4 layer because of the four types of blood cells in the data set. The FC64 is added between FC1000 and FC4 for reducing the dimensional differences.

### 3.3 Dropout

In the machine learning model, if the model's parameters are too many and the training samples are too few, the machine learning model is easy to produce the problem of overfitting. The overfitting problem is embodied in: the loss function of the model in the training data is small, and the prediction accuracy is high;

but, in the test data, the loss function is relatively large, and the prediction accuracy is low. Therefore, we use dropout to solve the overfitting problem in this paper. The dropout is shown in Fig. 4. Dropout can be used as a trick for training the deep neural network. The overfitting problem can be significantly reduced in each training batch by ignoring half of the feature detectors (making half of the hidden layer node value 0). We add the dropout layer between FC64 and FC4 to improve the classification performance.

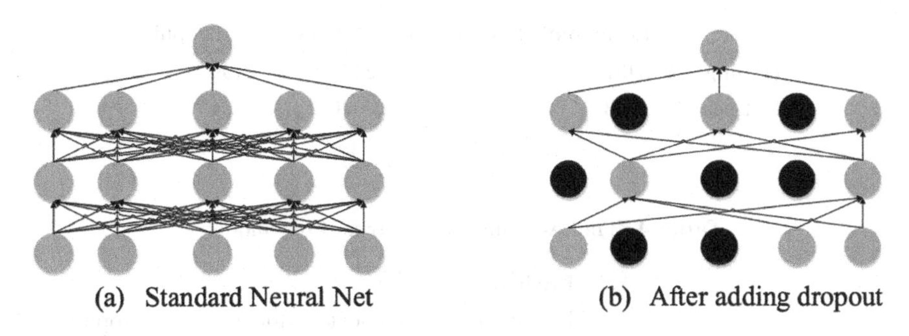


**Fig. 4.** Comparison between using dropout and not using dropout in network structure

# 4 Results

## 4.1 Experiment Settings

The hyper-parameter settings of the proposed model are modified. We set the minibatch size to 64 to overcome the problem of overfitting. The max-epoch is 1. Based on the experience, the learning rate is set as 1e-4. The hyper-parameter settings of the proposed model are given in Table 2.

**Table 2.** The hyper-parameter settings of the proposed model

| Hyper-parameter | Value |
|---|---|
| Minibatch size | 64 |
| Max-epoch | 1 |
| Learning rate | 1e–4 |

## 4.2 The Performance of the Proposed Model

In this paper, there are four types of blood cells: Eosinophil, Lymphocyte, Monocyte, and Neutrophil, respectively. Each of four different types of blood cells has about 3,000 images. The details of the data set are given in Table 3. The test confusion matrix of our model is shown in Table 4. The accuracy of our model is 86.53%. The specific calculation is:

$$accuracy(RDNet) = \frac{(492 + 618 + 474 + 568)}{(623 + 620 + 620 + 624)} = 86.53\% \tag{3}$$

**Table 3.** The details of the data set

| | Eosinophil | Lymphocyte | Monocyte | Neutrophil |
|---|---|---|---|---|
| Traning set | 2497 | 2483 | 2478 | 2499 |
| Test set | 623 | 620 | 620 | 624 |

**Table 4.** The test confusion matrix of our model

| | | Predicted class | | | |
|---|---|---|---|---|---|
| | | Eosinophil | Lymphocyte | Monocyte | Neutrophil |
| Actual class | Eosinophil | 492 | 1 | 0 | 130 |
| | Lymphocyte | 0 | 618 | 2 | 0 |
| | Monocyte | 60 | 0 | 474 | 86 |
| | Neutrophil | 56 | 0 | 0 | 568 |

### 4.3 Comparison of the Proposed Model with TRNet

To better show the superiority of our model, we compare the proposed RDNet with the TRNet. The test confusion matrix of the TRNet is demonstrated in Table 5. The accuracy of the TRNet is:

$$accuracy(TRNet) = \frac{(454 + 618 + 457 + 567)}{(623 + 620 + 620 + 624)} = 84.28\% \tag{4}$$

The proposed RDNet obtains better accuracy than the TRNet because we add dropout in RDNet.

## 5 Conclusion

The paper proposes a new model (RDNet) for the automatic classification of four types of blood cells. The proposed RDNet selects the pre-trained ResNet-18 as the backbone. We transfer the pre-trained ResNet-18 because of the difference between the blood cell data set with the ImageNet data set. The transferred pre-trained ResNet-18 is abbreviated as TRNet. We add dropout to improve the classification performance. The accuracy of the proposed RDNet is 86.53%. The proposed RDNet obtains better accuracy than the TRNet because we add

**Table 5.** The test confusion matrix of the TRNet

| | | Predicted class | | | |
|---|---|---|---|---|---|
| | | Eosinophil | Lymphocyte | Monocyte | Neutrophil |
| Actual class | Eosinophil | 454 | 4 | 5 | 160 |
| | Lymphocyte | 0 | 618 | 2 | 0 |
| | Monocyte | 5 | 0 | 457 | 158 |
| | Neutrophil | 57 | 0 | 0 | 567 |

dropout in RDNet. Based on the accuracy, the proposed model is an effective tool to classify blood cells.

Even though the proposed model gets good results, there are still some limitations. 1. In this paper, we only test single cells and do not test overlapping cells. 2. Although four kinds of blood cells are classified in this paper, there are many kinds of cells that we have not tested.

In future research, we will collect more kinds of blood cells for classification. What's more, we will classify single cells and include overlapping cells in the next paper.

## References

1. Banik, P.P., Saha, R., Kim, K.D.: An automatic nucleus segmentation and CNN model based classification method of white blood cell. Expert Syst. Appl. **149**, 113211 (2020)
2. Dhieb, N., Ghazzai, H., Besbes, H., Massoud, Y.: An automated blood cells counting and classification framework using mask R-CNN deep learning model. In: 2019 31st International Conference on Microelectronics (ICM), pp. 300–303. IEEE (2019)
3. Górriz, J.M., et al.: Artificial intelligence within the interplay between natural and artificial computation: advances in data science, trends and applications. Neurocomputing **410**, 237–270 (2020)
4. He, K., Zhang, X., Ren, S., Sun, J.: Deep residual learning for image recognition. In: Proceedings of the IEEE Conference on Computer Vision and Pattern Recognition, pp. 770–778 (2016)
5. Hegde, R.B., Prasad, K., Hebbar, H., Singh, B.M.K.: Comparison of traditional image processing and deep learning approaches for classification of white blood cells in peripheral blood smear images. Biocybern. Biomed. Eng. **39**(2), 382–392 (2019)
6. Kassim, Y.M., et al.: Clustering-based dual deep learning architecture for detecting red blood cells in malaria diagnostic smears. IEEE J. Biomed. Health Inform. **25**(5), 1735–1746 (2020)
7. Krizhevsky, A., Sutskever, I., Hinton, G.E.: Imagenet classification with deep convolutional neural networks. Adv. Neural. Inf. Process. Syst. **25**, 1097–1105 (2012)
8. Kutlu, H., Avci, E., Özyurt, F.: White blood cells detection and classification based on regional convolutional neural networks. Med. Hypotheses **135**, 109472 (2020)

9. Lamberti, W.F.: Blood cell classification using interpretable shape features: a comparative study of SVM models and CNN-based approaches. Comput. Methods Programs Biomed. Update **1**, 100023 (2021)
10. Long, F., Peng, J.J., Song, W., Xia, X., Sang, J.: BloodCaps: a capsule network based model for the multiclassification of human peripheral blood cells. Comput. Methods Programs Biomed. **202**, 105972 (2021)
11. Lu, S.Y., Satapathy, S.C., Wang, S.H., Zhang, Y.D.: PBTNet: a new computer-aided diagnosis system for detecting primary brain tumors. Front. Cell Dev. Biol. 2926 (2021)
12. Lu, S., Zhu, Z., Gorriz, J.M., Wang, S.H., Zhang, Y.D.: NAGNN: classification of Covid-19 based on neighboring aware representation from deep graph neural network. Int. J. Intell. Syst. **37**(2), 1572–1598 (2022)
13. Pasupa, K., Vatathanavaro, S., Tungjitnob, S.: Convolutional neural networks based focal loss for class imbalance problem: a case study of canine red blood cells morphology classification. J. Ambient Intell. Human. Comput. 1–17 (2020)
14. Patil, A., Patil, M., Birajdar, G.: White blood cells image classification using deep learning with canonical correlation analysis. IRBM **42**(5), 378–389 (2021)
15. Simonyan, K., Zisserman, A.: Very deep convolutional networks for large-scale image recognition. arXiv preprint arXiv:1409.1556 (2014)
16. Tiwari, P., et al.: Detection of subtype blood cells using deep learning. Cogn. Syst. Res. **52**, 1036–1044 (2018)

# Automatic Diagnosis of Myocarditis in Cardiac Magnetic Images Using CycleGAN and Deep PreTrained Models

Afshin Shoeibi[1](✉), Navid Ghassemi[2], Jonathan Heras[3], Mitra Rezaei[4], and Juan M. Gorriz[5]

[1] Faculty of Electrical Engineering, FPGA Lab, K. N. Toosi University of Technology, Tehran, Iran
afshin.shoeibi@gmail.com

[2] Computer Engineering Department, Ferdowsi University of Mashhad, Mashhad, Iran

[3] Department of Mathematics and Computer Science, University of La Rioja, La Rioja, Spain

[4] Electrical and Computer Engineering Department, Tarbiat Modaters University, Tehran, Iran

[5] Department of Signal Theory, Networking and Communications, Universidad de Granada, Granada, Spain

**Abstract.** Myocarditis is a cardiovascular disease caused by infectious agents, especially viruses. Compared to other cardiovascular diseases, myocarditis is very rare, occurring mainly due to chest pain or heart failure. Cardiac magnetic resonance (CMR) imaging is a popular technique for diagnosis of myocarditis. Factors such as low contrast, different noises, and high CMR slices of each patient cause many challenges when diagnosing myocarditis by specialist physicians. Therefore, it is necessary to introduce new artificial intelligence (AI) techniques for diagnosis of myocarditis from CMR images. This paper presents a new method to detect myocarditis in CMR images using deep learning (DL) models. First, the Z-Alizadeh Sani myocarditis dataset was used for simulations, which included CMR images of normal subjects and myocardial infarction patients. Next, preprocessing is performed on CMR images. CMR images are created with the help of the cycle generative adversarial network (GAN) model at this step. Finally, pretrained models including EfficientNet B3, EfficientNet V2, HrNet, ResNetrs50, ResNest50d, and ResNet 50d have been used to classify the input data. Among pretrained methods, the EfficientNet V2 model has achieved 99.33% accuracy.

**Keywords:** Myocarditis · Diagnosis · Deep learning · Pretrained · Cycle GAN

## 1 Introduction

Cardiovascular disease (CVD) is one of the most common causes of death worldwide today [17]. According to the World Health Organization (WHO),


J. M. Ferrández Vicente et al. (Eds.): IWINAC 2022, LNCS 13258, pp. 145–155, 2022.
https://doi.org/10.1007/978-3-031-06242-1_15

cardiovascular disease is the leading cause of death worldwide, with an estimated 17.9 million people dying from CVD in 2016 [8,12]. Coronary arteries [25], rheumatoid heart [39], and myocarditis [3] are the most prevalent types of CVD.

The myocardium is the main part of the heart responsible for contracting and resting in pumping blood to the heart, outside the heart, and other parts of the body [3]. Myocarditis is a heart disease caused by inflammation of the myocardium. Research has shown that viruses such as COVID-19, adenovirus, and AIDS can lead to myocardial infarctions [30]. This disease is very dangerous and in some cases leads to the death of the patient.

In recent years, significant advances have been made regarding the diagnosis and treatment of cardiovascular diseases and the reduction of deaths caused by these diseases [1]. Medical imaging techniques play a vital role in the diagnosis of CVDs, which include CMR [24], computed tomography (CT) [4], and ultrasound [36] modalities that accurately show the anatomical functions of the heart. The CMR method is a medical imaging technology for the non-invasive evaluation of the performance and structure of the cardiovascular system [24]. This imaging technique plays an essential role in myocarditis diagnosis [24].

CMR images often have low resolutions and different noises, making it difficult for physicians to diagnose CVDs accurately. In addition, to diagnose CVD, it is necessary to record various CMR data, and it can be challenging for physicians to review each of these images. To overcome these challenges, research has been conducted on the detection of CVDs using AI techniques [33]. AI is divided into machine learning methods (ML) [23,31] and deep learning (DL) [14], which are widely used in CVDs detection research [21].

In this paper, we present a new method for diagnosing myocarditis using the GAN cycle and pretrained models. The Z-Alizadeh Sani myocarditis dataset is selected for testing [30]. In the second step, preprocessing involves noise removal, image resizing, and data augumentation using cycle GAN [16]. Finally, various types of pretrained models including EfficientNet B3 [34], EfficientNet V2 [35], HrNet [37], ResNetrs50 [2], ResNest50d [40] and ResNet50d [15] are used to classify the input data.

In the following, other sections of the article are introduced. The third section is dedicated to materials and methods. In this section, details of the proposed method including datasets, preprocessing and DL models are introduced. The simulation results are in the third section. Finally, the discussion, conclusion, and future work are described in Sect. 4.

## 2 Material and Methods

In this section, the proposed method of diagnosing myocarditis using CMR imaging and DL models is presented. Figure 2 shows the block diagram of the proposed myocarditis diagnosis method, including datasheet steps, processing techniques, and pretrained models for classification. According to Fig. 1, the Z-Alizadeh Sani myocarditis dataset was selected for testing. Next, the noise removal and image size reduction are done in the pre-processing step. In addition, in this step, the Cycle GAN technique is used to increase the input data.

In the third step, a variety of pretrained models, including EfficientNet B3, EfficientNet V2, HrNet, ResNetrs50, ResNest50d, and ResNet 50d are tested to classify the input data.

### 2.1 Dataset

The data collection took place from September 2018 to September 2019 at the CMR department of OMID hospital in Tehran, IRAN. The data collection process was approved by the local ethical committee of the OMID hospital [30]. A 1.5-T system (MAGNETOM Aera Siemens, Erlangen Germany) was used for CMR examination. Dedicated body coils were used to scan all patients in the standard supine position. CINE-segmented images and pre-contrast T2-Weighted (trim) images in short and long-axis views were performed. Acquisition of pre contrast T1-Weighted Relative images were done in axial views of the myocardium [30]. The T1-Weighted relative sequence was repeated right after Gadolinium injection (DOTAREM 0/1 mmol/kg). Late Gadolinium Enhancements (LGE- high-resolution PSIR) sequences in short and long-axis views were performed after 10–15 min. The total number of images examined is 10425 [30]. The number of images representing healthy and Myocarditis patients was 7000 and 6000, respectively.

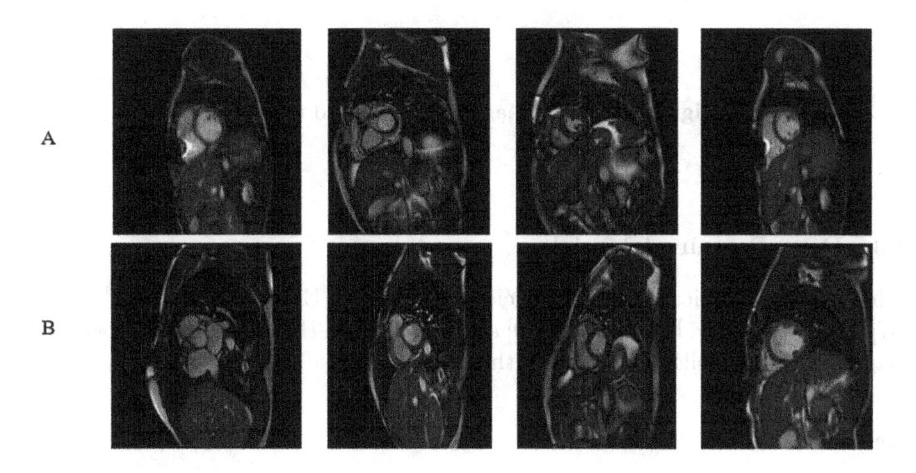


**Fig. 1.** Sample CMR images dataset A) Normal B) Abnormal.

### 2.2 Data Augmentation Using Cycle GAN

In this section, CMR image pre-processing techniques are presented, which include noise cancellation, image resizing, and DA using Cycle GAN. In DL research, the lack of medical data and its classification is always a challenge [16]. In recent years, several DA techniques have been proposed, among which

GAN models are highly efficient [16]. Cycle GAN architecture is one of the most important DA techniques based on image-to-image translation [16]. The cycle GAN architecture consists of two interconnected conditional GAN that function similarly to AE models [16]. In this model, cost functions play a significant role in discriminator sections [16]. A cycle GAN model similar to the one that is cited in the reference [13] is used in this research. Lastly, the block diagram of the proposed method is shown in Fig. 2.

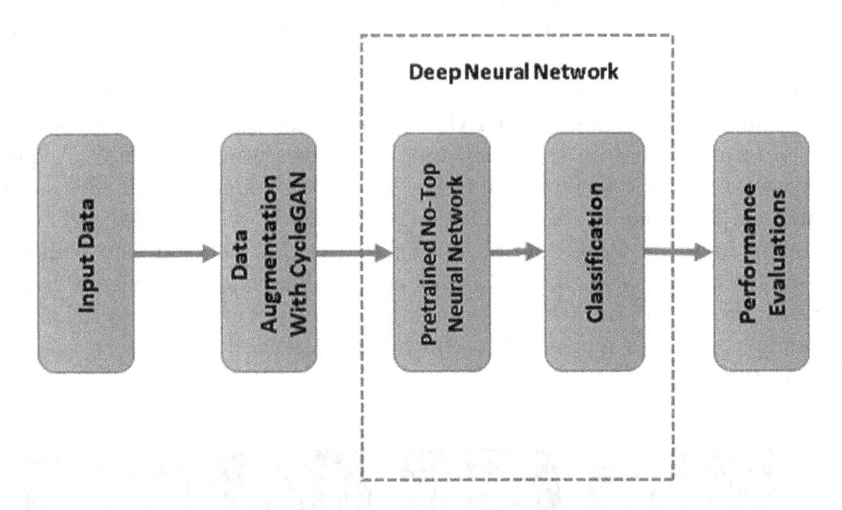


**Fig. 2.** The block diagram of proposed method.

### 2.3 Deep Pertained Models

This section is dedicated to the EfficientNet B3 [34], EfficientNet V2 [35], HrNet [37], ResNetrs50 [2], ResNest50d [40] and ResNet 50d [15] models for myocarditis diagnosis. In the following, each of these methods is discussed.

**ResNet**

ResNet [15] architecture was introduced in 2015, and it was the first convolutional architecture with more than 20 layers that was successfully trained. Such a landmark was achieved thanks to the usage of residual blocks that deal with the vanishing gradient problem. Since this architecture was released, it has been the main baseline for applying transfer learning to different contexts. The number of layers of the ResNet architecture might vary, but the most widespread version of the ResNet architecture has 50 layers, and this is the version employed in the present paper. The ResNet architecture has also been the basis for developing new architectures including ResNeSt and ResNetRS which are also considered in this paper.

### ResNeSt

ResNeSt [40] architecture was developed by researchers from Amazon and UC Davis, and it replaces the residual block of the ResNet architecture with a split-attention block. The split-attention block is a computational unit that consists of feature map group and split attention operations. Such operations capture cross-feature interactions from different layers of the architecture, and also learn diverse representations. As in the ResNet architecture, several versions of the ResNeSt architecture exist, and we use the version with 50 layers.

### ResNetRS

ResNetRS authors [2] did not design a new family of architectures, but instead revisited the canonical ResNet and studied new training and scaling strategies. In particular, the training strategy employs the cosine learning rate schedule and several regularization methods including weight decay, label smoothing, dropout, stochastic depth, and RandAugment. In addition, two new scaling strategies are applied by scaling model depth and increasing image resolution slowly. In our work, we have employed the checkpoint of the ResNetRS architecture with 50 layers.

### EfficientNet

EfficientNet [34] architecture was designed by applying Neural Architecture Search methods to find the optimal combination of depth, width, and resolution of the input images. In particular, the EfficientNet architecture was developed by using a step-by-step scheme that, as a first step, built a baseline network and then used compound scaling to increase the capacity of the network without increasing the number of parameters greatly. The scaling process produced a family of EfficientNet architectures that goes from B0 (the smallest) to B7 (the largest). For our experiments, we have used the B3 version of the EfficientNet family.

### EfficientNet v2

EfficientNet v2 [35] was developed exactly in the same way as the EfficientNet architecture but dealt with the limitations of such an architecture. From the architecture design, EfficientNetV2 architecture extensively utilizes both MBConv and Fused-MBConv layers. Moreover, the architecture was trained by using a progressive learning scheme that combines progressive resizing with a strong regularization scheme.

### HrNet

HrNet [37] is a general-purpose convolutional neural network for tasks like semantic segmentation, object detection and image classification. The HrNet is able to maintain high-resolution representations throughout the whole process. This is achieved by starting from a high-resolution convolution stream, gradually adding high-to-low resolution convolution streams one by one, and connecting the multi-resolution streams in parallel. The high-resolution representations learned from HrNet are not only semantically strong, but also spatially precise since this architecture connects high-to-low resolution convolution streams in parallel rather than in series, and multi-resolution fusions boost the high-resolution representations with the help of the low-resolution representations, and vice versa.

## 3 Experiment Results

In this section, the results of the proposed method for the detection of myocarditis from CMR images using DL techniques are reported. A hardware system with 16 GB RAM, CPU Core i7, and NVidia GeForce 1080 was used for simulation. Also, all implementations of the proposed method are in the Python environment using the TensorFlow 2 [10], Keras [19], and Skit-learn [27] tools. To evaluate the proposed method, the parameters of Acc, Sens, Spec, Prec, and F1-S have been calculated based on [20]. Table 1 presents the simulation results of the EfficientNet B3, EfficientNet V2, HrNet, ResNetrs50, ResNest50d, and ResNet 50d models without using Cycle GAN.

**Table 1.** Results for pertained models

| Models | Acc | Prec | Rec | Spec | F1 |
|---|---|---|---|---|---|
| efficientnet_b3 | 99,52 (3,48) | 99,47 (17,87) | 99,14 (30,96) | 99,72 (17,85) | 99,3 (24,07) |
| efficientnetv2 | 99,54 (2,58) | 99,19 (15,64) | 99,47 (20,46) | 99,57 (6,89) | 99,33 (21,2) |
| HrNet | 99,04 (0,28) | 98,11 (0,31) | 99,12 (0,51) | 99 (0,16) | 98,61 (0,41) |
| resnest50d | 99,23 (0,4) | 99,69 (0,44) | 98,06 (0,72) | 99,84 (0,23) | 98,87 (0,58) |
| resnet50d | 99,62 (0,14) | 99,51 (0,27) | 99,39 (0,38) | 99,74 (0,14) | 99,45 (0,2) |
| resnetrs50 | 99,42 (0,33) | 99,04 (0,69) | 99,29 (1,03) | 99,49 (0,37) | 99,16 (0,48) |

According to Table 1, it can be seen that the efficient net v2 model has a better performance compared to other pretrained models. The efficientnet_b3 model has also achieved successful results. In the following, the results of EfficientNet B3, EfficientNet V2, HrNet, ResNetrs50, ResNest50d, and ResNet 50d models are discussed along with the cycle GAN method and pretrained models. In Table 2, the results of pretrained models using the Cycle GAN technique are reported. According to Table 2, the efficientnetv2 model has been able to achieve successful results compared to other pretrained models. Also, the results of Table 2 show that the use of the cycle GAN has a significant role in increasing the efficiency and accuracy of myocarditis diagnosis from CMR images.

**Table 2.** Results for pertained models with Cycle GAN.

| Models | Acc | Prec | Rec | Spec | F1 |
|---|---|---|---|---|---|
| efficientnet_b3 | 99,33 (0,37) | 99,47 (0,34) | 98,57 (1,42) | 99,72 (0,18) | 99,01 (0,56) |
| efficientnetv2 | 99,33 (0,07) | 99,04 (0,7) | 99,03 (0,87) | 99,49 (0,37) | 99,03 (0,11) |
| HrNet | 86,1 (13,25) | 81,66 (21,32) | 89,47 (9,88) | 84,34 (22,45) | 83,33 (12,12) |
| resnest50d | 98,85 (0,68) | 98,49 (0,54) | 98,16 (1,95) | 99,21 (0,29) | 98,32 (1,01) |
| resnet50d | 99,27 (0,2) | 98,91 (0,78) | 98,98 (0,5) | 99,42 (0,42) | 98,94 (0,29) |
| resnetrs50 | 98,75 (0,61) | 97,16 (1,84) | 99,31 (0,7) | 98,46 (1,03) | 98,21 (0,86) |

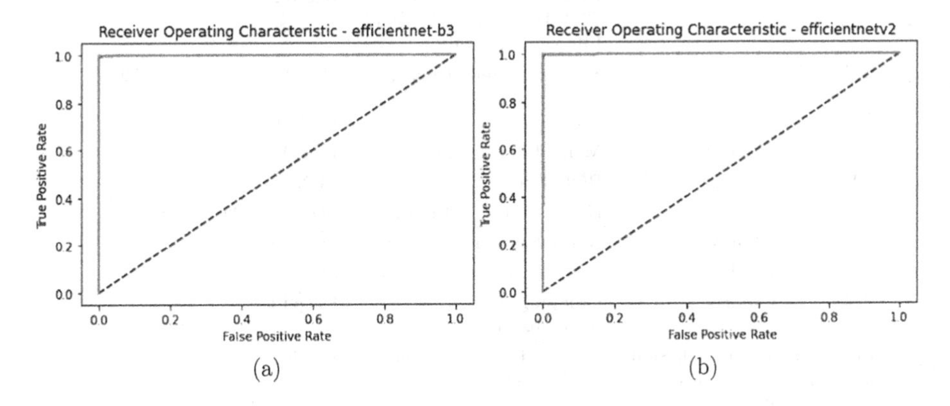


**Fig. 3.** ROC curves for efficientnetv2 and efficientnet_b3 models.

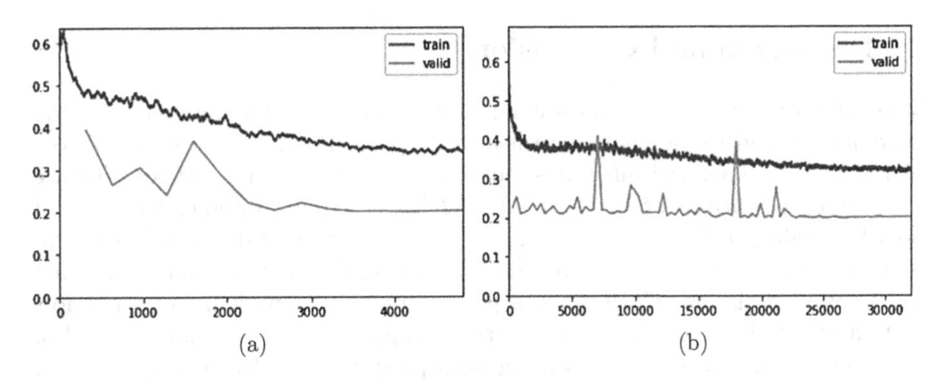


**Fig. 4.** Learning and loss curves for efficientnetv2 and efficientnet_b3 models.

Figure 3 shows the ROC curves for efficientnetv2 and efficientnet_b3 methods alongside the cycle GAN technique. In addition, learning curves and loss curves for efficientnetv2 and efficientnet_b3 are also described in Fig. 4.

The results of Tables 1 and 2 show that efficientnetv2 and efficientnet_b3 models have satisfactory results compared to a variety of pretrained methods. In this paper, for the first time, the cycle GAN technique and pretrained models are used to diagnose myocarditis from CMR images.

**Table 3.** Related works

| Works | Dataset | Preprocessing | DL Model | Acc (%) |
|---|---|---|---|---|
| [41] | MICCAI 2020 EMIDEC | Different Methods | U-Net | – |
| [29] | Clinical | Standard Preprocessing | 2D-CNN | 96.70 |
| [22] | MICCAI 2020 EMIDEC | Normalization, Resampling, Segmentation, Applying Three-order Spline Interpolation | 2D U-Net | 92.00 |
| [5] | Clinical | Resampling, Cropping, CLAHE, LV Localization, Filtering Methods | ResNet-56 | – |
| [26] | Clinical | Standard Preprocessing | GoogLeNet, AlexNet, ResNet-152 | 79.50 |
| [42] | Clinical | Manual Segmentation | LSTM | – |
| [11] | Clinical | ROIs Extraction, DA | FCN | – |
| [6] | Clinical | Normalization, filtering | SDAE+SVM | 87.60 |
| [7] | Clinical | Normalization | 2D-CNN | – |
| [38] | Clinical | Countering | 2D-CNN | 94.50 |
| [18] | Clinical | Cropping, IFT, Normalization | DeepT1 | – |
| Proposed method | Z-Alizadeh Sani myocarditis dataset | Standard Preprocessing, Cycle GAN | EfficientNet V2 | 99.33 |

# 4 Discussion and Conclusion

Myocarditis is a type of cardiovascular disease that causes inflammation of the heart muscle and threatens human health [3]. This disease is caused by factors such as viral or bacterial infections, heart surgery, rheumatic fever, and the use of certain medications or toxins [8,12,17,25]. Chest pain, fatigue, shortness of breath, bulging of the jugular veins, etc. are among the most important symptoms in patients with myocarditis [3,30]. Among the methods for diagnosing cardiovascular disease, CMR images provide physicians with important information about the structure of the heart. This paper presents a new method of intelligent myocarditis diagnosis using deep pretrained models and cycle GAN. In this work, the Z-Alizadeh Sani myocarditis dataset was used for simulations. Next, preprocessing includes noise cancellation, image size reduction, and the cycle GAN to produce artificial CMR images. Finally, pretrained models were tested to classify the input data. In this section, the results of the EfficientNet B3, EfficientNet V2, HrNet, ResNetrs50, ResNest50d, and ResNet 50d models were reviewed. EfficientNet B3 and EfficientNet V2 models were found to be more efficient than other DL techniques. Table 3 compares the results of the proposed method with those of several studies performed in the field of cardiovascular disease diagnosis.

Based on Table 3, this paper's method achieves higher accuracy than other similar studies. The method described in this paper can be applied in specialized centers for diagnosing cardiovascular disease as software for diagnosing myocarditis in the future. The following are some suggestions for future work in the field of myocarditis detection from CMR images using DL techniques. For future work, new GAN models such as inFoGAN can be used in the preprocessing section [9]. As another future work, the use of transformers [28] and attention [32] models can increase the accuracy of myocarditis detection from CMR images.

## References


1. Alizadehsani, R., et al.: Coronary artery disease detection using artificial intelligence techniques: a survey of trends, geographical differences and diagnostic features 1991–2020. Comput. Biol. Med. **128**, 104095 (2021)
2. Bello, I., et al.: Revisiting resnets: improved training and scaling strategies. Adv. Neural Inf. Process. Syst. **34** (2021)
3. Blauwet, L.A., Cooper, L.T.: Myocarditis. Progress Cardiovascular Dis. **52**(4), 274–288 (2010)
4. Borggreve, A.S., Goense, L., van Rossum, P.S., van Hillegersberg, R., de Jong, P.A., Ruurda, J.P.: Generalized cardiovascular disease on a preoperative CT scan is predictive for anastomotic leakage after esophagectomy. Eur. J. Surg. Oncol. **44**(5), 587–593 (2018)
5. Chen, A., et al.: Transfer learning for the fully automatic segmentation of left ventricle myocardium in porcine cardiac cine MR images. In: Pop, M., et al. (eds.) STACOM 2017. LNCS, vol. 10663, pp. 21–31. Springer, Cham (2018). https://doi.org/10.1007/978-3-319-75541-0_3
6. Chen, M., Fang, L., Zhuang, Q., Liu, H.: Deep learning assessment of myocardial infarction from MR image sequences. IEEE Access **7**, 5438–5446 (2019)
7. Chen, Z., et al.: Myocardial infarction segmentation from late gadolinium enhancement MRI by neural networks and prior information. In: 2020 International Joint Conference on Neural Networks (IJCNN), pp. 1–8. IEEE (2020)
8. Cosselman, K.E., Navas-Acien, A., Kaufman, J.D.: Environmental factors in cardiovascular disease. Nat. Rev. Cardiol. **12**(11), 627–642 (2015)
9. Creswell, A., White, T., Dumoulin, V., Arulkumaran, K., Sengupta, B., Bharath, A.A.: Generative adversarial networks: an overview. IEEE Signal Process. Mag. **35**(1), 53–65 (2018)
10. Dillon, J.V., et al.: Tensorflow distributions. arXiv preprint arXiv:1711.10604 (2017)
11. Fahmy, A.S., El-Rewaidy, H., Nezafat, M., Nakamori, S., Nezafat, R.: Automated analysis of cardiovascular magnetic resonance myocardial native t1 mapping images using fully convolutional neural networks. J. Cardiovascular Magn. Reson. **21**(1), 1–12 (2019)
12. Gaziano, T., Reddy, K.S., Paccaud, F., Horton, S., Chaturvedi, V.: Cardiovascular disease. Disease Control Priorities in Developing Countries. 2nd edition (2006)
13. Ghassemi, N., et al.: Automatic diagnosis of covid-19 from CT images using cyclegan and transfer learning. arXiv preprint arXiv:2104.11949 (2021)
14. Górriz, J.M., et al.: Artificial intelligence within the interplay between natural and artificial computation: advances in data science, trends and applications. Neurocomputing **410**, 237–270 (2020)
15. He, K., Zhang, X., Ren, S., Sun, J.: Proceedings of the IEEE conference on computer vision and pattern recognition (2016)

16. Huang, Z., et al.: Cagan: a cycle-consistent generative adversarial network with attention for low-dose CT imaging. IEEE Trans. Comput. Imaging **6**, 1203–1218 (2020)
17. Imes, C.C., Lewis, F.M.: Family history of cardiovascular disease (CVD), perceived CVD risk, and health-related behavior: a review of the literature. J. Cardiovascular Nursing **29**(2), 108 (2014)
18. Jeelani, H., Yang, Y., Zhou, R., Kramer, C.M., Salerno, M., Weller, D.S.: A myocardial t1-mapping framework with recurrent and u-net convolutional neural networks. In: 2020 IEEE 17th International Symposium on Biomedical Imaging (ISBI), pp. 1941–1944. IEEE (2020)
19. Ketkar, N.: Introduction to Keras. In: Deep Learning with Python, pp. 95–109. Apress, Berkeley, CA (2017). https://doi.org/10.1007/978-1-4842-2766-4_7
20. Khodatars, M., et al.: Deep learning for neuroimaging-based diagnosis and rehabilitation of autism spectrum disorder: a review. Comput. Biol. Med. **139**, 104949 (2021)
21. Liu, X., Wang, H., Li, Z., Qin, L.: Deep learning in ECG diagnosis: a review. Knowl.-Based Syst. **227**, 107187 (2021)
22. Ma, J.: Cascaded framework for automatic evaluation of myocardial infarction from delayed-enhancement cardiac MRI. arXiv preprint arXiv:2012.14556 (2020)
23. Mohammadpoor, M., Shoeibi, A., Shojaee, H., et al.: A hierarchical classification method for breast tumor detection. Iranian J. Med. Phys. **13**(4), 261–268 (2016)
24. Mouquet, F., et al.: Characterisation of peripartum cardiomyopathy by cardiac magnetic resonance imaging. Eur. Radiol. **18**(12), 2765–2769 (2008)
25. Ogden, J.A.: Congenital anomalies of the coronary arteries. Am. J. Cardiol. **25**(4), 474–479 (1970)
26. Ohta, Y., Yunaga, H., Kitao, S., Fukuda, T., Ogawa, T.: Detection and classification of myocardial delayed enhancement patterns on MR images with deep neural networks: a feasibility study. Radiol. Artif. Intell. **1**(3), e180061 (2019)
27. Pedregosa, F., et al.: Scikit-learn: machine learning in python. J. Mach. Learn. Res. **12**, 2825–2830 (2011)
28. Sadeghi, D., et al.: An overview on artificial intelligence techniques for diagnosis of schizophrenia based on magnetic resonance imaging modalities: methods, challenges, and future works. arXiv preprint arXiv:2103.03081 (2021)
29. Scannell, C.M., Veta, M., Villa, A.D., Sammut, E.C., Lee, J., Breeuwer, M., Chiribiri, A.: Deep-learning-based preprocessing for quantitative myocardial perfusion MRI. J. Magn. Reson. Imaging **51**(6), 1689–1696 (2020)
30. Sharifrazi, D., et al.: CNN-KCL: automatic myocarditis diagnosis using convolutional neural network combined with k-means clustering (2020)
31. Shoeibi, A., et al.: Detection of epileptic seizures on EEG signals using ANFIS classifier, autoencoders and fuzzy entropies. Biomed. Signal Process. Control **73**, 103417 (2022)
32. Shoeibi, A., et al.: Automated detection and forecasting of covid-19 using deep learning techniques: a review. arXiv preprint arXiv:2007.10785 (2020)
33. Suri, J.S., et al.: Understanding the bias in machine learning systems for cardiovascular disease risk assessment: the first of its kind review. Comput. Biol. Med. 105204 (2022)
34. Tan, M., Le, Q.: Efficientnet: rethinking model scaling for convolutional neural networks. In: International Conference on Machine Learning, pp. 6105–6114. PMLR (2019)
35. Tan, M., Le, Q.: Efficientnetv2: smaller models and faster training. In: International Conference on Machine Learning, pp. 10096–10106. PMLR (2021)

36. Villanueva, F.S., Wagner, W.R.: Ultrasound molecular imaging of cardiovascular disease. Nat. Clin. Pract. Cardiovascular Med. **5**(2), S26–S32 (2008)
37. Wang, J., et al.: Deep high-resolution representation learning for visual recognition. IEEE Trans. Patt. Anal. Mach. Intell. **43**(10), 3349–3364 (2020)
38. Wang, S.H., McCann, G., Tyukin, I.: Myocardial infarction detection and quantification based on a convolution neural network with online error correction capabilities. In: 2020 International Joint Conference on Neural Networks (IJCNN), pp. 1–8. IEEE (2020)
39. Wolfe, F., Michaud, K.: Heart failure in rheumatoid arthritis: rates, predictors, and the effect of anti-tumor necrosis factor therapy. Am. J. Med. **116**(5), 305–311 (2004)
40. Zhang, H., et al.: Resnest: split-attention networks. arXiv preprint arXiv:2004.08955 (2020)
41. Zhang, Y.: Cascaded convolutional neural network for automatic myocardial infarction segmentation from delayed-enhancement cardiac MRI. In: Puyol Anton, E., Pop, M., Sermesant, M., Campello, V., Lalande, A., Lekadir, K., Suinesiaputra, A., Camara, O., Young, A. (eds.) STACOM 2020. LNCS, vol. 12592, pp. 328–333. Springer, Cham (2021). https://doi.org/10.1007/978-3-030-68107-4_33
42. Zhou, H., et al.: Deep learning algorithm to improve hypertrophic cardiomyopathy mutation prediction using cardiac cine images. Eur. Radiol. **31**(6), 3931–3940 (2021)

# Quantifying Inter-hemispheric Differences in Parkinson's Disease Using Siamese Networks

Juan E. Arco[1,3](✉), A. Ortiz[1,3], Diego Castillo-Barnes[2,3], Juan M. Górriz[2,3], and Javier Ramírez[2,3]

[1] Communications Engineering Department, University of Málaga, 29004 Málaga, Spain
jearco@ic.uma.es

[2] Department of Signal Theory, Communications and Networking, University of Granada, 18060 Granada, Spain

[3] Andalusian Data Science and Computational Intelligence Institute (DaSCI), Granada, Spain

**Abstract.** Classification of medical imaging is one of the most popular application of intelligent systems. A crucial step is to find the features that are relevant for the subsequent classification. One possibility is to compute features derived from the morphology of the target region in order to check its role in the pathology under study. It is also possible to extract relevant features to evaluate the similarity between different regions, in addition to compute morphology-related measures. However, it can be much more useful to model the differences between regions. In this paper, we propose a method based on the principles of siamese neural networks to extract informative features from differences between two brain regions. The output of this network generates a latent space that characterizes differences between the two hemispheres. This output vector is then fed into a linear SVM classifier. The usefulness of this method has been assessed with images from the Parkinson's Progression Markers Initiative, demonstrating that differences between the dopaminergic regions of both hemispheres lead to a high performance when classifying controls *vs* Parkinson's disease patients.

**Keywords:** Deep learning · Siamese network · Parkinson's disease · SPECT images

## 1 Introduction

Current medical images provide an extremely useful information for the diagnosis of a wide range of diseases. Despite these images have a high quality, their correct interpretation and the subsequent diagnosis is not a straightforward task. The emergence of artificial intelligence has revolutionized the study of different pathologies, given the ability of this kind of techniques for being


J. M. Ferrández Vicente et al. (Eds.): IWINAC 2022, LNCS 13258, pp. 156–165, 2022.
https://doi.org/10.1007/978-3-031-06242-1_16

used within a computer aided diagnosis (CAD) system [12]. When applying to neuroimaging, this tool usually finds relevant patterns that are extremely useful for the identification of neurological disorders. In fact, a high number of studies have developed intelligent systems for this purpose. For instance, these methods have demonstrated a good performance when diagnosing Alzheimer's disease [1,2,9,16]. These works use information contained in magnetic resonance or positron emission tomograpy (PET) images to classify controls *vs* AD patients, in addition to detect the progression from mild cognitive impairment to a severe dementia. In a similar way, changes associated with Parkinson's disease have also been automatically detected by these alternatives [6,17,21]. They usually employ DaTSCAN neuroimages given their suitability for visualizing the dopamine deficiency.

Moreover, classification models focus their analysis on a specific brain region that characterizes the pathology. Besides, the definition of this region of interest considerably reduces the dimensionality of the data, addressing the curse of dimensionality problem associated with statistical classification. After that, the simplest alternative is to evaluate the intensity of the voxels contained in this region. However, it is unlikely that differences in intensity allow to interpret the cognitive state of the patient. Another option is to compute features based on the morphology of the region [11,26]. Thus, the region is characterized by a series of measurements such as size, position of the centroid, roundness, etc. Despite this alternative has been successfully employed in previous studies, it seems suboptimal for one crucial reason: the relevance of features varies for each individual classification context. This means that some features can be extremely informative in one scenario, but completely irrelevant in a different one. An interesting alternative is to directly model the differences between regions, instead of extracting features associated with each one of them. In this case, a siamese neural network is an excellent choice [8,14]. According to its name, this architecture consists of two identical neural networks sharing the same structure that compare their individual outputs at the end by using a distance metric. A global output is then generated from this resulting measure.

In this work, we propose an alternative for automatically computing differences between two regions that can be subsequently used for classification. Specifically, our proposal relies on the use of a siamese neural network with two inputs that extracts informative features from both regions. Once the model is trained, the latent space of the dense layer leads to a feature vector that is entered as input of a linear classifier. The proposed methodology has been evaluated using PET images from the Parkinson's Progression Markers Initiative [18]. Specifically, we aim at demonstrating that differences between the dopaminergic regions of both hemispheres can be used as discriminative features to distinguish between controls and patients suffering from Parkinson's disease.

The rest of the paper is organized as follows. Section 2.1 contains a description of the dataset used for the evaluation of the performance of the system. Section 2.2 details the preprocessing applied to this database before entering into the siamese neural network, which is explained in Sect. 2.3. Results are summarized in Sect. 3, whereas conclusions and future work are available in Sect. 4.

## 2 Material and Methods

### 2.1 Dataset

The method proposed in this work aims at measuring the asymmetry between different brain regions that can be relevant for classification purposes. The database employed in this work contains DaTSCAN SPECT images from 1413 subjects, 1218 from patients suffering Parkinson's disease (PD) and 195 controls (CN) from the PPMI dataset [19]. This repository contains data from an observational clinical study to verify the progression markers in Parkinson's disease. Raw projection data are acquired into a $128 \times 128$ matrix stepping each 3° for a total of 120 projection into two 20% symmetric photopeak windows centered on 159 KeV and 122 KeV with a total scan duration of approximately 30–45 min. Table 1 summarizes the demographics of the patients in the database. Thus, the goal in this specific context is to evaluate if differences in the shape of the dopaminergic regions can be used to distinguish between PD patients and CN (Fig. 1).

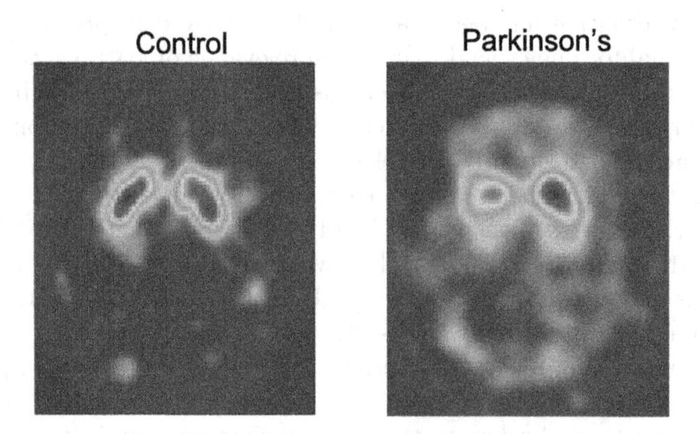


**Fig. 1.** Slice of the 3D images corresponding to a control (left) and a PD patient (right).

**Table 1.** Patient demographics

| Evaluation | Sex (M/F) | Mean age ± Std |
|---|---|---|
| NC | 130/65 | $61.02 \pm 11.25$ |
| PD | 798/420 | $62.93 \pm 9.92$ |

### 2.2 Preprocessing

DaTSCAN images from the PPMI database were spatially normalized according to the MNI152 template, which is based on the average of 152 scans from normal subjects. We further used SPM12 [27] to preprocess the images, applying affine and local deformations to achieve the best warping between the images with the DaTSCAN template defined in [22]. After that, the regions of interest were selected, which refers to those that reveal dopaminergic activity. As a result, this process led to a reduction in the size of the images, from the original (95, 69, 79) to the final (29,25,41). This final step allows a reduction in the computational cost of the classification system while preserving the information contained in the target regions.

Another crucial aspect is related to the intensity levels of the images. The idea behind these functional images is that the intensity in each pixel provides an indirect measure of the neurophysiological activity. This means that the same value in two different pixels should correspond to the same drug uptake, whereas abnormal differences in these values can reveal a wide range of pathologies [7, 24,25]. This paper employs Integral Normalization [13], in order to ensure that there is a clear relationship between intensity levels and drup uptakes, as follows:

$$\hat{\mathbf{I}}_i = \frac{\mathbf{I}_i}{I_{n,i}} \tag{1}$$

where $\mathbf{I}_i$ refers for the image of the *ith* subject in the database, $\hat{\mathbf{I}}_i$ is the resulting normalized image, and $I_{n,i}$ is the intensity normalization value. This is computed as the mean of the image for each independent subject. Then, the resulting values were standardized in the range [0, 1]. Finally, each resulting image is then partitioned into two subimages. The first one contains the dopaminergic region of the left hemisphere, whereas the second includes the dopaminergic region of the right one. These two images for each individual subject are entered into the classification system proposed in this work, which is fully described in next sections.

### 2.3 Siamese Neural Network

The siamese architecture was introduced in the 1990s within a signature verification system [5]. A siamese neural network consists of the union of two identical neural networks with exactly the same configuration. This means that they have the same parameters and even share common weights. During the training, each network processes the inputs as an individual feedforward network, processing information in only one direction. Briefly, each neuron of a specific layer processes the input, and sends the output to all the neurons of the following layer. Since both networks share the same weights, they are updated at the same moment through the error back-propagation process. Given that the siamese architecture is based on two individual networks, each one of them receives an input and produces an output in its final layer. The main aspect of this framework is that the outputs of both subnetworks are compared according to a distance measure. Based on this value, the final output of the siamese network is then used to assign a label to the data inputs. The output can be seen as the semantic difference between the projected representation of the inputs [8].

Despite this network has been widely used with the aim of evaluating the similarity between two inputs (e.g. fingerprints [3] or signatures [5]), it can be used as an intermediate stage within a different classification context. Figure 2 depicts a representation of the siamese neural network employed in this work. In our case, we trained the siamese network with the SPECT images of the database described in Sect. 2.1. The aim in this application context was to test that asymmetry between the dopaminergic regions of each hemisphere differs from PD patients and controls. In other words, our hypothesis is that differences between left and right striatum are relevant to distinguish people who suffer from Parkinson's disease and those that do not. This way, the left and right dopaminergic regions are entered into the two inputs of the siamese network. The model is then trained in order to learn the differences between the two inputs. To do so, the Hinge function [10] is used to compute the loss associated with the distance between the outputs of the two subnetworks, as follows:

$$l(y) = \begin{cases} 0 & t \cdot y \geq 1 \\ 1 - t \cdot y & \text{otherwise} \end{cases} \tag{2}$$

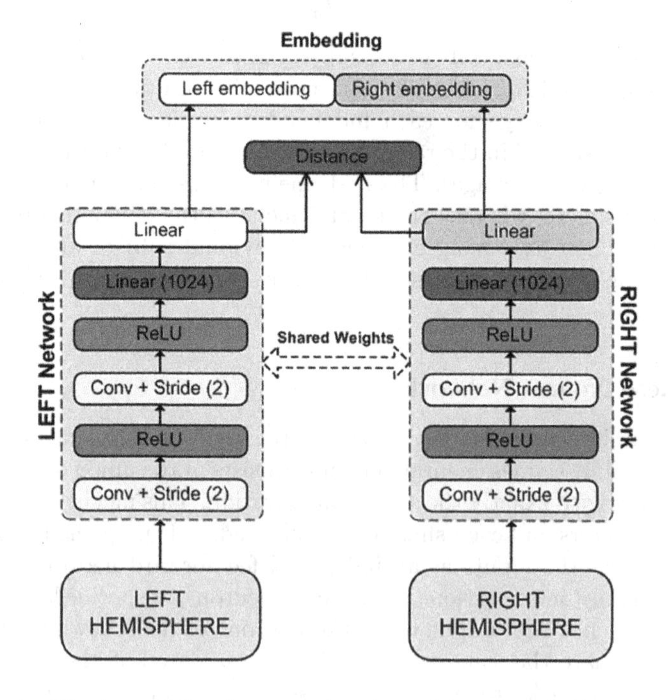


**Fig. 2.** Architecture of the siamese network used to compute the embeddings.

where $t = \{-1, 1\}$ corresponds to the actual label and $y$ is the output of the linear layers.

Once the training process is finished, the information contained in the final layer is retrieved. This is obtained as the combination of the output of the lineal layer of each individual branch. This embedding is then used as the feature vector within a classification framework. Specifically, the vectors associated with each individual sample are used to train a linear Support Vector Machine (SVM) classifier. Different metrics from the confusion matrix are employed to evaluate the performance of this scheme. Besides, a 5-fold cross-validation scheme was used to preserve the independence between training and test sets.

## 3 Results

In this section, we present the results obtained by the proposed method. First of all, we used the T-distributed Stochastic Neighbor Embedding (TSNE) for reducing the dimensionality of the embeddings to two dimensions in order to improve the visualization of data associated with the two classes: PD patients and controls. As Fig. 3 shows, the samples are grouped into two different clusters. It is worth highlighting that there is a clear separation between both classes since most of samples are correctly located. This demonstrates that the information contained in the embedding allows the distinction between the two classes.

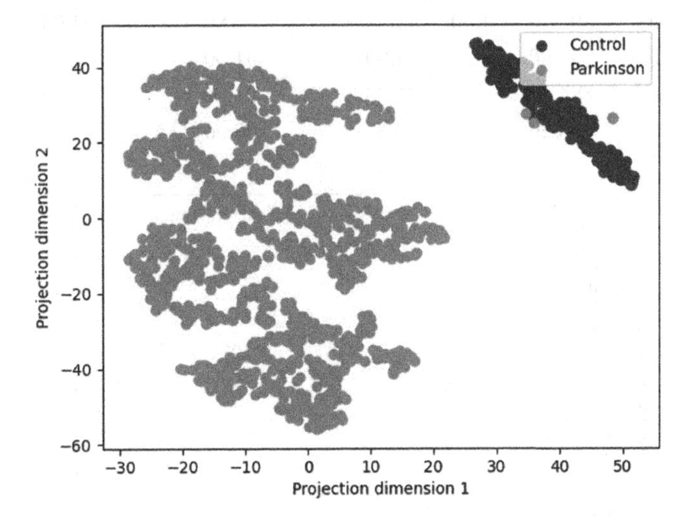


**Fig. 3.** Projection over the first two dimensions of the embeddings associated with controls (blue) and PD (red). (Color figure online)

Table 2 summarizes the performance in terms of different metrics. With reference to balanced accuracy, a 96.78% was obtained, whereas our proposal led to an AUC of 99.16%. This demonstrates the suitability of the proposed method for this classification context. Figure 4 depicts the ROC curve obtained by our method. This graphic provides a visual evidence of the large performance obtained by our proposal. In fact, the large AUC obtained manifests that most of samples are properly assigned to their corresponding class.

Figure 5 provides a crucial information that certifies the separability between the two classes. Figure 5a contains two different clusters, similar to the ones obtained in Fig. 3. However, in this case it is represented the pairwise distance between the vectors contained in the dense layers of both branches in the siamese network. It is important to note that this distance is much lower for controls than for PD patients, evidencing that the asymmetry between the dopaminergic regions of both hemispheres is higher in PD patients. Therefore, as Fig. 5 states, there is a much higher variability regarding the differences between both hemispheres in PD patients compared with controls. This means that PD patients are extremely different than controls, but they also vary one from each other, as the large variance in the sample shows.

**Table 2.** Classification results obtained by the proposed method and by other previous works.

| Method | Accuracy | Sensitivity | Specificity | AUC |
|---|---|---|---|---|
| **Ours** | **96.78 ± 1.99** | **97.66 ± 1.65** | **95.90 ± 3.84** | **99.16** |
| [25] | 95.00 ± 3.00 | 95.00 ± 5.00 | 95.00 ± 4.00 | 97.00 |
| [20] | 94.10 ± 4.50 | 93.30 ± 5.80 | 95.80 ± 7.90 | 94.60 |
| [4] | 92.00 | 94.00 | 91.00 | – |
| [23] | 84.60 | 71.60 | 97.50 | 85.90 |
| [15] | 95.20 | 97.50 | 90.90 | – |

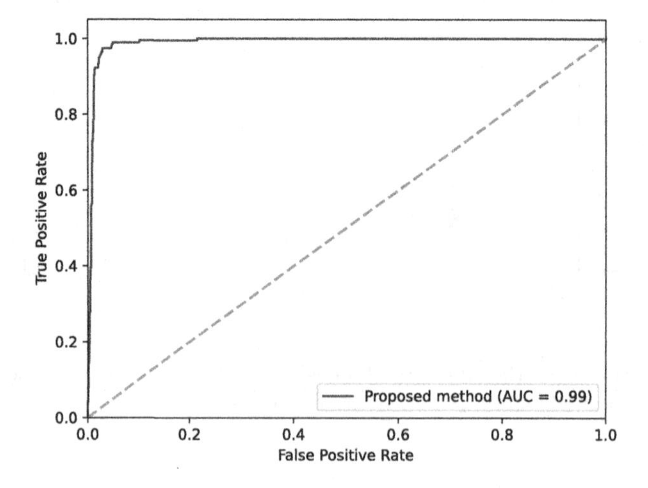


**Fig. 4.** ROC curves obtained for classification using PET, MRI and mixed mode images

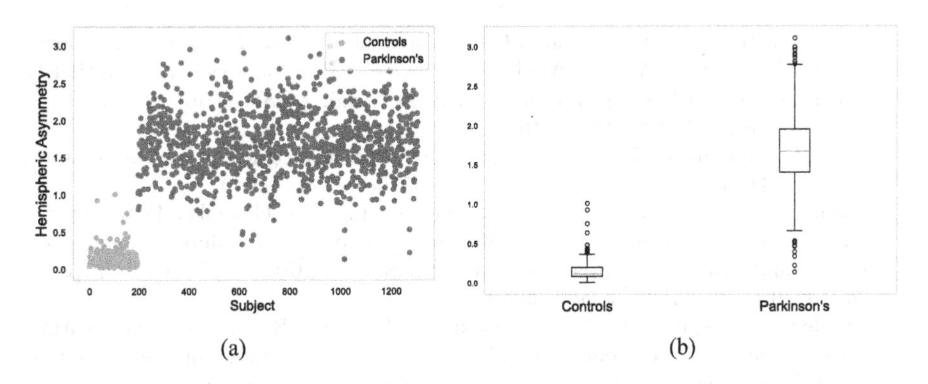


**Fig. 5.** (a) Distribution of the inter-hemispheric distances computed by the proposed model for Parkinson's patients and controls. (b) Boxplots of the distributions of distances in both classes.

## 4 Conclusions and Future Work

In this work, we propose a method based on siamese neural networks to assess anatomical differences in medical imaging. These differences are then evaluated in a classification context in order to quantify their relevance as a feature extractor. Once the network is optimized to compute similarities between two different inputs, vectors from the final dense layer are extracted. These features are then entered into a linear SVM classifier, leading to an accuracy of 96.78%

The excellent results manifest the usefulness of the method, paving the way to future research not only in brain imaging but also in other biomedical signals. Additionally, the inclusion of the cosine similarity in the loss function could help in a better computation of the similarities between the two inputs of the neural network. Finally, our findings reveal that differences between two structures can be even more informative that the nature of the regions itself, even when these differences had been considered as irrelevant in the context under study.

**Acknowledgments.** This work was supported by projects PGC2018-098813-B-C32 and RTI2018-098913-B100 (Spanish "Ministerio de Ciencia, Innovación y Universidades"), UMA20-FEDERJA-086, CV20-45250, A-TIC-080-UGR18 and P20 00525 (Consejería de economía y conocimiento, Junta de Andalucía) and by European Regional Development Funds (ERDF); and by Spanish "Ministerio de Universidades" through Margarita-Salas grant to J.E. Arco.

## References

1. Arco, J.E., Ramírez, J., Górriz, J.M., Ruz, M.: Data fusion based on searchlight analysis for the prediction of Alzheimer's disease. Expert Syst. Appl. **185**, 115549 (2021). https://doi.org/10.1016/j.eswa.2021.115549

2. Arco, J.E., Ramírez, J., Puntonet, C.G., Górriz, J.M., Ruz, M.: Improving short-term prediction from MCI to AD by applying searchlight analysis. In: 2016 IEEE 13th International Symposium on Biomedical Imaging (ISBI), pp. 10–13 (2016). https://doi.org/10.1109/ISBI.2016.7493199
3. Baldi, P., Chauvin, Y.: Neural networks for fingerprint recognition. Neural Comput. **5**(3), 402–418 (1993). https://doi.org/10.1162/neco.1993.5.3.402
4. Brahim, A., Ramírez, J., Górriz, J.M., Khedher, L., Salas-Gonzalez, D.: Comparison between different intensity normalization methods in 123i-ioflupane imaging for the automatic detection of parkinsonism. PLOS ONE **10**(6), 1–20 (2015). https://doi.org/10.1371/journal.pone.0130274
5. Bromley, J., Guyon, I., LeCun, Y., Säckinger, E., Shah, R.: Signature verification using a "Siamese" time delay neural network. In: Proceedings of the 6th International Conference on Neural Information Processing Systems, NIPS 1993, pp. 737–744. Morgan Kaufmann Publishers Inc., San Francisco (1993)
6. Castillo-Barnes, D., Martinez-Murcia, F.J., Ortiz, A., Salas-Gonzalez, D., Ramírez, J., Górriz, J.M.: Morphological characterization of functional brain imaging by isosurface analysis in Parkinson's disease. Int. J. Neural Syst. **30**(09), 2050044 (2020). https://doi.org/10.1142/S0129065720500446
7. Castillo-Barnes, D., Ramírez, J., Segovia, F., Martínez-Murcia, F.J., Salas-Gonzalez, D., Górriz, J.M.: Robust ensemble classification methodology for i123-ioflupane SPECT images and multiple heterogeneous biomarkers in the diagnosis of Parkinson's disease. Front. Neuroinform. **12**, 53 (2018). https://doi.org/10.3389/fninf.2018.00053
8. Chicco, D.: Siamese neural networks: an overview. Methods Mol. Biol. **2190**, 73–94 (2021)
9. Collazos-Huertas, D., Cárdenas-Peña, D., Castellanos-Dominguez, G.: Instance-based representation using multiple kernel learning for predicting conversion to Alzheimer disease. Int. J. Neural Syst. **29**(02), 1850042 (2019). https://doi.org/10.1142/S0129065718500429
10. Crammer, K., Singer, Y.: On the algorithmic implementation of multiclass kernel-based vector machines. J. Mach. Learn. Res. **2**, 265–292 (2002)
11. Gross, C.C., et al.: Classification of neurological diseases using multi-dimensional CSF analysis. Brain **144**(9), 2625–2634 (2021). https://doi.org/10.1093/brain/awab147
12. Górriz, J.M., et al.: Artificial intelligence within the interplay between natural and artificial computation: advances in data science, trends and applications. Neurocomputing **410**, 237–270 (2020). https://doi.org/10.1016/j.neucom.2020.05.078
13. Illán, I.Á., Gorrizz, J.M., Ramírez, J., Segovia, F., Jiménez-Hoyuela, J.M., Lozano, S.J.O.: Automatic assistance to Parkinson's disease diagnosis in DaTSCAN SPECT imaging. Med. Phys. **39**(10), 5971–80 (2012)
14. Koch, G.R.: Siamese neural networks for one-shot image recognition (2015)
15. Magesh, P., Myloth, R., Tom, R.: An explainable machine learning model for early detection of Parkinson's disease using lime on DaTSCAN imagery. Comput. Biol. Med. **126**, 104041 (2020). https://doi.org/10.1016/j.compbiomed.2020.104041
16. Mammone, N., et al.: Hierarchical clustering of the electroencephalogram spectral coherence to study the changes in brain connectivity in Alzheimer's disease. In: 2016 IEEE Congress on Evolutionary Computation (CEC), pp. 1241–1248 (2016). https://doi.org/10.1109/CEC.2016.7743929
17. Manzanera, O.M., et al.: Scaled subprofile modeling and convolutional neural networks for the identification of Parkinson's disease in 3d nuclear imaging data. Int. J. Neural Syst. **29**(09), 1950010 (2019). https://doi.org/10.1142/S0129065719500102

18. Marek, K., et al.: The Parkinson progression marker initiative (PPMI). Prog. Neurobiol. **95**(4), 629–635 (2011). https://doi.org/10.1016/j.pneurobio.2011.09.005. http://www.sciencedirect.com/science/article/pii/S0301008211001651
19. Marek, K., et al.: The Parkinson progression marker initiative (PPMI). Prog. Neurobiol. **95**(4), 629–635 (2011). https://doi.org/10.1016/j.pneurobio.2011.09.005
20. Martinez-Murcia, F.J., Górriz, J.M., Ramírez, J., Ortiz, A.: Convolutional neural networks for neuroimaging in Parkinson's disease: is preprocessing needed? Int. J. Neural Syst. **28**(10), 1850035 (2018). https://doi.org/10.1142/S0129065718500351
21. Martins, R., et al.: Automatic classification of idiopathic Parkinson's disease and atypical parkinsonian syndromes combining [11C]raclopride PET uptake and MRI grey matter morphometry. J. Neural Eng. **18**, 33848996 (2021). https://doi.org/10.1088/1741-2552/abf772
22. Mazziotta, J.C., et al.: A probabilistic atlas and reference system for the human brain: international consortium for brain mapping (ICBM). Philos. Trans. Royal Soc. Lond. Ser. B Biol. Sci. **356 1412**, 1293–322 (2001)
23. Modi, H., Hathaliya, J., Obaidiat, M.S., Gupta, R., Tanwar, S.: Deep learning-based Parkinson disease classification using PET scan imaging data. In: 2021 IEEE 6th International Conference on Computing, Communication and Automation (ICCCA), pp. 837–841 (2021). https://doi.org/10.1109/ICCCA52192.2021.9666251
24. Ortiz, A., Martínez Murcia, F.J., Munilla, J., Górriz, J.M., Ramírez, J.: Label aided deep ranking for the automatic diagnosis of parkinsonian syndromes. Neurocomputing **330**, 162–171 (2019). https://doi.org/10.1016/j.neucom.2018.10.074
25. Ortiz, A., et al.: Parkinson's disease detection using isosurfaces-based features and convolutional neural networks. Front. Neuroinform. **13**, 48 (2019). https://doi.org/10.3389/fninf.2019.00048
26. Segovia, F., Górriz, J.M., Ramírez, J., Martínez-Murcia, F.J., Castillo-Barnes, D.: Assisted diagnosis of parkinsonism based on the striatal morphology. Int. J. Neural Syst. **29**(09), 1950011 (2019). https://doi.org/10.1142/S0129065719500114
27. Wellcome Centre for Human Neuroimaging: Statistical Parametrical Mapping (2018). https://www.fil.ion.ucl.ac.uk/spm/software/spm12

# Analyzing Statistical Inference Maps Using MRI Images for Parkinson's Disease

C. Jimenez-Mesa[1(✉)], Diego Castillo-Barnes[1], Juan E. Arco[2], F. Segovia[1], J. Ramirez[1], and J. M. Górriz[1]

[1] Department of Signal Theory, Telematics and Communications, University of Granada, Periodista Daniel Saucedo Aranda, 18071 Granada, Spain
carmenj@ugr.es

[2] Department of Communications Engineering, University of Malaga, Blvr. Louis Pasteur 35, 29004 Malaga, Spain
http://sipba.ugr.es/

**Abstract.** Parkinson's Disease (PD) is one of the most prevalent and studied types of dementia. Traditionally, studies about this neurological disorder have made use of functional SPECT images. Nevertheless, to avoid some of its disadvantages with special focus on its expensive cost and low resolution, in the last years many studies have tried to use another imaging alternatives such as MRI scans able to evaluate subtle changes in the Grey Matter tissue. When analyzing the state of the art on this subject, we have found several shortcomings in the way of proceeding. Therefore, the work presented here presents a qualitative analysis of the regions of interest (ROIs) when using MRI for PD by the computation of statistical significance maps. For that, we have made use of a parametric and a non-parametric approaches using the widely known Statistical Parametric Mapping (SPM) package and the novel Statistical Agnostic Mapping (SAM) proposal. Results obtained suggest that there are no relevant ROIs in GM MRI imaging contrary to other modalities like the FP-CIT SPECT scans evaluated on the striatum region.

**Keywords:** Statistical Agnostic Mapping · Statistical Parametric Mapping · Neuroimaging · MRI · SPECT · Parkinson's Disease

## 1 Introduction

According to the World Health Organization (www.who.int), dementia is a syndrome that leads to a deterioration of cognitive function as a result of a variety of disorders that affect the brain. It is currently estimated that more than 55 million of people worldwide suffer from this disease, having a direct impact not only on the psychological, social and economic aspects for people living with dementia, but also for their relatives, carers and society in general.

Parkinson's Disease (PD) is considered the second most prevalent form of dementia. This neurodegenerative disorder results from a progressive loss


J. M. Ferrández Vicente et al. (Eds.): IWINAC 2022, LNCS 13258, pp. 166–175, 2022.
https://doi.org/10.1007/978-3-031-06242-1_17

of dopaminergic neurons in the nigrostriatal pathway by causes that are still unclear [17].

One common tool to diagnose PD is the use of FP-CIT SPECT brain scans. This kind of adquisitions make use of $I^{[123]}$-Ioflupane radioligand that presents a high binding affinity for Dopaminergic Transporters (DT) in the brain. As DT distribution gives us a quantitative measure of dopaminergic neuronal loss, we can use this information to develop Computer-Aided-Diagnosis (CAD) systems for PD [15,19].

Although FP-CIT SPECT brain scans are really useful to diagnose PD, their adquisition presents some disadvantages such as its high cost, the unreliable supply of radiotracer, and potential adverse effects like disorders of the heart beat (including cardiac arrest) [1]. For this reason, some recent works, such as [2,14,18], have proposed the use of MRI scans to try diagnosing PD only focusing on subtle patterns in Grey Matter (GM) atrophy. When analyzing these proposals, we have found that some of them try to identify some regions of interest (ROI) to reduce the amount of input features for classification. Although this practice is right, some of them are attempting to determine these regions outside the cross-validation loop. This is incorrect, because it leads to a potential bias and, consequently, false classification results.

Statistical brain mapping is a simple approach to detect ROIs when comparing two groups of subjects, i.e. healthy control (HC) and PD subjects. A group analysis can be helpful to delineate the ROIs of the brain for a specific condition/pathology [12]. Thus, when an exploratory analysis is required, statistical inference maps based on null-hypothesis ($H_0$) are commonly used [5]. This allows to detect relevant patterns between classes. Nevertheless, more and more studies are proposing new approaches due to its dependence on a variety of assumptions [3]. Data-driven statistical learning theory (SLT) would be one of those proposals [10,13] and, as an example of non-parametric approaches based on Machine Learning (ML) we could highlight the use of the Statistical Agnostic Mapping (SAM) tool [9]. This proposal is able to work with small sample sizes detecting (large and subtle) relevant effects.

In this context, our work aims to assess the reliability of the results generated by neuroimaging software tools to identify the most relevant ROIs when comparing Healthy Control subjects (HC) and patients with Parkinsonism using MRI and FP-CIT SPECT images. The tools analysed are the standard Statistical Parametric Mapping (SPM) package [4] and the SAM proposal [9].

## 2 Materials and Methods

### 2.1 Parkinson's Progression Markers Initiative

Data used for this work was obtained from the Parkinson's Progression Markers Initiative (PPMI). The PPMI is a public-private partnership funded by the Michael J. Fox Foundation for Parkinson's Research and the funding partners listed on ppmi-info.org/about-ppmi/who-we-are/study-sponsors. For more information on the study, please visit: ppmi-info.org.

For this contribution, we have compared a balanced dataset including a total of 40 HC subjects and 40 patients with PD. In all cases, the time between MRI and FP-CIT SPECT imaging adquisitions were no longer than 15 days. Demographics of the participants included in this work are depicted in Table 1.

**Table 1.** Demographics (age given in terms of its mean ± and standard deviation).

| | | Male | Female | Both |
|---|---|---|---|---|
| HC | # | 28 | 12 | 40 |
| | Age | $62.64 \pm 11.61$ | $61.25 \pm 7.48$ | $62.22 \pm 10.56$ |
| PD | # | 23 | 17 | 40 |
| | Age | $61.56 \pm 8.48$ | $60.76 \pm 7.00$ | $61.22 \pm 7.89$ |

### 2.2 Image Preprocessing

All the images used in this work have been spatially normalized (non-linear transformations) to a reference space defined by the Montreal Neurological Institute (MNI). For that, we have made use of the software tool `SPM12`[1] as follows:

- For MRI images, we have spatially registered each subject's scan to the `MNI152` template included in `SPM12` [11].
- In the case of FP-CIT SPECT scans, we first generated a functional template following [21], and then registered each sample to this new template (in the same position as the `MNI152` template).

Once spatially normalized, both MRI and FP-CIT SPECT scans presented the same size ($121 \times 145 \times 121$ voxels) and a voxel-size of $1.5 \times 1.5 \times 1.5$ mm.

Apart from the spatial registration, we have also included an intensity normalization for functional images using $\alpha$-stable distributions [20]. This method applies the linear transformation in Eq. (1) to each input sample, $X$, where $\gamma^*$ represents the mean of $\gamma$ (dispersion) parameters from all the input scans, and $\mu^*$ is calculated as the average of $\mu$ (location).

$$Y = \frac{\gamma^*}{\gamma} X + (\mu^* - \frac{\gamma^*}{\gamma} \mu) \quad (1)$$

As depicted in Fig. 1, this procedure reduces the differences between scans due to external factors such as the amount of radioligand injected to each patient, their absorption rate or the calibration of the acquisition equipment, among others.

[1] www.fil.ion.ucl.ac.uk/spm/software/spm12/.

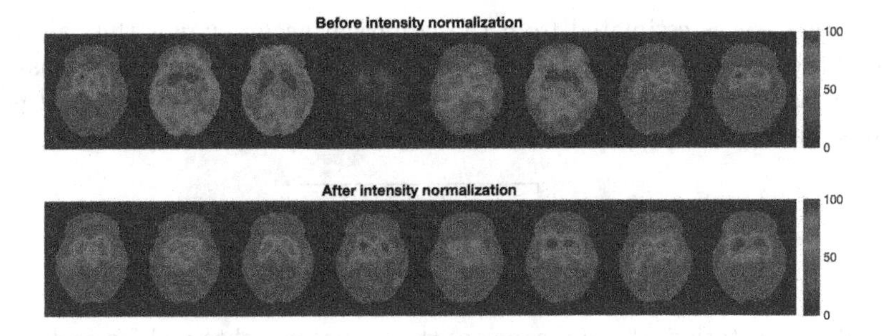


**Fig. 1.** Montage showing differences between FP-CIT SPECT scans randomly selected before/after intensity normalization using $\alpha$-stable distributions.

## 2.3 Statistical Parametric Mapping (SPM)

SPM constitutes an statistical neuroimaging tool commonly used in neuroimaging to carry out procedures such as realignment, segmentation, normalization and smoothing, among others. It allows to perform statistical analyses based on the General Linear Model (GLM) and Mixed Random Effects [6]. In our case, we have made use of SPM to compute an inference model by a two-sample t-test. For that, once the parameters of the model are estimated, we have generated significance maps using different configurations available in SPM. Both a $p_{\text{value}}$ corrected for multiple comparisons, named as FWE $p_{\text{value}}$, and an uncorrected $p_{\text{value}}$ are selected. The latter generates much less conservative significance maps with a high false positive (FP) rate [3]. Thus, the general recommendation is to add an extent threshold to limit the amount of significant voxels by requiring a minimum number of voxels to be clustered together.

## 2.4 Statistical Agnostic Mapping (SAM)

In contrast to SPM, SAM is based on data-driven approach based on concentration inequalities. Therefore, the main difference with the former is that the latter is a non-parametric method. As Gorriz et al. stated in [8], pattern classification by ML can be used as statistical inference approach. That is because the confidence intervals established during classification can be seen as statistical significance maps [9].

The procedure implemented in SAM is as follows. Firstly, the design matrix of the experiment and the data parcellation model are generated. To detect ROIs, a parcellation of the brain images by means of the Automated Anatomical Labeling (`AAL116`) atlas, [23], is applied. Then, a feature extraction and selection (FES) stage is also conducted prior to a classification step. For example, a t-test during the FES step can be applied together with Partial Least Squares (PLS) [24]. Later, a classification algorithm using Support Vector Machine (SVM) with linear kernel [22] is implemented. Then, using the whole dataset as training and test set (resubstitution), the actual accuracy under the worst case with probability $1-\delta$ is estimated from the empirical accuracy. For this, an upper bound of the actual error based on concentration inequalities for linear classifiers is applied as shown in Eq. (2) where $n$ is the size of the training set, $d$ is the feature's dimension and $\eta$ is the significance level established [7].

$$\mu \leq \sqrt{\frac{1}{2n} \ln \frac{\sum_{k=0}^{d-1} \binom{n-1}{k}}{0.5\eta}} \tag{2}$$

Finally, the statistical significance of the accuracies obtained for each region are evaluated by calculating the z-test statistic.

## 3 Results

To generate and evaluate the significance maps associated with each software tool, three experiments were proposed. The main one was a group analysis comparing HC vs PD whereas the other two confronted samples within the same class arbitrarily separated into two groups: i.e. HC vs HC and PD vs PD. In any case, the number of input samples per class were 40 (balanced classes).

Once the images are preprocessed, they were given as inputs for the statistical mapping tools. When using SPM, a standard two-sample t-test was conducted. Firstly, a factorial design along its design matrix was generated. Then, we performed the parameters estimation of GLM. Finally, a t-contrast test with $[1\ -1]$ as weight vector was defined and applied to the data. Here, two different inference models were studied: a voxelwise inference approach, where a FWE $p_{\text{value}} \leq 0.05$ was applied; and a clusterwise approach where an uncorrected for multiple comparisons using $p_{\text{value}} \leq 0.001$ along with an extent threshold of 10 voxels. In case of using SAM, the images were parcellated according to the `AAL116` atlas. The feature selection method chosen was t-test and the feature extraction method was PLS using 1 component. Then, the features were classified using a linear kernel SVM classifier and its accuracy was determined under the worst case with probability 95%. Finally, significance of each region was tested by means of a z-test statistic applied to each actual accuracy. In contrast to the previous methods, this would be a region-wise inference. Table 2 and Table 3 illustrate the results obtained for each experiment and image modality.

Figure 2 depicts the significance maps generated by the proposed inferences for the FP-CIT SPECT scans. The colour blue is associated to the significance map obtained using SAM. Colours red and green are related to the clusterwise and voxelwise inferences conducted in SPM. The same range of colours is used in the three experiments and in Fig. 3, where the significance maps for the GM MRI scans are illustrated. It should be noted that the significance maps of experiment HC vs HC is practically null for all the inferences analysed, so its inclusion has been discarded.

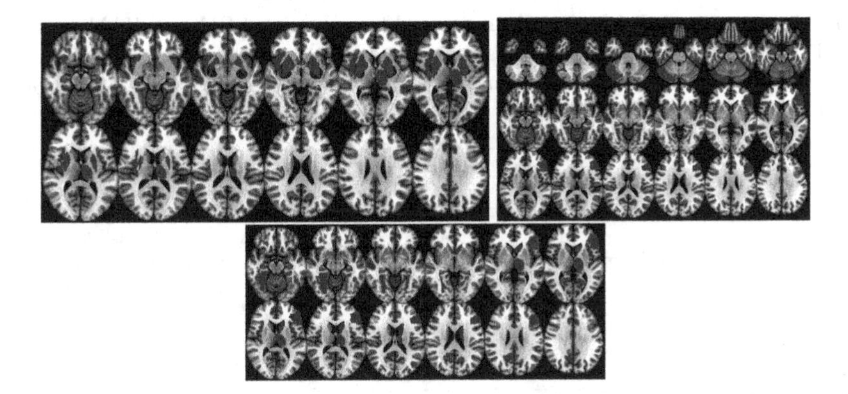

**Fig. 2.** Significance maps obtained for FP-CIT SPECT scans. Voxelwise SPM is represented in green, clusterwise SPM in red and SAM is blue. HC vs PD experiment ($n = 80$) is illustrated on the up left, PD vs PD ($n = 40$), on the up right and HC vs HC ($n = 40$) is located at the bottom. Voxelwise SPM is only non-null in the experiment HC vs PD. The ROIs obtained are overlapped over 59.50% between SAM and voxelwise SPM and 57.67% between SAM and clusterwise SPM. (Color figure online)

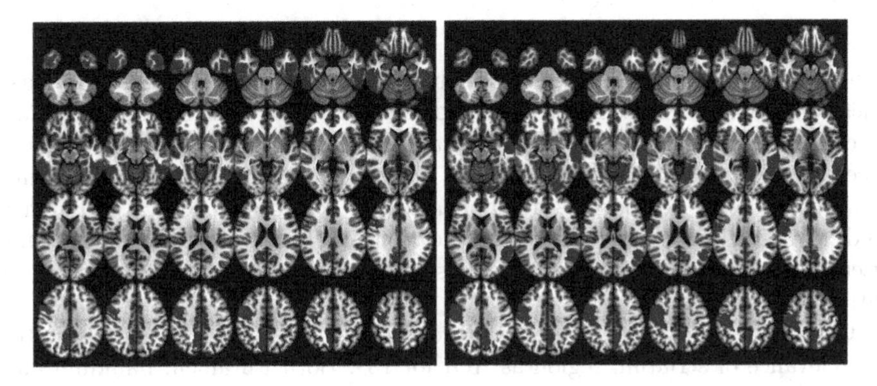

**Fig. 3.** Significance maps obtained for GM MRI scans. Clusterwise SPM is presented in red and SAM is blue. Voxelwise SPM is null in both experiments. HC vs PD experiment ($n = 80$) is illustrated on the left, PD vs PD ($n = 40$), on the right. In the HC vs HC experiment ($n = 40$), only clusterwise SPM is non-null but with very few voxels. No relevant overlapping was detected. (Color figure online)

## 4 Discussion

In this work, a qualitative analysis of significance maps of Parkinson's Disease is conducted. To do so, two different approaches were performed, a parametric approach (SPM) and a non-parametric approach (SAM).

The main experiment (HC vs PD) assesses the possibility of characterization of PD by the imaging modality under study. This kind of experiment indicates relevant regions to perform a two-class (binary) classification. In MRI scans,

**Table 2.** Summary of the statistical maps obtained using SPM and SAM tools for FP-CIT SPECT images. Experiments highlight regions of interest among different or same classes (HC and PD). The names of the regions are based on the AAL116 atlas nomenclature.

| Exp | Method | # Voxels | # ROIs | List of ROIs |
|---|---|---|---|---|
| HC vs PD | SPM (voxel) | 4583 | 12 | Insula$^L$, Insula$^R$, Hippocampus$^L$, Hippocampus$^R$, Amygdala$^L$, Amygdala$^R$, Putamen$^L$, Putamen$^R$, Pallidum$^L$, Pallidum$^R$, Thalamus$^L$, Thalamus$^R$ |
| | SPM (cluster) | 20662 | 46 | Frontal_Sup_Orb$^L$, Frontal_Inf_Oper$^L$, Hippocampus$^L$, ParaHippocampal$^R$, Amygdala$^R$, Caudate$^R$, Putamen$^R$ or Thalamus$^R$ among others |
| | SAM | 12933 | 6 | Insula$^R$, Putamen$^L$, Putamen$^R$, Pallidum$^L$, Pallidum$^R$, Thalamus$^L$ |
| HC vs HC | SPM (voxel) | 0 | 0 | |
| | SPM (cluster) | 327 | 7 | Frontal_Sup_Orb$^R$, Frontal_Mid_Orb$^R$, Sup_Motor_Area$^L$, Cingulum_Post$^L$, Calcarine$^L$, Precuneus$^L$, Vermis_4_5 |
| | SAM | 24526 | 5 | Frontal_Inf_Tri$^L$, Fusiform$^R$, Precuneus$^R$, Caudate$^L$, Thalamus$^L$ |
| PD vs PD | SPM (voxel) | 0 | 0 | |
| | SPM (cluster) | 53 | 2 | Calcarine$^R$, Thalamus$^R$ |
| | SAM | 12202 | 2 | Frontal_Inf_Tri$^L$, Cerebelum_Crus1$^R$ |

results do not suggest that there are any highly relevant ROIs to address the classification problem. As seen in Table 3 and Fig. 3 (left), SPM significant maps are practically empty. The only region that appears in both SPM clusterwise inference and SAM was the Precuneus (right hemisphere). Nevertheless, when using SAM, the Inferior Temporal Gyrus appears as ROI as described in [16]. Further study is needed to understand the implications of these regions for Parkinson's disease, especially as they are more closely associated with the dementia that patients may develop.

Contrary to MRI, significance maps related to FP-CIT SPECT clearly states the relevance of striatum region as ROI for PD. Both Putamen, Pallidum and Thalamus appear in all the inference maps as seen in Table 2. Indeed, the importance of this area can also be appreciated in Fig. 2 (up left) where significance maps of SAM and voxelwise SPM have an overlap of 59.50% and the ones of clusterwise SPM and SAM, 57.67%. Similar results are obtained in [9], where SPECT scans from PPMI and a dataset from "Virgen de la Victoria" Hospital (Malaga, Spain) were analysed.

At this point two more experiments were conducted selecting samples of the same class arbitrarily divided into two groups to analyze their contrast, i.e. HC vs HC and PD vs PD. These experiments allowed us to deduce the reliability of classification: occurrence rates of ROIs indicate if input data is highly heterogeneous and even if it could be separated among them. In any of the experiments involving SPM voxelwise inference maps point out several ROIs for both image modalities. Though the clusterwise also does, its number of detected voxels is small and they seem to have been randomly choosen (see Table 2 and Table 3). Similarly, for the SAM approach it also seems that detected regions

**Table 3.** Summary of the statistical maps obtained using SPM and SAM tools for GM MRI. Experiments highlight regions of interest among different or same classes (HC and PD). The names of the regions are based on the AAL116 atlas nomenclature.

| Exp | Method | # Voxels | # ROIs | List of ROIs |
|---|---|---|---|---|
| HC vs PD | SPM (voxel) | 0 | 0 | |
| | SPM (cluster) | 98 | 10 | Frontal_Sup$^{R}$, Frontal_Mid$^{R}$, Frontal_Inf_Tri$^{L}$, Fusiform$^{R}$, Postcentral$^{R}$, Parietal_Sup$^{R}$, Precuneus$^{R}$, Putamen$^{R}$, Cerebelum_Crus1$^{R}$, Cerebelum_6$^{R}$ |
| | SAM | 31525 | 4 | Precentral$^{R}$, Precuneus$^{R}$, Temporal_Inf$^{L}$, Temporal_Inf$^{R}$ |
| HC vs HC | SPM (voxel) | 0 | 0 | |
| | SPM (cluster) | 83 | 11 | Frontal_Inf_Oper$^{L}$, Frontal_Inf_Tri$^{L}$, Olfactory$^{R}$, Insula$^{R}$, ParaHippocampal$^{L}$, Cuneus$^{L}$, Lingual$^{L}$ Occipital_Mid$^{R}$, Paracentral_Lobule$^{L}$, Temporal_Sup$^{R}$, Temporal_Mid$^{R}$ |
| | SAM | 0 | 0 | |
| PD vs PD | SPM (voxel) | 0 | 0 | |
| | SPM (cluster) | 102 | 10 | Frontal_Sup$^{R}$, Frontal_Mid$^{L}$, Frontal_Mid$^{R}$, Supp_Motor_Area$^{L}$, Cingulum_Mid$^{L}$, Cingulum_Mid$^{R}$, Precuneus$^{L}$, Paracentral_Lobule$^{R}$, Temporal_Sup$^{L}$, Cerebelum_Crus1$^{R}$ |
| | SAM | 42163 | 5 | Precentral$^{R}$, Lingual$^{L}$, Postcentral$^{R}$, Precuneus$^{L}$, Temporal_Mid$^{L}$ |

have no relevant implications for the study. Only the results derived from HC vs HC experiment using FP-CIT SPECT should be further analyzed (especially if we focus on the Fusiform Gyrus).

Regarding the methods analysed, it is clear that the use of FWE $p_{\text{values}}$ generates highly conservative significance maps [3]. In fact, when a true effect appears, it can be seen in Fig. 2 (top left) how the least conservative significance map is the one that uses an uncorrected $p_{\text{value}}$ for multiple comparisons. However, this method increases the number of false positives. Therefore, its use is not advisable [3]. An intermediate term would be the significance maps generated using SAM. Indeed, as many other works in the current state of art where the small sample size problem appears, the classical statistics creates a challenging scenario where the generalization of the results is unclear. In this context, proposals like SAM should be stated as gold standard due to its reliability when working with small sample sizes as it is based on concentration inequalities [9].

After analyzing our results, we can state that MRI scans is not a much reliable source of information for PD diagnosis even despite the classification results reported in related works such as [2]. Though few other studies have proposed the use of MRI imaging markers as main inputs in CAD systems for PD, they might be falling into some inadvertencies. For example, they work in conjunction with SPECT and MRI, whereby the former is the main contributor to a high accuracy (biased) [14], or the feature selection is done outside the cross-validation loop [18].

In any case, to evaluate the use of MRI for PD, research projects using larger sample sizes are needed. And even then, it will be necessary to evaluate the effect of sample size on the inference maps and to analyze the effect of false positives.

## 5 Conclusions

The analysis of significance maps performed in this work concludes that with the studies conducted so far, it is not possible to identify ROIs for the study of PD when only evaluating T1-MRI scans. In contrast, FP-CIT SPECT scans clearly highlight the relevance of the striatum region for Parkinson's monitoring. A more in-depth study with a larger sample size is needed to confirm these findings.

**Acknowledgments.** This work was supported by the MCIN/ AEI/10.13039/501100011033/ and FEDER "Una manera de hacer Europa" under the RTI2018-098913-B100 project; by the Consejería de Economía, Innovación, Ciencia y Empleo (Junta de Andalucía) and FEDER under CV20-45250, A-TIC-080-UGR18, B-TIC-586-UGR20 and P20-00525 projects; and by the Ministerio de Universidades under the FPU18/04902 grant given to C. Jimenez-Mesa and the Margarita-Salas grant to J.E. Arco.

## References

1. Bateman, T.: Advantages and disadvantages of PET and SPECT in a busy clinical practice. J. Nucl. Cardiol. **19**(S1), 3–11 (2012). https://doi.org/10.1007/s12350-011-9490-9
2. Cigdem, O., Beheshti, I., Demirel, H.: Effects of different covariates and contrasts on classification of Parkinson's disease using structural MRI. Comput. Biol. Med. **99**, 173–181 (2018). https://doi.org/10.1016/j.compbiomed.2018.05.006
3. Eklund, A., Nichols, T.E., Knutsson, H.: Cluster failure: why fMRI inferences for spatial extent have inflated false-positive rates. Proc. Natl. Acad. Sci. **113**(28), 7900–7905 (2016). https://doi.org/10.1073/pnas.1602413113
4. Friston, K.: Statistical Parametric Mapping: The Analysis of Functional Brain Images. Elsevier/Academic Press, Amsterdam (2007)
5. Friston, K.: Sample size and the fallacies of classical inference. Neuroimage **81**, 503–504 (2013). https://doi.org/10.1016/j.neuroimage.2013.02.057
6. Friston, K., et al.: Classical and Bayesian inference in neuroimaging: applications. Neuroimage **16**(2), 484–512 (2002). https://doi.org/10.1006/nimg.2002.1091
7. Górriz, J., Ramírez, J., Suckling, J.: On the computation of distribution-free performance bounds: application to small sample sizes in neuroimaging. Pattern Recogn. **93**, 1–13 (2019). https://doi.org/10.1016/j.patcog.2019.03.032
8. Gorriz, J., et al.: A connection between pattern classification by machine learning and statistical inference with the general linear model. IEEE J. Biomed. Health Inform, 1 (2021). https://doi.org/10.1109/jbhi.2021.3101662
9. Górriz, J., et al.: Statistical agnostic mapping: a framework in neuroimaging based on concentration inequalities. Inf. Fusion **66**, 198–212 (2021). https://doi.org/10.1016/j.inffus.2020.09.008
10. Górriz, J.M., et al.: Artificial intelligence within the interplay between natural and artificial computation: advances in data science, trends and applications. Neurocomputing **410**, 237–270 (2020). https://doi.org/10.1016/j.neucom.2020.05.078
11. Grabner, G., et al.: Symmetric atlasing and model based segmentation: an application to the hippocampus in older adults. In: Larsen, R., Nielsen, M., Sporring, J. (eds.) MICCAI 2006. LNCS, vol. 4191, pp. 58–66. Springer, Heidelberg (2006). https://doi.org/10.1007/11866763_8

12. Jeong, Y., et al.: 18F-FDG PET findings in frontotemporal dementia: an SPM analysis of 29 patients. J. Nucl. Med. Official Publ. Soc. Nucl. Med. **46**, 233–239 (2005)
13. Kim, I., et al.: Classification accuracy as a proxy for two sample testing. Ann. Stat. **49**(1), 411–434 (2021)
14. Martins, R., et al.: Automatic classification of idiopathic Parkinson's disease and atypical parkinsonian syndromes combining [11c]raclopride PET uptake and MRI grey matter morphometry. J. Neural Eng. **18**(4), 046037 (2021). https://doi.org/10.1088/1741-2552/abf772
15. Palumbo, B., Bianconi, F., Nuvoli, S., Spanu, A., Fravolini, M.L.: Artificial intelligence techniques support nuclear medicine modalities to improve the diagnosis of Parkinson's disease and Parkinsonian syndromes. Clin. Transl. Imaging **9**(1), 19–35 (2020). https://doi.org/10.1007/s40336-020-00404-x
16. Pan, P., et al.: Abnormalities of regional brain function in Parkinson's disease: a meta-analysis of resting state functional magnetic resonance imaging studies. Sci. Rep. **7**(1) (2017). https://doi.org/10.1038/srep40469
17. Poewe, W., et al.: Parkinson's disease. Nat. Rev. Dis. Primers **3**(1) (2017). https://doi.org/10.1038/nrdp.2017.13
18. Rana, B., et al.: Relevant 3d local binary pattern based features from fused feature descriptor for differential diagnosis of Parkinson's disease using structural MRI. Biomed. Signal Process. Control **34**, 134–143 (2017). https://doi.org/10.1016/j.bspc.2017.01.007
19. Sakai, K., Yamada, K.: Machine learning studies on major brain diseases: 5-year trends of 2014–2018. Jpn. J. Radiol. **37**(1), 34–72 (2018). https://doi.org/10.1007/s11604-018-0794-4
20. Salas-Gonzalez, D., et al.: Linear intensity normalization of FP-CIT SPECT brain images using the $\alpha$-stable distribution. Neuroimage **65**, 449–455 (2013). https://doi.org/10.1016/j.neuroimage.2012.10.005
21. Salas-Gonzalez, D., et al.: Building a FP-CIT SPECT brain template using a posterization approach. Neuroinformatics **13**(4), 391–402 (2015). https://doi.org/10.1007/s12021-015-9262-9
22. Schoölkopf, B., et al.: Learning With Kernels - Support Vector Machines, Regularization, Optimization, and Beyond. MIT Press, Cambridge (2002)
23. Tzourio-Mazoyer, N., et al.: Automated anatomical labeling of activations in SPM using a macroscopic anatomical parcellation of the MNI MRI single-subject brain. Neuroimage **15**(1), 273–289 (2002). https://doi.org/10.1006/nimg.2001.0978
24. Wold, S., Ruhe, A., Wold, H., Dunn, W.: The collinearity problem in linear regression. The partial least squares (PLS) approach to generalized inverses. SIAM J. Sci. Stat. Comput. **5**(3), 735–743 (1984). https://doi.org/10.1137/0905052

# Evaluating Intensity Concentrations During the Spatial Normalization of Functional Images for Parkinson's Disease

Diego Castillo-Barnes[1](✉), Juan E. Arco[2], C. Jimenez-Mesa[1], J. Ramirez[1], J. M. Górriz[1], and D. Salas-Gonzalez[1]

[1] Department of Signal Theory, Telematics and Communications, University of Granada, Periodista Daniel Saucedo Aranda, 18071 Granada, Spain
diegoc@ugr.es

[2] Department of Communications Engineering, University of Malaga, Blvr. Louis Pasteur 35, 29004 Malaga, Spain
http://sipba.ugr.es/

**Abstract.** The aim of this study is to assess the changes in imaging of dopamine transporters using [123]I-FP-CIT SPECT when applying a deformation of the striatum using a linear and/or a non-linear registration to a reference template. For that, the deformation has been indirectly measured studying the changes in the intensity values in two different scenarios when, during the interpolation stage, the amount or the concentrations of intensity values are preserved. As showed by our results, the degree of deformation is greater in images from patients with Parkinson's Disease than in healthy control subjects.

**Keywords:** Spatial normalization · FP-CIT SPECT · Neuroimaging · Parkinson's Disease

## 1 Introduction

[123]I-FP-CIT SPECT (DaTSCAN) is a complementary tool in the differential diagnosis of patients with Parkinsonism (PKS). This functional imaging modality allows the evaluation of synaptic parts of the dopaminergic system in the striatum region [7].

When a Computer-Aided-Diagnosis (CAD) system is being developed, generally, the image acquisition is followed by a preprocessing step that consists in the spatial normalization of the input brain scans. This registration allows us to perform a direct voxel-by-voxel comparison between brain scans from different patients and/or different acquisitions from the same subject.

When we evaluate [123]I-FP-CIT SPECT scans from patients with PKS, we can observe an accentuated loss of dopaminergic neurons in the nigrostriatal pathway, with a certain degree of asymmetry and a rounded shape. On the


J. M. Ferrández Vicente et al. (Eds.): IWINAC 2022, LNCS 13258, pp. 176–186, 2022.
https://doi.org/10.1007/978-3-031-06242-1_18

contrary, images from healthy control subjects (NOR) are, in general, symmetric and with a well defined c-shape highly illuminated in the striatum [5]. As shown in Fig. 1, these patterns help us to discriminate between NOR and PKS [8].

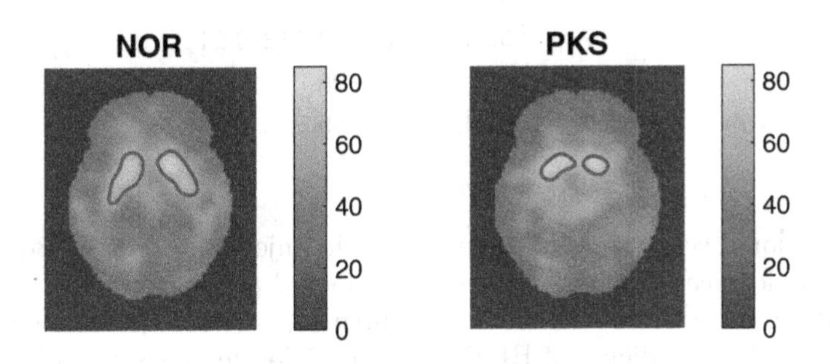


**Fig. 1.** Examples of [123]I-FP-CIT SPECT scans from NOR and PKS subjects in HUVN database.

Nevertheless, when the registration process is applied, usually by using a reference template built from a control subjects database, scans from PKS patients exhibit greater changes in their shape after the warping process in comparison with scans from control (NOR) subjects. This also applies to intensity preservation before/after the spatial normalization and it might alter artificially the interclass separation between the classes, which leads classification models based on Machine Learning to false optimistic results [4]. In this work we propose a comparison between intensity preservation of the concentration and intensity preservation of the amount during the spatial registration of [123]I-FP-CIT SPECT scans and determined their potential use when differentiating PKS and NOR subjects.

## 2 Materials and Methods

### 2.1 HUVN Dataset

Dataset used for this work was supplied by the Service of Nuclear Medicine from Hospital Universitario Virgen de las Nieves de Granada (https://www.huvn.es/). It comprises a total of 118 [123]I-FP-CIT SPECT scans from NOR and PKS patients recruited between 2007 and 2012 whose demographics have been summarized in Table 1.

**Table 1.** Demographics (age given in terms of mean ± standard deviation) from HUVN dataset.

| | | Male | Female |
|---|---|---|---|
| HC | # | 25 | 20 |
| | Age | 69.24 ± 10.80 | 73.20 ± 9.24 |
| PD | # | 45 | 28 |
| | Age | 69.34 ± 9.38 | 69.57 ± 8.81 |

All functional scans were obtained after the injection of 185 MBq of I123-Ioflupane radiotracer on previously thyroid-blocked participants.

Diagnostic labels were established by three experienced specialists from the Service of Nuclear Medicine of HUVN. All the informed consents are available through the institutional review boards from HUVN and approved by their local ethics committee local ethics committee.

### 2.2 Image Preprocessing: Spatial Registration

For this work we have compared a balanced subset of 20 NOR and 20 PKS scans randomly selected from patients included in HUVN database. During the pre-processing step, all the input scans have been spatially registered to a reference template generated ad-hoc following the procedure explained in [10]. For that, we have made use of the Statistical Parametric Mapping package available at https://www.fil.ion.ucl.ac.uk/spm/.

When using MRI, the spatial normalization procedure on SPM is based on segmentation [3]. On the contrary, [123]I-FP-CIT SPECT scans require having a reference MRI image at their same position. If this condition is not matched, an alternative method based on minimizing the mean squared difference between the template and a warped version of the image [9]. In any case, the finding of the deformation that better fits with the given data can be perform in two different ways:

1. Following an affine registration using a 12-parameter transformation that places each scan on the template applying translations, rotations, scaling and trimming. To optimize these parameters, one commonly used method is to minimize the mean squared differences between the image and the template [6].
2. Using a non-linear registration that also includes non-linear deformations to correct the residual differences between the input scan and the template by, for example, applying a linear combination of smooth cosine transform basis functions, as explained in [2].

During spatial normalization, the intensities of some brain regions are redistributed to compensate for the deformation applied by the spatial transformation. To this end, SPM proposes two models of intensity preservation [1]:

- **Intensity preservation of the concentration**. Where the spatially normalized images represent a weighted average of the signal under the smoothed kernel, roughly preserving the intensity of the original scans.
- **Intensity preservation of the amount**. In this case, the normalized scans preserve the total amount of signal for each region so intensities from expanded areas during warping are reduced.

Figure 2 includes an schema explaining how intensity preservation of the concentration and intensity preservation of the amount work.

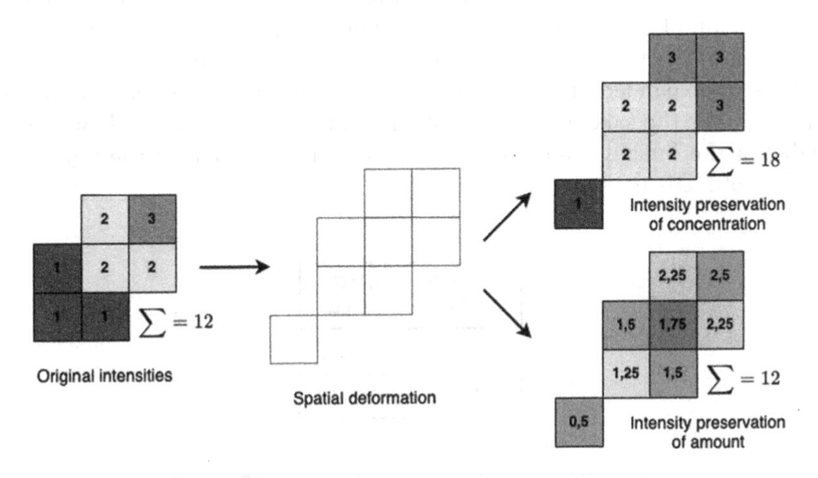


**Fig. 2.** Schema depicting the two models of intensity preservation included in SPM12.

### 2.3 Overview

For this work, we have normalized a total of 40 I[123]-FP-CIT SPECT scans (20 NOR and 20 PKS) to a reference space calculated following [10], for both affine and non-linear spatial registrations procedures and using both intensity preservation techniques (concentration or amount) in SPM12. Once the spatial registration was performed, we have selected a common reference Region Of Interest (ROI) given by the Automated Anatomical Labeling atlas [11], co-registered to our reference template, to focus our analysis on the striatum region [7]. Then, we have analyzed the distribution of intensity values within this region and compared all scenarios to determine the conditions where the interclass separation between NOR and PKS subjects is largest. Figure 3 shows a diagram with an overview of the procedure followed in this work.

## 3 Results

For this work, [123]I-FP-CIT SPECT scans from HUVN database have been spatially registered to the same reference space to evaluate differences between spatial registration preserving intensity concentration or preserving the amount of intensity for both affine and non-linear transformations. Figure 4 depicts the average of all the registered brain images and how intensities from NOR and PKS scans are altered as a function of the intensity preservation during the normalization when applying non-linear spatial deformations using SPM12.

In Fig. 5 we have included the histograms of the intensity values within the striatum region of each scan after the non-linear registration procedure using a preservation of the total amount of the signal in the brain images. As expected, areas that are expanded during the warping process are correspondingly reduced in intensity. For the sake of clarity, histograms have been ordered depending on the label of the images (NOR or PKS, having the first half of the figure filled with plots of NOR data).

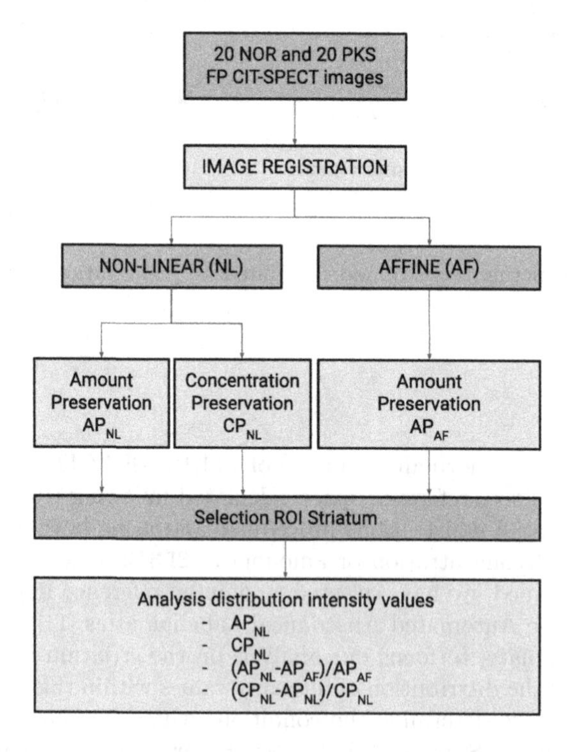


**Fig. 3.** General diagram.

Analogously, Fig. 6 shows the distribution of intensity values in the striatum when the database is registered and the intensity concentrations of the original images are preserved. In that case, due to the fact that originally PKS images present a smaller striatum area according to the functional information provided by the [123]I-FP-CIT radiotracer, we obtain lower mean values.

To compare both methods, we have included in Fig. 7 a representation of the distribution differences between the intensity values after preservation of the amount and concentration divided by the latter. This dimensionless factor provides a parsimonious way to measure the differences between NOR and PKS images.

And finally, Fig. 8 shows the histogram of the difference for the intensity values in the striatum between the registered database using both affine and non-linear spatial registration referred to the preservation of the amount of intensity.

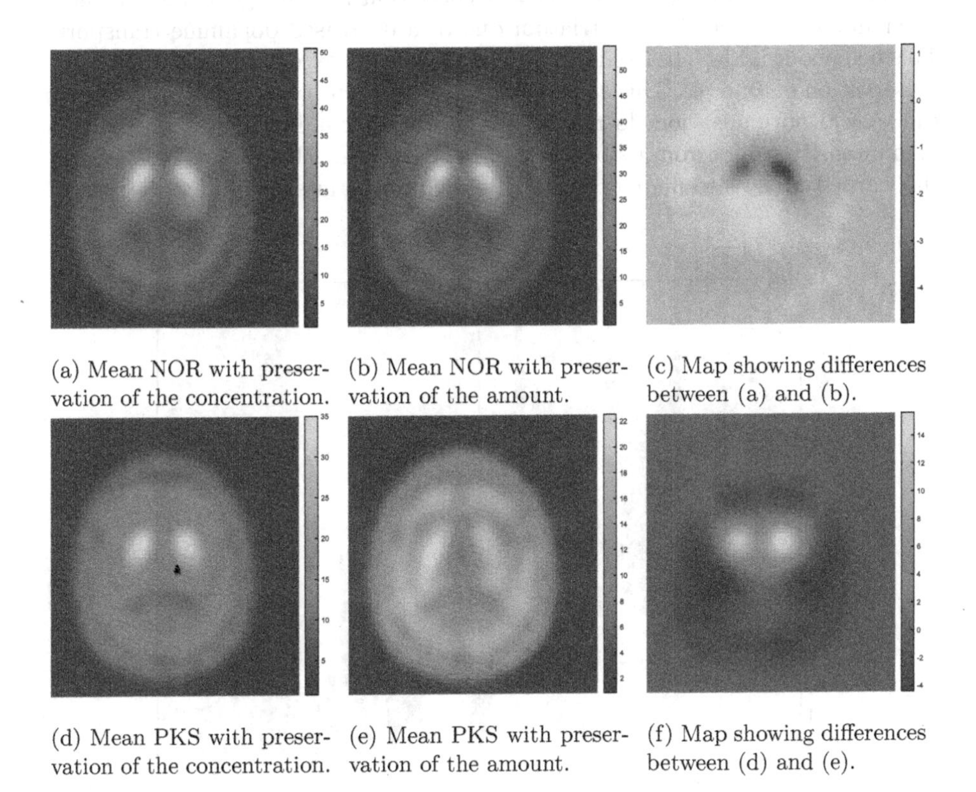

(a) Mean NOR with preservation of the concentration. (b) Mean NOR with preservation of the amount. (c) Map showing differences between (a) and (b).

(d) Mean PKS with preservation of the concentration. (e) Mean PKS with preservation of the amount. (f) Map showing differences between (d) and (e).

**Fig. 4.** Transaxial slices showing the average of HUVN dataset after the non-linear spatial registration. Note that affine registration scenario has not been included in order not to overload the figure.

## 4 Discussion

As shown in Fig. 4, the image registration process affects NOR and PKS images very differently with special emphasis on the striatum region. Whereas the NOR images remain approximately the same, when we focus on PKS scans, we observe that their striatum are highly illuminated when we choose concentration preservation in the registration process. This is due to the fact that, in the case of this type of images, the striatum region and its  have been deformed (enlarged) in excess to fit with the size of the reference template built only using NOR images [4,10]. Conversely, striatum region in NOR images is similar regardless using concentration preservation or amount preservation. In any case, on average, the final image is slightly more compressed in the striatum, so that the amount preservation shows slightly higher intensity values.

When focusing on Fig. 5, we can easily check that PKS images exhibit slightly lower intensity values in the striatum due to a decreased dopamine transport. This behaviour is also observed in Fig. 6 but to a lesser extent due to the intensity preservation of amount. This can be explained as intensity interpolations during the spatial normalization do not fill the space surrounding the striatum with high intensity values from its borders. Note that as the PKS data is increased in the warped process to match the NOR template, the intensity values decrease.

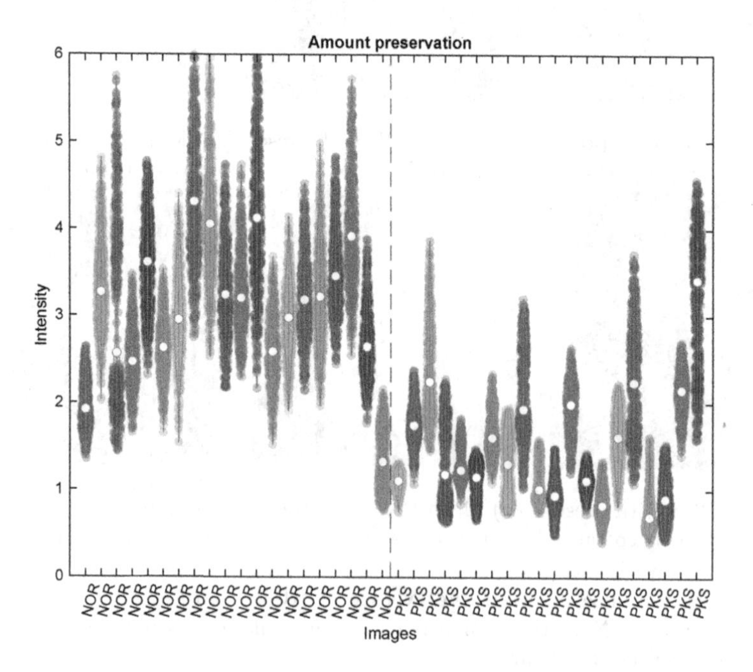


**Fig. 5.** Intensity distribution at striatum region for each image in HUVN dataset when referring to the amount preservation approach.

Many CAD systems performing any morphological analysis of the brain, e.g. [5], may take advantage of a better delimitation of the striatum borders due to the intensity preservation of amount approach, which will probably lead to better classification rates. In any case, regardless being based on affine transformations or nonlinear transformations, these CAD models might also exploit the use of the dimensionless features computed when comparing the preservation of the amount and the preservation of the concentration in Fig. 7 and Fig. 8. As can be seen in both representations, there are large differences not only at the level of average intensity levels but also at the level of variability of the results when comparing NOR and PKS subjects.

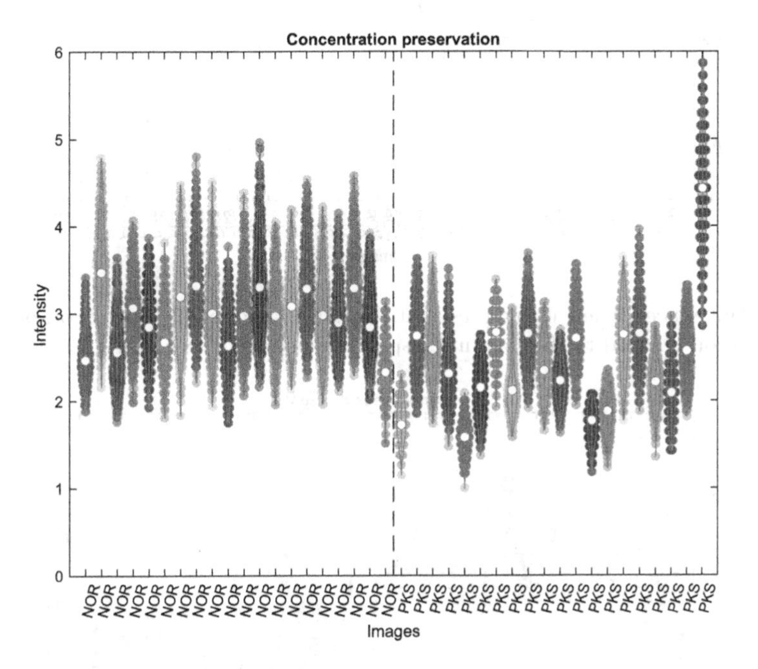


**Fig. 6.** Intensity distribution at striatum region for each image in HUVN dataset when referring to the concentration preservation approach.

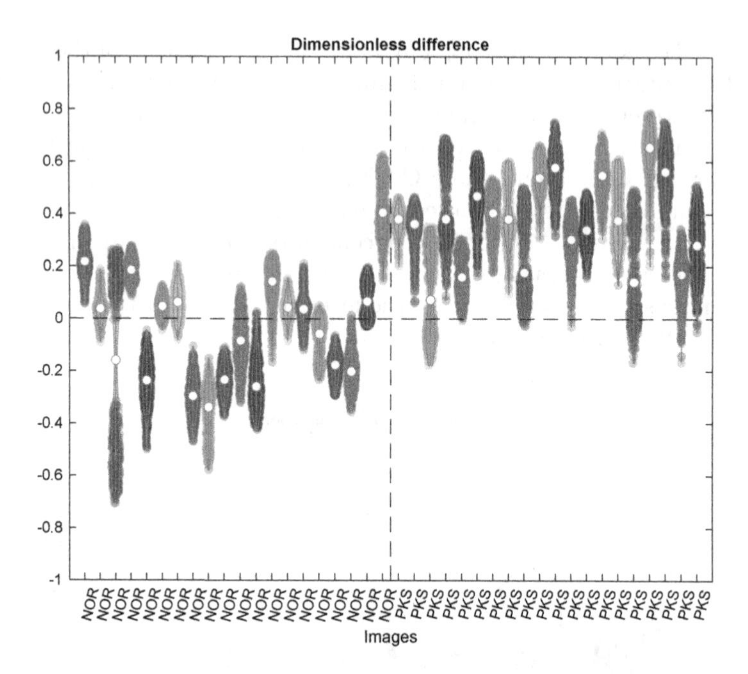


**Fig. 7.** Differences when using preservation of the amount and preservation of the concentration referred to the non-linear spatial registration scenario.

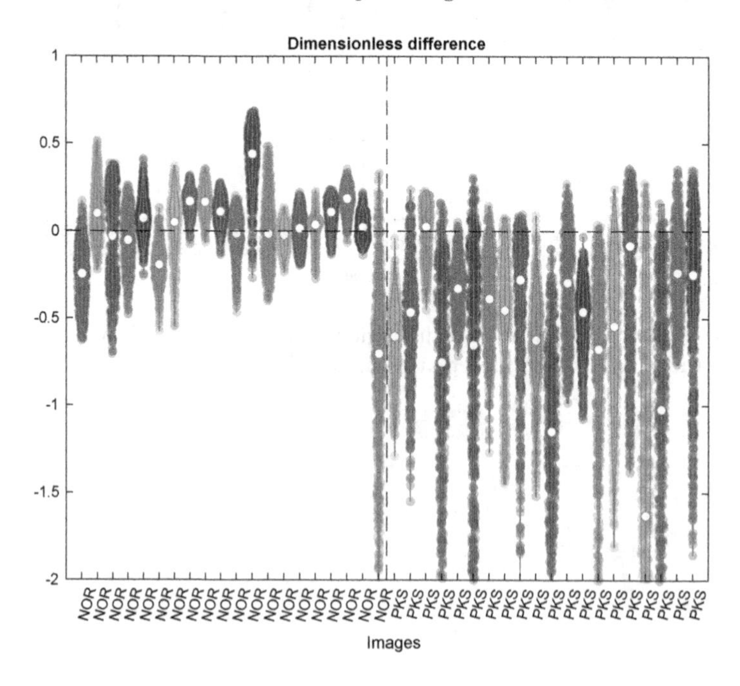


**Fig. 8.** Differences between affine and non-linear spatial registration referred the intensity preservation of the amount.

## 5 Conclusions

Although intensity preservation of concentration is the default practice when normalizing brain scans to a reference template, its use for both affine and non-linear spatial registration should be taken with care. In this work we have shown that the preservation of the amount is more effective when comparing I[123]-FP-CIT SPECT scans for Parkinson's Disease due to the way it fills the space surrounding the striatum borders, specially in patients with Parkinsonism. This fact increases the interclass separation and help Computer-Aided-Diagnosis systems to differentiate between healthy controls and patients with Parkinson's. Apart from this, our proposal has identified a dimensionless imaging biomarker that increases this interclass separation when comparing both affine and non-linear spatial transformations referred to the intensity preservation of the amount.

**Acknowledgments.** This work was supported by the MCIN/AEI/10.13039/5011000 11033/ and FEDER "Una manera de hacer Europa" under the RTI2018-098913 -B100 project; by the Consejería de Economía, Innovación, Ciencia y Empleo (Junta de Andalucía) and FEDER under CV20-45250, A-TIC-080-UGR18, B-TIC-586-UGR20 and P20-00525 projects; and by the Ministerio de Universidades under the FPU18/04902 grant given to C. Jimenez-Mesa and the Margarita-Salas grant to J.E. Arco.

## References

1. Ashburner, J., et al.: SPM12 Manual. UCL Queen Square Institute of Neurology, London, UK, October 2021. https://www.fil.ion.ucl.ac.uk/spm/doc/spm12_manual.pdf
2. Ashburner, J., Friston, K.J.: Nonlinear spatial normalization using basis functions. Hum. Brain Mapp. **7**(4), 254–266 (1999)
3. Ashburner, J., Friston, K.J.: Unified segmentation. Neuroimage **26**(3), 839–851 (2005). https://doi.org/10.1016/j.neuroimage.2005.02.018
4. Castillo-Barnes, D., et al.: Quantifying differences between affine and non-linear spatial normalization of FP-CIT SPECT images. Int. J. Neural Syst. (2022)
5. Castillo-Barnes, D., et al.: Morphological characterization of functional brain imaging by isosurface analysis in Parkinson's disease. Int. J. Neural Syst. **30**(09), 2050044 (2020). https://doi.org/10.1142/s0129065720500446
6. Friston, K.J., et al.: Spatial registration and normalization of images. Hum. Brain Mapp. **3**(3), 165–189 (1995). https://doi.org/10.1002/hbm.460030303
7. Ikeda, K., Ebina, J., Kawabe, K., Iwasaki, Y.: Dopamine transporter imaging in Parkinson disease: progressive changes and therapeutic modification after anti-parkinsonian medications. Intern. Med. **58**(12), 1665–1672 (2019). https://doi.org/10.2169/internalmedicine.2489-18
8. Marek, K., et al.: [123i]-CIT SPECT imaging assessment of the rate of Parkinson's disease progression. Neurology **57**(11), 2089–2094 (2001). https://doi.org/10.1212/wnl.57.11.2089
9. Palumbo, L., et al.: Evaluation of the intra- and inter-method agreement of brain MRI segmentation software packages: a comparison between SPM12 and FreeSurfer v6.0. Physica Medica **64**, 261–272 (2019). https://doi.org/10.1016/j.ejmp.2019.07.016

10. Salas-Gonzalez, D., et al.: Building a FP-CIT SPECT brain template using a posterization approach. Neuroinformatics **13**(4), 391–402 (2015). https://doi.org/10.1007/s12021-015-9262-9
11. Tzourio-Mazoyer, N., et al.: Automated anatomical labeling of activations in SPM using a macroscopic anatomical parcellation of the MNI MRI single-subject brain. NeuroImage **15**(1), 273–289 (2002). https://doi.org/10.1006/nimg.2001.0978

# Neuromotor and Cognitive Disorders

# Monitoring Motor Symptoms in Parkinson's Disease Under Long Term Acoustic Stimulation

L. Sigcha[1], David Gonzalez Calleja[1], I. Pavón[1], J.M. López[2], and G. de Arcas[1](✉)

[1] Grupo de Investigación en Instrumentación y Acústica Aplicada, ETSI Industriales, Departamento de Ingeniería Mecánica, Universidad Politécnica de Madrid, 28010 Madrid, Spain
g.dearcas@upm.es

[2] Grupo de Investigación en Instrumentación y Acústica Aplicada, Departamento de Ingeniería Telemática y Electrónica, ETSIS Telecomunicacion, Universidad Politécnica de Madrid, 28031 Madrid, Spain

**Abstract.** Parkinson's disease (PD) is a progressive neurodegenerative disease that presents motor (tremor, rigidity, bradykinesia, and instability) and non-motor dysfunctions. In this study we present a methodology to monitor the evolution of motor symptoms of Parkinson disease patients remotely, and its application to asses the effects of an experimental therapeutic intervention based on acoustic stimulation in a two months' study. Monitoring is based on wearing a commercial smartwatch while performing a set of exercises extracted from the Unified Parkinson's Disease Rating Scale. Three indicators are extracted from the triaxial accelerometer's signals to monitor tremor and bradykinesia. Results of the evolution of these indicators over a 2 months' study are shown for two PD patients following a therapeutic intervention based on acoustic stimulation and a healthy control group of similar age and gender. The feasibility of using consumer smartwatches for remote monitoring of motor symptom's is discussed together with its limitations, specially regarding long-term assessment of acoustic stimulation interventions.

**Keywords:** Acoustic stimulation · Parkinson disease · Motor symptoms · Wearables · Pervasive technologies

## 1 Introduction

Parkinson's disease (PD) is a progressive neurodegenerative disease (ND) characterized by the presence of motor dysfunctions and non-motor symptoms because of low dopamine levels in the brain. PD incidence increases with age reaching rates of 1–2% in people over 60 years of age. In the early stages, PD presents

Supported by Instrumentation and applied acoustics research group.


J. M. Ferrández Vicente et al. (Eds.): IWINAC 2022, LNCS 13258, pp. 189–198, 2022.
https://doi.org/10.1007/978-3-031-06242-1_19

cognitive deficits, while in the late stages, PD presents motor dysfunctions such as tremor, rigidity, bradykinesia and instability [11]. Currently, PD has no cure. Pharmacotherapy and surgery are the most common therapies used in clinical practice in combination with non-pharmacological treatments to ameliorate the symptoms. The most used drug is the metabolic precursor of dopamine (levodopa). In the early stages of PD, levodopa is very effective in ameliorating hypokinesia (reduced movement speed) and tremor. However, within a therapeutic window of 5 to 10 years, PD patients can develop dyskinesias (involuntary muscle movements) [3].

In this context, non-pharmacological treatments play a critical role as a complement to medication. They are targeted to improve the patient's quality of life by reducing the impact, and/or by slowing down the progression of specific symptoms. Experimental therapies under research include non-invasive neurostimulation techniques or magnetic and acoustic stimulation. These therapeutic interventions have attracted great interest in the scientific and clinical communities in the last few years [2].

Among them, acoustic stimulation in the audio range has shown a great potential to produce motor and non-motor effects that could benefit PD patients. In previous works [6] we observed positive effects in brain activity in the short term, which could contribute to improve motor symptom's in the long term.

In order to increase the sample size and duration of these studies we developed a methodology to monitor the evolution of the motor symptoms of PD patients using wearables under long term acoustic stimulation. This work presents this methodology and preliminary results obtained in a double blinded randomized controlled study performed during 3 months.

## 2 Fundamentals

Auditory stimulation at frequencies >30 Hz is proven to trigger the so call entrainment effect, which is the coupling between the external stimulus and internal oscillator in brain activity [18]. The frequency of the stimulus is "mirrored" at the auditory cortex (AC), and the entrained neural oscillations at the AC spread around the brain through neural pathways. These oscillations, in turn, elicit others at different brain locations. For example, auditory stimuli relate to motor functions [5] as well as to cognitive functions [17].

The fact that the brain connectivity during entrainment occurs via the thalamus suggests that auditory stimuli may compensate the lack of endogenous oscillations at the basal ganglia due to neurodegeneration [18]. PD patients show decreased dominant (peak) frequency and increased theta band power, both correlated with cognitive decline. The oscillations in the basal ganglia of PD patients typically shift down to frequencies in the beta band, 14–30 Hz, characteristic of akinetic states or dopamine deficiency, as well as to <10 Hz frequencies, associated with tremor, dystonia and sleep; while the gamma band, >60 Hz, is restored during voluntary movements and dopamine-based treatment [7].

Based on this, different auditory stimulation protocols have been tested in preclinical and clinical studies. Among these protocols we find those based on binaural beats, a specific type of auditory stimulation which has been studied in healthy subjects [4]. In a previous work we obtained positive results on electroencephalogram power, functional connectivity and working memory in a group of 14 PD patients after two sessions of 10 min separated by a week [6]. Although these results are promising and in line with the theoretical background, they have important limitations for scaling the sample size and duration of the studies. Even more, from a practical point of view positive changes in brain activity are not enough if they do not result in improvements in motor or non-motor symptom's.

Since the evaluation of motor symptoms is the basis of PD diagnosis, the assessment of motor symptoms using sensors can represent a major advance in objective assessment by reducing the absence of inter-observer variability and allowing remote evaluations [13]. Objective symptom monitoring can be used for research purposes, allowing a better understanding of the patient's condition and response to treatments and therefore improving the quality of life of patients [8].

The portability of wearable sensors, combined with technological maturity in artificial intelligence (AI), provides the opportunity to improve data acquisition methodologies for objective monitoring of motor symptoms [14]. In addition, these technologies can enable continuous monitoring of motor activity in their free-living environment, which may allow a representative monitoring of the evolution of the disease, and work as a complement to traditional assessment methods [10].

## 3 Materials and Methods

This study was carried out using data from PD patients who participated in the TECA-PARK project [1], where a custom-built mHealth mobile and wearable application was developed and used to monitor the evolution of motor and non-motor symptom's of 29 patients during three months while they listened either to an experimental or a control acoustic stimulation twice a day. Both experimental (A) and control (B) simulations were the same as in our previous study [6] and participants were randomly assigned to each group. All participants had been diagnosed previously with similar values (2–3) in the Hoehn and Yahr scale.

In order to track motor symptom's a consumer smartwatch was used for data acquisition and the custom mobile application was used to guide the participants to perform a set of exercises [15] under the supervision of a therapist once a day as shown in Fig. 1. Data was collected always on the same day of the week and in a similar schedule to consider the possible variations produced by pharmacodynamics.

The exercises were based on the Unified Parkinson's Disease Rating Scale (UPDRS) parts 3.6 (pronation and supination movements), 3.17 (rest tremor amplitude) and 3.18 (constancy of rest tremor). The device was placed on the wrist of the most affected side of each patient, and the smartwatch was previously analyzed to identify its frequency response by using a methodology described in

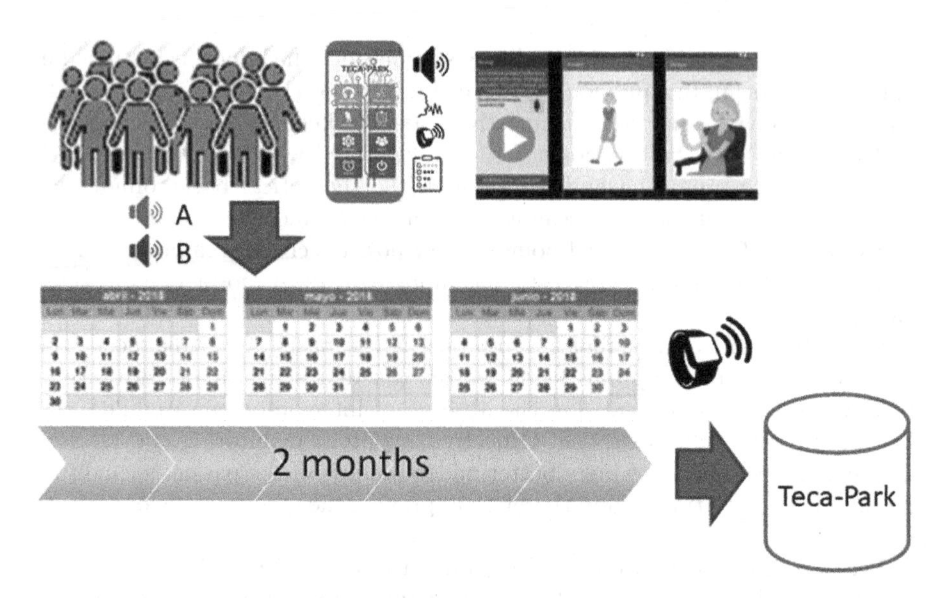


**Fig. 1.** Methodology for long-term monitoring of motor symptoms based on smartwatches.

a previous work [16]. In addition, the same protocol was followed by a group of 5 healthy participants of similar age/gender for 8 weeks.

In this work we will show and discuss results for the following three indicators of motor activity that had been extracted using data from the triaxial accelerometer signals while performing some of these exercises:

1) An indicator of tremor magnitude, obtained as the 75th percentile of the power in the 3.5–7.5 Hz frequency band extracted from the acceleration magnitude signal for windows corresponding to a situation of "hand-rest and presence of tremors".
2) An indicator of tremor constancy, obtained as the total time of tremor presence during the resting periods between the above-mentioned exercises by direct observation of the tests performed.
3) An indicator of bradykinesia obtained as the power expressed in decibels of the signal for the frequency band between 0.5 and 3.5 Hz during the pronation and supination exercise.

The methodology to obtain the two tremor indicators is described in [15], and the methodology used to extract the bradykinesia indicators is described in [9].

On the one hand, the validity of the tremor indicators was assessed in [15] by comparing the results of constancy and amplitude obtained from a multitask convolutional neural network (CNN) with the results of the clinical assessment using the MDS-UPDRS scale. To extract this indicator, the raw triaxial signal of the entire sequence of exercises (duration 7 min) was filtered with a hi-pass butterworth third-order filter (cut-off 0.25 Hz) to eliminate the effects of gravity, then the acceleration magnitude was calculated. The resulting signal was split using temporal windows of 128 samples (2.56 s) to feed the CNN. The outcomes of the multitask CNN were the window segments detected as tremor and resting periods. The tremor amplitude indicator was extracted using the power of the tremor-band in each window (with the presence of tremor) and calculating the 75th percentile to report a single value as indicator. While the information with the number of windows detected as resting times and tremor was used to calculate the tremor constancy as percentage.

On the other hand, the indicator of bradykinesia was calculated using the accelerometer signal of the segment corresponding to the pronation and supination exercise (duration 10 s) of each test. For this, the acceleration magnitude was calculated from the triaxial signal and filtered (butterworth third-order) in the movement band (0.5 and 3.5 Hz) proposed in [9]. Finally, the power of the resulting signal was calculated and expressed in acceleration level (dB) to be used a indicator of bradykinesia.

## 4 Results

In this work we present the time evolution of the above mentioned indicators for one participant of each acoustic stimulation group (labelled PD-A and PD-B) and for those of the healthy control group (labelled CTL1-5). Table 1 shows age and gender information for each individual.

**Table 1.** Gender and age data of each participant.

| Participant ID | Gender | Age |
|---|---|---|
| PD-A | M | 52 |
| PD-B | F | 60 |
| CTRL1 | F | 63 |
| CTRL2 | F | 61 |
| CTRL3 | M | 65 |
| CTRL4 | M | 59 |
| CTRL5 | M | 58 |

PD-A performed the tests always in the mornings (approximately at 11 am), while subject PD-B performed them in the middle of the day (approximately at 1 pm). From these data, the following indicators were obtained:

### 4.1 Tremor Indicators

Two indicators have been extracted to monitor the evolution of tremor: one representing the amplitude of the vibration signal, shown in Fig. 2, and the other representing the constancy of tremors, shown in Fig. 3.

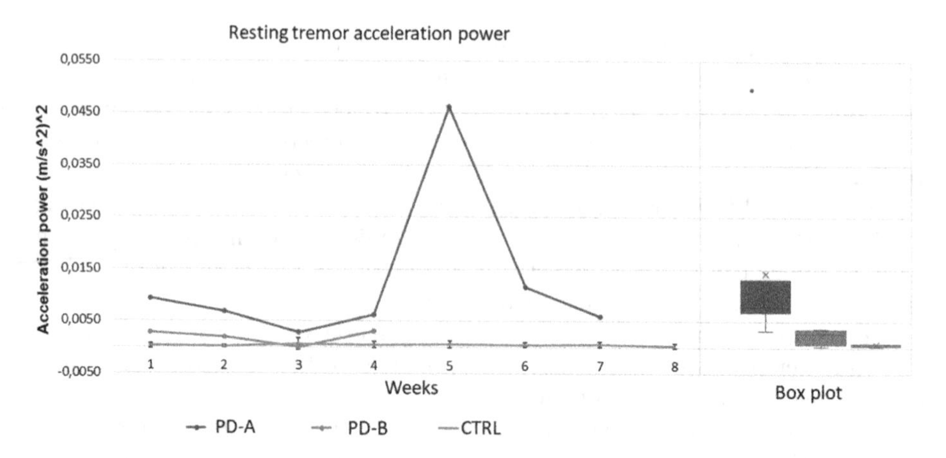


**Fig. 2.** Resting tremor - magnitude indicator. The left part shows the evolution of this indicator over several weeks for two PD subjects and five healthy subjects (mean $\pm$ std). The right part of the figure shows a blox plot with the data of all weeks for PD subjects and all control subjects.

Figure 2 shows higher magnitude values for PD-A and PD-B users compared to the average magnitude values of the control subjects, which present average values close to 0 with low variability (inter-patient) because these subjects did not present Parkinsonian tremors.

Tests performed in the first weeks for PD patients (PD-A and PD-B) show variability in magnitude in the tremor band, while in the fifth week a peak in magnitude is identified for patient PD-A. Meanwhile, the results for patient PD-B show a lower tremor magnitude than PD-A in all weeks, however, in most weeks it shows higher magnitude values than the control subjects.

The results of resting tremor constancy (see Fig. 3) show a very low presence (close to 0%) of resting tremor in the control group (as expected), with low inter-subject variability. PD patients show a presence of tremor of up to 70%, however, PD-B shows a lower percentage of tremor than PD-A throughout all weeks, as well as a lower dispersion in the data due to a marked reduction in the constancy in weeks 3 and 4.

### 4.2 Bradykinesia Indicator

The magnitude indicator for bradykinesia Fig. 4 shows higher average values in the control group data, although with similar values in PD patients in the first weeks. Most of the tests performed by PD patients show lower acceleration levels, these results are due to the difficulties they may present when performing exercises of a repetitive nature such as those proposed in the MDS-UPDRS scale. Similarly, to the tremor results, a greater variability is identified in the data of subject PD-A compared to PD-B, this behavior could be caused by the effects of medication and its effectiveness in controlling symptoms (motor fluctuations) and have a significant influence on the results obtained.

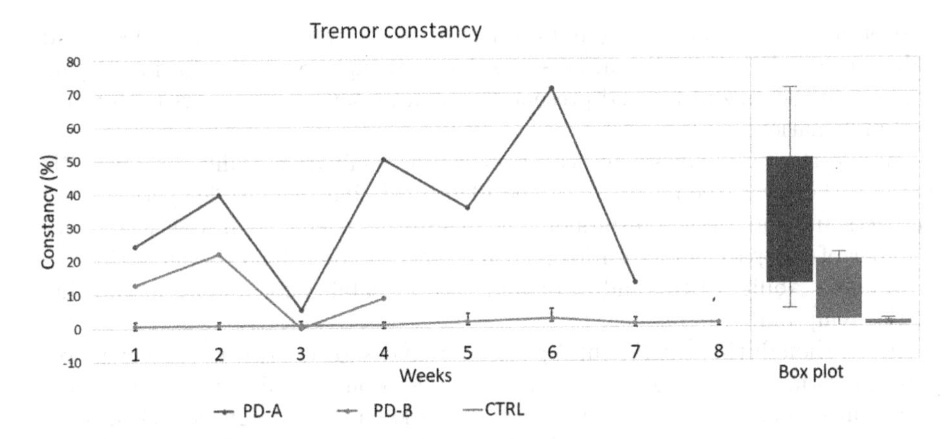


**Fig. 3.** Resting tremor - constancy indicator: The left part shows the evolution of this indicator over several weeks for two PD subjects and five healthy subjects (mean ± std). The right part of the figure shows a blox plot with the data of all weeks for PD subjects and all control subjects.

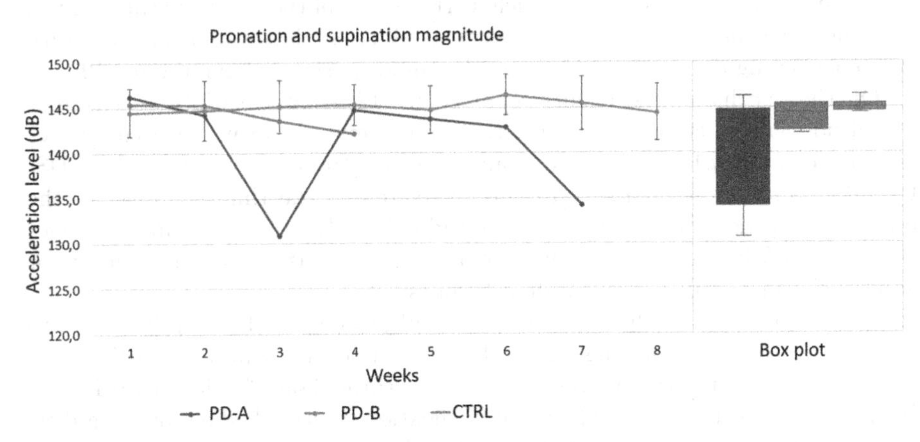


**Fig. 4.** Bradykinesia indicator: The left part shows the evolution of this indicator over several weeks for two PD subjects and five healthy subjects (mean ± std). The right part of the figure shows a blox plot with the data of all weeks for PD subjects and all control subjects.

The results show that the proposed indicator could be considered as an option for monitoring bradykinesia, however, a combined assessment including exercises of a similar nature (e.g., toe or foot-tapping) could provide a broader view of the symptom.

From a technical perspective, the results in the weekly assessment of the amplitude and constancy of resting tremors suggest that the current configuration of the acceleration sensor in the amplitude range (±2 g factory set), and in the sampling frequency (50 Hz, programmatically set) can provide relevant information for tremor monitoring and discrimination between healthy subjects and subjects with PD. In addition, these results show high sensitivity for identifying changes in the magnitude and presence (or non-presence) of tremors both intra- and inter-patient.

While the bradykinesia indicator shows discriminative ability between PD patients, however, the results obtained identify similar magnitudes between PD subjects and the average results of control subjects over several weeks. A larger number of PD patients and a longer time interval could help to verify the discriminative ability of this indicator and its applicability for remote monitoring of the symptom.

Even though the experimental protocol was designed to ensure the exercises were performed at the same time of the day and on a similar day of the week, the results show a high variability over the weeks in the data of the PD subjects (PD-A and PD-B), compared to the data of the healthy control subjects for the three proposed indicators.

The variability identified in the results is consistent with those identified in related studies, where significant changes have been observed in the indicators used to monitor motor symptoms such as tremor amplitude, both within a given day (intra-day pattern) as well as between different days (inter-day pattern), even considering data from subjects with a mild presence of symptoms [12].

On the one hand, the results reported in [12] show a variation in tremor magnitude with fluctuating patterns that can change, for example, 4 times over the course of a day for patients with an amplitude tremor score of 1 to 3 (MDS-UPDRS), and whose changes are closely related to the time of intake of the prescribed medication. Whereas a longitudinal analysis has identified changes of up to 20% in tremor constancy at 40-day intervals in patients with poor adherence to prescribed medication schedules.

The variability identified in the data of subjects with PD shows the need to consider objective monitoring with a higher temporal resolution, which would allow identifying changes in motor competence throughout the day or week, considering among other factors the usual medication intake schedules or the adherence to pharmacological treatments, to provide useful information to increase knowledge about the disease, and the development of personalized therapies. This will require to move towards a free-living environment in order to reduce the burden of the patient.

Unfortunately, due to the limitations abovementioned it has not been possible to asses if the acoustic stimulation has had any effect in the evolution of the motor symptom's of these individuals. Our goal was to check if the effects that we had seen in electroencephalogram power in the motor area after a single stimulation session would produce changes in the motor activity in the long term. A longitudinal study of such phenomena using electrophysiology is required to confirm if the results obtained here are due to the limitations of the monitoring technique or a consequence of the type of stimulation.

## 5 Conclusions

Although the results obtained in this study are preliminary, they provide several guidelines to be considered in future studies using a remote data collection methodology. For example, it is necessary to consider variables that may influence the motor capacity of patients, such as, the schedule of the evaluation test and its relationship with the effects and duration of the medication throughout the day, the mood of the patients or the presence of motor fluctuations (On-Off state) and dyskinesias in patients in advanced stages of the disease. In addition, it should be considered the bias produced by procedural problems such as the proper placement of the monitoring devices or by errors in the performance of the experimental protocols.

Future studies could include a larger number of subjects, assessed over an extended period (several months) and with higher temporal resolution, to identify with more certainty the validity and sensitivity of the indicators for discriminating change in intra and inter-patient motor competence. In addition, studies under free-living conditions may be considered in conjunction with periodic clinical evaluations to provide reference points to assess the validity of objective indicators such as those proposed in this work.

Finally, these results also show that these methods might not be robust enough yet as to detect the possible effects of experimental interventions such as this type of acoustic stimulation paradigm, requiring other more traditional approaches based on imaging and/or electrophisiology.

**Acknowledgements.** This work has been funded by grant TECA-PARK_55_01 under the POCTEP 0348_CIE_6_E program of Fundacion General del CSIC. The authors would like to thank the Parkinson's associations of Madrid, Valladolid, Burgos, Asturias, Jovellanos and Hogar Santa Estefânia de Guimaraes for their participation in this study, and also to Dr. J. C. Martiinez-Castrillo, Dr. M. Gago and Dra. M. Blazquez for their contributions and continuous support.

## References

1. Tecapark. https://www.i2a2.upm.es/tecapark. Accessed 7 Jan 2022
2. Bloem, B.R., de Vries, N.M., Ebersbach, G.: Nonpharmacological treatments for patients with Parkinson's disease. Mov. Disord. **30**(11), 1504–1520 (2015)
3. Díez-Cirarda, M., Ibarretxe-Bilbao, N., Peña, J., Ojeda, N.: Neurorehabilitation in Parkinson's disease: a critical review of cognitive rehabilitation effects on cognition and brain. Neural Plast. **2018**, 2651918 (2018)
4. Garcia-Argibay, M., Santed, M.A., Reales, J.M.: Efficacy of binaural auditory beats in cognition, anxiety, and pain perception: a meta-analysis. Psychol. Res. **83**(2), 357–372 (2018). https://doi.org/10.1007/s00426-018-1066-8
5. García-Casares, N., Martín-Colom, J.E., García-Arnés, J.A.: Music therapy in Parkinson's disease. J. Am. Med. Dir. Assoc. **19**(12), 1054–1062 (2018)
6. Gálvez, G., Recuero, M., Canuet, L., Del-Pozo, F.: Short-term effects of binaural beats on EEG power, functional connectivity, cognition, gait and anxiety in Parkinson's disease. Int. J. Neural Syst. **28**(5), 1750055 (2018)

7. Johns, P.: Chapter 13-Parkinson's disease. In: Clinical Neuroscience, pp. 163–179 (2014)
8. Luis-Martínez, R., Monje, M.H.G., Antonini, A., Sánchez-Ferro, A., Mestre, T.A.: Technology-enabled care: integrating multidisciplinary care in Parkinson's disease through digital technology. Front. Neurol. **11**, 575975 (2020)
9. Mahadevan, N., et al.: Development of digital biomarkers for resting tremor and bradykinesia using a wrist-worn wearable device. NPJ Digit. Med. **3**, 5 (2020)
10. Monje, M.H.G., Foffani, G., Obeso, J., Sánchez-Ferro, Á.: New sensor and wearable technologies to aid in the diagnosis and treatment monitoring of Parkinson's disease. Annu. Rev. Biomed. Eng. **21**, 111–143 (2019)
11. Poewe, W.: Parkinson disease Primer - a true team effort. Nat. Rev. Dis. Primers **6**(1), 31 (2020)
12. Powers, R., et al.: Smartwatch inertial sensors continuously monitor real-world motor fluctuations in Parkinson's disease. Sci. Transl. Med. **13**(579), eabd7865 (2021)
13. Prasad, R., Babu, S., Siddaiah, N., Rao, K.: A review on techniques for diagnosing and monitoring patients with Parkinson's disease. J. Biosens. Bioelectron. **7**(203), 2 (2016)
14. Rovini, E., Maremmani, C., Cavallo, F.: How wearable sensors can support Parkinson's disease diagnosis and treatment: a systematic review. Front. Neurosci. **11**, 555 (2017)
15. Sigcha, L., et al.: Automatic resting tremor assessment in Parkinson's disease using smartwatches and multitask convolutional neural networks. Sensors (Basel) **21**(1), 291 (2021)
16. Sigcha, L., Pavón, I., Arezes, P., Costa, N., De Arcas, G., López, J.M.: Occupational risk prevention through smartwatches: precision and uncertainty effects of the built-in accelerometer. Sensors **18**(11) (2018). https://www.mdpi.com/1424-8220/18/11/3805
17. Sihvonen, A.J., Särkämö, T., Leo, V., Tervaniemi, M., Altenmüller, E., Soinila, S.: Music-based interventions in neurological rehabilitation. Lancet Neurol. **16**(8), 648–660 (2017)
18. Swerdlow, N.R., Bhakta, S.G., Light, G.A.: Room to move: plasticity in early auditory information processing and auditory learning in schizophrenia revealed by acute pharmacological challenge. Schizophr. Res. **199**, 285–291 (2018)

# Evaluation of TMS Effects on the Phonation of Parkinson's Disease Patients

Andrés Gómez-Rodellar[1], Jiri Mekyska[2], Pedro Gómez-Vilda[3](✉), Lubos Brabenec[4], Patrik Simko[4], and Irena Rektorova[4,5]

[1] Usher Institute, Faculty of Medicine, University of Edinburgh, Edinburgh, UK
a.gomezrodellar@ed.ac.uk
[2] Department of Telecommunications, Brno University of Technology, Brno, Czech Republic
mekyska@vut.cz
[3] NeuSpeLab, CTB, Universidad Politécnica de Madrid, 28220 Pozuelo de Alarcón, Madrid, Spain
pedro.gomezv@upm.es
[4] Applied Neuroscience Research Group, Central European Institute of Technology - CEITEC, Masaryk University, Brno, Czech Republic
{lubos.brabenec,patrik.simko,irena.rektorova}@ceitec.muni.cz
[5] First Department of Neurology, Faculty of Medicine and St. Anne's University Hospital, Masaryk University, Brno, Czech Republic

**Abstract.** Repetitive Transcranial Magnetic Stimulation (rTMS) is a non-invasive technique which is known to produce modifications in cortical brain activity. This paper is devoted to describe potential beneficial effects of rTMS on the phonation stability of Parkinson's Disease Patients (PDPs). To this end, several measurements derived from phonation have been studied. The stability of phonation is evaluated on sustained emisions of certain open vowels, as [a:]. Using vocal tract inversion, a correlate of the glottal source (pressure on the supraglottal rim of the vocal folds) is estimated from vowel emissions. The glottal source power spectral density is used to indirectly estimate the biomechanical tension of the vocal folds. The neuromotor instabilities experienced by PDPs, affecting the vocal fold tension are used as perturbation features related to tremor bands. A longitudinal analysis of the features from an active rTMS case can be compared in different time laps after rTMS, and tested against those from a similar study on a sham rTMS case. Relevant improvements on phonation stability may be appreciated on the active rTMS case compared to the sham one, which are reflected on several features as biomechanical tremor bands. These results open

This research received funding from the European Union's Horizon 2020 research and innovation program under the Marie Skłodowska-Curie grant agreement no. 734718 (CoBeN) and from a grant from the Czech Ministry of Health, 16-30805A, and from grants TEC2016-77791-C4-4-R (Ministry of Economic Affairs and Competitiveness of Spain), and Teca-Park-MonParLoc FGCSIC-CENIE 0348-CIE-6-E (InterReg Programme).


J. M. Ferrández Vicente et al. (Eds.): IWINAC 2022, LNCS 13258, pp. 199–208, 2022.
https://doi.org/10.1007/978-3-031-06242-1_20

a new non-invasive, costless and remote methodology for PD functional neuromotor evaluation.

**Keywords:** Repetitive Transcranial Magnetic Stimulation · Neuromotor diseases · Voice production · e-Health

# 1 Introduction

## 1.1 Background

Parkinson's Disease is a neurodegenerative illness first described by James Parkinson in 1817. Recent studies quantify its incidence in 15 cases per 100,000, with a prevalence ranging from 100 to 200 cases per 100,000 [1]. The impact of the disease in the quality of life of PDPs is associated to motor symptoms as bad neuromotor control, difficulty in walking and handling objects, resting tremor, facial rigidity, etc., [2], and other non-motor symptoms which are also challenging PDPs' ability to carry an independent life. PDP also experience alterations in respiration, phonation, articulation, and prosody, collectively referred to as hypokinetic dysarthria (HD), which is a complex motor speech disorder characterized by manifestations such as monopitch and monoloudness, imprecise articulation, impaired speech rate and rhythm, and irregular pitch fluctuations [2]. PD symptoms may be mitigated by medication, neurostimulation and rehabilitation, and stabilized temporarily to facilitate patients' motor functions [3]. Transcranial magnetic stimulation (TMS) is a non-invasive method for the stimulation of neural tissue, including cerebral cortex, spinal roots, and cranial and peripheral nerves [4]. In this sense, repetitive TMS (rTMS) has been proposed as a therapy to improve patients' neuromotor conditions [4]. The rest of Sect. 1 presents the working hypotheses of the study. Section 2 is devoted to describe the estimation procedures of the features used, the databases from which the active and sham rTMS cases have been derived, as well as the methodological procedures used. The results of both cases are presented in Sect. 3, and discussed in Sect. 4. Conclusions and findings are commented in Sect. 5.

## 1.2 Working Hypotheses

The aim of the present study is to explore the possibility of using motor features estimated on phonation biomechanics to assess the extent to which rTMS affects the phonation motor function, and for how long this effect can be observed. For such, phonation tremor features from a specific case of active rTMS stimulation have been compared with similar ones from a case of sham rTMS. For the comparisons, the timely evolution model, which this effect is expected to follow, has been considered, as shown in Fig. 1. It assumes that rTMS has an effect on phonation and that it could be observed by examining certain features derived from biomechanical behavior of the vocal folds, which are assumed to take the following behavior:

- Improvement phase (I): distortion and instability will be reduced, features will approach to normative, and separate from baseline.
- Stabilization phase (II): distortion and instability will remain reduced, features will stay closer to normative than baseline
- Regression phase (III): distortion and instability are expected to resume pre-stimulus values closer to baseline than to normative

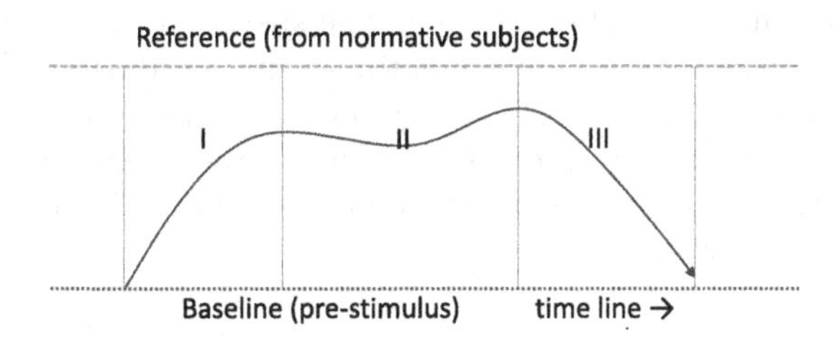


**Fig. 1.** Expected effect of rTMS on an assumed instability-related feature: Phase I-Improvement (estimations separate from a baseline pre-stimulus level: lower dot-red line) and approach a normative level observed on non-pathological participants (upper dash-green line); Phase II-Stabilization (features keep closer to normative than to baseline); Phase III-Regression (features separate from normative and regress to baseline as the stimulation effects fade away). (Color figure online)

## 2 Materials and Methods

### 2.1 Experimental Protocol

The present study is based upon the extraction of tremor features in phonation, estimated from the vocal fold tension. It is a well established fact that vocal fold vibration produces an excitation at the glottis known as glottal source, which may be seen as the pressure wave measured at the supraglottal ridge of the vocal folds. A reliable estimation of the vocal fold tension may be obtained from sustained open vowels as [a:], that allows for a better separation of the vocal tract effects from the glottal source, compared to other vowels [5].

This study has been conducted on two cases selected from 39 participants with clinically diagnosed PD showing mild to moderate HD, all of them right handed, and native speakers of Czech. All were on a stable dopaminergic medication during the whole study. The patients were tested in the ON state. They were informed of the nature of the research and gave their written consent.

All participants went through a baseline assessment (pre-stimulus at T0), i.e. 10 stimulation sessions (within two weeks); a follow-up assessment right after stimulation (post-stimulus at T1); additional follow-up evaluations at 6 weeks (post-stimulus at T2) and 10 weeks (post-stimulus at T3). In some cases, as the ones reported in the present study, after the baseline assessment, an additional evaluation was conducted (post-stimulus at T4). A multiple stimulation session protocol was conducted on each participant, randomly assigned to active or sham rTMS groups. Each session lasted 40 min, using a figure-eight-shaped coil, producing pulses 1 Hz at 100% intensity, applying 1800 pulses/session. The sham stimulation followed identic steps, settings and sounds, although no magnetic field was applied. A perceptual assessment was conducted by a speech therapist rating speech performance, faciokinesis, phonorespiration, and phonetic competence at each evaluation step. The speech exercises demanded from each participant consisted in 2 short emotion-neuter readings in Czech; 1 short emission of vowels [a:], [i:], [u:] (of ~1.5 s); one long emission of a sustained [a:] (of around ~15 s); 1 long emission of a diadochokinetic exercise, consisting in the repetition of [pataka] (lasting >10 s), and one single emission of ten different selected trisyllabic words in Czech.

### 2.2 Feature Estimation

The study was conducted on two cases, which were used for comparison purposes to illustrate the procedural concepts. The data produced by a PD participant (male, 58 years-old, UPDRS-III = 9, LED = 990 mg) consisting in long sustained vowel emissions of [a:] before (baseline at T0, day 0) and after active rTMS (T1 at 15, T2 at 26, T3 at 78 and T4 at 109 days) were used in the study. Similarly, vowel emissions of [a:] from a PD participant (male, 70 years-old, UPDRS-III = 16, LED = 1000 mg) before (T0 at day 0) and after sham rTMS (T1 at, T2 at, T3 at, and T4 at days) were used for comparison purposes. The features estimated from these two cases were confronted against similar ones from a subset of 16 male subjects who were selected for reference from HUGM (a database built of sustained vowel emissions from 100 gender-balanced normative participants recorded at the ENT services of Hospital Universitario Gregorio Marañón, Madrid, Spain). Recordings of vowel [a:] from the five evaluations of the active- and sham-rTMS cases were low-pass filtered (anti-aliasing) and down-sampled to 16 kHz, as well as the nomative ones. The vocal tract transfer function of each utterance was estimated by a 24-pole iterative adaptive inverse filtering (IAIF) algorithm [6], implemented as a lattice-ladder, to be removed from the spectral contents of the speech signal (inversion), producing a filtering residual, which once integrated produced an estimation of the glottal source [7].

### 2.3 Feature Assessment

Each feature distribution estimated from emissions of [a:] at the post-stimulation instants T1–T4 was tested against equivalent features estimated from pre-stimulation emissions at T0 (baseline), and against emissions from the 16 normative participants selected from HUGM. The probability density distributions (pdd) of these features were compared using Jensen-Shannon Divergence (JSD). Let these pdds be $p_i(x)$ and $p_j(x)$, then JSD would be calculated as:

$$D_{JSij}\{p_i(x), p_j(x)\} = D_{KLim}\{p_i(x), p_m(x)\} \quad (1)$$

where

$$p_m(x) = (p_i(x) + p_j(x))/2 \quad (2)$$

and

$$D_{KLij}\{p_i(x), p_j(x)\} = -\int_{\zeta=0}^{\infty} p_i(\zeta) ln[p_i(\zeta)/p_j(\zeta)]d\zeta \quad (3)$$

The features used in the comparisons were the relative frequency and amplitude perturbations Jitter (Jt) and Shimmer (Sh), the Cepstral Peak Prominence (CPP) and the tremor of the vocal fold tension splitted in the frequency bands $\delta$ (f $\leq$ 4 Hz), $\vartheta$ (4 Hz $<$ f $\leq$ 8 Hz), $\alpha$ (8 Hz $<$ f $\leq$ 16 Hz), $\beta$ (16 Hz $<$ f $\leq$ 32 Hz), $\gamma$ (32 Hz $<$ f), and $\mu$ (8 Hz $<$ f $\leq$ 12 Hz), corresponding to the ones classically used in Electroencephalography (EEG). The following procedure was used to estimate the JSDs between each sample feature at T1–T4 with respect to T0, and between T0–T4 with respect to the normative set from HUGM:

- Recordings of the vowel [a:] from the PD participants (active and sham stimulated) were down-sampled to 16 KHz.
- The recordings of the vowel [a:] from the normative subset of male participants from HUGM were also down-sampled at 16 kHz.
- The glottal source and the vocal tract transfer function from all emissions were evaluated by a 24-pole adaptive inverse lattice-ladder filter.
- The glottal source was segmented into glottal cycles between neighbour Minimum Flow Declension Ratio instants (MFDR, see [8]).
- The spectral power distribution of the glottal source was estimated using Fourier methods on each glottal cycle.
- The spectral power distribution was matched against a functional (trans-admittance) of a 2-mass vocal fold model.
- Estimates of the biomechanical features of the vocal fold body tension were obtained for each glottal cycle.
- A bank of zero-phase digital filters of order 4 was used to split the timely evolution of the vocal fold tension in the EEG-corresponding bands $\delta$, $\vartheta$, $\alpha$, $\beta$, $\gamma$, and $\mu$.
- Estimations of Jt, Sh and CPP were included following [9].
- The amplitude probability distributions of each feature was evaluated from the normalized 50-bin amplitude histogram of the feature considered as $p_i(k) = h_{ik}/\Sigma_k h_{ik}$, where $h_{ik}$ is the amplitude histogram of feature $i$ and $k$ is the bin number.
- The JSD from feature at T1–T4 was estimated with respect to its corresponding baseline at T0. Similarly, the JSD from each vowel emission at T0–T4 was estimated with respect to the normative set.

## 3 Results

An example of the vocal fold tension obtained from a segment of 4 s corresponding to the T0 emission of vowel [a:] from the active tTMS participant, its splitting in EEG-compatible tremor bands, and the resulting band distributions in amplitude and frequency are presented in Fig. 2 to illustrate the procedures involved in feature estimation and assessment. For each tremor band distributions in amplitude and frequency are included. This set of figures illustrates the build up process used in the evaluation and assessment of feature data.

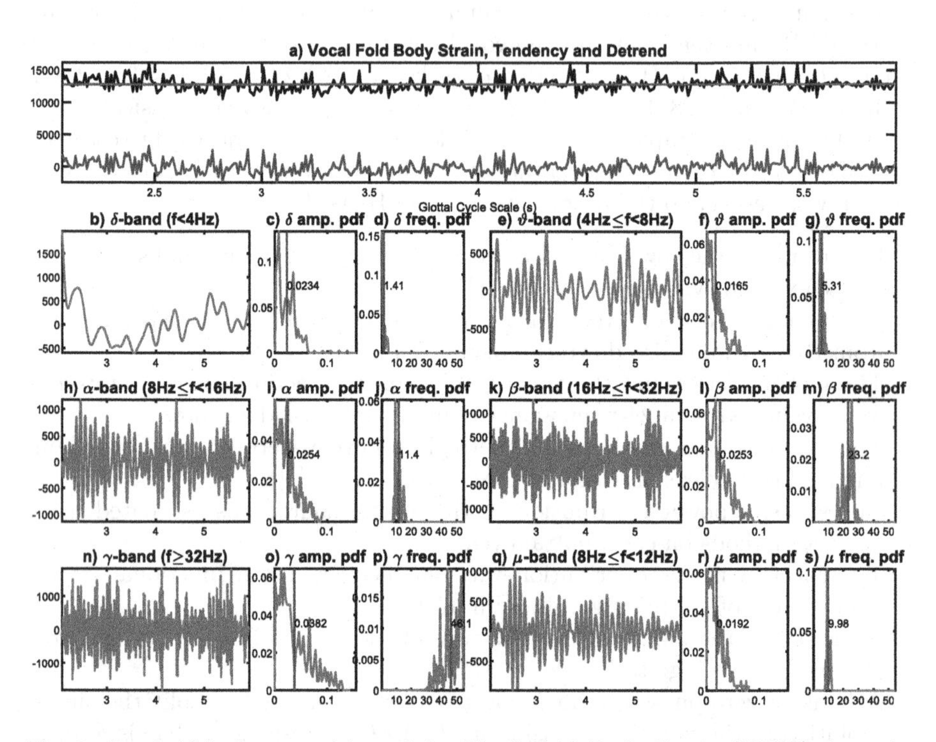


**Fig. 2.** Bandwidth decomposition of the Vocal Fold Body Strain (VFBS) from the active rTMS PD patient at T0 (medians of distributions signaled by vertical red lines): a) VFBS in g.s$^{-2}$ (black), its detrended tendency (red) and unbiased estimate (low blue); b) VFBS component on the $\delta$-band, its amplitude distribution and median (c), and its frequency distribution and median (d); e, f, g) Id. on the $\vartheta$-band; h, i, j) Id. on the $\alpha$-band; k, l, m) Id. on the $\beta$-band; n, o, p) Id. on the $\gamma$-band; q, r, s) Id. on the $\mu$-band. (Color figure online)

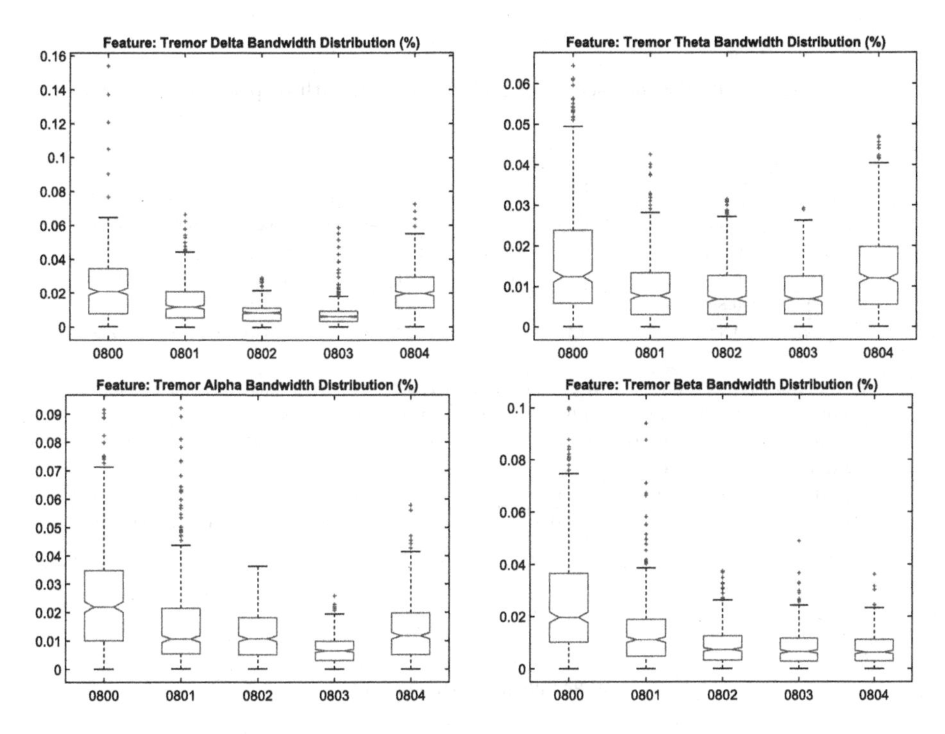


**Fig. 3.** Boxplots of the pre- (T0) and post-evaluations (T1, T2, T3, and T4) on the $\delta$-, $\theta$-, $\alpha$-, and $\beta$-band distributions.

The amplitude distributions of the active rTMS study corresponding to the four most interesting bands ($\delta, \vartheta, \alpha$, and $\beta$) are summarized in Fig. 3.

The divergences from each sample to its baseline or to the normative reference subset may be combined in a single Feature Equidistance Score (FES), defined as $\lambda = (JSD_{iN} - JSD_{iB})/JSD_{BN}$, where $\mathrm{JSD}_{iB}$ and $\mathrm{JSD}_{iN}$ are the JSDs from sample $i$ to the baseline or the normative reference, and $JSD_{BN}$ is the divergence between the baseline and the normative reference of each feature. $\lambda$ may be interpreted as the likelihood of the feature under test being closer to the baseline ($\lambda < 0$) or to the normative dataset ($\lambda > 0$), as JSD may be interpreted as the square of a distance [10]. The FES ($\lambda$) for Jt, Sh, CPP, and the EEG-related features from the active and sham data are given in Table 1, and in Fig. 4.

**Table 1.** Feature equidistance scores of each evaluation with respect to normative and baseline references

| Active rTMS | | | | | | | | | |
|---|---|---|---|---|---|---|---|---|---|
| Eval (days) | Jt | Sh | CPP | $\delta$-band | $\vartheta$-band | $\alpha$-band | $\beta$-band | $\gamma$-band | $\mu$-band |
| T0 (0) | −1.000 | −1.000 | −1.000 | −1.000 | −1.000 | −1.000 | −1.000 | −1.000 | −1.000 |
| T1 (15) | −0.500 | −0.785 | 0.561 | 0.429 | 0.288 | 0.017 | 0.191 | −0.121 | −0.274 |
| T2 (26) | 0.525 | −0.858 | 0.319 | 0.603 | 0.321 | 0.434 | 0.375 | 0.289 | 0.674 |
| T3 (78) | 0.805 | −0.884 | 0.621 | −0.347 | 0.172 | 0.475 | 0.310 | 0.294 | 0.719 |
| T4 (109) | −0.328 | −0.866 | 0.470 | 0.442 | 0.227 | 0.428 | 0.415 | −0.128 | 0.639 |
| Sham rTMS | | | | | | | | | |
| Eval (days) | Jt | Sh | CPP | $\delta$-band | $\vartheta$-band | $\alpha$-band | $\beta$-band | $\gamma$-band | $\mu$-band |
| T0 (0) | −1.000 | −1.000 | −1.000 | −1.000 | −1.000 | −1.000 | −1.000 | −1.000 | −1.000 |
| T1 (24) | −0.222 | −0.747 | −0.644 | 0.471 | −0.272 | 0.154 | 0.325 | 0.000 | −0.322 |
| T2 (54) | −0.214 | −0.716 | −0.588 | 0.569 | −0.131 | 0.395 | −0.004 | −0.808 | 0.353 |
| T3 (81) | −0.151 | 0.001 | −0.539 | 0.700 | 0.113 | 0.311 | −0.263 | −0.610 | 0.209 |
| T4 (119) | −0.863 | −0.336 | −0.291 | 0.554 | 0.466 | 0.369 | 0.230 | −0.341 | −0.074 |

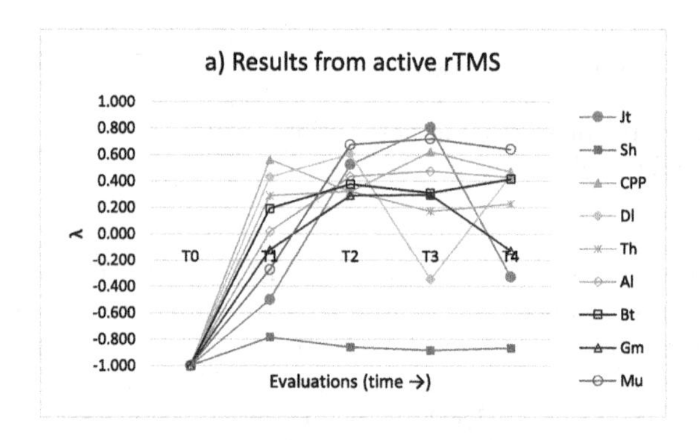


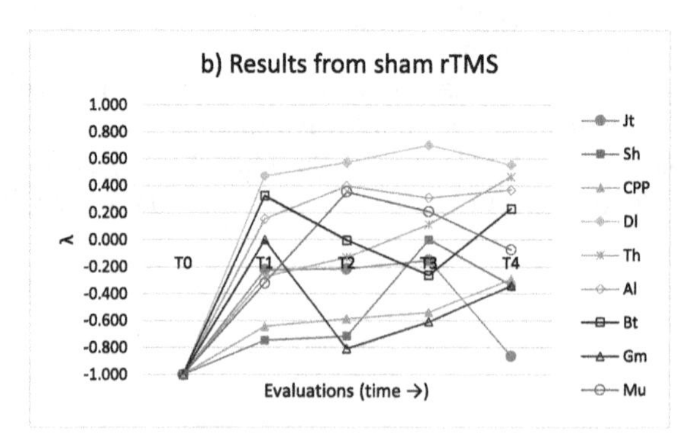


**Fig. 4.** Timely evolution of $\lambda$: a) active rTMS case; b) sham rTMS case

# 4 Discussion

From the examination of the results in Fig. 2 it may be seen that important vocal fold tension instabilities may be observed in all the bands explored, ranging approximately between $-1$ N.m$^{-1}$ and 1 N.m$^{-1}$ (where 1 N.m$^{-1}$ is equivalent to 1000 g.s$^{-2}$), which appear as amplitude modulations on the average band frequencies ($\delta$, $\vartheta$, $\alpha$, $\beta$, and $\gamma$) at 1.41, 5.31, 11.4, 23.2, and 46.1 Hz, respectively (9.98 Hz for the $\mu$-band). Besides, it may be observed that the amplitude distributions behave as $\chi^2$ distributions. The results in Fig. 3 show that the tremor associated to the $\delta$-band follow the pattern assumed in Fig. 1, showing first an amplitude decay (T1) followed by a stabilization at T2 and T3, and by a regression (increment) in T4. This same behavior, somewhat less clear can be seen also in the $\vartheta$- and $\alpha$-bands. The behavior in the $\beta$-band is much more interesting, as it shows a progressive decay in stability in all the post-evaluation means, and a reduction of dispersion, which is another interesting indicator of the therapeutic effects of rTMS. Regarding the results in Table 1 and Fig. 4, jitter (Jt) after active rTMS shows a possible improvement in T2 and T3 ($\lambda > 0$) with regression at T4, whereas in the sham case all the evaluations are negative ($\lambda < 0$). On its turn, shimmer (Sh) does not seem to be affected by stimulation in the active case. Cepstral Peak Prominence (CPP) shows beneficial effects in the active case ($\lambda > 0$), which are quite mild in the sham case. The analysis of the $\delta$-band results do not show a clear cut between active and sham, as interestingly, a beneficial effect may be observed in both cases. The $\lambda$ of the $\vartheta$-band show a relative modest but persistent improvement for the active case, and a progressive improvement for the sham case. The scores for the $\alpha$-band show also improvements for both cases, which are more notorious in the active case. Those of $\beta$-band show clear improvements in the active case, and less clear in the sham one. The results of $\gamma$-band show moderate improvements and final regression in the active case, and no clear beneficial effects in the sham case. The tremor in the $\mu$-band experience a beneficial effect in both cases, and less clear in the sham one. Now, the burning question is that besides observing beneficial effects after active rTMS, why apparently sham rTMS may produce improvements in the stability of the vocal fold tension in certain features as CPP, $\alpha$ and $\mu$? Although the present study cannot give a conclusive response to this question, a possible hypothetical explanation may be behind the beneficial effects of vocal exercise on PD patients, as it is well known that speech rehabilitation may produce these effects [11], because participants are exposed to voice exercising during recordings, which may have a beneficial effect per se. Of course, if it were the case, these effects would add up to those experienced by active rTMS as well. The limitations of the present study are due to its very specific character, centered on the longitudinal examination of only two cases (active and sham). It seems that the likelihood score defined $\lambda$ may be a good indicator of the stimulation effects, although the confounding factor of the rehabilitative vocal exercise may mix up with the real effects of rTMS stimulation.

## 5 Conclusions

Certain potentially beneficial effects may be observed in some of the features monitored when comparing active vs sham stimulation. These effects may also be perceived on certain features in the sham stimulation case studied, although not as clear or intense. Whether this may be due to the improvements in phonation after recording sessions progress is a matter which is to be investigated. More research is needed studying other cases and introducing some changes in the evaluation protocol to assess the possible effects of speech exercises on the improvement of phonation. As future lines, the relationship between phonation tremor bands and EEG activity measured proximity of speech neuromotor areas is to be investigated. However, this line presents the problem associated to electromyographic activity developed during articulation, especially on the masseter, which may contaminate EEG recording. It is expected that the systemic chain behind the relationship linking EEG and tremor could provide relevant information on the neuromotor behavior behind speech and phonation.

## References


1. Tysnes, O.-B., Storstein, A.: Epidemiology of Parkinson's disease. J. Neural Transm. **124**(8), 901–905 (2017). https://doi.org/10.1007/s00702-017-1686-y
2. Duffy, J.R.: Motor Speech Disorders: Substrates, Differential Diagnosis, and Management, 3rd edn. Elsevier, St. Louis (2013)
3. Brabenec, L., Klobusiakova, P., Simko, P., Kostalova, M., Mekyska, J., Rektorova, I.: Non-invasive brain stimulation for speech in Parkinson's disease: a randomized controlled trial. Brain Stimul. **14**, 571–578 (2021). https://doi.org/10.1016/j.brs.2021.03.010
4. Hallett, M.: Transcranial magnetic stimulation: a primer. Neuron **55**(2), 187–199 (2007). https://doi.org/10.1016/j.neuron.2007.06.026
5. Titze, I.: Nonlinear source-filter coupling in phonation: theory. J. Acoust. Soc. Am. **123**, 2733–2749 (2008). https://doi.org/10.1121/1.2832337
6. Alku, P., et al.: OPENGLOT-an open environment for the evaluation of glottal inverse filtering. Speech Commun. **107**, 38–47 (2019). https://doi.org/10.1016/j.specom.2019.01.005
7. Gómez, P., et al.: Glottal Source biometrical signature for voice pathology detection. Speech Commun. **51**, 759–781 (2009). https://doi.org/10.1016/j.specom.2008.09.005
8. Titze, I.R.: Theoretical analysis of maximum flow declination rate versus maximum area declination rate in phonation. J. Speech Lang. Hear. Res. **49**, 439–447 (2006). https://doi.org/10.1044/1092-4388(2006/034)
9. Mekyska, J., et al.: Robust and complex approach of pathological speech signal analysis. Neurocomputing **167**, 94–111 (2015). https://doi.org/10.1016/j.neucom.2015.02.085
10. Connor, R., Cardillo, F.A., Moss, R., Rabitti, F.: Evaluation of Jensen-Shannon distance over sparse data. In: Brisaboa, N., Pedreira, O., Zezula, P. (eds.) SISAP 2013. LNCS, vol. 8199, pp. 163–168. Springer, Heidelberg (2013). https://doi.org/10.1007/978-3-642-41062-8_16
11. Ramig, L.O., Fox, C., Sapir, S.: Speech treatment for Parkinson's disease. Expert Rev. Neurother. **8**, 297–309 (2008). https://doi.org/10.1586/14737175.8.2.297

# Effects of Neuroacoustic Stimulation on Two Study Cases of Parkinson's Disease Dysarthria

Pedro Gómez-Vilda[1](✉), Andrés Gómez-Rodellar[2], Daniel Palacios-Alonso[3], and Agustín Álvarez-Marquina[1]

[1] NeuSpeLab, CTB, Universidad Politécnica de Madrid, 28220 Pozuelo de Alarcón, Madrid, Spain
pedro.gomezv@upm.es, aalvarez@fi.upm.es

[2] Usher Institute, Faculty of Medicine, University of Edinburgh, Edinburgh, UK
a.gomezrodellar@ed.ac.uk

[3] E.T.S. de Ingeniería Informática - Universidad Rey Juan Carlos, Campus de Móstoles, Tulipán, s/n, 28933 Móstoles, Madrid, Spain
daniel.palacios@urjc.es

**Abstract.** The first prevailing Neuromotor Disorder (ND) is Parkinson's Disease (PD) with steadily increasing incidence rates, lacking a definitive cure. Nevertheless, dopaminergic medication and rehabilitation may improve the living conditions of people affected by PD. Neuroacoustical stimulation is a non-invasive, which may improve some motor symptoms associated to PD, as Hypokinetic Dysartrhia (HD). The aim of this research is to extend previous findings in phonation features before and after neuroacoustical binaural stimulation using two-tone combinations. The study evaluated two study cases (one male and one female) from a database collected on smartphone terminals, consisting in utterances of sequences of vowels peripheral to the vowel triangle. The longitudinal evolution of the participants under an ongoing weekly neuroacoustical stimulation stood on the energy and F0 profiles, the Vowel Space Area, Formant Centralization Ratio, Vowel Articulation Index, and Dynamic Formant Spans. The two cases comprised four recordings, and the comparisons were referred to the first recording. The results reported mixed behavior on the traits studied. The male case showed improvements with respect to phonation quality and tremor, whereas the female case showed improvements with respect to vowel space features, energy and fundamental frequency dispersion.

**Keywords:** Absolute Kinematic Velocity · Parkinson's Disease · Vowel Space Area · Formant Centralization Ratio · Hypokinetic dysarthria

---

This research received funding from grants TEC2016-77791-C4-4-R (Ministry of Economic Affairs and Competitiveness of Spain), and Teca-Park-MonParLoc FGCSIC-CENIE 0348-CIE-6-E (InterReg Programme).


J. M. Ferrández Vicente et al. (Eds.): IWINAC 2022, LNCS 13258, pp. 209–218, 2022.
https://doi.org/10.1007/978-3-031-06242-1_21

# 1 Introduction

Parkinson's Disease (PD) is the most prevalent neuromotor disorder, with a rate of 100-200 cases, and an incidence rate of 15 cases, per 100,000 individuals [1]. PD neuromotor symptoms are shown as rigidity, tremor, and general progressive loss of motor control [2]. People with Parkinson (PwP) might exhibit other non-motor symptoms, as sleep problems, depression or cognitive decline. As neuromotor symptoms are reflected in speech, this makes speech signal a very adequate multi-trait vehicle for PD characterization, monitoring, and rehabilitation, using correlates derived from respiration, phonation, articulation, prosody and fluency marks [3]. The aim of this research is to extend previous findings from the analysis of phonation features before and after neuroacoustical binaural stimulation of PwP using two-tone combinations [4–6]. It is well known that neuroacustical stimulation based on binaural beats may modify the brain activity measured in the cortex by electroencephalography (EEG) as a frequency following response (FFR) [7], influencing cognition, mental states and motor performance. Stimulation by binaural beats responds to the effect of hearing two pure tones separately through each ear [8] resulting in the perception of beats corresponding to the two-tone frequency differences. In this study the neuroacoustic stimulation was the result of using a sine wave 154 Hz through the left ear, and 168 Hz through the right ear, inducing in a binaural perceived tone 14 Hz. The specific objective of this study is to test if neuroacoustical stimulation is able of modifying the speech articulation in PwP characterizing post-stimulation estimations of articulation features from the acoustic analysis of PwP's speech. The working hypothesis assumes that binaural beat stimulation will produce an improvement in the articulation of sequences of vowels as a consequence of better-regulated articulation neuromotor activity. The present research is conceived to capitalize on previous studies conducted on the examination pre-and post stimulation responses on EEG and phonation signals, where positive effects on the participants' neuromotor function were detected [4–6].

The paper is organized as follows: Sect. 2 is devoted to describe the data recording protocols from the PwP participants in the experiment, as well as the features used in the study and the data processing methods to extract them. Section 3 describes the results derived from the application of the methodology expressed in Sect. 2 to the data recruited. Section 4 is intended to comment and criticize the results shown, and to expose the limitations of the research. Section 5 summarizes the main contributions, findings, and conclusions derived from the study.

# 2 Materials and Methods

The study stands on the examination of post-stimulation results on two specific PwP (a male and a female, ages 52 and 60 years old). These participants were a part of a wider experiment within the project Teca-Park, involving the recruitment and recording limb neuromotor and speech signals produced during the execution of a battery of tests, which in the case of speech consisted in the utterance of sustained vowels, dyphthongs, tripthongs, and diadochokinetic exercises.

The recordings were collected on a weekly basis using standard smartphones equiped with a specific app to record speech [9]. The neuroacoustical stimulation was conducted following the protocol described in [5]. The participants were recruited on different PwP patients' associations in nortwest Spain and northern Portugal. The recording protocol was approved by the Ethical Committee of UPM (MonParLoc, 18/06/2018). The voluntary participants were informed about the experiments to be conducted, the protection of personal data, and signed an informed consent form. The methodology was strictly aligned with the Declaration of Helsinki. A brief participants' description is given in Table 1.

**Table 1.** Participants' biometrical data. MP1: male PD participant; FP1: female PD participant; H&Y: Hoehn and Yahr PD rating scale; State: medication state (ON: under the effects of dopaminergic medication).

| Code | Gender | Age | Condition | H&Y | State |
|---|---|---|---|---|---|
| MP1 | M | 52 | PD | 2 | ON |
| FP1 | F | 60 | PD | 2 | ON |

The specific articulation test used in the present study consisted in the sustained utterance of the five cardinal vowels [a:, e:, i:, o:, and u:] in sequence. Four recordings were collected from each participant at different dates, as reported in Table 2. The recordings were taken on general-purpose low-cost smartphones with a sampling frequency 48000 Hz and 16 bits. Speech recordings were stored in the smartphone internal memory and transferred via a wifi link to a central repository for their bach processing [9]. The neuroacoustical stimulation and the speech tests were conducted under the supervision of a personal assistant.

**Table 2.** Recording table from the utterance of the sustained vowel suite by a male PD participant (MP1, four recordings: MP11-14) and a female PD participant (FP1, four recordings: FP11-14). Recording date and time are given using the format YYYY-MM-DD-HH,mm,SS (YYYY: year; MM: month; DD: day; HH: hour; mm: minute; SS: Second)

| Code | Date | Code | Date |
|---|---|---|---|
| MP11 | 2019-12-05-11-06-24 | FP11 | 2019-10-10-12-17-18 |
| MP12 | 2019-12-13-10-22-42 | FP12 | 2019-10-15-12-51-00 |
| MP13 | 2019-12-20-10-18-54 | FP13 | 2019-10-22-12-54-00 |
| MP14 | 2020-01-03-09-44-05 | FP14 | 2019-12-03-12-47-04 |

The study of the articulation kinematics in PD followed the evolution of representations on the vowel triangle measuring the vocal span, and dynamics features expressing the jaw-tongue joint performance in diadochokinetic tests.

Vowel triangle representation tests using suites of peripherical vowels to the vowel triangle are based on the capability shown by participants to produce regular and ample movements in the jaw-tongue joint, which are estimated using the Vowel Space Area (VSA) [10] and the Formant Centralization Ratio (FCR) [11,12]. Vowel space exploring tests are based on peripheral vowel sequences if the aim of the test is to check sustained articulation stability. In this last case, the Absolute Kinematic Velocity (AKV) [13] may help in quantifying undesired and uncontrolled jaw-tongue movement instabilities. In the present research, the following features derived from the utterance of the peripheral five-vowel test [a:→e:→i:→o:→u:] will be used:

- logVSA: decimal logarithm of the Vowel Space Area, as defined in [10,11].
- FCR: Formant Centralization Ratio, as defined in [11,12].
- VAI: Vowel Articulation Index, the reciprocal of FCR.
- $F_{2s} = F_{2i}/F_{2u}$: High vowel second-formant span (range of the second formant positions from [i:] to [u:].
- $\Delta F_{1N} = \Delta F_1/F_{1\mu}$: Normalized first-formant span (first formant range with respect to the vowel triangle first formant center ($F_{1\mu} = mean\{F_{1i}\}$).
- $\Delta F_{2N} = \Delta F_2/F_{2\mu}$: Normalized first-formant span (first formant range with respect to the vowel triangle first formant center ($F_{2\mu} = mean\{F_{2i}\}$).
- $|\Delta F_{12N}| = \sqrt{\Delta F_{1N}^2 + \Delta F_{2N}^2}$: Modulus of Normalized first- and second-formant spans.
- AKV: Absolute Kinematic Velocity of the jaw-tongue joint, as defined in [13]. In vowel suites, as it is the case, a low value may be an indication of hypokinetic speech.

The evaluation of the above mentioned peripheral five-vowel test features relies on the accurate estimation of the first two formants, as described in [13].

The phonation quality features used in the present study were estimated from probability density functions of the different correlates listed below:

- $E_n$: Energy profile in the time domain, estimated from the Teager-Kaiser Energy Operator [14]. The most relevant feature is its dispersion, expected to manifest instabilities on articulation and phonation due to PD.
- $F_0$: Fundamental Frequency profile, estimated from the autocorrelation function [15]. Similarly, its dispersion is its most relevant statistics, due to PD speech instability.
- HNR: Harmonic-Noise Ratio, expressing the proportion of the power spectrum contributed by pseudo-periodical phonation components with respect to turbulent components [16]. A low value may indicate the presence of airy voice due to weak phonation, typical in PD speech.
- CPP: Cepstral Peak Prominence. It is also related to the proportion of harmonic contents with respect to turbulent ones, although being estimated in the cepstral domain [17]. The larger its value, the better phonation quality.
- $T_\vartheta$: Phonation Tremor in the $\vartheta$ band. A large value may indicate phonation instability associated to neurological tremor, which is estimated from the band-filtered vocal fold biomechanical strain [18]. In the present study, the tremor aligned with the $\vartheta$ band is used.

The peripheral five-vowel space and phonation quality features are to be confronted in a longitudinal basis with respect to the first estimation available using Jensen-Shannon Divergence (JSD), a measurement of the mutual information contents between two feature probability density functions (PDFs) evaluated from feature amplitude histograms. JSD is bound between 0 (low divergence, meaning large common information contents between objective distributions) and 1 (maximum divergence, low common information contents). As there is no indication of the divergence sense, a Log-Likelihood Ratio (LLR) between both PDFs is also evaluated [19], which may take a positive value if the distribution under test is more disperse that the reference distribution, or negative otherwise.

## 3 Results

An example of the peripheral five-vowel test [a:→e:→i:→o:→u:] from a female PD participant (FP11) is presented in Fig. 1 for reference. The total duration of the test is 10 s, which corresponds roughly to 2 s per vowel. Templates c) and d) present the projection of the first two formant estimations on a cartesian plot and on a linguistic plot (Formant axes swapping in reverse ordering).

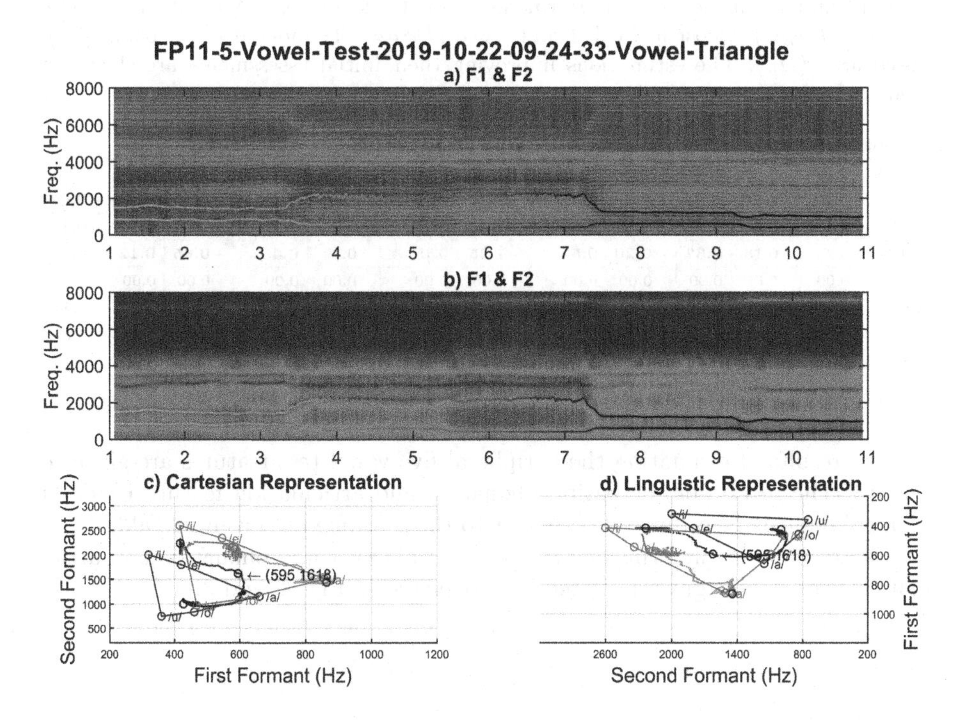


Fig. 1. Fourier and LPC spectrograms from the peripheral five-vowel test [a:→e:→i:→o:→u:]. The first two formants are superimposed in time-color coding (cyan to red symbolize the earlier to latter instants): a) Fourier Spectrogram; b) LPC Spectrogram; c) Cartesian Vowel Plot; d) Linguistic Vowel Plot. (Color figure online)

**Table 3.** Vowel Space features considered in the longitudinal evaluation: decimal logarithm of the Vowel Space Area (logVSA); Formant Centralization Ratio (FCR); Vowel Articulation Index (VAI); Ratio between the vertices of the second formant values of [i:] and [u:] ($F_{2s}$); First Formant Normalized Span ($\Delta F_{1N}$); Second Formant Normalized Span ($\Delta F_{2N}$); Modulus of the Normalized Span ($|\Delta F_{12N}|$). The estimations improving their initial assessments are shown in bold.

| Record | logVSA | FCR | VAI | $F_{2s}$ | $\Delta F_{1N}$ | $\Delta F_{2N}$ | $\lvert\Delta F_{12N}\rvert$ |
|---|---|---|---|---|---|---|---|
| MP11 | 5.61 | 0.70 | 1.42 | 3.24 | 1.24 | 1.32 | 1.81 |
| MP12 | 5.38 | 0.80 | 1.25 | **3.43** | 0.63 | 1.27 | 1.41 |
| MP13 | 5.52 | 0.80 | 1.26 | **3.91** | 0.90 | 1.31 | 1.59 |
| MP14 | 5.45 | 0.83 | 1.21 | **3.40** | 0.78 | 1.15 | 1.39 |
| FP11 | 5.44 | 1.06 | 0.94 | 2.25 | 0.75 | 0.78 | 1.08 |
| FP12 | 5.36 | **1.04** | **0.96** | **2.69** | 0.58 | **0.94** | **1.10** |
| FP13 | **5.47** | **1.00** | **1.00** | **2.43** | **0.80** | **0.89** | **1.17** |
| FP14 | **5.50** | **1.00** | **1.01** | **2.71** | 0.71 | **0.89** | **1.14** |

**Table 4.** Kinematic features considered in the longitudinal evaluation: Jensen-Shannon Divergence (D) and Log-Likelihood Ratio estimated on the Energy Profile ($D_{En}$), Fundamental Frequency ($D_{F0}$), Harmonic-Noise Ratio ($D_{HN}$), Absolute Kinematic Velocity ($D_{AKV}$), Cepstral Peak Prominence ($D_{CPP}$), Tremor on the Z-band ($D_{\vartheta}$); Fused JSD ($D_F$). The estimations improving their initial assessments are shown in bold.

| Record | $D_{En}$ | $\lambda_{En}$ | $D_{F0}$ | $\lambda_{F0}$ | $D_{HNR}$ | $\lambda_{HNR}$ | $D_{AKV}$ | $\lambda_{AKV}$ | $D_{CPP}$ | $\lambda_{CPP}$ | $D_{\vartheta}$ | $\lambda_{\vartheta}$ |
|---|---|---|---|---|---|---|---|---|---|---|---|---|
| MP11 | 0.00 | 0.00 | 0.00 | 0.00 | 0.00 | 0.00 | 0.00 | 0.00 | 0.00 | 0.00 | 0.00 | 0.00 |
| MP12 | **0.21** | **−0.19** | **0.21** | **−0.02** | **0.61** | **−1.46** | 0.22 | 0.28 | **0.28** | **−0.19** | **0.13** | **−0.07** |
| MP13 | 0.19 | 0.05 | **0.23** | **−0.14** | **0.58** | **−1.46** | 0.24 | 0.30 | **0.17** | **−0.08** | **0.16** | **−0.12** |
| MP14 | 0.23 | 0.13 | **0.32** | **−0.20** | **0.54** | **−1.25** | 0.15 | 0.19 | **0.48** | **−0.65** | **0.12** | **−0.05** |
| FP11 | 0.00 | 0.00 | 0.00 | 0.00 | 0.00 | 0.00 | 0.00 | 0.00 | 0.00 | 0.00 | 0.00 | 0.00 |
| FP12 | **0.22** | **−0.03** | **0.27** | **−0.20** | **0.26** | **−0.40** | 0.16 | 0.20 | **0.15** | **−0.19** | 0.34 | 0.24 |
| FP13 | **0.16** | **−0.32** | **0.44** | **−0.72** | **0.38** | **−0.69** | **0.06** | **−0.03** | **0.36** | **−0.44** | 0.30 | 0.21 |
| FP14 | **0.18** | **−0.39** | **0.38** | **−0.57** | **0.50** | **−1.21** | 0.14 | 0.16 | **0.11** | 0.11 | 0.16 | 0.13 |

The results of evaluating the peripheral five-vowel test features are given in Table 3. The results of comparing phonation and articulation feature PDFs of each subsequent recording with respect to the first one are given in Table 4.

The results of comparing the PDFs of peripheral vowel phonation and articulation features are given in Table 4 are depicted in Figs. 2 and 3.

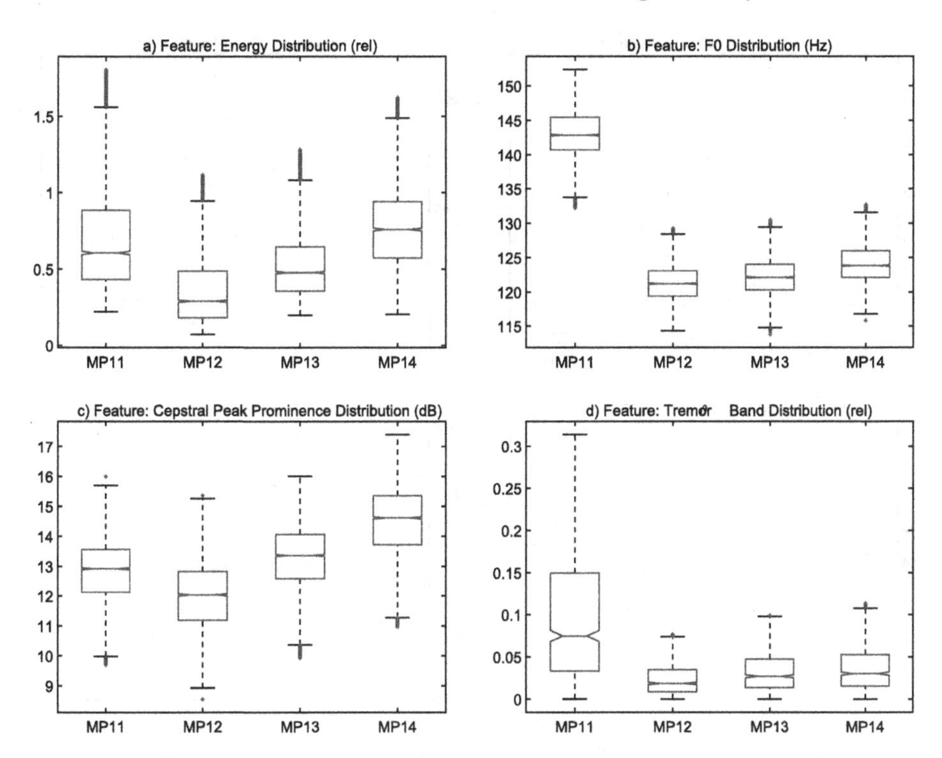


**Fig. 2.** Longitudinal evolution of selected feature distributions, male participant's (MP1) four recordings: a) Relative Energy Profile; b) Fundamental Frequency (F0); c) Cepstral Peak Prominence (CPP); d) Tremor on the $\vartheta$-Band.

# 4 Discussion

It may be seen in Fig. 1 that the strongest dynamic activity from the utterance of the peripheral five-vowel test sequence is to be expected in the transitions [a:→e:] and [i:→o:] as a result of strong jumps on the second formant, and to a lesser level, on the first formant. These jumps will affect mainly the features $D_{AKV}$ and $\lambda_{AKV}$. The utterance is well inscribed within the average female vowel triangle (in purple). The feature estimations corresponding to this recordig are given in Tables 3 and 4 under the horizontal entry FP11. Table 3 presents results from the male and female cases, which show important differences between them. The only feature from the male participant expressing a potential improvement is the second formant span ratio $F_{2s}$ (larger in samples MP12, MP13 and MP14 with respect to MP11). Samples from the female case show improvements in logVSA (FP13 and FP14), FCR and VAI (this last feature being the more normal the closer to 1), $F_{2s}$, and $|\Delta F_{12N}|$. The results presented in Table 4 correspond to scores from longitudinal comparisons between each first recording quality feature with respect to subsequent recordings. One of the comparisons is done in terms of JSDs, and the second one is done in terms of LLRs. The JSD is propor-

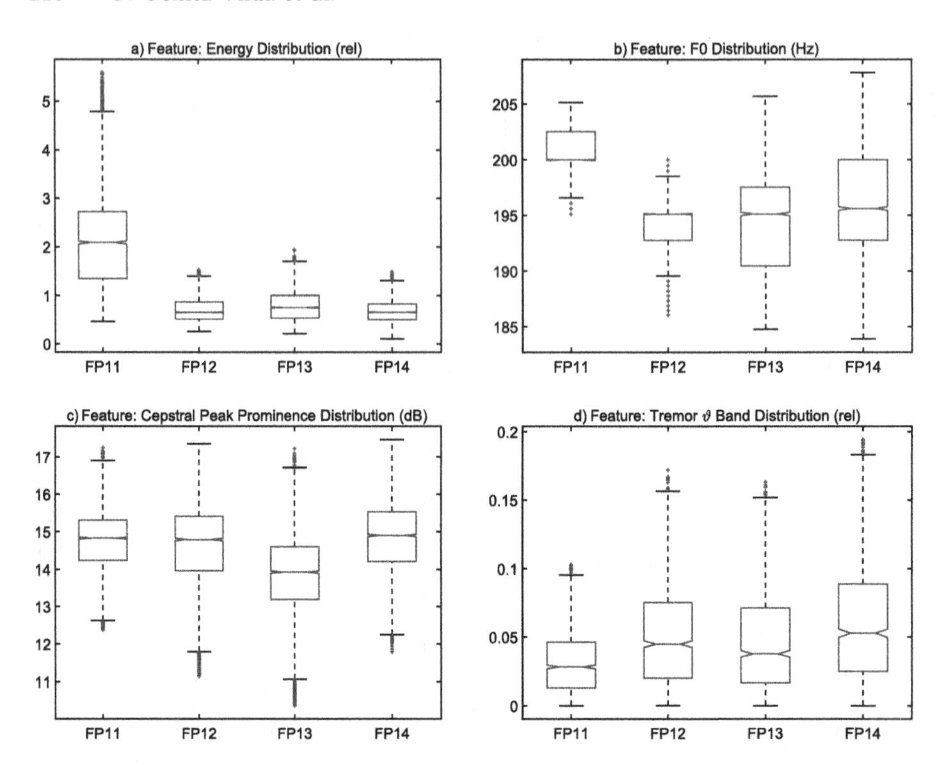


**Fig. 3.** Longitudinal evolution of selected feature distributions, female participant's (FP1) four recordings: a) Relative Energy Profile; b) Fundamental Frequency (F0); c) Cepstral Peak Prominence (CPP); d) Tremor on the $\vartheta$ Band.

tional in absolute terms to the divergence between each sample feature PDF with respect to the initial sample. The LLR is either positive or negative depending on the sense of the comparison. In general, positive values might indicate that the PDF under test is more disperse (larger standard deviation) than the PDF of reference (the initial sample), and vice versa. In general, it is expected that the energy and fundamental frequency profiles ($E_n$ and $F_0$) will show more dispersion from unstable dysfunction alterations due to PD. Therefore, large JSD and negative LLR would indicate an improved behavior. The results in Table 4 are also presented in Figs. 2 and 3 for a better understanding of underlying comparisons. The features presented in Fig. 2 (male participant) reveal the following behavior:

- The $E_n$ profile shows a reduction on the value of the energy, and a slight reduction on dispersion.
- The profile of $F_0$ shows a reduction on the value of the fundamental frequency, which may be taken as an indication of a less tense fonation.
- The CPP is larger in the last two evaluations (MP13 and MP14), which implies a better phonation quality.

- The tremor in the $\vartheta$ band is clearly much smaller and less disperse, this being a clear indication of a less unstable phonation.

On its turn, the features presented in Fig. 3 (female participant) show the following behavior:

- The profile of $E_n$ expresses a reduction in the value of the energy, which is a relative magnitude. More significantly, the dispersion of the profiles is clearly smaller in FP12-14, which is an indication of a more stable phonation.
- The CPP does not change much in sessions FP12-14.
- The tremor in the $\vartheta$ band does not show important improvements, on the contrary, it shows worsening conditions for utterances FP12-14, both in the average value and in dispersion.

## 5 Conclusions

From what has been exposed in the previous tables and figures, it may be concluded that the potential benefits of neuroacoustic stimulation are not equally present in both cases studied, where certain features express positive effects in the male case ($F_2s$, $F_0$, HNR, CPP and $T_\vartheta$), whereas in the female case the most positive effects are seen in VAI, $F_2s$, $|\Delta F_{12N}|$, $E_n$, $F_0$, HNR, and CPP (in a lesser intensity). Whether this behavioral difference is due to recording conditions, phonation quality, or stimulation protocol vulnerability, is a question which will need a further study extending the examination of results to other participants. Substantially, these limitations hamper the relevance of the present study, which is planned to inspect wider databases, including also HC and PD participants and other diadochokinetic exercises. The results of the study may help in the design of fast and accurate characterization methods for PD speech in developing application-specific monitoring for clinical purposes.

## References


1. Tysnes, O.-B., Storstein, A.: Epidemiology of Parkinson's disease. J. Neural Transmission **124**(8), 901–905 (2017). https://doi.org/10.1007/s00702-017-1686-y
2. Duffy, J.R.: Motor Speech Disorders: Substrates, Differential Diagnosis, and Management, 3rd edn., Elsevier (2013)
3. Hlavnička, J., Čmejla, R., Tykalová, T., Šonka, K., Růžička, E., Rusz, J.: Automated analysis of connected speech reveals early biomarkers of Parkinson's disease in patients with rapid eye movement sleep behaviour disorder. Sci. Rep. **7**(1), 1–13 (2017). https://doi.org/10.1038/s41598-017-00047-5
4. Gálvez, G., Recuero, M., Canuet, L., Del-Pozo, F.: Short-term effects of binaural beats on EEG power, functional connectivity, cognition, gait and anxiety in Parkinson's disease. Int. J. Neural Syst. **28**(5), 1750055 (2018). https://doi.org/10.1142/S0129065717500551

5. Gálvez-García, G., Gómez-Rodellar, A., Palacios-Alonso, D., de Arcas-Castro, G., Gómez-Vilda, P.: Neuroacoustical stimulation of Parkinson's disease patients: a case study. In: Ferrández Vicente, J.M., Álvarez-Sánchez, J.R., de la Paz López, F., Toledo Moreo, J., Adeli, H. (eds.) IWINAC 2019. LNCS, vol. 11487, pp. 329–339. Springer, Cham (2019). https://doi.org/10.1007/978-3-030-19651-6_32
6. Gálvez, G., Gómez, A., Palacios, D.: Temporal reversion of phonation instability in Parkinson's disease by neuroacoustical stimulation. In Models and Analysis of Vocal Emissions for Biomedical Applications, MAVEBA 11th Int. Workshop, Firenze University Press, pp. 21–24 (2019)
7. Hink, R.F., Kodera, K., Yamada, O., Kaga, K., Suzuki, J.: Binaural interaction of a beating frequency-following response. Audiology **19**(1), 36–43 (1980). https://doi.org/10.3109/00206098009072647
8. Oster, G.: Auditory beats in the brain. Sci. Am. **229**(4), 94–102 (1973)
9. Palacios, D., Meléndez, G., López, A., Lázaro, C., Gómez, A., Gómez, P.: MonParLoc: a speech-based system for Parkinson's disease analysis and monitoring. IEEE Access **8**, 188243–188255 (2020). https://doi.org/10.1109/ACCESS.2020.3031646
10. Kent, R., Kim, Y.: Toward an acoustic typology of motor speech disorders. Clin. Linguistics Phonetics **17**, 427–445 (2003). https://doi.org/10.1080/0269920031000086248
11. Sapir, S., Ramig, L.O., Spielman, J.L., Cynthia Fox, C.: Formant centralization ratio: a proposal for a new acoustic measure of dysarthric speech. J. Speech, Lang. Hearing Res. **53**(1), 114–125 (2010). https://doi.org/10.1044/1092-4388(2009/08-0184)
12. Skodda, S., Visser, W., Schlegel, U.: Vowel articulation in Parkinson's disease. J. Voice **25**(4), 467–472 (2011). https://doi.org/10.1016/j.jvoice.2010.01.009
13. Gómez, A., Tsanas, A., Gómez, P., Palacios-Alonso, D., Rodellar, V., Álvarez, A.: Acoustic to kinematic projection in Parkinson's Disease Dysarthria. Biomed. Signal Process. Control **66** (2021). https://doi.org/10.1016/j.bspc.2021.102422
14. Gómez, P., Gómez, A., Palacios, D., Tsanas, A.: Performance of Monosyllabic vs Multisyllabic Diadochokinetic exercises in evaluating Parkinson's Disease Hypokinetic Dysarthria from fluency distributions. Proc. BIOSTEC **2021**(4), 114–123 (2021). https://doi.org/10.5220/0010380301140123
15. Tsanas, A., Zañartu, M., Little, M.A., Fox, C., Ramig, L.O., Clifford, G.D.: Robust fundamental frequency estimation in sustained vowels: detailed algorithmic comparisons and information fusion with adaptive Kalman filtering. J. Acoustical Soc. Am. **135**(5), 2885–2901 (2014). https://doi.org/10.1121/1.4870484
16. Qi, Y., Hillman, R.E.: Temporal and spectral estimations of harmonics-to-noise ratio in human voice signals. J. Acoustical Soc. Am. **102**, 537 (1997) https://doi.org/10.1121/1.419726
17. Heman-Ackah, Y.D., Michael, D.D., Jr Goding, G.S.: The relationship between cepstral peak prominence and selected parameters of dysphonia. J. Voice **16**(1), 20–27 (2002). https://doi.org/10.1016/S0892-1997(02)00067-X
18. Gómez-Vilda, P., et al.: Parkinson's disease monitoring from phonation biomechanics. In: Ferrández Vicente, J.M., Álvarez-Sánchez, J.R., de la Paz López, F., Toledo-Moreo, F.J., Adeli, H. (eds.) IWINAC 2015. LNCS, vol. 9107, pp. 238–248. Springer, Cham (2015). https://doi.org/10.1007/978-3-319-18914-7_25
19. Staude, G.H.: Precise onset detection of human motor responses using a whitening filter and the log-likelihood-ratio test. IEEE Trans. Biomed. Eng. **48**(11), 1292–1305 (2001). https://doi.org/10.1109/10.959325

# Characterizing Masseter Surface Electromyography on EEG-Related Frequency Bands in Parkinson's Disease Neuromotor Dysarthria

Andrés Gómez-Rodellar[1], Pedro Gómez-Vilda[2(✉)], José Manuel Ferrández-Vicente[3], and Athanasios Tsanas[1]

[1] Faculty of Medicine, Usher Institute, University of Edinburgh, Edinburgh, UK
{a.gomezrodellar,athanasios.tsanas}@ed.ac.uk

[2] NeuSpeLab, CTB, Universidad Politécnica de Madrid, 28220 Pozuelo de Alarcón, Madrid, Spain
pedro.gomezv@upm.es

[3] Universidad Politécnica de Cartagena, Campus Muralla del Mar, Pza. Hospital 1, 30202 Cartagena, Spain
jm.ferrandez@upct.es

**Abstract.** Speech has proven to be an effective neuromotor biomarker, capitalizing on the capabilities of contact-free technology. This study aims to evaluate the behavior of facial muscles' activity estimating the entropy of their surface electromyographic (sEMG) activity during the production of diadochokinetic speech tests. The study explores the entropic behavior of the sEMG signal in certain frequency bands associated to EEG activity comparing participants affected by neuromotor diseases than in age-matched normative participants. Using recordings from two PD vs two HC participants on 5 EEG bands $(\delta, \vartheta, \alpha, \beta, \gamma)$, the maximum entropy estimated on the HC group was 5.70 $10^{-5}$, whereas the minimum entropy on the PD group was 7.25 $10^{-5}$. A hypothesis test rejected the similarity between the PD and HC results with a p-value under 0.0003. This different behavior might open the way to a wider study in characterizing neuromotor disease alterations from neuromotor origin.

**Keywords:** Entropy · EEG · Surface electromyography · Neuromotor diseases · Hypokinetic dysarthria · Parkinson's Disease

## 1 Introduction

Neurological diseases are rated third among health disorders resulting in disability and premature death within the European Union [1], Parkinson's Disease

This research received funding from grants TEC2016-77791-C4-4-R (Ministry of Economic Affairs and Competitiveness of Spain), and Teca-Park-MonParLoc FGCSIC-CENIE 0348-CIE-6-E (InterReg Programme). The authors wish to thank Víctor Lorente for his inspiring thoughts (School of Veterinary, UCM, Spain).


J. M. Ferrández Vicente et al. (Eds.): IWINAC 2022, LNCS 13258, pp. 219–228, 2022.
https://doi.org/10.1007/978-3-031-06242-1_22

(PD) rated as the most predominant neuromotor disorder, with a prevalence rate of 105 cases per 100,000 individuals, an incidence rate of 13 cases, per 100,000 individuals, and an average number of Years Living with Disability (YLDs rate) of 15 years. Characteristic PD symptoms include rigidity, tremor, and general progressive loss of motor control to the point that people with PD (PwP) require continuous assistance in their daily life activity during the late stages of the disorder. Additionally, they often exhibit other non-motor symptoms which may hamper considerably their life quality. Speech is one of the motor symptoms most altered by PD, dysarthria developing at some point during the disease course in about 90% of people with PD (PwP) [2]. Speech involves the respiration, phonation, articulation, and their associated premotor activation systems. Each one of those key physiological mechanisms may be affected in different degrees, therefore, correlates of PD-induced alterations may be found in the analysis of electroencephalography (EEG) frequency bands related with the activity of premotor areas [3].

The aim of the present research is to provide insights into the use of entropy estimations from surface electromyograpy (sEMG) on the masseter on EEG-related frequency bands to characterize PD dysarhria by building on previous work [4] relating some speech physiological mechanisms involved in the articulation gestures of certain diadochokinetic tests with sEMG. It is well known that sEMG is the signal produced by group discharges following the activation of neuromotor units on muscle fibers [5]. Although sEMG reflects groupal activity of multiple fibers, there is some informed opinion on the possibility of finding meaningful correlates with the groupal activity of neuromotor cortical areas related with the activation of the specific muscles involved, especially relevant in the field of neuromotor rehabilitation [3]. In this sense, it is thought that the masseter sEMG could show some features specific of EEG activity in the same frequency bands regarding cortico-muscular coupling, as EEG-EMG coherence showing functional connectivity in limb movement was clearly observed [6] and is being used for rehabilitation purposes [7].

The fact that the movement of the jaw-tongue biomechanical joint is very much conditioned by the activity of the masseter can be used to establish clear relationships between acoustical features as formants, and the neuromotor activity of the masseter by means of inverse projection methods [8]. Therefore, a clear connection may be established between speech acoustics and EEG cortical activity by the interplay of sEMG. Thus, a description of sEMG in EEG frequency bands could offer some insight onto coherence and causality of cortical oscillations and movement.

On its turn, the cortico-muscular relationship between neuromotor activity and limb movement has been quite clearly established by studies on Brain-Computer Interfaces (BCI), and the most relevant EEG-frequency bands associated have been determined [9–12]. A similar relationship is likely expected between cortico-facial, glossal and laryngeal systems, responsible for speech phonation and articulation [13,14].

It is well known that entropy is strongly related with information contents of a given stochastic process. A possible characterization of the sEMG signal could be provided in terms of its entropy description, using different approximations, as Renyi's [15], approximate and sample entropy [16], multiscale [17], or transfer entropy [18], to estimate causality relationships among bands' activity. In the present research simple entropy will be used on an exploratory study.

The main hypothesis to be treated in the present research is the possibility of successfully distinguishing PD from HC speech articulation from the utterance of a sequence of five peripheral vowels in terms of the EEG-related frequency bands' entropy from the surface myoelectrical signal measured on the masseter.

The paper is organized as follows: A description of the Information Theory methods based on Entropy contents proposed in the study is provided in Sect. 2, explaining the influence of neuromotor disorders in the statistical properties of amplitude Probability Density Functions (PDFs) of EEG-related frequency bands found in the sEMG signals. The data set used in the research is also described in Sect. 2. Section 3 is devoted to present the results of processing the recordings from the four participants in the study in terms of their PDFs, and in their entropy contents, which are compared graphically and in tables. Section 4 is intended to discuss the implications of the comparisons, analyze their possible clinical applications, and remark the study limitations and its possible future extension. Section 5 summarizes the main contributions, findings and conclusions.

## 2 Experimental Framework

### 2.1 Methods

An experimental framework has been devised to test the relative effects of HD by means of the recording of the sEMG signal produced by the participants during the utterance of a fast repetition of the dyphthong [a → i]. The main features considered are the EEG-equivalent frequency bands in the sEMG signal ($\delta, \vartheta, \alpha, \beta, \gamma$, and $\mu$). The working hypothesis is that these characteristic sEMG bands might be considered associated to EEG activity in the motor cortex when inducing speech related neuromotor discharges on the masseter. The methodology used in the study is based on the estimation of the EEG-related bands by the following procedures:

- The sEMG signal was recorded on a Biopac MP150 EMG100 platform at 2 kHz and 16 bits. The fixture to record sEMG from the masseter is shown in Fig. 1.
- A notch filter tuned 50 Hz and its higher harmonics was used for electric power artifact removing.
- EEG-related band filtering was implemented on the notch-filtered unbiased (zero-mean) sEMG signal, $s_n$ ($n$ being the discrete time index) using 6-order bandpass filters $F_i\{\cdot\}$ following (1), tuned to each one EEG band:
  i = 1, $\delta$: $0 \leq f < 4$ Hz;

i = 2, $\vartheta$: 4 ≤ f < 8 Hz;
i = 3 $\alpha$: 8 ≤ f < 16 Hz;
i = 4, $\beta$: 16 ≤ f < 32 Hz;
i = 5, $\gamma$: 32 ≤ f < 64 Hz;
i = 6, $\mu$: 8 ≤ f < 12 Hz).

- The amplitude histogram $h_i$ of EEG-related band $i$ is built on the filtered sEMG signal in band $i$ $s_{i,n}$, following (2), where $b_k = k\delta_b$ and $b_{k-1} = (k-1)\delta_b$ are the k-th bin limits for the bin index $1 \leq k < K$ and the bin size $\delta_b$, $W$ being the time window considered.
- The probability density function $p_{i,k}$ of band $i$, is estimated from its normalized histogram $h_{i,k}$ following expression
- The entropy of band $i$ is estimated following expression (4).

$$s_{i,n} = F_i\{s_n\} \tag{1}$$

$$\forall n \in W \Rightarrow \quad h_{i,k} = \begin{cases} h_{i,k-1} + 1 & b_{k-1} \leq s_{i,n} < b_k; \\ h_{i,k-1} & otherwise \end{cases} \tag{2}$$

$$p_{i,k} = h_{i,k} / \sum_{k=1}^{K} h_{i,k} \tag{3}$$

$$E_i = -\sum_{k=1}^{K} p_{i,k} log_{10}(p_{i,k}) \tag{4}$$

The band entropy $E_i$ is estimated by a normalized 400-bin amplitude histogram of frequency band signals $s_{i,n}$, where i is the band index. Smooth approximations of the band PDFs $p_{i,k}$ may be produced using the Kolmogorov-Smirnov fit [19].

### 2.2 Materials

The behavior in terms of EEG frequency band entropy contents was assessed on stable vowel production (minialterations due to poor or unstable articulation) and on continuous gliding diadochokinetic voice production.

The specific articulation test used in the present study because it fulfills both conditions consisted in the sustained utterance of the peripheral five-vowel sequence test [a: → e: → i: → o: → u:]. Results from recordings by two PD participants (a male 69 year-old and a female 70 year-old, respectively) from the APARKAM association of PD patients were analyzed following the methods described in expressions (1–4). The results from these recordings were compared to similar tests against age- and gender-matched normative HC participants (69 and 62 years-old). The recording protocol was approved by the Ethical Committee of UPM (MonParLoc, 18/06/2018). The voluntary participants were informed about the experiments to be conducted, the protection of personal data,

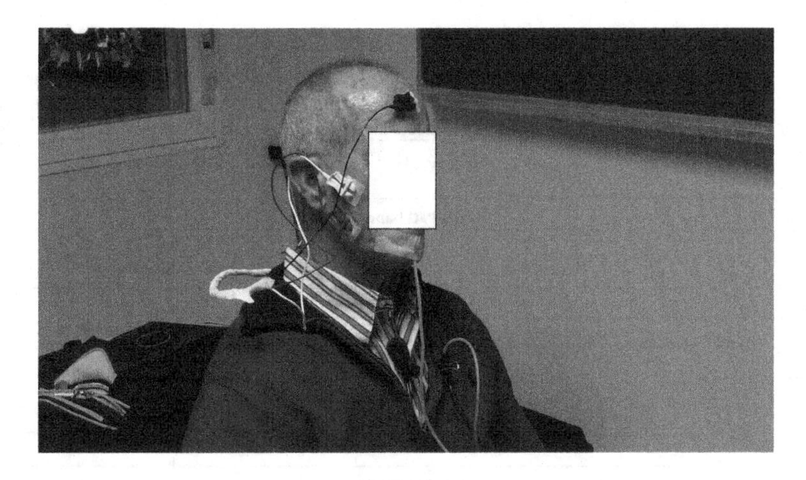

**Fig. 1.** Recording fixture of sEMG on the masseter, accelerometry, and speech from a male PD participant (MP1).

and signed an informed consent form. The methodology was strictly aligned with the Declaration of Helsinki. The biometrical data of the participants are given in Table 1.

**Table 1.** Participants' biometrical data. MC: male control participants; MP: male PD participants; FC: female control participants; FP: female PD participants; H&Y Hoehn and Yahr PD rating scale; State: medication state (on: under medication; –: not applicable)

| Code | Gender | Age | Condition | H &Y | State |
|---|---|---|---|---|---|
| MC1 | M | 69 | – | – | – |
| FC1 | M | 62 | – | – | – |
| MP1 | M | 69 | PD | 2 | ON |
| FP1 | M | 70 | PD | 2 | ON |

## 3 Results

The utterance recordings of the peripheral vowel sequence test [a: → e: → i: → o: → u:] from each participant were processed to obtain the frequency band contents described in Subsect. 2.1. An example of one such recording produced by the male HC participant MC1 is shown in Fig. 2.

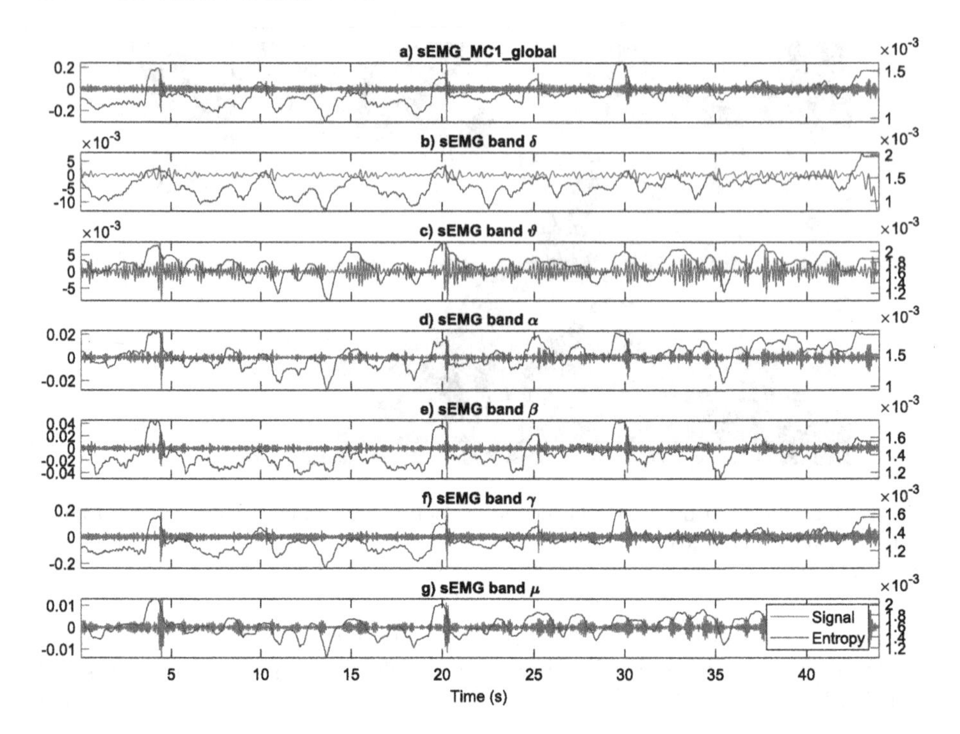


**Fig. 2.** Example of sEMG contents in terms of EEG-related frequency bands during the utterance of the peripheral vowel sequence test [a: → e: → i: → o: → u:] by the HC male participant (MC1): a) sEMG signal; b) $\delta$ band; c) $\vartheta$ band; d) $\alpha$ band; e) $\beta$ band; f) $\gamma$ band; g) $\delta$ band; (blue: sEMG band component, red: short time entropy). (Color figure online)

The PDFs of each frequency band from the four participants was estimated accordingly to expressions (2–3) on 400 bin amplitude histograms over the whole band window, and are presented for reference in Fig. 3.

The entropy associated to each PDF in Fig. 3 is estimated accordingly to expression (4). The result of these estimations are presented in Table 2.

For the sake of a better interpretability the band entropies per participant are also depicted in Fig. 4.

**Table 2.** Entropy of sEMG activity of participants in the different EEG-Bands.

| Participant | Global | $\delta$ | $\theta$ | $\alpha$ | $\beta$ | $\gamma$ | $\mu$ |
|---|---|---|---|---|---|---|---|
| MC1 | 4.56E−05 | 5.05E−05 | 5.70E−05 | 5.13E−05 | 4.88E−05 | 4.67E−05 | 5.40E−05 |
| FC1 | 4.21E−05 | 3.42E−05 | 5.16E−05 | 5.29E−05 | 4.84E−05 | 4.24E−05 | 5.53E−05 |
| MP1 | 7.49E−05 | 7.72E−05 | 7.58E−05 | 7.28E−05 | 7.25E−05 | 7.88E−05 | 7.55E−05 |
| FP1 | 1.17E−04 | 1.23E−04 | 1.20E−04 | 1.22E−04 | 1.17E−04 | 1.14E−04 | 1.22E−04 |

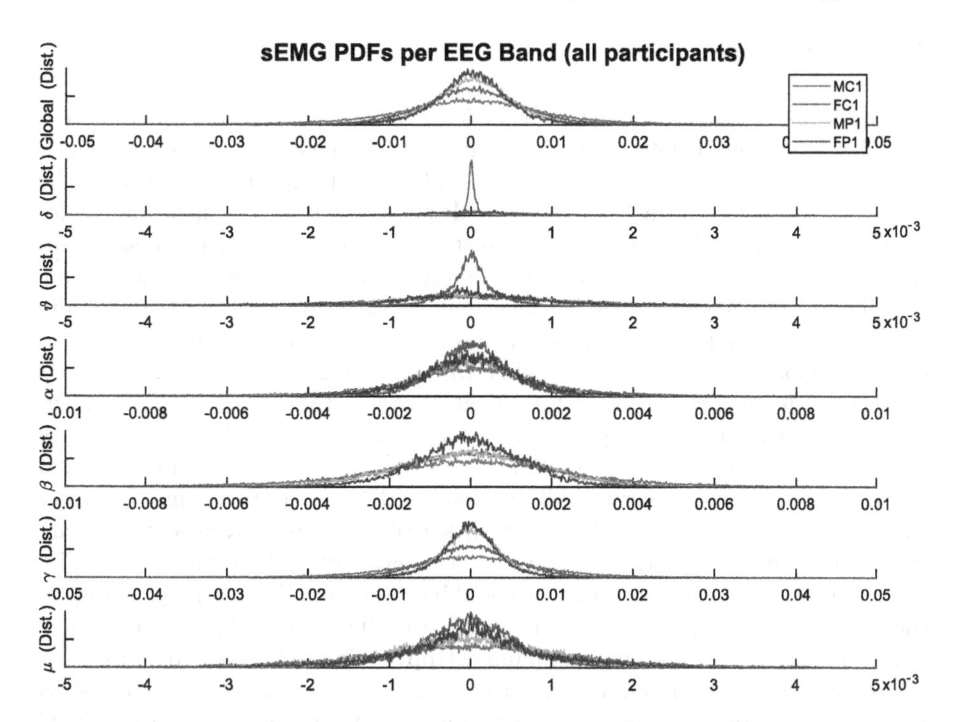


**Fig. 3.** Probability distributions of the sEMG amplitude in the EEG-related bands from the participants: HC male (MC1: blue) and female (FC1: red), and PD male (MP1: orange) and female (FP1: purple). PDFs of the different bands from top to bottom: whole unbiased notch-filtered sEMG; $\delta$; $\vartheta$; $\alpha$; $\beta$; $\gamma$; $\mu$. (Color figure online)

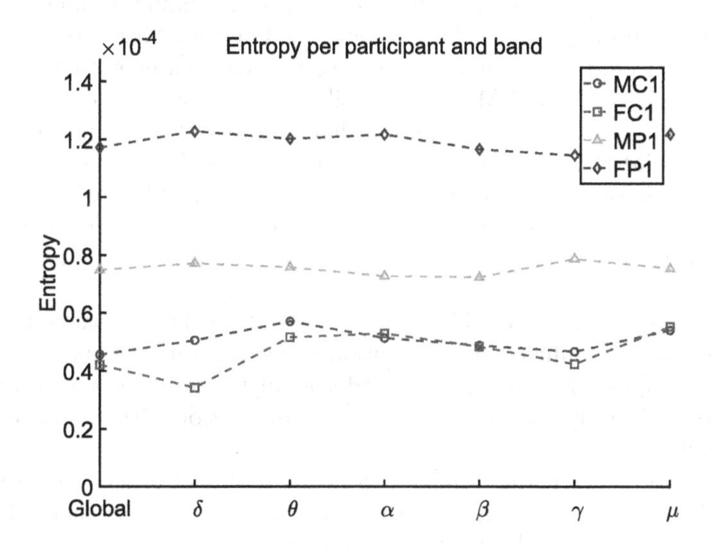


**Fig. 4.** Entropy of sEMG activity in the EEG-related bands from the participants: HC male (MC1: blue circles) and female (FC1: red squares), and PD male (MP1: orange triangles) and female (FP1: purple diamonds). (Color figure online)

## 4 Discussion

An important feature of normative behavior is dispersion. As the frequency band sEMG signals have been unbiased to remove their means, it is expected them to be centered with respect to their amplitude axes (horizontal). As it might be expected, the distributions showing more dispersion will spread their amplitudes over a larger span. The more stable a distribution is, the lower dispersion, and the smaller its associated entropy. This is the case of FC1, which shows a large stability on the $\delta$ and $\vartheta$ bands. The analysis of entropy by participant and band given in Table 2 and Fig. 3 give also a general idea on the behavior of each participant's neuromotor stability through the dispersion of their sEMG activity on the masseter. The first observation is that there is a strong alignment between the two normative participants (MC1 and FC1), which repeat a very similar and stable pattern. The minimum entropy is estimated for FC1 in the $\delta$ band, associated with a very small dispersion. MP1, on its turn, shows a larger entropy in all the examined bands, with a really stable pattern among them. At a farther distance, corresponding to extremely large values, participant FP1 shows the worst results in terms of instability (maximum entropy), also in all the bands considered. Comparing the results the PD vs the HC participants on 5 EEG bands $(\delta, \vartheta, \alpha, \beta, \gamma)$, the maximum entropy estimated on the HC group was 5.705E−05, whereas the minimum entropy on the PD group was 7.247E−05. A hypothesis test rejected the similarity between both groups on the same EEG bands with a p-value under 0.0003.

Obviously, the results reported lack any statistical relevance, given the small size of the sample studied. This is due to the exploratory nature of the research, which has considered a minimum number of cases including normative and PD participants of both genres. If these results could be generalized to a larger sample size, they would open a new perspective to analyze a neuromotor symptom severity associated with sEMG as a possible diagnostic tool in terms of information theory contents. An interesting characteristic of this kind of feature as entropy, associated with information contents, is that it could open the study using mutual information measurements, as Kulback-Leibler or Jensen-Shanon divergence, as well as other entropy definitions, as Renyi's, sample and multiscale Entropy, or transfer entropy, to estimate causality relationships among bands' activity.

Given the limitations posed by the small size of the dataset studied, a further study is needed to confirm this assumption by using larger size databases of HC and PD participants including, other diadochokinetic exercises, and studying the use of these estimations in pathology severity evaluation with clinical enhanced explainability.

The results of the research may help in the design of other characterization methods for PD speech, essential in the design of application-specific PD speech monitoring approaches, also for clinical purposes.

## 5 Conclusions

The differential behavior of sEMG band entropy contents between PD with respect to HC participants is a most inspiring insight, as it could lead to open new possibilities of adding meaningful features acceptable for clinical application. As the semantics behind entropy is quite clear under the point of view of interpretability in terms of instability in functional behavior, it can be easily accepted under the clinical point of view. One important challenge for its direct applicability is its relatively complex protocol, as it requires fixing electrodes on the participants' face, which raises concerns about its acceptability. Nevertheless, if a direct association may be established between facial sEMG and recorded speech, showing reliable resolution and correlation, an interesting methodology could be open in the application of EEG-related frequency band contents through the vehicular use of speech, as a much more ubiquitous and simpler recording procedure.

## References

1. Deuschl, G., et al.: The burden of neurological diseases in Europe: an analysis for the Global Burden of Disease Study 2017. Lancet Public Health **5**, e551–e567 (2020). https://doi.org/10.1016/S2468-2667(20)30190-0
2. Duffy, J.R.: Motor Speech Disorders: Substrates, Differential Diagnosis, and Management, 3rd edn. Elsevier, St. Louis (2013)
3. Brambilla, C., Pirovano, I., Mira, R.M., Rizzo, G., Scano, A., Mastropietro, A.: Combined use of EMG and EEG techniques for neuromotor assessment in rehabilitative applications: a systematic review. Sensors **21**(21), 7014 (2021). https://doi.org/10.3390/s21217014
4. Gómez, P., et al.: Neuromechanical modelling of articulatory movements from surface electromyography and speech formants. Int. J. Neural Syst. **29**(02), 1850039 (2019). https://doi.org/10.1142/S0129065718500399
5. Cram, J.R.: The history of surface electromyography. Appl. Psychophys. Biofeedback **28**(2), 89–91 (2003). https://doi.org/10.1023/A:1023802407132
6. Baker, S.N., Olivier, E., Lemon, R.N.: Coherent oscillations in monkey motor cortex and hand muscle EMG show task-dependent modulation. J. Physiol. **501**, 225–41 (1997). https://doi.org/10.1111/j.1469-7793.1997.225bo.x
7. Krauth, R., et al.: Cortico-muscular coherence is reduced acutely post-stroke and increases bilaterally during motor recovery: a pilot study. Front. Neurol. **10**, 126 (2019). https://doi.org/10.3389/fneur.2019.00126
8. Gómez, A., Tsanas, A., Gómez, P., Palacios-Alonso, D., Rodellar, V., Álvarez, A.: Acoustic to kinematic projection in Parkinson's Disease Dysarthria. Biomed. Signal Process. Control **66**, 102422 (2021). https://doi.org/10.1016/j.bspc.2021.102422
9. Palmer, S.J., Lee, P.W.H., Wang, Z.J., Au, W.L., McKeown, M.J.: $\vartheta$, $\beta$ But not $\alpha$-band EEG connectivity has implications for dual task performance in Parkinson's disease. Park. Relat. Disord. **16**(6), 393–397 (2010). https://doi.org/10.1016/j.parkreldis.2010.03.001

10. Hülsdünker, T., Mierau, A., Neeb, C., Kleinöder, H., Strüder, H.K.: Cortical processes associated with continuous balance control as revealed by EEG spectral power. Neurosci. Lett. **592**, 1–5 (2015). https://doi.org/10.1016/j.neulet.2015.02.049
11. Ofner, P., Schwarz, A., Pereira, J., Müller-Putz, G.R.: Upper limb movements can be decoded from the time-domain of low-frequency EEG. PLoS ONE **12**(8), e0182578 (2017). https://doi.org/10.1371/journal.pone.0182578
12. Liu, J., Sheng, Y., Liu, H.: Corticomuscular coherence and its applications: a review. Front. Hum. Neurosci. **13**, 100 (2019). https://doi.org/10.3389/fnhum.2019.00100
13. Hickok, G.: A cortical circuit for voluntary laryngeal control: implications for the evolution language. Psychonomic Bull. Rev. **24**(1), 56–63 (2016). https://doi.org/10.3758/s13423-016-1100-z
14. Kent, R.D., Duffy, J.R., Slama, A., Kent, J.F., Clift, A.: Clinicoanatomic studies in dysarthria. J. Speech Lang. Hear. Res. **44**(3), 535–551 (2001). https://doi.org/10.1044/1092-4388(2001/042)
15. Cover, T.M., Thomas, J.A.: Elements of Information Theory, 2nd edn. Wiley, Hoboken (2006)
16. Delgado-Bonal, A., Marshak, A.: Approximate entropy and sample entropy: a comprehensive tutorial. Entropy **21**(6), 541 (2019). https://doi.org/10.3390/e21060541
17. Costa, M., Goldberger, A.L., Peng, C.K.: Multiscale entropy analysis of biological signals. Phys. Rev. E **71**(2), 021906 (2005). https://doi.org/10.1103/PhysRevE.71.021906
18. Vicente, R., Wibral, M., Lindner, M., Pipa, G.: Transfer entropy-a model-free measure of effective connectivity for the neurosciences. J. Comput. Neurosci. **30**, 45–67 (2011). https://doi.org/10.1007/s10827-010-0262-3
19. Bowman, A.W., Azzalini, A.: Computational aspects of nonparametric smoothing with illustrations from the SM library. Comp. Stat. Data Anal. **42**, 545–560 (2003). https://doi.org/10.1016/S0167-9473(02)00118-4

# Acquisition of Relevant Hand-Wrist Features Using Leap Motion Controller: A Case of Study

Carlos Rodrigo-Rivero[1], Carlos Garre del Olmo[1], Agustín Álvarez-Marquina[2], Pedro Gómez-Vilda[2], Francisco Domínguez-Mateos[1], and Daniel Palacios-Alonso[1,2](✉)

[1] Escuela Técnica Superior de Ingeniería Informática, Universidad Rey Juan Carlos, Campus de Móstoles, Tulipán, s/n, 28933 Móstoles, Madrid, Spain
daniel.palacios@urjc.es

[2] Neuromorphic Speech Processing Lab, Center for Biomedical Technology, Universidad Politécnica de Madrid, Campus de Montegancedo, 28223 Pozuelo de Alarcón, Madrid, Spain

**Abstract.** All the gestures and movements we make are influenced by our psychomotor abilities. This mobility deteriorates over the years. It is logical to think that an older individual has worse mobility than a younger one if they do not suffer from other pathologies. On this premise, the main aim of this research work is based on the detection of semantic biomechanical features, using a hand-tracking controller (Leap Motion) through three calibration tests. The data collected by this hand-tracker has allowed to measure hand-wrist movements. Likewise, this paper aims to highlight different tasks based on those used by neurologists customarily based on the Hoehn Yahr [11] and UPDRS scales [9]. Indeed, this study intends to visualize the differences between healthy participants. The manuscript provides some promising findings that will help tailoring biometric indicators for non-normative participants in future research using this technology.

**Keywords:** Hand-tracking · Leap motion controller · Hand-wrist movement · Parkinson

## 1 Introduction

Computer graphics have always sought ways to make visual information more realistic and accessible to the user. With this objective in mind, its use in the research aims to provide accurate and high quality virtual feedback. Indeed,

---

This research work was partly funded by one intramural project of Rey Juan Carlos University and a contract with the Spanish Defense Ministry (2022/00004/004 and 2021/00168/001, respectively).


J. M. Ferrández Vicente et al. (Eds.): IWINAC 2022, LNCS 13258, pp. 229–238, 2022.
https://doi.org/10.1007/978-3-031-06242-1_23

technological advances have allowed us to have more and more powerful processors and graphics. This enables a much greater computing and rendering capacity. Likewise, auxiliary technological resources such as motion tracking devices have been improving in parallel, creating branches of development with a strong impact on today's world, such as virtual reality (VR). These add-ons allow us to take our graphical perception to another level, complementing our graphical experience and improving it enormously. The types of interaction with the virtual environment through these devices is very varied and depends on the device we are using. There exist devices capable of capturing different degrees of freedom of human movement, depending on the application. Motion tracking can also be combined with haptic feedback for more realistic interaction. Various implementations of this class of devices can be found in the automotive sector, art, aviation, medicine, mobile devices, puzzles, computers, robotics, space, tactile electronic devices, teleoperators, simulators, neurorehabilitation, video games and virtual reality [15].

Considering computer graphics from a research point of view and focusing on motion tracking, multiple useful devices of interest are found. The choice of the device depends mainly on the degrees of freedom (DoFs) that needs to be tracked for each specific application. In some of them, it is enough to capture six DoFs for global hand position and orientation. For instance, Braz et al. perform a systematic review of the literature on the effectiveness of Nintendo Wii in the improvement of functional and health results of people with Parkinson's disease [3]. Their results suggest that the use of Wii seems effective to improve functional outcomes (balance, mobility, motor performance and independence) and health (reducing the risk of falls), the results being more consistent for the improvement of balance. When a reconstruction of full body skeletal DoFs is needed, devices such as Kinect are more appropriate. In [13], a comprehensive study with healthy individuals using Kinect 2 was carried out. During this study three objectives were addressed, the first one was based on determining the sensitivity for detecting impairments and gait, as well as to determine the age of the users. The second aim which pursued was to detect concurrent validation of users with stroke. Finally, the last one was intended to detect the possibility of falling after stroke.

Even though Kinect skeletal reconstrution can include the hands, there exists devices specifically designed for applications where the focus is only on tracking hand motion, such as the Leap Motion Controller (LMC) from Ultraleap Company [1]. This device tracks the user hands with optical sensors and infrared light, and do an internal reconstruction of the hands' skeleton. This reconstruction takes into account joint constraints and past information in order to provide output on every frame, even if some fingers are not clearly visible. The hand DoFs can be expressed in reduced coordinates when taking into account the different joint constraints (hinge between the phalanxes, and saddle in the knuckles), but LMC provides maximal coordinates' output (position and orientation of each bone) for easier integration in video games and virtual reality applications. Through its application programming interface (API) it is very easy to get a real time visual feedback of the hand with its representation of bones or rigging on the screen as shown in Fig 1.

Multiple predecessor studies have been carried on using the LMC, both technical and practical, in different areas such as sign language [7,16], writing [12,17,20], biomedicine [5,6,8,18], education [2,4,14] or training [21]. Taking into account the present research, it is based on the biomechanical branch using this device.

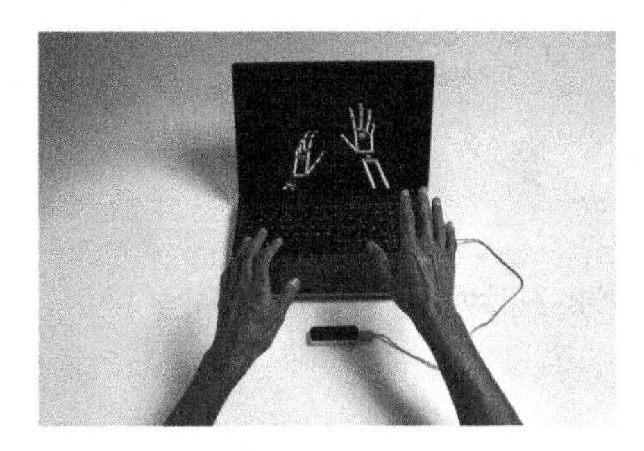

**Fig. 1.** Example of screenshot of Leap motion controller interface (source: Ultraleap webpage).

With this objective, the research group has developed three exercises that confirms this device as a powerful tool to monitor normative subjects hand activity.

The paper is organized as follows: Sect. 2 is devoted to describe the hand-tracking LMC device, objectives, methodology, corpus, technical description of the tests and the framework and hardware which is used in the study. Section 3 describes the results produced by the graphical analysis of the data collected. Finally, Sect. 4 summarizes the main findings and conclusions.

# 2 Methods and Materials

## 2.1 Leap Motion Controller

Leap motion controller (LMC) has optical sensors with infrared light capable of capturing hand motion with a field of 150 °C, with an optimal range of detection between 25 and 600 mm above the device sensors (which is considered to be the Y-axis in the LMC coordinate system) [19]. LMC uses a three-dimensional Cartesian coordinate system. The origin is centered at the upper middle point of the device, the X and Z axis are placed on the base of the device, the X axis exiting on the right side, the Z axis towards the user's position and, finally, the Y axis in the vertical, taking positive values above the tracker (see Fig. 2) [10]. Likewise, the API of LMC provides the developer with detailed information about the hand position, orientation and gesture of each detected hand by frame. The device is capable of detecting multiple hands, but it is optimized for only one or two hands (single user). Global motion output is provided (such as corresponding arm, palm position and velocity or pinch strength) together with detailed bone positions and orientations for every finger.

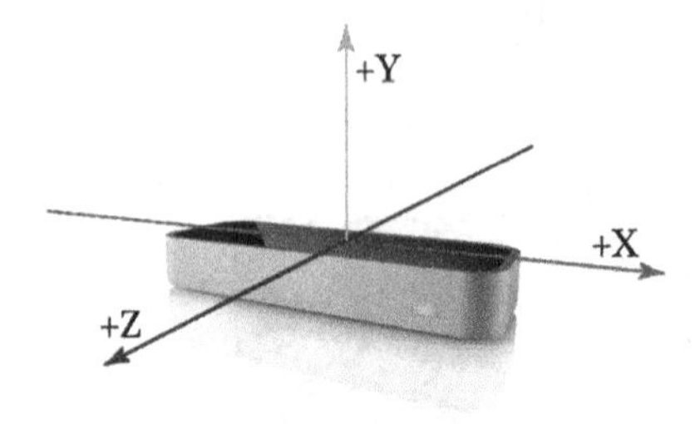


**Fig. 2.** Example of Gizmo 3D in LMC.

### 2.2 Objectives

As mentioned above, the main objective of this research work is aimed at obtaining as much information as possible that will allow to determine the biometrical characteristics of the user who performs each of the tasks. One of the main approaches is the detection of small gestures, details or patterns. To facilitate this task, the following age groups were established: Cluster A from 20 to 30 years-old, Cluster B from 31 to 50 years-old, and Cluster C from 51 to 75 years-old. Likewise, the aim is to capture the motor skills of individuals, such as mobility, precision, or neuromotor reaction in one hand against to the other. For this reason, detecting multiple hands allows a more reliable approximation of the state of both hands, or detects any anomaly movement in them.

### 2.3 Methodology

The study is based on three calibration tests focusing on biometric values of hands (see Fig. 3). The first screen requests personal data such as a nickname, age and gender. As it follows, the first test out three consists on holding the hands in prone position at rest parallel to the floor. The second one requires bending the indicated finger towards the inside of the hand as fast as possible keeping the hand at rest. The final test requires opening and closing the hand as fast as possible creating a fist and counting the number of fists done. All tests will be activated sequentially, requiring to clap in order to show the next text. While executing them, the position, grip strength and speed of the hands are being measured by the LMC providing 10 features by frame. Once each test is finished, it is saved in the database selected previously. When the three tests are completed, the research team can select those they want to be presented graphically, which are used to compare the metrics accomplished. According the results, the most interesting values are speed and position in these three tests.

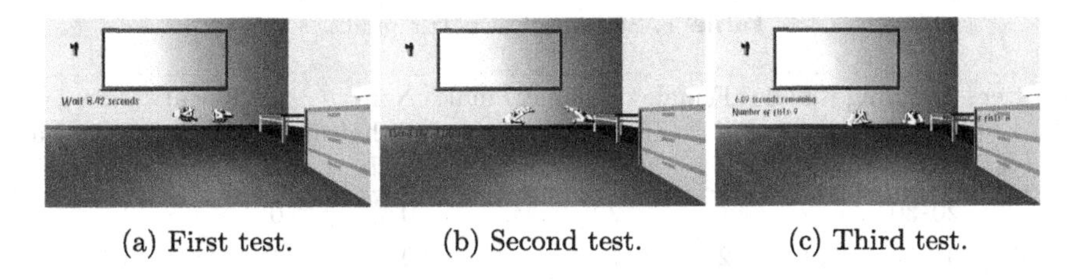

(a) First test. (b) Second test. (c) Third test.

**Fig. 3.** Graphical summary about the three tests.

*Test 1.* It involves placing the hands in parallel, creating a fist in both hands. Its purpose goes beyond the detection of movement in the X or Y axis by the user. It is also aimed at getting both the maximum degree of fist closure and acceleration when performing this gesture, denominated *grip strength.* Throughout this test, the position of the hand in the X, Y and Z axis and its speed will be evaluated (see Fig. 3a).

*Test 2.* In it, the fingers to be bent will be indicated sequentially. Whereas the user is focused on performing, the research group gathers information about the vertical speed of the hand and tremor in low-motion performance (see Fig. 3b).

*Test 3.* It is aimed at completing the information collected from previous tests on involuntary tremors and hand speed. This is accomplished by detecting the speed of closing and opening the hands as fast as the user can perform. The faster the opening and closing speed of the hands, the denser the mass of dots on the resulting graph. This can be seen in the results' section on comparing among different participants (see Fig. 3c).

### 2.4 Framework and Hardware

It is worth describing the XAMPP 3.2.2 database management system, which has allowed to save the data. Regarding the graphical section, Matlab R2020b has been used to plot the recorded information. Unity 3D (version 2019.4.21f21) is the video game engine chosen to perform the different tests. A Lenovo laptop with i5-7200U CPU 2.50 GHz and 12 GB RAM has been used as a supporting platform.

**Table 1.** Summary of participants.

| Cluster | Age range | Male | Female | Technical male | Non-technical male | Technical female | Non-technical female |
|---|---|---|---|---|---|---|---|
| A | 20–30 | 2 | 1 | 2 | 0 | 0 | 1 |
| B | 31–50 | 1 | 2 | 1 | 0 | 1 | 1 |
| C | 51–75 | 2 | 2 | 2 | 0 | 0 | 2 |

### 2.5 Corpus

The selected cohort of participants is composed of 10 normative individuals (five males and five females). The distribution is shown in Table 1.

Notice that technical users are considered to be all individuals who have studied technological disciplines such as computer science, video games, industrial or telecommunication engineering, etc. This factor can influence their experience with the device and framework, offering a better adaptation for performing the tests.

## 3 Results

After testing and collecting 10 features by frame in each test, the most representative examples and statistics are shown in what follows.

This figure depicts the speed graphics of three different participants. This picture is divided into two rows. The first one (a, b and c) illustrates the three raw samples for the three participants. It is mandatory to focus the attention on the abrupt changes in the intermediate zone (between 100 and 500 frames captured by LMC). For this reason, it was ignored the beginning and the end of these samples where the participant is closing and opening the fist with both hands. In the second row (d, e and f), a zoom was carried out. Thus, the user's movement is depicted in the intermediate area where the maximum of hand instability is shown. Finally, notice that the user of cluster C presents small dyskinesias in both hands concerning other users. To demonstrate this fact, a parametric test (t-test) was performed between these three participants of different ages as shown in Table 2.

**Table 2.** Statistical test of different participants for the Test 1.

| Comparison (FHS) | *p-value* |
|---|---|
| CRR vs DPA | 0.96 |
| CRR vs AJRA | **$\ll$0.00** |
| DPA vs AJRA | **$\ll$0.00** |

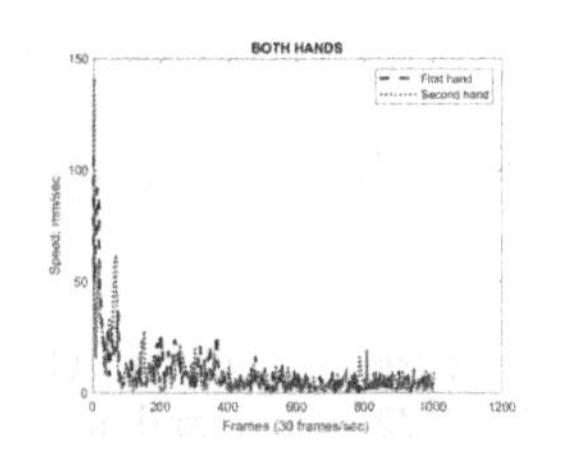


(a) CRR - Cluster A. Male, 25 years, and technical profile

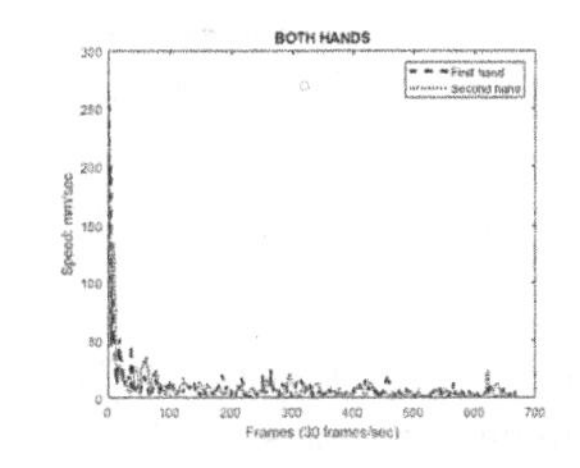


(b) DPA - Cluster B. Male, 39 years, and technical profile

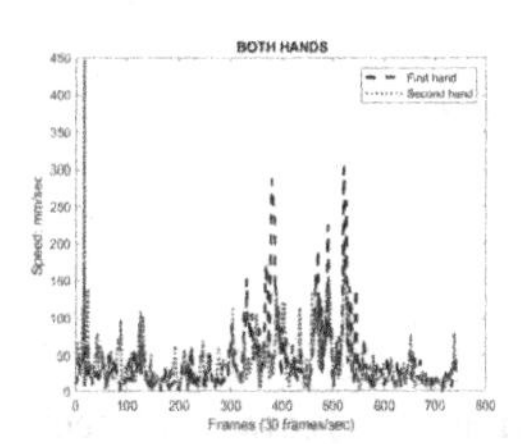


(c) AJRA - Cluster C. Female, 54 years, and non technical profile

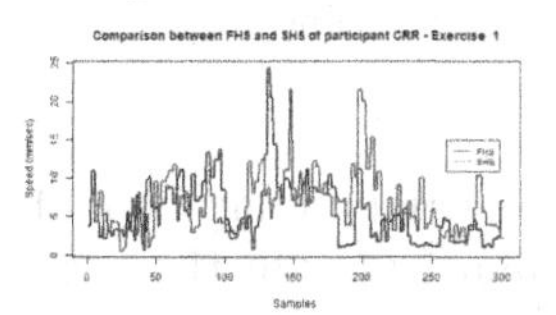

(d) CRR - Cluster A. Male, 25 years, and technical profile

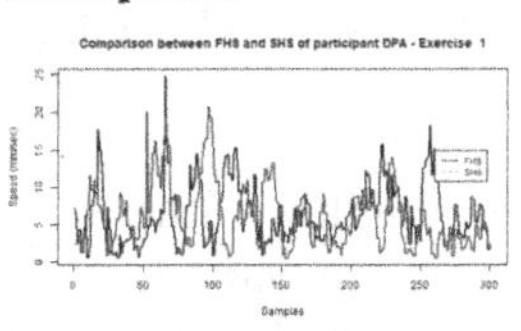

(e) DPA - Cluster B. Male, 39 years, and technical profile

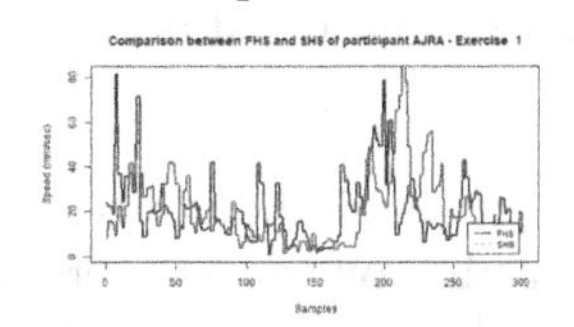

(f) AJRA - Cluster C. Female, 54 years, and non technical profile

**Fig. 4.** Graphical abstract about the first test. Notice that blue color depicts the first hand activity and red color depicts the second hand activity. (Color figure online)

The null hypothesis of this test is based on testing whether or not the two distributions belong to the same individual. If the hypothesis is rejected ($p - value \leq 0.05$), it indicates that there are indications that the two distributions are different. In any other case, there is insufficient evidence to indicate that both distributions do not belong to the same individual. This test was performed with the distributions obtained from the dominant hand (FHS). As it can be seen in the Table 2, the comparisons between the younger participants (Cluster A and B) do not reject the null hypothesis. However, comparisons of the two first clusters with the third cluster do reject the null hypothesis, returning an infinitesimal p-value.

The values obtained by three other different individuals for test 2 are presented below in Fig. 5. Note that the most influential sections of the graph have been plotted and not the raw sample.

Due to the lack of homoscedasticity of the data obtained for the samples of test four, it was decided to use a nonparametric statistic equivalent to the Student's test, the Wilcoxon test. The results obtained are shown in Table 3.

**Table 3.** Statistical test of different participants for the Test 2.

| Comparison (FHS) | *p-value* |
|---|---|
| CRR vs MFA | 0.33 |
| CRR vs BPA | $\ll$**0.00** |
| MFA vs BPA | $\ll$**0.00** |

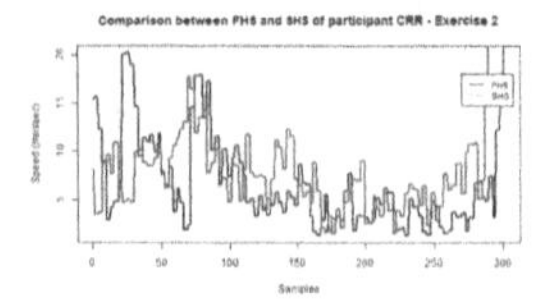

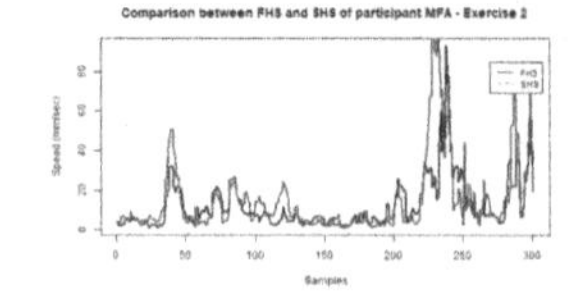

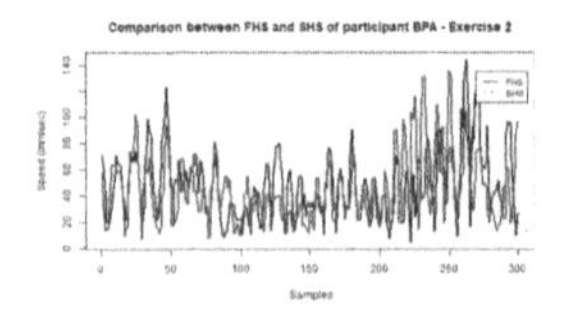

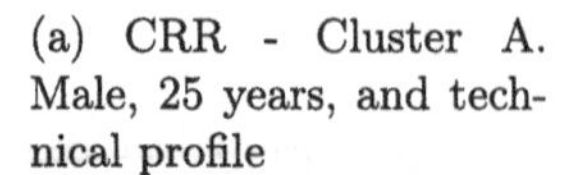

(a) CRR - Cluster A. Male, 25 years, and technical profile

(b) MFA - Cluster B. Female, 46 years, and non technical profile

(c) BPA - Cluster C. Male, 72 years, and non technical profile

**Fig. 5.** Graphical abstract about the second exercise. Notice that blue color depicts the first hand and red color depicts the second hand. (Color figure online)

It should be noticed how this null hypothesis is not rejected among individuals in younger clusters, where the *p-value* is greater than 0.05. On the other hand, comparison between the older individual and the participants in Cluster A and B provides an infinitesimal *p-value*, which not show any correlation between the individual Cluster C and the younger participants data.

Finally, regarding the third exercise (see Fig. 6), the difference in speed between the subjects in cluster C and the rest of the individuals can be verified. It is worth noting the density of points in the first two graphs and a lower number of points in the last one. This means that the third user opened and closed with greater difficulty than the two previous users.

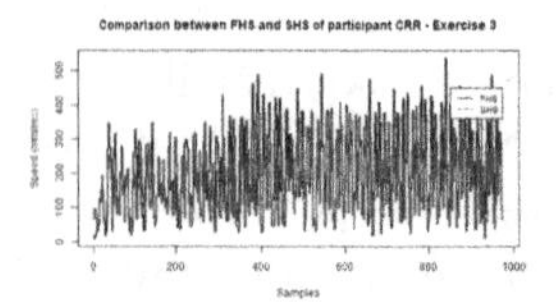

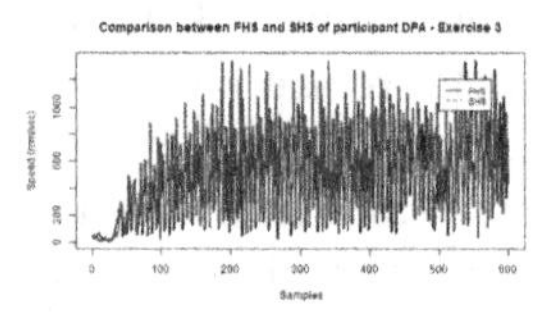

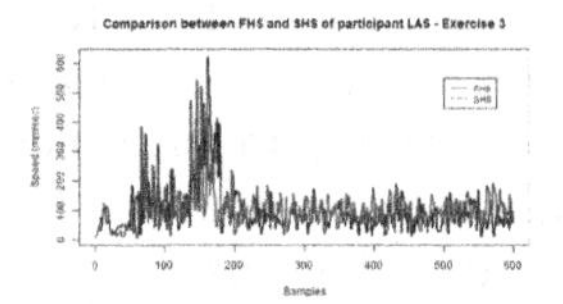

(a) CRR - Cluster A. Male, 25 years, and technical profile

(b) DPA - Cluster B. Male, 39 years, and technical profile

(c) LAS - Cluster C. Female, 72 years, and non technical profile

**Fig. 6.** Graphical abstract about the third exercise. Notice that blue color depicts the first hand and red color depicts the second hand. (Color figure online)

According the third test (see Table 4), the null hypothesis is not rejected by the comparison between participants of Cluster A and B. However, as it can be seen in previous tests the data recorded from Cluster C individuals is not correlated with Cluster A and B participants' measures.

**Table 4.** Statistical test of different participants for the Test 3.

| Comparison (FHS) | *p-value* |
|---|---|
| CRR vs DPA | 0.08 |
| CRR vs LAS | **0.01** |
| DPA vs LAS | $\ll$**0.00** |

## 4 Conclusions

The aims that have been successfully achieved are the following:

- **Predominance between hands.** Most participants show one hand more skilled than the other. This detail can be noticed in some tests and can be the key point of a predominance study.
- **Detecting motion patterns (highlighting the differences between participants).** This has been the main objective to achieve during this research work which has been accomplished.
- **Involuntary tremors.** LMC should be a remarkable device to detect small fluctuations. These movements in this preliminary study cannot be conclusive because all participants are healthy people. However, for the following steps in this research work, it can be a powerful tool to monitor and study non-normative participants such as Parkinson's disease patients. The values of position and speed show key data to identify neuromotor disorders.
- **Correct prediction of the performance of most tests.** In all of them, it was possible to successfully predict the interaction of the participants. Enabling to classify the actions of the participants according to the biometric data collected from their hands' activity.

This research work provides some promising findings that will help tailor biometric indicators for non-normative participants in future works with this technology.

## References

1. Ultraleap Company. https://www.ultraleap.com/. Accessed 2 Mar 2022
2. Al-Khalifa, H.S.: CHEMOTION: a gesture based chemistry virtual laboratory with leap motion. Comput. Appl. Eng. Educ. **25**(6), 961–976 (2017)
3. Braz, N.F.T., Dutra, L.R., Medeiros, P.E.S., Scianni, A.A., Faria, C.D.C.D.M.: Effectiveness of Nintendo Wii in functional and health outcomes of individuals with Parkinson's disease: a systematic review. Fisioterapia e Pesquisa **25**, 100–106 (2018)
4. Brown, D., Renney, N., Stark, A., Nash, C., Mitchell, T.: Leimu: gloveless music interaction using a wrist mounted leap motion (2016)
5. Butt, A.H., et al.: Objective and automatic classification of Parkinson disease with leap motion controller. Biomed. Eng. Online **17**(1), 1–21 (2018)

6. Butt, A.H., Rovini, E., Dolciotti, C., Bongioanni, P., De Petris, G., Cavallo, F.: Leap motion evaluation for assessment of upper limb motor skills in Parkinson's disease. In: 2017 International Conference on Rehabilitation Robotics (ICORR), pp. 116–121. IEEE (2017)
7. Elons, A., Ahmed, M., Shedid, H., Tolba, M.: Arabic sign language recognition using leap motion sensor. In: 2014 9th International Conference on Computer Engineering & Systems (ICCES), pp. 368–373. IEEE (2014)
8. Fernández-González, P., et al.: Leap motion controlled video game-based therapy for upper limb rehabilitation in patients with Parkinson's disease: a feasibility study. J. Neuroeng. Rehabil. **16**(1), 1–10 (2019)
9. Goetz, C.G., et al.: Movement disorder society-sponsored revision of the unified Parkinson's disease rating scale (MDS-UPDRS): scale presentation and clinimetric testing results. Mov. Disord. Off. J. Mov. Disord. Soc. **23**(15), 2129–2170 (2008)
10. Guna, J., Jakus, G., Pogačnik, M., Tomažič, S., Sodnik, J.: An analysis of the precision and reliability of the leap motion sensor and its suitability for static and dynamic tracking. Sensors **14**(2), 3702–3720 (2014)
11. Hoehn, M.M., Yahr, M.D., et al.: Parkinsonism: onset, progression, and mortality. Neurology **50**(2), 318 (1998)
12. Kumar, P., Saini, R., Roy, P.P., Dogra, D.P.: 3D text segmentation and recognition using leap motion. Multimed. Tools Appl. **76**(15), 16491–16510 (2017)
13. Latorre, J., Colomer, C., Alcañiz, M., Llorens, R.: Gait analysis with the Kinect v2: normative study with healthy individuals and comprehensive study of its sensitivity, validity, and reliability in individuals with stroke. J. Neuro Eng. Rehabil. **16**(1), 1–11 (2019)
14. Nicola, S., Stoicu-Tivadar, L., Virag, I., Crişan-Vida, M.: Leap motion supporting medical education. In: 2016 12th IEEE International Symposium on Electronics and Telecommunications (ISETC), pp. 153–156. IEEE (2016)
15. Piggott, L., Wagner, S., Ziat, M.: Haptic neurorehabilitation and virtual reality for upper limb paralysis: a review. Crit. Rev. TM Biomed. Eng. **44**(1–2) (2016)
16. Potter, L.E., Araullo, J., Carter, L.: The leap motion controller: a view on sign language. In: Proceedings of the 25th Australian Computer-Human Interaction Conference: Augmentation, Application, Innovation, Collaboration, pp. 175–178 (2013)
17. Roy, P.P., Kumar, P., Patidar, S., Saini, R.: 3D word spotting using leap motion sensor. Multimed. Tools Appl. **80**(8), 11671–11689 (2021)
18. Solari, F., Chessa, M., Chinellato, E., Bresciani, J.P.: Advances in human-computer interactions: methods, algorithms, and applications (2018)
19. Weichert, F., Bachmann, D., Rudak, B., Fisseler, D.: Analysis of the accuracy and robustness of the leap motion controller. Sensors **13**(5), 6380–6393 (2013)
20. Xu, N., Wang, W., Qu, X.: Recognition of in-air handwritten Chinese character based on leap motion controller. In: Zhang, Y.-J. (ed.) ICIG 2015. LNCS, vol. 9219, pp. 160–168. Springer, Cham (2015). https://doi.org/10.1007/978-3-319-21969-1_14
21. Yang, B., Xia, X., Wang, S., Ye, L.: Development of flight simulation system based on leap motion controller. Proced. Comput. Sci. **183**, 794–800 (2021)

# A Pilot and Feasibility Study of Virtual Reality as Gamified Monitoring Tool for Neurorehabilitation

Daniel Palacios-Alonso[1,2](✉), Agustín López-Arribas[1,2], Guillermo Meléndez-Morales[1,2], Esther Núñez-Vidal[1], Andrés Gómez-Rodellar[2,4], José Manuel Ferrández-Vicente[3], and Pedro Gómez-Vilda[2]

[1] Escuela Técnica Superior de Ingeniería Informática, Universidad Rey Juan Carlos, Campus de Móstoles, Tulipán, s/n, 28933 Móstoles, Madrid, Spain
daniel.palacios@urjc.es

[2] Neuromorphic Speech Processing Lab, Center for Biomedical Technology, Universidad Politécnica de Madrid, Campus de Montegancedo, 28223 Pozuelo de Alarcón, Madrid, Spain

[3] Universidad Politécnica de Cartagena, Campus Universitario Muralla del Mar Pza. Hospital 1, 30202 Cartagena, Spain

[4] Usher Institute, Faculty of Medicine, University of Edinburgh, Edinburgh, UK

**Abstract.** This research work focuses on the collection of indicators (biomarkers) related to the use of virtual reality among participants with different age ranges. The key aim is to monitor participants' psychomotor skills, especially wrist, elbow, and shoulder movements. All the obtained data have been collected and stored in bimodal form. It is worth mentioning a strong gamification work to achieve a total immersion effect for participants. Likewise, the scenarios must be fun, entertaining to the participant. To obtain these features, two serious games based on medieval sports games such as archery and javelin throw have been developed. To apply these two serious games, the participant uses two wireless actuators, called motion controllers, whose control is based on movement, and virtual reality goggles. The data set is composed of eight normative participants of different ages. Also, these participants at the end of the session answered a complete questionnaire. This survey provides valuable and extrapolate information on the use of virtual reality as a vehicle for monitoring the participant's psychomotor skills. For this reason, future research works will attempt to adapt it with patients with neurodegenerative diseases such as Parkinson's, Alzheimer's, or ALS to carry out longitudinal studies.

**Keywords:** Monitoring · Virtual reality · Serious games · Parkinson disease

This research received funding from grants TEC2016-77791-C4-4-R (Ministry of Economic Affairs and Competitiveness of Spain), and Teca-Park-MonParLoc FGCSIC-CENIE 0348-CIE-6-E (InterReg Programme).


J. M. Ferrández Vicente et al. (Eds.): IWINAC 2022, LNCS 13258, pp. 239–248, 2022.
https://doi.org/10.1007/978-3-031-06242-1_24

## 1 Introduction

Today, new devices are coming to the market as entertainment tools such as virtual reality goggles, trackers, cell phones and tablets. However, these devices can be used not only for gaming but also for rehabilitative functions. Likewise, the number of virtual reality devices sold in the last five years has increased considerably [12]. In addition, large multinational social networking companies such as Facebook, also known as Meta, have acquired leading companies in the manufacture of virtual reality googles (e.g., Oculus). This fact favors a democratization of prices and that these devices, little by little, reach all households. This is a favorable point for the development of new longitudinal monitoring applications based on these new devices.

Applications using virtual reality and several trackers have begun to stand out in recent years [9,13]. What started out as an exclusive and very expensive technology, due to the high cost of both the virtual reality googles and the high-performance personal computer (PC) that these applications need to work, is now becoming cheaper and better, allowing many more people to start developing using this type of device. However, the use of this technology is not new and many researchers have used it in recent years. In the following, we will present some of the most interesting research works for the monitoring of neurodegenerative diseases such as Parkinson's, using these kinds of devices.

Docks et al. [3] propose a review was to determine the effectiveness of virtual reality (VR) exercise interventions for rehabilitation in Parkinson's disease (PD). The authors determine that VR interventions may lead to greater improvements in step and stride length compared with physiotherapy interventions. Fernández-González et al. [4] also use Leap Motion Controller® (LMC) in order to test its effectiveness to PD patients. The preliminary findings show the amelioration of coordination and rapidness of motion in those patients. Latorre et al. [7] uses the Kinect v2-based system to complement gait assessment in clinical settings. They conclude that is could be an affordable alternative.

Studies on the use of devices such as 3D Oculus Rift CV1 or Leap motion controller to improve patients' quality of life are growing. Cikajlo and Peterlin Potisk [2] present the use of these devices to study functional improvements, motivation aspects and clinical effectiveness. Indeed, Campo-Prieto et al. [1] examine if the VR games can be used to rehabilitate PD patients. They conclude that games improve the clinical situation of patients without having adverse effects.They also indicate that patients welcome the games positively. Likewise, Imbimbo et al.[6] test that VR rehabilitation improve walking and balance of patients affected by idiopathic PD. The authors conclude that there is a correlation between the use of VR and the patient improvement.

On its turn, Lina et al. [8] and Wang et al. [14] propose a review of the effectiveness of virtual reality interventions in PD patients. Those studies take into account balance and gait in people with Parkinson's disease. Both of them conclude that there are benefits but further studies are necessary as the results are not conclusive in terms of gait. Other research work in which VR rehabilitation is used is the Pazzaglia et al. work [10]. They resolve that VR rehabilitation is more feasible than the conventional rehabilitation. On the other hand, the results of the use of VR are coming slowly. Severiano et al. [11] study the rehabilitation with virtual reality games.The authors finds that the Tightrope Walk and Ski Slalom virtual games benefices patients. Finally, Hawkins et al. [5] present a study in which they analyze how PD interacts using VR comparing to healthy controls participants.

## 2 Objectives

The main objective of this case study focuses on capturing a set of features related to the locomotor capacity of the participant's upper and lower trunk, using two serious games developed for virtual reality. In order to evaluate the effectiveness of these two games a comprehensive questionnaire has been answered by the participants, asking about ease of use, intelligibility of the exercises, motion sickness, entertainment, degree of immersion, graphic quality, among others. In this way, with the feedback obtained, longitudinal studies can be carried out with future patients with neurodegenerative diseases.

## 3 Methods and Materials

### 3.1 Participants

The number of participants in this case study was eight. Three males ($30 \pm 8$ years old) and five females ($30.4 \pm 6.31$ years old). This meager number of participants was due to the impossibility of conducting such a study in Parkinson's associations because of the active state of alarm due to the global pandemic by COVID-19. Note that technical participants are considered to be all those who have studied technological disciplines such as computer science, video games, industrial or telecommunications engineering, etc. This factor may influence their experience and handling with the device, offering a better adaptation for the performance of the tests. It should be noted that the number of females is higher than that of males, an irregular situation for this type of studies.

### 3.2 Frameworks and Hardware

The VR goggles used in this research work were the Lenovo Explorer googles that use Windows Mixed Reality technology. This kind of technology combines virtual reality with augmented reality. In addition, this technology is compatible with most platforms and libraries for video games such as SteamVR. It is a component that facilitates the use of virtual reality by providing both an interface for selecting games and playing them, as well as a series of libraries that can be used free of charge to program new applications that automatically integrate with Steam platform. This library has been used throughout the development of the video game to handle all the built-in VR elements. It has a version compatible with the Unity 3D game engine.

As aforementioned, VR technology requires a very high-performance computer to work properly. This means the computer should be able to maintain high refresh rates at high resolutions (90 frames per second at least). On the contrary, the participant could experiment with what is called motion sickness which provokes dizziness, disorientation, or loss of balance. The computer used while performing the test was a desktop computer that contained a motherboard AMD Ryzen 2700X with 8 cores, running at 3.6 GHz, and it was powered by 16 Gigabytes of RAM DDR4. About the graphics card, it was an Nvidia RTX 2080 which was one of the top graphic cards available at the moment.

Finally, the data collected by the video game were stored in two different databases, one local and the other remote. This option was developed because an Internet connection could not be guaranteed during the performance of tests.

### 3.3 Description of Two Scenarios

The study case consists of two complete scenarios, each one of them designed keeping in mind the kind of movement desired for the participant to collect high-quality locomotor biomarkers. During each level, the participant has to complete different challenges by performing several actions across all scenarios, using the hands to interact with the virtual world. Both of them involve the entire arm of the participant (i.e., from the hand to the shoulder). The first test focuses the attention on performing a horizontal pull (i.e., noak the bow with an arrow and shoot it). The second test is based on a horizontal push (i.e., throw a spear as far as possible). These two scenarios are discussed as follows.

**Kyūdō.** In this level, the participant spawns inside the parade ground of a medieval castle, looking towards a bulls-eye as is shown in Fig. 1a. On the right side, there is a table with a bow on top that has to be grabbed following the instructions displayed on the controller. This first action allows the participant to interact with the environment and gradually get used to it. The GUI shows different indications to perform the exercises of the level.

Once the participant has grasped the bow placed on the table (see Fig. 1a), the bow and arrow will appear in the hands automatically. The tutorial also teaches the participant how to shoot an arrow and achieve the bull-eye without the help of the researchers. To perform this task, the participant has to pull the trigger of the controller and place the arrow next to the bow. Next, the arm holding the arrow will pull the arm taut, performing the effect of loading the bow (see Fig. 1b). Finally, when the user releases the finger from the controller, the arrow will be shot. Consequently, the participant performs a horizontal pulling movement. It is logical to think that the ultimate aim of this test is to hit the bull-eye as often as possible. Once three hits have been achieved, the bull-eye will be moved to another position on the stage but at a greater distance. This forces participants to raise and lower their arms to perform a parabolic shot. Finally, the participant can end the level at any time by leaving the bow on the table.

(a) Virtual scenario. (b) Real scenario.

**Fig. 1.** Point of view of the participant at the Kyūdō scene.

As shown in Fig. 2, the main indicators allow the collection of the coordination between wrist, elbow, and shoulder, as well as the time spent in performing the movements. It is noteworthy how these features provide information about relevant movements such as the circumduction movement produced when the bow is caught and the pronation of the forearm when the bowstring is pulled. Likewise, secondary data such as hits, impact location, and head mobility are gathered too.

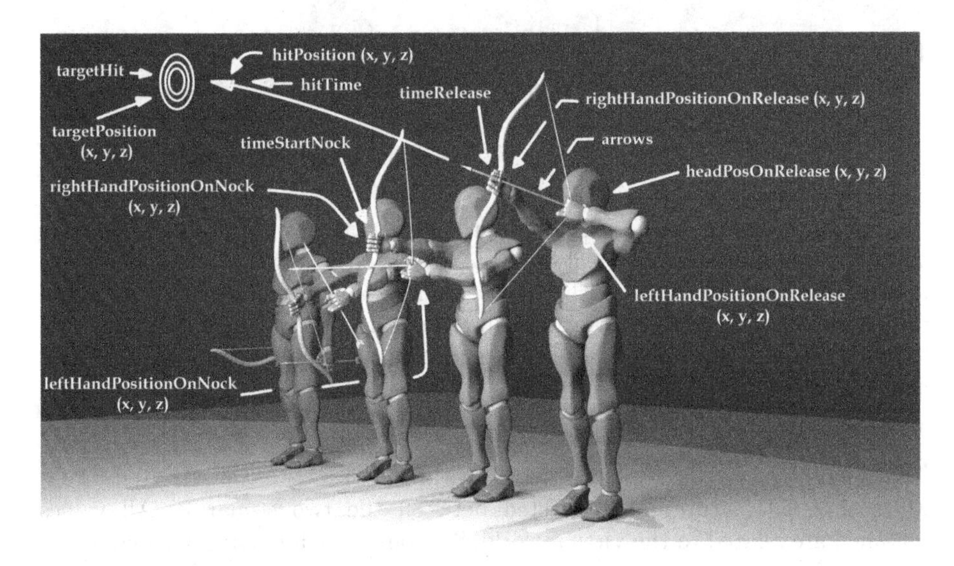


**Fig. 2.** Representation of movement and gathering of main indices collected, using the Kyūdō level.

**Gungnir.** This level takes place on a dock where the participant spawns looking at the sea with a bunch of spears, standing on one side as can be seen in Fig. 3.

Following the instructions of the tutorial about the motion controller, the participant has to grab a spear and throw it into the sea. Thus, the participant carries out a throwing movement based on the horizontal push. As soon as the spear collides with the sea, a score will appear and show the total distance made by the spear. The objective is to throw each spear the furthest the participant can do in order to make the highest score. This level will finish after the participant has thrown six spears.

**Fig. 3.** Point of view of the participant at the Gungnir scene.

As shown in Fig. 4, the coordination between the arm joints and the time spent in the execution of the movements are saved in database. As aforementioned in the Kyūdō level, both the circumduction movement and the pronation of the forearm are performed when the spear is picked up, positioned and throwed. In addition, secondary data such as maximum height reached and throwing distance or number of throws are collected too.

### 3.4 Indices Collected

More than 60 features are collected between the two scenarios. All values obtained are related to the virtual world. The only way a participant may interact with the virtual environment is by using motion controllers, VR goggles, and headphones. The position of each device is recorded in all axes of the virtual world (i.e., X, Y, Z). Note that the position of the controllers is relative to that of the position of the goggles. However, other characteristics, referring to the gamification aspect, are collected such as the number of hits on the bull-eye, distance reached with spears, etc. These data are not binding for the study, however, they offer a more immersive experience for the participants.

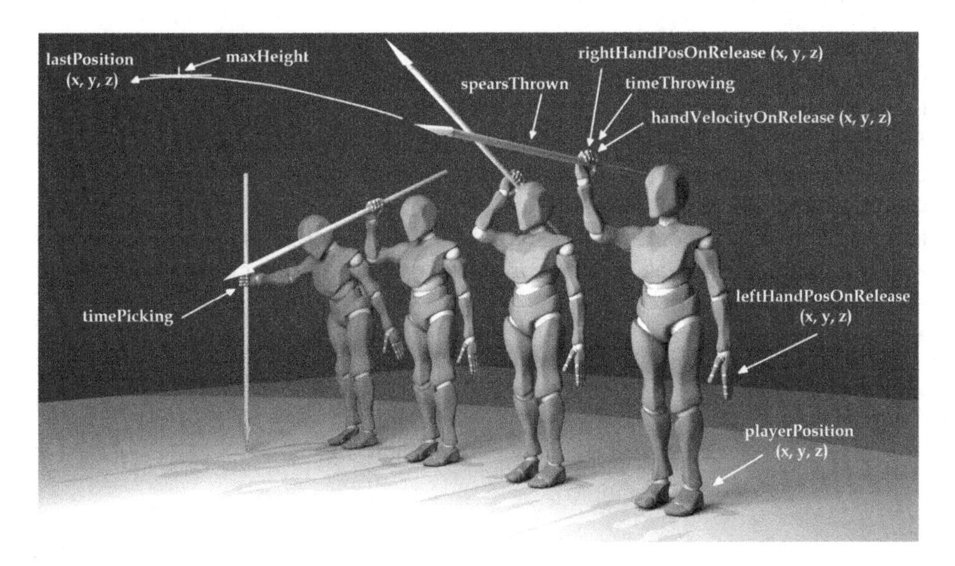


**Fig. 4.** Representation of movement and gathering of main indices collected, using the Gungnir level.

### 3.5 Questionnaire

In order to quantify the degree of achievement of the objectives, a questionnaire was elaborated taking into account some of the most important points in the development of VR simulators such as level design, font size, listening to music while using VR goggles, lighting, texturing. All these questions were directed to avoid the symptoms of motion sickness in the participants. On the other hand, other questions about usability, user-friendliness, entertainment were also asked to the participants. Finally, the participants had the opportunity to rate the scenarios with Likert scale.

## 4 Results

Once the introduction to the scenarios and the questionnaire have been completed, the main results obtained from the responses of the study participants will be presented.

First, five out of the eight participants were not regular gamers. This was considered very relevant due to the possible feedback obtained on the ease of use for future trials. Also, three-quarters of the participants had used VR goggles before.

Next, they were asked several questions about the countermeasures taken to avoid motion sickness. On the one hand, they found the music to be in keeping with the game presented in both scenarios. Also, the font size was considered adequate and did not require modification. On the other hand, the open level design and the refresh rate achieved were highly rated.

The following section of the questionnaire focused on direct questions about each of the scenarios presented. Firstly, they found the tutorials for both levels to be concise and generally easy to follow. However, 25% of the participants found the controls difficult in the spear scenario (Gungnir) and 12.5% in the bow scenario (Kyūdō). The most applauded game among the participants was the bow (Kyūdō) game, obtaining on six occasions a five-point rating and on two occasions a four-point rating on a Likert scale (see Fig. 5a). The game of the spear (Gungnir) obtained a 50% score between four and five points on a Likert scale (see Fig. 5b).

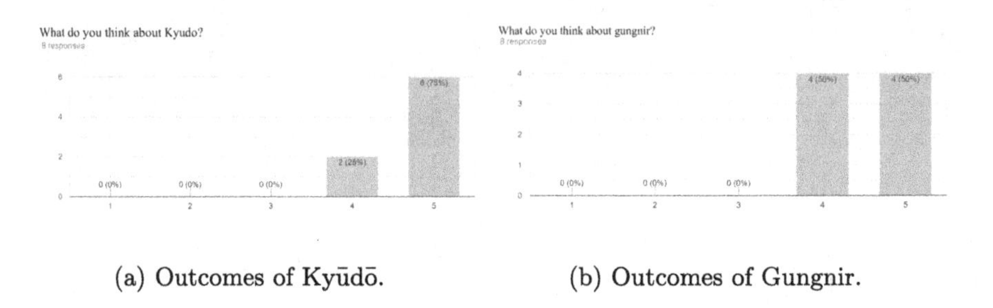


(a) Outcomes of Kyūdō. (b) Outcomes of Gungnir.

**Fig. 5.** Graphical abstract about the results accomplished of participants.

However, when it comes to the levels of difficulty, the opinions of the participants were more disparate. For example, in the case of the archery game, five out of the eight participants scored four out of five for the levels of difficulty, two out of eight scored three, and finally, one participant rated two out of five, i.e., the participant found the levels difficult to achieve (see Fig. 6a). In turn, in the case of the spears, the opinions were mostly neutral, with 62.5% of the participants assessing with three points out of five. The rest of the participants gave an excellent score to the level design of this scenario (see Fig. 6b).

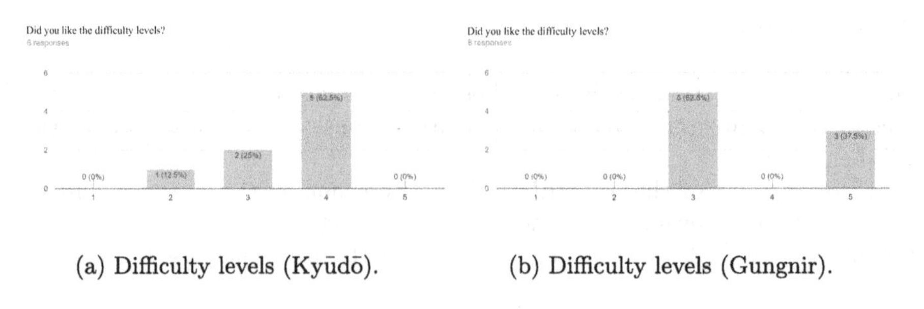


(a) Difficulty levels (Kyūdō). (b) Difficulty levels (Gungnir).

**Fig. 6.** Graphical abstract about the results accomplished of participants.

## 5 Conclusion

This study aims to present two serious games based on virtual reality technology as a longitudinal monitoring tool. To observe their feasibility, a small corpus was carried out with a set of close participants. The reason lies in the impossibility of access to neurodegenerative disease associations due to the pandemic and the active alarm state at the time of this study.

It appears that the questionnaire data provide positive feedback on the design and development of the two scenarios. The various levels seem easy to understand and use. Therefore, it seems logical to think that future the target group of participants (non normative) will also be of the same opinion. Likewise, one of the key points of this research work is that none of the participants who collaborated in the various tests suffered from motion sickness. It should be noted that this type of reaction is very detrimental and dangerous for the future work we want to carry out with the target population.

Given the outcomes accomplished, it seems to be adequate to develop more scenarios to gather different information about the locomotor system of participants. In addition, more in-depth work on the indicators collected and their potential use for monitoring should be carried out in future research.

Finally, under the current circumstances, we could say that the first step of the research work has been achieved with relative success.

**Acknowledgments.** The authors would like to thank all participants of this study case for their dedication and patience in these tough moments.

## References

1. Campo-Prieto, P., Rodríguez-Fuentes, G., Cancela-Carral, J.M.: Can immersive virtual reality videogames help Parkinson's disease patients? A case study. Sensors **21**(14), 4825 (2021)
2. Cikajlo, I., Peterlin Potisk, K.: Advantages of using 3d virtual reality based training in persons with Parkinson's disease: a parallel study. J. Neuroeng. Rehabil. **16**(1), 1–14 (2019)
3. Dockx, K., et al.: Virtual reality for rehabilitation in Parkinson's disease. Cochrane Database Syst. Rev. (12) (2016)
4. Fernández-González, P., et al.: Leap motion controlled video game-based therapy for upper limb rehabilitation in patients with Parkinson's disease: a feasibility study. J. Neuroeng. Rehabil. **16**(1), 1–10 (2019)
5. Hawkins, K.E., Paul, S.S., Chiarovano, E., Curthoys, I.S.: Using virtual reality to assess vestibulo-visual interaction in people with Parkinson's disease compared to healthy controls. Exp. Brain Res. **239**(12), 3553–3564 (2021)
6. Imbimbo, I., et al.: Parkinson's disease and virtual reality rehabilitation: cognitive reserve influences the walking and balance outcome. Neurol. Sci. **42**(11), 4615–4621 (2021)
7. Latorre, J., Colomer, C., Alcañiz, M., Llorens, R.: Gait analysis with the Kinect V2: normative study with healthy individuals and comprehensive study of its sensitivity, validity, and reliability in individuals with stroke. J. Neuroeng. Rehabil. **16**(1), 1–11 (2019)

8. Lina, C., et al.: The effect of virtual reality on the ability to perform activities of daily living, balance during gait, and motor function in Parkinson disease patients: a systematic review and meta-analysis. Am. J. Phys. Med. Rehabil. **99**(10), 917–924 (2020)
9. O'Neil, O., et al.: Virtual reality for neurorehabilitation: insights from 3 European clinics. PM&R **10**(9), S198–S206 (2018)
10. Pazzaglia, C., et al.: Comparison of virtual reality rehabilitation and conventional rehabilitation in Parkinson's disease: a randomised controlled trial. Physiotherapy **106**, 36–42 (2020)
11. Severiano, M.I.R., Zeigelboim, B.S., Teive, H.A.G., Santos, G.J.B., Fonseca, V.R.: Effect of virtual reality in Parkinson's disease: a prospective observational study. Arq. Neuropsiquiatr. **76**, 78–84 (2018)
12. Statista Company: https://www.statista.com/study/29689/virtual-reality-vr-statista-dossier/ (January 2021)
13. Turner, T.H., Atkins, A., Keefe, R.S.: Virtual reality functional capacity assessment tool (VRFCAT-SL) in Parkinson's disease. J. Parkinson's Dis. (Preprint), 1–9 (2021)
14. Wang, B., Shen, M., Wang, Y.X., He, Z.W., Chi, S.Q., Yang, Z.H.: Effect of virtual reality on balance and gait ability in patients with Parkinson's disease: a systematic review and meta-analysis. Clin. Rehabil. **33**(7), 1130–1138 (2019)

# Pairing of Visual and Auditory Stimuli: A Study in Musicians on the Multisensory Processing of the Dimensions of Articulation and Coherence

Octavio de Juan-Ayala[1,4], Vicente Caruana[2], José Javier Campos-Bueno[3], Jose Manuel Ferrández[2], and Eduardo Fernández[4,5](✉)

[1] Professional Conservatory of Music, Alicante, Spain
[2] Department of Electronics, Computer Architecture and Projects, Technical University of Cartagena, Cartagena, Spain
[3] Universidad Complutense, Madrid, Spain
[4] Bioengineering Institute, Universidad Miguel Hernandez, Elche, Spain
e.fernandez@umh.es
[5] CIBER Research Center on Bioengineering, Biomaterials and Nanomedicine (CIBER BBN), Madrid, Spain

**Abstract.** There are stimuli than can evoke us a feeling of connection between them. For example, we can easily appreciate differences when a musical piece is played in a continuous (legato) or discontinued way (martellato). This is called articulation, and this feature can be also applied to other sensorial modalities such as paintings or food images. In this framework, we wonder if the brain processing of a musical piece played in a discontinued way (i.e., martellato) could be similar to the brain processing of images with analogous features. We used Functional Magnetic Resonance Imaging (fMRI) to assess how the brain processes the dimensions of articulation in relation to discontinuity (time or space) and coherence. To avoid any potential bias due to the lack of knowledge of the articulations in music, all the participants were professional musicians. Although more studies are still needed, our results suggest that different areas of auditory and visual cortex are specialized in processing several types of articulations and that brain activity is greater when coherent stimuli were used.

**Keywords:** fMRI · Music · Brain processing · Paintings · Neurograstronomy

## 1 Introduction

Stimuli can be classified according to different properties.The physical properties are multiple and refer to the sensory nature of the stimulus; in addition, the

O. de Juan-Ayala and V. Caruana—Contributed equally to this paper.


J. M. Ferrández Vicente et al. (Eds.): IWINAC 2022, LNCS 13258, pp. 249–258, 2022.
https://doi.org/10.1007/978-3-031-06242-1_25

stimuli also have basic affective properties -pleasant and unpleasant or neutral- on which emotions and feelings are elaborated [1–3]. A special category of stimuli refers to objects with aesthetic value, whether natural or man-made. Art, constituted by these artistic stimuli, has been considered a useful tool to provide valuable scientific knowledge of the neural processes that underlie aesthetic behaviour [4,7,9,10].

Many experiments using fMRI, MEG, EEG have shown that some stimuli (music, visual art, films) provoke emotional responses but there are only a few studies regarding how different senses can interact. Another interesting issue is how stimuli of different sensory modalities (e.g. pictorial, musical and gastronomic [12]) are processed according to their shared sensory dimensions. In this framework, the main aim this study is to study how the brain processes different auditory and visual stimuli. We have focused on examining how the brain processes the dimensions of articulation in relation to discontinuity (time or space) and coherence (similar or different elements shared by the stimuli) that could hinder or facilitate the multisensory processing of different stimuli.

## 2 Methods

### 2.1 Subjects

The study was approved by the Ethics committee of Hospital General Universitario Morales Meseguer, Murcia, Spain. Twelve people participated in the study. All the subjects were professional musicians. Before enrolling in the study, all the subjects completed the consent information and performed several tests to assess their mental state, including: Hamilton Anxiety Rating Scale, Barratt Impulsiveness Scale, CES-D scale and Oldfield test.

### 2.2 Stimuli Used

We used auditory and visual stimuli with different articulations and coherence [3,5,6]. The term articulation in an auditory stimulus refers to the temporal dimension of sound presentation. Thus, when a fragment of a musical piece is played, the interpreter must produce the sounds indicated in the score according to the written notes and the rhythm of the piece. On bowed string instruments (for example, the viola, which was the instrument used in all the experiments) the player inevitably has to introduce temporary cuts (technically called: Legato, Martellato or Spiccato). Our hypothesis is that these sound cuts in time, typical of auditory stimuli, could also have their equivalent counterpart in the representation of visual stimuli. To emulate these auditory articulations, we used images from several paintings using different pictorial techniques (painting with palette, pointillism, etc.) and also images related with the sense of taste, such as meat with several cuts (thin steaks (carpaccio), meat diçes, mince (tartar) or different presentations of chocolate (liquid, ounces, pearls) (Figs. 1, 2).

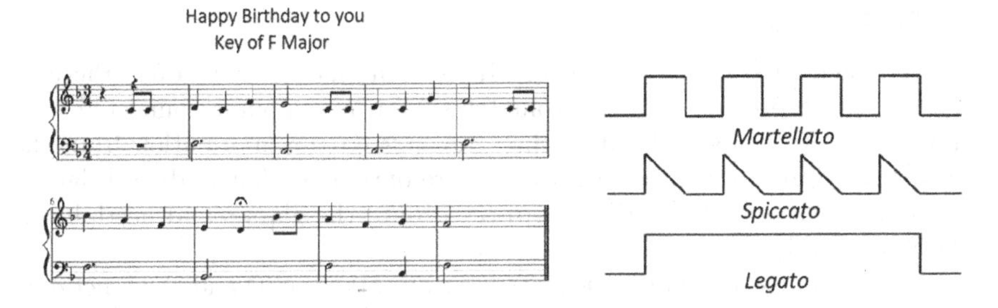


**Fig. 1.** "The figure shows the representation of the temporal cut of the sound matter when a piece of music is performed on a bowed string instrument in the three modalities of articulation with the bow: legato, martellato, spiccato.

The dimension that we have called coherence refers to the greater or lesser number of elements shared by the stimuli when the temporal and spatial dimensions are simultaneously confronted. We hypothesize that a match of these features in terms of greater or lesser degree of coherence-referring to articulation continuity/discontinuity-would be reflected in the brain processing these stimuli if the subjects were using common processing pathways (e.g. auditory stimuli played with Martellato with visual stimuli resembling the spatial Martellato - squares of chocolate).

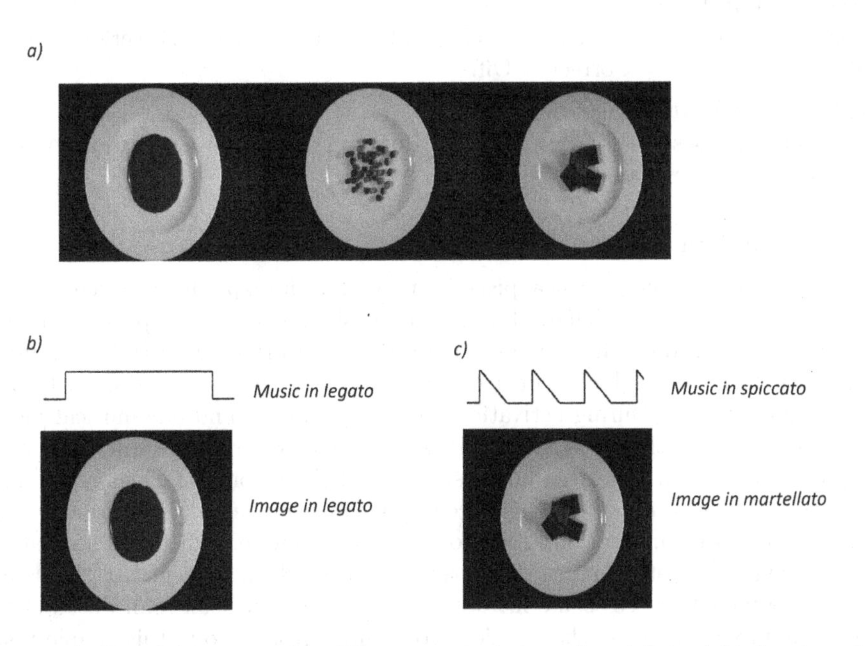


**Fig. 2.** a) Chocolate presented in differents articulations, from left to right: legato, spicatto and martellato. b) Example of coherent articulated stimuli (legato). c) Example of incoherent articulated stimuli (spiccato and martellato).

## 2.3 fMRI Scanning

The experiments were done with a GE MRI system, model HDxt 1.5T. Pictures and music were presented to the volunteers from a computer connected to a stimuli device (VisuaStim Digital) that transmitted the information from the operator room to the scan room through fibre optic lines. Then, digital data is converted to image and sound and projected onto the glasses and headphones worn by the subject.

The protocol used includes a localizer, a calibration, a 3D and the functional sequence. Parameters for the functional sequence (BOLD) are the following:

**Patient Position**
Patient Entry = Head First, Patient Position = Supine, Coil = 8 Ch Brain Hi Res.

**Image Parameters**
Plane = Axial, Mode = 2D, Pulse Family = EPI, Pulse Sequence = GRE-EPI, Imaging Options = FC, VBw, Multi-phase.

**Scan Times**
Number of shots = 1, TE = 60, TR = 2000, Flip angle = 90°.

**Acquisition Range**
FOV = 240 mm, Slice Thickness = 5 mm, Spacing = 3.75 mm, Number of slices = 23

**Acquisition Times**
Frequency = 64, Phase = 64, NEX = 1, PFOV = 1, Frequency Direction = R/L, Shim = Auto, Phase Correct = Off.

**Additional Multiphases Parameters**
Number of phases = 23, Delay Time = Minimum delay after acquisition, Acquisition order = Descending.

## 2.4 Paradigm Design

9 different images and 6 musical pieces were used in the experiment articulated in legato, spiccato and martellato. Images and music were created expressly for this experiment. Based on the type of stimuli we were working with [11], paradigm was structured in 20 blocks of activation and rest with a duration of 6 and 10 s respectively 3. During activation blocks, one image and one musical piece were presented at the same time, during the rest blocks the screen is black and no music is played. Pictures and music with all these considerations of timing, number of blocks, etc. were sequenced in a video made in Movie Maker and played from a computer connected to the stimuli device. Functional sequences started with "ghost" samples to warm up the gradient system. At the same time, the volunteer sees some introductory messages to relax him. Just before the stimuli starts, he is asked to do a countdown from 5 to 0, this simple task tries to help the participant clear his mind and start the data collection focused on the stimuli. Up until now, we have done 2 experiments trying to find which area in the brain relates to different articulations.

**First Experiment.** 4 volunteers participated in the first experiment, but one of them was discarded because there wasn't activation in occipital and temporal lobes, that means something went wrong during the data collection. The rest of volunteers were 2 females and 1 male (ages 40–52, mean 44). For this experiment we had 9 images (painting, chocolate and meat) and 6 popular musical pieces (Happy Birthday and Septimino op.20 by Beethoven). For each image and each piece, we had 3 articulations (legato, spicatto and martellato). That means we had 15 different stimuli in total. All of them were combined in 20 different blocks giving a total duration for the functional sequence of 5 min 30 s (Fig. 3).

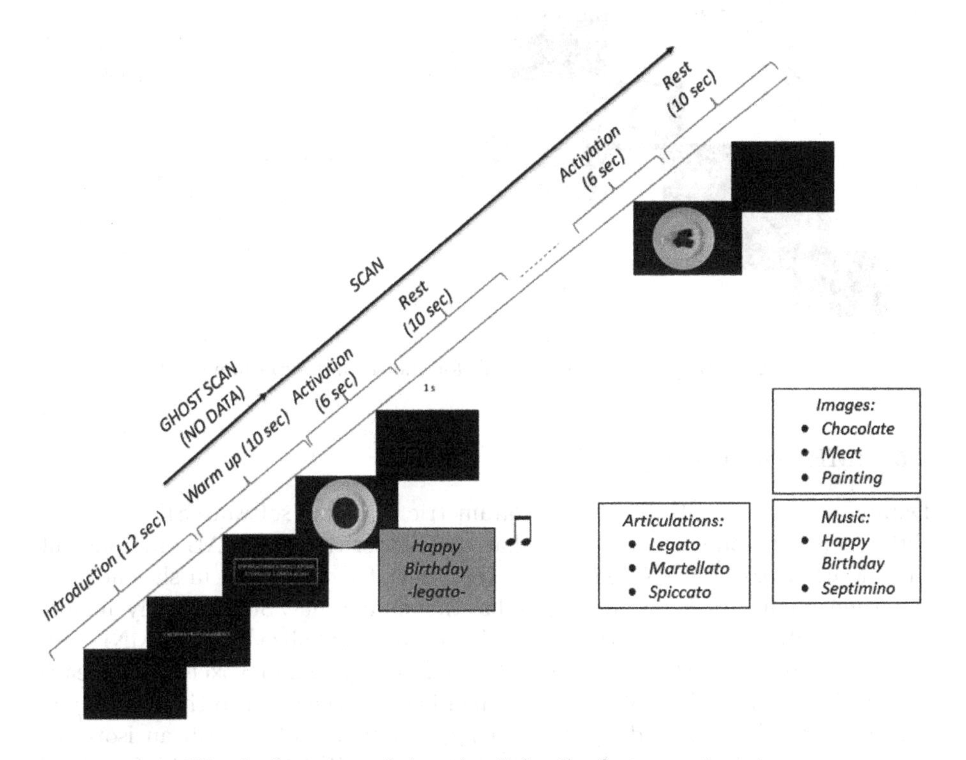


**Fig. 3.** Sequence of the stimuli for the first experiment.

**Second Experiment.** As the results of the first experiment were not conclusive, we decided to simplify the experiment using only two articulations (martellato and legato) and increased number of blocks to 28, scan time (*). 7 volunteers participated in this experiment (ages 15–52, mean 36). For the second experiment we discarded the piece of Septimino and the images of meat and painting (Fig. 4).

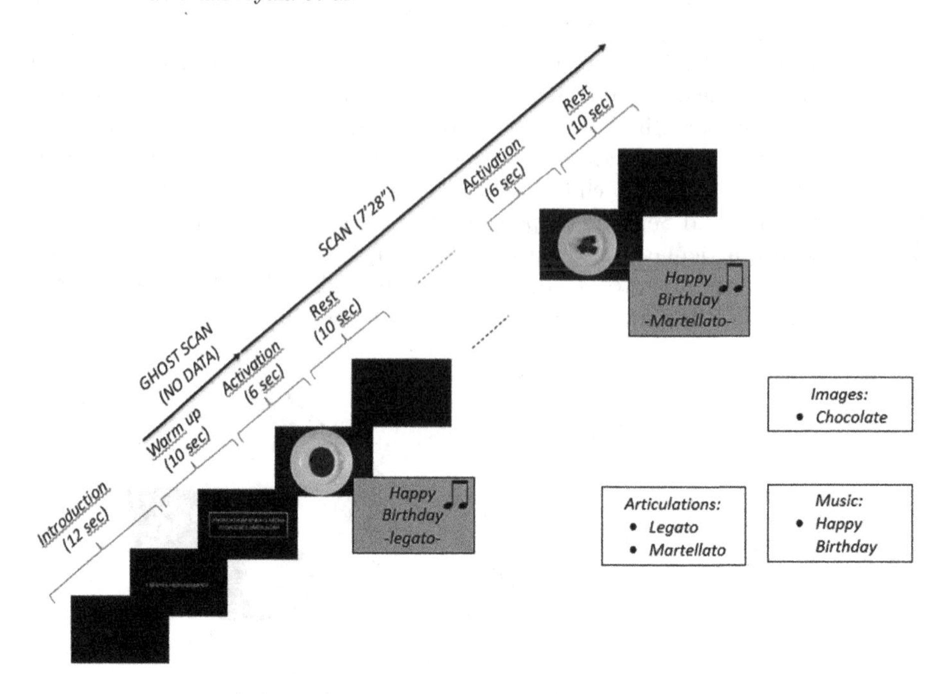


**Fig. 4.** Sequence of the stimuli for the second experiment.

### 2.5 fMRI Data Analysis

Data was analyzed with Statistical Parametric Mapping software SPM12 (http://www.fil.ion.ucl.ac.uk/spm) that runs under the MATLAB environment (The Math-Works Inc., Natick, MA). To correct for differences in slice acquisition time, all images were synchronized to the middle slice. Subsequently, images were spatially realigned to the first volume and normalized to the MNI template supplied with SPM12 (resampled to $2 \times 2 \times 2$ mm voxels) using each subjects' anatomical T1 volume, which had been coregistered to the mean functional image. Functional data were then spatially smoothed with an isotropic Gaussian Kernel of 8-mm fullwidth half-maximum (FWHM). In the first level analysis (intra-subject) all stimuli-functions were convolved with the SPM12 canonical hemodynamic response function (HRF). The time series at each voxel for each subject were high-pass filtered at 128 s to remove low-frequency artifacts and temporally corrected for autocorrelations using the AR(1) model in SPM12. Interscan interval was set to 2 s, microtime resolution to 16 and microtime onset to 1 s. To take into account "random effects" a second level analysis was performed (inter-subjects). Within this ANOVA, we isolated brain activation that resulted from different contrasts depending of the correlations we were studying [8].

# 3 Results

## 3.1 Differences Between Coherent Articulations

Activation for coherent stimuli MM (image in Martellato; music in Martellato) versus LL (image in Legato; music in legato) shows clearly a focused response in auditive and visual cortex in both hemispheres (LH and RH) for the martellato articulation. When we studied the opposite, LL versus MM the activation was more widespread. Notice the p value was $p < 0.05$ not corrected and threshold limited to 20 voxels to detect the differences in activation for so few participants.

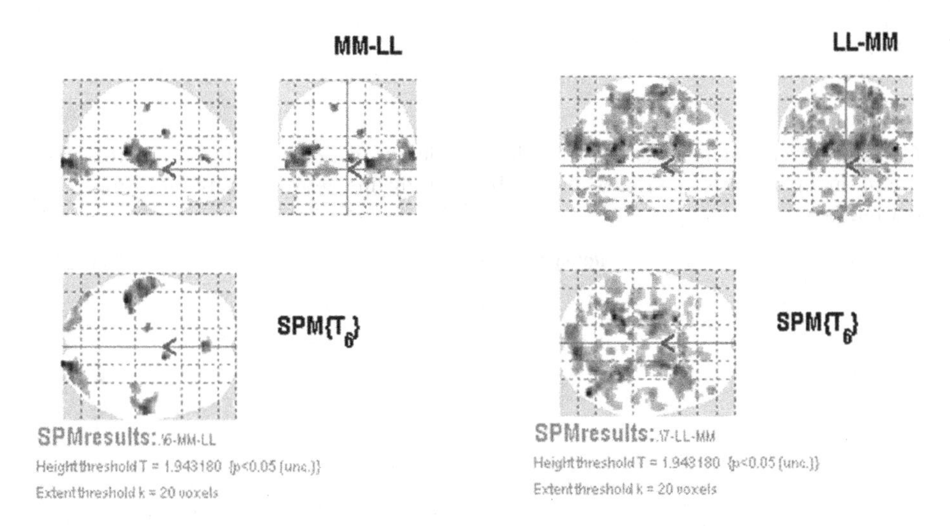


**Fig. 5.** Activation for coherent articulations MM versus LL and viceversa. Notice first letter correspond to image articulation and second to music articulation.

We have also studied what happened if we grouped the activation of musical stimuli independently to the image presented (Fig. 6). Similarly, as we saw in Fig. 5, activation for martellato shows a stronger activation in superior temporal lobe (LH ad RH) and anterior cingulate cortex versus legato.

In Fig. 7 we grouped same image articulations independently to the piece of music, in the same way as we had done in the previous analysis. This time, neural activity is concentrated in the cuneus (LH and RH) and superior motor area (LH and RH) when legato is weighted versus martellato. In the opposite case, martellato had more activity versus legato mainly in the middle occipital gyrus (LH and RH), the pons and the precentral gyrus (LH).

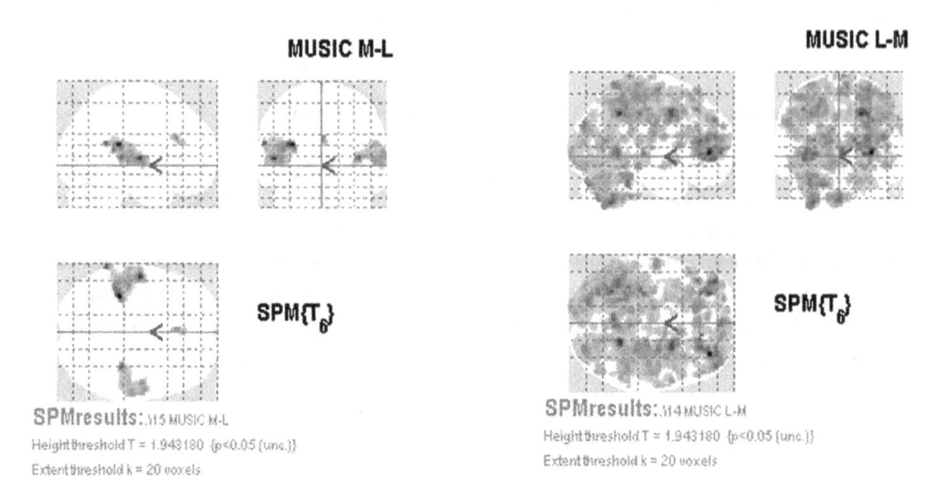


**Fig. 6.** Left image shows activation when music was articulated in martellato versus legato, independently the image shown. In the right image we weighted legato against martellato.

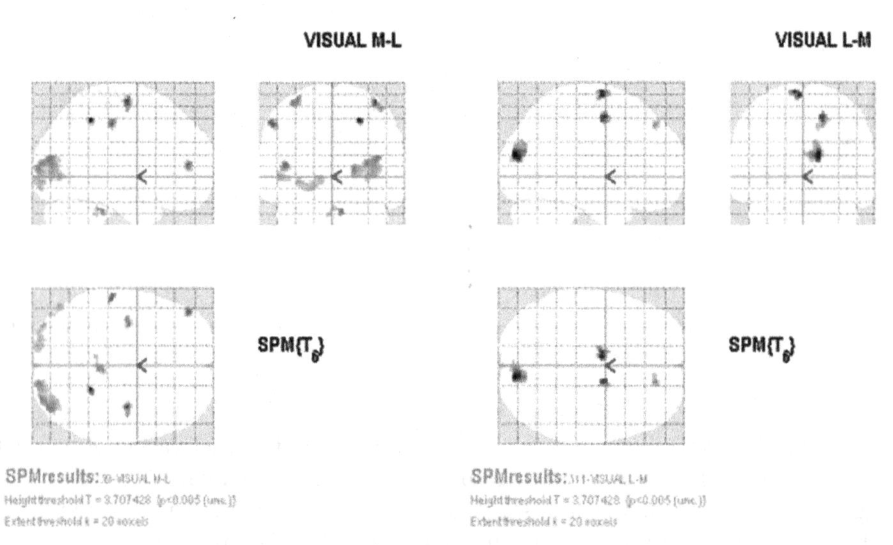


**Fig. 7.** Left image shows activation when images were articulated in martellato versus legato, independently to the music heard. The image on the right shows the opposite case and it is independent to the music playing. Notice $p < 0.005$ and threshold = 20 voxels.

## 3.2 Differences Between Coherent and Incoherent Articulations

If we compare when stimuli articulation is coherent versus incoherent, we see more activation for coherent stimuli in posterior cingulate, thalamus (LH and RH), Heschl's gyrus (RH) and caudate nucleus (LH) (Fig. 8), it doesn't matter if they are in legato, martellato or both together. For a p value of $p < 0.001$ threshold limited to 20 voxels there are not any voxel surviving when the incoherent activation is weighted versus coherent.

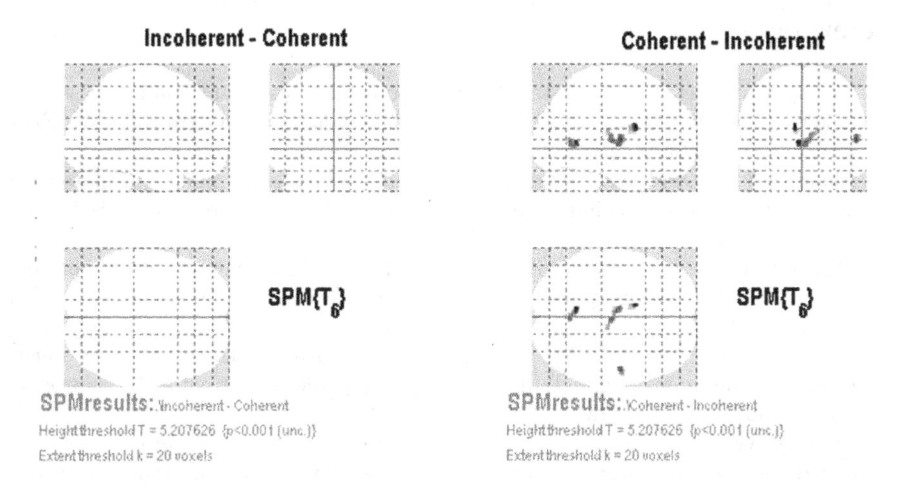


**Fig. 8.** Comparison between Incoherent and coherent articulations.

## 4 Conclusions and Future Development

Different areas of the auditory and visual cortex seem to be specialized in processing different types of articulations in music and images respectively. For auditory stimuli, activation is much more concentrated in the superior temporal lobe (both hemispheres) and anterior cingulate for martellato than for legato. It may be due to the task of "connecting of dots" we unconsciously do when we hear a piece of music articulated in martellato.

Brain association of different stimuli with same or different affective modality is hardly differentiated at this point as we saw in the coherent-incoherent comparison. Nevertheless, brain activity was greater when coherent stimuli were being presented to the subjects versus incoherent stimuli. Locating the area in the cortex where this association is processed may require some modification in the experiment where the participants could express at what moment they are aware of this coherence.

As the results for differences between articulations are promising, it would also be interesting to add other non-musicians subjects to the study and analyse if they are so sensitive and process these stimuli in the same manner.

**Acknowledgements.** This project has received funding by grant RTI2018-098969-B-100 from the Spanish Ministerio de Ciencia Innovación y Universidades and by grant PROMETEO/2019/119 from the Generalitat Valenciana (Spain).

We also want to thank Jose María García Santos, Head of the Radiology Department. University Hospital Morales Meseguer (Spain) for his support in this project, as well as to Francisco Fuentes, Chef of the Palacete La Seda Restaurant in Murcia.

## References

1. Arkhipova, A., et al.: Changes in brain responses to music and non-music sounds following creativity training within the "different hearing" program. Front. Neurosci. **15**, 703620 (2021)
2. Blood, A.J., Zatorre, R.J.: Intensely pleasurable responses to music correlate with activity in brain regions implicated in reward and emotion. Proc. Natl. Acad. Sci. U S A **98**(20), 11818–23 (2001)
3. Campos-Bueno, J.J., DeJuan-Ayala, O., Montoya, P., Birbaumer, N.: Emotional dimensions of music and painting and their interaction. Span J. Psychol. **18**, E54 (2015)
4. Chan, M.M.Y., Han, Y.M.Y.: The functional brain networks activated by music listening: a neuroimaging meta-analysis and implications for treatment. Neuropsychology **36**(1), 4–22 (2022)
5. De Juan Ayala, O.: La interrelación música pintura: un análisis comparativo actualizado de sus principales fundamentos técnicos y expresivos, pp. 163–164 (2010). http://hdl.handle.net/10201/19757
6. De Juan Ayala, O.: Beethoven y Goya en tu cerebro, pp. 22,156. Compobel (2017)
7. Greenberg, D.M., Decety, J., Gordon, I.: The social neuroscience of music: understanding the social brain through human song. Am. Psychol. **76**(7), 1172–1185 (2021)
8. Jahn, A., et al.: justbennet: andrewjahn/andysbrainbook: (Jan 2022). https://doi.org/10.5281/zenodo.5879294
9. Pearce, M.T., et al: Neuroaesthetics: the cognitive neuroscience of aesthetic experience. Perspect. Psychol. Sci. **11**(2), 265–79 (2016)
10. Peretz, I., Zatorre, R.J.: Brain organization for music processing. Annu. Rev. Psychol. **56**, 89–114 (2005)
11. Poldrack, R., Mumford, J., Nichols, T.: Handbook of Functional MRI Data Analysis. Cambridge University Press (2011). https://books.google.es/books?id=OywOnwEACAAJ
12. Shepherd, G.: Neurogastronomy: How the Brain Creates Flavor and Why It Matters. Columbia University Press (2011). https://books.google.es/books?id=gEigoDUBvA4C

# Design of Educational Scenarios with BigFoot Walking Robot: A Cyber-physical System Perspective to Pedagogical Rehabilitation

Valentin Nikolov[1,2], Maya Dimitrova[2(✉)], Ivan Chavdarov[2,3], Aleksandar Krastev[2], and Hiroaki Wagatsuma[4]

[1] Sensata Technologies, 1528 Sofia, Bulgaria
vnn@sensata.com
[2] Institute of Robotics, Bulgarian Academy of Sciences, 1113 Sofia, Bulgaria
m.dimitrova@ir.bas.bg
[3] Faculty of Mathematics and Informatics, Sofia University, 1164 Sofia, Bulgaria
ivannc@fmi.uni-sofia.bg
[4] Graduate School of Life Science and Systems Engineering, Kyushu Institute of Technology (KYUTECH), Kitakyushu 808-0196, Japan
waga@brain.kyutech.ac.jp

**Abstract.** The currently designed novel educational scenarios with the walking robot BigFoot from a cyberphysical system perspective to pedagogical rehabilitation is described in the paper. The sensor system of the robot is presented, which is being developed further in order to adequately apply it to two newly formulated educational scenarios. The results of a pilot study are discussed.

**Keywords:** Walking robot · Cyber-physical system · Pedagogical rehabilitation · Special education

## 1 Cyber-physical Systems for Pedagogical Rehabilitation

The concept of Cyber-Physical Systems (CPS) has been introduced recently to account for technical devices with certain adaptive, sensing and reasoning abilities with a varying degree of autonomous behaviour within networked environments (i.e. internet-of-things) - with or without the human in the information and control loop [1–5]. The CPS approach to education and pedagogical rehabilitation is of special interest for its potential to become a ubiquitous educational tool (also in times of pandemics), as currently developed within the H2020 funded project "CybSPEED: Cyber-physical Systems for Pedagogical Rehabilitation in Special Education" (2017–2022) [6]. Pedagogical rehabilitation is understood by the authors of the paper as a set of behavioural methods for teaching new learning and social skills, resembling games and classroom activities (rather than therapeutic approaches), encompassing the widest range of possible corrections of


J. M. Ferrández Vicente et al. (Eds.): IWINAC 2022, LNCS 13258, pp. 259–269, 2022.
https://doi.org/10.1007/978-3-031-06242-1_26

neurodevelopmental disorders like speech therapy, or focusing on minimal brain dysfunction, delays in acquisition of learning abilities, hyperactivity, attention deficit, etc. in an educational system's framework.

The currently designed novel educational scenarios with the walking robot BigFoot - patented at IR-BAS [6–10] - from a CPS perspective to pedagogical rehabilitation - are described in the present paper. The non-humanoid robot BigFoot is small in size, like a toy, controlled via laptop or joystick (or eye-gaze control in the future) and does not physically interact with the child.

### 1.1 Background Studies of the Appropriateness of Using Toy-like Robots in the Pedagogical Rehabilitation of Children with Autism

Robots interacting with children are being considered in several EU funded projects, both completed and ongoing. The project ALIZ-E designed scenarios for long-term interaction with social robots of children with diabetes in hospital settings [11]. The project MOnarCH focuses on modelling mixed human-robot societies [12]. The project DREAM is using NAO and Probo (elephant robot) to communicate and help teach social skills to children with autism [13]. These projects emphasise the clinical relevance of robotic technology whereas we place the technology in the broader context of pedagogical and social communication in standard and special education - hence in inclusive education settings.

Intrinsic motivations [14] are being modelled in the FP7 project IM-CLeVeR, where agents - animals, humans and robots - are guided by internal drives for entertainment and socialization being more sophisticated than the basic survival drives. The intrinsic motivation is the attraction of a cognitive system to novelty, making a difference between novelty and surprise. IM-CLeVeR is embodying in i-Cub robot the intrinsic motivation of higher-level brains to seek new knowledge [15], thereby sustaining learning and self-improvement in the course of life.

CybSPEED Action emphasizes a similar to the intrinsic-motivation approach to learning by designing human-robot situations (games, pedagogical cases, and artistic performances) and advanced interfaces where children and students interact with the novel technology to enhance the underlying self-compensation and complementarity of brain encoding during learning.

## 2 Functionalities of the BigFoot Robot

One of the main elements of the proposed CPS is a 3D printed walking robot which was called BigFoot in games with children, built on the basis of a minimalistic principle via 3D printing technology [6,7,10]. 3D printing technology is entering many fields today, including robotics, because the technology is affordable and allows the rapid creation of functional models [16,17]. Children with learning difficulties enjoyed playing with both humanoid and nonhumanoid robots such as NAO and BigFoot within the METEMSS project [18,19]. Here we present the further development of BigFoot as a tool in education towards acquiring novel cognitive and social skills.

FigFoot has only two engines, but at the same time several functionalities - it can walk, turn at 360° and even overcome obstacles. The robot consists of a round base 1 on which the body 2 is mounted (Fig. 1). The drive elements and electronic components are located in the body 2 of the robot.

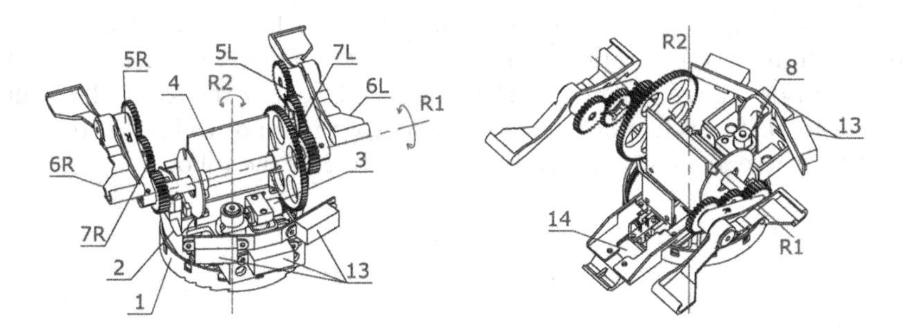


**Fig. 1.** Two views of the BigFoot robot.

The possible movements of the robot are of two types - moving forward or back-ward with stepping and changing orientation:

- The stepping is performed by a DC motor 3, which drives the gear shaft 4, the arms 5L, 5R and the steps 6L, 6R by means of a gear. The steps move in an arc from a circle and, thanks to the gears 7L and 7R, maintain a constant orientation relative to the round base 1 (Fig. 2 left). Thus, the robot goes through two main phases: a fixed circular base 1, in which the steps are moved, and fixed steps, in which all the other elements are moved.
- The change of orientation is carried out by a DC motor 8, which rotates by means of the gear 9 the body of the robot relative to the fixed circular base 1 (Fig. 2 right). This movement is possible only in the first phase in which the robot has stepped with its round base on the ground.

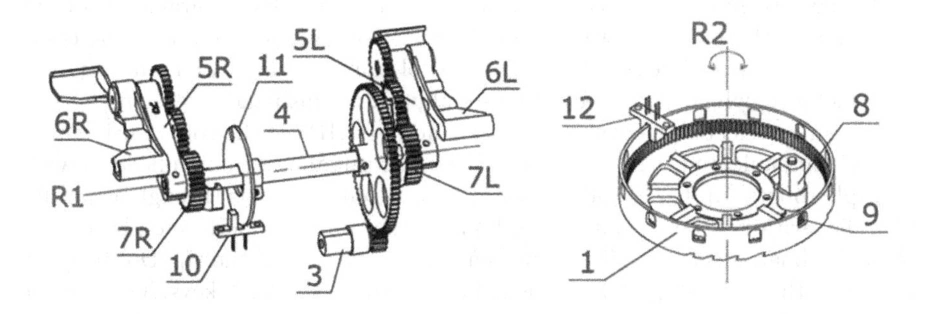


**Fig. 2.** Robot actuators. Left - walking mechanism; Right - rotation mechanism.

The main mechanical components of the robot are made with a 3D printer. Detailed information on the movements, kinematics and overcoming obstacles by the robot is given in [6–9]. The dynamics of the robot is studied in [10]. Here the sensor system of the robot is presented, which is being developed further in order to adequately apply it to the newly formulated educational scenarios.

**Sensor System of BigFoot.** A system of optocoupler 10 and disk 11 with a hole is used to count the steps. A second optocoupler 12 detects the rotation of the robot body 2 relative to the circular base 1. There are 12 holes in the base, which allow the rotation to be read by 30°. Three analog infrared (IR) distance sensors 13 are mounted in the designated front of the robot. They allow distances to obstacles in front, or to the sides, of the robot to be registered and measured. There is a color sensor 14 in the body 2 of the robot.

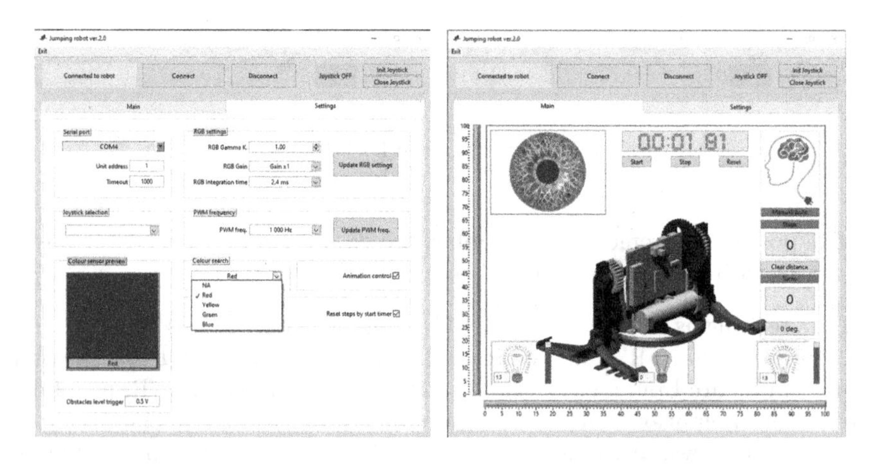


**Fig. 3.** Software management dialogs. Left - settings panel; Right - main panel.

Interactive software has been developed to manage the robot movement, based on the LabView language [20]. Figure 3 shows the two dialog boxes of the platform.

In the Setting panel, Fig. 3 (left), the user can change basic parameters of the operation of the robot, as well as to add and disable functionalities. The communication between the robot and the control computer is carried out remotely via Bluetooth, which creates a virtual serial port. In this panel, a specific "Serial port" can be selected, as well as a specific address (ID) of the robot with which it will communicate ("Unit address"). The platform allows communication with multiple robots (up to 9), and the access to a specific robot is through a specific identifier with numbers from 1 to 9. Usually the control is performed by means of the keyboard, as the Up/Down movements are done with the Up/Down arrow keys, and the Left/Right rotation is done with the Left/Right keys. The control also allows the use of a joystick. If one is enabled, it can be selected and activated from the corresponding menu. As the platform is designed to work with different robots, which in some cases work with different types (e.g. parameters) of electric motors, there is space for the option to change the frequency factor of the pulse-width (PWM) modulation ("PWM freq.") in the Settings panel.

The selected infrared distance sensors allow the registration of objects (obstacles) in the range of 10–30 cm, and proportionally the output voltage of the sensor varies in the range of 0.2–2.5 V at a supply voltage of 3.3 V. In order to set the maximum permissible safety distance to an obstacle, the voltage (respectively the distance) is set in the field "Obstacles level trigger", below which an alarm signal must be generated and a response should follow. In other words, this parameter binaryizes the analog input of the IR sensor and generates an event after the voltage rises above a certain value.

The color sensor mounted in the robot base allows a wide range of colors to be registered, but for training purposes and easier interaction with the environment, the platform is programmed to recognize the three primary colors (Red, Green, Blue) plus Yellow. Due to the sensitivity of the sensor to the surrounding (parasitic) light sources, the sensor allows it to be adjusted by changing the Gama coefficient, Gain and integration time. Each of these parameters has its own control and can be changed by the user. To activate the recognition algorithm, from the drop-down menu "Color search", the user selects a specific color. The currently registered color is presented in the large square field of the "Color sensor preview", and the name of the recognized color (if any) is represented by the field at the base of the square.

The "Main" panel in Fig. 3 (right) represents the main interface in which the interaction with the environment takes place. All the sensory information coming from the robot is presented with the animation adapted according to the functionality, as well as the movements according to the selected direction. On the left and bottom of the window there are two sliders in yellow color, with values 0–100% setting the filling factor of PWM, which can adjust the rotational speeds of both engines, respectively, for walking and lifting the legs. At the base of the Main window the status of each of the three IR distance sensors are visualized as electric lamps. When an obstacle appears below the limit set in the Setting menu, the lights turn yellow and start flashing. The value of the measured distance is visible in a window inside the indicator icon as well as by a linear indicator on the side.

The result of the currently registered color is presented in the form of a large human eye that dynamically changes its color. When the specific color recognition function is activated and the robot steps on such a colored surface, the background of the panel of normally pure white is dotted with a color picture. When the robot is controlled and moves in a certain direction, its movements are animated, and the animation changes depending on the direction of movement. As mentioned, the steps and angles of rotation of the robot are registered by means of optocouplers, and their values are presented in the respective fields "Steps" and "Turns".

For the implementation of different game scenarios the panel contains a timer counting down the time to perform a certain task. The function for automatic timer stop is provided in cases when color recognition is activated. In this way, the time from a certain starting position to reaching a target position marked with a certain color can be accurately measured. By starting the timer, the number of steps can be reset. At the end of the interval the complexity of the movements is taken into account and the approach to the target point is analyzed. A function is in the process of development where each step is recorded as a control algorithm. Subsequently, this algorithm can be used to automatically execute the recorded trajectory. So, the user will be able to set programs for subsequent self-control of the robot.

To demonstrate autonomous control, the software implements an algorithm for automatic control with bypassing obstacles without user intervention. When the function is activated (via the "Manual/Auto" button), the robot starts moving in the right direction until obstacles appear in its path. Then it turns in a certain direction or steps back to overcome it.

## 3 Educational Scenarios with BigFoot

The concept of the script is based on the interaction between two children and the walking robot. A simple scene is created in which the starting position of the robot is marked. There are several possible positions of the target to which the robot controlled by the user should move. The positions are hexagonal holes Pi ($i = 1 \div 6$) in the rectangular area (Fig. 4).

Two children take part in the game - robot operator and goal setter. The goal setter (3) uses colored hexagrams (4) which s/he puts in position Pi. The hypothesis is that the following elements are learned: directions, distance, color, teamwork. Socialization. The robot can measure: time to complete the task, finding the right color, number of steps (distance), and direction. A task can be formulated: go with the robot to a hexagram of a certain color. The hexagrams are of the four main colors. The user (5a) controls the walking robot by means of a laptop (6) or a tablet in order to reach the goal (Fig. 5).

**Educational Game Scenario with BigFoot.** A game scenario has been developed with the participation of two children and a teacher. A child's goal is to place a hexagram tile of one of the following colors yellow, red, blue or green on a board with holes. The other child's goal is to control the robot via a computer to make it step on the tile and recognizes the color with the color sensor (Fig. 5). When the goal is reached, the child, who placed the tile says "Bravo, you succeeded!", and in case of failure s/he says "Try again!". This is an incentive to encourage the child to play, to enjoy the game and not to give up in case of failure.

Intrinsic motivation is an important factor in the acquisition of new knowledge and skills by children. During the game, the teacher observes the children's actions and, if necessary, mentors or helps them. At the discretion of the teacher, the two children change roles. The child, who controlled the robot, will place a colored hexagon tile on the board, and the other will control the robot. The

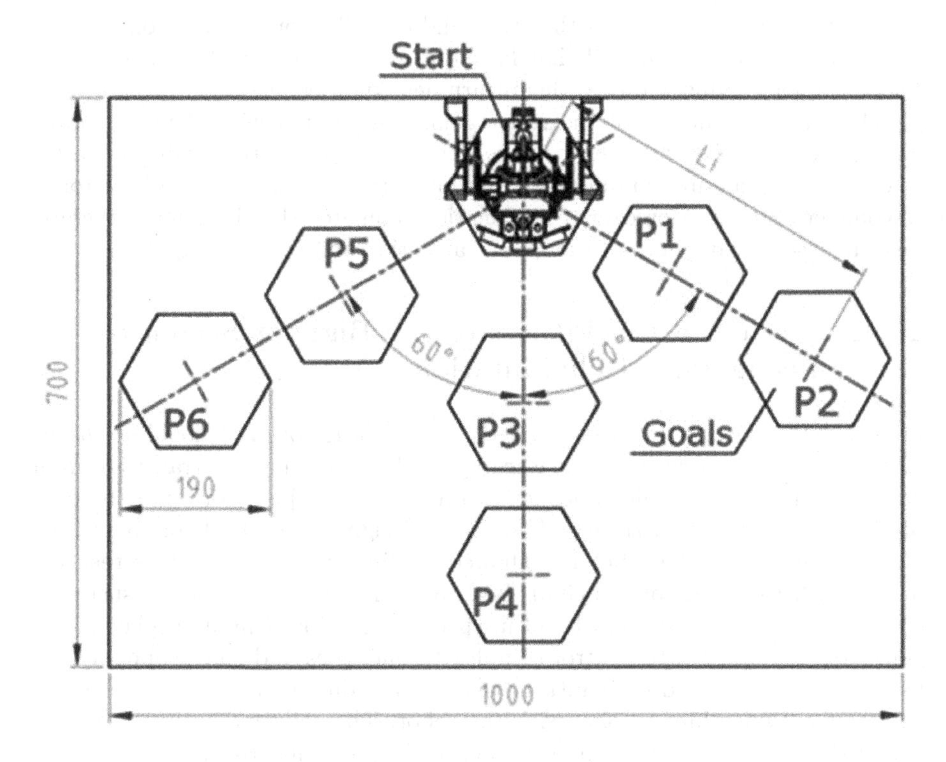


**Fig. 4.** Scenarios for robot movement.

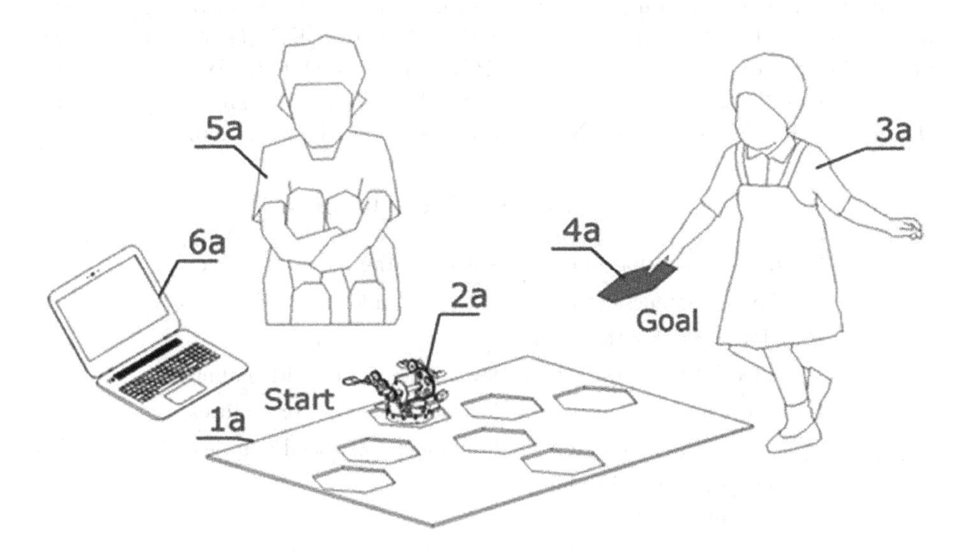


**Fig. 5.** A concept for interaction between two children and the robot.

expected result of the game is that the children will improve their communication with each other and, or if they have difficulty recognizing the directions or colors, the joint game will help them learn new knowledge.

The main advantages of the robot game are the following: Cheap, simple management, climbs obstacles, collects information about the quality of tasks. The game puts children in different situations - to set tasks or control the robot to complete a task. Moreover, the scenario stimulates the three main therapy tasks in cases of autism, imitation, joint attention and turn taking [21–24].

## 4 Evaluation of the Fitness of the BigFoot Scenarios for Pedagogical Rehabilitation

From a systems perspective the so called evolving design of educational situations (games) for children with special learning needs, describing the transition from one experiment in real-life conditions to another, not just from pilot to real-life testing was tested with one of the early designs of the BigFoot robot [18]. The architecture of the robot has allowed modifications requested by teachers in line with the needs of the child. BigFoot scenario was one of the successes with children both in standard and in special education. Children, playing the game, feel empowered to control complex technologies and solve sophisticated tasks. Moreover, a child with autism, who was avoiding human presence, let the therapist convince him to play with the robot. The child enjoyed the sessions with the robot. In the present work we are investigating the potential of the BigFoot to enhance social communications of children with autism, playing in turns.

At the moment, the empirical evaluation of the novel scenario is being performed. The Ethics Committee for Scientific Research (ECSR) of IR-BAS gave permission to conduct the study with Protocol 4/10.02.2022. We present here the data from 2 cases - 2 children with high-functioning autism (ASC) playing the game, and 2 neurotypical (NT) children, playing the same game.

**Table 1.** Teacher's ratings of the game

| No | Likert scales for game assessment by the parent/teacher |
|---|---|
| 1 | Appropriateness of the game for the child |
| 2 | Motivating for the child |
| 3 | Effect of the game for the development of cognitive abilities |
| 4 | Effect of the game for the development of motor abilities |
| 5 | Effect of the game for the development of social abilities |
| 6 | Interesting for the child |
| 7 | Difficulty of the game |
| 8 | Appropriateness of collective participation in the game |
| 9 | Role of the game for formation of novel policies in education |
| 10 | Number of self-initiates social contacts |

Table 1 presents the Likert scales, which parents/teachers filled in during and after the game. Figure 6 present some basic comparison of the obtained mean scores.

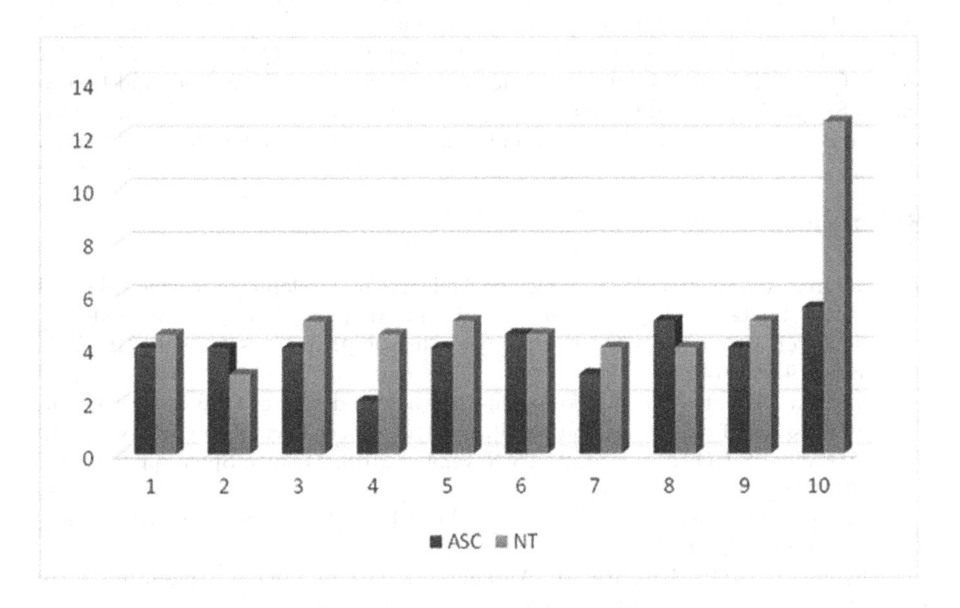


**Fig. 6.** A concept for interaction between two children and the robot.

The results can be interpreted as follows: All parents/teachers consider the game as highly appropriate for the overall development of their children. We believe that it can be commercialized and standardized for use in education. Difference is observed in the item on child's motor development. Parents of NT children does not rate it high for child's motor development since children are probably more involved in other activities, unlike children with ASC. The main difference is in the score for self-initiated social contact (SISC). Parents/teachers declare lower rate of SISC in ASC, yet they believe the game addresses the issue in a very appropriate way. The game will be further tested with other groups of children as well.

## 5 Conclusions

The paper presented the currently designed novel educational scenarios with the walking robot BigFoot from a cyber-physical system perspective to pedagogical rehabilitation. The sensor system of the robot, developed towards following child's strategy in the game and performing high level analysis of the observed child-robot interaction is presented. The results of the pilot study demonstrate a significant potential to improve the imitation, eye-contact and turn taking skills of children in a positive, encouraging learning educational environment.

**Acknowledgement.** We express our gratitude to the team of the Medical Center "Children with Developmental Problems" in Sofia for our long-term collaboration, and in particular to Dr. Daniela Milanova and Ms. Snejina Mihailova. This research work has been partially supported by the National Science Fund of Bulgaria (scientific project "Digital Accessibility for People with Special Needs: Methodology, Conceptual Models and Innovative Ecosystems"), Grant Number KP-06-N42/4, 08.12.2020; EC for project CybSPEED, № 777720, H2020-MSCA-RISE-2017 and OP Science and Education for Smart Growth (2014–2020) for project Competence Center "Intelligent mechatronic, eco- and energy saving sytems and technologies" № BG05M2OP001-1.002-0023.

## References

1. Törngren, M., Sellgren, U.: Complexity challenges in development of cyber-physical systems. In: Lohstroh, M., Derler, P., Sirjani, M. (eds.) Principles of Modeling. LNCS, vol. 10760, pp. 478–503. Springer, Cham (2018). https://doi.org/10.1007/978-3-319-95246-8_27
2. Dimitrova, M., et al.: A multi-domain approach to design of CPS in special education: issues of evaluation and adaptation. In: Proceedings of the 5th Workshop of MPM4CPS COST Action, 24–25 November 2016, Malaga, Spain, pp. 196–205 (2016)
3. Dimitrova, M., Wagatsuma, H. (eds.): Cyber-Physical Systems for Social Applications. IGI Global, (2019)
4. Khargonekar, P.P., Sampath, M.: A framework for ethics in cyber-physical-human systems. IFAC-PapersOnLine **53**(2), 17008–17015 (2020)
5. Nisiotis, L., Alboul, L., Beer, M.: A prototype that fuses virtual reality, robots, and social networks to create a new cyber-physical-social eco-society system for cultural heritage. Sustainability **12**(2), 645 (2020)
6. Chavdarov, I.: 3D Printed Walking Robot Based on a Minimalist Approach, in book: collaborative and Humanoid Robots. InteachOpen. https://www.intechopen.com/chapters/76475 (2021)
7. Patent name: "Walking Robot", Reg. No 111362/05.12.2012, Inventors. In: Chavdarov, I., Tanev, T., Pavlov, V. (eds.) Official Bulletin No 6, 30.06.2014. p. 11. (in Bulgarian)
8. Chavdarov, I., Naydenov, B..: Design and kinematics of a 3-D printed walking robot "Big Foot", overcoming obstacles. Int. J. Adv. Robot. Syst., 1–12 (2019)
9. Chavdarov, I., Krastev, A., Naydenov, B., Pavlova, G.: Analysis and experiments with a 3D printed walking robot to improve climbing obstacle. Int. J. Adv. Robot. Syst., 1–13 (2020)
10. Stefanov, A., Chavdarov, I., Nedanovski, D.: Detailed dynamical model of a simple 3D print-ed walking robot. In: AIP Conference Proceedings, vol. 2321, p. 030031 (2021)
11. The ALIZ-E Project: Adaptive Strategies for Sustainable Long-Term Social Interaction. www.dfki.de/KI2012/PosterDemoTrack/ki2012pd09.pdf
12. MONARCH. http://monarch-fp7.eu/
13. DREAM. http://www.dream2020.eu
14. Baldassarre, G., Stafford, T., Mirolli, M., Redgrave, P., Ryan, R., Barto, A.: Intrinsic motivations and open-ended development in animals, humans, and robots: an overview. Front. Psychol. **5**, 985 (2014). https://doi.org/10.3389/fpsyg.2014.00985
15. IM-CLeVeR. http://www.im-clever.eu/

16. Kim, J., Kang, T., Song, D., Yi, S.-J.: Design and control of an open-source, low cost, 3D printed dynamic quadruped robot. Appl. Sci. **11**, 3762 (2021)
17. Lin, S., Su, J., Song, S., Zhang, J.: An event-triggered low-cost tactile perception system for social robot's whole body interaction. IEEE Access **9**, 80986–80995 (2021). https://doi.org/10.1109/ACCESS.2021.3053117
18. Dimitrova, M., et al: Cyber-physical systems for pedagogical rehabilitation from an inclusive education perspective. BRAIN. Broad Res. Artif. Intell. Neurosci. **11**(2Sup1), 186–207 (2019)
19. Methodologies and technologies for enhancing the motor and social skills of children with developmental problems-METEMSS. http://www.iser.bas.bg/metemss/en/index.html
20. Jeffrey, T.: LabVIEW for everyon : graphical programming made easy and fun. Kring, Jim. (3rd ed.). Prentice Hall, Upper Saddle River, NJ (2006)
21. Richardson, K., et al.: Robot enhanced therapy for children with autism (DREAM): a social model of autism. IEEE Technol. Soc. Mag. **37**(1), 30–39 (2018)
22. Huskens, B., Verschuur, R., Gillesen, J., Didden, R., Barakova, E.: Promoting question-asking in school-aged children with autism spectrum disorders: effectiveness of a robot intervention compared to a human-trainer intervention. Dev. Neurorehabil. **16**(5), 345–356 (2013)
23. Happe, F.: Autism: cognitive deficit or cognitive style? Trends Cogn. Sci. **3**(6), 216–222 (1999)
24. Dimitrova, M.: Gestalt processing in human-robot interaction: a novel account for autism research. BRAIN Broad Res. Artif. Intell. Neurosci. **6**(1–2), 30–42 (2015)

# Feasibility Study of a ML-Based ASD Monitoring System

José María Vicente-Samper[1(✉)], Ernesto Ávila-Navarro[2], and José María Sabater-Navarro[1]

[1] Neuroengineering Biomedical Research Group, Miguel Hernandez University of Elche, 03202 Elche, Spain
{jose.vicentes,j.sabater}@umh.es

[2] Department of Materials Science, Optics and Electronic Technology, Miguel Hernández University of Elche, 03202 Elche, Spain
eavila@umh.es

**Abstract.** People with Autism Spectrum Disorder (ASD) show a great heterogeneity in their atypical sensory behaviours due to they often suffer from a Sensory Processing Disorder (SPD). This nervous system condition is associated with the social interaction, learning and behavioural problems experienced by people with ASD. This work shows a study conducted in a clinical scenario where the participating users carry out tasks to improve their skills. During the execution of these activities, data about the physiological state of the user and the stimuli present in the environment is recorded using an electronic monitoring platform for people with ASD. From the acquired signals by the devices, different features are created, and a dataset is built with them. Finally, information is extracted from the dataset using machine learning techniques to try to relate changes in the user's state with environmental stimuli.

**Keywords:** Autism spectrum disorder · Artificial intelligence · Electronics

## 1 Introduction

Sensory integration can be defined as the organisation of the information from the senses to be used and it allows the individual to interact with the environment around himself [1]. However, most people with Autism Spectrum Disorder (ASD) have an associated Sensory Processing Disorder (SPD) which affects the way in which the nervous system processes the information coming from the senses [2,3]. For people with ASD, these problems are related to difficulties in social interaction, learning and the behaviour. The SPD affects people with ASD

This work was partially funded by Spanish Research State Agency and European Regional Development Fund through "Race" Project (PID2019-111023RB-C32). The work of J.M.V.-S. is supported by the Conselleria d'Educació, Investigació, Cultura i Esport (GVA) through FDGENT/2018/015 project.


J. M. Ferrández Vicente et al. (Eds.): IWINAC 2022, LNCS 13258, pp. 270–280, 2022.
https://doi.org/10.1007/978-3-031-06242-1_27

differently for visual, tactile, sound, smell and taste stimuli. For example, a person can be very sensitive to a certain sound, which is considered soft by the majority.

The purpose of this work is to present the details of a test performed in a clinical setting where the environmental stimuli and the physiological signals of the user are monitored during the course of skill-enhancing sessions for people with ASD.

The paper is organized as follows, in the materials and methods section, the used devices are presented, the setup of the environment and the data acquisition protocol are briefly described, and the make up of the generated dataset is explained. Section III presents of the obtained results during the experimentation with one user. In addition, the process to extract information from the data by using machine learning techniques is shown. Finally, the conclusions are presented.

## 2 Materials and Methods

This section describes the purpose of this study, that is to monitor the sensory activity in the environment during the development of the designed sessions. The user's physiological signals are also monitored during the session, so that a relationship between the changes in the person's state and the stimuli present in the environment can be established. First, the used devices for the signal acquisition are introduced. Then, the configuration of the environment where the test is performed with users and the followed protocol for the data acquisition are shown. Finally, the dataset with the generated features from the collected signals is assembled.

### 2.1 Monitoring System

The used monitoring devices are part of an acquisition platform especially developed for people with ASD. The development of these devices can be found in [4,5] and Fig. 1 shows a picture of the different devices that make up the platform. On one hand, a soft wristband to measure different physiological parameters of the user (heart rate, body temperature and motor activity). On the other hand, there are three devices in charge of acquiring stimuli from the environment; the first of them is a small device that collects the environmental conditions of the room (luminosity, environmental temperature, relative humidity and atmospheric pressure). There is also a device with a 360-degree camera that measures visual stimuli (number of people and optical flow). Finally, an Android smartphone which manages the platform, shows relevant information in the interface and also acts as a sound analysis sensor. All the information collected by the platform is stored in a remote database.

**Fig. 1.** Picture of the monitoring platform's devices.

## 2.2 Setup of the Clinical Environment

Given that it is difficult to perform tests with people with ASD due to their reluctance to the use of foreign bodies, it is important to choose the environment in order to try to facilitate their adaptation to the system. For this reason, a controlled environment is proposed as a starting point, where users can come periodically and the tasks they carry out can be planned. Many people with ASD, especially in childhood, require work sessions with specialised professionals to help them to manage the problems derived from their pathology, for example aggressiveness, phobias to objects or tasks such as hygiene or hyperactivity problems, among many others. The professionals prepare tasks that are adjusted to the level of development of the person and always looking for an evolution in the problem to be treated. For this reason, the work described in this paper has been carried out in collaboration with the University Clinic. It is a centre for applied research and healthcare transfer of the Miguel Hernández University of Elche, where there is a specialised ASD unit.

In order to perform the experimentation, one of the clinic's rooms has been set up where the professionals carry out the sessions with the children. The monitoring platform devices have been permanently installed in this room to facilitate the work of the therapists and the adaptation of the users, who are already habituated to working in it. Figure 2 shows some pictures of the room where the experimentation is conducted. The room has several working areas where users can develop activities of different purposes. There is also a multitude of available materials for the tasks, such as mats, trampolines, pilates balls, materials for general gross motor skills tasks, blackboards, etc. In addition, the lighting in the room can also be controlled to regulate the intensity or the colour according to the moment or the child's needs, which allows different chromatic ambiences to be created. For this purpose, the windows are fitted with opaque panels that prevent the entry of outside light when it is required. The room also has a background music system to play pleasant music for the user or to help them to relax if it is necessary, as well as a projector to display digital content.

**Fig. 2.** Pictures of the room set up for the experimentation at the University Clinic.

The visual stimulus acquisition device has been placed in the wall area between the windows, on a piece of furniture and in a half-height position. In this way, the 360-degree camera will be able to capture an image of the whole room, regardless of the area where the user is working. This device will remain fixed during all the sessions of the experimentation in the same place. However, both the ambient conditions acquisition device and the sound spectrum management and analysis device will vary their location in the room depending on the working area where the user is located, so that they will be close to the user without having to be next to him/her. Finally, the personal monitoring device is placed on the user's wrist during the entire session as tightly as the user is able to tolerate it.

### 2.3 Protocol for Data Acquisition

Users were recruited by clinic's specialists. The legal guardians of each of the users have given their approval to participate in the study. Researchers managed the collected data anonymously and the Ethical Principles for Medical Research Involving Human Subjects of the Declaration of Helsinki was respected during the experimentation [6].

For each session, a series of tasks were defined focused on improving some aspect of the user's daily life, in a way that the user feels comfortable working on it. Before starting with the tasks, the recording of signals with the devices was initiated. The start-up procedure was carried out by the clinic's specialists at all times, so that there were no outsiders to interfere with the normal work in the session.

First, the monitoring bracelet was fitted to the user. In the first session, it was given to them to handle it in order to help them to tolerate its use. Once the wristband had been attached as firmly as the user allows, the other devices were powered on and the recording of signals begins. From that moment on, the therapists start the defined tasks to work with the user during the session.

**Table 1.** Summary report of a performed session with an user in the clinic.

| Session date | 8/3/2021 |
|---|---|
| Session aims | 1. Tolerate scissor stimulation<br>2. Increase autonomy in daily life activities (shower and washing):<br>- Recognise stains<br>- Hand washing with only verbal support |
| Session schedule | a) [17:00 - 17:35] Circuit with trampoline, cones, plasticine and plastic knife. At the end user will have cut the paper and make the mandala<br>b) [17:35 - 17:42] Show the pictograms for hand washing, and then go to the toilet with the material. Let her perform the action herself with visual support |
| Session progress | Tolerate scissors stimuli: When cutting the mandala, she presses to cut, and one of the times she has made the up-down movement. We have put tape on her nails, but she has not let us cut it. She has tolerated seeing it on someone else, and cut the tape herself<br>Increasing autonomy in activities of daily living (showering and hand washing): She looks at the pictograms and understands them, but finds it difficult to carry them out |

At the end of the session, all devices were stopped and the monitoring bracelet was removed from the user. Finally, the therapists generated a session report specifying the performed tasks, their start times and the user's behaviour for each of them to facilitate the subsequent processsing and to understand possible system failures. In those sessions where an extra person, such as the user's parent, participates, this is also specified in the session report. Table 1 shows as an example a report of one of the sessions carried out with a user. The development of the session shown in the table does not contain major incidents, however, there are some sessions where major difficulties appear, either with the devices or with the user's behaviour.

Thanks to the platform's working architecture, where all the collected information by the different monitoring devices is stored in a remote database, the recorded data can be accessed during the sessions in real time. This allows to follow the sessions from a different place so as not to interfere with the work of the therapists in the room. In addition, it allows working with the data after the end of each session without the need to bother the clinic's professionals with downloading or sending data.

### 2.4 Signal Processing and Dataset Generation

The signal processing applied to the recorded signals during the sessions is very similar to the one described in [5], where the different generated attributes and the construction of the dataset were also described. However, in addition to the selected attributes in [5], the second derivative in time, the dominant frequency and the accumulated energy of the frequency response were also calculated (defined as Eq. 1), where FFT is the Fast Fourier Transform, *conj* it's the conjugate of a complex number and $signal_i$ it's the values array of the window

i. As a result, the dataset consist of 182 different attributes and an extra column with a reference label. This label identifies the type of activity of the corresponding entry among four different groups. These groups are the result of the classification of all the tasks carried out by the user during the different sessions according to their level of intensity and nature, which have been extracted from the generated reports by the professionals of the clinic.

$$Energy = \sum_{i=1}^{n} FFT(signal_i) * conj(FFT(signal_i)) \quad (1)$$

The activities in **Group A** correspond to the stimulation of the user in front of objects and tasks that cause them phobia. In addition, changes in the light intensity and colour of the light in the room are used to see if the user's tolerance to the objects improves. Regarding to the **Group B**, there are activities where the user combines the development of gross psychomotor tasks, such as physical activity, and fine psychomotor tasks such as the use of modelling clay. On the other hand, activities in **Group C** are focused on learning, e.g. working with pictograms related to the bathroom. Activities related to learning to tolerate healthcare equipment or objects that can be found in a doctor's room or hospital are also classified in this group. Finally, in **Group D** there are activities where physical tasks such as obstacle courses are carried out, and the activity is often accompanied by music.

Once the dataset with the 183 columns has been composed, all the features, except the people quantification and the main sound frequency band, are normalized so that the mean of each group is 0 and the standard deviation is 1. In addition, a *one-hot* coding is applied to people quantification and the main sound frequency band. Finally, the dataset entries are mixed and split to obtain the training dataset (80%) and the test dataset (20%).

## 3 Results

This section shows the obtained results of the experimentation with one participant user as an example. This user was selected because she obtained positive results from a small number of sessions. She is a 12 years old girl with a high degree of severity in ASD. The total recording period for this user was just over four months, with one session every week and approximately 35 min of recordings per session. In total, the user participated in the study for 16 sessions, 10 of which were valid for experimentation. For the remaining 6 sessions the records were incomplete due to problems with the user during the session or failures with one of the monitoring devices. Therefore, the total recording time was 5 h and 38 min of valid data.

If the value of the different generated attributes for the different dataset entries is analysed according to the group of activities in which they have been classified, it can be observed that there are differences between them according to the group of activities to which they belong. Figure 3 shows a graph with the

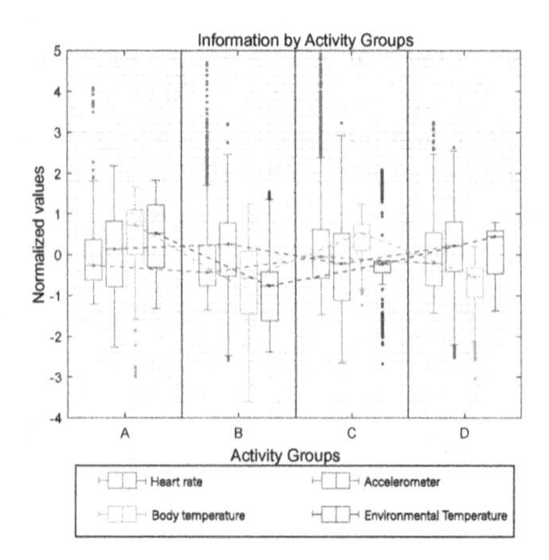


**Fig. 3.** Some of the generated attributes with the obtained signals during the experimentation. The normalised values for the different groups of activities are represented by the median value and the error box whose extremes represent the first and third quartiles.

result of grouping by the activity group of each input some of the generated attributes from different measured signals.

By looking at the heart rate (HR) results it can be seen that the range of the majority of values remains very similar in the four groups of activities. However, it can be seen that in activity groups B and C the number of outliers is significantly higher than in the other two groups. This is because in these groups (B and C), the majority of activities where the user performs constant manual tasks, such as trimming, and with continuous movement of the fingers and wrist, more motion artifacts are produced than in the other two groups. However, if we look at the results of the accelerometer signal, it can be seen that the user constantly performs arm movements during the sessions, although the intensity values are higher for groups B and D, whose activities have a higher intensity of movements, as these groups include gross psychomotor tasks, such as the obstacle courses. In line with this, this higher intensity of movements is also seen in the activities of groups B and D if looking at the results of the body temperature signal. It can be seen how body temperature values are lower in these two groups (B and D), as body temperature increases with inactivity, and decreases with increasing motor activity. Finally, the room temperature result is also included as a sample of a signal measured in the environment. It is observed for example in group C that the body temperature is higher than in group D even though the ambient temperature is lower in group C. This strengthens the relationship between body temperature and motor activity.

### 3.1 Use of ML Algorithms for Information Extraction

As a result of signal processing and attribute extraction, a dataset of 9942 entries is generated. Once the dataset is built, underlying information is extracted from

**Table 2.** Variance of the original data captured by different number of principal components of the PCA result.

| Number of principal components | Captured variance (%) |
|---|---|
| 10 | 66,86 |
| 20 | 77,56 |
| 50 | 94,95 |
| 100 | 99,99 |
| 182 | 100 |

these data using ML algorithms. Unsupervised learning methods will be used for this task, as no information about the user or his behaviour is available beyond the one provided by the reports of the sessions performed. Therefore, the generated labels with the group of activities to which the dataset entries belong will only be used as a guide to analyse the obtained results of the used algorithms.

First, dimensionality reduction methods are applied to the dataset, which will allow subsequent algorithms to work faster and also prevent some of them not providing valuable information. The first of the applied methods was the PCA. The results of applying this algorithm are shown in Table 2. It can be seen that with the first 50 calculated principal components, almost 95% of the variance of the original data is retained. Using only the 10 first components would conserve more than 65% of the variance in the data, and the first two components account more than 30% of the variance data. Figure 4 shows the space distribution using the first two components of the PCA with the group labels of the training dataset. It can be seen how with only two components it is able to separate the different inputs of the dataset. Other algorithms for dimensionality reduction such as incremental PCA, sparse PCA, Singular Value Decomposition and Random Gaussian Projection have also been applied. The result of these algorithms has been very similar with the PCA results. Therefore, we have continued to work with the obtained results with PCA, as it is simpler and therefore the execution times are shorter than with other more complex.

The next step was to apply clustering algorithms to try to identify those entries in the dataset that are similar to each other. First, the K-Means algorithm is applied. For this, it is necessary to specify the number of clusters into which the algorithm will group the entries of the dataset. The algorithm will try to optimise the model by minimising the variation of the data within each cluster, known as inertia. Initially, a sweep of a range of clusters has been performed to see how inertia decreases as the number of clusters increases. Figure 5a a shows this. Another way to check how increasing the number of clusters improves the result, i.e. the different groups will be more homogeneous, is by calculating the accuracy using the generated labels with the different groups of activities as

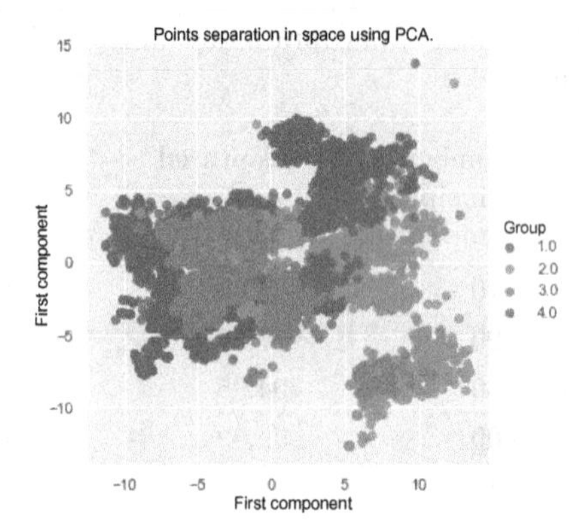


**Fig. 4.** Spatial distribution of the task group labels using the two first components resulting from the PCA algorithm.

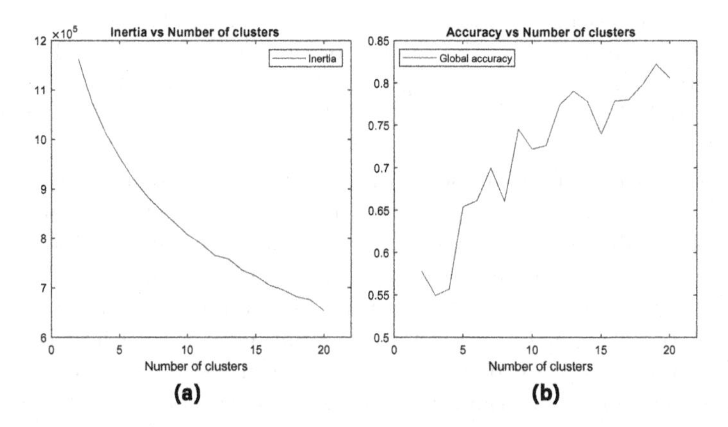


**Fig. 5.** **(a)** Graph of the K-Means algorithm inertia results according to the number of clusters used in the training. **(b)** Graph of the overall accuracy of the model as a function of the number of the used clusters.

a reference. Figure 5b shows the overall accuracy of the model for a different number of clusters. It can be seen that the accuracy increases with increasing the number of clusters. It should be noted that the model has not been trained with the activity group labels, so it is the model which has extracted an underlying structure from the dataset.

On the other hand, the hierarchical clustering algorithm has also been tested. On this occasion it is not necessary to indicate the number of clusters for training, but rather the algorithm will generate a dendogram where each of the entries are in the base and these are grouped into clusters until the final cluster is reached, which would be the complete dataset. Then, it is decided where to cut and this fixes the number of clusters. To do this, it must be determined what euclidean distance will be the maximum allowed between two clusters. To compare the results with the previous model, the distance for which the number of clusters will set to 20 are selected. For the obtained results, the maximum distance between clusters should be set to 151.5. The overall accuracy of the model for this choice is 81.57%, very similar with the previous one (80%). Finally, two other clustering algorithms, DBSCAN and HDBSCAN, a hierarchical version of DBSCAN, have also been applied. However, the achieved results with these algorithms were worse than with the previous algorithms (49.31% and 68.86% accuracy, respectively).

In order to continue extracting information about the user's response to different stimuli, it will be necessary to consider as future work the inclusion of more information about the performed activities and the user's behaviour during these activities. This will allow to apply algorithms for anomaly detection, to combine semi-supervised learning and to establish situations that should be detected.

## 4 Conclusions

The conducted study has allowed to evaluate the use of a monitoring platform for people with ASD in a clinical scenario. The obtained results are presented with a participant user. In addition, from the collected signals during the sessions, it has been possible to extract information about the environment and the state of the person using machine learning techniques. However, it will be necessary to improve the acquisition protocol and the information about the user during the performance of the tasks in order to continue working with the collected signals. This would allow to establish a relationship between changes or anomalies in the user with the present stimuli in the environment.

**Acknowledgements.** The authors would like to thank the participants and their families for their participation in the study. In addition, they would like to thank the collaboration of the professionals of the Clínica Universitaria.

## References

1. Ayres, A.J.: Sensory integration and learning disorders. Western Psychological Services (1972)
2. Pfeiffer, B.A., Koenig, K., Kinnealey, M., Sheppard, M., Henderson, L.: Effectiveness of sensory integration interventions in children with autism spectrum disorders: a pilot study. Nat. Instit. Health **65**(1), 76–85 (2013)
3. Baker, A.E., Lane, A., Angley, M.T., Young, R.L.: The relationship between sensory processing patterns and behavioural responsiveness in autistic disorder: a pilot study. J. Autism Dev. Disord. **38**(5), 867–875 (2008). https://doi.org/10.1007/s10803-007-0459-0
4. Vicente-Samper, J.M., Ávila-Navarro, E., Sabater-Navarro, J.M.: Data acquisition devices towards a system for monitoring sensory processing disorders. IEEE Access **8**, 183596–183605 (2020). https://doi.org/10.1109/ACCESS.2020.3029692

5. Vicente-Samper, J.M., Avila-Navarro, E., Esteve, V., Sabater-Navarro, J.M.: Intelligent monitoring platform to evaluate the overall state of people with neurological disorders. Appl. Sci. **11**(6), 2789 (2021). https://doi.org/10.3390/app11062789
6. World Medical Association: World medical association declaration of Helsinki: ethical principles for medical research involving human subjects. JAMA **310**(20), 2191–2194 (2013). https://doi.org/10.1001/jama.2013.281053

# ApEn: A Stress-Aware Pen for Children with Autism Spectrum Disorder

Jing Li[1(✉)], Emilia Barakova[1], Jun Hu[1], Wouter Staal[2], and Martine van Dongen-Boomsma[2]

[1] Eindhoven University of Technology, 5612 AZ Eindhoven, The Netherlands
{j.li2,e.i.barakova,J.Hu}@tue.nl

[2] Karakter academisch centrum, Horalaan 5, Ede 6717 LX, The Netherlands
{w.staal,m.vandongen-boomsma}@karakter.com
https://www.karakter.com/

**Abstract.** Children with Autism Spectrum disorder (ASD) often experience high levels of anxiety and stress. Many children with ASD have difficulty in being aware of their stress and communicating distress to family and caregivers. Stress detection and regulation are vital for their mental well-being. This paper presents a stress-aware pen (ApEn) that detects real-time stress-related behaviors and interacts with users with vibrotactile and light as a feedback indication of interpreted stress levels. ApEn is a context-aware tool for collecting behavioral data related to the expression of stress and can increase users' stress awareness. A pilot test was conducted with typical developed children to investigate how to detect stress in their daily environment. The pilot test results indicate that ApEn is a promising tool for detecting stress-related behaviors and can attend the user about the detected stress through designed sensory feedback.

**Keywords:** Smart everyday objects · Children with ASD · Stress detection

## 1 Introduction

Children with Autism Spectrum Disorder (ASD) have a higher prevalence of anxiety disorders and poorer stress management than in the general population [16]. They have difficulty in having sufficient awareness of stress and communicating their distress to family and caregivers [28]. Stress is extremely dangerous for children with ASD, since it can result in restrictive, repetitive behaviors, and even self-injurious behavior [28]. Therefore, timely stress detection and regulation are highly required for enhancing the well-being in children with ASD.

Devices for monitoring physiological signals (e.g. smart wristbands, and chest bands) are widely applied in stress research by children with ASD [2,6,12,28]. These devices often do not embody a potential of behavior change. In addition, the existing devices mostly monitor physiological responses that provide limited


J. M. Ferrández Vicente et al. (Eds.): IWINAC 2022, LNCS 13258, pp. 281–290, 2022.
https://doi.org/10.1007/978-3-031-06242-1_28

contextual information [19]. We argue that everyday objects that are already used in the daily life of children with ASD, can help unobtrusively detect stress from psychological, behavioral and environmental variables [7] and allow continuous and more accurate estimation and indication of stress [15,21,29].

Children with ASD often have a hard time staying focused on learning tasks, and thus stress is more noticeable during moments of learning. Therefore, stress detection during studying is important for children with ASD and their caregivers. Apen is therefore designed for monitoring stress in the natural learning environment. We explored which types of interactions with everyday objects were potential indicators of stress under the context. A stress-aware pen was introduced since there was a predicted correlation between the writing behavior and stress. Data were collected through a design probe - ApEn to find out "How an everyday object as a pen should detect and respond to stressful behaviors to increase the awareness of stress in children with ASD? ".

We conducted an exploratory experiment in typical developing children to investigate if ApEn can detect stress behavior and provide awareness of stress. It reveals the correlation between stress and pressure applied in writing and holding the pen. ApEn provided vibrotactile and light feedback to children's stress levels. The data from the pressure sensors was collected and the designed feedback was evaluated with children and their families.

## 2 Related Work

**Stress Measurement.** Stress can be classified into physiological stress and perceived stress. Physiological stress can be detected mainly in the physiological response of an individual, and perceived stress is based on mental appraisal and psychological response of an individual [25]. Physiological stress can be quantified from several biomarkers such as heart rate (HR), and heart rate variability(HRV) [22]. Perceived stress is often measured through periodical self-report collected from the individuals [25]. Perceived and physiological stress were shown to be correlated [6].

People respond to stress in different ways, physiologically, psychologically, and behaviorally [5]. Children exhibit a clear behavioral response under stress. The anxious behaviors can be observed in object manipulation(e.g. playing with an object) [5,8]. Research reports results on stress detection from behavioral data such as movements or activities [9,19] with sensors embedded in objects.

**Smart and Interactive Everyday Objects.** Previous research has identified several requirements for designing stress detection devices for children with ASD [12]. The device has to be a non-invasive [2]; it should be easy to use and easy to learn [20]; and, the cost of purchasing and maintaining the device should be affordable for families [28].

Everyday objects that we are comfortable and familiar to live with, such as a cup, a lightbulb, a table, etc. The way we interact with them does not require a new set of skills, and we don't have to learn new gestures, icons, color codes, and button combinations [23]. Those pervasive everyday objects can be used to collect stress related information through human-product interactions [14,17,18] in an unobtrusive way [23], to further help recognize stress situation and identify stressful behavior [19].

**Behavior Study.** Research has been conducted on evaluating which behaviors are associated with a pen when one feels anxious and stressed. The work of Bruns [5] found that pressure or force could indicate stress, and squeezing and pressing are predictors of stress while one interacts with a pen. Other studies [10,11] reported that more pressure was detected on handwriting when people are under stress. The above evidence leads us to measure the handwriting pressure, and the pressure applied on the body of the pen.

**Feedback Exploration.** Light effects are widely applied in the interaction design, with the purpose of informing, alerting also influencing human behaviors. Different light color and light patterns especially breathing light often applied in stress-related design [27,31]. The breathing light pattern is also implemented in the final design of an ApEn, in consonance with the levels of the detected pressure from the pen, for increasing users' stress awareness and influencing their behaviours. Vibrotactile effects has great value in passing on information [24]. A library called VibViz1 [24] has been built to reveal the possible interpretation from users depending on the intensity, duration, rhythm, and location of vibration signals. Based on the library, we selected and tested the vibration patterns for drawing attention and influencing behaviours for the final design.

## 3 Method and Design Process

We conducted interview with parent couples having a child with ASD. Three families with children with autism severity 1–2 [26] and age between 8–10 years old took part in the research. The interviews indicated that stress in children with autism is often noticed during studying. Sitting for a long time and focusing on study can be challenging and increase their stress level. The parents mentioned some typical stressful behaviors in children with ASD such as mumbling, shouting and some repetitive body movements. These stressful behaviors are often associated with some objects, such as a pen and an eraser. In the interviews the parents reported that children with ASD tend to put harder force on writing, and bend or squeeze the pen when they are under pressure. Parents think it is important but challenging for them to accompany children during studying activities while coping with children's stress timely, since it requires a great attention and a huge amount of time from the parents. Thus, we focus on detecting and regulating stress in children with ASD during their focusing time.

We followed a Research-through-Design process [13] that involved creation of a design probe, testing, data collection and analysis and a corresponding redesign of the design probe towards the final ApEn design.

## 4 Final Design

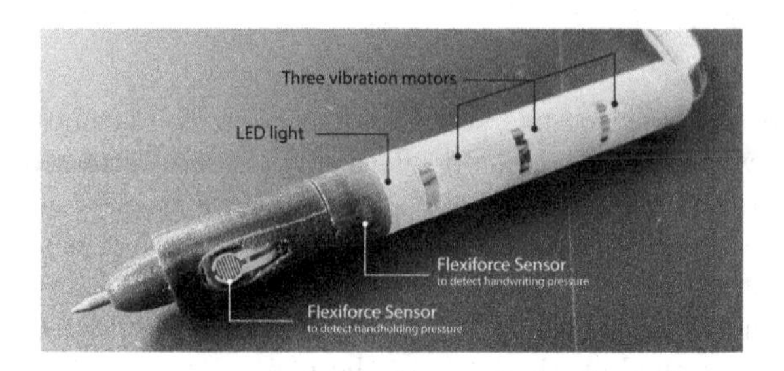


**Fig. 1.** Apen: the stress-aware pen.

The stress-aware pen(ApEn, as shown in Fig. 1) is designed to detect stress-related behaviors by sensing the handwriting and hand-holding pressure. Two Flexiforce sensors are embedded to detect pressure through the pen lead and the pen body. In order to draw children's attention to their stress-related behaviors, three vibration motors and one LED light are used to provide feedback, as Fig. 2(a) left side shows. In the ***feedback mode***, there are three thresholds set for handwriting and hand-holding pressure. When the average handwriting pressure is above threshold 1 for 10 s, the LED light turns on with blue color, and the vibration is growing gradually from low to high in one second. When the average handwriting pressure is above threshold 2 for 10 s or the hand-holding pressure is above threshold 3, the LED light turns red, and one vibration motor vibrates 3 times with a short duration and a strong intensity. The thresholds are calibrated according to the handwriting and hand-pressing habits of each user.

The other mode is the ***feedforward mode***, as Fig. 2(b) right side shows. It is designed not only for helping users be aware of stress but also for evaluating if ApEn can intervene and cope with stress. The breathing blue light changes the breathing rate depending on the different ranges of handwriting pressure. The higher the average pressure of handwriting is, the faster the light breathes. The maximum breathing rate of LED is once each two seconds that is within the average respiration rate of a child. In this way, the breathing rate of LED can not only indicate the stress levels but also be expected to influence users' respiration when users start to breath with it. In the meantime, when handwriting force is above the threshold 3, the vibration starts from the bottom vibration motor, moving to the middle one and then the top motor. With the vibrotactile guide on bottom-up direction, the users are expected to adapt their handwriting behavior, for example, lifting hands from writing when the high pressure is detected. The pen also reacts to handholding pressure in real-time when above threshold 4, with a short and intensive vibration to alert users.

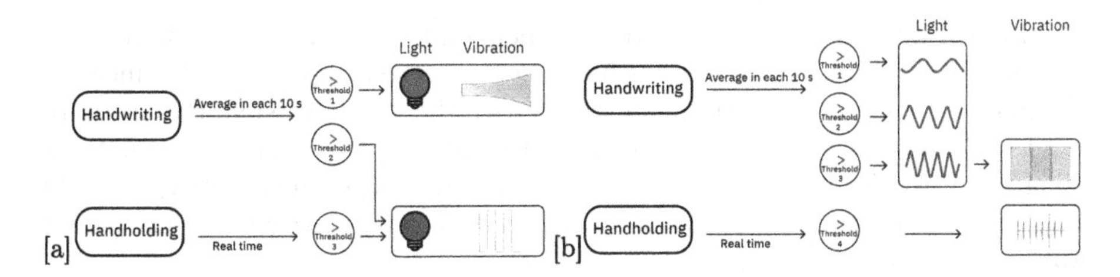


**Fig. 2.** The feedback mode of the stress-aware pen (a). The feedforward mode of the stress-aware pen (b).

## 5 Experiments

**Participants.** Three parent couples of in total four typical developing children (age 7–10 years, two girls and two boys) were included. The child participant with one of their parents joined in the experiment. The children from these three families all experienced stressful situations in studying.

**Experiment Design and Data Collection.** Stress detection experiments are often designed in the laboratory with inducing stress deliberately [9]. Since stress is less common and distinct in real life [9] to be detected in short time experiments and the physiological measurement is easy to be disturbed in the field experiments [9]. Therefore, our experiments in the field were designed similar to laboratory experiments, with structured context containing potential stressors.

The experiments were done in the context of studying at participants' home. In order to conduct experiments in the field, the context of the experiment needs to be personalized depending on each child's educational background in order to induce potential stress. To do so, studying tasks consisting of mathematic and linguistic questions for each child were prepared from easy to hard level depending on their educational progress. For motivation and pressure consideration, a reward was promised to all children if they finished the tasks on time with a certain accuracy.

The data collected in the experiments consist of psychological, physiological, and behavioural data. The physiological data were collected by parents via self-assessment manikin(SAM) [4]; physiological data, including HR and BVP, was captured using an E4 wristband equipped with sensors. ApEn detected the real-time behavioural data in terms of handwriting and hand-holding pressure. Additionally, video was recorded during each experiment to provide additional insights of the user behavior and the contextual information.

**Procedure.** Before the experiment, each child participant was asked to try out ApEn with their normal and high pressure to calibrate the thresholds. Each experiment consisted of two sessions: the feedback and the feedforward sessions. When the experiment started, the children were asked to wear the E4 wristband

and use ApEn to finish the prepared tasks independently. Each session took up to 20 mins. The children had a 10 mins break between the two sessions to reduce the possibility of accumulated stress interfering with the second session. Parents were requested to stay beside the children only to annotate their children's emotions using the SAM form once every 5 mins. When the experiment ended, Parents were interviewed by the facilitator to reflect on the experiment and the use of ApEn.

## 6 Results

**Stress and the Corresponding Behaviors.** The experiments were carried out without fully controlled stressors, however, the study needed to have an objective evidence of aroused stress. Physiological data specially HRV allows the reliable identification of stressful events [3]. Thus, HRV was calculated. It was analyzed from photoplethysmography (PPG) data from the E4 wristband using Kubios [1] in MatLab.

The detected stress from HRV of one of the children during feedback session shows an abrupt change in mean HR between 7 to 13 minuses after the experiment started. Correspondingly, Fig. 3 shows that pressure data in handwriting and hand-holding had an elevation, and feedback (light and vibration) was given by ApEn a few times during the same time slot. Simultaneously, parent marked a raise (from level one to level two) in arousal index in SAM form during 5–10 minuses and it stayed at level two till the end of the next time slot. There was only one stress event detected among all participants' physiological data. Moreover, the pressure signals from force sensors dropped every time after feedback given by the pen, as shown in Fig. 3.

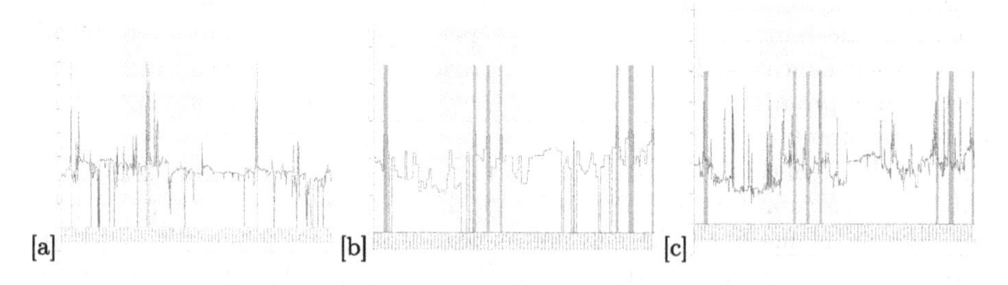


**Fig. 3.** Handholding (a), Handwriting average every 10 s (b), and real time (c) pressure of one of the children in feedback session (in blue). Orange column is when feedback is given. Light yellow space is a marked growth in arousal (SAM). (Color figure online)

**Observed Stress and the Corresponding Behaviors.** Besides detected stress in physiological data, stress was also marked in one of the parents' annotation and observed from video footage and the handwriting of her child. Figure 4

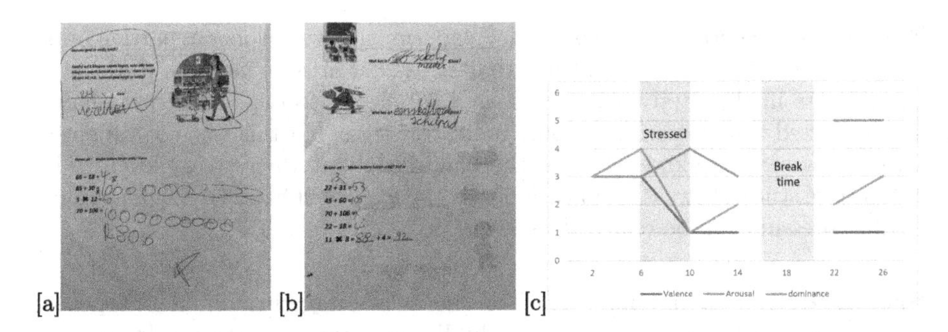


**Fig. 4.** Handwriting at stress moments (a) as observed by the parents and Handwriting without noticed stress (b). Data Visualization of one participant's SAM (c), the red period was "stressed" moment between 4 to 10 mins after feedback session started, and the empty area was during the break without annotation data.

(a) and (b) shows the comparison between the child's handwriting with and without stress. The detected stress moments are based on the stress assessment by the parent in (c) Fig. 4. The handwriting shows more aggression and strength on (a) comparing to (b).

**User Experience with ApEn.** We collected user experience feedback from the parents and children. First, the vibration feedback was sensed well by the children, also noticed by parents in general. For example, a child mentioned: "After vibration happens, I relaxed my fingers or re-positioned my fingers on the pen." Second, regarding the light feedback, both parents and children could notice the light and the color change in the feedback session without putting too much attention on it.

## 7 Discussion

**Potential Difference in Stress Responses of Children with ASD.** Physiological measures such as skin conductance and heart rate variability are common indication for stress [28]. Therefore, we can expect a lower HRV by the children with ASD during testing the ApEn. The expressions of stress behaviors in children without ASD in this study could be different from children with ASD. The reason to conduct experiments with typically developing children is to avoid potential risks to the vulnerable group of children with ASD at this early stage of research. We expected that even behavioral expression of stress could differ by children with and without ASD, physiological expression are similar and it is possible to gain valuable insights on the research questions and provide input and benchmark for the next steps. Further research needs to be done with the target group.

**Exploratory Study.** The exploratory experiments were done at participants' home. There was no direct evidence that stress did arise in children during the experiments in this study since stress is a subjective feeling. All stress related data collected from experiments only indicates stress, but this was not validated since we were not allowed to ask the children about their stress levels. The explorative study shows the indication that the change of handwriting and hand-holding pressure can be related to stress.

**Bias in User Experience.** The thresholds of triggering different types of feedback were set only based on each child's short time calibration. In general, to make sure giving the same user experience to every participant and study the effect of the feedback, it requires more calibration for better thresholds, or instead of thresholding, more advanced machine learning to detect stress patterns. This study is exploratory and the focus is not user experience or stress reduction, but the forms of sensory feedback in raising awareness of stress in children. Nevertheless, we should still consider the potential bias in users' feedback.

**Future Work.** Firstly, ASD is a heterogeneous condition. The profile of each child or adult with ASD is unique. Thus, a personalized system needs to be built to learn behavioral patterns from daily life. Also, physiological data can be used not only for detecting stress events but also for designing biofeedback [30] to increase stress awareness to cope with stress. The ultimate goal of this research is to intervene and help children's stress reduction. Social support from family and people around children with ASD is essential to cope with their stress, so their stress can be managed with external assistance in time. Enhancing ApEn to a connected object can contribute to data collection for the daily-based or clinical assessment on stress. Ultimately, involving ApEn to an IoT system that enables children, caregivers and clinics can work together on data collection and stress coping strategy is a promising direction for stress detection and regulation in children with ASD.

## 8 Conclusion

This research aims to study stress-related behaviors in the natural environment and explore how to enhance everyday objects for stress detection and regulation. The results indicate that the relation of children's interaction with Apen with their stress. There is a great potential to learn children's stress from their interaction with everyday objects. Although the design focus is on children with ASD, ApEn can be applied for different scenarios. For example, ApEn can be used for clinical research in assessing children's stress level. ApEn can also support children with stress regulation when they have heavy study load at school. This paper offers an example of enhancing everyday objects to support stress detection and insights on intervening in stress. Based on the preliminary results, further design can explore how to make the pen as a connected object to better support stress detection and reduction.

## References


1. Kubios hrv analysis software. https://www.kubios.com/
2. Airij, A.G., Bakhteri, R., Khalil-Hani, M.: Smart wearable stress monitoring device for autistic children. Jurnal Teknologi **78**(7–5) (2016)
3. Azzaoui, N., et al.: Classifying heartrate by change detection and wavelet methods for emergency physicians. ESAIM: Proc. Surv. **45**, 48–57 (2014)
4. Bradley, M.M., Lang, P.J.: Measuring emotion: the self-assessment maninkin and semantic differential. J. Behav. Ther. Exp. Psychiatr. **25**, 49–59 (1994)
5. Bruns, M.: Relax!: inherent feedback during product interaction to reduce stress (2010)
6. Can, Y.S., Arnrich, B., Ersoy, C.: Stress detection in daily life scenarios using smart phones and wearable sensors: a survey. J. Biomed. Inform. **92**, 103139 (2019)
7. Carr, E.G., Durand, V.M.: Reducing behavior problems through functional communication training. J. Appl. Behav. Anal. **18**(2), 111–126 (1985)
8. Fleege, P.O., Charlesworth, R., Burts, D.C., Hart, C.H.: Stress begins in kindergarten: a look at behavior during standardized testing. J. Res. Child. Educ. **7**(1), 20–26 (1992)
9. Gjoreski, M., Gjoreski, H., Luštrek, M., Gams, M.: Continuous stress detection using a wrist device: in laboratory and real life. In: proceedings of the 2016 ACM International Joint Conference on Pervasive and Ubiquitous Computing: Adjunct, pp. 1185–1193 (2016)
10. Halder-Sinn, P., Enkelmann, C., Funsch, K.: Handwriting and emotional stress. Percept. Mot. Skills **87**(2), 457–458 (1998)
11. Keinan, G., Eilat-Greenberg, S.: Can stress be measured by handwriting analysis? The effectiveness of the analytic method. Appl. Psychol. Int. Rev. **42**(2), 153–170 (1993)
12. Koo, S.H., Gaul, K., Rivera, S., Pan, T., Fong, D.: Wearable technology design for autism spectrum disorders. Arch. Des. Res. **31**(1), 37–55 (2018)
13. Koskinen, I., Zimmerman, J., Binder, T., Redstrom, J., Wensveen, S.: Design research through practice: from the lab, field, and showroom. IEEE Trans. Prof. Commun. **56**(3), 262–263 (2013)
14. Liang, Y., Zheng, X., Zeng, D.D.: A survey on big data-driven digital phenotyping of mental health. Inf. Fus. **52**, 290–307 (2019)
15. Marvar, P.J., et al.: T lymphocytes and vascular inflammation contribute to stress-dependent hypertension. Biol. Psychiat. **71**(9), 774–782 (2012)
16. Merikangas, K.R., et al.: Lifetime prevalence of mental disorders in us adolescents: results from the national comorbidity survey replication-adolescent supplement (NCS-A). J. Am. Acad. Child Adolescent Psychiatr. **49**(10), 980–989 (2010)
17. Mohr, D.C., Zhang, M., Schueller, S.M.: Personal sensing: understanding mental health using ubiquitous sensors and machine learning. Annu. Rev. Clin. Psychol. **13**, 23–47 (2017)
18. Moura, I., et al.: Mental health ubiquitous monitoring supported by social situation awareness: a systematic review. J. Biomed. Inform. **107**, 103454 (2020)
19. de Moura, I.R., Teles, A.S., Endler, M., Coutinho, L.R., da Silva E Silva, F.J.: Recognizing context-aware human sociability patterns using pervasive monitoring for supporting mental health professionals. Sensors **21**(1), 86 (2021)
20. Pantelopoulos, A., Bourbakis, N.G.: A survey on wearable sensor-based systems for health monitoring and prognosis. IEEE Trans. Syst. Man Cybern. Part C (Appl. Rev.) **40**(1), 1–12 (2009)

21. Parlak, O., Keene, S.T., Marais, A., Curto, V.F., Salleo, A.: Molecularly selective nanoporous membrane-based wearable organic electrochemical device for noninvasive cortisol sensing. Sci. Adv. **4**(7), eaar2904 (2018)
22. Picard, R.W.: Affective computing: challenges. Int. J. Hum Comput Stud. **59**(1–2), 55–64 (2003)
23. Rose, D.: Enchanted objects: Design, human desire, and the Internet of things. Simon and Schuster (2014)
24. Seifi, H., Zhang, K., MacLean, K.E.: VibViz: organizing, visualizing and navigating vibration libraries. In: 2015 IEEE World Haptics Conference (WHC), pp. 254–259. IEEE (2015)
25. Sharawi, M.S., Shibli, M., Sharawi, M.I.: Design and implementation of a human stress detection system: a biomechanics approach. In: 2008 5th International Symposium on Mechatronics and Its Applications, pp. 1–5. IEEE (2008)
26. Speaks, A.: DSM-5 diagnostic criteria. Retrieved from (2014)
27. Ståhl, A., Jonsson, M., Mercurio, J., Karlsson, A., Höök, K., Johnson, E.C.B.: The soma mat and breathing light. In: Proceedings of the 2016 CHI Conference Extended Abstracts on Human Factors in Computing Systems, pp. 305–308 (2016)
28. Taj-Eldin, M., Ryan, C., O'Flynn, B., Galvin, P.: A review of wearable solutions for physiological and emotional monitoring for use by people with autism spectrum disorder and their caregivers. Sensors **18**(12), 4271 (2018)
29. Tonhajzerova, I., Mestanik, M., Mestanikova, A., Jurko, A.: Respiratory sinus arrhythmia as a non-invasive index of 'brain-heart' interaction in stress. Indian J. Med. Res. **144**(6), 815 (2016)
30. Yu, B., Funk, M., Hu, J., Wang, Q., Feijs, L.: Biofeedback for everyday stress management: a systematic review. Front. ICT **5**, 23 (2018)
31. Yu, B., Hu, J., Funk, M., Feijs, L.: Delight: biofeedback through ambient light for stress intervention and relaxation assistance. Pers. Ubiquit. Comput. **22**(4), 787–805 (2018)

# Anxiety Monitoring in Autistic Disabled People During Voice Recording Sessions

Marina Jodra-Chuan[1,2], Paula Maestro-Domingo[3], and Victoria Rodellar-Biarge[3,4](✉)

[1] Department of Personality, Assessment and Clinical Psychology, Faculty of Education, Complutense University of Madrid, C/ Rector Royo Villanova 1, 28040 Madrid, Spain
majodra@ucm.es

[2] Asociación Nuevo Horizonte, Comunidad de Madrid 43, Las Rozas de Madrid, 28231 Madrid, Spain

[3] Escuela Técnica Superior de Ingeniería Informática, Universidad Politécnica de Madrid, Campos de Montegancedo, Boadilla del Monte, 28660 Madrid, Spain

[4] NeuSpeLab, Center for Biomedical Technology, Universidad Politécnica de Madrid, Campus de Montegancedo, 28220 Pozuelo de Alarcón, Madrid, Spain
mariavictoria.rodellar@upm.es

**Abstract.** The objective of the present study is to define potential biomarkers from the phonation of People with Autistic Spectrum Disorder (PwASD), with the purpose of better understanding the syndrome functional behavior regarding excitment and anxiety arousal when conducting phonation exercises, in order to be able of developing specific rehabilitation protocols contributing to improve their quality of life. The study of sustained vowel utterances from a male participant with autism and intellectual disability has allowed obtaining longitudinal estimations of vocal fold tremor, potentially associated with neurological excitment in performing vocalization tests. Relative important correlations have been found between the neurological and flutter tremor bands and surface skin conductance.

**Keywords:** Autistic disorder · Actigraphy technology · Electrodermal activity · Accelerometry · Phonation tremor

## 1 Introduction

This work is part of a research project whose objective is to find digital biomarkers present in the phonation of People with Autistic Disorder (PwASD), and intellectual disability, with the purpose of better understanding the syndrome,

This research received funding from grants TEC2016-77791-C4-4-R (Ministry of Economic Affairs and Competitiveness of Spain), and Teca-Park-MonParLoc FGCSIC-CENIE 0348-CIE-6-E (InterReg Programme). The authors wish to thank Asociación Nuevo Horizonte participants for their valuable collaboration.


J. M. Ferrández Vicente et al. (Eds.): IWINAC 2022, LNCS 13258, pp. 291–300, 2022.
https://doi.org/10.1007/978-3-031-06242-1_29

and to be able to develop specific tools that contribute to improve their quality of life [1].

Nowadays, there is an increasing intense activity in the search for ASD biomarkers capable of obtaining biochemical data from urine and blood and other biomedical and clinical data, collected during periodic check-ups of PwASD. The procedures associated to these biomarker acquisition methods are often invasive and expensive, and present an incomplete view of the nature, dynamical and functional behavior of ASD due to the limited number of measurements that can be performed on a given session and over a specific timeline, having in mind the specificities and conduct limitations of PwASD. The fast surge of the Internet of the Things (IoT), the pervasive capacity of modern smart devices (phones, tablets, and others), and the development of Actigraphy based on sensor wearable devices have pushed a major revolution in the acquisition of biological data, and in the proposal of new digital biomarkers [2]. These new markers are generally non-invasive, easy, and cheap to record, can produce qualitative and quantitative measurements, and allow continuous data acquisition, thus facilitating longitudinal studies outside the clinics. Typical highly-semantic measurements that these devices allow recording of biomedical interest are heart and respiratory rate, blood oxygen level, electrocardiographic activity, blood pressure, glucose level, temperature, electrodermal activity (EDA), etc., as well as geo-positioning, movement detection (X, Y, Z axes), eye movements, pace rhythm and stride, etc.

Speech and voice are specific and highly semantic correlates conveying important information on emotional activation and neurological alterations (cognitive or neuromotor). Dysfunctions and alterations that can occur in speech planning, in the activation of the neuromotor cortex and in the deterioration of the extrapyramidal and bulbar systems leave complex and semantic-rich marks on voice and speech. ASD people show emotional and neurological alterations, therefore their voice and speech seem to be good candidates to look for biomarkers related to autism, conveying important advantages, as ubiquity, simplicity, and low cost. On the other hand, a digital biomarker that is of high significance in ASD is Electro-Dermal Activity (EDA), also known as skin conductance or galvanic response. It is typically used to detect levels of stress and to prevent behavioral problems, seizures, as it tends to increase with emotional or physical arousal [3].

One of the main problems limiting progress in using biomarkers from speech and phonation is the lack of data collections or open databases of biometrical speech signals or biomarkers that allow comparing results between different studies. This research is focused on the problem of voice data acquisition and other digital biomarkers in ASD, and on the generation of a database that might be shared among researchers.

In this sense, the objectives of this work are focused on proposing a methodology for the synchronized recording of voice signals and EDA in PwASD, and in showing how to analyze the anxiety that the recording sessions may produce in the participants. The study has been carried out on three male and three female participants during three consecutive recording sessions, although detailed results from only a male participant have been reported, for the sake of the manuscript space limits.

The paper is organized as follows: A description of the fundamentals supporting the study is provided in Sect. 2. The data set used in the research is also described in Sect. 3, as well as the methodology used in the study. Section 4 is devoted to present the results of recordings obtained from a study case in terms of timely evolution and correlation contents, which are compared graphically and in tables. Section 5 is intended to discuss the implications of the comparisons, analyze their possible application, and remark the study limitations and its possible future extension. Section 6 summarizes the main contributions, findings and conclusions.

## 2 Fundamentals

People with neurological disorders may manifest functional altered behavior in speech, as in fluency and prosody with slow and difficult language, in dysarthrias, manifested as dyskinetic facial, jaw, or lingual movements, or as dysphonias, showing perturbations in the fundamental frequency. These manifestations may appear as a consequence of cerebrovascular accidents, Parkinson's Disease, or Multiple or Amyotrophic Lateral Sclerosis, and many other examples.

ASD manifests neurological disorders, therefore it is expected that speech alterations may appear as a consequence. The compilation work by Fusaroli [4] capitalizing on whether the voice is a marker of the autism spectrum offers a very interesting overview and exposes that most of the studies of the voice in people with ASD have been carried out standing mainly on prosody analysis. The characteristics that have classically been analyzed in these studies are: the mean and the variation of the fundamental frequency, and of the intensity profile of the sound wave, the duration, the pauses and the speed of speech and the quality of the voice described on simple perturbation features (jitter and shimmer). The main finding of Fusaroli's work is that there is no conclusive evidence of acoustic markers of ASD that can predict the clinical characteristics of the syndrome.

Quite a different approach in the search for markers derived from ASD voice is the use of biomechanical features estimated using an inverse model to process speech in the reconstruction of the glottal source [5]. The inverse problems start from observable data produced by a given process or model, and look for the causes or excitations that produced those observables. This is the approach being used in the present research, in which an inverse model reconstructs the glotal source, and estimate the parameters of a biomechanical model of phonation to search for markers in voice. It starts from the voice signal, removing the radiation effects of the lips and the vocal tract transfer function, reconstructing the glottal source [6]. The inversion of a simple mass-and-spring biomechanical model the biomechanical strain of the musculus vocalis is estimated, from which, phonation tremor may be characterized. This estimate is to be correlated with the emotional and neurological states of the speaker.

The ultimate goal of this approach is to use tremor features as biomarkers of possible altered neurological states of PwASD, as voice is a vehicular pervasive and easy to record signal, although the relatively complex inversion process has to be supported by some kind of *ground truth* observable. It is suggested that this supporting trait migth be provided by EDA, as it is well known that this observable is highly related to emotionally altered neurological states [3].

There are different devices to record EDA, although considering the behavioral characteristics of PwASD, it has been decided that the device to be used in this work be the Empatica E4 bracelet [7], because it is oriented towards research, and was one of the first initiatives in the development of portable sensors for the acquisition of physiological data in real time, and there is ample information on its application to ASD. The E4 bracelet is designed to capture good quality signals and has the advantage of not needing the application of contact gel on the sensors or the participants' skin for its operation.

EDA is made up of two main components. One is the Skin Conductance Level (SCL), or tonic level, which varies slowly and depends on hydration, skin dryness, or autonomic regulation, and can differ significantly between individuals. The other component is the Skin Conductance Response (SCR) which normally occurs in response to an emotionally arousing stimulus. This response adds to the tonic changes and shows significantly more rapid alterations.

The argumental line of this research is to show that there is a strong connection between skin conductance alterations and phonation tremor in the studied cases of PwASD.

## 3 Experimental Framework

### 3.1 Materials

This research is part of a project on the longitudinal study of six PwASD (three males and three females), speakers of Spanish, between 40 and 50 years old, suffering from ASD and intellectual disability. It is being conducted on the association *Nuevo Horizonte* in the municipality of Las Rozas de Madrid, Spain. A team composed of psychological and pedagogical specialists are taking part in it. This teams was responsible for the selection of participants. Their demographical data is presented in Table 1, listing the gender, age, and co-morbidities present in each participant, as well as their Childhood Autism Rating Scale (CARS) [8] and Dysexecutive Questionnaire (DEX) [9] scores. CARS is a test to identify the severity of ASD symptoms, and although being initially conceived for children, it may also be applied to adult population [10]. DEX is intended to evaluate real-life executive dysfunctions. The selected participants show relatively well-matched CARS and DEX scores.

The association *Nuevo Horizonte* authorized the signed consent for the experimental recordings in alignment of Helsinki Declaration Ethical Reccommendations.

**Table 1.** Participants' demographical data. M-: male participants; F-: female participants; CARS: Childhood Autism Rating Scale; DEX: Dysexecutive Questionnaire; SID: Severe Intellectual Disability; PE: Psychotic Episodes; E: Epilepsy; OCD: Obsesive-Compulsive Disorders; D: Dysthymia

| Code | Gender | Birth Year | Co-morbidities | CARS | DEX |
|---|---|---|---|---|---|
| F1 | F | 1974 | SID | 41 | 51 |
| F2 | F | 1976 | SID | 41 | 35 |
| F3 | F | 1974 | SID | 45 | 45 |
| M1 | M | 1973 | SID, PE, E | 40 | 29 |
| M2 | M | 1981 | SID, PE, E | 41 | 44 |
| M3 | M | 1970 | SID, OCD, D | 42 | 32 |

The study is based on two types of signals to be collected from participants on different recording sessions: speech and skin conductance. Signal recording requires the use of specific instrumentation depending on its nature. Some PwASD tend to reject the use of measurement devices on their body or clothes and tend to take off and throw them away, so the instrumentation used must be as robust and non-invasive or non-perturbing, and must pass as unnoticed to them as possible. The participants in the study are people who show a certain degree of intellectual disability, and sometimes have difficulty understanding the instructions on how to perform the tests (what sounds they must utter and how they have to). This prevents using complicate voice tests, therefore the voice recording protocol has been made very simple, and consists of uttering the vowel [a:] for more than one second in a sustained manner. There are two important reasons favoring the use of [a:] instead of other vowels: being the most open vowel, it is the best suited phonation regarding vocal tract inversion, and for the same reason, its sustaining till flow exhaustion is a semantic duration mark of respiratory capacity.

**Recording Speech.** The recordings were taken with the Sennheiser SK 300 G2A wireless cardioid Lavalier microphone/transmitter pair, located 15 cm from the mouth, a Sennheiser EM 300 G2 receiver and Adobe Audition® software to manage data acquisition and vowel emission selection). Speech was sampled at 44.1 kHz and encoded at 16-bit resolution in uncompressed .wav format. Further signal processing was carried on using BioMet®Phon, a specific app to estimate biomechanical tremor [11].

The recording sessions took a duration of 3 to 4 min per person; if the duration was longer, the participants might show signs of fatigue and anxiety. Another challenging fact to be taken into account is that the original recording sessions may contain undesirable sounds, as caregivers' speech, noise, screaming, laughter, etc., and sometimes phonations are too short to support robust biomarker estimation.

The presence of caregivers is essential to improve recording speed and quality, as they are usually the persons they interact with daily, therefore it is easier for them to transmit correct instructions to participants. Even though, recordings lasting more than 2 s were seldom. Recordings were taken in a relaxing environment within a quiet and comfortable room, which participants were familiarized with.

**Recording Skin Conductance.** The use of the wristband E4 to obtain skin conductance recordings is straightforward. The main difficulty found in this respect was how to synchronize it with voice recordings within a same session. The instrumentation used for both types of signals only allow independent and asynchronous recording. The method used to make synchronization feasible was to take advantage of the signals provided by the 3D accelerometer included in E4. The participant is asked to raise the arm the E4 is tied to, and lower it suddenly. The caregiver must produce an acoustic mark at the same time, which is later on used to associate the acoustic recording with the skin conductance counting on the accelerometry output showing the arm lowering event. Once the event occurence and duration have been identified the skin conductance from E4 and the voice recordings may be aligned to select the phonation segments of interest for the study and the evolution of the skin conductance.

The recorded data on E4 are collected through a Bluetooth Low Energy connection and the E4 Realtime app, which must be installed on a smartphone or a tablet. The data are transfered to an E4 platform (Empatica E4 connect) where they must be downloaded from for batch-processing.

### 3.2 Methods

The selected vowel segments with a duration of 400 ms were processed with the App BioMet®Phon, which allows extracting up to 72 features from each phonation segment, including perturbation features as jitter, shimmer, and harmonic-noise ratio, as well as a cepstral description of the glottal source. Another relevant feature set are the biomechanical mass and strain associated to the *musculus vocalis*, as well as their unbalance between neighbour phonation cycles. A third set of features are the contact, aduction and abduction gaps, related to the open and close segments of the phonation cycle. Finally, a set of tremors, estimated from the oscillations present in the vocal fold strain, is also evaluated. This is the set of features used in the present study:

- PTA: Physiological Tremor Amplitude in the band $0 < f \leq 4$ Hz.
- NTA: Neurological Tremor Frequency in the band $4 < f \leq 8$ Hz.
- FTA: Flutter Tremor Amplitude in the band $8 < f \leq 12$ Hz.
- GTA: Global Tremor Amplitudes, summarizing mean square root of tremor in all bands.

Tremor features provide information on the presence of defects, instabilities, or feedback problems in the neuromotor system linked to the activation of the *musculus vocalis*.

# 4 Results

**Study Case.** The results shown in detail are based on three samples from participant M1, corresponding to a male born in 1973 (48 years old a the time the recordings took place), who presents an intellectual disability, psychotic episodes, and epilepsy, with a CARS of 40 and a DEX of 29, separated on a week interval, taken on November 19 and 26, and December 3, 2021. Valid utterances of a sustained [a:] lasting more than 400 ms were selected from the recordings, corresponding to 12 valid segments during the two first sessions, and 18 valid segments during the third session. These segments were processed as described in Sect. 3 to estimate the tremor in the PTA, NTA, FTA, and GTA bands. These estimations were compared with the normalized EDA value recorded by the wristband E4 using correlation. The results presented in Fig. 1 correspond to the three recording sessions.

**Tremor and EDA.** The plots in Fig. 1 show some joint evolution of tremor bands and skin conductance. To assess possible relationships, the estimated values of Pearson's correlation coefficient are given in Table 2. Similarly, the evolution of the skin conductance on the three sessions (S1: 2021.11.19; S2: 2021.11.26; S3: 2021.12.03) has been also correlated. The results are $\rho_{12} = 0.95$; $\rho_{13} = 0.88$; $\rho_{23} = 0.97$;

**Table 2.** Pearson's correlation coefficient between EDA and tremor bands. $\rho_{EP}$: EDA vs PTA; $\rho_{EN}$: EDA vs NTA; $\rho_{EF}$: EDA vs FTA; $\rho_{EG}$: EDA vs GTA.

| Session | $\rho_{EP}$ | $\rho_{EN}$ | $\rho_{EF}$ | $\rho_{EG}$ |
|---|---|---|---|---|
| S1: 2021.11.19 | −0.02 | −0.13 | 0.06 | 0.08 |
| S2: 2021.11.26 | −0.37 | −0.58 | −0.64 | −0.33 |
| S3: 2021.12.03 | 0.36 | 0.46 | 0.30 | 0.38 |

# 5 Discussion

The results presented in Fig. 1 and Table 2 allow to extract interesting conclusions. On the one hand, it seems that EDA follows a clear ascend in the three sessions being exposed. This is confirmed by the high values of their Pearson's coefficients. This fact could be interpreted in the sense that the level of excitation in the participant followed a clear progress. Nevertheless, the degree of excitation might not be associated to an increment in anxiety, accordingly to what will be discussed next. On the other hand, tremor seemed to descend in S1 except for segment #9, where the maximum value of all tremors was attained (the series were normalized for an easier comparison). A clearer descent was observed in S2 between the first seven segments with respect to the five latter ones. In S3, there was a steady descent in the average, with sudden jump ups in segments #15–17.

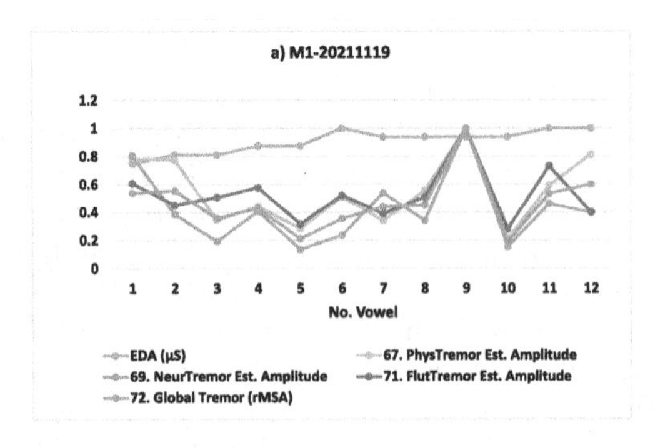


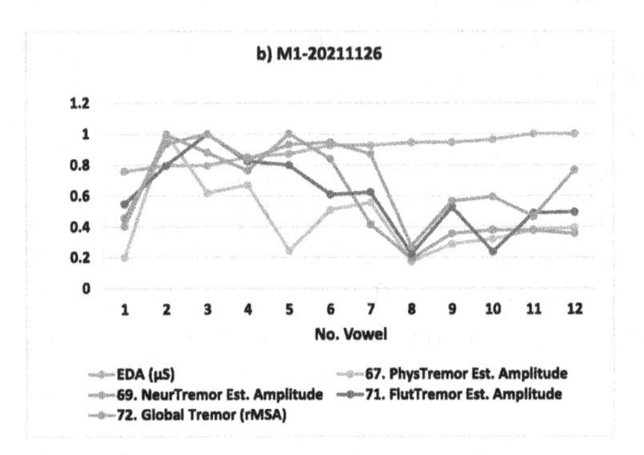


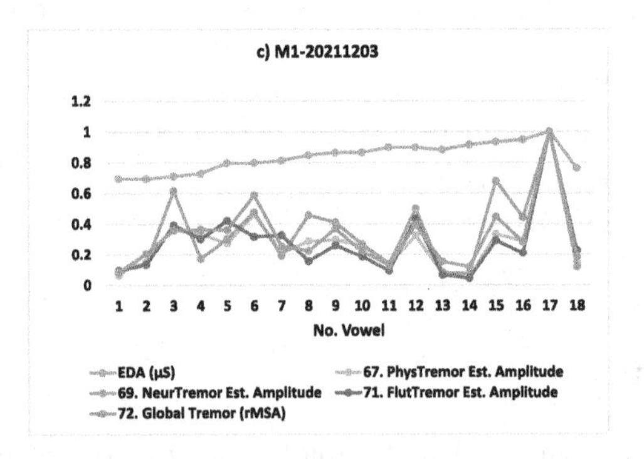


**Fig. 1.** Longitudinal evolution of tremor features and EDA from male participant M1: a) Session S1-2021.11.19; b) Session S2-2021.11.26, 2021; c) Session S3-2021.12.03.

As a consequence, some controversial behavior may be observed in the correlation values shown in Table 2 between EDA and the tremor bands. Session S1 shows almost no correlation between both types of signals, although this might be a consequence of the singular behavior of segment #9. This is a question which should be further investigated, as a single salience may alter the sense of the correlation. Session S2 shows clear although moderate negative correlation in all bands, especially in NTA and FTA, as neurological tremor descended with ascending skin conductance. Could this be interpreted in the sense that EDA is recording excitation, whether this be due to positive excitment or enthusiasm in following the tests, or on the contrary to fatigue or hastening? Necessarily more experiments using other correlates to assess valence should be proposed to find a clear response to this question. Session S3, on its turn, shows lower values of positive correlation in this case. Nevertheless, the general tendency shows a decline trend between segments #3–14, which the exception of a jump up in segment #12, that could be considered an exception. But the most intriguing pattern is the strong increment in tremor activity shown by segments #15–17, responsible for a sing change in the correlation values. Could this factor be due to a longer recording and exhausting session? As with comments regarding S1, new observables should be taken into account to assess valence during the recording sessions. Needless to say, these results lack any statistical relevance, as they come from the single participant studied. If these results could be generalized including the whole set of participants is a matter of a wider study still pending, as the processing of available samples is cumbersome, and will require extra efforts. Of course, new results would open a wider perspective to analyze neurological alterations due to emotional changes during tests. A very important task pending of study is the association of tremor bands related to the FTA within the frequency range associated to the EEG $\beta$ and $\mu$ bands, as it is established that alterations of activity in these bands might be strongly associated with mirror neuron disfunction, thought to be behind ASD [12]. A relevant restraint of the study is its specific scope limited to a single study case. The results from the participants not included in the present study are still under research and discussion. Another limitation is the reduced number of tremor frequency bands included, which barely cover the equivalent $\delta$. $\vartheta$, and $\alpha$ EEG bands. A new version of the App used in the estimation of the frequency band contents of the vocal fold strain will allow sheding more light on possible anomalies in the $\alpha$ and $\mu$ bands. It is expected that the study results may open new possibilities for designing more sensitive protocols in the stimulation and speech rehabilitation of PwASD.

## 6 Conclusions

The study of sustained vowel utterances from an ASD participant has allowed obtaining longitudinal estimations of vocal fold tremor, potentially associated with neurological excitment in performing vocalization tests. Relative relevant correlations have been found between NTA and FTA band tremor and surface skin conductance. The apparently controversial correlation results from the three recording sessions studied pose an important challenge in determining the valence of increasing neurological excitment produced during test performance.

## References

1. Bridgemohan, C., et al.: Investigating Potential Biomarkers in Autism Spectrum Disorder, Frontiers in Integrative Neuroscience, 13, Art. 31 (2019). https://doi.org/10.3389/fnint.2019.00031
2. Babrak, L.M., et al.: Traditional and digital biomarkers: two words apart? Digital Biomarkers **3**(2), 92–102 (2019). https://doi.org/10.1159/000502000
3. Greco, A., et al.: Acute stress state classification based on electrodermal activity modeling. IEEE Trans. Aff. Comput. (2021). https://doi.org/10.1109/TAFFC.2021.3055294
4. Fusaroli, R., Lambrechts, A., Bang, D., Bowler, D.M., Gaigg, S.B.: Is voice a marker for Autism spectrum disorder? A systematic review and meta-analysis. Autism Res. **10**(3), 384–407 (2017). https://doi.org/10.1002/aur.1678
5. Gómez, P., et al.: Glottal Source biometrical signature for voice pathology detection. Speech Commun. **51**(9), 759–781 (2009). https://doi.org/10.1016/j.specom.2008.09.005
6. Alku, P., et al.: OPENGLOT-an open environment for the evaluation of glottal inverse filtering. Speech Commun. **107**, 38–47 (2019)
7. E4 wristband. https://www.empatica.com/en-eu/research/e4/. Accessed 21 Jan 2022
8. Schopler, E., Reichler, R.J., DeVellis, R.F., Daly, K.: Toward objective classification of childhood autism: Childhood Autism Rating Scale (CARS). J. Autism Dev. Disord. **10**(1), 91–103 (1980). https://doi.org/10.1007/BF02408436
9. Pedrero-Pérez, E.J., Ruiz-Sánchez-de-León, J.M., Winpenny-Tejedor, C.: Dysexecutive Questionnaire (DEX): unrestricted structural analysis in large clinical and non-clinical samples. Neuropsychol. Rehabil. **25**(6), 879–894 (2015). https://doi.org/10.1080/09602011.2014.993659
10. García-Villamisar, D., Muela, C.: Psychometric properties of the Childhood Autism Rating Scale (CARS) as a diagnostic tool for autistic adults in the workplace. Revista de psicología general y aplicada **53**, 515–521 (2000)
11. Gómez P., et al.: A system to monitor phonation in clinics. In: The Fifth International Conference in eHealth, Telemedicine and Social Medicine (eTELEMED), falyan paginas y lugar 253–258 (2013)
12. Dumas, G., Soussignan, R., Hugueville, L., Martinerie, J., Nadel, J.: Revisiting mu suppression in autism spectrum disorder. Brain Res. **1585**, 108–119 (2014). https://doi.org/10.1016/j.brainres.2014.08.035

# What Can Technology Do for Autistic Spectrum Disorder People?

Marina Jodra[1,2](✉) and Victoria Rodellar[3]

[1] Department of Personality, Assessment and Clinical Psychology, Faculty of Education, Complutense University of Madrid, C/ Rector Royo Villanova 1, 28040 Madrid, Spain
majodra@ucm.es

[2] Asociación Nuevo Horizonte, Comunidad de Madrid 43, Las Rozas de Madrid, 28231 Madrid, Spain

[3] NeuSpeLab, Center for Biomedical Technology, Universidad Politécnica de Madrid, Campus de Montegancedo, 28220 Pozuelo de Alarcón, Madrid, Spain
mariavictoria.rodellar@upm.es

**Abstract.** People with Autism Spectrum Disorders (ASD) need specialized support through-out their lives. New technologies are an opportunity to complement psychoeducational interventions, diagnosis and medical care for these individuals, but it is important to design technologies tailored to each individual and validate their use in these programs. This paper reviews the assistive technologies currently in use and raises new challenges to respond to the central problems of this population such as early diagnosis, support in aging processes or access to the world of work and housing.

**Keywords:** Autism Spectrum Disorders · Actigraphy technology · Quality of life

## 1 Introduction

Autism Spectrum Disorders (ASD) affect the area of social communication of individuals, and are characterized by stereotyped behaviors, restricted interests and sensory disturbances. In addition to these core symptoms, people with ASD often present comorbidity with other diseases and psychopathologies, such as epilepsy, intellectual disability, depression, anxiety, etc. [1].

The diagnosis of ASD is currently made through behavioral observation, since to date there are no biomarkers to help us in this process. The persons affected by ASD will need support and accompaniment throughout their life. One of the support modalities that in recent years is expanding and evolving to a greater extent are the Information and Communication Technologies (ICTs) [2–5].

The paper is organized as follows: A description of ASD disorders is provided in Sect. 2, explaining the neurological and behavioral characteristics of

The authors would like to thank Asociación Nuevo Horizonte for unconditional support.


J. M. Ferrández Vicente et al. (Eds.): IWINAC 2022, LNCS 13258, pp. 301–309, 2022.
https://doi.org/10.1007/978-3-031-06242-1_30

ASD affecting language, intellectual disability, anxiety, depression, epilepsy, and sleep problems. Section 3 overviews supportive solutions for ASD. Section 4 is intended to summarize needs not yet covered currently. The main insights and conclusions derived from the study are given in Sect. 5.

## 2 Description of Autism Spectrum Disorders

Both the severity of core and comorbid symptomatology will determine the level and type of supports that each person needs. In ASD the most significant symptomatology can be summarized in what follows.

Nuclear symptoms:

1. Socio-communicative disturbances
   One of the core characteristics of this disorder is socio-communicative disturbances, which translate into a range of atypical behaviors. There may appear as low interest in social relationships, inappropriate approaches (such as not respecting physical space), deficits in the recognition of emotions or in communication (total absence of oral language or difficulties with the pragmatic aspects of language that make it difficult for them to understand jokes, irony, sarcasm, etc.). This range of behaviors sometimes translates into negative conducts, increased anxiety in social situations, problematic comportment when facing difficulties in communicating needs, etc.
2. Stereotyped behaviors, restricted interests, and sensory disturbances.
   Characteristic symptoms in this area are closely related to the occurrence of behavioral problems. For example, if there is hypersensitivity to a particular sensory stimulus and the person is exposed to it for a long time or with great intensity, the chances of a behavioral problem inevitably increase. Another example is the inflexibility that many people with ASD exhibit when faced with small changes in their environment or routines. This inflexibility often also results in increased anxiety and possible behavioral problems.
   Repetitive behaviors and sensory hyper- or hypo-responsivity are also common in people with ASD. The need to adapt the environment and offer support for this symptomatology is nuclear to improve the quality of life of these individuals.

On the other hand, there are many comorbid symptoms of ASD that can sometimes also influence an increase problematic behaviors or distress in this population [6]:

1. Language disorders
   Language difficulties and disorders are a common feature in many people with ASD. These difficulties lead to behavioral problems when they are unable to communicate their needs, discomfort and intentions to others or fail to interpret environmental cues appropriately.
2. Intellectual disability
   ASD often show an associated intellectual disability, which means that they cannot compensate with cognitive strategies, many difficulties caused by the symptoms associated to the disorder.

3. Anxiety
   Anxiety accompanies many individuals with autism throughout their development and in numerous day-to-day experiences. This, together with the existing difficulties in communicating their emotions and needs, leads to behavioral problems due to inadequate self-management of these situations.
4. Depression
   There are numerous studies on the association between depression and ASD [50]. The depressive symptoms observed in this population are usually irritability and behavioral disturbances.
5. Epilepsy
   Sometimes, the onset of epileptic seizures is accompanied by behavioral seizures or generalized behavior worsening.
6. Sleep problems
   The prevalence of sleep problems in people with ASD is 50–80% [7]. The shortage and problems during sleep cause tiredness, mood swings and behavioral problems, being one of the variables most associated with the appearance of these behaviors in people with autism.

## 3 Overview of Supportive Applications for ASD

The above described characteristic symptoms mean that ASD individuals need specialized supports. In the recent decades, new technologies are playing a fundamental role in the design of materials, adaptation of environments and development of interventions, since they facilitate more predictable environments, eliminate social complexity and provide stimuli through various sensory channels, giving special attention to visual stimuli. An estimated 33% of people with ASD better retain information presented through computers or tablets [8]. In addition, digital tablets improved communication and social language skills in students with ASD by 25%. Tables 1, 2 and 3 show a summary of the ICTs that have been used in recent years to try to improve the quality of life of people with ASD.

**Table 1.** Assistive technologies used in the diagnostic of ASD individuals

| Area | Technology |
|---|---|
| Diagnosis | Eye-tracking [22,23] |
| | Robotics [24] |
| | Biological markers (sensors in watches, clothing, cameras, etc.) |
| | – Voice [25] |
| | – EDA, EEG, EMG, kinematics and peripheral temperature [26–29] |

**Table 2.** Assistive technologies used in the intervention in nuclear areas with people with ASD

| Areas | Technology |
|---|---|
| Social disturbances | Robotics [9–13]:<br>– "Nao"; verbal communication and imitation<br>– "Zeno R-30" - imitation, "Robokind" emotion recognition<br>Apps:<br>– Open Autism Software [14]<br>– Virtual reality [15–17]<br>Social interaction assistance:<br>– Cameras - MOSOCO; [18,19])<br>– Glasses [20]<br>Theory of Mind Learning Materials: Apps and eLearning programs [21] |
| Communication and language disorders | Robots: Nao [30]<br>Communicators and alternative materials:<br>– Proloquo2Go [51]<br>– Día a D'ia [52]<br>– AbaPlanet [53]<br>– ARASACC [54]<br>– MeCalendar [31] |
| Self-regulation and stereotyped behaviors | Apps:<br>– Berrinche [55]<br>Watches [35–37] |
| Learning of daily living skills and academic content | Adapted materials on mobile devices, computers, tablets, digital whiteboards, robots, etc.<br>Attention, categorization, memory, etc.<br>– MOTIVATEME [38]<br>– PICAA2 [56]<br>– TINY TAP [57]<br>– Virtual reality environments [39,40]<br>– Pictoaplications [58]<br>– NeuronUp [59]<br>– Robot "Lucy" [41]<br>– Robot "Kaspar" [42] |

**Table 3.** Assistive technologies used in intervention in comorbid symptoms in ASD people

| Areas | Technology |
|---|---|
| Epilepsy | Watches [60] |
| Anxiety, depression, and sleep and behavioral problems | Robotics, to analyze the person's emotions and adapt the environment and tasks [32]<br>Watches [33]<br>– E4 Empatica [61]<br>– Taimun Watch [62]<br>Therapeutic clothing with sensors [34] |

# 4 Needs for New Applications

For the design of technological tools in the coming years, it is important to detect what problems people with ASD currently are facing, to help through the design of assistive technologies to make society a more inclusive environment, increasing research towards the adult stage, since most studies focus on childhood [49]. Some of these challenges we currently have to deal with in the field of psycho-educational intervention and in adult life are:

- Access to the world of work (training job skills through apps, virtual reality, robotics, etc.).
- Access to housing (support through home automation with domotic devices, cognitive accessibility of environments, specialized psychological and therapeutic support, etc.).
- Early aging (assistive technologies for health care, fall prevention, support to internal medicine, detection of early aging through markers, design of psycho-educational materials for cognitive reserve and adult education, etc.).
- Improve early diagnosis through biological variables and biomarkers (detection of these markers).

# 5 Conclusions

New technologies are part of everyone's daily life, and have come to play a leading role in the therapies and diagnosis of people with ASD. It is important to develop multidisciplinary projects that take into account different professional perspectives and the opinion of people with ASD, since within this common spectrum there is a great heterogeneity of cases. The aim is to design ICTs that are flexible and adapted to each person so that they can provide real support. In addition to involving different professionals in the design of these technologies, it is necessary to scientifically validate the interventions, since at present we do not have enough empirical studies on their efficacy. Current research depends on small samples, mostly including children and people without intellectual disabilities. Therefore, it is important to increase the number of people evaluated in each study and to include adults and contemplate the intellectual disability that so often accompanies ASD [43,44].

The cognitive profile of people with ASD conditions that in many occasions the approach to new technologies does not occur quickly and naturally, so any technology to be designed for these people has to contemplate a first familiarization phase [45]. On the other hand, the sensory profile forces us to take into account that the devices used for ASD have to be light, small, non-intrusive, with the minimum number of distractors (alarms, vibrations, lights, etc.), so that the person feels comfortable and accepts their use [46]. On many occasions the person must be familiarized with the use of ICTs relying on the help of specialized behavioral therapists who know the person.

Along with this, assistive technologies for people with ASD must be designed to withstand sometimes inappropriate use due to repeated touching, being turned off on more occasions, being bumped or removed when they should not be interrupted, etc. Therefore, they must be rugged and support much more frequent data transmission [37,47].

Robotics has its limitations, as it is an expensive resource and requires supervision of the intervention by specialized professionals [48]. In addition, it is necessary to assess how to generalize the learnings and to take precautions regarding the attachment that the person with ASD establishes with the robot. It could be counterproductive for us to develop socioemotional skills through robotics and enhance in the person a refusal to interact with other people, since through robots we display a fictitious, predictable, less hostile and pleasant but artificial social environment and that can lead to greater social isolation.

Finally, it is important to know that the monitoring of a person's physiological variables has repercussions with respect to their privacy, so these data must be treated with the utmost confidentiality and only for therapeutic purposes. The selection of which physiological variables we are going to measure for each objective and the appropriate and longitudinal measurement and treatment of these data is also an issue that needs to be studied and further improved in the future, since the massive accumulation of data without adequate post-processing can sometimes make therapeutic interpretation inadequate, leaving us with a lot of meaningless data [49].

## References

1. American Psychiatric Association.: Diagnostic and Statistical Manual of Mental Disorders DSM-5 (5th. ed.). Arlington, VA, USA (2013)
2. Abdo, M., Al Osman, H.: Technology impact on reading and writing skills of children with autism: a systematic literature review. Heal. Technol. **9**(5), 725–735 (2019). https://doi.org/10.1007/s12553-019-00317-4
3. Knight, V., McKissick, B., Saunders, A.: A review of technology-based interventions to teach academic skills to students with autism spectrum disorder. J. Autism Dev. Disord. **43**(11), 2628–48 (2013)
4. Leung, P.W.S., Li, S.X., Tsang, C.S.O., Chow, B.L.C., Wong, W.C.W.: Effectiveness of using mobile technology to improve cognitive and social skills among individuals with autism spectrum disorder: systematic literature review. JMIR Ment. Health **8**(9), e20892 (2021)
5. Valencia, K., Rusu, C., Quiñones, D., Jamet, E.: The impact of technology on people with autism spectrum disorder: a systematic literature review. Sensors **19**(20), 4485 (2019)
6. Mazurek, M., Kanne, S., Wodka, E.L.: Physical aggression in children and adolescents with autism spectrum disorders. Res. Autism Spectr. Disord. **7**, 455–465 (2013)
7. Souders, M.C., et al.: Sleep behaviors and sleep quality in children with autism spectrum disorders. Sleep **32**(12), 1566–78 (2009)
8. McEwen, R.: Mediating sociality: the use of iPod Touch™ devices in the classrooms of students with autism in Canada. Inf. Commun. Soc. **17**(10), 1264–1279 (2014)

9. Coeckelbergh, M., et al.: A survey of expectations about the role of robots in robot-assisted therapy for children with ASD: ethical acceptability, trust, sociability, appearance, and attachment. Sci. Eng. Ethics **22**, 47–65 (2016)
10. Dautenhahn, K., Werry, I.: Towards interactive robots in autism therapy: background, motivation and challenges. Pragmat. Cognit. **12**, 1–35 (2004)
11. Liu, X., Wu, Q., Zhao, W., Luo, X.: Technology-facilitated diagnosis and treatment of individuals with autism spectrum disorder: an engineering perspective. Appl. Sci. **7**, 1051 (2017)
12. Robins, B., Dickerson, P., Stribling, P., Dautenhahn, K.: Robot-mediated joint attention in children with autism: a case study in robot-human interaction. Interact. Stud. **5**(2), 151–198 (2004)
13. Van den Berk-Smeekens, I., et al.: Pivotal response treatment with and without robot-assistance for children with autism: a randomized controlled trial. Eur. Child Adolesc. Psychiatry (2021). PMID: 34106357
14. Hourcade, J.P., Bullock-Rest, N.E., Hansen, T.E.: Multitouch tablet applications and activities to enhance the social skills of children with autism spectrum disorders. Pers. Ubiquit. Comput. **16**(2), 157–168 (2012)
15. Halabi, O., El-Seoud, S.A., Alja'am, J., Alpona, H., Al-Hemadi, M., Al-Hassan, D.: Design of immersive virtual reality system to improve communication skills in individuals with autism. IJET **12**(05), 50–64 (2017)
16. Kuriakose, S., Lahiri, U.: Design of a physiology-sensitive VR-based social communication platform for children with autism. IEEE Trans. Neural Syst. Rehabil. Eng. **25**(8), 1180–1191 (2017)
17. Lahiri, U.: A Computational View of Autism. Using Virtual Reality Technologies in Autism Intervention. Springer, Switzerland (2020). https://doi.org/10.1007/978-3-030-40237-2
18. Escobedo, L., et al.: MOSOCO: a mobile assistive tool to support children with autism practicing social skills in real-life situations. In: Proceedings of the 2012 ACM Annual Conference on Human Factors in Computing Systems (CHI 2012), pp. 2589–2598. ACM, New York, NY, USA (2012)
19. Madsen, M., El Kaliouby, R., Goodwin, M., Picard, R.: Technology for just-in-time in-situ learning of facial affect for persons diagnosed with an autism spectrum disorder. In: Proceedings of Assets 2008, pp. 19–26 (2008)
20. Kinsella, B.G., Chow, S., Kushki, A.: Evaluating the usability of a wearable social skills training technology for children with autism spectrum disorder. Front. Robot. AI **4**, 31 (2017)
21. Garc'ia-Villamisar, D., Jodra, M., Muela, C.: Programa eLearning para el reconocimiento de emociones en personas adultas con autismo. In: Congreso AETAPI. San sebastian, Spain (2008)
22. Anderson, C.J., Colombo, J., Shaddy, D.J.: Visual scanning and pupillary response in young children with autism spectrum disorder. J. Clin. Exp. Neuropsychol. **28**, 1238–1256 (2006)
23. Wan, G., Kong, X., Sun, B., Yu, S., Tu, Y., Park, J.: Applying eye tracking to identify autism spectrum disorder in children. J. Autism Dev. Disord. **49**(1), 209–215 (2019)
24. Golliot, J., et al.: A tool to diagnose autism in children aged between two to five old: an exploratory study with the robot que-ball. ACM/IEEE Human-Robot Interaction Extended Abstracts. In: HRI 2015, pp. 61–62 (2015)
25. Rodellar-Biarge, V., Jodra-Chuan, M.: A longitudinal study of voice tremor in intellectually impaired autistic persons. In: MAVEBA, 12th International Workshop, pp. 67–70. Firenze University Press, Firenze, Italy (2021)

26. Aparicio Betancourt., M., Dethorne, L.S., Karahalios, K., Kim, J.G.: Skin conductance as an in situ marker for emotional arousal in children with neurodevelopmental communication impairments: methodological considerations and clinical implications. ACM Trans. Access. Comput. **9**(3), 1–29 (2017)
27. Chaspari, T., Tsiartas, A., Stein Duker, L.I., Cermak, S.A., Narayanan, S.S.: EDA-gram: designing electrodermal activity fingerprints for visualization and feature extraction. In: IEEE EMBC, pp. 403–406 (2016)
28. Nehme, B., Youness, R., Hanna, T.A., Hleihel, W., Serhan, R.: Developing a skin conductance device for early Autism Spectrum Disorder diagnosis. In: 3rd Middle East Conference on Biomedical Engineering (MECBME), pp. 139–142 (2016)
29. Spiel, K., Makhaeva, J., Frauenberger, C.: Embodied companion technologies for autistic children. Tangible, Embedded & Embodied Interaction. In: TEI 2016, pp. 245–252 (2016)
30. Alemia, M., Mahboub, B.N.: Exploring social robots as a tool for special education to teach English to Iranian kids with autism. Int. J. Robot. **4**(4), 30–41 (2016)
31. Wilson, C., Brereton, M., Ploderer, B., Sitbon, L., Saggers, B.: Digital strategies for supporting strengths- and interests- based learning with children with autism. In: ASSETS 2017, Baltimore, MD, USA (2017)
32. Liu, C., Conn, K., Sarkar, N., Stone, W.: Online affect detection and robot behavior adaptation for intervention of children with autism. IEEE Trans. Robot. **24**, 883–896 (2008)
33. Chen, J., Abbod, M., Shieh, J.S.: Pain and stress detection using wearable sensors and devices-a review. Sensors **21**, 1030 (2021)
34. Warren, S., et al.: Design projects motivated and informed by the needs of severely disabled autistic children. In: IEEE EMBC, pp. 3015–3018 (2016)
35. Chuah, M., Diblasio, M.: Smartphone based autism social alert system. In: Mobile Ad-hoc and Sensor Networks (MSN), pp. 6–13 (2012)
36. Gonçalves, N., Rodrigues, J.L., Costa, S., Soares, F.: Automatic detection of stereotyped hand flapping movements: two different approaches. In: 2012 IEEE RO-MAN: IEEE Robot and Human Interactive Communication, pp. 392–397. Paris (2012)
37. Zakaria, C., Davis, R.C., Walker, Z.: Seeking independent management of problem behavior: a proof-of-concept study with children and their teachers. In: IDC 2016, pp. 196–205 (2016)
38. Bakhai, A., Constantin, A., Alexandru, C.A.: MOTIVATEME!: an alexa skill to support higher education students with autism. In: International Conferences Interfaces and Human Computer Interaction and Game and Entertainment Technologies (2020)
39. Herrera, G., et al.: Pictogram room: natural interaction technologies to aid in the development of children with autism. Annuary Clin. Health Psychol. **8**, 39–44 (2012)
40. Liu, R., Salisbury, J.P., Vahabzadeh, A., Sahin, N.T.: Feasibility of an autism-focused augmented reality smartglasses system for social communication and behavioral coaching. Front. Pediatr. **5**(145) (2017)
41. Khosla, R., Nguyen, K., Chu, M.T.: Socially assistive robot enabled home-based care for supporting people with autism. In: Proceedings of the Pacific Asia Conference on Information Systems, p. 12. Singapore (2015)
42. Huijnen, C.A.G.J., Lexis, M.A.S., Jansens, R., de Witte, L.P.: How to implement robots in interventions for children with autism? A co-creation study involving people with autism, parents and professionals. J. Autism Dev. Disord. **47**(10), 3079–3096 (2017). https://doi.org/10.1007/s10803-017-3235-9

43. DiPietro, J., Kelemen, A., Liang, Y., Sik-Lanyi, C.: Computer- and robot-assisted therapies to aid social and intellectual functioning of children with autism spectrum disorder. Medicina **55**, 440 (2019)
44. Jaliaawala, M.S., Khan, R.A.: Can autism be catered with artificial intelligence-assisted intervention technology? A comprehensive survey. Artif. Intell. Rev. **53**, 1039–1069 (2020)
45. Jiang, X., Boyd, L.A.E., Chen, Y., Hayes, G.R.: ProCom: designing a mobile and wearable system to support proximity awareness for people with autism. In: UbiComp 2016, pp. 93–96 (2014)
46. Weisberg, O., et al.: TangiPlan: designing an assistive technology to enhance executive functioning among children with ADHD. In: IDC 2014, pp. 293–296 (2014)
47. Washington, P., et al.: A wearable social interaction aid for children with autism. In: SIGCHI EA 2016, pp. 2348–2354 (2016)
48. Miskam, M.A., Shamsuddin, S., Yussof, H., Ariffin, I.M., Omar, A.R.: A questionnaire-based survey: therapist's response on emotions gestures using humanoid robot for autism. In: Proceedings of the International Symposium on Micro-Nano Mechatronics and Human Science (MHS), pp. 1–7 (2015)
49. Sharmin, M., Hossain, M.M., Saha, A., Das, M., Maxwell, M., Ahmed, S.: From re-search to practice: informing the design of autism support smart technology. In: CHI 2018, Montréal, QC, Canada (2018)
50. Rai, D., et al.: Association of autistic traits with depression from childhood to age 18 years. JAMA Psychiatry **75**(8), 835–843 (2018). https://doi.org/10.1001/jamapsychiatry.2018.1323
51. Proloquo2Go. https://www.assistiveware.com/es/productos/proloquo2go. Accessed 15 Nov 2021
52. D'ia a D'ia. https://www.fundacionorange.es/aplicaciones/dia-a-dia/. Accessed 15 Nov 2021
53. AbaPlanet. https://apps.apple.com/es/app/abaplanet/id989142096. Accessed 20 Nov 2021
54. ARASACC. https://arasaac.org. Accessed 20 Nov 2021
55. Berrinche. https://apps.apple.com/us/app/berrinche/id1479562280. Accessed 25 Oct 2021
56. Picaa2. https://apps.apple.com/es/app/picaa-2/id938321978. Accessed 12 Jan 2022
57. Tiny Tap. https://www.tinytap.com. Accessed 12 Jan 2022
58. Pictoaplicaciones. https://www.pictoaplicaciones.com. Accessed 12 Jan 2022
59. NeuronUp. https://www.neuronup.com. Accessed 15 Jan 2022
60. Tang, J., et al.: Seizure detection using wearable sensors and machine learning: setting a benchmark. Epilepsia **62**(8), 1807–1819 (2021). https://doi.org/10.1111/epi.16967
61. Empatica E4. https://www.empatica.com/en-eu/research/e4/. Accessed 11 Jan 2022
62. TAIMUN watch. https://www.tecnobility.com/es/noticia/una-app-para-relojes-ayuda-autorregularse-personas-con-tea-que-sufran-una-crisis. Accessed 11 Jan 2022

# Autism Spectrum Disorder (ASD): Emotional Intervention Protocol

Gema Benedicto[1,2(✉)], Mikel Val[2], Eduardo Fernández[2], Francisco Sánchez Ferrer[1], and José Manuel Ferrández[1]

[1] Universidad Politécnica de Cartagena, 30202 Murcia, Spain
benedicto.gema@gmail.com
[2] Instituto de Bioingeniería, Universidad de Miguel Hernández, 03202 Alicante, Spain

**Abstract.** The main aim of this study is to show an intervention protocol for emotional deficits through the use of robotics as a therapeutic tool in children with Autism Spectrum Disorder (ASD). Pepper, the humanoid robot which is capable of detecting emotions and is designed to interact with people, was used together with the Empatica E4 bracelet, a portable device during interactive and invasive states, capable of collecting physiological data such as electrodermal activity that reflects changes in the skin's electrical resistance (conductance) due to different types of emotions. This intervention takes place through the relationship between the psychological state originated by therapy sessions and the galvanic skin response throughout the sessions based on pivotal area training, complemented with a set of psychological tests that verify the diagnosis of ASD and the evolution of the disorder during therapy.

**Keywords:** Autism spectrum disorder · Robotic therapy · PRT · GSR

## 1 Introduction

Autism Spectrum Disorder (ASD) is one of the neurodevelopmental disorders whose prevalence is increasing over the years. It is more common in young men than in young women in a ratio of 4:1 and its origin or specific cause is still unknown, although genetic (to a greater extent) and environmental factors are know to be involved [1]. ASD is often characterised by persistent deficits in social communication-interaction including problems in social empathy and perseveration [2], and involves restrictive and repetitive behaviour patterns, interests or activities [24]. ASD patients show deficits in executive functions, such as cognitive flexibility, planning ability and inhibition. Working memory is also affected and thus patients perform poorly in go-no-go tasks.

Social competence is understood as using information about society in order to recognise and interpret social facts and conflicts in order to make a decision and to be able to make a response. Social competence takes place thanks to two fundamental pillars, cognition and social behaviour, both independent but


J. M. Ferrández Vicente et al. (Eds.): IWINAC 2022, LNCS 13258, pp. 310–322, 2022.
https://doi.org/10.1007/978-3-031-06242-1_31

linked, with cognitive processes being the basis of social behaviour [21]. Individuals with ASD present deficits in social cognition and difficulties in the process of empathy, which includes alterations in communication and theory of mind, denying them understanding not only other people's behaviour, but also their emotions, intentions or knowledge. Other theories that also help us to understand some of the points that make up the neuropsychological basis of individuals with ASD are the empathy theory (deficit of social affect, theory of mind and communication disorder) and systemisation (over-expression in repetitive skills and behaviours), thus constituting a system of compensation between two brain types: social (empathy) and analytical (systemisation). This triad of theories could be the cause of the symptoms presented by patients with ASD [20].

In this sense, the problems they present in imitation, a behaviour thanks to which bodily expressions are recognised (the basis of social interaction considered by some), are understandable [1]. This is carried out naturally and with greater speed and precision in neurotypical children, but this does not happen in the same way in children with ASD, who show difficulties in recognising different emotions exposed for a very short time [6] (alexithymia between 33.3%-63%) [9], even in individuals who show high functioning.

Social communication takes place through the recognition of emotions [19], thanks to the range of facial expressions we possess [9]. One of the innate and universal qualities of living beings is emotion. It is a physiological reaction to any stimulus that leads to a correct adaptation to the surrounding environment. Because of this, there is a rapid growth of social and emotional areas from birth. Emotional areas, including the amygdala and the limbic system, are responsible for interpreting these stimuli and responding to them, playing a major role in the emotional regulation of the individual [13].

Until now, attempts have been made to compensate these emotional deficits in autism from the field of psychology, as they have focused on behavioural interventions such as ABA (Applied Behaviour Analysis), whose main objective is to reduce inappropriate behaviours and increase appropriate ones, based on three main principles: Analysis (registering progress), Behaviour and Applied (from the observation of certain behaviours).

PRT is a type of training based on ABA and developed by Robert I. Koegel, L.K. Koegel, and L. Shreibman, characterised by an increase in the child's motivation, responsiveness to multiple cues and self-management, as well as self-initiations [2].

An alternative to the treatment of patients with ASD is to combine the field of psychology and technology, resulting in an improvement in their social and communication skills, which is one of the most affected abilities of patients with ASD. To this end, studies combining different types of training with social robots, so called because their main objective is to maintain communication and interaction with all human beings around them and they are characterised by flexibility in their programming, adapting more and more to the patient [1], are becoming more and more frequent.

In an analysis conducted by De Korte et al. in 2020, on the effect of PRT treatment versus PRT treatment with the participation of a robot on self-initiatives by children with ASD, there is an increase in self-initiatives in both treatment groups, with a greater impact on the interventions in which the robot collaborates.

Although there are few studies that demonstrate the long-term effects of this type of therapy, the use of a social robot offers a number of advantages, such as multimodal interaction through facial expressions, gestures..., or an improvement in the frequency of eye contact by patients, thus being able to obtain benefits in other areas of their lives [1].

More information is needed on emotional development and for this, a physiological signal is monitored, which is the galvanic skin response (GSR). Thanks to the GSR, information can be obtained about the emotional state of an individual, as it produces a response to environmental stimuli such as sounds, stress, odour... [8]. Skin conductance is due to the variability in the activity of the sweat glands, i.e., the galvanic response is under the control of the sympathetic autonomic nervous system and this regulates temperature, blood pressure, heart rate, etc. and in times of danger or stress triggered by emotional stimuli (positive and negative), it is responsible for regulating a totally involuntary response: change in galvanic conductance. Therefore, this alteration in the galvanic response is related to the environment around us, acting in its increase as a representation of the intensity of the different emotions (arousal) [3].

The dimensions measured through GSR are arousal (resulting from positive or negative emotion), valence (positive/negative emotion) and emotional dominance/impact (changes during emotional arousal) (Chayo-Dichy et al. 2003) [4].

Skin conductance is measured in micro Siemens ($\mu$S) [12] and is composed of two components: skin conductance level (SCL) and skin conductance response (SCRs). SCL reflects the skin tonic level, i.e., changes in slow-acting autonomic activation controlled by sympathetic inputs and involves the background of the GSR signal, as opposed to SCR (phasic level), associated with rapid changes of the sympathetic nervous system and the elicitation of measurable peaks of the signal [15].

Individuals with autism spectrum disorder seem to present stress invisibly, giving an outward appearance of calm that does not reflect reality: high arousal of the autonomic system, so that, thanks to the control of physiological signals, we are faced with an indicator of arousal that can provide more information about emotions, helping in their understanding.

Skin conductance not only provides information on the psycho-physiological state of the individual, but also helps to ascertain the child's level of engagement with the therapy, considering engagement as the psychological state of proactive involvement with an object or agent that has a positive affective nuance [16]. That is why, throughout this protocol, two measures of engagement with therapy are addressed: galvanic skin conductance (GSR) and a series of psychological tests to parents, in addition to intervening in the regulation of emotional deficits with the use of robots during pivotal area training (PRT).

## 2 Objectives

In order to carry out the therapeutic intervention it is necessary to identify the emotions that need to be regulated and to implement a series of emotional regulation strategies to increase or decrease these emotions. Because of this, an experimental study is conducted in which a multimodal database (video recordings, electrodermal activity signals) will be collected while individuals with ASD participate in different game scenarios through PRT training with a social robot, Pepper.

The impact of the robotic therapy, the type of training from psychological tests to parents and the influence of the use of the E4 Empathic from which approximate information about the emotional state of the individual can be obtained from the characteristics extracted from the physiological signals of tonic GSR (SCL) and phasic GSR (SCR), as follows, are studied: mean value of the GSR signal in a stage (MEAN GSR), standard deviation of EDA (STD GSR), harmonic summation of EDA (SUM H GSR), number of SCR events in each of the different stages (EVE SCR), frequency of occurrence of SCR in a stage (FRC SCR), average amplitude value of SCR responses in a stage (AMP SCR), GSR signal magnitude area (SMA GSR) and number of GSR peaks (NPR GSR).

## 3 Materials

To carry out this study it is necessary to have a webcam placed in the upper body area of the Pepper robot, a laptop computer to control the Pepper robot whose main function is to identify non-verbal language, gestures and emotions through a 3D camera, different sensors such as tactile and a screen on which visual stimuli are displayed.

Children also participate in a series of games and activities such as Tangram of objects, numbers and animals, Labyrinths of letters, fruits and geometric shapes, and play with plastic, wooden and dental building blocks. Each of the different play scenarios are divided into four levels of complexity.

The galvanic skin response (GSR) is measured by the E4 empathic bracelet consisting of two non-polarised silver (Ag) electrodes that measure the potential that has been generated. Finally, the whole experiment is accompanied by a set of psychological tests and questionnaires to validate the clinical diagnosis of the patient and the evolution of the disorder throughout the robotic therapy.

## 4 Participants

This research is intended for children with Autism Spectrum Disorder (ASD) who meet the DSM-V criteria for ASD, and may have comorbidities as long as the primary diagnosis is Autism Spectrum Disorder (ASD), equal/similar level of severity of impairments, and if uncertain, is obtained from the CSS scores of the ADOS. Must have the ability to speak at least one word accompanied by utterances and not have received prior pivotal response training (PRT).

The activities that are part of the present design are focused on children aged 4–6 years, and according to the inclusion and exclusion criteria, patients must meet requirements such as an updated clinical diagnosis of ASD together with ADOS 2 (Autism Diagnostic Observation Scale) or ADI-R (Autism Diagnostic Interview) test. Normal IQ by the Wechsler scale (WISC-III), equal or higher than 70 points and must have been obtained in the last two years, otherwise the estimation must be carried out through the Wechsler Infant Intelligence Scale (from 6 years to 16 years and 11 months), Wechsler Preschool and Primary Intelligence Scale (2 years and 6 months to 7 years and 7 months) or Mullen Early Learning Scale (from birth to 68 months of age).

# 5 Measures

## 5.1 IQ

Drawn from the Wechsler Infant Intelligence Scale (6 years to 16 years and 11 months), Wechsler Preschool and Primary Intelligence Scale (2 years and 6 months to 7 years and 7 months) or Mullen Early Learning (birth to 68 months of age).

## 5.2 Attitude Towards the Robotic Therapy

By means of a questionnaire, following a Likert scale ranging from 1 to 5, parents indicate whether they have a positive attitude towards the robot and the methodology of each of the sessions, just as the patient completes a questionnaire at the end of the programme of sessions to find out whether they have enjoyed Pepper's participation. In addition, there is an Emodiana (dartboard of emotions) for the child at the beginning and end of each session to rate with a maximum of 3 points on a Likert scale how he/she feels.

## 5.3 Engagement Level

Provides more information about the subject's involvement in the therapy (activities, robot, therapist, etc.). An approximation of the level of engagement is obtained from the results obtained for each of the different characteristics extracted from the galvanic skin conductance (GSR).

## 5.4 Evolution of the Disorder

It is essential to check whether a number of changes in patients' behaviour have occurred before and after the intervention programme. This is verified by means of tests such as ADOS-2 and ADI-R.

Before starting the first session, parents are offered a Social Communication Questionnaire (SCQ) (Screening for Autistic Spectrum Disorders with an approximate duration of 10 min) to be completed at the beginning and end of each session, to check if there is an evolution in the individual's social behaviour.

The children will have been previously diagnosed of autism through the ADI-R (Autism Diagnostic Interview) or ADOS 2 (Autism Diagnostic Observation Scale) test and will be tested once the intervention programme is finished in order to evaluate the possibility of reduction of ASD symptoms.

Other parameters such as medication are of great relevance in the study as they may be implicated in the effects of the treatment carried out with patients, causing them harm [17]. These can be classified into banal errors (almost no or no harm), adverse events (serious or mild harm) which in turn include preventable and non-preventable adverse events and potential adverse events (serious harm) [14]. Therefore, it is necessary for parents to continuously report a change in their medication that poses a risk to research.

## 6 Procedure

Participants are randomly divided into two groups following these conditions: PRT training + Robot Pepper + GSR band (first group), and PRT training + GSR band (second group).

Each of these conditions has a duration of two months, each month being divided into four sessions of 30 min each, giving a total of 16 sessions (8 sessions with the robot and 8 without it). Each condition has a total duration of 240 min (4 h), with a total time investment for children and parents (initial questionnaire, treatment, final questionnaire) of approximately 1 h.

During the sessions of each of the two established groups, the parents remain all the time observing through a glass the type of training that is being carried out. If, during each of the first sessions, an individual has problems adapting to the experiment, the parents can be present for the first 15 min (Fig. 1).

**Fig. 1.** Play scenario from the child's point of view. It can see the robot Pepper, the researcher who controls the robot and the glass through which the parents can observe.

## Intervention Protocol

Parents are informed by means of an informative letter and are given informed consent to authorise the therapy programme, in addition to analysing the possible subjects who meet the inclusion requirements. After selection, four days before the start of the experiment, the subject will use the Even Better virtual platform where the most basic emotions and how to identify them will be explained. After the four days, at the start of each session, the child will be given a 3-point score to evaluate their mood, choosing between happiness, sadness and anger (represented by a character on the virtual platform) and their intensity, by means of an Emodiana. They must also do so at the end of the session. Meanwhile, the development of the programme is carried out by a researcher who holds a PRT certificate.

PRT training follows basic motivational procedures: child choice, maintenance tasks, natural rewards and reward attempts. It has been used to improve language skills, play and social behaviour in children with ASD through the use of natural reinforcement by offering as reinforcement that which is related to the behaviour being taught, and by offering a reward when it is attempted even though it is not performed perfectly, except in the case of a lack of effort or a chance hit [11]. The development of PRT focuses on specific areas that in turn produce generalised positive improvements and the production of a consequence after a given behaviour, since in children with ASD the term learned helplessness (they do not understand the connection between response and consequence) occurs [10].

Each of the eight sessions take place once a week, the first session being dedicated to the presentation of the project to the parents and the first contact of the children with the robot, controlled by the researcher through the computer, which models the techniques and actions of the robot according to the behaviour of the patient and the programming of the games is adaptable, i.e. the robot can act in different ways: choice of the game by the subject, after the patient's self-initiation and after the response of the individual [5].

The hierarchy of the elements involved in training are 1) movement of the robot providing an indirect cue to the child that does not deviate from the target behaviour, 2) open-ended question asked by the robot, 3) the robot waits for a response after having created the opportunity to respond, and after the response, positive reinforcement occurs. During the moments of reinforcement, dismissal and the rest of the game, three different types of emoticons appear on Pepper's screen.

At the beginning of the session, the child is offered a choice of game, and all activities are characterised by a variety of visual elements, and the sessions are recorded for subsequent analysis of the patient's behaviour, in order to synchronise the different behaviours and times of the game with the data obtained from the GSR sensor.

Once the subject is comfortable with the GSR band, the session begins, with the data acquisition process being divided into three stages: conditioning phase, i.e. habituation to the device, data collection phase from the sensors, which are subjected to further processing, and baseline phase [12]. For this purpose, proprietary software is developed from which the results are inferred with the features extracted from the filtered GSR signal, specifically from SCR and SCL, but mainly by putting the focus of the study on two main characteristics of a signal: latency and threshold.

The GSR analysis starts by dividing the data collection into a series of phases:

1. noting down all session characteristics such as date, type of activity, participant's initial, or ID number of both the session and the wristband.
2. Use Matlab/Python to analyse the signal, i.e. plot the different signals, remove possible artefacts...
3. Find and eliminate those signals that are not valid.
4. GSR signals that are considered valid will be synchronised with the video recording, exported from Kinovea.

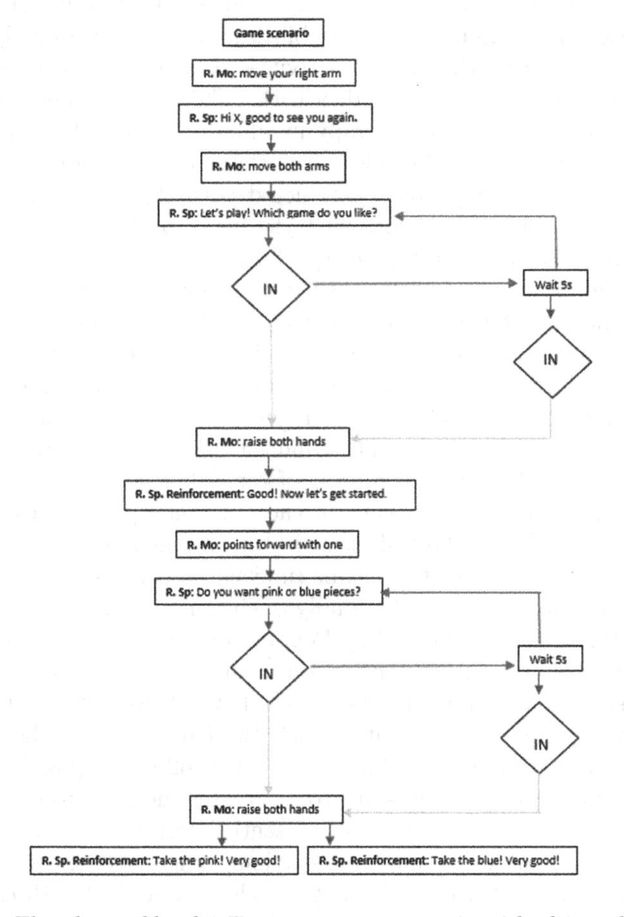


**Fig. 2.** Flowchart of level 1 Tangram game scenario with object theme.

5. Align the exact timestamp with the video recording (video timelines + signal timelines).
6. Identify the different stages in which the activity or game is divided.
7. Label the stages to extract useful features about the mechanism of the GSR signal.
8. Finally, signal pre-processing and feature extraction of the activity windows takes place [16] (Fig. 2).

## 7 Discussion

With the aim of a correct socio-communicative development of patients with Autism Spectrum Disorder (ASD), a mixed intervention takes place in which the field of psychology (therapy with PRT) and robotics (Pepper) participate. Through therapy, attempts have been made to compensate for the deficits presented by patients with ASD, but the expected results have not been obtained in some cases, with a number of possibilities such as a variety in the IQ of the individuals involved, the type of activities they perform or cognitive differences [6]. However, in previous studies with PRT training, high effect ratings were obtained for a part of the therapy in which the robot did not participate, but the parents played with the child using the PRT training methods, and less positive effect ratings for therapy conducted with the robot alone, making it necessary to alternate PRT sessions with the participation of the robot in them to maintain the motivation of the subject [22].

Therapeutic interventions are classified into several groups such as naturalistic, evolutionary, or behavioural interventions. PRT is one of the therapeutic intervention treatments with a more naturalistic approach that follows Applied Behavioural Analysis (ABA), i.e. a global model of behavioural intervention with high social validity, in which certain behaviours are learned and are under the influence of external (environmental) stimuli. Unlike other therapeutic interventions, with this method there is a sharing of games between the therapist and the child, adapting to their needs, taking into account their preferences and focusing on several pivotal areas of the child that in turn influence other collateral areas, instead of going one by one, at the same time as the parents participate and can extrapolate the information to the family environment [2].

This treatment is implemented by Pepper, a humanoid social robot that has certain advantages over NAO (a robot commonly used in educational therapies) such as presenting a screen in the chest area that can display visual stimuli that keep the child's attention and with which the child can establish interaction (communication) with the robot during therapy. It offers the possibility of longer sessions compared to other robots due to its greater energy autonomy, and it can recognise the voice and face, as well as presenting certain movements that make its behaviour more similar to human behaviour. The visual tracking during the activity is the most distinctive and suitable element to deal with children with autism (ASD), as it maintains a continuous visual contact with the child, thus improving the state of engagement with the task [7].

To date, high ratings of the robot by both parents and children have been observed, and there may be a difference between ratings by parents of one sex compared to those of the other sex, although these ratings of social behaviour are derived from questionnaires and tests that may be influenced by environment and caregiver bias, and the social deficits they present are not exclusively necessary for a diagnosis of Autism Spectrum Disorder (ASD) [2]. However, there are still no validated self-report scales and tests that measure the level of agreeableness and affection in children with ASD.

An improvement is expected in the adaptation of the game scenarios to children, as the chosen activities have a multitude of completely different vocabulary and at each of the levels of complexity of each of the game scenarios, being games such as Tangram that can be adapted to different abilities and ages.

In the last study conducted on robotic therapy using PRT training with children, up to nine game scenarios and seven levels of complexity of games such as Lego or puzzles were created, proving too easy or too difficult for the subjects, and they stopped playing because they found the behaviour of the NAO robot predictable [22]. But in this case, an improvement is expected in the adaptation of the game scenarios to the children, as the chosen activities have a multitude of completely different vocabulary and at each of the levels of complexity of each of the game scenarios, being games such as Tangram that can be adapted to different abilities and ages, without having to create a larger library of games with higher levels of difficulty.

Moreover, the reinforcement is direct and contingent, without a large loss of time (2–4 s) between the subject's action and the robot's response. However, if the individual shows a kind of spontaneous initiative and wants the robot to do a certain movement or action, he or she will not receive the desired response, as only the game scenarios are programmed [5].

On the other hand, the Empathic E4 wristband is placed on the wrist area of the individuals, although ideally in the future it would be ideal to conduct this study with a wristband that could measure GSR in the lower calf area of the subject, as in that area data is recorded more accurately in younger children while they play or perform an activity comfortably [18]. From the results to be obtained, although the GSR signal does not provide data on the severity of autism [23], we expect to find changes in SCL and SCR that vary depending on the age and severity of autism of the individual, an increase in SCL at times that demand greater attention during play scenarios and corroborate that the galvanic conductance response is an appropriate peripheral-physiological measure to assess engagement.

## 8 Conclusion

Thanks to this type of combined therapy, professionals such as therapists or psychologists will be able to intervene in the resolution of socio-emotional problems even during the infancy stage of children with Autism Spectrum Disorder (ASD) if they have a better understanding of their emotional development, and thus a more effective intervention. Learning to regulate different emotions and impulses is crucial for social success in the environment [13], in addition to the existing difficulty of understanding the needs of an individual for whom communication is an obstacle.

Much information and quantitative data is needed to understand how people with autism reflect their psychological states. The level of engagement can help to a better knowledge in this field, for example to discern what kind of activities are the most appealing for them, to build other technologies more attractive and able to detect the psychological state and level of engagement of the subject, but, above all, that can be combined with different psychological treatments, making the therapy a more effective and impacting method.

Although there are still limitations, such as the acquisition of signals that do not occur online, but rather a collection of data during each of the sessions with which the physiological state of the child can be subsequently determined and intervened. In addition, other more naturalistic or evolutionary interventions might have better results in combination with Pepper, although their use in research is still limited, even though they have a number of advantages over NAO that are clearly suitable for certain types of subjects, as in this case.

It is important to determine whether this type of intervention with physiological signal recording shows impactful results in the long term and the level of engagement is maintained throughout therapy, as well as on a larger scale, with a larger number of individuals. In addition, it may be of greater benefit to use a robot with facial expression and gestures such as the Moxie or QT robots, or to obtain more information about the level of engagement in the intervention through a system of coding the individual's behaviour throughout the sessions, complemented by real-time image processing.

**Acknowledgements.** This project has received funding by grant PID2020-115220RB-C22 funded by MCIN/AEI/ 10.13039/501100011033 and, as appropriate, by "ERDF A way of making Europe", by the "European Union" or by the "European Union NextGenerationEU/PRTR", and by grant RTI2018-098969-B-100 from the Spanish Ministerio de Ciencia Innovación y Universidades and by grant PROMETEO/2019/119 from the Generalitat Valenciana (Spain).

## References

1. Adrover-Roig, D., Rendón, L., Pinel, V.: Los robots sociales como promotores de la comunicación en los trastornos del Espectro Autista (TEA). Letras de Hoje **53**, 39 (2018)
2. Álvarez, R., et al.: Manual didáctico para la intervención en atención temprana en trastorno del espectro del autismo. (2018)
3. Bhumika, B., Sivakumar, R.: Galvanic skin response for detecting emotional arousal and stress. In: 10th International Conference on Application of Information and Communication Technology and Statistics in Economy and Education (ICAITSEE-2020) (2020)

4. Chayo-Dichy, R., García, A.E.V., García, N.A., Castillo-Parra, G., Ostrosky-Solis, F.: Valencia, activación, dominancia y contenido moral, ante estímulos visuales con contenido emocional y moral: Un estudio en población mexicana, vol. 13 (2003)
5. De Korte, M.W., et al.: Self-initiations in young children with autism during pivotal response treatment with and without robot assistance. Autism Int. J. Res. Pract. **24**(8), 2117–2128 (2020)
6. Drimalla, H., Baskow, I., Behnia, B., Roepke, S., Dziobek, I.: Imitation and recognition of facial emotions in autism: a computer vision approach. Mol. Autism **12**(1), 27 (2021)
7. Efstratiou, R., et al.: Teaching daily life skills in autism spectrum disorder (ASD) interventions using the social robot pepper. In: Lepuschitz, W., Merdan, M., Koppensteiner, G., Balogh, R., Obdržálek, D. (eds.) RiE 2020. AISC, vol. 1316, pp. 86–97. Springer, Cham (2021). https://doi.org/10.1007/978-3-030-67411-3_8
8. Hernández, M.M.: Procesado y Análisis de Señales Fisiológicas en Tareas de Rehabilitación con Robots. Universidad de Valladolid, Escuela de Ingenierías Industriales (2016)
9. Keluskar, J., Reicher, D., Gorecki, A., Mazefsky, C., Crowell, J.A.: Understanding, assessing, and intervening with emotion dysregulation in autism spectrum disorder. Child Adolesc. Psychiatr. Clin. N. Am. **30**(2), 335–348 (2021)
10. Koegel, R.L., Mentis, M.: Motivation in childhood autism: can they or won't they? J. Child Psychol. Psychiatry **26**(2), 185–191 (1985)
11. Koegel, L., Koegel, R., Harrower, J., Carter, C.: Pivotal response intervention I: overview of approach. J. Assoc. Persons Severe Handicaps **24**, 174–185 (1999)
12. Krupa, N., Anantharam, K., Sanker, M., Datta, S., Sagar, J.V.: Recognition of emotions in autistic children using physiological signals. Heal. Technol. **6**(2), 137–147 (2016). https://doi.org/10.1007/s12553-016-0129-3
13. Malik, F., Marwaha, R.: Developmental Stages of Social Emotional Development in Children. StatPearls Publishing, StatPearls (2022)
14. de Murcia, C.d.S.R.: Recuperado 31 de enero de 2022. Información al profesional para mejorar la seguridad en el uso de los medicamentos, murciasalud. (s.f.)
15. Novak, D., et al.: Psychophysiological responses to robotic rehabilitation tasks in stroke. IEEE Trans. Neural Syst. Rehabil. Eng. **18**(4), 351–361 (2010)
16. Perugia, G., Diaz-Boladeras, M., Catala-Mallofre, A., Barakova, E.I., Rauterberg, M.: ENGAGE-DEM: a model of engagement of people with dementia. IEEE Trans. Affect. Comput. 1 (2020)
17. Preventing Medication Errors in Hospitals. ASHP Guidelines on Preventing Medication Errors in Hospitals, Medication Safety-Guidelines, pp. 267–289 (2018)
18. Prince, E.B., et al.: Relationship between autism symptoms and arousal level in toddlers with autism spectrum disorder, as measured by electrodermal activity. Autism **21**(4), 504–508 (2017)
19. Ríos-Hernández, I.: El lenguaje: herramienta de reconstrucción del pensamiento. Razón y Palabra **72** (2010)
20. Ruggieri, V.L.: Empatía, cognición social y trastornos del espectro autista. Revista de Neurología **56**(S01), 13 (2013)
21. Simmons, G.L., Ioannou, S., Smith, J.V., Corbett, B.A., Lerner, M.D., White, S.W.: Utility of an observational social skill assessment as a measure of social cognition in autism. Autism Res. **14**(4), 709–719 (2021)
22. Van den Berk-Smeekens, I., et al.: Adherence and acceptability of a robot-assisted pivotal response treatment protocol for children with autism spectrum disorder. Sci. Rep. **10**(1), 8110 (2020)

23. Vernetti, A., et al.: Atypical emotional electrodermal activity in toddlers with autism spectrum disorder. Autism Res. **13**(9), 1476–1488 (2020)
24. Zúñiga, A.H., Balmaña, N., Salgado, M.: Los trastornos del espectro autista (TEA). Pediatría Integr. **21**(2), 92–108 (2017)

# Creating Vignettes for a Robot-Supported Education Solution for Children with Autism Spectrum Disorder

Trenton Schulz(✉) and Kristin Skeide Fuglerud

Norwegian Computing Center/Norsk Regnesentral,
Postbox 414, Blindern, 0314 Oslo, Norway
{trenton,kristins}@nr.no

**Abstract.** As part of the work in developing a robotic toolkit to help children with ASD develop their social and communication skills, we have adapted a method for creating vignettes that has been fruitful in other studies. We present a brief background of earlier work that has been done with robots and children with ASD, the idea behind our toolkit, and how we have involved the different stakeholders so far. We then present our updated method for creating vignettes and how we plan to run our workshops. We close with discussing some current challenges and next steps.

**Keywords:** Robot · Human-robot interaction · Autism spectrum disorder · Children · Education

## 1 Introduction

Autism Spectrum Disorder (ASD) is characterized by poor nonverbal conversation skills, uneven language development, repetitive or rigid language, and narrow inter-ests in specific areas. Many children with ASD have difficulty understanding body language and the meaning and rhythm of words and sentences. Developing social interaction and communication skills can be challenging, but these skills form the basis for the children's future opportunities, degree of independence, and quality of life. Language skills are vital for education, expressing needs, and participating in society and work life [18].

The current recommendations for programs targeting communication skills of children with ASD should: (*a*) begin at preschool and continue through school; (*b*) be tailored to the child's age and interests; (*c*) address communication and behavior; and (*d*) offer regular reinforcement of positive actions [6,17]. These programs often require special educators and teacher aids. Information and communication technology (ICT) resources may increase the quality of the children's support and reach additional children with ASD. We are currently working on a tool to support language development of children with ASD using social robots.


J. M. Ferrández Vicente et al. (Eds.): IWINAC 2022, LNCS 13258, pp. 323–331, 2022.
https://doi.org/10.1007/978-3-031-06242-1_32

Social robots interact with people in a natural, interpersonal way, and socially assistive robotics (SAR) assist people by using a robot for social interaction (speech, gestures, and body language) [15].

There were several good reasons for selecting robots to help children with ASD. Robots can elicit motivation and provide physical presence and a more tailored experience than other ICT solutions [1]. Robots can provide teachers with new tools [12] and deliver predictable behaviors and repetitive feedback. In addition, a robot can help build social behavior skills, teach, or demonstrate socially desirable behaviors to children with ASD who have trouble expressing themselves. Robots do not get angry, tired, or stressed, and they can be tailored to the needs of a specific child and used repetitively [10]. A child-sized or smaller robot is less intimidating than adults, and many children with ASD therefore feel safer interacting with social robots [15]. Children with ASD who had trained with robots paid closer attention during interactions with adults long after the robot training ended [15], and children with ASD were more likely to complete a treatment session when the session included a robot [23]. Other studies reported improved social skills, increased involvement, more positive behavior, and better social interaction [5,10,14,20].

A review of robots in ASD interventions defined four categories of intervention goals: social, communication, maladaptive behavior, and academic skills [1]. Most studies, however, target only one of these goals, and they normally target only one kind of social robot. Research is needed on how social robots in general can meet the challenge of targeting all or a combination of the goals, in particular combining supporting social skills with language learning. Robots have been shown to be effective in teaching knowledge and skill-based topics, but research is needed on how effectively they teach language [2,13]. To our knowledge, there are no studies of robot-supported development of primary language skills for children with ASD nor any attempt to make the lessons work on multiple kinds of robots. To develop robots in this field, technological and multidisciplinary research is needed in human-robot interaction (HRI), human-computer interaction (HCI), robot-assisted learning, privacy, and ethics.

Our overall objective is to use social robots to improve language, social, and communication skills for children with ASD. We plan to do this by researching how to best apply a robot for this activity by involving teachers, parents, and children in the design process, and to develop a toolbox that the teachers can use to personalize lessons for children with ASD. To meet this objective, we need to understand what possible scenarios work well for teaching children in this diverse group using a robot.

This paper introduces the ROSA toolbox (Sect. 2), discusses some other studies in the area (Sect. 3) and describes on our present research activity of determining the use cases for teaching, and how to involve the children, parents, and teachers (Sect. 4). We also document the process for finding vignettes in our current workshop and how it has been adapted it from other contexts to the context of a robot in the school (Sect. 5). We end with discussing challenges and next steps in our research (Sect. 6).

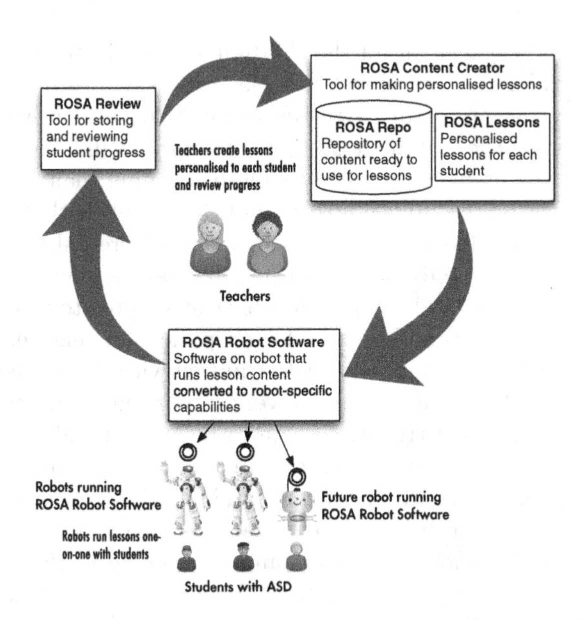


**Fig. 1.** The ROSA toolbox consists of three parts: a content creator, software that runs on the robot for interpreting the lessons, and a review panel for teachers.

## 2 The ROSA Toolbox

To help meet our objective, we are working on building the RObot Supported Education for children with ASD (ROSA) toolbox. The toolbox consists of three parts (Fig. 1): (*a*) ROSA Content Creator, a tool for easily creating tailored one-on-one lessons for children with ASD; (*b*) ROSA Robot Software reads the lessons and runs lesson content customized to the robots capabilities; and (*c*) ROSA Review, a tool for following lesson progress and input for the next lesson. The goal of the toolbox is to make teachers more effective by providing tailor-made education plans for children with ASD and easy to follow progress. For children with ASD, the toolbox lessons will be tailored to their unique needs, increase the children's motivation for learning and result in children developing better language, social, and communication skills. The robot will present content customized to the robot's capabilities. We posit the ROSA toolbox can provide tailored, motivating educational and communication support by exploring and exploiting the unique affordances of a social robot as an expressive medium and educational tool for children with ASD.

## 3 Other Studies Using Robots with Children with ASD

The ROSA toolbox is not the first study to look at how to incorporate a robot with teaching children with ASD. A current question we are examining is the role of the robots in the lessons and what sort of motivation mechanisms that can be used to hold engagement of the children and motivate them to continue. We have examined how other research has used robots. For example, robots have been used to help develop sensory experiences for children with ASD [11]. In this study, a robot helped the children experience a senses at different stations. This study found that the humanoid robot was more engaging than a wheeled robot.

An earlier review on the clinical use of robots for people with ASD was critical about the use of robots because many of the studies were only exploratory and the results normally were only ever published in robotics journals and not journals about ASD [7]. Some studies since then have attempted to be more grounded in ASD research (e.g., [23]) and be over longer periods of time (e.g., [22]).

Others have used robust robots to help children with ASD experience things through play [3]. In this situation, it was very important to create a robot that could be built cheaply, robustly, big enough to be hugged and touched, and run for long periods without maintenance. The created robot (TeoG) was successful and able to stand up to children tackling and hugging the robot in the play sessions. Others have gone further and developed qualitative and quantitative requirements for using touch with robots and children with ASD [4]. Care must be taken, however, as some children with ASD seek out touch, while others avoid it. Regardless, the therapists involved in designing the guidelines argue that touch would be useful regardless of the role the robot played (e.g., teacher, friend, companion); although there was no agreement on the shape of the robot [4].

Some studies have used educational games [19,23,24] that have been good for teaching concepts and improving skills, such as turn-taking, social interaction, self-initiation, and understanding other people's perspective. In one of the studies, the robot was only used for part of a therapy session, but the children reported liking the *whole* session better (i.e., not just the parts including the robots) [23]. Regardless, of how the robot is used, technical issues need to be addressed as well. Some solutions can be having a second backup robot [19] or ensuring that the robot is responsive to the actions that happen from the child [23]. Using *Wizard of Oz* techniques may address these issues in the short-term, but are likely expensive in the long term.

## 4 Involving Children, Parents, and Teachers in the Creation Process

To ensure success of the ROSA toolbox, we are working closely with a school that specializes working with children with ASD and other cognitive disabilities. We have visited the school on multiple occasions and observed how different classes teach the children. These observations were useful for understanding the range of abilities of the children in the school.

We have also held meetings with teachers where we have presented the goals of the project and previous work that has been done in the area (e.g., from Mengoni et al. [16]). We also asked teachers to anonymously answer a questionnaire about what they thought about possibilities for using a robot in the classroom. This included questions like what feelings would be raised in the children, if the robot can help children with developing language or communication skills, other social skills, and if the robot could help in maintaining motivation, attention, joint attention, and concentration. We also asked the teachers if the robot should be used in groups or individually.

We also had the robot come to visit a couple of classes. There were also a couple of sessions where the robot ran a program that was developed for a different target group [8], but could still be useful for communication in some classes. The visits were useful from a technical standpoint as it highlighted a variety of technical issues that would need to be solved to make the robot work seamlessly in the classroom.

One activity we devised for understanding the context was to have the teachers work together with the researchers to develop vignettes of activities that could serve as the basis of one or more lessons.

## 5 Finding Vignettes

Vignettes are small, self-contained, reusable, temporarily ordered set of events that can be put into multiple scenarios [21]. The scenarios here in this case would be the in the potential lessons created by the ROSA content creator and run in the ROSA robot software. This requires cooperation from teachers as the experts in understanding the children's needs.

Our method for finding vignettes is based on a technique that we have used in other projects [9]. The goal is to gather groups of local experts in a workshop with the goal of vignettes as an output from the workshop. The local experts work together on a worksheet that list the necessary bits that form the building blocks of the vignette. From experience, having experts that understand the context and the goal has led to developing many vignettes in the very short time the workshop was run [9].

Since this technique can be used in different context, the different items on the worksheet are dependent on the context being worked on. Since the context here is children with ASD in school and the local experts were teachers, the items in the worksheet were adjusted to use pedagogical and work terms that the teachers were familiar with.

Our categories of items included:

- Purpose
- Selected goals and tasks from previous work
- Student's or students' skill level
- Learning prerequisites
- Digital learning environment
- Amplifier or reward
- Prompts that can be given
- Learning outcome
- Other

The workshop will be run as part of a larger two-day pedagogical workshop at the school. The workshop itself is split into a section for both days. On the first day, the ROSA toolbox will be explained and some of the different technologies that we are working with will also be explained. We will then introduce them to the basics of the workshop, introduce the categories, and show the groups for the activities for the next day. We have worked with the administration before the pedagogical in the school to split up the teachers into groups where they had similar levels of students. In addition, each group will have an explicit area of focus (for example, part of a specific curriculum or technology) to help anchor the vignette finding.

On the second day, the teachers will work in the groups introduced from the previous day to come up with as many vignettes as they can in a 90-min period. Since there will be more groups than researchers, researchers will switch between groups to help keep the groupwork on track. After the 90-min period is up, everyone will return and summarize their results from each group.

## 6 Challenges and Next Steps

The vignettes will provide a starting point on creating scenarios and content that can be added to the ROSA toolbox. We plan on putting together a early prototype that can be used before summer or in the start of the next school year.

The teachers that answered our survey indicated that the children would probably be likely excited and skeptical to a robot in the classroom. They also were unsure how well the robot would work in the classroom, but were positive to finding out how this could work.

The robot visits in the classroom drew interest from the children. The robot seemed to capture the children's attention, although some were uncertain of the robot. Unfortunately, due to illness we were unable to complete the visits to all the classes we wanted, but we hope to resume this later, perhaps with some of the vignettes we are working on.

The visiting robot also uncovered many technical issues with the robot, network connections, and the software. The biggest issues was connectivity; it turns out that the connection between the robot, the software, and offsite cloud service was more fragile than anticipated. This led to many issues in having the robot not function as expected, but it was good feedback to use for future iterations of the software. Making the technical issues disappear is important as our observations and feedback from the teachers indicates that if the robot does not work as expected, children will not be draw motivation from the robot, and it won't be used.

Also, the COVID-19 pandemic still causes issues in schools with unexpected sick leave and potential closing of schools. Our partner school has been very flexible and robust, so this has been less of an issue so far than was initially anticipated. We continue monitoring the situation to ensure that risks of infection remain low.

**Acknowledgments.** This work is partly supported by the Research Council of Norway as part of the RObot Supported Education for children with ASD (ROSA) project, under grant agreement 321821. We would also like to thank the other project members, and a special thank you to the teachers and students at *Frydenhaug skole* for their participation in the project.

## References

1. Begum, M., Serna, R.W., Yanco, H.A.: Are robots ready to deliver autism interventions? A comprehensive review. Int. J. Soc. Robot. **8**(2), 157–181 (2016). ISSN: 1875-4805. https://doi.org/10.1007/s12369-016-0346-y
2. Belpaeme, T., et al.: Guidelines for designing social robots as second language tutors. Int. J. Soc. Robot. (2018). ISSN: 1875-4791, 1875-4805. https://doi.org/10.1007/s12369-018-0467-6. Accessed 26 Jan 2018
3. Brivio, A., Rogacheva, K., Lucchelli, M., Bonarini, A.: A soft, mobile, autonomous robot to develop skills through play in autistic children. Paladyn, J. Behav. Robot. **12**(1), 187–198 (2021). ISSN: 2081-4836. https://doi.org/10.1515/pjbr-2021-0015.https://www.degruyter.com/document/doi/10.1515/pjbr-2021-0015/html. Accessed 29 06 2021
4. Burns, R.B., Seifi, H., Lee, H., Kuchenbecker, K.J.: Getting in touch with children with autism: specialist guidelines for a touch-perceiving robot. Paladyn, J. Behav. Robot. **12**(1), 115–135 (2021). ISSN: 2081-4836. https://doi.org/10.1515/pjbr-2021-0010. https://www.degruyter.com/document/doi/10.1515/pjbr-2021-0010/html. Accessed 29 Jun 2021
5. Dautenhahn, K., Werry, I.: Towards interactive robots in autism therapy: background. Motiv. Challenges Pragmatics Cogn. **12**(1), 1–35 (2004). ISSN: 0929-0907, 1569-9943. https://doi.org/10.1075/pc.12.1.03dau. https://www.jbe-platform.com/content/journals/10.1075/pc.12.1.03dau. Accessed 23 June 2020
6. Deafness and Other Communication Disorders (NIDCD), N.I. on: Autism Spectrum Disorder: Communication Problems in Children, NIDCD. (2015). https://www.nidcd.nih.gov/health/autism-spectrum-disorder-communication-problemschildren. Accessed 12 Nov 2020
7. Diehl, J.J., Schmitt, L.M., Villano, M., Crowell, C.R.: The clinical use of robots for individuals with autism spectrum disorders: a critical review. Res. Autism Spectrum Disord. **6**(1), 249–262 (2012). ISSN: 17509467. https://doi.org/10.1016/j.rasd.2011.05.006. https://linkinghub.elsevier.com/retrieve/pii/S1750946711000894. Accessed 12 Nov 2020
8. Halbach, T., Schulz, T., Leister, W., Solheim, I.: Robot-enhanced language learning for children in norwegian day-care centers. MTI **5**(12), 74 (2021). ISSN: 2414-4088. https://doi.org/10.3390/mti5120074. https://www.mdpi.com/2414-4088/5/12/74. Accessed 13 Jan 2022
9. Hannay, J.E., Fuglerud, K.S., Schulz, T., Leister, W.: Scenario design for healthcare collaboration training under suboptimal conditions. In: 13th International Conference on Digital Human Modeling and Applications in Health, Safety, Ergonomics and Risk Management. LNCS. Springer, Gothenburg (2022, inpress)
10. Huijnen, C.A.G.J., Lexis, M.A.S., Jansens, R., de Witte, L.P.: How to implement robots in interventions for children with autism? A co-creation study involving people with autism, parents and professionals. J. Autism Dev. Disord. **47**(10), 3079–3096 (2017). ISSN: 1573-3432. https://doi.org/10.1007/s10803-017-3235-9. https://doi.org/10.1007/s10803-017-3235-9. Accessed 12 Nov 2020

11. Javed, H., Burns, R., Jeon, M., Howard, A.M., Park, C.H.: A robotic framework to facilitate sensory experiences for children with autism spectrum disorder: a preliminary study. J. Hum.-Robot Interact. **9**(1), 3:1–3:26 (2019). https://doi.org/10.1145/3359613. Accessed 04 Feb 2020
12. Jones, A., Castellano, G.: Adaptive robotic tutors that support self-regulated learning: a longer-term investigation with primary school children. Int. J. Soc. Robot. **10**(3), 357–370 (2018). https://doi.org/10.1007/s12369-017-0458-z. Accessed 12 Nov 2020
13. Kanero, J., Geçkin, V., Oranç, C., Mamus, E., Küntay, A.C., Göksun, T.: Social robots for early language learning: current evidence and future directions. Child Dev. Perspect. **12**(3), 146–151 (2018). ISSN: 17508592. https://doi.org/10.1111/cdep.12277. http://doi.wiley.com/10.1111/cdep.12277. Accessed 26 Mar 2020
14. Lee, C.-H.J., Kim, K., Breazeal, C., Picard, R.: ShyBot: friend-stranger interaction for children living with autism. In: CHI 2008 Extended Abstracts on Human Factors in Computing Systems. CHI EA 2008, pp. 3375–3380. Association for Computing Machinery, New York (2008). ISBN: 978-1-60558-012-8. https://doi.org/10.1145/1358628.1358860. Accessed 12 Jan 2020
15. Matarić, M.J., Scassellati, B.: Socially assistive robotics. In: Siciliano, B., Khatib, O. (eds.) Springer Handbook of Robotics, pp. 19731–1994. Springer, Cham (2016). https://doi.org/10.1007/978-3-319-32552-1_73. Accessed 12 Nov 2016
16. Mengoni, S.E., et al.: Feasibility study of a randomised controlled trial to investigate the effectiveness of using a humanoid robot to improve the social skills of children with autism spectrum disorder (KasparRCT): a study protocol. BMJ Open **7**(6), e017376 (2017). ISSN 2044-6055, 2044-6055. https://doi.org/10.1136/bmjopen-2017-017376.pmid:28645986. https://bmjopen.bmj.com/content/7/6/e017376. Accessed 25 Feb 2022
17. Olaff, H.S., Eikeseth, S.: Variabler som kan påvirke effekter av tidlig og intensiv opplaring basert på anvendt atferdsanalyse (EIBI/TIOBA) (2015). ISSN: 0809-781X. https://oda-hioa.archive.knowledgearc.net/handle/10642/3068. Accessed 12 Nov 2020
18. Parmenter, T.: Promoting training and employment opportunities for people with intellectual disabilities: international experience (2011). https://ecommons.cornell.edu/handle/1813/76807. Accessed 26 Jan 2022
19. Sandygulova, A., et al.: Interaction design and methodology of robot-assisted therapy for children with severe ASD and ADHD. Paladyn, J. Behav. Robot. **10**(1), 330–345 (2019). https://doi.org/10.1515/pjbr-2019-0027. https://www.degruyter.com/view/journals/pjbr/10/1/article-p330.xml. Accessed 22 Jun 2020
20. Simm, W., Ferrario, M.A., Gradinar, A., Whittle, J.: Prototyping 'clasp': implications for designing digital technology for and with adults with autism. In: Proceedings of the 2014 Conference on Designing Interactive Systems. DIS 2014, pp. 345–354. Association for Computing Machinery, New York (2014). ISBN: 978-1-4503-2902-6. https://doi.org/10.1145/2598510.2600880. Accessed 12 Nov 2020
21. Simulation Interoperability Standards Organization: SISO-GUIDE-006-2018 - Guideline on Scenario Development for Simulation Environments (2018)
22. Taheri, A., Shariati, A., Heidari, R., Shahab, M., Alemi, M., Meghdari, A.: Impacts of using a social robot to teach music to children with low-functioning autism. Paladyn, J. Behav. Robot. **12**(1), 256–275 (2021). ISSN: 2081-4836. https://doi.org/10.1515/pjbr-2021-0018. https://www.degruyter.com/document/doi/10.1515/pjbr-2021-0018/html. Accessed 29 Jun 2021

23. Van den Berk-Smeekens, I., et al.: Adherence and acceptability of a robot-assisted pivotal response treatment protocol for children with autism spectrum disorder. Sci. Rep. **10**(1), 8110 (2020). ISSN: 2045-2322. https://doi.org/10.1038/s41598-020-65048-3. https://www.nature.com/articles/s41598-020-65048-3. Accessed 10 Jul 2020
24. Wood, L.J., Robins, B., Lakatos, G., Syrdal, D.S., Zaraki, A., Dautenhahn, K.: Developing a protocol and experimental setup for using a humanoid robot to assist children with autism to develop visual perspective taking skills. Paladyn, J. Behav. Robot. **10**(1), 167–179 (2019). ISSN: 2081-4836. https://doi.org/10.1515/pjbr-2019-0013. https://www.degruyter.com/view/j/pjbr.2019.10.issue-1/pjbr-2019-0013/pjbr-2019-0013.xml?format=INT. Accessed 27 May 2019

# Identification of Parkinson's Disease from Speech Using CNNs and Formant Measures

Agustín Álvarez-Marquina[1(✉)], Andrés Gómez-Rodellar[2], Pedro Gómez-Vilda[1], Daniel Palacios-Alonso[3], and Francisco Díaz-Pérez[4]

[1] NeuSpeLab, CTB, Universidad Politécnica de Madrid, 28223 Pozuelo de Alarcón, Madrid, Spain
{agustin.alvarez,pedro.gomezv}@upm.es

[2] Faculty of Medicine, Usher Institute, University of Edinburgh, Edinburgh, UK
a.gomezrodellar@ed.ac.uk

[3] E.T.S. de Ingeniería Informática - Universidad Rey Juan Carlos, Campus de Móstoles, Tulipán, s/n, 28933 Móstoles, Madrid, Spain
daniel.palacios@urjc.es

[4] E.T.S. de Ingeniería de Sistemas Informáticos - Universidad Politécnica de Madrid, Campus Sur, Alang Turing, s/n, 28031 Madrid, Spain
francisco.diazp@upm.es

**Abstract.** Parkinson's Disease (PD) is a neurodegenerative disorder that severely impacts the motor capabilities of patients. Dysarthria is one of the symptoms that can be accurately characterized using speech analysis, tracking the deterioration associated with the evolution of the disease. Through the present work the use of machine learning-based technologies, more specifically the Convolutional Neural Networks (CNNs) and the direct application of formant features extracted form sustained phonations of vowel /a/ are proposed. The main goal is to investigate the effects of the speech articulatory movements affected by hypokinetic dysarthria in Parkinson's Disease as this would allow to use speech as a reliable monitoring tool. The study employs voice recording of 593 subjects form the Patient Voice Analysis dataset (PVA) and 687 health controls from the Saarbrücken Voice Database (SVD). The k-fold cross-validation trials provided the best results when the length of the utterances is limited to 2 s, achieving a sensibility of 0.96 and a specificity of 0.99.

**Keywords:** Convolutional neural networks · Formant measurement · Sustained vowel phonation · Parkinson's disease · Hypokinetic dysarthria

This research received funding from grants TEC2016-77791-C4-4-R (Ministry of Economic Affairs and Competitiveness of Spain), and Teca-Park-MonParLoc FGCSICCE-NIE 0348-CIE-6-E (InterReg Programme). PVA datasets were generated through collaboration between Sage Bionetworks, PatientsLikeMe and Dr. Max Little as part of the Patient Voice Analysis study (PVA). They were obtained through Synapse ID [syn2321745].



J. M. Ferrández Vicente et al. (Eds.): IWINAC 2022, LNCS 13258, pp. 332–342, 2022.
https://doi.org/10.1007/978-3-031-06242-1_33

# 1 Introduction

Parkinson's disease (PD) affects the cells responsible of producing dopamine in the brain. Symptoms include an increased muscle tonus (rigidity), diminished and slow movements (bradykinesia), the inability of voluntary movement (akinesis), and shaking movement in the resting position (tremor) [1]. Current studies show that it is the second most prevalent neurodegenerative disease after Alzheimer's [2] being reported as one of the fastest-growing neurological disorders in terms of prevalence and deaths [3]. Impairment of voice and speech appears to be a common overall affectation with the progression of PD [4]. Over 90% of patients with PD experience some form of speech-motor impairment, namely, hypokinetic dysarthria characterized by reduced pitch and loudness [5,6]. In that sense, acoustic voice analysis may be considered as an important non-invasive tool for screening and monitoring this disorder. PD features derived from speech may be comprised of voice tremor, imprecise articulation, monotonous speech and, hoarseness. Latency in response accompanied by rushes of speech may also observed [7]. Additionally, changes in formant frequencies during the vowel phonation due to dysarthria have also been reported in studies [8].

Artificial intelligence techniques using the information extracted from speech may be helpful in identifying Parkinson's disease, assessing the extent of the progression, and suggesting important cues about the acoustic units that are more affected by the disease [9]. The underlaying goal is obtaining new biomarkers that are directly linked with the articulatory dysfunctions introduced by the disease as early as its initial stages [10]. It might prove useful to extend the use of features devoted to general voice disorder detection, typically using sustained vowels to detect dysphonia associated to PD [11]. However, it is also possible to use state-of-the-art speaker identification techniques such as GMM-UBM and i-Vectors [12] or x-Vectors [13] by means of diadochokinetic tests and text dependent utterances with the application of Perceptual Linear Prediction coefficients (PLP) or Mel-Frequency Cepstral Coefficients (MFCC) as descriptors. Recent advances using these descriptors have allowed the use of deep learning algorithms to the estimation of PD through speech, by employing the application of artificial neural networks [14]. Solutions originally proposed for image processing as Convolutional Neural Networks (CNN) [15] and transfer learning strategy [16] may also be applied for the present task. Besides, Long Short-Term Memory (LSTM) recurrent neural networks have been successfully applied for this purpose [17].

Regarding the identification and separation of people with Parkinson's (PwP) from healthy controls, it usually ranges from 85% to 99% but it is dependent on the task and the databases considered [18]. Recent studies using sustained vocalizations show remarkable differences. In [19] detection rates of 95.66% (90.70% of sensibility and 98.1% of specificity) are achieved for 26 PD patients and 16 subject controls using distributions of vocal fold biomechanical and tremor estimations with Random Forests (RF). However, [20] reports only 67.43% sensitivity and 67.25% specificity for a much larger population of 1078 PD and 5453 control participants using a selected set of 304 dysphonia measures to train a Support Vector Machine (SVM). Finally, a latest study using a CNN-based approach from direct spectrogram feature extraction of sustained phonation of vowels yields a sensitivity of 86.2% and specificity of 93.3% [21].

The objective of this study is to extend a previous work where formant acoustics were translated to neuromotor activity through speech articulatory movements [22].

The paper is organized as follows: A description of the features and CNN architecture proposed in the study is given in Sect. 2. Section 3 describes the dataset and the experimental protocol to test the functionality of the CNN from PD and healthy controls (HC). Section 4 summarizes the experimental results and their discussion. The conclusions derived from the study are given in Sect. 5.

## 2 Materials and Methods

### 2.1 Formant Features

Figure 1 shows an example of the measures estimated for a female PD patient and a female control. The normalized features extracted every 5 ms are then used to feed the CNN input.

The features extracted in the present study are six formant measures obtained directly from the FFT spectrogram of the speech signals. Measures consist of the first three formant frequency values $F_1$, $F_2$ and $F_3$, and the energy calculated associated to the previous estimations: $eF_1$, $eF_2$ and $eF_3$. Those values are estimated from the log-power spectrum in the frequency domain performing the following steps:

- High quality SVD recordings are downsampled to 8 kHz for PVA telephone-channel compatibility.
- A simple energy VAD is applied for the detection of the vocalizations. The length of speech fragments considered ranges from 250 ms to 4 s.
- A zero-padding mechanism is used for speech files shorter than the required segment length.
- Spectrograms are generated by means of the FFT using a Hamming sliding window of 25 ms with a 5 ms step (20 ms overlap).
- Formant tracking is obtained frame by frame from the peaks of the FFT envelope without applying a smoothing procedure.
- Finally, every single formant estimation, being frequency or energy, is mean subtracted and divided by the standard deviation calculated using whole speech segments.

### 2.2 CNN Architecture

The neural network proposed in this study is a CNN architecture as described in Table 1.

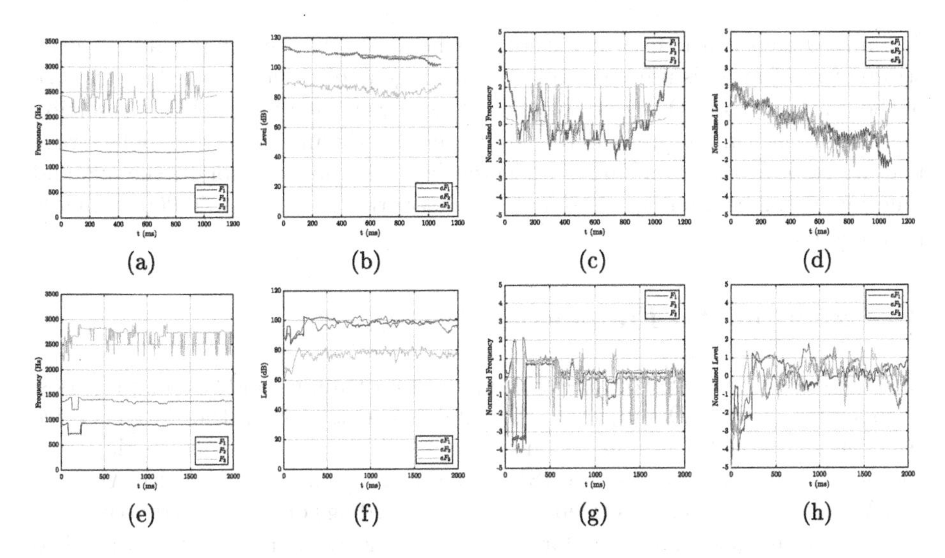


**Fig. 1.** Formant features of a control female subject: a) $F_1$, $F_2$ and $F_3$ frequencies in Hz; b) Log-power of $F_1$, $F_2$ and $F_3$ in dB; c) and d) Mean and standard deviation normalization of features in a) and b) respectively. Formant features of a PD female patient: e) $F_1$, $F_2$ and $F_3$ frequencies in Hz, f) Log-power of $F_1$, $F_2$, $F_3$ in dB, g) and h) Mean and standard deviation normalization of features in e) and f) respectively.

**Table 1.** Architecture of the proposed convolutional neural network.

| Layer | Configuration | Output size |
|---|---|---|
| $Input$ | | $(M, N)$ |
| $1D\_Convolutional + BN + ReLU + Drop_{(0.1)}$ | $f = 16, k = 5, s = 2$ | $(\frac{M}{2}, 16)$ |
| $1D\_Convolutional + BN + ReLU + Drop_{(0.1)}$ | $f = 32, k = 3, s = 1$ | $(\frac{M}{2}, 32)$ |
| $1D\_AveragePooling$ | $k = 2$ | $(\frac{M}{4}, 32)$ |
| $1D\_Convolutional + BN + ReLU + Drop_{(0.1)}$ | $f = 64, k = 3, s = 1$ | $(\frac{M}{4}, 64)$ |
| $1D\_AveragePooling$ | $k = 2$ | $(\frac{M}{8}, 64)$ |
| $1D\_Convolutional + BN + ReLU + Drop_{(0.1)}$ | $f = 128, k = 3, s = 1$ | $(\frac{M}{8}, 128)$ |
| $1D\_AveragePooling$ | $k = 2$ | $(\frac{M}{16}, 128)$ |
| $1D\_Convolutional + BN + ReLU + Drop_{(0.1)}$ | $f = 256, k = 3, s = 1$ | $(\frac{M}{16}, 256)$ |
| $GlobalAveragePooling$ | | 256 |
| $Drop_{(0.5)} + Dense_{(64)} + Drop_{(0.5)}$ | | 64 |
| $Dense_{(1)}$ | $Sigmoid$ | 1 |

The input $M$ corresponds to the number of 5 ms frames extracted from the speech signal, and $N$ is the number of formant features considered. The network comprises five convolutional layers followed by a batch normalization (BN) step and a rectified linear unit (ReLU) as a non-linear function. Every convolution matrix is of the form $3 \times f$, where $f$ starts with 16 and ends at 128. These values correspond to the number of filters applied in the previous convolution operation.

The initial stage constitutes an exception where a $5 \times N$ convolving matrix with a stride of two frames is used instead of an average pooling layer to reduce the size of the temporal axis. Dropout layers are extensively used for randomly resetting inputs during training time in order to prevent overfitting. After a global average pooling step, a 64-neuron fully-connected layer is applied prior to the sigmoid function acting as a binary classifier. The network has around 150,000 trainable parameters.

## 3 Experimental Framework

In this work, two speech corpora are employed for training and testing purposes. A collection of 593 PD subjects (289 male, 304 female) from the Patient Voice Analysis dataset (PVA) and 687 health controls (259 male, 428 female) from the Saarbrücken Voice Database (SVD) is selected. The Patient Voice Analysis (PVA) dataset [23] contains telephone voice recordings of a sustained vowel /a/, hold as long as possible at a comfortable level. Patients are ranged from stage 1 to 5 of the Hoehn and Yahr scale and mean age is 62.17 years old (maximum 84 and minimum 34). On the other hand, the Saarbrücken Voice Database (SVD) [24] contains recordings of sustained vowels /a/, /i/ and /u/ with normal, higher, lower, rising and falling pitch, together with a short phrase. A total of 1002 speakers exhibiting a wide range of voice disorders and 851 controls (ages varies from 9–84 years) are included. However, only normal pith /a/ utterances from healthy subjects are considered in this study thus reducing the final number of available speakers. In both speech databases, a small portion of the speakers contributes with more than a single session. Also, for the SVD data a telephone filtered version of all recordings is also added.

As female and male speakers are not separated into distinct groups, a total amount of 740 PD and 1408 control samples are applied to the network using a k-fold cross-validation approach. Following this methodology, the whole data is divided into 10 equal subsets. During each iteration of the cross-validation procedure eight partitions are used for training, one partition is used for development and the remaining one is applied to test the system. That development partition is used to prevent overfitting and to provide an early stopping mechanism during the training. A maximum of 100 epochs is applied to the network. However, when the validation score does not improve for 20 successive iterations the training process is aborted. The network providing the best validation score is then recovered to apply the test subset. Initially, the learning rate is 0.001 with a five-epoch exponential decay of 0.95. To measure the system performance, the Mathews correlation coefficient (MCC) [25] is used along the process as the dataset exhibit some imbalance. MCC is given by:

$$MCC = \frac{TP \cdot TN - FP \cdot FN}{\sqrt{(TP+FP)(TP+FN)(TN+FP)(TN+FN)}} \tag{1}$$

## 4 Results and Discussion

The results of the identification of Parkinson's disease patients and control subjects are summarized for different feature combinations in Table 2 for speech segments of 250 ms, Table 3 (500 ms), Table 4 (1 s), Table 5 (2 s) and Table 6 (4 s). In addition to MCC, other binary classification metrics as sensibility, specificity, accuracy, F1, and the area under the curve (AUC) are calculated.

As a rule, it may be appreciated that the increase in the portion of speech analysed by the neural network is linked to a better separation between patients and control groups. However, that clear trend reaches its limit for 2 s segments.

**Table 2.** Identification results for segment size of 250 ms *($M = 50$)*.

| Feature set | MCC | Sensibility | Specificity | Accuracy | F1 score | AUC |
|---|---|---|---|---|---|---|
| $F_1$ | 0.6913 | 0.7797 | 0.9048 | 0.8617 | 0.7953 | 0.9284 |
| $F_2$ | 0.5492 | 0.6689 | 0.8686 | 0.7998 | 0.6972 | 0.8440 |
| $F_3$ | 0.4005 | 0.6554 | 0.7550 | 0.7207 | 0.6178 | 0.7723 |
| $F_1$, $F_2$ | 0.6629 | 0.7514 | 0.9013 | 0.8496 | 0.7749 | 0.9113 |
| $F_1$, $F_3$ | 0.5982 | 0.6973 | 0.8871 | 0.8217 | 0.7293 | 0.8777 |
| $F_2$, $F_3$ | 0.5300 | 0.6608 | 0.8594 | 0.7910 | 0.6854 | 0.8560 |
| $F_1$, $F_2$, $F_3$ | 0.6923 | 0.7446 | 0.9261 | 0.8636 | 0.7900 | 0.9166 |
| $eF_1$ | 0.8667 | 0.9068 | 0.9574 | 0.9399 | 0.9123 | 0.9781 |
| $eF_2$ | 0.8698 | 0.9095 | 0.9581 | 0.9413 | 0.9144 | 0.9770 |
| $eF_3$ | 0.6981 | 0.8432 | 0.8686 | 0.8599 | 0.8057 | 0.9241 |
| $eF_1$, $eF_2$ | 0.9120 | 0.9162 | 0.9837 | 0.9604 | 0.9410 | 0.9874 |
| $eF_1$, $eF_3$ | 0.8578 | 0.9081 | 0.9503 | 0.9358 | 0.9069 | 0.9788 |
| $eF_2$, $eF_3$ | 0.8798 | 0.9054 | 0.9673 | 0.9460 | 0.9203 | 0.9775 |
| $eF_1$, $eF_2$, $eF_3$ | 0.9058 | 0.9270 | 0.9737 | 0.9576 | 0.9378 | 0.9850 |
| $F_1$, $eF_1$ | 0.8997 | 0.9257 | 0.9702 | 0.9548 | 0.9339 | 0.9889 |
| $F_2$, $eF_2$ | 0.8719 | 0.9108 | 0.9588 | 0.9423 | 0.9158 | 0.9774 |
| $F_3$, $eF_3$ | 0.7468 | 0.8189 | 0.9219 | 0.8864 | 0.8324 | 0.9437 |
| $F_1$, $F_2$, $eF_1$, $eF_2$ | 0.9057 | 0.9176 | 0.9787 | 0.9576 | 0.9372 | 0.9839 |
| $F_1$, $F_3$, $eF_1$, $eF_3$ | 0.8671 | 0.8757 | 0.9744 | 0.9404 | 0.9101 | 0.9809 |
| $F_2$, $F_3$, $eF_2$, $eF_3$ | 0.8587 | 0.8784 | 0.9673 | 0.9367 | 0.9053 | 0.9784 |
| $F_1$, $F_2$, $F_3$, $eF_1$, $eF_2$, $eF_3$ | 0.8819 | 0.8851 | 0.9794 | 0.9469 | 0.9199 | 0.9817 |

The examination of the results shows a better behaviour for energy derived features ($eF_1$, $eF_2$ and $eF_3$) compared to the frequency set ($F_1$, $F_2$ and $F_3$) when considered in isolation or in groups. Moreover, that performance differences are increased when shorter segment lengths are considered. Besides, joining a formant frequency and its energy values provides an improvement in the system detection capabilities specially for the first two formants. Nevertheless, the concatenation of several frequency and energy features (e.g., $F_1$, $F_2$, $eF_1$ and $eF_2$) usually does not lead to a further enhancement.

**Table 3.** Identification results for segment size of 500 ms *(M = 100)*.

| Feature set | MCC | Sensibility | Specificity | Accuracy | F1 score | AUC |
|---|---|---|---|---|---|---|
| $F_1$ | 0.7451 | 0.8216 | 0.9190 | 0.8855 | 0.8317 | 0.9486 |
| $F_2$ | 0.6551 | 0.7324 | 0.9070 | 0.8468 | 0.7672 | 0.9212 |
| $F_3$ | 0.4576 | 0.6405 | 0.8161 | 0.7556 | 0.6436 | 0.8113 |
| $F_1$, $F_2$ | 0.7272 | 0.7514 | 0.9460 | 0.8790 | 0.8105 | 0.9263 |
| $F_1$, $F_3$ | 0.7068 | 0.7784 | 0.9169 | 0.8692 | 0.8039 | 0.9331 |
| $F_2$, $F_3$ | 0.6103 | 0.7284 | 0.8764 | 0.8254 | 0.7419 | 0.8943 |
| $F_1$, $F_2$, $F_3$ | 0.7658 | 0.7851 | 0.9538 | 0.8957 | 0.8384 | 0.9452 |
| $eF_1$ | 0.9008 | 0.9297 | 0.9688 | 0.9553 | 0.9348 | 0.9866 |
| $eF_2$ | 0.8841 | 0.9324 | 0.9553 | 0.9474 | 0.9243 | 0.9816 |
| $eF_3$ | 0.7571 | 0.8635 | 0.9020 | 0.8887 | 0.8425 | 0.9499 |
| $eF_1$, $eF_2$ | 0.9317 | 0.9405 | 0.9844 | 0.9693 | 0.9547 | 0.9912 |
| $eF_1$, $eF_3$ | 0.8995 | 0.9149 | 0.9759 | 0.9548 | 0.9332 | 0.9851 |
| $eF_2$, $eF_3$ | 0.9016 | 0.9203 | 0.9744 | 0.9558 | 0.9348 | 0.9869 |
| $eF_1$, $eF_2$, $eF_3$ | 0.9183 | 0.9216 | 0.9851 | 0.9632 | 0.9453 | 0.9892 |
| $F_1$, $eF_1$ | 0.9195 | 0.9459 | 0.9730 | 0.9637 | 0.9472 | 0.9924 |
| $F_2$, $eF_2$ | 0.9039 | 0.9311 | 0.9702 | 0.9567 | 0.9368 | 0.9865 |
| $F_3$, $eF_3$ | 0.8006 | 0.8514 | 0.9418 | 0.9106 | 0.8678 | 0.9653 |
| $F_1$, $F_2$, $eF_1$, $eF_2$ | 0.9142 | 0.9351 | 0.9751 | 0.9614 | 0.9434 | 0.9927 |
| $F_1$, $F_3$, $eF_1$, $eF_3$ | 0.8891 | 0.8946 | 0.9794 | 0.9502 | 0.9252 | 0.9854 |
| $F_2$, $F_3$, $eF_2$, $eF_3$ | 0.9006 | 0.9230 | 0.9723 | 0.9553 | 0.9343 | 0.9870 |
| $F_1$, $F_2$, $F_3$, $eF_1$, $eF_2$, $eF_3$ | 0.9101 | 0.9068 | 0.9872 | 0.9595 | 0.9391 | 0.9847 |

**Table 4.** Identification results for segment size of 1 s *(M = 200)*.

| Feature set | MCC | Sensibility | Specificity | Accuracy | F1 score | AUC |
|---|---|---|---|---|---|---|
| $F_1$ | 0.8020 | 0.8351 | 0.9517 | 0.9115 | 0.8668 | 0.9618 |
| $F_2$ | 0.7839 | 0.8514 | 0.9297 | 0.9027 | 0.8577 | 0.9574 |
| $F_3$ | 0.6900 | 0.7919 | 0.8963 | 0.8603 | 0.7962 | 0.9168 |
| $F_1$, $F_2$ | 0.8335 | 0.8595 | 0.9602 | 0.9255 | 0.8883 | 0.9668 |
| $F_1$, $F_3$ | 0.8341 | 0.8757 | 0.9517 | 0.9255 | 0.8901 | 0.9707 |
| $F_2$, $F_3$ | 0.7808 | 0.8162 | 0.9474 | 0.9022 | 0.8519 | 0.9430 |
| $F_1$, $F_2$, $F_3$ | 0.8525 | 0.8770 | 0.9638 | 0.9339 | 0.9014 | 0.9749 |
| $eF_1$ | 0.9297 | 0.9392 | 0.9837 | 0.9683 | 0.9534 | 0.9914 |
| $eF_2$ | 0.9235 | 0.9405 | 0.9787 | 0.9655 | 0.9495 | 0.9913 |
| $eF_3$ | 0.8577 | 0.8703 | 0.9709 | 0.9362 | 0.9039 | 0.9709 |
| $eF_1$, $eF_2$ | 0.9514 | 0.9595 | 0.9879 | 0.9781 | 0.9680 | 0.9972 |
| $eF_1$, $eF_3$ | 0.9328 | 0.9446 | 0.9830 | 0.9697 | 0.9556 | 0.9905 |
| $eF_2$, $eF_3$ | 0.9276 | 0.9297 | 0.9872 | 0.9674 | 0.9516 | 0.9882 |
| $eF_1$, $eF_2$, $eF_3$ | 0.9380 | 0.9446 | 0.9865 | 0.9721 | 0.9588 | 0.9941 |
| $F_1$, $eF_1$ | 0.9390 | 0.9514 | 0.9837 | 0.9725 | 0.9598 | 0.9932 |
| $F_2$, $eF_2$ | 0.9484 | 0.9500 | 0.9908 | 0.9767 | 0.9657 | 0.9939 |
| $F_3$, $eF_3$ | 0.8723 | 0.8919 | 0.9695 | 0.9427 | 0.9148 | 0.9748 |
| $F_1$, $F_2$, $eF_1$, $eF_2$ | 0.9381 | 0.9554 | 0.9808 | 0.9721 | 0.9593 | 0.9941 |
| $F_1$, $F_3$, $eF_1$, $eF_3$ | 0.9267 | 0.9257 | 0.9886 | 0.9669 | 0.9507 | 0.9880 |
| $F_2$, $F_3$, $eF_2$, $eF_3$ | 0.9110 | 0.9176 | 0.9822 | 0.9600 | 0.9404 | 0.9819 |
| $F_1$, $F_2$, $F_3$, $eF_1$, $eF_2$, $eF_3$ | 0.9172 | 0.9297 | 0.9801 | 0.9628 | 0.9451 | 0.9856 |

**Table 5.** Identification results for segment size of 2 s *(M = 400)*.

| Feature set | MCC | Sensibility | Specificity | Accuracy | F1 score | AUC |
|---|---|---|---|---|---|---|
| $F_1$ | 0.9464 | 0.9432 | 0.9929 | 0.9758 | 0.9641 | 0.9883 |
| $F_2$ | 0.9348 | 0.9419 | 0.9858 | 0.9707 | 0.9568 | 0.9759 |
| $F_3$ | 0.9265 | 0.9351 | 0.9837 | 0.9669 | 0.9512 | 0.9785 |
| $F_1$, $F_2$ | 0.9383 | 0.9284 | 0.9950 | 0.9721 | 0.9582 | 0.9871 |
| $F_1$, $F_3$ | 0.9319 | 0.9284 | 0.9908 | 0.9693 | 0.9542 | 0.9824 |
| $F_2$, $F_3$ | 0.9317 | 0.9392 | 0.9851 | 0.9693 | 0.9547 | 0.9861 |
| $F_1$, $F_2$, $F_3$ | 0.9370 | 0.9351 | 0.9908 | 0.9716 | 0.9578 | 0.9832 |
| $eF_1$ | 0.9556 | 0.9581 | 0.9915 | 0.9800 | 0.9706 | 0.9952 |
| $eF_2$ | 0.9587 | 0.9581 | 0.9936 | 0.9814 | 0.9726 | 0.9964 |
| $eF_3$ | 0.9422 | 0.9432 | 0.9901 | 0.9739 | 0.9614 | 0.9872 |
| $eF_1$, $eF_2$ | 0.9546 | 0.9554 | 0.9922 | 0.9795 | 0.9698 | 0.9950 |
| $eF_1$, $eF_3$ | 0.9453 | 0.9446 | 0.9915 | 0.9753 | 0.9635 | 0.9881 |
| $eF_2$, $eF_3$ | 0.9608 | 0.9608 | 0.9936 | 0.9823 | 0.9740 | 0.9942 |
| $eF_1$, $eF_2$, $eF_3$ | 0.9525 | 0.9554 | 0.9908 | 0.9786 | 0.9685 | 0.9911 |
| $F_1$, $eF_1$ | 0.9545 | 0.9595 | 0.9901 | 0.9795 | 0.9699 | 0.9962 |
| $F_2$, $eF_2$ | 0.9529 | 0.9405 | 0.9986 | 0.9786 | 0.9680 | 0.9946 |
| $F_3$, $eF_3$ | 0.9412 | 0.9392 | 0.9915 | 0.9735 | 0.9606 | 0.9840 |
| $F_1$, $F_2$, $eF_1$, $eF_2$ | 0.9547 | 0.9486 | 0.9957 | 0.9795 | 0.9696 | 0.9894 |
| $F_1$, $F_3$, $eF_1$, $eF_3$ | 0.9475 | 0.9419 | 0.9943 | 0.9763 | 0.9647 | 0.9891 |
| $F_2$, $F_3$, $eF_2$, $eF_3$ | 0.9454 | 0.9392 | 0.9943 | 0.9753 | 0.9633 | 0.9859 |
| $F_1$, $F_2$, $F_3$, $eF_1$, $eF_2$, $eF_3$ | 0.9475 | 0.9419 | 0.9943 | 0.9763 | 0.9647 | 0.9859 |

**Table 6.** Identification results for segment size of 4 s *(M = 800)*.

| Feature set | MCC | Sensibility | Specificity | Accuracy | F1 score | AUC |
|---|---|---|---|---|---|---|
| $F_1$ | 0.9364 | 0.9230 | 0.9964 | 0.9711 | 0.9566 | 0.9755 |
| $F_2$ | 0.9426 | 0.9284 | 0.9979 | 0.9739 | 0.9608 | 0.9712 |
| $F_3$ | 0.9455 | 0.9378 | 0.9950 | 0.9753 | 0.9632 | 0.9841 |
| $F_1$, $F_2$ | 0.9488 | 0.9351 | 0.9986 | 0.9767 | 0.9651 | 0.9834 |
| $F_1$, $F_3$ | 0.9487 | 0.9365 | 0.9979 | 0.9767 | 0.9652 | 0.9778 |
| $F_2$, $F_3$ | 0.9385 | 0.9243 | 0.9972 | 0.9721 | 0.9580 | 0.9783 |
| $F_1$, $F_2$, $F_3$ | 0.9415 | 0.9297 | 0.9964 | 0.9735 | 0.9602 | 0.9767 |
| $eF_1$ | 0.9504 | 0.9527 | 0.9908 | 0.9777 | 0.9671 | 0.9796 |
| $eF_2$ | 0.9504 | 0.9514 | 0.9915 | 0.9777 | 0.9670 | 0.9884 |
| $eF_3$ | 0.9436 | 0.9311 | 0.9972 | 0.9744 | 0.9616 | 0.9718 |
| $eF_1$, $eF_2$ | 0.9568 | 0.9486 | 0.9972 | 0.9804 | 0.9710 | 0.9841 |
| $eF_1$, $eF_3$ | 0.9507 | 0.9405 | 0.9972 | 0.9777 | 0.9667 | 0.9815 |
| $eF_2$, $eF_3$ | 0.9477 | 0.9365 | 0.9972 | 0.9763 | 0.9645 | 0.9820 |
| $eF_1$, $eF_2$, $eF_3$ | 0.9518 | 0.9405 | 0.9979 | 0.9781 | 0.9673 | 0.9828 |
| $F_1$, $eF_1$ | 0.9505 | 0.9473 | 0.9936 | 0.9777 | 0.9669 | 0.9851 |
| $F_2$, $eF_2$ | 0.9527 | 0.9432 | 0.9972 | 0.9786 | 0.9681 | 0.9892 |
| $F_3$, $eF_3$ | 0.9495 | 0.9473 | 0.9929 | 0.9772 | 0.9662 | 0.9860 |
| $F_1$, $F_2$, $eF_1$, $eF_2$ | 0.9579 | 0.9500 | 0.9972 | 0.9809 | 0.9717 | 0.9929 |
| $F_1$, $F_3$, $eF_1$, $eF_3$ | 0.9485 | 0.9419 | 0.9950 | 0.9767 | 0.9654 | 0.9855 |
| $F_2$, $F_3$, $eF_2$, $eF_3$ | 0.9508 | 0.9392 | 0.9979 | 0.9777 | 0.9666 | 0.9761 |
| $F_1$, $F_2$, $F_3$, $eF_1$, $eF_2$, $eF_3$ | 0.9529 | 0.9405 | 0.9986 | 0.9786 | 0.9680 | 0.9814 |

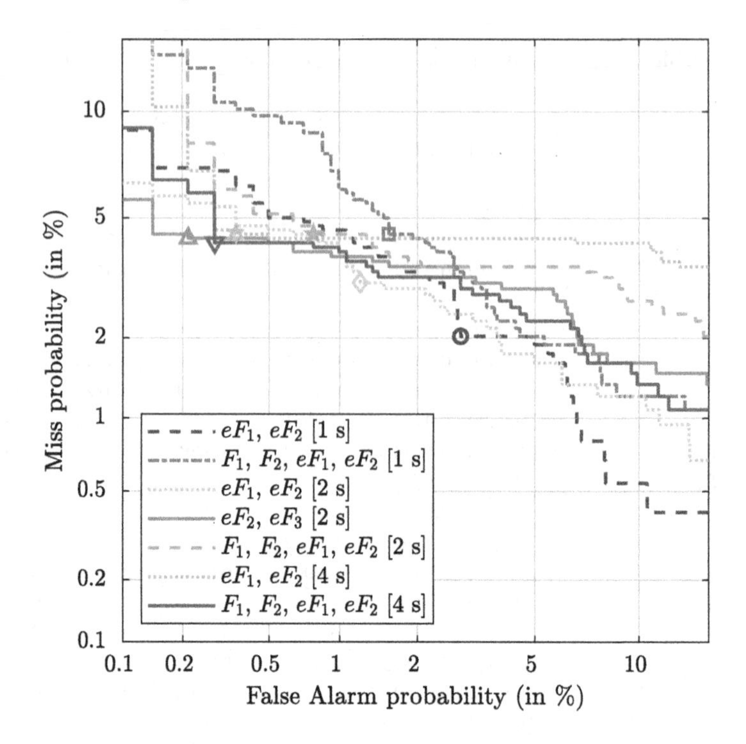


**Fig. 2.** DET curves for system configurations providing best performance.

The best results for speech segments from 250 ms to 1 s is achieved by $eF_1$, $eF_2$ where MCC ranges from 0.9120 to 0.9514, sensibility from 0.9162 to 0.9595, and specificity from 0.9837 to 0.9879. Moreover, the same feature set reaches the top scores for a length of 4 s with a MCC of 0.9568, sensibility of 0.9486, and specificity of 0.9972. However, for a window of 2 s the set $eF_2$, $eF_3$ provides the highest scores, being MCC=0.9608 (sensibility is 0.9608 and specificity is 0.9936).

Figure 2 shows the Detection Error Trade-off (DET) curves for a selection of the best results considering different costs for a PD missing detection and a false-alarm in the two-class separation problem. Markers correspond to the situation where $C_{miss} = C_{fa} = 1.0$ with a $P_{target} = 0.5$. As it may be seen, there is not a unique feature dataset providing the best results for every working conditions.

## 5 Conclusions

A lightweight CNN model has been proposed and tested for identifying PD dysarthria by means of sustained /a/ vowel utterances. A total of six input features are extracted using measures from formants $F_1$, $F_2$ and $F_3$. Various combinations of several normalized formant frequencies and energies have been studied for different speech segment lengths, ranging from to 250 ms to 4 s. The designed CNN model has less than 150k parameters and achieves a sensibility of 0.96 and a specificity of 0.99 for the set $eF_2$, $eF_3$.

## References


1. Jankovic, J.: Parkinson's disease: clinical features and diagnosis. J. Neurol. Neurosurg. Psychiatry **79**, 368–376 (2008). https://doi.org/10.1136/jnnp.2007.131045
2. de Rijk, M.C., et al.: Prevalence of Parkinson's disease in Europe: a collaborative study of population-based cohorts. Neurol. Diseases Elderly Res. Group Neurol. **54**(11 Suppl. 5), S21–S23 (2000). PMID: 10854357. https://doi.org/10.1136/jnnp.62.1.10
3. Ray Dorsey, E., et al.: Global, regional, and national burden of Parkinson's disease, 1990–2016: a systematic analysis for the global burden of disease study 2016. Lancet Neurol. **17**(11), 939–953 (2018). https://doi.org/10.1016/S1474-4422(18)30295-3
4. Skodda, S., Grönheit, W., Mancinelli, N., Schlegel, U.: Progression of voice and speech impairment in the course of Parkinson's disease: a longitudinal study. Parkinson's Disease **2013**, 389195 (2013). https://doi.org/10.1155/2013/389195
5. New, A.B., et al.: The intrinsic resting state voice network in Parkinson's disease. Hum. Brain Mapp. **36**, 1951–1962 (2015). https://doi.org/10.1002/hbm.22748
6. Sapir, S.: Multiple factors are involved in the dysarthria associated with Parkinson's disease: a review with implications for clinical practice and research. J. Speech Lang. Hear. Res. **57**, 1330–1343 (2014). https://doi.org/10.1044/2014JSLHR-S-13-0039
7. Sureshbabu, S.: Clinical speech impairment in Parkinson's disease, progressive supranuclear palsy, and multiple system atrophy. Neurol. India **56**(2), 122–126 (2008). https://doi.org/10.4103/0028-3886.41987
8. Kent, R.D., et al.: Acoustic studies of dysarthric speech: methods, progress, and potential. J. Commun. Disorders **32**(3), 141–186 (1999). https://doi.org/10.1016/s0021-9924(99)00004-0
9. Godino-Llorente, J.I., Moro-Velázquez, L., Gómez-García, J.A., Choi, J.-Y., Dehak, N., Shattuck-Hufnagel, S.: Approaches to evaluate parkinsonian speech using artificial models. In: Godino-Llorente, J.I. (ed.) AAPS 2019. CCIS, vol. 1295, pp. 77–99. Springer, Cham (2020). https://doi.org/10.1007/978-3-030-65654-6_5
10. Godino-Llorente, J.I., et al.: Towards the identification of idiopathic Parkinson's disease from the speech. New articulatory kinetic biomarkers. PLoS ONE **12**(12) (2017). https://doi.org/10.1371/journal.pone.0189583
11. Little, M.A., McSharry, P.E., Hunter, E.J., Spielman, J., Ramig, L.O.: Suitability of dysphonia measurements for telemonitoring of Parkinson's disease. IEEE Trans. Biomed. Eng. **56**, 1015–1022 (2009). https://doi.org/10.1109/TBME.2008.2005954
12. Moro-Velázquez, L., et al.: Analysis of speaker recognition methodologies and the influence of kinetic changes to automatically detect Parkinson's Disease. Appl. Soft Comput. **62**, 649–666 (2018). https://doi.org/10.1016/j.asoc.2017.11.001
13. Moro-Velazquez, L., Villalba, J., Dehak, N.: Using x-vectors to automatically detect Parkinson's disease from speech. In: Proceedings of ICASSP 2020, pp. 1155–1159. https://doi.org/10.1109/ICASSP40776.2020.9053770
14. Berus, L., Klancnik, S., Brezocnik, M., Ficko, M.: Classifying Parkinson's disease based on acoustic measures using artificial neural networks. Sensors (2019). https://doi.org/10.3390/s19010016

15. Vaiciukynas, E., Gelzinis, A., Verikas, A., Bacauskiene, M.: Parkinson's disease detection from speech using convolutional neural networks. In: Guidi, B., Ricci, L., Calafate, C., Gaggi, O., Marquez-Barja, J. (eds.) GOODTECHS 2017. LNICST, vol. 233, pp. 206–215. Springer, Cham (2018). https://doi.org/10.1007/978-3-319-76111-4_21
16. Rios-Urrego, C.D., Vásquez-Correa, J.C., Orozco-Arroyave, J.R., Nöth, E.: Transfer learning to detect Parkinson's disease from speech in different languages using convolutional neural networks with layer freezing. In: Sojka, P., Kopeček, I., Pala, K., Horák, A. (eds.) TSD 2020. LNCS (LNAI), vol. 12284, pp. 331–339. Springer, Cham (2020). https://doi.org/10.1007/978-3-030-58323-1_36
17. Bhati, S., Moro-Velázquez, L., Villalba, J., Dehak, N.: LSTM Siamese network for Parkinson's disease detection from speech. In: 2019 IEEE Global Conference on Signal and Information Processing (GlobalSIP), pp. 1–5 (2019). https://doi.org/10.1109/GlobalSIP45357.2019.8969430
18. Orozco-Arroyave, J.R., et al.: Automatic detection of Parkinson's disease in running speech spoken in three different languages. J. Acoust. Soc. Am. **139**(1), 481–500 (2016). https://doi.org/10.1121/1.4939739
19. Meghraoui, D., Boudraa, B., Merazi-Meksen, T., Gómez-Vilda, P.: A novel preprocessing technique in pathologic voice detection: application to Parkinson's disease phonation. Biomed. Sig. Process. Control **68** (2021). https://doi.org/10.1016/j.bspc.2021.102604
20. Arora, S., Tsanas, A.: Assessing Parkinson's disease at scale using telephone-recorded speech: insights from the Parkinson's voice initiative. Diagnostics. **11**(10), 1892 (2021). https://doi.org/10.3390/diagnostics11101892
21. Hireš, M., et al.: Convolutional neural network ensemble for Parkinson's disease detection from voice recordings. Comput. Biol. Med. **141**, 105021 (2022). https://doi.org/10.1016/j.compbiomed.2021.105021
22. Gómez, A., et al.: Acoustic to kinematic projection in Parkinson's disease dysarthria. Biomed. Sig. Process. Control **66**, 102422 (2021). https://doi.org/10.1016/j.bspc.2021.102422
23. Tsanas, A., et al.: Novel speech signal processing algorithms for high-accuracy classification of Parkinson's disease. IEEE Trans. Biomed. Eng. **59**(5), 1264–1271 (2012). https://doi.org/10.1109/TBME.2012.2183367
24. Putzer, M., Barry, W.: Saarbrucken voice database, Institute of Phonetics, University of Saarland. http://www.stimmdatenbank.coli.uni-saarland.de/. Accessed 15 Feb 2022
25. Chicco, D., Tötsch, N., Jurman, G.: The Matthews correlation coefficient (MCC) is more reliable than balanced accuracy, bookmaker informedness, and markedness in two-class confusion matrix evaluation. BioData Min. **14**, 13 (2021). https://doi.org/10.1186/s13040-021-00244-z

# Characterization of Hypokinetic Dysarthria by a CNN Based on Auditory Receptive Fields

Pedro Gómez-Vilda[1](✉), Andrés Gómez-Rodellar[2], Daniel Palacios-Alonso[3], Agustín Álvarez-Marquina[1], and Athanasios Tsanas[2]

[1] NeuSpeLab, CTB, Universidad Politécnica de Madrid, 28220 Pozuelo de Alarcón, Madrid, Spain
pedro.gomezv@upm.es, aalvarez@fi.upm.es

[2] Faculty of Medicine, Usher Institute, University of Edinburgh, Edinburgh, UK
{a.gomezrodellar,athanasios.tsanas}@ed.ac.uk

[3] E.T.S. de Ingeniería Informática - Universidad Rey Juan Carlos, Campus de Móstoles, Tulipán, s/n, 28933 Móstoles, Madrid, Spain
daniel.palacios@urjc.es

**Abstract.** Parkinson's Disease (PD) is a major neurodegenerative disorder with steadily increasing incidence rates, demanding overgrowing resources from national health systems and imposing considerable burden on caregivers. Cost-effective and efficient turn-around time monitoring methods are required to facilitate regular, longitudinal, accurate clinical assessment and symptom management. Speech has proven to be an effective neuromotor biomarker, capitalizing on the capabilities of contact-free technology. This study aims to evaluate processing speech from people diagnosed with Parkinson's Disease using Convolutional Neural Networks (CNN) towards characterizing speech articulation kinematics to explore differences between Healthy Controls (HC) and PD participants with Hypokinetic Dysarthria (HD), using Auditory Receptive Fields (ARFs) in the convolutional layers. The proposed proof of concept is based on a CNN described in detail, using an Extreme Learning Machine (ELM) at the output projection layer. This structure is evaluated on speech recordings from 6 PD and 6 HC participants. The performance of the approach is evaluated in terms of correlation and the log-likelihood ratio on the softmax output, showing the efficiency and retrieving properties of the CNN on speech auditory images, towards providing new insights on the pathophysiology of PD speech.

This research received funding from grants TEC2016-77791-C4-4-R (Ministry of Economic Affairs and Competitiveness of Spain), and Teca-Park-MonParLoc FGCSIC-CENIE 0348-CIE-6-E (InterReg Programme). The authors want to thank the APARKAM association of Parkinson's Disease patients of Alcorcón and Leganés in Madrid, and the voluntary participants for contributing to this initiative.


J. M. Ferrández Vicente et al. (Eds.): IWINAC 2022, LNCS 13258, pp. 343–352, 2022.
https://doi.org/10.1007/978-3-031-06242-1_34

**Keywords:** Convolutional neural networks · Parkinson's disease · Auditory receptive fields · Extreme learning machines · Hypokinetic dysarthria kinematics

## 1 Introduction

Parkinson's Disease (PD) is the second most predominant neurodegenerative disorder, with a prevalence rate of 100–200 cases, and an incidence rate of 15 cases, per 100,000 individuals, respectively [1]. Characteristic PD symptoms include rigidity, tremor, and general progressive loss of motor control to the point that people with PD (PwP) require continuous assistance in their daily life activity during the late stages of the disorder [2]. Additionally, they often exhibit other non-motor symptoms which may hamper considerably their life quality. Speech is one of the motor symptoms most altered by PD, as it was commented in the first known study of the disease, by J. Parkinson [3]. Speech involves the respiration, phonation, articulation, and their associated premotor activation systems. Each one of those key physiological mechanisms may be affected in different degrees, therefore, correlates of PD-induced alterations may be found in the analysis of PD speech [4]. The features derived from these speech correlates are based on their parametric algorithmic expressions which often rely on the use of standard statistical descriptors, such as the mean, standard deviation, quartiles, skewness, and kurtosis, as well as maximin range. Feature classification relies on an ample set of machine learning methods, such as k-nearest neighbors (kNNs), hierarchical trees, random forests, Gaussian mixture models (GMMs), support vector machines (SVMs), and artificial neural networks (ANNs), among others. A wide and comprehensive view of both conceptual fields is presented in [5]. In this sense, it must be said that although important progress has been achieved in both objectives, the clinical translation of the results produced, is yet far from being accepted by the medical community [6]. The set-up of the present research is oriented to include articulation dynamics from PD speech with advanced machine learning methods to add a bioinspired semantics to the characterization of speech dynamics. The aim of this study is to investigate the properties of speech articulation in PwP employing CNNs, to propose a new paradigm in modeling speech kinematics by CNNs using ARFs as an alternative to Visual Receptive Fields, using an Extreme Learning Machine (ELM) implementation for the output projection layer, to study how ELM dynamics may offer new insights to characterize Hypokinetic Dysarthria (HD) in PwP, reduce overfitting, and improving training computational costs.

The paper is organized as follows: A description of the ML methods proposed in the study is provided in Sect. 2, explaining the neurological fundamentals of ARFs in speech perception, the architecture of the convolutional part, as well as the classification layer of the CNN. Section 3 describes the experimental protocol designed to test the functionality of the CNN on a size-limited speech database from PD and healthy control (HC) participants (proof of concept). Section 4 is intended to summarize the validation results. The discussion of the results is extended in Sect. 5. The main insights and conclusions derived from the study are given in Sect. 6.

## 2 Methods

Conventional CNNs use convolutional Visual Receptive Fields (VRFs) inspired by the work of Hubel and Wiesel [7] to capture basic visual patterns, such as straight lines, contours, and curves relevant to image representations [8]. In the case of audio and speech perception, similar receptive fields in the auditory cortical areas of the mustached bat were extensively studied by N. Suga [9] and may be used with the same purpose. These time-frequency receptive fields are specialized in retrieving static and dynamic displacement of frequency in the time domain. Inspired by these VRFs, a CNN has been designed whose first convolutional layers use ARFs, which are a simplified version of Suga's CF and FM units. Its purpose is to detect and encode the complex dynamic time-domain scenarios [10] produced by speech articulation to characterize HD in PD.

The CNN proposed in the present study is a classical convolution-max-pooling architecture using a fully-connected backpropagation-output classifier [11], modified to introduce ARFs inspired in human speech perception and an Extreme Learning Machine (ELM) known as a Random Least-Squares Feedforward Network (RLSFN) substituting the output classifier. The proposed CNN architecture is described in Fig. 1.

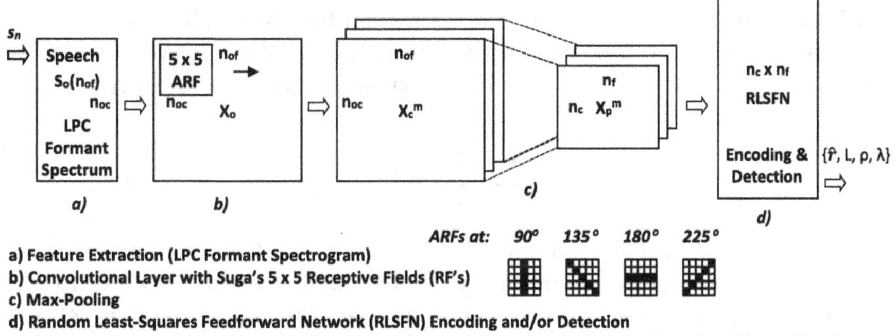


**Fig. 1.** Architecture of the proposed CNN

The input layer for data preprocessing (a) is based on an adaptive lattice-ladder Linear Prediction Coding (LPC) algorithm [12] to estimate the strongest frequency bands (formants) in the time-frequency domain. The LPC preprocessing stage produces a formant spectrogram in the following steps:

- Remove the low frequency artifacts and offset, the recordings are high-pass filtered 20 Hz. High quality speech samples are recorded at 44.1 kHz, although for telephone-channel compatibility these recordings are down-sampled at 8 kHz.

- A segment of 2 s long is selected from the middle part of each recording (onset and decay are neglected).
- Windows of 512 samples are used, equivalent to 64 ms at 8 kHz. An overlapping stride of 10 ms is used to produce an LPC spectrogram 100 times per second. The discrete-time index of each spectrogram frame will be $1 \leq n_{of} \leq 280$, where $n_{of}$ is the frame time index.
- A high-pass filter is used to compensate radiation effects.
- A first-order LPC filter is used to remove the glottal formant.
- A nine-pole lattice-ladder is used to implement an Iterative Adaptive Inverse Filter (IAIF) [13] on the speech signal $s_n$, to estimate an optimal LPC set of weights $\{a_{io}\}$ (weights of the predictor-error filter). From this set of weights an LPC spectrum $\mathbf{S}_o$ is evaluated at $n_{of}$ as

$$\mathbf{S}_O(n_{of}) = \frac{1}{1 - \sum_{i=1}^{K} a_{i,n_{of}} e^{-jn_{oc}\Omega}} \tag{1}$$

where $1 \leq n_{oc} \leq 800$ is the frequency channel index and $\Omega = 5\,\text{Hz}$ is the frequency resolution of the LPC spectrogram. A full description of the lattice-ladder inversion filter details can be found in [14].

For memory-saving reasons, a size reduction of the spectrogram is carried on by a watershed algorithm [15] and ReLU units. The result will be a formantgram $\mathbf{X}_o$ matrix of $n_{oc} \times n_{of}$ to be seen as an auditory image. The matrices $\mathbf{X}_c^m$ are the result of a time-frequency convolution of $\mathbf{X}_o$ with a set of $5 \times 5$ ARFs, corresponding to 90°, 135°, 180°, and 225° time-frequency lines expressed as

$$x^c_{n_c,n_f} = f\Big(\sum_{k=-2}^{2} \sum_{i=-2}^{2} w_{k,i} x^o_{n_c+k,n_f-i}\Big) \tag{2}$$

where $x^o_{i,j}$ and $x^c_{i,j}$ are the generic elements of $\mathbf{X}_o$ and $\mathbf{X}_c$, respectively, and $w_{k,i}$ the weights of the corresponding ARF. The nonlinear function $f$ is a ReLU unit with a threshold fixed at half the weight sum of the corresponding RF unit (2.5 for this particular case). A max-pooling of 8:2 is applied to the set $\mathbf{X}_c^m$ to reduce their dimensionality to $100 \times 140$, therefore the new frequency resolution will be $\Omega = 80\,\text{Hz}$ and the time frame stride will be $\Delta = 20\,\text{ms}$. The results of max-pooling are a set of $n_{oc}$ x $n_{of}$ matrices denoted as $\mathbf{X}_p^m$. An output neural network (RLSFN) is used for detection, encoding, or classification.

The classifier proposed is a multiple-layer ANN where the weights of the input and output layers are fixed by stochastic methods (input layer) and least-squares learning (output layer), known as Random Vector Functional-Link Networks (RVFLN) [16]. In the case under study it will be assumed (see Fig. 1) that the input is an $n_{oc} \times n_{of}$ matrix $\mathbf{X}_c \in \Re^{n_c \times n_f}$. The target elements will be defined as $t_s \in \{-1, +1\}$, depending on HC (−1), or PD (+1) participants. The first layer of the RLSFN is defined as $\mathbf{W}_1 \in \Re^{n_c \times n_h}$, where $n_h$ is the number of hyperplanes projecting the input matrices $\mathbf{X}$ to the hidden layers, as

$$\mathbf{Y} = \sigma_1\{\mathbf{W}_1 \mathbf{X}\} \tag{3}$$

where $\mathbf{W}_1$ is filled with real random values following a normal distribution $N\{\mu = 0, \sigma = 1\}$, $\sigma_1\{\cdot\}$ is a sigmoid mapping to the interval $[-1, +1]$, and $\mathbf{Y}$ is the hidden layer activation. The output is an approximation to the expected target vector $\mathbf{t}$

$$\mathbf{z} = \sigma_2\{\mathbf{W}_2\mathbf{Y}\} \tag{4}$$

$\mathbf{z} \in \Re^{n_c}$, is the output vector, where $n_c$ is the number of channels in the frequency domain, $\sigma_2\{\cdot\}$ being also a sigmoid centered in the origin mapping the output to the interval $[-1, +1]$. $\mathbf{W}_2$ is the output mapping, defined as $\mathbf{W}_2 = \mathbf{Pt}$, where $\mathbf{P}$ is Moore-Penrose's pseudoinverse $\mathbf{P} = \mathbf{Y}^T(\mathbf{YY}^T)^{-1}$. As the training error $\mathbf{e} = \mathbf{t} - \mathbf{z}$ has to consider all the error components $n_c$, a normalized overall score will be defined

$$\hat{\mathbf{r}} = \frac{\mathbf{r} - \mathbf{r}_{min}}{\mathbf{r}_{max} - \mathbf{r}_{min}}; \quad \mathbf{r} = \boldsymbol{f}\left(\sum_{n_c} \mathbf{e}\right) \tag{5}$$

$\boldsymbol{f}(\cdot)$ being a softmax function. Defining $\hat{\mathbf{r}}_m$ as the average of $\hat{\mathbf{r}}$ on the max-pooling set for the $n_s$ participants, the following performance scores may defined

$$L = \frac{1}{n_s}\sum_{i=1}^{n_s} \parallel \mathbf{e}_i \parallel; \quad \rho = c_p\{\mathbf{t}, \hat{\mathbf{r}}\}; \quad \lambda = log\left\{\frac{\sum_{i=1}^{n_s} \sigma_m(\hat{\mathbf{r}}_m, t_i = +1)}{\sum_{i=1}^{n_s} \sigma_m(\hat{\mathbf{r}}_m, t_i = -1)}\right\} \tag{6}$$

where $L$ is the overall error function, $c_p\{\cdot, \Delta\cdot\}$ is Pearson's correlation coefficient ($\rho$), and $\lambda$ is the log-likelihood between the softmax ($\sigma_m$) of the normalized errors from the PD participants ($t_i = +1$) and the HC participants ($t_i = -1$). The operations of training and testing are as follows:

- **Feedforward dataflow.** It consists of the successive applications of expressions (3–4) on the input matrix ($\mathbf{X}_p^m$).
- **Training dataflow.** During the training phase, the hidden layer activation $\mathbf{Y}$ is used to estimate $\mathbf{P}$ and the optimal output layer weight matrix $\mathbf{W}_2 = \mathbf{Pt}$.
- **Testing dataflow.** During the testing phase, the previously estimated matrices $\mathbf{W}_1$ and $\mathbf{W}_2$ are used to produce the results in terms of $\{\hat{\mathbf{r}}_m, L, \rho$, and $\lambda\}$, following expressions (3)–(6).

A time-domain variant of the aforementioned frequency-domain training method may be carried on simply presenting the transposes of each matrix in the participant training set $(\mathbf{X}_p^m)^T$.

## 3 Experimental Framework

The study of the articulation kinematics in PD is based on the capability shown by patients to produce regular and ample movements in the jaw-tongue joint, among other diadochokinetic exercises. These include regular and fast repetitions of the diphthong $[a \rightarrow i]$. This diadochokinetic exercise is used for its easy utterance by all patients and its capability for testing the fast jaw movements in the frequency and time domain. Participants were asked to utter this sequence for a minimum duration of 2 s. The study cohort comprised 6 PD participants (three male, three female, all Spanish native speakers, stage 2 in the H&Y scale) who were recruited from a PwP association in the metropolitan area of Madrid (Asociación de Pacientes de Parkinson de Alcorcón y Móstoles, APARKAM). For comparison purposes, three male and three female HC age-paired participants have been also included in the study. The research protocol was approved by the Ethical Committee of UPM (MonParLoc, 18/06/2018). The voluntary participants were informed about the experiments to be conducted, the protection of personal data, and signed an informed consent form. The methodology was strictly aligned with the Declaration of Helsinki. The demographical data of the participants are given in Table 1.

**Table 1.** Biometrical profile of participants

| Code | Gender | Age | Condition | Medication | H&Y |
|---|---|---|---|---|---|
| PMa | M | 73 | PD | ON | 2 |
| PMb | M | 76 | PD | ON | 2 |
| PMc | M | 69 | PD | ON | 2 |
| CMa | M | 69 | HC | – | – |
| CMb | M | 70 | HC | – | – |
| CMc | M | 68 | HC | – | – |
| PFa | F | 78 | PD | ON | 2 |
| PFb | F | 73 | PD | ON | 2 |
| PFc | F | 71 | PD | ON | 2 |
| CFa | F | 72 | HC | – | – |
| CFb | F | 65 | HC | – | – |
| CFc | F | 66 | HC | – | – |

The experimental protocol has considered conducting PD vs HC detection separately by gender, to study the effects of convolving auditory images separately with ascending RFs (135°) and descending RFs (225°), and to compare the PD vs HC detection using auditory images on the frequency domain ($\mathbf{X}_p^m$) or on the time domain $(\mathbf{X}_p^m)^T$. The processing steps are:

- The steps described in Sect. 2 are followed to generate an 800 × 200 LPC spectrogram (resolution 5 Hz) on a 64 ms sliding window with a stride of 10 ms (auditory image). A watershed formant detector and a ReLU stage are applied to the spectrogram, which is then convolved with a 5 × 5 ascending or descending ARF filter (2), to produce the convolved matrices.
- Max-pooling (8:2) is applied to the convolved matrices to produce a pooled dynamic matrix (100 × 140). Either the pooled dynamic matrix (frequency domain) or its transpose (time-domain) are processed by an RLSFN with $n_h = 1600$ input hyperplanes.

# 4 Results

Two examples, one from an HC (CMa) and one from a PD participant (PMb) are presented in Fig. 2 showing input speech data for comparison purposes. Their corresponding RLSFN data are shown in Fig. 3. It may be seen that the utterance by the HC participant is more stable, the evolution of harmonics and formants showing a quasi-periodic behavior. The sequence produced by the PD participant is more irregular in frequency. These differences are particularly evident in the respective LPC spectrograms. The comparison between the different formant bands to the output matrices **Z** (Fig. 3.f left and rightside) shows a better definition of the band contributions in the HC case with respect to the PD case.

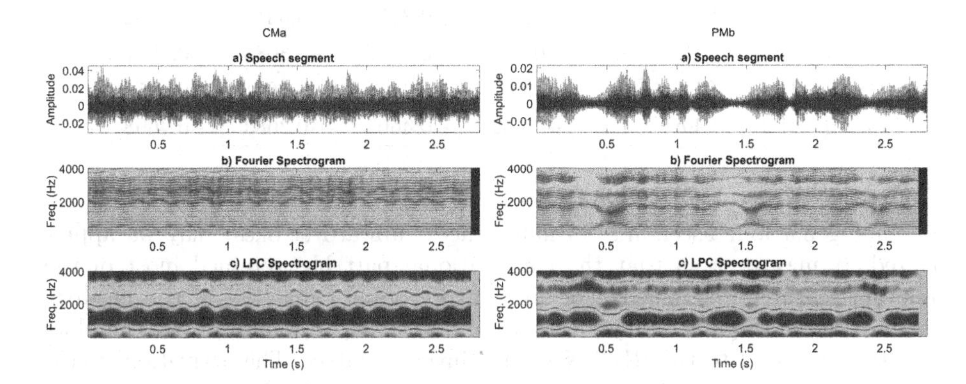


**Fig. 2.** Input speech and spectrograms. The left templates show the results from the HC participant (CMa). The right templates show those from the PD participant (PMb). a) Speech signal corresponding to sequence $[\ldots a \rightarrow i \rightarrow a \ldots]$; b) Fourier Spectrogram; c) LPC Spectrogram.

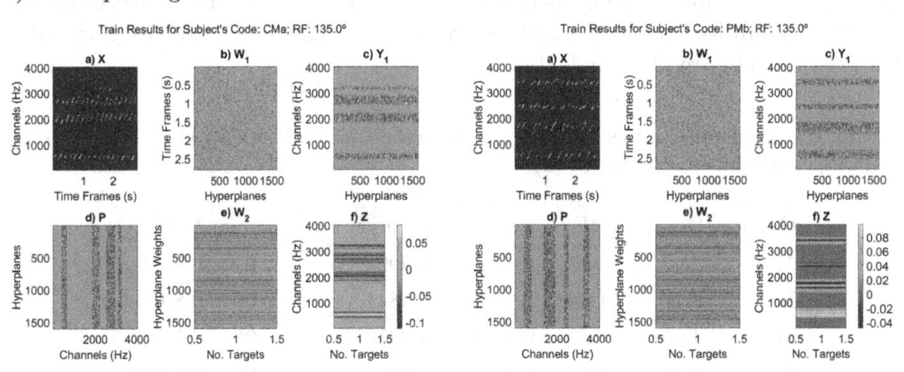


**Fig. 3.** RLSFN data structures from the two examples shown in 2 (frequency domain, ascending ARFs). The a–f) templates on the left correspond to the HC participant (CMa). The a–f) templates on the right correspond to the PD participant (PMb). a) Input data matrix (**X**) after ReLU, and 8:2 max-pooling; b) Input projection matrix ($\mathbf{W}_1$); c) Hidden layer activation ($\mathbf{Y}_1$); d) Pseudoinverse **P**; e) Output projection layer ($\mathbf{W}_2$); f) Output vector (**Z**).

**Table 2.** Participants' scores $\hat{\mathbf{r}}_m$ (targets are PD:+1 and HC:−1).

| Gender | Cond. | Code | Target | Freq/Asc | Freq/Desc | Time/Asc | Time/Desc |
|---|---|---|---|---|---|---|---|
| M | PD | PMa | +1 | 0.4828 | 0.4660 | 0.3153 | 0.3590 |
| M | PD | PMb | +1 | 0.4584 | 0.3422 | 0.3151 | 0.3590 |
| M | PD | PMc | +1 | 0.5027 | 0.4100 | 0.3140 | 0.3432 |
| M | HC | CMa | −1 | −0.4973 | −0.3378 | −0.6847 | −0.6410 |
| M | HC | CMb | −1 | −0.4606 | −0.3474 | 0.0553 | −0.0097 |
| M | HC | CMc | −1 | −0.4859 | −0.5340 | −0.3148 | −0.4104 |
| F | PD | PFa | +1 | 0.4521 | 0.5007 | 0.3603 | 0.5255 |
| F | PD | PFa | +1 | 0.4995 | 0.5212 | 0.1127 | 0.3391 |
| F | PD | PFa | +1 | 0.4439 | 0.3692 | 0.6317 | 0.5302 |
| F | HC | CFa | −1 | −0.4628 | −0.4517 | −0.3683 | −0.4697 |
| F | HC | CFb | −1 | −0.5005 | −0.4788 | −0.3681 | −0.4554 |
| F | HC | CFc | −1 | −0.4323 | −0.4605 | −0.3683 | −0.4698 |

In Fig. 3.a four ascending formant traces (upward strokes) may be appreciated. It may be seen that the convolutional part has retained most of the relevant features mentioned. The formant encoding capability of the network can be seen in the hidden layer activation matrix $\mathbf{Y}_1$ (c). On its turn, the plots in (d) correspond to the RLSFN pseudo-inverse matrix. The activation vector given in (f) shows the relevance of each frequency band contribution to the final score by position, sign, and intensity (see the side scale bar). The scores of each speaker in the dataset in differentiating HC from PD auditory images after a training-testing process are given in Table 2, taking into consideration the frequency or time domain approach and using two different ARFs (ascending or descending). In its turn, the quality scores for each gender set including HC and PD participants are given in Table 3.

**Table 3.** Performance scores by gender, domain and ARF ($L, \rho$, and $\lambda\}$) under different training conditions according to (6).

| Gender | Domain | RF | Loss ($L$) | Pearson ($\rho$) | p-value | LLR ($\lambda$) |
|---|---|---|---|---|---|---|
| M | Freq. | Asc. | 0.1347 | 0.9994 | <0.001 | 0.9626 |
| M | Freq. | Desc. | 0.1791 | 0.9841 | <0.001 | 0.8094 |
| M | Time | Asc. | 0.2576 | 0.8274 | <0.001 | 0.5844 |
| M | Time | Desc. | 0.2259 | 0.8866 | <0.001 | 0.6728 |
| F | Freq. | Asc. | 0.1434 | 0.9984 | <0.001 | 0.9303 |
| F | Freq. | Desc. | 0.1450 | 0.9946 | <0.001 | 0.9296 |
| F | Time | Asc. | 0.2108 | 0.9262 | <0.001 | 0.7589 |
| F | Time | Desc. | 0.1451 | 0.9909 | <0.001 | 0.9338 |

## 5 Discussion

An examination of the results allows us to derive interesting insights from the present study. The first one is the capability of ARFs to track formant dynamics, as well as the ability of the RLSFN to capture and encode relevant formant band information in its hidden layer ($\mathbf{Y}_1$), from the examination of Table 2. The first three columns give the gender, condition and code of each participant. The fourth column gives the target value for supervised training (−1 for HC and +1 for PD). The fifth column gives the results using frequency domain profiles and ascending ARFs. The sixth column gives the scores for frequency profiles and descending ARFs. The rightmost columns give the scores for time profiles and ascending or descending ARFs. Positive scores are expected for PDs and negative ones for HCs. These expected outcomes are fulfilled by frequency profiles, but not that clearly by time profiles, where CMb produces a low but positive score for time-ascend configuration (in bold) and a weak negative with the time-descent one. The results from other participants are acceptable, except for PFb and time-ascend configuration, producing a low positive result. According to quality measurements given in Table 3, the best results (in terms of $L, \rho$, and $\lambda\}$) are produced by the male dataset combining the frequency profile and the ascend ARF. The second rightmost column corresponds to the p-value associated to $\rho$. The worst behavior, also for the male dataset corresponds to the time profile and ascending ARF. In the case of the female dataset, the best performance is obtained with the frequency profile and ascending ARF, and the worst one is produced with the time profile and the ascending ARF. It seems that frequency profiles perform better than time profiles, possibly because formant bands may better encode the most relevant information from speech dynamics. It must be stressed that the results from the present ARF-CNN approach seem to be competitive enough to open a comparison with hand-crafted features used by other competing approaches which have been widely used in this field. The automatically extracted frequency or time domain patterns produced from input data by ARFs may gather the adequate information contents on speech dynamics in a completely data-driven. This does not mean that this approach is necessarily better at this stage, however we envisage it is worth exploring it in further detail and potentially directly in comparison with hand-engineered features. Given the limitations posed by the small size of the dataset studied, a further study is needed to confirm this assumption by using larger size databases of HC and PD participants including other diadochokinetic exercises and the evaluation of performance in comparison with handcrafted approaches. The results of the study may help in the design of fast and accurate characterization methods for PD speech, essential in the design of application-specific PD speech monitoring approaches for clinical purposes.

## 6 Conclusions

An ARF-based CNN has been proposed and tested for the processing of speech diadochokinetic exercises in the study of PD dyskinesia. The insights derived from the study presented point to that the convolutional processing using ARFs seems to capture the essence of speech formant dynamics. Besides, RLSFN used in the detection and evaluation of the convolved auditory images is capable of storing relevant characteristic information from the participant samples presented during the training phase. It seems also that relevant information on the evaluation of each gender set appears to be tentatively encoded in the quality factors. The whole train-test process on the same reference set seems to perform well given the quality factors considered. Finally, the best performing configuration results are produced combining frequency profile matrices with either ascend or descend ARFs.

## References

1. Tysnes, B., Storstein, A.: Epidemiology of Parkinson's disease. J. Neural Transm. **124**, 901–905 (2017)
2. Duffy, J.R.: Motor Speech Disorders: Substrates, Differential Diagnosis, and Management, 3rd edn. Elsevier, Amsterdam (2013)
3. Parkinson. J.: An essay on the shaking palsy, Sherwood, Neely and Jones, London, 1817. J. Neuropsychiatry Clin. Neurosci. **12**(2), 223–236 (2002)
4. Tsanas, A.: Accurate telemonitoring of Parkinson's disease symptom severity using nonlinear speech signal processing and statistical machine learning. Ph.D. thesis, University of Oxford, UK, June 2012
5. Hedge, H., et al.: A survey on machine learning approaches for automatic detection of voice disorders. J. Voice **33**(6), 947.E11–E33 (2019)
6. Cerasa, A.: Machine learning on Parkinson's disease? Let's translate into clinical practice. J. Neurosci. Meth. **266**, 161–162 (2016)
7. Hubel, D.H., Wiesel, T.N.: Receptive fields and functional architecture of monkey striate cortex. J. Physiol. **195**, 215–243 (1968)
8. LeCun, Y., Bengio, Y., Hinton, G.: Deep learning. Nature **521**(7553), 436–444 (2015)
9. Suga, N.: Basic acoustic patterns and neural mechanisms shared by humans and animals for auditory perception. In: Greenberg, S., et al. (eds.) Speech Processing in the Auditory System, pp. 159–181. Springer, New York (2004)
10. Greenberg, S., Ainsworth, W.A.: Speech processing in the auditory system: an overview. In: Greenberg, S., et al. (eds.) Speech Processing in the Auditory System, vol. 18, pp. 1–62. Springer, New York (2004). https://doi.org/10.1007/0-387-21575-1_1
11. Forsyth, D.: Applied Machine Learning, pp. 401–419. Springer, Cham (2019). https://doi.org/10.1007/978-3-030-18114-7
12. Deller, J.R., et al.: Discrete-Time Processing of Speech Signals. Macmillan, New York (1993)
13. Alku, P., et al.: OPENGLOT - an open environment for the evaluation of glottal inverse filtering. Speech Commun. **107**, 38–47 (2019)
14. Gómez, P., et al.: Glottal source biometrical signature for voice pathology detection. Speech Commun. **51**(9), 759–781 (2009)
15. Osma, V., et al.: An improved watershed algorithm based on efficient computation of shortest paths. Pattern Recogn. **40**(3), 1078–1090 (2007)
16. Huang, G-B., Siew, C.-K.: Extreme learning machine: RBF network case. In: Proceedings of the ICARCV, pp. 1029–1033 (2004)

# Evaluation of the Presence of Subharmonics in the Phonation of Children with Smith Magenis Syndrome

Rafael Martínez-Olalla[2(✉)], Daniel Palacios-Alonso[1,2(✉)], Irene Hidalgo-delaGuía[3], Elena Garayzabal-Heinze[3], and Pedro Gómez-Vilda[2]

[1] Escuela Técnica Superior de Ingeniería Informática, Universidad Rey Juan Carlos, Campus de Móstoles, Tulipán, s/n, 28933 Móstoles, Madrid, Spain
daniel.palacios@urjc.es

[2] Neuromorphic Speech Processing Lab, Center for Biomedical Technology, Universidad Politécnica de Madrid, Campus de Montegancedo, 28223 Pozuelo de Alarcón, Madrid, Spain
rmolalla@fi.upm.es

[3] Universidad Autónoma de Madrid, Ciudad Universitaria de Cantoblanco, 28049 Madrid, Spain

**Abstract.** Smith-Magenis syndrome SMS is a rare disease relatively unknown and underdiagnosed. This genetic disorder involves among others, neurodevelopmental deficits, which affect the complex mechanisms implicated in the production of speech. In this way, all levels of speech production are impacted. Individuals diagnosed with SMS present a deep, hoarse voice, dysphonia, excess of vocal muscle stiffness, disfluencies, tachylalia, high unintelligibility. This study investigates the origin of one of these characteristics (deep, hoarse voice) and compares de voice of a group of children with SMS (N = 12) with a healthy normative control group of children (N = 12) matched in age and gender. To observe the possible effects of age and gender, four groups were considered: boys and girls, both genders divided in two groups of age: 5 to 7 years old, and 8 to 12 years old. As the study is centred in the cause of the characteristic deep, hoarse voice of patients with SMS, acoustic analysis of a sustained vowel /a/ was performed, thus avoiding coarticulation effects. The present study is based in the search of subharmonic components of voice in the cepstrum domain that may explain that characteristic of SMS voices.

**Keywords:** Smith-Magenis syndrome · Speech · Cepstrum analysis · Phonation stability · Children

This research received funding from grants TEC2016-77791-C4-4-R (Ministry of Economic Affairs and Competitiveness of Spain), and Teca-Park-MonParLoc FGCSIC-CENIE 0348-CIE-6-E (InterReg Programme).


J. M. Ferrández Vicente et al. (Eds.): IWINAC 2022, LNCS 13258, pp. 353–362, 2022.
https://doi.org/10.1007/978-3-031-06242-1_35

# 1 Introduction

Speech production is a complex neurocognitive process that involves numerous processes, from message encoding to phonoarticulatory realization. For the explanation of the set of mechanisms necessary for oral language production, different approaches have been followed [5,8], but in all cases the neurological foundations for the production of speech and phonation coincide in a set of cortical and subcortical areas specialized in the different stages, from the organization of the message, to the neuromotor planning of the articulatory organs involved in speech. In the process of speech production, a series of mechanisms occur that are intimately linked to the subject's clinical situation. Thus, in healthy subjects, this neurocognitive activity occurs successfully, however, in individuals with pathologies of neurological origin, some of the cortical areas involved in the process may be affected. This can impact on various stages of the speech production processes from message production to phonation itself. The condition of the neuromotor profile of speech associated with numerous pathologies such as Parkinson's, amyotrophic lateral sclerosis, cerebral palsy, etc. has been extensively studied [1,9]. However, there are numerous pathologies that, due to their low prevalence or their recent description, are less known and have not yet been studied in depth. This is the case of the Smith-Magenis syndrome (SMS), [11,12]. This syndrome is characterized by its low prevalence (estimated at 1 in 15,000–25,000 cases) and it is underdiagnosed. It is a genetic disorder that involves intellectual deficits, sleep disorders, craniofacial and skeletal abnormalities, psychiatric disorders, and motor and speech delay [6,7,10]. With regard to speech production, velopharyngeal insufficiency, a deep, hoarse voice, and vocal cord nodules and polyps are common features [4,6,10]. In [3], the vocal quality of most children with SMS was found to be lower than that of a normative control group. This vocal quality was evaluated by measuring the Cepstral Peak Prominence (CPP). This result has been described with other pathologies, non-tremor-dominant phenotipe [2], cognitive impairment [13], cerebral palsy [1], or Williams syndrome [14], which, like SMS, is a neurodevelopmental disorder. Nevertheless, although CPP is related to speech quality, it does not explain some characteristics of SMS voices such as the deep and hoarse quality of voice. The hypothesis in the present study is that the neuromotor conditions of SMS affect the vibration of the vocal folds which can be noticed through irregularities in the speech periodicity. These irregularities will show up as subharmonics due to the momentary loss of the glottic arches, which will explain the deep and rough voice characteristics that are observed.

# 2 Materials and Methods

## 2.1 Participants

SMS is a neurodevelopmental disorder with a very low prevalence, and therefore little is known about it and the reason why it is also underdiagnosed. There are currently 72 cases diagnosed with SMS in the Spanish Association of Smith-Magenis Syndrome (ASME) of various ages. In this study, the voices of 12 children with SMS (six boys and six girls), grouped in two age ranges 5–7 years and 8–12 years, were recorded and analysed. The control group (Normative Group, NG), is a set of 12 children matched in age and sex to the SMS group.

This control group came from the Public School "Maria Luisa Cañas" (Ciudad Real). Legal tutors and teachers were informed about this research. The speech therapist of the school verified the inexistence of vocal pathology in the subjects of the NG. The information of both the SMS and the control groups is summarized in Table 1.

An informed consent was given to parents and legal tutors of the SMS participants who signed it. This research does not violate any rights of minors and complies with all the ethical principles set out in the Declaration of Helsinki by the World Medical Association in 1964 (Povl Riis, 2003).

**Table 1.** Total number of normative group (NG) and Smith-Magenis Syndrome (SMS) group participants distributed by gender and age.

| Population | Gender | Group | Number of cases/IDs |
|---|---|---|---|
| NG | Female | 5–7 years | 3 (517A, 612A, 637A) |
| | | 8–12 years | 3 (10AGPC, 11AAZM, 12109A) |
| | Male | 5–7 years | 3 (511O, 618O, 743O) |
| | | 8–12 years | 3 (819O, 842O, 11OADS) |
| SMS | Female | 5–7 years | 3 (SMS4, SMS5, SMS6) |
| | | 8–12 years | 3 (SMS10, SMS11, SMS12) |
| | Male | 5–7 years | 3 (SMS1, SMS2, SMS3) |
| | | 8–12 years | 3 (SMS7, SMS8, SMS9) |

### 2.2 Recording Procedure

The nature of this work is exploratory and aims to determine the manifestations of neuromotor deficits in the voice and phonation of people with SMS. For this purpose, the recording of sustained vowels lasting approximately 1 s was made. The use of sustained vowels was preferred over whole-word utterances because the work focuses on studying the characteristics of the glottic source and thus its effect on voice production can be better isolated from the set of muscles involved in the articulation of sounds. The study considered the emission of the vowel /a/. The use of the phoneme /a/ is due to phonological reasons: it is the most open vowel in Spanish and favors a more natural phonation.

Given the age of the participants and their behavioral phenotype, it was decided to use a clip-on cardioid microphone (Audio-Technica ATR-3350). The microphone was placed at a distance of approximately 20 cm from the participants' mouths. For both groups, a room with similar characteristics was available: small size, and good acoustic characteristics. To achieve naturally emitted recordings, the examiner played and practiced with each child for at least half an hour before making the recordings. In this way the subject became familiar with the environment and with the materials used in the recording. The recordings were made with a sampling frequency of 22050 Hz (SMS set) and 48000 Hz (normative set), stereo 16 bits.

## 3 Acoustic Processing

### 3.1 Preprocessing of Data

To avoid possible DC component introductions in the registration process, all the files were filtered to ensure that they had zero mean. After that they were normalized in amplitude to avoid possible overflows. In addition, the sampling frequency was unified by resampling the SMS set at 48000 Hz.

### 3.2 Subharmonic Detection

The algorithm used for the detection of subharmonics was based on cepstrum. This algorithm requires a detailed calculation of the fundamental frequency f0, which can be especially complicated in the case of pathological voices. Furthermore, in certain cases, f0 can actually be the value of the subharmonic. The search for f0 value was limited to a range between 40 Hz and 705.9 Hz. These limits may seem exaggerated for a child population, but it was preferred to maintain this range, given the presence of pathological voices and the fact that subharmonics and artifacts in voice were sought. The period in samples in the cepstral domain corresponding to these frequencies was between $fs/40 = 1200$ samples (Tmax) and $fs/705.9 = 68$ samples (Tmin). The files were block-processed with a window length of 64 ms (3,072 samples) and a displacement of 8 ms (384 samples). That is, with an overlap between windows of 87.5%. Each frame was multiplied by a Hamming window and the real cepstrum was calculated. Consecutive cepstrum values were collected into a matrix, with the cepstrum values placed in the columns of the matrix. The cepstrum matrix thus obtained was filtered using a 2D low-pass filter of size $19 \times 9$ (19 rows and 9 columns). The filter was obtained by the point-to-point product of the values of two matrices: the first consisted of 19 rows of Blackman window vectors of length 9, and the second of 9 columns of Blackman window vectors of length 19. The frequency response of that filter can be seen in Fig. 1. This filter had a low-pass characteristic, more pronounced in the direction of each cepstral vector (y axis) than between consecutive frames of the cepstrum (x axis or temporal direction). This filtering reinforced the positions of the true energy maxima, at the same time that smoothed out spurious peaks. In addition, since it affected the temporal dimension as well as the cepstral dimension, it reduced the possible undesired effects of the windowing process as it involved successive frames.

After this filtering, the cepstral vectors were compensated to enhance the low frequencies. To calculate the fundamental period, the position of the maximum of each column of the filtered matrix between the samples Tmin and Tmax was searched. To verify the possible existence of subharmonics, the presence of amplitude maxima located in the middle or third part of the position in samples of the global maximum was sought. The presence of a local maximum in a position half the number of samples from that of the global maximum and with an amplitude close to that maximum, revealed the possible presence of a subharmonic of half frequency. If the position of that local maximum was in the third part of the position in samples of the global maximum, then the global maximum corresponded to a subharmonic of frequency the third part of the fundamental frequency. Figure 2 shows an example of the cepstral matrix with local maxima at positions one half and one third of global maxima. Figure 3 show a detail of the waveform corresponding to the positions where these half (Fig. 3a) and third (Fig. 3b) frequency subharmonics were present. The results of this processing were only considered when voiced speech with a sufficient periodic quality was detected. For the evaluation of voice quality, Cepstral Peak Prominence (CPP) was calculated for every frame.

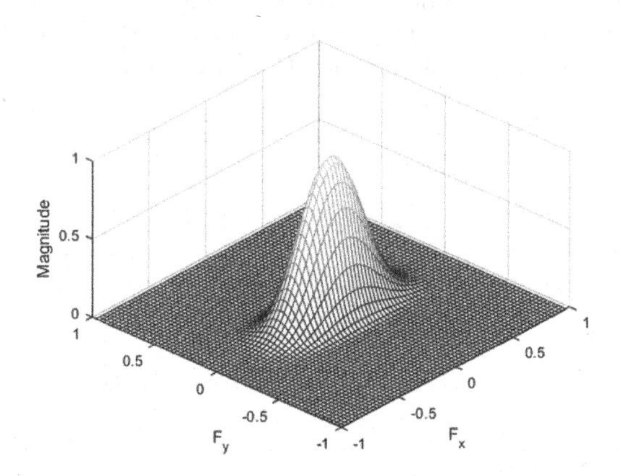


**Fig. 1.** Frequency response of 2D filter used to smooth the cepstrum matrix.

## 4 Results

The normative and the SMS database were processed, and the possible presence of subharmonics analysed as described in the previous section. In all frames of the normative group, the first peak of the cepstrum was the highest one, so no subharmonic was found in those recordings. A graphic of the subharmonic detection is presented in Fig. 4a as a function of the speaker. All values are either 0 (no periodic voice detected) or 1 (periodic voice without subharmonics is detected). The case of the SMS group is completely different: only in the voice of two children no subharmonic contents were found. In the other ten cases, the higher level of the second or even the third cepstral peak revealed the presence of subharmonics. These results are shown in Fig. 4b. It can be noticed that

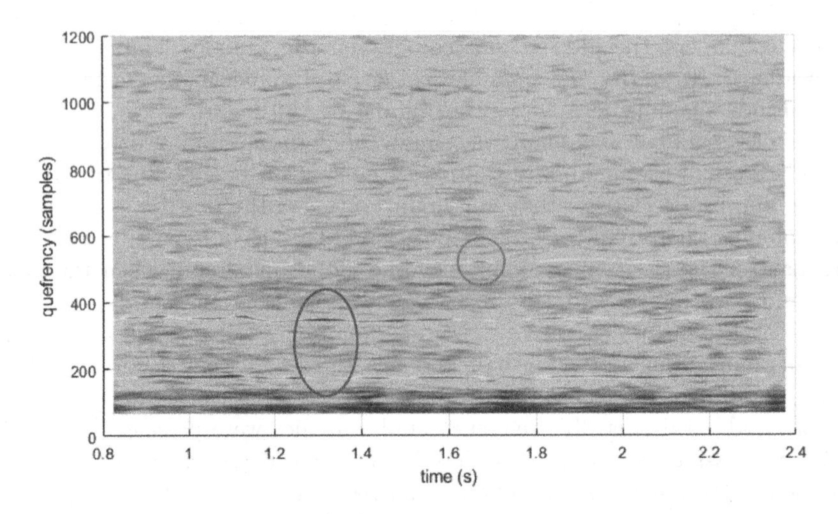


**Fig. 2.** Cepstrum matrix of an utterance of the /a/ vowel produced by a SMS patient. Red oval shows the second peak of the cepstrum being higher than the first one (subharmonic of half the frequency). The green circle shows the global maximum in the third peak of the cepstrum (subharmonic of one third the frequency). (Color figure online)

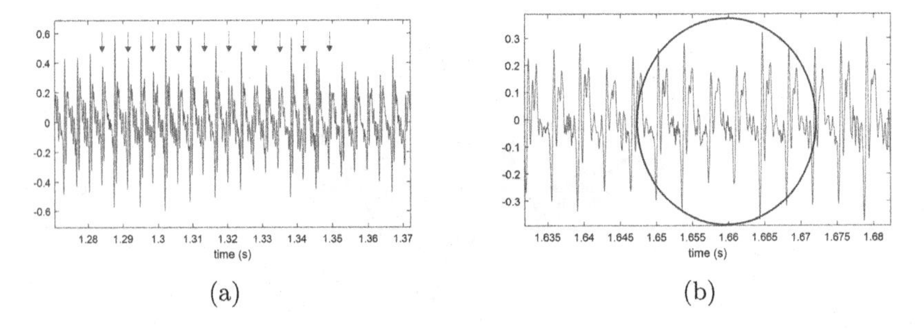


**Fig. 3.** (a) Detail of the speech waveform corresponding to the area identified in Fig. 2 with the red oval. Arrows point to the glottal arches that decrease in amplitude and almost disappear giving place to the presence of a subharmonic with half the frequency. (b) Detail of the speech waveform corresponding to the area identified in Fig. 2 with the green circle (Color figure online)

for almost every speaker (all but SMS6 and SMS7), the graphic takes values higher than 1, indicating that the second (value 2) or the third (value 3) peak of cepstrum is the global maximum and subharmonic contents are present.

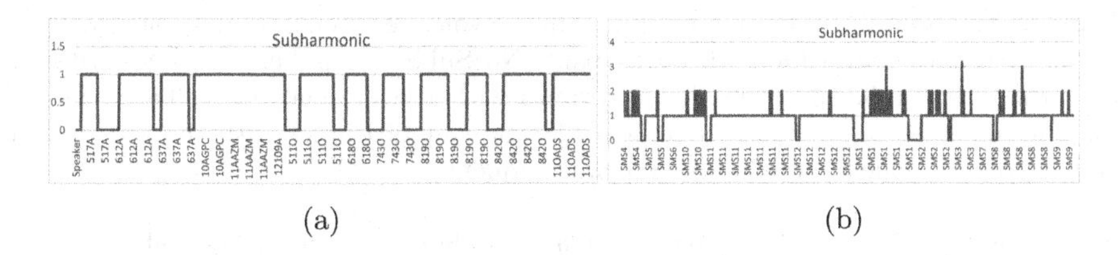


(a) (b)

**Fig. 4.** Subharmonic detection for the normative population: 0: No periodic signal; 1: No subharmonic; 2: Subharmonic with half the frequency; 3: subharmonic with one third the frequency; 3.2: Both subharmonics with half the frequency and one third the frequency are present. The speaker identification is indicated in the x axis.

Table 2 summarises the results of the database analysis. No subharmonics were found in the NG. Nevertheless, in the case of the SMS group subharmonics were found in ten of the twelve participants. In some cases (for example speaker SMS10), almost one third of the frames presented subharmonics. The presence of subharmonics does not follow a pattern of gender and age: the voices of 5 boys and 5 girls present subharmonics and the only two voices of the SMS group without subharmonics correspond to a girl from the first age group (5–7 years) and to a boy of the second age group (8–12 years).

Some cases are very illustrative. In Fig. 5a the case of the speaker SMS10 is shown. This participant, a 10 year old girl, not only presents a great number of frames with subharmonic contents (see green and purple rectangles, and the corresponding waveform in Fig. 5b), but also has a very low fundamental frequency (about 144 Hz). The frequency of highest energy in the waveform of Fig. 5b is 72 Hz. This low frequency content gives to this recording in particular a roughness characteristic.

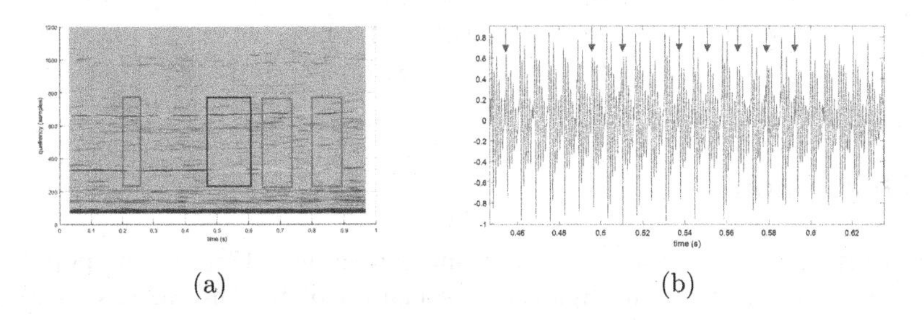

(a) (b)

**Fig. 5.** Left side - cepstral matrix of the recording of speaker SMS10. Rectangles show the places where the second cepstral peak is higher than the first one. The waveform correspondent to the signal in the purple rectangle is showed in (b). Rigtht side - detail of the waveform of the emission produced by speaker SMS10 correspondent to the zone marked with the purple rectangle in (a). Red arrows point to the arches that are decreased and give place to the presence of a subharmonic with half the frequency. (Color figure online)

**Table 2.** Summary of subharmonics found in the voice registers. Frames: total number of voiced frames found for each participant; No Subhar: number of frames without subharmonics; Subhar 1/2: number of frames with the presence of subharmonics of half the fundamental frequency; Subhar 1/3: number of frames with the presence of subharmonics of one third the fundamental frequency

| | Gender | Group | Speaker ID | Frames | No Subhar | Subhar 1/2 | Subhar 1/3 |
|---|---|---|---|---|---|---|---|
| NG | Female | 5–7 years | 517A | 66 | 66 | 0 | 0 |
| | | | 612A | 94 | 94 | 0 | 0 |
| | | | 637A | 77 | 77 | 0 | 0 |
| | | 8–12 years | 10AGPC | 94 | 94 | 0 | 0 |
| | | | 11AAZM | 94 | 94 | 0 | 0 |
| | | | 12109A | 50 | 50 | 0 | 0 |
| | Male | 5–7 years | 511O | 108 | 108 | 0 | 0 |
| | | | 618O | 57 | 57 | 0 | 0 |
| | | | 743O | 62 | 62 | 0 | 0 |
| | | 8–12 years | 819O | 129 | 129 | 0 | 0 |
| | | | 842O | 109 | 109 | 0 | 0 |
| | | | 11OADS | 94 | 94 | 0 | 0 |
| SMS | Female | 5–7 years | SMS4 | 78 | 55 (70.5%) | 23 (29.5%) | 0 |
| | | | SMS5 | 57 | 51 (89.5%) | 6 (10.5%) | 0 |
| | | | SMS6 | 94 | 94 | 0 | 0 |
| | | 8–12 years | SMS10 | 94 | 64 (68%) | 30 (32%) | 0 |
| | | | SMS11 | 368 | 359 (97.5%) | 9 (2.5%) | 0 |
| | | | SMS12 | 239 | 232 (97%) | 7 (3%) | 0 |
| | Male | 5–7 years | SMS1 | 197 | 149 (75.6%) | 46 (23.4%) | 2 (1%) |
| | | | SMS2 | 120 | 95 (79.2%) | 25 (20.8%) | 0 |
| | | | SMS3 | 94 | 86 (91.5%) | 6 (6.4%) | 2 (2,1%) |
| | | 8–12 years | SMS7 | 94 | 94 | 0 | 0 |
| | | | SMS8 | 237 | 223 (94.1%) | 12 (5.1%) | 2 (0.8%) |
| | | | SMS9 | 94 | 87 (92.6%) | 7 (7.4%) | 0 |

## 5 Discussion

This study encompasses a very significant percentage (17%) of the population diagnosed with SMS in the Spanish Association of Smith-Magenis Syndrome (ASME). Even so, this only comprises 12 patients, who have been chosen in a balanced way in terms of sex and age: 6 boys and 6 girls divided into 2 age groups 5–7 and 8–12 years. The voices of these children have been compared to the 12 participants of the control group whose gender and ages coincided with those of the children with SMS. The nature of the work is exploratory and aims to search for acoustic characteristics that are present in the population with SMS and that may correspond to neurological disorders. In [3], statistically significant differences were established between the Cepstral Peak Prominence (CPP) values of the compared SMS and normative populations. The CPP is considered one of the most reliable acoustic parameters for the detection of dysphonia. In the current work, the exploration of the presence of subharmonics in the voice of patients with SMS has been carried out to try to explain the harsh characteristic of some of these voices. The phonatory and articulatory characteristics of these voices could be related to the neuromotor and behavioral profile of the syndrome. In particular, the clear loss of glottic arches that is observed in some areas in which the second or third peak of the cepstrum is predominant, could be related to a neuromotor deficit in patients with SMS. It has to be noticed that this loss occurs in the central part of the emission, when it should remain stable and it cannot be attributed to instabilities such as the initial explosion of the phonation or due to the loss of pressure as it happens at the end of the emission.

## 6 Conclusion

The main contribution of this work is that, despite the small size of the sample (12 normative speakers and 12 speakers with SMS), we have been able to identify a characteristic that seems typical of the SMS group: the presence of subharmonics in the emission in areas where the phonation should be stable. This characteristic is present in 10 of the 12 patients and does not appear in any of the NG participants. This subharmonic content also explains the deep, hoarse voice that is characteristic of SMS patients. Our prevention in adopting an analysis window long enough to observe relatively low-frequency phenomena and widening the search range for f0 between 40 Hz and 706 Hz has been supported by the results obtained. In Figs. 5a and 5b, corresponding to the SMS10 speaker, it is observed that both abnormally low f0 for his gender and age, as well as the clear presence of subharmonics (at a frequency of 72 Hz), made it necessary to extend that search range in the cepstral domain. It should be noted that, in this example, the possible presence of a third frequency subharmonic would have a frequency of 48 Hz and would therefore correspond to a position with a period of 1000 samples in the cepstral domain. See the light line that appears in Fig. 5a at the height of the value 1000, corresponding to the third peak of the cepstrum. It is also important to point out some limitations of the present study. The sample size is small. This is a common situation in the case of rare and underdiagnosed diseases. The work is also limited to the study of the sustained vowel /a/ and avoids coarticulation mechanisms to focus on the part of the emission corresponding to the vocal cords and isolate it as much as possible from the effects of the set of muscles involved in the process of articulation. In this way, it would be interesting to make several recordings that include aspects related to articulation. Given the limited nature of the sample, it has not been possible to observe differences in phonation between the gender and age groups, but with only 3 subjects in each group (gender x age), it is not possible to know whether age or gender may be factors that affect the neuromotor profile and therefore to the phonation of patients with SMS.

## References

1. van Brenk, F., Kuschmann, A.: Acoustic markers of dysarthria in children with cerebral palsy: comparison of speech tasks. In: Annual ASHA convention 2018 (2018)
2. Burk, B.R., Watts, C.R.: The effect of parkinson disease tremor phenotype on cepstral peak prominence and transglottal airflow in vowels and speech. J. Voice **33**(4), 580–e11 (2019)
3. Garayzábal-Heinze, E., Gómez-Vilda, P., Martínez-Olalla, R., Palacios-Alonso, D., et al.: Acoustic analysis of phonation in children with Smith-Magenis syndrome. Front. Hum. Neurosci. **15**, 259 (2021)
4. Garayzábal-Heinze, E., Gómez-Vilda, P., et al.: Voice characteristics in Smith-Magenis syndrome: an acoustic study of laryngeal biomechanics. Languages **5**(3), 31 (2020)
5. Garrett, M.: Levels of processing in sentence production. In: Language Production, Vol. 1: Speech and Talk, pp. 177–220. Academic Press, Cambridge (1980)
6. Greenberg, F., et al.: Multi-disciplinary clinical study of Smith-Magenis syndrome (deletion 17p11. 2). Am. J. Med. Genet. **62**(3), 247–254 (1996)
7. Madduri, N., Peters, S.U., Voigt, R.G., Llorente, A.M., Lupski, J.R., Potocki, L.: Cognitive and adaptive behavior profiles in Smith-Magenis syndrome. J. Dev. Behav. Pediatrics **27**(3), 188–192 (2006)
8. McClelland, J.L., Rumelhart, D.E., Hinton, G.E.: The Appeal of Parallel Distributed Processing, pp. 3–44. MIT Press, Cambridge (1986)
9. Palacios-Alonso, D., Meléndez-Morales, G., López-Arribas, A., Lázaro-Carrascosa, C., Gómez-Rodellar, A., Gómez-Vilda, P.: Monparloc: a speech-based system for Parkinson's disease analysis and monitoring. IEEE Access **8**, 188243–188255 (2020)
10. Smith, A.C., Gropman, A.L.: Smith-Magenis syndrome. In: Cassidy and Allanson's Management of Genetic Syndromes, pp. 863–893 (2021)
11. Smith, A.C., et al.: Interstitial deletion of (17)(p11. 2p11. 2) in nine patients. Am. J. Med. Genet. **24**(3), 393–414 (1986)
12. Smith ,A.C.M.M.L., Waldstein, G.: Deletion of the 17 short arm in two patients with facial clefts. Am. J. Hum. Genet. **34** (1982)
13. Themistocleous, C., Eckerström, M., Kokkinakis, D.: Voice quality and speech fluency distinguish individuals with mild cognitive impairment from healthy controls. PLoS ONE **15**(7), e0236009 (2020)
14. Watts, C.R., Awan, S.N., Marler, J.A.: An investigation of voice quality in individuals with inherited elastin gene abnormalities. Clin. Linguist. Phonet. **22**(3), 199–213 (2008)

# Speech Analysis in Preclinical Identification of Alzheimer's Disease

Olga Ivanova[1,2(✉)] and Juan José García Meilán[1,2]

[1] Universidad de Salamanca, 37008 Salamanca, Spain
olga.ivanova@usal.es
[2] Instituto de Neurociencias de Castilla y León, 37007 Salamanca, Spain

**Abstract.** In this paper we overview the role of speech variables in the preclinical discrimination of Alzheimer's Disease (AD). As a neurodegenerative process, AD displays important cognitive and language impairments, which affect speech production from early, even preclinical stages of the disease. In this work, we review speech affection in different neurodegenerative diseases in order to focus on specific speech traits of AD. We define speech production in individuals with AD and highlight how their speech changes in comparison with healthy elderly. Finally, we briefly talk about automatic speech analysis (ASA) as one of the most promising techniques for non-invasive and reliable assessment of speech for preclinical identification of AD.

**Keywords:** Speech · Alzheimer's disease · Automatic speech analysis · Cognitive impairment · Language

## 1 Introduction

Speech is one of the most affected functions in neurodegenerative diseases. The term 'speech' can refer to two language-related phenomena. On the one hand, 'speech' can be used as synonym to 'discourse' or 'communication sequence' and in such cases it is usually accompanied by the modifier 'connected' (e.g. connected speech). On the other hand, in the strict sense 'speech' refers to a motor-physiological and cognitive-driven process through which humans are able to produce articulated sounds of language. In a narrow sense, speech is defined as the process of perception and production of sounds conveying phonetic structure [1]. As such, speech is a dynamic process, which realization requires both the activation of specific anatomical organs (mainly known as speech organs: larynx, vocal cords, tongue, jaws, etc.) and of the language-cognitive structure. Most models of speech production recognize that speech is, at least, a two-stage process involving (i.) the lemma stage (codification of word) and (ii.) the phonological stage (codification of the word phonological form), which may be referred to as conceptual and articulatory stages, respectively. Thus, disorders or impairments at both motor and cognitive levels can drive changes in human speech production.


J. M. Ferrández Vicente et al. (Eds.): IWINAC 2022, LNCS 13258, pp. 363–368, 2022.
https://doi.org/10.1007/978-3-031-06242-1_36

Speech changes due to motor and cognitive impairments have been extensively studied in different aging-related clinical disorders. Neurodegenerative diseases are among the most common causes of speech alterations, since speech production strongly relies on specific brain areas that can be affected by degenerative processes. Changes in speech can be driven by the disruptions in the brain areas responsible for either motor, cognitive or motor-cognitive control. Thus, for example, degenerative processes can lead to the weakening of nerve connection of muscles involved in speech or negatively affect the coordination between the conceptual (message) stage and the phonological (articulation) stage [2].

Motor-driven speech alterations are usually due to impoverished muscular control of phonatory and articulatory organs. Motor-driven speech impairments typically define such neurodegenerative diseases as motor neuron disease [3], non-fluent primary progressive aphasia, progressive apraxia of speech or primary progressive apraxia of speech [4]. Parkinson's disease (PD), one of the most well-known motor neurodegenerative diseases, presents with important speech and voice impairments, which lead to perceptual hypophonia (reduced loudness), monotony (reduced pitch inflection), dysarthria (imprecise and difficult articulation) and hoarse voice [5,6]. In view of this, current studies recur to the speech analysis in order to improve accuracy classifiers for detecting patients with PD [7] or to predict the itinerary of cognitive decline in speakers suffering from this neurodegeneration [8].

Yet, there is also a strong cognitive background of impaired speech in neurodegenerative diseases. Some speech changes may be the direct result of cognitive, rather than motor dysfunction and impairment in people suffering from neurodegenerative diseases [4]. Speech alterations have been reported for different diseases with cognitive affectation. The neural background behind speech alterations within the framework of dementia-driven cognitive impairment is related to neurodegenerative processes affecting language-related areas, mainly, the frontal lobe and the temporal lobe in the left hemisphere [9]. Evidence from speech-language-directed studies on people with different dementia variants proves a direct relationship between conceptual language knowledge and speech production. Consequently, cognitive-based speech changes in dementia may drive for the language dysfunction itself [4]. For example, impaired semantic knowledge gives rise to increased phonological errors in speakers with semantic dementia [10]. Speech deficits in frontotemporal dementia are consistent with language impairments at lexical-semantic level and in written language use (reading and writing) [11], and are systematically evidenced as temporal and prosodic disturbances, mainly increased intervals between words, reduced speech rate and regularity, and reduced stress [12].

One of the most promising research lines in psycholinguistics and clinical neurolinguistics focuses on the analysis of speech traits in Alzheimer's Disease, the most common cause of dementia Worldwide. Clinically, Alzheimer's Disease (AD) is characterized by the impairment of memory and one another superior cognitive function, which is frequently the language function. Language impairment is considered as a prominent clinical symptom of AD [13]. Yet, one of the most important challenges in diagnosing AD is that its clinical manifestations can be frequently shared with other cognitive disorders, for example, the Major Depression Disorder (MDD). Furthermore, pronounced cognitive symptoms of AD are not easily identifiable until early-moderate or even moderate stages of dementia.

Still, recent experimental research has shown that cognitive decline in AD can be already detected at early, even preclinical stages, through the analysis of speech traits. AD-driven speech changes are a consequence of the language dysfunction in dementia: speakers with AD present word-finding difficulties, which cause such temporal changes in their discourse production as salient pauses and slow pace [13]. Communicative deficits in AD also affect speech performance during such controlled tasks as oral reading. While orally reading, speakers with AD dementia exhibit reduced speech rate, reduced articulation rate, phonation time with low effectiveness, as well as increased number and proportion of pauses [14].

In the following section we describe the most outstanding speech features of AD and their value for the preclinical discrimination of this neurodegeneration from other clinical pictures.

## 2 Speech Traits of Alzheimer's Disease (AD)

Speech changes in AD can be considered as a preclinical marker of this dementia condition. Changes in speech in AD mainly belong to acoustic, temporal, and prosodic properties. Acoustic properties of speech refer to the structure and the nature of language sounds, that is, to their configuration as fluctuations in air pressure from the vocal tract [15]. Acoustic properties are responsible for phonemic differentiation between language sounds and are usually assessed in clinical speech through such variables as fundamental frequency (f0), formants or spectral characteristics. Temporal properties refer to the duration with which sound or syllabic segments are articulated and produced [16]. Finally, prosodic features of speech are an intrinsic element of message generation. Frequently classified within the suprasegmental level, that is, as features not limited to segments, prosodic cues are responsible for meaning modulation of longer combinations of segments, like syllables, words or utterances. Prosodic cues include intonation, rhythm, pitch contour, duration, stress, juncture or tone, and their role in the meaning modulation of the message can vary across languages. Yet, in all languages prosodic cues can add both linguistic and emotional information to the message: on the one hand, they can mark the mode or the type of the utterance (e.g. declarative vs. interrogative vs. exclamative), and, on the other hand, they can also convey information about speaker's attitude towards the uttered message [17]. The nature of the functions that prosodic cues perform in meaning modulation allows to define prosody as a cognitive function [9]. Indeed, prosody can mark semantic and syntactic relationship through rhythmic grouping, emphasis or other non-segmental cues, being able to reflect attitudinal and affective state of the utterer [18].

Altogether, acoustic, temporal, and prosodic traits of speech are responsible for organizing and structuring speech sounds from both cognitive and sensorimotor control centers [16]. Importantly, their specific alterations because of AD-driven neurodegeneration allow to sensitively distinguish between healthy older adults and older adults with mild AD. AD speakers produce a more temporally slow and interrupted speech: they involve much more time on phonation specifically and on speech production generally, and their articulation and elocution rates are slower than in healthy aging. In addition, speech production in AD includes more pauses and more irregular duration of syllabic intervals, which are much more variable across AD group than across healthy aging group. Finally, speakers with AD tend to show a higher proportion of voice breaks, a diminished ratio of noise-to-harmonics, and a general decrease in speech intensity and energy. Their voice sounds, in such a way, as highly dysphonic (Table 1).

**Table 1.** Discriminating speech variables of Alzheimer's Disease.

| Speech variable | Manifestation in AD |
|---|---|
| Speech time | *Increased* |
| Speech rate | *Decreased* |
| Phonation time | *Increased* |
| Syllabic duration | *Altered* |
| Number of pauses | *Increased* |
| Proportion of pauses | *Increased* |
| Normalized PVI | *Increased* |
| Voice breaks | *Increased* |
| Voiceless segments (percentage) | *Increased* |
| Articulation rate | *Decreased* |
| Formants | *Distortion* |
| Acoustic Voice Quality | *Decreased* |
| Skewness | *Decreased* |

As suggested by a recent systematic review [19], it may be difficult to establish which of the described variables can have a more significant role in clinical discrimination of AD from healthy aging. By the moment, combinatory approaches based on the joint analysis of several speech parameters, particularly of prosodic and temporal variables, seem to be the best discriminating option.

In the next, and final section of this paper we focus on the application of automatic speech analysis as one of the most promising techniques for discriminating assessment of speech in preclinical AD.

## 3 Automatic Speech Analysis of AD

Automatic Speech Analysis (ASA) is one of the most promising non-invasive techniques allowing to assess and analyze extensive databases of speech. At present day, ASA is extensively used in testing speech assessment as a unique and novel method for preclinical discriminating between healthy aging, Mild Cognitive Impairment (MCI), AD and other possible clinical pictures with cognitive affection (MDD, other types of dementia, etc.). Indeed, as a recent systematic review [19] highlights, ASA is highly sensitive even to subtle changes in speech of speakers with AD, allowing for a precise discrimination of dementia picture from healthy aging and MCI.

At present, most experimental research on preclinical speech assessment of AD is based on the development of ASA protocols, experimental designs, and algorithms of analysis. Experimental designs may be based on different types of stimuli, and we suggest that selected tasks can have a direct influence on the reliability of ASA assessment of preclinical AD. For example, studies based on ASA evaluation of spontaneous speech allow to discriminate healthy aging from pathological aging at 78.8% [20]. ASA evaluation of picture-description task can vary from 81% [21] to 90% [22] accuracy level of discrimination. To the best of own knowledge, the highest levels of accuracy have been achieved so far by ASA evaluation of reading tasks, with accuracy levels of discrimination between healthy speakers and AD speakers reaching from 87% [23] to 91,2% [24].

All in all, ASA is currently the most promising technique for assessing speech as preclinical marker of AD and more insightful studies would be needed in order to test the highest levels of accuracy of different experimental and protocol designs.

## References

1. Liberman, A.M., Whalen, D.H.: On the relation of speech to language. Trends Cogn. Sci. **4**(5), 187–196 (2000)
2. Vizza, P., et al.: Methodologies of speech analysis for neurodegenerative diseases evaluation. Int. J. Med. Inform. **122**, 45–54 (2019)
3. Bak, T.H., O'Donovan, D.G., Xuereb, J.H., Boniface, S., Hodges, J.R.: Selective impairment of verb processing associated with pathological changes in Brodmann areas 44 and 45 in the motor neurone disease-dementia-aphasia syndrome. Brain **124**(1), 103–120 (2001)
4. Poole, M.L., Brodtmann, A., Darby, D., Vogel, A.P.: Motor speech phenotypes of frontotemporal dementia, primary progressive aphasia, and progressive apraxia of speech. J. Speech Lang. Hear. Res. **60**(4), 897–911 (2017)
5. Olson Ramig, L., Fox, C., Sapir, S.: Parkinson's disease: speech and voice disorders and their treatment with the Lee Silverman voice treatment. Thieme: Sem. Speech Lang. **25**(2), 169–180 (2004)
6. Sakar, N.E., et al.: Collection and analysis of a Parkinson speech dataset with multiple types of sound recordings. IEEE J. Biomed. Health Inform. **17**(4), 828–834 (2013)

7. Gómez-Vilda, P., et al.: Parkinson disease detection from speech articulation neuromechanics. Front. Neuroinform. **11**, 56 (2017)
8. Rektorova, I., et al.: Speech prosody impairment predicts cognitive decline in Parkinson's disease. Parkinsonism Relat. Disord. **29**, 90–95 (2016)
9. De Stefano, A., et al.: Changes in speech range profile are associated with cognitive impairment. Dementia Neurocogn. Disord. **20**(4), 89–98 (2021)
10. Meteyard, L., Patterson, K.: The relation between content and structure in language production: an analysis of speech errors in semantic dementia. Brain Lang. **110**(3), 121–134 (2009)
11. Geraudie, A., et al.: Speech and language impairments in behavioral variant frontotemporal dementia: a systematic review. Neurosci. Behav. Rev. **131**, 1076–1095 (2021)
12. Vogel, A.P., et al.: Motor speech signature of behavioral variant frontotemporal dementia. Refining the phenotype. Neurology **89**(8), 837–844 (2017)
13. Lee, H., Gayraud, F., Hirsh, F., Barkat-Defradas, M.: Speech dysfluencies in normal and pathological aging: a comparison between Alzheimer patients and healthy elderly subjects. In: ICPhS, vol. XVII, pp. 1174–1177 (2011)
14. Martínez-Sánchez, F., Meilán, J.J.G., García-Sevilla, J., Carro, J., Arana, J.M.: Oral reading fluency analysis in patients with Alzheimer disease and asymptomatic control subjects. Neurologia **28**(6), 325–331 (2013)
15. Tiwari, M.: Speech acoustics: how much science? J. Nat. Sci. Biol. Med. **3**(1), 24–31 (2012)
16. Ivanova, O., Meilán, J.J.G., Martínez-Sánchez, F., Martínez-Nicolás, I., Llorente, T.E., Carcavilla González, N.: Discriminating speech traits of Alzheimer's disease assessed through a corpus of reading task for Spanish language. Comput. Speech Lang. **73**, 101341 (2022)
17. Wartenburger, I., Steinbrink, J., Telkemeyer, S., Friedrich, M., Friederici, A.D., Obrig, H.: The processing of prosody: evidence of interhemispheric specialization at the age of four. Neuroimage **34**(1), 416–425 (2007)
18. Wagner, M., Watson, D.G.: Experimental and theoretical advances in prosody: a review. Lang. Cognit. Process. **25**(7–9), 905–945 (2010)
19. Martínez-Nicolás, I., Llorente, T.E., Martínez-Sánchez, F., Meilán, J.J.G.: Ten years of research on automatic voice and speech analysis of people with Alzheimer's disease and mild cognitive impairment: a systematic review article. Front. Psychol. **12**, 620251 (2021)
20. Toth, L., et al.: A speech recognition-based solution for the automatic detection of mild cognitive impairment from spontaneous speech. Curr. Alzheimer Res. **15**(2), 130–138 (2018)
21. Fraser, K.C., Meltzer, J.A., Rudzicz, F.: Linguistic features identify Alzheimer's disease in narrative speech. J. Alzheimers Dis. **49**, 407–422 (2016)
22. Wahlforss, A., Aslaksen Jonasson, A.: Early dementia diagnosis from spoken language using a transformer approach. Alzheimer's dement. J. Alzheimer's Assoc. **16**(S11), e043445 (2020)
23. Martínez-Sánchez, F., Meilán, J.J.G., Carro, J., Ivanova, O.: Prototype for the voice analysis of Alzheimer's disease. J. Alzheimer Disease **64**, 473–481 (2018)
24. Meilán, J.J.G., Martínez-Sánchez, F., Carro, J., Carcavilla, N., Ivanova, O.: Voice markers of lexical Access in mild cognitive impairment and Alzheimer's disease. Curr. Alzheimer Res. **15**, 111–119 (2018)

# The Effect of Breathing Maneuvers on the Interaction Between Pulse Fluctuation and Heart Rate Variability

Nicolás Alberto Posteguillo[1(✉)] and María Paula Bonomini[1,2]

[1] Instituto Tecnológico de Buenos Aires (ITBA), Buenos Aires, Argentina
nposteguillo@itba.edu.ar

[2] Instituto Argentino de Matemática "Alberto P. Calderón" (IAM), CONICET, Buenos Aires, Argentina
paula.bonomini@conicet.gov.ar

**Abstract.** Respiration has appeared as a useful knob to turn up and down the autonomic nervous system (ANS). In particular, several studies have reported plenty of benefits on emotional and physiological health from slow paced breathing, even though little is known about paced breathing at normal or even fast rates. In this work, we have systematically investigated the physiological interactions between the blood volume pressure signal (Bvp) and the heart rate (HR) during spontaneous and paced breathing at normal (NB), fast (FB) and slow (SB) rates. Such interactions were quantified by the cross-amplitude spectrum, by detecting peak amplitude and peak frequency at every phase. The spontaneous phase presented no structured interaction, while the SB showed, as expected, clearly defined peaks at the breathing frequency (at about 0.1 Hz). The NB phase presented peaks at both, under and above 0.1 Hz, revealing energy at both sympatho-vagal bands. The FB phase showed a vagal withdrawal, with peaks located at about 0.05 Hz; very similar to those interactions arisen during the stressor test, which consisted of the N-Back test (cognitive load). In conclusion, the cross amplitude spectrum satisfactorily separated four different ANS scenarios, which might have potential for emotion recognition.

**Keywords:** ANS · Paced breathing · Cross amplitude spectrum · Bvp-HR interactions

## 1 Introduction

In the intact organism, variations in the heart rate (HR) and in the blood pressure (BP) are interconnected by means of the baroreceptor reflex, which acts as a rapid mechanism for blood pressure control. Baroreceptors are nerve cells specialized to sense changes in blood pressure. When barorecptors sense an increase in blood pressure, the heart rate will, through a negative feedback loop, decrease to compensate. If, on the other hand, a decrease in blood pressure is sensed,


J. M. Ferrández Vicente et al. (Eds.): IWINAC 2022, LNCS 13258, pp. 369–379, 2022.
https://doi.org/10.1007/978-3-031-06242-1_37

the heart rate will increase. Thus, it seems quite obvious that the variabilities observed in heart rate and blood pressure are highly correlated [1,4].

Similarly, respiration is also linked to blood pressure and heart rate by the Respiration Sinus Arrhythmia (RSA) effect, a cyclic variation that accelerates the heart rate during inhalation and decelerates it during exhalation [13]. This mechanism produces a frequency modulation effect on heart rate, impacting on the Bvp signal as well. Finally, and due to the possibility of conscious control, respiration has proved useful to produce specific autonomic conditions to better understand the heart rate and blood pressure interactions [2]. In this line, paced breathing, in particular slow paced breathing, has gained an increasingly body of evidence reporting all kinds of benefits [3,9,10]; although little is described about pacing breath at other frequencies, such as fast and normal rates, to the best of our knowledge.

With the advent of new technologies, it is now possible to record blood volume pressure (Bvp) with wearable devices. From the Bvp signal, the HR series can be obtained. Therefore, it is quite natural to address the nature of the interaction between the Bvp and HR variabilities on spontaneous and paced breathing systematically. In this case, paced breathing spanned normal, fast and slow rates. We also compared these patterns of interaction to those produced in stressor tests, under cognitive load.

## 2 Materials and Methods

**Study Population and Experimental Paradigm.** Twenty three young healthy subjects were enrolled aged $34.4 \pm 7.2$ years old (12 male). From this population, two subjects were discarded due to noisy records.

Groups were defined according to three respiratory frequencies; normal breathing at about 12 breaths per minute (NB), fast breathing at about 20 breaths per minute (FB), and slow breathing below 6 breaths per minute (SB). In addition, a control group without breath pacing (spontaneous breath) was included. During respiratory phases, NB, FB and SB subjects were asked to close their eyes; except for control, where they remained with their eyes open. All experiments were accomplished in the morning, in the same room. Blood volume pulse (Bvp) was obtained from photoplethysmography using the wereable device E4 Empatica wristband. After the control period and every breathing phase, subjects were asked to complete a cognitive task consisting in the N-Back task with $N = 2$. Briefly, in the 2-Back task the participants were presented with a sequence of stimuli one-by-one. For each stimulus, they had to decide if the current stimulus was equal to the one presented two trials ago. The 2-Back test is a particular implementation of the N-Back test, which aims to assess working memory [8]. The order of the respiratory sessions was randomized to avoid bias due to training.

**Pulse and Heart Rate Variability Interactions. Cross-spectrum Definitions.** In order to quantify the interactions between the Bvp and HR variabilities, we computed their cross and phase spectra. The sample cross spectrum between Bvp and HR ($C_{Bvp,hr}$) is defined as the Fourier Transform of the cross-correlation estimate $c_{Bvp,hr}$:

$$C_{Bvp,hr}(f) = \int_{-T}^{T} c_{Bvp,hr}(u)e^{-j2\pi fu}\,du \tag{1}$$

This is an asintotically biased estimator, since its mean value equals:

$$E[(C_{Bvp,hr}(f))] = \int_{-T}^{T} (1 - \frac{u}{T})\gamma_{Bvp,hr}(u)e^{-j2\pi fu}\,du \tag{2}$$

and as T tends to infinity, this mean value tends to the theoretical cross spectrum $\Gamma_{Bvp,hr}(f)$:

$$\lim_{T\to\infty} E[C_{Bvp,hr}(f)] = \Gamma_{Bvp,hr}(f) = \int_{-\infty}^{\infty} \gamma_{Bvp,hr}(u)e^{-j2\pi fu}\,du \tag{3}$$

The co and quadrature spectra can be obtained from the cross spectrum definition by writing $\gamma_{Bvp,hr}(u)$ as the sum of an even part $\lambda_{Bvp,hr}(u)$ and an odd part $\psi_{Bvp,hr}(u)$:

$$\lambda_{Bvp,hr}(u) = \frac{1}{2}(\gamma_{Bvp,hr}(u) + \gamma_{Bvp,hr}(-u)) \tag{4}$$

$$\psi_{Bvp,hr}(u) = \frac{1}{2}(\gamma_{Bvp,hr}(u) - \gamma_{Bvp,hr}(-u)) \tag{5}$$

from which the co spectrum $\Lambda_{Bvp,hr}(f)$ and the quadrature spectrum $\Psi_{Bvp,hr}(f)$ are defined as:

$$\Lambda_{Bvp,hr}(f) = \int_{-\infty}^{\infty} \lambda_{Bvp,hr}(u)cos(2\pi fu)\,du \tag{6}$$

$$\Psi_{Bvp,hr}(f) = \int_{-\infty}^{\infty} \psi_{Bvp,hr}(u)sin(2\pi fu)\,du \tag{7}$$

With these definitions, the cross spectrum $\Gamma_{Bvp,hr}(f)$ can now be written in terms of the co and quadrature spectra, which in turn has a complex representation expressed by an amplitude $\alpha_{Bvp,hr}(f)$ times a phase $\phi_{Bvp,hr}(f)$, both functions of frequency:

$$\Gamma_{Bvp,hr}(f) = \Lambda_{Bvp,hr}(f) + j\Psi_{Bvp,hr}(f) = \alpha_{Bvp,hr}(f)e^{-j\phi_{Bvp,hr}(f)} \tag{8}$$

From the above, $\alpha_{Bvp,hr}(f)$ is called the cross amplitude spectrum and $\phi_{Bvp,hr}(f)$ is the phase spectrum, mathematically defined as:

$$\alpha_{Bvp,hr}(f) = \sqrt{\Lambda^2_{Bvp,hr}(f) + \Psi^2_{Bvp,hr}(f)} \tag{9}$$

$$\phi_{Bvp,hr}(f) = \arctan \frac{-\Psi_{Bvp,hr}(f)}{\Lambda_{Bvp,hr}(f)} \quad (10)$$

We computed the amplitude spectrum $\alpha_{Bvp,hr}(f)$ for every experimental group (Control, NB, FB, and SB) in order to characterize the Bvp-HR interaction in different breathing phases and study whether those representations changed under cognitive load (the test phases). Also, we used the amplitude and frequency of the dominant peak in $\alpha_{Bvp,hr}(f)$ as features that will feed a classification algorithm in order to separate the different breathing phases.

**Statistical Analysis and Classification.** Kruskal-Wallis ANOVA was used to compare within and between group comparisons, followed by Mann-Whitney post-hoc comparisons. Statistical significance was defined for $p<0.05$. Two separate analysis were accomplished: one for the breathing phases, and one for the cognitive phases following the former.

In order to check whether amplitude and frequency of the main peaks in the amplitude spectrum presented sufficiently separated patterns for every phase, we fed a kmeans algorithm and compared the means of the resulting clusters against those displayed in Fig. 1.

## 3 Results

The Bvp-HR cross amplitude spectrum $\alpha_{Bvp,hr}(f)$ showed significantly different interaction patterns during the breathing phases. Figure 1 shows the interval and blood pulse interactions for one paradigmatic subject for control, NB, FB and SB phases. The top panels display the temporal evolution of both the NN and Bvp series. The NN series is the corrected interval series, where RR intervals belonging to ectopic beats were corrected with the linear interpolation method [11]. Notice that RR and NN intervals are the inverse of instantaneous heart rate. Below the temporal series evolution, the cross amplitude spectrum is shown for every phase.

Notice the synchronous oscillatory pattern at around 0.1 Hz (the respiratory frequency) emerging, as expected, on the SB phase (see Fig. 1-d). However, during FB, all Bvp-HR interactions shifted towards the very low and low frequencies; suggesting a dominant sympathetic drive (see Fig. 1-c). On the other hand, the NB phase tended to present activation at both sympathetic and vagal drive, with one peak under 0.1 Hz and another one beyond 0.1 Hz most of the time (Fig. 1-b). Finally, the control phase presented a rather unstructured interaction, with low activity at almost all the frequency bands (Fig. 1-a).

Despite the clear structure found in the paced breathing phases, we failed to find clear interactions when analysing the subsequent phases where the subjects responded to the N-Back test (cognitive phases), besides a stronger activity in the low frequencies for every phase regardless of the respiratory phase that preceded the actual cognitive phase. Figure 2 shows the same analysis as that from Fig. 1, but accounting for the cognitive tasks. Note that in this case no Cross amplitude spectrum showed energy above 0.15 Hz, making evident a suppressed vagal drive, which tells that the cognitive load acts as an effective stressor.

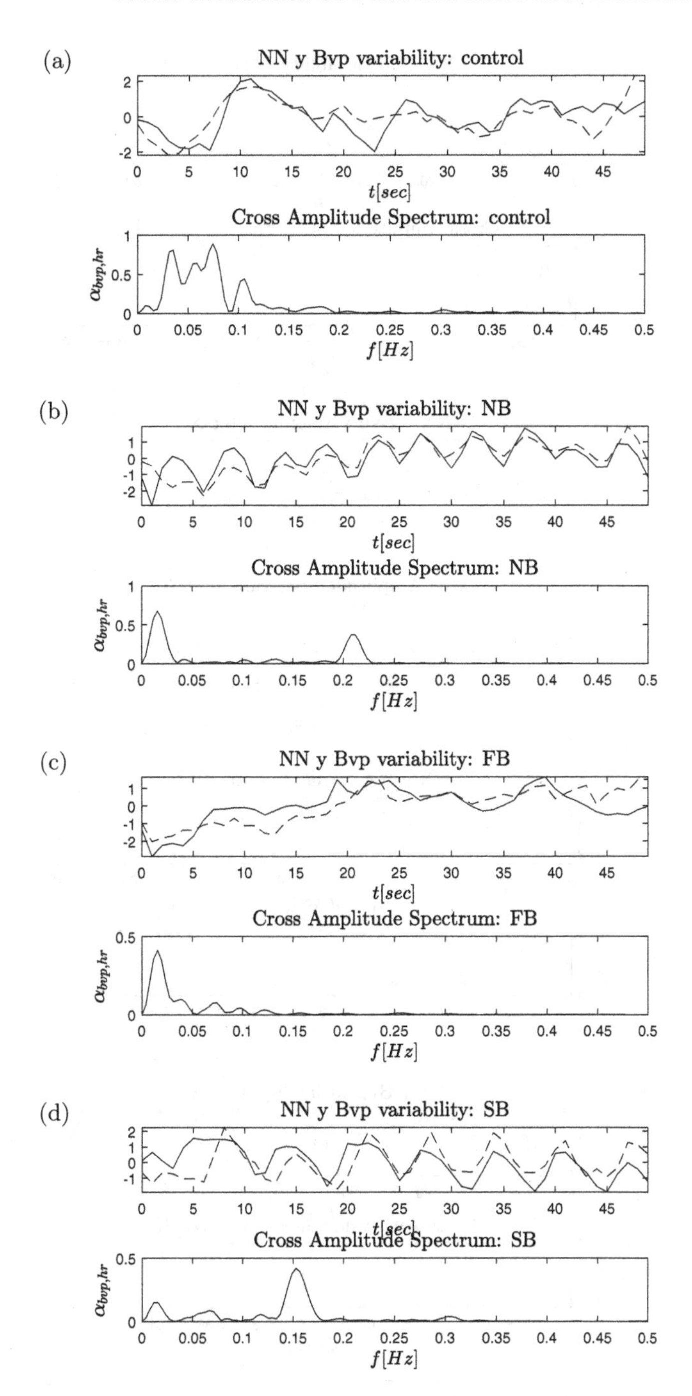


**Fig. 1.** Interaction between Bvp (dashed) and NN (solid) for Control, NB, FB and SB phases. Top panels: NN and Bvp variability series. Bottom panels: Bvp-NN cross amplitude spectrum.

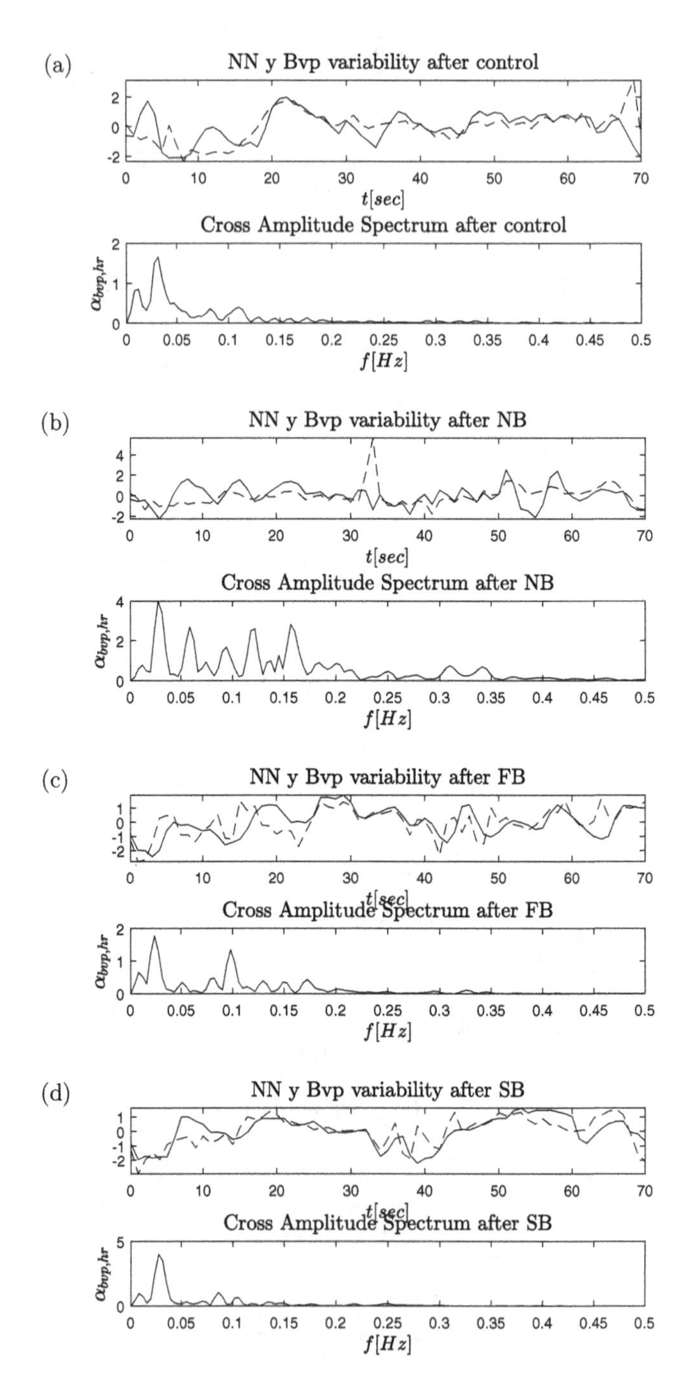


**Fig. 2.** Interaction between Bvp (dashed) and NN (solid) for cognitive phases following Control, NB, FB and SB respiratory phases. Top panels: NN and Bvp variability series. Bottom panels: Bvp-NN cross amplitude spectrum.

The trends shown in Fig. 1 for the breathing maneuvers and Fig. 2 for the cognitive load, were confirmed on the entire population. Figure 3 (left) shows boxplots for the peak amplitudes and dominant frequencies of the cross amplitude spectrum $\alpha_{Bvp,hr}(f)$ for the breathing phases; while its counterpart for the cognitive phases are shown in Fig. 3 (right). Note the increasing peak amplitude towards breathing phases with more defined peaks, such as SB and FB; while the frequency of the dominant peak in $\alpha_{Bvp,hr}(f)$ changes only at FB and SB, with an increase in SB above 0.1 Hz and a decrease at FB about 0.05 Hz.

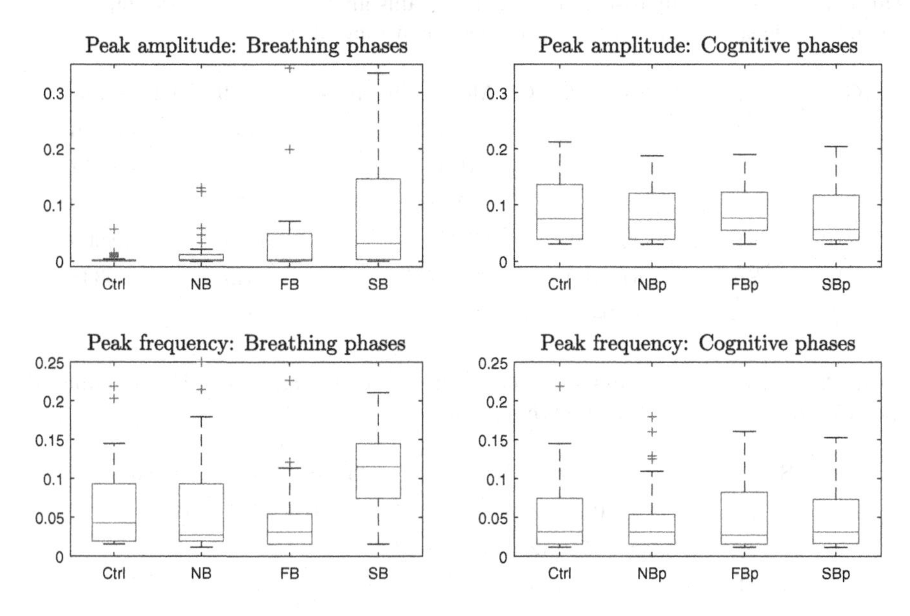


**Fig. 3.** Boxplots for the peak amplitude and peak frequency for Breathing and Cognitive phases. Note the increasing amplitude from NB to SB and the clear peak frequency separation in FB and SB for the breathing phases, telling about pace of breath and structure. Also, in the Cognitive phases there is lack of structure throughout the phases.

Table 1 shows the Kruskalwallis analysis for the peak amplitudes in $\alpha_{Bvp,hr}(f)$ on the different breathing groups, followed by a Tuckey-Kramer post-hoc comparison (Table 2). Note that all respiratory phases significantly increased their peak amplitudes with respect to Control, suggesting a progressive structure in the cross amplitude peaks, and therefore a more synchronous interaction between Bvp and HR throughout the respiratory maneuvers. Similarly, Table 3 and Table 4 show the Kruskalwallis analysis and post-hoc comparisions for the dominant frequency in $\alpha_{Bvp,hr}(f)$, respectively. Herein, it is evident a significant separation between the dominant frequencies in the cross amplitude spectrum at FB and SB, accounting for sympathetic-driven or vagal-driven interactions respectively.

**Table 1.** Kruskalwallis analysis for the peak amplitude in the amplitude spectrum $\alpha_{Bvp,hr}(f)$ at every breathing phase.

| Source | SS | df | MS | Chi-sq | Prob>Chi-sq |
|---|---|---|---|---|---|
| Groups | 5.1505e+04 | 3 | 1.7168e+04 | 28.7978 | 2.4696e-06 |
| Error | 2.0783e+05 | 142 | 1.4636e+03 | | |
| Total | 2.5933e+05 | 145 | | | |

**Table 2.** Post-hoc comparisons from Kruskalwallis analysis for the peak amplitude in the amplitude spectrum $\alpha_{Bvp,hr}(f)$ at every breathing phase.

| Group 1 | Group 2 | Lower 95% CI | Mean differences | Upper 95% CI | p-value |
|---|---|---|---|---|---|
| Ctrl | NB | –48.9006 | –24.5380 | –0.1754 | 0.0476 |
| Ctrl | FB | –56.1704 | –31.6266 | –7.0828 | 0.0052 |
| Ctrl | SB | –78.5607 | –52.7155 | –26.8703 | 0.0000 |
| NB | FB | –32.5231 | –7.0886 | 18.3459 | 0.8908 |
| NB | SB | –54.8700 | –28.1775 | –1.4849 | 0.0338 |
| FB | SB | –47.9469 | –21.0889 | 5.7691 | 0.1815 |

**Table 3.** Kruskalwallis analysis for the frequency of the main peak in the amplitude spectrum $\alpha_{Bvp,hr}(f)$ at every breathing phase.

| Source | SS | df | MS | Chi-sq | Prob > Chi-sq |
|---|---|---|---|---|---|
| Groups | 3.7144e+04 | 3 | 1.2381e+04 | 20.9516 | 1.0774e-04 |
| Error | 2.1992e+05 | 142 | 1.5487e+03 | | |
| Total | 2.5706e+05 | 145 | | | |

**Table 4.** Post-hoc comparisons from Kruskalwallis analysis for the dominant frequency in the amplitude spectrum $\alpha_{Bvp,hr}(f)$ at every breathing phase.

| Group 1 | Group 2 | Lower 95% CI | Mean differences | Upper 95% CI | p-value |
|---|---|---|---|---|---|
| Ctrl | NB | –22.4625 | 1.7932 | 26.0489 | 0.9976 |
| Ctrl | FB | –10.0637 | 14.3724 | 38.8085 | 0.4308 |
| Ctrl | SB | –57.6400 | –31.9081 | –6.1763 | 0.0079 |
| NB | FB | –12.7437 | 12.5792 | 37.9021 | 0.5782 |
| NB | SB | –60.2768 | –33.7014 | –7.1259 | 0.0062 |
| FB | SB | –73.0207 | –46.2806 | –19.5404 | 0.0001 |

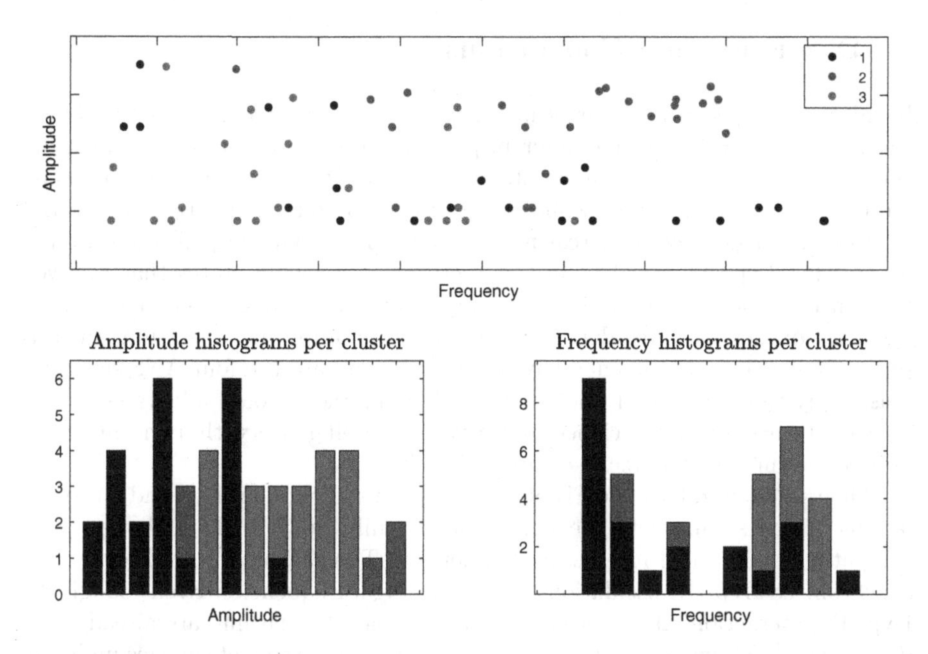


**Fig. 4.** Cluster separation. Top: scatter of the two-featured elements of the k-means output. Clusters were set to three (K = 3) in order to keep the complexity of the classifier low due to a small amount of observations. Bottom: Histograms per cluster for peak amplitude (left) and peak frequency (right).

Finally, a clustering algorithm (kmeans, k = 3) was carried out to assess the performance of using 'amplitude' and 'dominant frequency' as features to separate into the different breathing phases. Note that we have ruled out the NB phase, since the amount of data was too small for clustering. Figure 4 shows the clusters for control, FB and SB phases in a log-log scale for better representation.

In order to check the accuracy of the non-supervised clustering algorithm, we computed the mean values for every cluster produced by the kmeans algorithm (K=3). Table 5 shows the mean for amplitude and dominant frequency for every cluster arisen at Fig. 4. Notice that every group falls into the median values for Control, FB and SB given in Fig. 2 for both features (Amplitude and Frequency).

**Table 5.** Mean amplitude and dominant frequency of every cluster at Fig. 4.

| Label | GroupCount | mean_Amplitude | mean_Frequency |
|---|---|---|---|
| 1 | 22 | 0.0067 | 0.0481 |
| 2 | 22 | 0.1152 | 0.1012 |
| 3 | 22 | 0.2073 | 0.0435 |

## 4 Discussion and Conclusions

In this work, the Bvp-HR interaction, measured by means of the cross spectrum, was revisited in order to see if different patterns arose for four different breathing conditions. We have seen that spontaneous (automatic) respiration generates no structured interaction at all, while paced respiration builds up a tighter Bvp-HR interaction, progressively increasing from NB to SB. This organization around more defined spectral peaks can be described as resonant states that are well differentiated according to the breathing maneuver that originated them. As a matter of fact, features derived from the cross amplitude spectrum, such as peak amplitude and frequency have proved to separate the four ANS scenarios reasonably well (see Fig. 4 and Table 5). This markers would add value to the recent advances in data science and artificial intelligence within the interplay between natural and artificial computation [7].

Then we showed that the FB phase produced synchronous Bvp and HR oscillations at the sympathetic frequency band, while the NB phase showed both sympathetic and vagal interaction frequencies. The SB, as expected, generated synchronized Bvp-HR oscillations about the vagal frequencies. Even though the Bvp-HR interaction effects produced by slow paced breathing are already well described [6], no one has systematically described these sort of changes under all kind of respiratory frequencies, including normal and fast breathing, to the best of our knowledge. This systematic analysis allows us to quantify the capability of respiration to induce a discrete number of quite well defined Bvp-HR states [2]. In the future, we could therefore address whether this breath-induced ANS modulation is alike that induced by emotions themselves; so that the cross spectrum would go further than autonomic states and also help in emotion recognition. Under this light, we could regard not only the SB as a down-regulator of emotions, which is quite well established already, but also the FB as an up-regulator. Furthermore, by taking advantage of brain plasticity, we could speculate with some training based on respiratory maneuvers to prepare for cognitive load or to cope with certain stressful situations [5]. In this sense, respiration would act by anticipating emotions, as in Stearns et al., who utilized slow and simple motor movements (like tapping) as an explicit process of emotion regulation, successfully attenuating autonomic arousal [12].

## References

1. Baselli, G., et al.: Spectral and cross-spectral analysis of heart rate and arterial blood pressure variability signals. Comput. Biomed. Res. **19**, 520–534 (1986)
2. Bonomini, M.P., Calvo, M.V., Morcillo, A.D., Segovia, F., Vicente, J.M.F., Fernandez-Jover, E.: The effect of breath pacing on task switching and working memory. Int. J. Neural Syst. **30**(06), 2050028 (2020)
3. Chen, S., Sun, P., Wang, S., Lin, G., Wang, T.: Effects of heart rate variability biofeedback on cardiovascular responses and autonomic sympathovagal modulation following stressor tasks in prehypertensives. J. Hum. Hypertens. **30**, 105–11 (2015)

4. DeBoer, R.W., Karemaker, J., Strackee, J.: Beat-to-beat variability of heart interval and blood pressure. Automedica **4**, 217–222 (1983)
5. Ferrandez, J., et al.: Brain plasticity: feasibility of a cortical visual prosthesis for the blind. In: Proceedings of the 25th Annual International Conference of the IEEE Engineering in Medicine and Biology Society (IEEE Cat. No.03CH37439), vol. 3, pp. 2027–2030 (2003). https://doi.org/10.1109/IEMBS.2003.1280133
6. Folschweiller, S., Sauer, J.: Respiration-driven brain oscillations in emotional cognition. Front. Neural Circuits **15**, 761812 (2021)
7. Górriz, J.M., et al.: Artificial intelligence within the interplay between natural and artificial computation: advances in data science, trends and applications. Neurocomputing **410**, 237–270 (2020)
8. Jaeggi, S., Buschkuehl, M., Perrig, W., Meier, B.: The concurrent validity of the n-back task as a working memory measure. Memory **18**, 394–412 (2010)
9. Park, Y.J., Park, Y.B.: Clinical utility of paced breathing as a concentration meditation practice. Complement. Ther. Med. **20**, 393–399 (2012)
10. Rijken, N., Soer, R., de Maar, E., et al.: Increasing performance of professional soccer players and elite track and field athletes with peak performance training and biofeedback: a pilot study. Appl. Psychophysiol. Biofeedback **41**, 421–430 (2016)
11. Rincon Soler, A., Silva, L., Fazan, R.J., Murta, L.J.: The impact of artifact correction methods of RR series on heart rate variability parameters. J. Appl. Physiol. **124**, 646–652 (2018)
12. Stearns, S., Fleming, R., Fero, L.: Physiological arousal through the manipulation of simple hand movements. Int. J. Neural Syst. **42**, 39–50 (2017)
13. Travaglini, A., Lamberti, C., DeBie, J., Ferri, M.: Respiratory signal derived from eight-lead ECG. In: Computers in Cardiology 1998, vol. 25, pp. 65–68 (Cat. No.98CH36292) (1998). https://doi.org/10.1109/CIC.1998.731718

# Horizon Cyber-Vision: A Cybernetic Approach for a Cortical Visual Prosthesis

Mikel Val Calvo[1(✉)], Roberto Morollón Ruiz[1], Leili Soo[1], Dorota Wacławczyk[1], Fabrizio Grani[1], José Manuel Ferrández[3], and Eduardo Fernández Jover[1,2]

[1] Instituto de Bioingeniería, Univ. Miguel Hernández, Alicante, Spain
mikel1982mail@gmail.com
[2] CIBER-BBN, Madrid, Spain
[3] Dpto. de Electrónica, Tecnología de Computadoras y Proyectos, Univ. Politécnica de Cartagena, Cartagena, Spain

**Abstract.** The way towards the next generation of visual cortical prosthesis is visualised through an engineering cycle based on a cybernetic paradigm. Our proposal is to develop a configurable and wearable system that will generate simulated prosthetic vision, while on the other hand, perform intracortical stimulation when applied to blind patients, so that it is expected that improvements with sighted volunteers, in combination with transformed reality strategies, will correlate with similar improvements in blind patients. The resulting cybernetic model involves modelling from stimuli to visual percepts, and in parallel, developing the best suited transformed reality strategy leading to a better perception of the environment. Deep learning approaches for object detection, monocular depth estimation, or structural edge detection, in combination with the use of an eye-tracking system, will lead to an integrated system that has proved to be wearable, optimised, modular, and computationally lightweight. To assess the cybernetic approach, behavioural experiments are proposed using two different scenarios. Firstly, a corridor with a series of obstacles and a controlled but more complex environment that resembles a city square, called StreetLab.

**Keywords:** Cybernetic · Bio-inspired · Prosthesis · Cortical · Deep learning

## 1 Introduction

Cybernetics can be understood as the methodology for understanding the qualitative aspects of the interrelationships between the various components of a system and the resulting synthetic behaviour in interaction with the environment [10,26,27,36,43]. The visual cortical prostheses (VCPs) acts as an interplay between natural and artificial computation, where the environment feeds the artificial system with images; the input, and the VCP encodes that information; the output, to the natural computing system; the human, who will carry


J. M. Ferrández Vicente et al. (Eds.): IWINAC 2022, LNCS 13258, pp. 380–394, 2022.
https://doi.org/10.1007/978-3-031-06242-1_38

out actions based on the transformed reality(TR) provided by the cybernetic vision system [24,25]. It is the aim of the present work to materialise the next generation of VCPs through a cybernetic paradigm engineering cycle.

Studies concerning the visual system have been performed since early the 19th century, Schultze [38] investigated the different types of neurons in the retina characterising the rods and cones using the Golgi method. Hartile [18] described the discharge of impulses in single optic nerves fibbers in response to illumination of the retina, and later on, Brindley [3–5], Dobelle and Mladjilovsky [11,12,14], showed that visual percepts; called phosphenes, could be evoked by direct stimulation of the visual cortex. But still, there are a number of limitations to be faced, such as the target area of the visual pathway to be stimulated, which can range from the retina, the visual nerve, the lateral geniculate nucleus, or the different cortical processing areas, V1, V2, V3,... For the present project, V1 is the target area in which the electrodes will be implanted. Schiller and Tehovnic [37] suggested that the implantation of area V1 is very promising since it is a large volume area and therefore can accommodate large arrays of electrodes. Furthermore, receptive fields are smaller in V1 than in extra-extriate cortical areas, being the cortex in that area more uniform in density and thickness. Besides, visual fields are laid out in near topographical order so the spatial integrity of the images is more manageable here than in other areas.

Currently, research efforts are focused on the description of evoked phosphenes [8] since the VCP must be able to produce functional vision and evoking phosphenes has shown have a direct relationship with stimulation of the visual cortex. This process is directly related to how hardware, connected to the brain, can stimulate populations of neurons to generate a TR. An appropriate stimulation strategy must be performed and consequently the search for optimal electrical parameters leading is necessary [33]. Such a study should also take into account the mapping between the stimulation electrodes and their location on the visuotopic map with respect to the shape and location of the perceived phosphenes in the visual field [16,34]. Therefore, the magnification factor and eccentricity concerning the foveal visuotopic area must be taken into account [32]. In addition, stimulating the electrodes generates a lot of variability with respect to the positions of the perceived phosphenes, i.e., it generates spatial noise. Electrode dropouts leading to inconsistent performance of the resulting artificial vision may occur. Aside, any cybernetic vision(CV) system based on cortical implants should take gaze direction into account, as it influences the resulting navigation strategies in terms of head scanning [6], while coping with nystagmus, an uncontrollable and involuntary eye movement typically experienced by blind people [17]. Finally, the terms CV and TR, together emphasise a different modality of vision, focusing on being able to capacitate the blind and not visual restoration, the latter of which is not possible today given the technology available in terms of the electrical stimulation of the visual cortex and phosphene perception [33]. Therefore, VCP should nowadays be considered as the additional source of information that provides the blind with adequate information when the cane is no sufficient.

Inducing a TR is a branch of science in itself and there are in fact different approaches to achieve this. Bio-inspired strategies constitute a broad field of study in which the visual pathway, from the retina to the cortical areas related to visual processing, has been modelled using a set of varied paradigms [15,28,31,39]. Spiking neural networks or computational models of biological neural networks, based on the mathematical definition of a single neuronal cell, represent a research area in such a way [9,19,23,41,42]. However, deep learning models [21] have rapidly improved our ability to automate processes and are the most widely used in industry. Deep learning models computationally mimic the down-up strategy of abstracting structural information from images. Layers of artificial neurons in deep convolutional networks progressively capture the intrinsic properties of images at different levels of granularity, from individual edge detection to basic shape compositions, and finally association to one or more categorical artificial neurons. Several task-oriented deep learning models have been developed, such as structural edge detection [22,35], monocular depth estimation [45], object detection [44] or segmentation strategies [29]. All these could be combined to generate a VCP, taking into account that each of them is best suited for specific contexts.

Our proposal is to develop a configurable and wearable system that will generate simulated prosthetic vision (SPV) on sighted volunteers, while on the other hand, perform intracortical stimulation when applied to blind patients, using the VCP. The main aim is to generate an engineering cycle in which it is expected that improvements with sighted volunteers, in combination with TR strategies, will correlate with similar improvements in blind patients. The proposed system allows the implementation of a cybernetic engineering approach where theory can go to reality using "human-in-the-loop" paradigms. Lets explain this approach better as exposed in [30] page 32 (Fig. 1):

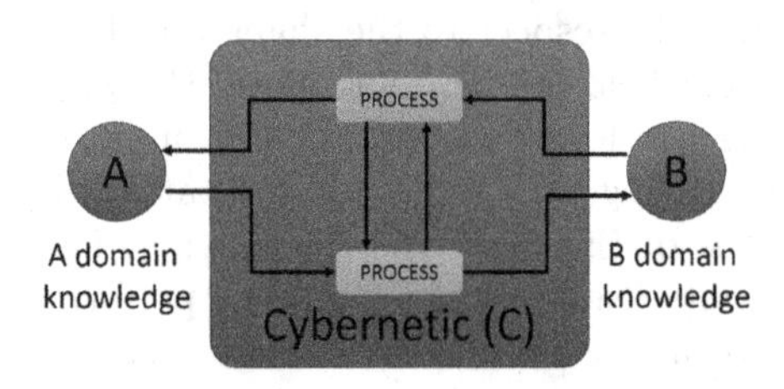


**Fig. 1.** Graphical description of cybernetic action strategy. Adapted from [30] page 32.

1. A phenomenon or process of A is transferred to C.
2. An attempt is made to identify it with one or more other representations of processes of B in C.
3. The knowledge of A is transferred to B, while the codifier is modified and a doctrine, common to both domains, is elaborated.
4. The new cybernetic model-version of the starting process is transferred to domain A and the validity of the results provided by the analogy established in C is verified.

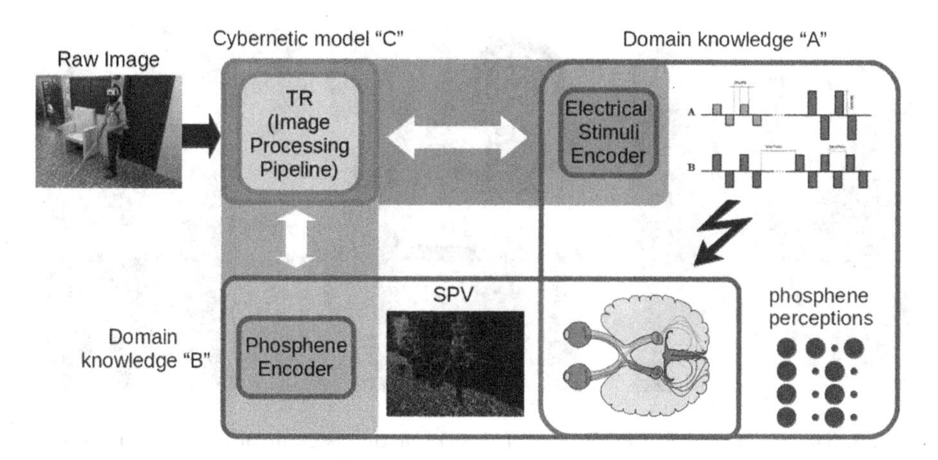


**Fig. 2.** Cybernetic action applied to the present work.

In the present study, see Fig. 2, domain A refers to the method of stimulation of the visual cortex in order to generate a series of punctate phosphenes that resemble the source image to some extent. This knowledge can be achieved experimentally by stimulating the visual cortex using different stimuli configurations while asking the patient to draw the perceived phosphenes [6]. Alternatively, domain B refers to improvements in functional vision when using TR strategies. That is, image processing techniques can be applied in many different ways resulting in the highlight of certain properties of the source image [20], but with varied outcomes in terms of functional vision. The resulting cybernetic model C, allows for the design of the inverse model with regards to domain A. That is, checking which stimuli configurations are needed when looking for a specific perception of phosphenes, and also, reversing the engineering process of designing TR image processing strategies to allow for a better perception of the environment, regarding domain B.

The engineering process of VCP cannot be understood without a proper evaluation methodology which can objectively measure the improvement of blind development in activities of daily life [1], To accomplish these tasks, two different scenarios are proposed. Firstly, a corridor with a series of obstacles is used to measure navigation, and secondly, a controlled but more complex environment which resembles a city square, called StreetLab, both shown in Fig. 3.

Finally, not many devices have been developed that fully address the process of image processing up to phosphene generation in the visual cortex [13,16], although there exist diferent SPV systems [7,20,40]. The present work must be seen as the starting point of the development of a neuro-prosthesis device for intracortical stimulation using a cybernetic approach. Flexibility in design and software modularity is the upmost priority, but also optimisation of the vision processing algorithms with the use of graphical processing units and optimisation libraries. Eye-tracking will also be performed as it gives the system an attention mechanism, which can also be used in order to minimise the amount

**Fig. 3.** Environments used in the present work for the assessment of functional vision. Corridor on the left, StreetLab on the right

of visual information to encode. In order to analyse the possible limitations of the prosthetic system a minimal battery of tests is provided.

## 2 Materials and Methods

The present paper aims to define the VCP design approach. The following subsections explain how in the same device, both strategies, SPV and VCP, can be merged in the same device.

### 2.1 Hardware

To have a system that allows for the different configurations, SPV and VCP, the hardware required is listed below.

1. Logitech c920 Camera for image acquisition.
2. Virtual reality headset (VR).
3. High-resolution screen (4K) embedded in the VR headset.
4. Eye tracking system TObii Pro3 for gaze2d estimation in real-time.
5. Intracortical 96-channel Microelectrode UTAH Array.
6. Stimuli generator and neural recording system, Summit Neuroprocessor, from the Ripple Neuromed Company.
7. Jetson Tx2 embedded AI computing device from NVIDIA.
8. Backpack for portable use.
9. 19 V Battery to power all the system.

Figure 4 shows how each of the key hardware components is connected. The system is comfortable, safe, and allows for complete mobility.

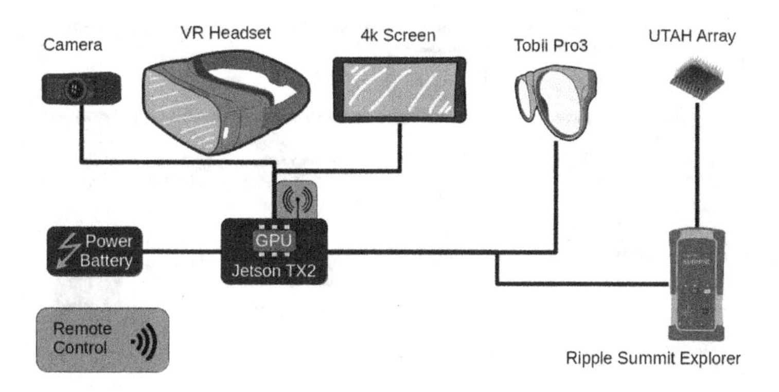


**Fig. 4.** Graphical description of cybernetic action strategy for the present work.

### 2.2 Simulated Prosthetic Vision

Figure 5 shows hardware components required by the SPV system:

There are four main algorithm components for the SPV. First, images are obtained from the frontal camera on the VR headset. Second, eye-tracking position(gaze2D) is computed from the eye-tracking system and a configurable region of interest(ROI) is set around the gaze2D coordinates. Third, the processed image is combined with the related ROI. Image preprocessing is performed depending on the expected outcome(further details in Sect. 2.5). Fourth, phosphene simulation is performed that yields to the SPV to display the outcome on the screen.

### 2.3 Visual Cortical Prosthesis

Figure 5 shows hardware and software components required by the stimuli generator system:

The proposed algorithm follows the SPV strategy although it substitutes the phosphene simulation step with the stimuli generator one.

### 2.4 Graphical Interface Software

Figure 6 shows the graphical interface to set up experiments. It is composed of four main areas. The area labelled "Controls" is used to allow the experimenter to set up the filename of all the data generated, start and stop visualisations and recordings, and to set the source data to be visualised, which can be; raw camera images, monocular depth data, augmented reality from object detection algorithm, or SPV generator. The area labelled "Drivers" is used to activate each of the processes that run in parallel and are used by the underlying algorithms. In this area, webcam, eye tracking, monocular depth, ROI selection, or stimuli generator processes can be activated independently. The third area, labelled "Retina", focuses on the object detection models, as this area requires special

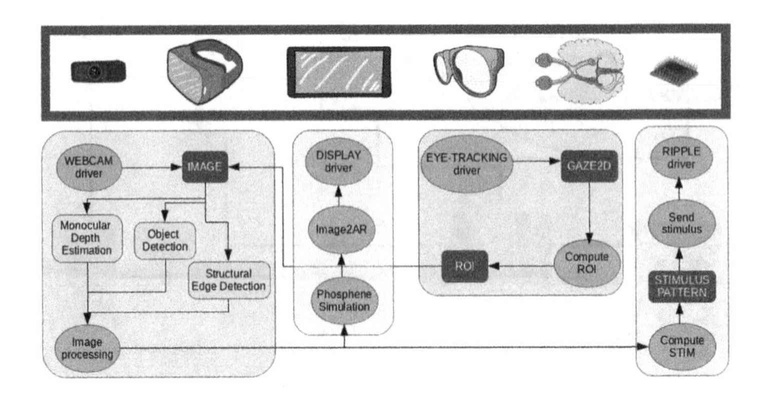


**Fig. 5.** Visualization of the high-level design of the software. It describes the different processes that interact to produce SPV or VCP. This software is intended to be modular and therefore highly modifiable.

consideration, each of the detected objects can be linked with a specific stimuli pattern. Finally, the "Logger" area helps to follow all the internal states the software is performing.

All the code has been written using python 3.6. The PyQt5 library is used for GUI design, the Jetson-inference library for deep learning models optimisation and the Pytorch library for deep learning models definition and optimisation of convolution operation in the GPU for the case of SPV.

## 2.5 Technical Description of Image Processing Strategies

The device is capable of processing simultaneously: raw image acquisition, monocular depth estimation, object detection tasks and gaze2D data acquisition, with the proper use of python multiprocessing tools, to produce the resulting SPV.

SPV is generated from source images obtained with the attached webcam to the headset; (320*240) resolution. Canny edge detection with $threshold = 100$ is computed and the resulting image is convolved ($convolvestep = 8$) with a 2D Gaussian filter; (10*10) resolution and $radius = 3$.

Through the software GUI, source image visualisation is performed at 37.78($\pm$2.84) Frames/s. Monocular depth estimation processing and visualisation is performed in 42.21($\pm$4.54) Frames/s. Object detection processing and visualisation takes 23.99($\pm$2.15). SPV performs in 45.92($\pm$8.32) Frames/s. Eye-tracking system generates data at 100 Hz rate. All this processes running in parallel potentially can be combined to generate different TR strategies without a compromise of the real-time constraints (Fig. 7).

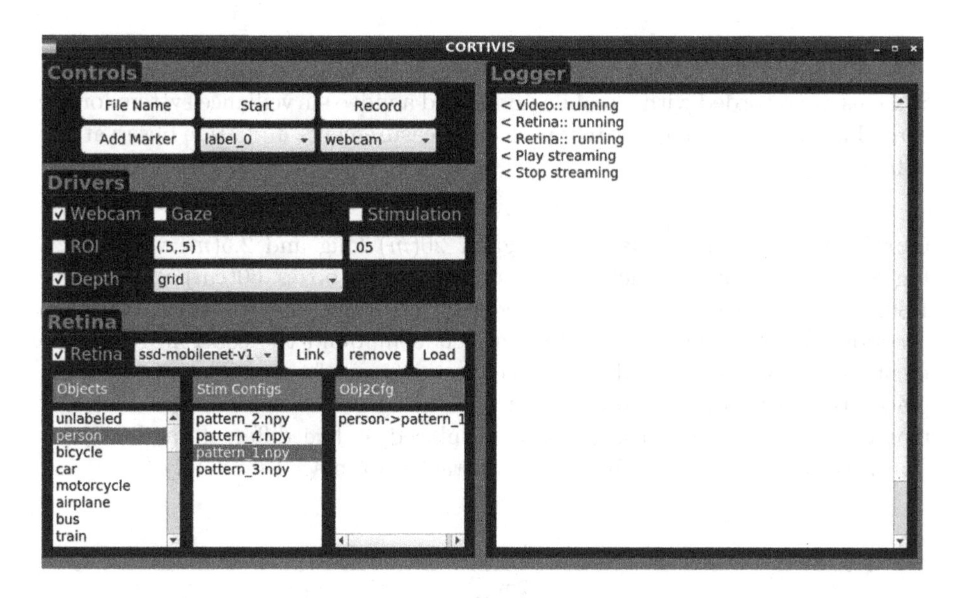


**Fig. 6.** Graphical user interphase developed to set up experiments.

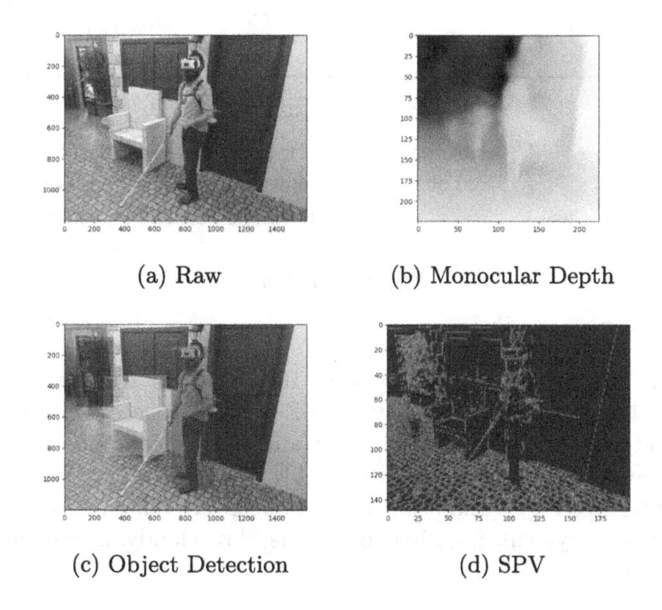

(a) Raw (b) Monocular Depth

(c) Object Detection (d) SPV

**Fig. 7.** a) Raw image from the camera attached to the headset. b) Monocular Depth Estimation processed image. c) Augmented Reality using 'ssd-mobilenet-v1' Object Detection deep learning model. d) SPV image.

### 2.6 Environments

Sessions are recorded with a mobile phone and a video surveillance system for the StreetLab, in order to accurately perform measurements and aside observations offline.

**Corridor.** The corridor is a rectangular 20($m$) long and 2.5($m$) wide space, Fig. 8. Thirteen big obstacles made of cardboard boxes 60($cm$) × 40($cm$) × 1.80($cm$) are placed. Three closed doors on the right and left sides, one open corridor and two safety showers on the left compound the controlled environment. The showers are hidden by obstacles in front of them. The obstacles are placed two meters away from one another along the corridor. As the corridor measures 2.5($m$) wide, obstacles can be placed at five different random positions. At 6($m$), 14($m$) and 20($m$) two obstacles are placed randomly.

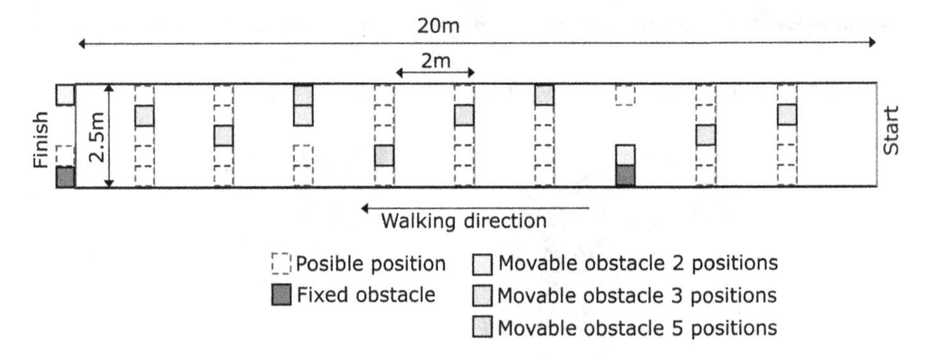


**Fig. 8.** Corridor.

**StreetLab.** The StreetLab is a space of 100($m2$) that mimics a city square, designed for the assessment of activities of daily life. Decorative walls, doors, windows, and obstacles of different heights and shapes, compound this scenario. Besides, three movable wooden boxes of 2.5($m$) × 1($m$) allow the generation of different routes, see Fig. 9. Experiments without interference are recorded with a video surveillance system. Lighting conditions; day, cloudy, afternoon, or night, can also be set.

### 2.7 Test Battery

The experimental paradigm followed for the assessment of mobility and orientation is performed with five volunteers to measure their adaptability. Two conditions in two different scenarios are tested: First, blindfolded and with a cane, and second, using the SPV with a $32 \times 32$ phosphene grid; both conditions are performed five times per volunteer both in the corridor and in the StreetLab.

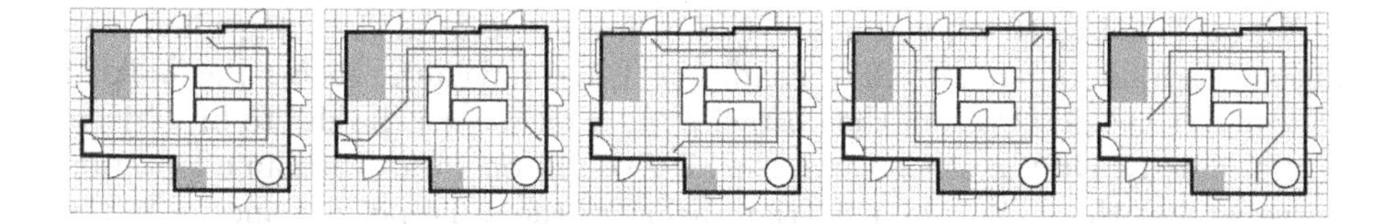

**Fig. 9.** StreetLab.

In the corridor, obstacles are placed randomly in the set of selected areas. For the StreetLab, five different round-trip routes are tested. The blind trials are performed before the SPV trials.

Instructions provided to the participants:

1. Do not use your hands, use the walking cane instead when needed.
2. The researcher will only interfere with the volunteer if safety is not warrantied.
3. Do not interact with the researcher, focus on completing the task.
4. Walk at a pace that you find comfortable.
   - Corridor:
     (a) Reach the end of the corridor.
     (b) Avoid hitting the obstacles.
   - StreetLab:
     (a) Route pathway is explained, i.e.: "Walk straight. Then, turn left and walk until the en of the corridor. Turn left and go ahead. Find the cash machine on your right-hand side against the wall".
     (b) Try to walk down the middle of the corridor.
5. Do you have any questions before starting the route?
6. Let's get started! I will count to three, then you can start walking.
7. Touch the objective to finish the route.
8. Are you ready? So ... one, two, three!

## 3 Results

Functional vision assessment requires a wider range of tests than those offered in the present work. However, in order to provide a proof of concept and a glimpse at the issues to be faced later, the proposed test battery is analysed. The expected outcome should make it possible to assess the adaptive learning curve that occurs when a volunteer learns to use the SPV system. The results will allow to improve the determination of the final battery of tests needed to adequately assess functional vision.

Two basic behavioural measurements, walking time and number of collisions, have been performed. As shown in Fig. 10 a), the use of 5 trials appears to be sufficient to learn to navigate through the simpler corridor environment. However, the SPV condition requires more than five trials since, as can be seen in the figure, the plateau has not been reached.

Regarding Fig. 10 b), the number of collisions performed during blind condition is stable, but not in SPV condition. Blind navigation using the walking cane allows the detection of the obstacles. As a result, it is harder to hit objects with the body. On the other hand, the SPV condition varies with an adaptation curve, and at the end of the five trials, achieves the same performance as in the blind condition.

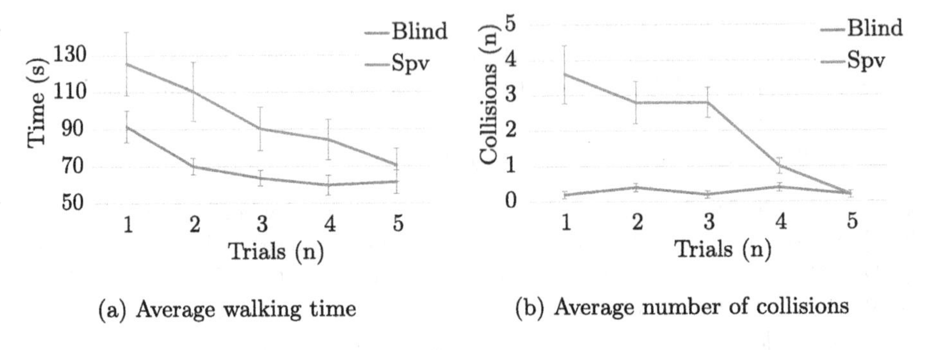


(a) Average walking time

(b) Average number of collisions

**Fig. 10.** Assessment in the corridor for the two conditions: (Blue) Blind with a walking cane, (orange) SPV. a) Average walking time. b) Average number of collisions. (Color figure online)

As shown in Fig. 11 a), walking time in the StreetLab achieves similar performance in both conditions, which is a positive result with regards to the use of the SPV system. About the number of collisions, Fig. 11 b) shows that more collisions are measured depending on the route, making clear that in a realistic scenario, the complexity of each route biases in different ways navigability. There are more collisions for the SPV condition, so TR improvements are required.

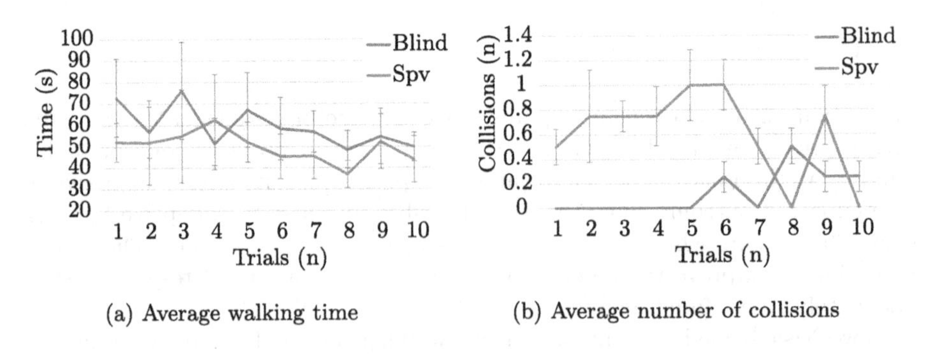


(a) Average walking time

(b) Average number of collisions

**Fig. 11.** a) Mean walking time performance in seconds of the StreetLab assessment. b) Mean number of collisions for each route in the StreetLab.

## 4 Discussion

The term cybernetics, as José Mira explains in [30] page 37, is linked to a series of quasi-axioms of which the most relevant to this work will be listed. First, any system behaviour, the Environment-VCP-Blind system in this case, can be described in mathematical form based on its inputs, internal states, and outputs. Second, the system takes information from the environment, from effectors, the VCP; and from a feedback mechanism that informs it about the efficacy of the action, the behavioural success. Although the development circle has not been completed and only tests have been performed with the SPV system, it is shown that a whole series of TR strategies can potentially be performed in the CVP with the aim of improving the functional vision.

Experiments in very controlled conditions do not represent the scenarios that a blind person will face when performing daily-living activities [1]. Therefore, more realistic scenarios, such as the StreetLab are needed in order properly to evaluate the impact of the VCP. Each environment, in interaction with the TR strategy, provides a different bias effect regarding the adaptation and learning phase [40]. As shown with the gait time measurements in the StreetLab, performance during learning appears to be similar for both conditions, whereas this is not true in the corridor. As a counterpart, the success of VCP in improving the living conditions of the blind is related to the development of brain-computer interface technologies [33]. This issue severely conditions the TR strategies that are to be applied. In conclusion, the field is still making its first steps.

For the present work, canny edge detection followed by convolution using a 2D Gaussian filter was performed. Such a strategy involves harsh and overly simplistic assumptions taking into account the perception that can be achieved with the use of intracortical electrodes attached to cortical area V1. The visuotopic map, along with the magnification factor as a function of the eccentricity [2], must be taken into account to achieve a realistic TR. Eye tracking can alleviate computing constraints by decreasing the amount of information to be computed, using the ROI around the gaze position.

The optimisation of the full device in order to achieve a discrete but useful system that the blind would be comfortable using and carrying could be achieved using neuromorphic architectures instead of Von-Neuman architectures.

Finally, the proposed prototype for the VCP development could be easily adapted to any other visual modality, such as virtual reality, augmented reality, and other transformative reality approaches.

## 5 Conclusions

This work has introduced the concept of cybernetics for engineering and research process of a VCP and SPV system. The use of deep learning libraries optimised for the Jetson Tx2 computer in conjunction with wearable sensors and effectors could enable the development of the VCP in more realistic scenarios. Combined with the set of environments, the corridor, and StreetLab, the VCP system will come closer to the real needs that a blind person requires. Finally, the SPV processing strategy used in the presented work is the starting point, but the system has proven to meet real-time constraints and be highly configurable. In future iterations, more complex TR strategies will be performed and measured concerning functional vision.

**Acknowledgements.** This project has received funding by grant RTI2018-098969-B-100 from the Spanish Ministerio de Ciencia Innovación y Universidades, by grant PROMETEO/2019/119 from the Generalitat Valenciana (Spain), by the European Union's Horizon 2020 research and innovation programme under the Marie Skłodowska-Curie grant agreement No 861423 (enTRAIN Vision) and by grant agreement No. 899287 (project NeuraViPer).

## References

1. Ayton, L.N., et al.: Harmonization of outcomes and vision endpoints in vision restoration trials: recommendations from the international hover taskforce. Transl. Vis. Sci. Technol. **9**(8), 25 (2020)
2. Benson, N.C., Kupers, E.R., Barbot, A., Carrasco, M., Winawer, J.: Cortical magnification in human visual cortex parallels task performance around the visual field. Elife **10**, e67685 (2021)
3. Brindley, G., Lewin, W.: Short-and long-term stability of cortical electrical phosphenes. J. Physiol. **196**, 479–493 (1968)
4. Brindley, G.S., Lewin, W.S.: The sensations produced by electrical stimulation of the visual cortex. J. Physiol. **196**(2), 479–493 (1968)
5. Brindley, G.: Effects of electrical stimulation of the visual cortex. Hum. Neurobiol. **1**, 281–283 (1982)
6. Caspi, A., et al.: Eye movements and the perceived location of phosphenes generated by intracranial primary visual cortex stimulation in the blind. Brain Stimul. **14**(4), 851–860 (2021)
7. Chen, S.C., Suaning, G.J., Morley, J.W., Lovell, N.H.: Simulating prosthetic vision: I. visual models of phosphenes. Vis. Res. **49**(12), 1493–1506 (2009)
8. Chen, X., Wang, F., Fernandez, E., Roelfsema, P.R.: Shape perception via a high-channel-count neuroprosthesis in monkey visual cortex. Science **370**(6521), 1191–1196 (2020)
9. Chou, T.S., et al.: Carlsim 4: an open source library for large scale, biologically detailed spiking neural network simulation using heterogeneous clusters. In: 2018 International Joint Conference on Neural Networks (IJCNN), pp. 1–8. IEEE (2018)
10. Craik, K.J.W.: The Nature of Explanation, vol. 445. CUP Archive, Cambridge (1952)
11. Dobelle, W., Mladejovsky, M.: Phosphenes produced by electrical stimulation of human occipital cortex, and their application to the development of a prosthesis for the blind. J. Physiol. **243**(2), 553–576 (1974)
12. Dobelle, W.H., Mladejovsky, M.G., Girvin, J.: Artificial vision for the blind: electrical stimulation of visual cortex offers hope for a functional prosthesis. Science **183**(4123), 440–444 (1974)
13. Dobelle, W.H.: Artificial vision for the blind by connecting a television camera to the visual cortex. ASAIO J. **46**(1), 3–9 (2000)

14. Dobelle, W.H., Mladejovsky, M.G., Evans, J.R., Roberts, T., Girvin, J.: Braille reading by a blind volunteer by visual cortex stimulation. Nature **259**(5539), 111–112 (1976)
15. Eshraghian, J.K., et al.: Formulation and implementation of nonlinear integral equations to model neural dynamics within the vertebrate retina. Int. J. Neural Syst. **28**(07), 1850004 (2018)
16. Fernández, E., et al.: Visual percepts evoked with an intracortical 96-channel microelectrode array inserted in human occipital cortex. J. Clin. Investig. **131**(23), e151331 (2021). https://doi.org/10.1172/JCI151331
17. Hall, E.C., Gordon, J., Abel, L.A., Hainline, L., Abramov, I.: Nystagmus waveforms in blindness. Vis. Impair. Res. **2**(2), 65–73 (2000)
18. Hartline, H.K.: The response of single optic nerve fibers of the vertebrate eye to illumination of the retina. Am. J. Physiol. Leg. Content **121**(2), 400–415 (1938)
19. Kumarasinghe, K., Kasabov, N., Taylor, D.: Deep learning and deep knowledge representation in spiking neural networks for brain-computer interfaces. Neural Netw. **121**, 169–185 (2020)
20. Li, W.: Wearable computer vision systems for a cortical visual prosthesis. In: Proceedings of the IEEE International Conference on Computer Vision Workshops, pp. 428–435 (2013)
21. Liu, L., et al.: Deep learning for generic object detection: a survey. Int. J. Comput. Vis. **128**(2), 261–318 (2020)
22. Liu, Y., Cheng, M.M., Fan, D.P., Zhang, L., Bian, J.W., Tao, D.: Semantic edge detection with diverse deep supervision. Int. J. Comput. Vis. **130**(1), 179–198 (2022)
23. Lozano, A., Soto-Sanchez, C., Garrigos, J., Martínez, J.J., Ferrández, J.M., Fernandez, E.: A 3d convolutional neural network to model retinal ganglion cell's responses to light patterns in mice. Int. J. Neural Syst. **28**(10), 1850043 (2018)
24. Lui, W.L.D., Browne, D., Kleeman, L., Drummond, T., Li, W.H.: Transformative reality: augmented reality for visual prostheses. In: 2011 10th IEEE International Symposium on Mixed and Augmented Reality, pp. 253–254. IEEE (2011)
25. Lui, W.L.D., Browne, D., Kleeman, L., Drummond, T., Li, W.H.: Transformative reality: improving bionic vision with robotic sensing. In: 2012 Annual International Conference of the IEEE Engineering in Medicine and Biology Society, pp. 304–307. IEEE (2012)
26. McCulloch, W.S.: Embodiments of Mind. MIT press, Cambridge (2016)
27. McCulloch, W.S., Pitts, W.: A logical calculus of the ideas immanent in nervous activity. Bull. Math. Biophys. **5**(4), 115–133 (1943)
28. Melanitis, N., Nikita, K.S.: Biologically-inspired image processing in computational retina models. Comput. Biol. Med. **113**, 103399 (2019)
29. Minaee, S., Boykov, Y.Y., Porikli, F., Plaza, A.J., Kehtarnavaz, N., Terzopoulos, D.: Image segmentation using deep learning: a survey. IEEE Trans. Pattern Anal. Mach. Intell. (2021)
30. Mira, J.: Entre lo biológico y lo artificial: selección de publicaciones: con 24 escritos originales de colaboradores del autor. Red Temática en Tecnologías de Computación Artificial/Natural (RTNAC) (2014)
31. Muratore, D.G., Chichilnisky, E.J.: Artificial retina: a future cellular-resolution brain-machine interface. In: Murmann, B., Hoefflinger, B. (eds.) NANO-CHIPS 2030. TFC, pp. 443–465. Springer, Cham (2020). https://doi.org/10.1007/978-3-030-18338-7_24

32. Oswalt, D., et al.: Multi-electrode stimulation evokes consistent spatial patterns of phosphenes and improves phosphene mapping in blind subjects. Brain Stimul. **14**(5), 1356–1372 (2021)
33. Pio-Lopez, L., Poulkouras, R., Depannemaecker, D.: Visual cortical prosthesis: an electrical perspective. J. Med. Eng. Technol. **45**(5), 394–407 (2021)
34. Polimeni, J.R., Balasubramanian, M., Schwartz, E.L.: Multi-area visuotopic map complexes in macaque striate and extra-striate cortex. Vis. Res. **46**(20), 3336–3359 (2006)
35. Poma, X.S., Riba, E., Sappa, A.: Dense extreme inception network: towards a robust CNN model for edge detection. In: Proceedings of the IEEE/CVF Winter Conference on Applications of Computer Vision (WACV) (March 2020)
36. Rosenblueth, A., Wiener, N., Bigelow, J.: Behavior, purpose and teleology. Philos. Sci. **10**(1), 18–24 (1943)
37. Schiller, P.H., Tehovnik, E.J.: Visual prosthesis. Perception **37**(10), 1529–1559 (2008)
38. Schultze, M.: The retina is the membrane-like terminal expansion of the. Man. Hum. Comparat. Histol. **3**, 218 (1873)
39. Shah, N.P., Chichilnisky, E.: Computational challenges and opportunities for a bi-directional artificial retina. J. Neural Eng. **17**(5), 055002 (2020)
40. Van Steveninck, J.D.R., et al.: Real-world indoor mobility with simulated prosthetic vision: the benefits and feasibility of contour-based scene simplification at different phosphene resolutions. J. Vis. **22**(2), 1 (2022)
41. Subbulakshmi Radhakrishnan, S., Sebastian, A., Oberoi, A., Das, S., Das, S.: A biomimetic neural encoder for spiking neural network. Nat. Commun. **12**(1), 1–10 (2021)
42. Tavanaei, A., Ghodrati, M., Kheradpisheh, S.R., Masquelier, T., Maida, A.: Deep learning in spiking neural networks. Neural Netw. **111**, 47–63 (2019)
43. Wiener, N.: Cybernetics or Control and Communication in the Animal and the Machine. Technology Press, Cambridge (1948)
44. Zaidi, S.S.A., Ansari, M.S., Aslam, A., Kanwal, N., Asghar, M., Lee, B.: A survey of modern deep learning based object detection models. Digital Signal Process. **126**, 103514 (2022)
45. Zhao, C.Q., Sun, Q.Y., Zhang, C.Z., Tang, Y., Qian, F.: Monocular depth estimation based on deep learning: an overview. Sci. China Technol. Sci. **63**(9), 1612–1627 (2020). https://doi.org/10.1007/s11431-020-1582-8

# The Assessment of Activities of Daily Living Skills Using Visual Prosthesis

Dorota Waclawczyk(✉), Leili Soo, Mikel Val, Roberto Morollon, Fabrizio Grani, and Eduardo Fernandez

Bioengineering Institute, University Miguel Hernández, 03202 Elche, Spain
dwaclawczyk@umh.es

**Abstract.** In the field of prosthetic vision, functional assessment refers to tests that capture a person's ability to use vision to perform everyday tasks. That includes assessments ranging from basic psychophysical tests of light perception and discrimination to performance-based tests. Visual prosthesis devices are not able to provide high-resolution visual acuity, hence standard vision tests are not sufficient to measure post-intervention improvements in vision. The lack of validation of assessment and reporting of patient wellbeing challenge interpreting the outcomes of clinical studies. Here we review the techniques of Activities of Daily Living assessment as well as the post-implantation training techniques. We explain why there is a need for an update and for a standard method which will enable the comparison of the results between research groups.

**Keywords:** Prosthetic vision · Activities of daily living

## 1 Introduction

The artificial restoration of vision in blind people by direct electrical stimulation has been the subject of research since the early 1970s [1,2]. Electrical stimulation of visual pathways induces discrete spots of light, called phosphenes. Phosphenes have a potential to be used as building blocks to compose a visual scene in blind, previously sighted patients. The ideal device should capture real time images and electrically stimulate the visual cortex to create dotted perceptions that are representing the scene accurately. The site of stimulation depends on the pathology of the blindness and is chosen in a way to bypass the interrupted segment of the visual pathway in normal signal flow [3]. Current visual prosthesis approaches include the electrical stimulation of the retina (subretinal, epiretinal, suprachoroidal), optic nerve, lateral geniculate nucleus of the thalamus, optic radiations, and cortical neurons [4]. Regardless of the differences in strategies for the acquisition of optical information and image processing, most visual prostheses share a common set of components. A head mounted camera that provides information of the visual space in front of the blind user, and a processor that transforms it into patterns of electrical stimulation and sends them to multiple microelectrodes situated at chosen part of the visual pathway


J. M. Ferrández Vicente et al. (Eds.): IWINAC 2022, LNCS 13258, pp. 395–404, 2022.
https://doi.org/10.1007/978-3-031-06242-1_39

[3]. Device that utilises phosphenes appears as a demanding but feasible way to sight restoration; visual prostheses have achieved significant development in recent years. Current research has advanced to a point where there are already commercially available devices [5,6]. However at the present stage of development they are able to restore only a very limited vision, with relatively low spatial resolution, lack of stereoscopic depth perception, and they are implanted only in one eye.

**Assessing the Benefit of the Prosthesis.** Blindness is one of the most debilitating disorders as quality of life is seriously deteriorated. Beginning from the physical limitations as reduced mobility, to psychological perspective, like insomnia, social isolation, or even suicidal thoughts [7]. The overall goal of developing a visual prosthesis is to enhance the quality of recipients' life by making previously difficult tasks easier, giving them at least some useful vision [8]. The degree to which the subjects can effectively use artificial vision in everyday life should be the determining factor in the success of a visual prosthesis [9]. The challenge is also learning how to use this kind of vision and how to quantify the improvement to differentiate the results between different groups and device approaches. The need for post-implantation rehabilitative strategies that will teach the subject how to use the prosthetic vision is pressing. The more visual prostheses devices are tested in clinical trials, the more important is the need for the standardised assessment to evaluate the effectiveness of vision restoration [10]. Prognosis of visual functions provided by such devices on the basis of technical specifications alone is not possible; novel test strategies are needed to comprehensively describe visual performance. It is indispensable to develop a specialised set of tests for evaluating visual functions in patients with artificial vision [11,12]. The goal should be to focus not only on technical specifications, but to reflect the patients benefits in daily living and provide a training that will help understanding of the functioning of the device.

**Assessment Provided by Existing Methods.** Vision provided by the electrical input to the visual cortex differs considerably from natural vision experienced prior to blindness. Hence the tests used for assessing natural vision are not suitable. The specialised tests for the assessment of very low vision, such as Snellen acuity, IADL-VLV [13], LoVE [14], have poor prediction in the context of artificial vision. This is mainly due to the inability of the prosthesis to provide high-resolution visual acuity [15]. The companies producing commercially-approved visual prosthesis claim that the visual prostheses can provide useful vision by allowing recipients to distinguish and interpret patterns, recognize outlines of people, basic shapes and navigate more independently through the world [5,6]. However, the literature reports that no visual prostheses recipient has achieved projected acuity goals, and even the best performing recipients have not improved to the level of "legal blindness" on standard measures of acuity [16]. The efficacy of commercially available Alpha IMS subretinal implant and the Second Sight Argus II epiretinal implant was evaluated using The Basic Assessment of Light and Motion (BaLM). BaLM is a standardised clinical method which assesses light perception, visual temporal resolution, object localization, and movement detection [17] providing a quantitative assessment of visual function in the ultra-low-vision range. Nonetheless, lacking measurement of improvement recipients quality of life, BaLM is not sufficient for assessing the visual prostheses.

Effectiveness of visual prostheses is measured by assessing the visual function, orientation and mobility, and activities of daily living (ADL). ADL refers to a series of self-care tasks that are essential for maintaining independence [17], and forms part of quality-of-life assessment. ADLs may be measured by a combination of self-report, researcher report, and direct observation. A combination of self-report and performance-based measures of ADL may be the best way to fully capture the picture of disability for a given individual [29]. In clinical practice the most commonly used test of activities of daily living is Katz Index of Independence [19]. This evaluation is designed for older adults with severely reduced physical mobility, hence does not correspond with ultra-low vision cases. Melbourne Low Vision ADL Index MLVAI [20,21] was developed to investigate the relationship between vision impairment and performance of activities of daily living, though is not adjusted for visual prostheses users. For the subjects whose vision has been partially restored with the Argus II, a pilot assessment was developed: FLORA (the Functional Low-Vision Observer Rated Assessment) [22]. FLORA focuses on subjects everyday lives and captures the functional vision ability and well-being of subjects. Although it is time-consuming, it includes tasks related not only to vision, and it is not suitable for the vision provided by different device approaches. Different groups are trying to develop novel test methods for ultra-low vision, including perception of light, light localization, direction of motion, as well as real-world functional tasks such as object localization and recognition, sock sorting, sidewalk tracking and walking direction tests [5,6,23]. The question which measures of visual function should be used to assess the outcomes of these early-stage devices also remains open.

**The Need of Standardised Tool for Assessment and Training.** Comparative analysis between devices' outcome is difficult given the novelty of the tasks and no consistency between the groups. Several guidelines have been released [17,24,25] in order to address the issue of heterogeneity in outcomes. To generate consensus on the methods of testing and reporting outcomes in vision restoration trials the HOVER taskforce [17] was developed. The following table summarises the most important components that should be included in the assessment.

Based on that recommendation, the test battery should be designed on three pillars. The first pillar is based on classical, highly standardised psychophysical testing, which contributes to good reliability, validity, and objectiveness. The second pillar consists of a testing performance in everyday tasks, which helps to estimate the efficacy of the device. The third one is meant to get the impression of a patient's own feelings. It enhances the usefulness of the device by providing authentic and quantifiable evidence how the visual prosthesis improves visual

**Table 1.** Summary of the most important component of ADL assessment

| | |
|---|---|
| Pillars of the test [25] | – Standard psychophysical testing<br>– Performance in task that are used in real-life situations<br>– Questionnaires that assess the patient's own feelings<br>– Clinical feasibility (limiting the duration of the test) |
| Items that should be tested [26] | – Standardisation of ambient illumination<br>– Correlation between individual items<br>– Instruction and item presentation should be as similar as possible<br>– Standardisation of ambient illumination<br>– Criterion-free test procedures (forced choice testing)<br>– Allowing patients use visual aids they customarily use |
| General prescriptions [17] | – Task performance must be recorded<br>– Task performance in terms of both speed and accuracy<br>– Performance measure on a continuous scale (if not quantization step as small as possible)<br>– Assessment of all tasks with and without input of the prosthetic device |

function to make a difference to patients' lives and provides a high reliability of the test [25,26]. To enhance reliability of the test, Lauren et al. [17] proposed comparing all the results with the device ON and OFF. Sharon et al. [20] suggested determining the normal time taken to perform each observed item for a group of age-matched subjects with normal vision. This should enhance scale of the performance by assigning an appropriate range of times to each rating level and later on to enhance the reliability. The usefulness of validated questionnaires to assess the psychosocial characteristics of potential visual prosthesis recipients has not been investigated to date. Such research could also facilitate the development of a test protocol to identify individuals who might be most suitable for retinal prosthesis trials [24]. To be suitable for evaluating the psychological condition in blind patients during the first clinical trial, possibly facing tremendous expectations and disappointments, such questionnaires should meet special requirements. For that Wilke et al. [11] proposed using the Brief Symptom Inventory test [27].

To minimise the effect of the experimenter, the instructions given to the participants should be clear and well repeated to all participants. The open questions may affect the given answer. A forced-choice test requires the participant to choose among two or more alternatives, only one of which is the correct answer. Without it is difficult to distinguish changes in visual function from changes in motivation (willingness to guess the correct answer). Forced-choice testing procedures should be used whenever possible to reduce the influence of criterion effects, as it minimizes biases due to differences among participants and their willingness to guess [17,28].

Goals of any kind of therapy should be: specific, measurable, achievable, relevant, and time limited. Additionally, clinical assessment needs to be reliable and valid. The test should be meaningful with a high test-retest reproducibility and as objective as possible. It has to be sensitive enough to detect even small changes in visual performance. The assessment should not only measure the technical specifications, but also reflect the patients benefit in daily living and help the experimenters to gain the better ability of understanding artificial vision devices for the future. Tests should be carried under controlled conditions to detect even slight improvements. Phosphene perception differs significantly between recipients - each recipients' qualitative experience is distinct. Each process should be tailored to the individual recipient and suitable for generic use to facilitate comparison of the functional results of different technical approaches in humans [11].

**Activities of Daily Living.** Recipients of the visual prostheses, during years of being completely blind, have developed strategies to manage the various tasks of day-to-day living to remain as independent as possible. Even though some participants are much more able to cope with severe vision loss than others. Furthermore, activities of daily living are intertwined with medical, psychological and social well-being, which makes independent measurement exceedingly difficult [21]. Restoring basic visual functions by visual prostheses will not necessarily improve these patients' ability to master everyday tasks. Nevertheless, improvement in the ability to pursue such activities would be a valuable endpoint in determining the efficacy of a visual prosthesis and could turn out to be more relevant than mere visual resolution values [11]. When designing a new ADL method, the first question that needs to be answered is what tasks or activities should be included. Many ADLs, including pouring water, can be performed with little if any visual input [17]. Many tasks, such as clock reading, can be solved by using other devices. Testing performance in daily activities by their nature is subjective and influenced by habituation [11].

**Ecological Validity.** Developing a test to measure performance on ADLs puts restrictions between the need to standardise testing conditions and ecological validity [17]. It's obvious that functional vision tests needs to be carefully standardised, but it is less obvious that functional tests can be over standardised [28]. Ecological validity refers to the link between the laboratory measurement and the participant's performance of similar real-world tasks, to ensure that the test is truly representative of everyday activities. A well-constructed performance-based task may be administered under carefully standardised lighting conditions on a clean, dry, level surface, or composed of uniform black and white components, but to be ecologically valid it should predict performance in the real world [28]. "Minimally Important Difference" (MID), specifies the smallest difference in a test score that is still sufficient to make a difference to the participant, either beneficial or harmful [30]. It is important to find a balance between standardisation, to ensure that the tests are reproducible across sites, and natural conditions, to ensure that the tests reflect performance in real-world conditions is a demanding task. To detect even slight improvements tests should be carried out under controlled conditions. Observation of the patient using their device in their home plays an important role in outcome assessment [22] but controlled laboratory-based testing is pivotal in order to fully understand the capabilities and limitations of a visual prostheses.

For the object recognition task, the Alpha-IMS group used a black table surface as a backdrop for geometric shapes, dining objects (such as cups and cutlery) [23]. All of the objects are uniformly white, which severely limits the validity of the test. It is questionable whether the object recognition test is generalizable to objects in the real world; whether it is ecologically valid. Even the rudimentary shape of vision can help a previously blind subject in being independent. Thorn et al. [31] explains how localisation of a single light source can be used by a person with ultra-low vision to orient and navigate more efficiently than if totally blind. However, it is questionable how much the requirements of the task can be simplified by increasing the size and contrast of visual stimuli. Second sight implanted recipients [23] were asked to walk along a bold white stripe on a black floor. The tests were designed in the way possible for a patient to accomplish. It is questionable if a strongly modified version of the real-world task still represents a realistic assessment of daily life activities. An outcome on that type of specialised test may be useful for rehabilitation tasks of visually guided behaviour, though should not be used to make an assertion about the ability of a patient with prosthetic vision to perform tasks of daily living.

**Post-implantation Training Program.** The standard procedure is needed in order to fully realise the potential of visual prosthesis devices and to get the ability to extract the most relevant visual features from a complex and dynamic visual environment. The post-implantation training should focus on teaching the recipients to understand what they are seeing and remodel new viewing habits to maximise the utility of the device [32]. Users need extensive education in regards to what prosthetic vision intervention can and cannot do. Having realistic expectations is vital to the success of rehabilitation [11]. The degree to which the subjects can effectively use artificial vision in everyday life should be the determining factor in the success of a visual prosthesis [9]. The relationship between vision and performance is complex, because both entities are multidimensional. In performance-based measures the information obtained can inform understanding of measures presently used to assess patients, such as visual acuity, visual field and instruments designed to study quality of life [26]. The proposed tasks should be performance-based, as they will be the crucial part of the process of adaptation to prosthetic vision. Time is required for patients to complete their adjustment to the prosthesis and reintegrate all the sensory input that it provides. Future vision rehabilitation programs should also emphasise the integration of prosthetics with other sensory information to reduce the mental effort [31].

**Virtual Reality.** First and foremost, limitation in building the standardised assessment is the number of recipients limited by small study populations. More complete accounts of recipients' perceptual experience can inform training and rehabilitation strategies, as well as aid in the development of a more accurate model of artificial vision. The better vision rehabilitation specialists are able to understand the recipient's qualitative experience, the more they are able to assist them in learning to use and live with the device [16]. To improve the performance of the participant and better understand the sensation of prosthetic vision the pilot test can be performed on sighted, simulated subjects [8,33]. The simulated vision implanted in a virtual reality environment provides a real-life experience. Subject is wearing a head-mounted low vision system that generates images similar to the one we expect in visual prostheses recipients. Virtual reality (VR) has advantages over other forms of ADL rehabilitation offering the potential to develop a human performance testing and training environment. Using VR technology, we can produce applications to test and assess patients in ways relevant to daily living, which provides a level of realism elusive by other techniques, and which have the potential to teach skills of practical relevance [8]. Complete control over content is possible, and performance data may be stored in a database. Moreover, VR provides patients with added motivation by adding gaming factors in a safe virtual environment that eliminates risks caused by errors [34]. As disussed above, functional assessment must strike a balance between standardisation, to ensure that the tests are reproducible across sites, and natural conditions, to ensure that the tests reflect performance in real-world conditions [15]. Validate standardised test batteries of functional vision assessment for ADL and Orientation and Mobility can be designed in VR and easily distributed between laboratories approaching in visual prosthesis.

**General Recommendations.** The selected set of tasks should be based and limited to activities that could be observed, standardised, timed, pilot tested, randomised and prospective. They should be achievable for cortical VP users after some training and show the different difficulty levels. Should be neither too difficult nor too easy, designed to mimic the real-world scenario and encompass aspects of everyday life which aims to make VP users more independent. The design of experiments should show if volunteers can learn to integrate the electrical stimulation of brain visual areas into meaningful percept. Proposed task should minimise the tactile input, sound and proprioception that may determine performance of these tasks much more than visual function.

The post-implantation training and assessment test should start as soon as possible after implantation, however it is important to consider that initially the subject has difficulties in the discrimination between spontaneous and evoked phosphenes. Any future experiment should take into account the initial learning period. It takes approximately two months to overcome this impediment and begin the meaningful stimulation experiments [35]. The intervention must last for multiple sessions so that older adults have sufficient time to adopt new knowledge and skills into daily activities.

A popular strategy for designing a functional assessment is to take a hierarchical approach, beginning with the simplest and most basic visual abilities - light detection and localisation - and moving up through more complex vision tasks - motion detection, and resolution (acuity) before moving on to everyday visual activities such as navigation and object recognition [28].

## 2 Conclusion

The need for an ADL test battery for persons with severe vision loss has become urgent over the past few years, with the commencement of clinical trials for vision restoration techniques. Albeit for the moment it is hard to compare the results between different laboratories. In this paper, we reviewed the existing methods of assessment of functional vision and pointed out that there is a need to update assessment of Activities of Daily Living and develop a standard method of testing performance in everyday activities. The subject-tailored methodology of an assessment should capture the activities of daily living at the level of vision provided by visual prostheses. Functional assessment must strike a balance between standardisation - to ensure that the tests are relevant and reproducible across sites, and natural conditions - to ensure that the tests reflect performance in real-world conditions. In addition to testing specifications and performances of visual functions with visual prostheses, it is important to evaluate the subject's well-being and its implications for test performances. The battery test should be appropriate to the patient, include replicable tasks from day-to-day life and make possible checking if performance of the task is easier or faster with the implant. As there is no direct physical correlation that can be measured, the test has to rely on indirect measuring techniques. The test battery should be built on three pillars: Activities of Daily Living, Psychophysical testing, Questionnaire. Performance-based functional outcomes should be considered in conjunction with patient-reported outcomes. There is a need for the standardised tool that can be used in clinical trials of devices and blindness treatments, to demonstrate the impact of the treatment, and to identify areas where rehabilitation might increase the functional skills at the level of vision provided by the visual prostheses. In the near future the standardised assessment for orientation and mobility for visual prostheses users should also be developed.

**Acknowledgements.** We would like to thank B.G. and her husband for their extraordinary commitment to this study. This project has received funding by grant RTI2018-098969-B-100 from the Spanish Ministerio de Ciencia Innovación y Universidades, by grant PROMETEO/2019/119 from the Generalitat Valenciana (Spain), by the European Union's Horizon 2020 research and innovation programme under the Marie Skłodowska-Curie grant agreement No 861423 (enTRAIN Vision) and by grant agreement No. 899287 (project NeuraViPer).

## References

1. Brindley, G.S., Lewin, W.S.: The sensations produced by electrical stimulation of the visual cortex. J. Physiol. (Lond.) **196**, 479–93 (1968)
2. Dobelle, W.H., Mladejovsky, M.G., Girvin, J.P.: Artificial vision for the blind: electrical stimulation of visual cortex offers hope for a functional prosthesis. Science **183**, 440–4 (1974)
3. Fernandez, E.: Development of visual neuroprostheses: trends and challenges. Bioelectron. Med. **4**, 12 (2018)
4. Niketeghad, S., Pouratian, N.: Brain machine interfaces for vision restoration: the current state of cortical visual prosthetics. Neurotherapeutics **16**(1), 134–143 (2019)
5. Humayun, M.S., et al.: Argus II study group.: interim results from the international trial of second sight's visual prosthesis. Ophthalmology **119**(4), 779–788 (2012)
6. Stingl, K., et al.: Artificial vision with wirelessly powered subretinal electronic implant alpha-IMS. Proc. R. Soc. B Biol. Sci. **280**(1757), 20130077 (2022)
7. Barriga-Rivera, A., Bareket, L., Goding, J., Aregueta-Robles, U.A., Suaning, G.J.: Visual prosthesis: interfacing stimulating electrodes with retinal neurons to restore vision. Front. Neurosci. **14**(11), 620 (2017)
8. Srivastava, N.R., Troyk, P.R., Dagnelie, G.: Detection, eye-hand coordination and virtual mobility performance in simulated vision for a cortical visual prosthesis device. J Neural Eng. **6**(3), 035008 (2009)
9. Titchener, S.A.: Oculomotor Behavior and Perceptual Localization in Retina Prostheses, PhD thesis (2020)
10. Merabet, L.B., Rizzo, J.F., Pascual-Leone, A., Fernandez, E.: Who is the ideal candidate?: decisions and issues relating to visual neuroprosthesis development, patient testing and neuroplasticity. J Neural Eng. **4**(1), S130-5 (2007)
11. Wilke, R., Bach, M., Wilhelm, B., Durst, W., Trauzettel-Klosinski, S., Zrenner, E.: Testing visual functions in patients with visual prostheses. In: Humayun, M.S., Weiland, J.D., Chader, G., Greenbaum, E. (eds.) Artificial Sight. Biological and Medical Physics, Biomedical Engineering, pp. 91–110. Springer, New York, NY (2007). https://doi.org/10.1007/978-0-387-49331-2_5
12. Stelmack, J.A., Stelmack, T.R., Massof, R.W.: Measuring low-vision rehabilitation outcomes with the NEI VFQ-25. Invest. Ophthalmol. Vis. Sci. **43**(9), 2859–68 (2002)
13. Finger, R.P., et al.: Developing an instrumental activities of daily living tool as part of the low vision assessment of daily activities protocol. Invest. Ophthalmol. Vis. Sci. **55**(12), 8458–8466 (2014)
14. Tamai, M., et al.: An instrument capable of grading visual function: results from patients with retinitis pigmentosa. Tohoku J. Exp. Med. **203**(4), 305–312 (2004)
15. Ayton, L.N., Rizzo, J.: Assessing patient suitability and outcome measures in vision restoration trials. In: Gabel, V.P. (ed.) Artificial Vision, pp. 3–8. Springer, Cham (2017). https://doi.org/10.1007/978-3-319-41876-6_1

16. Erickson-Davis, C., Korzybska, H.: What do blind people see with retinal prostheses? Observations and qualitative reports of epiretinal implant users. PLoS One **16**(2), e0229189 (2021)
17. Ayton, L.N., et al.: HOVER international taskforce. harmonization of outcomes and vision endpoints in vision restoration trials: recommendations from the international hover taskforce. Transl. Vis. Sci. Technol. **9**(8), 25 (2020)
18. Bach, M., Wilke, M., Wilhelm, B., Zrenner, E., Wilke, R.: Basic quantitative assessment of visual performance in patients with very low vision. Invest. Ophthalmol. Vis. Sci. **51**(2), 1255–60 (2010)
19. Shelkey, M., Wallace, M.: Katz index of independence in activities of daily living. Home Healthc. Nurse **19**, 323–3240 (2001)
20. Haymes, S.A., Johnston, A.W., Heyes, A.D.: The development of the melbourne low-vision ADL index: a measure of vision disability. Invest. Ophthalmol. Vis. Sci. **42**(6), 1215–1225 (2001)
21. Haymes, S.A., Johnston, A.W., Heyes, A.D.: Relationship between vision impairment and ability to perform activities of daily living. Ophthalmic Physiol. Opt. **22**, 79–91 (2002)
22. Stingl, K., et al.: Subretinal visual implant alpha ims-clinical trial interim report. Vis. Res. **111**(Pt B), 149–60 (2015)
23. Bentley, S.A., et al.: Psychosocial assessment of potential retinal prosthesis trial participants. Clin. Exp. Optom. **102**(5), 506–512 (2019)
24. Chader, G.J., Weiland, J., Humayun, M.S.: Artificial vision: needs, functioning, and testing of a retinal electronic prosthesis. Prog. Brain Res. **175**, 317–32 (2009)
25. Wei, H., et al.: A clinical method to assess the effect of visual loss on the ability to perform activities of daily living. Br. J. Ophthalmol. **96**(5), 735–41 (2012)
26. Boulet, J.: Reliability and validity of the brief symptom inventory. Psychol. Assess. **3**(3), 433–437 (1991)
27. Rubin, G.S.: Functional assessment of artificial vision. In: Gabel, V.P. (ed.) Artificial Vision, pp. 9–19. Springer, Cham (2017). https://doi.org/10.1007/978-3-319-41876-6_2
28. Chader, G.J.: Artificial vision. Prog Brain (2009)
29. Bravell, M.E., Zarit, S.H., Johansson, B.: Self-reported activities of daily living and performance-based functional ability: a study of congruence among the oldest old. Eur. J. Ageing **8**(3), 199–209 (2022)
30. Guyatt, G.H., Osoba, D., Wu, A.W., Wyrwich, K.W., Norman, G.R.: Clinical significance consensus meeting group.: methods to explain the clinical significance of health status measures. Mayo Clin. Proc. **77**(4), 371–83 (2002)
31. Geruschat, D.R., Deremeik, J.: Activities of daily living and rehabilitation with prosthetic vision. In: Dagnelie, G. (eds.) Visual Prosthetics, pp. 413–425. Springer, Boston, MA (2011). https://doi.org/10.1007/978-1-4419-0754-7_21
32. Chen, S.C., Suaning, G.J., Morley, J.W., Lovell, N.H.: Rehabilitation regimes based upon psychophysical studies of prosthetic vision. J. Neural Eng. **6**(3), 035009 (2009)
33. Thorn, J.T., Migliorini, E., Ghezzi, D.: Virtual reality simulation of epiretinal stimulation highlights the relevance of the visual angle in prosthetic vision. J. Neural Eng. **17**(5), 056019 (2020)
34. Lee, J.H., et al.: A virtual reality system for the assessment and rehabilitation of the activities of daily living. Cyberpsychol. Behav. **6**(4), 383–8 (2003)
35. Fernández, E., et al.: Visual percepts evoked with an intracortical 96-channel microelectrode array inserted in human occipital cortex. J. Clin. Invest. **131**(23), e151331 (2021)

# Affective Analysis

# Artificial Intelligence Applied to Spatial Cognition Assessment

Michela Ponticorvo[1(✉)], Mario Coccorese[1], Onofrio Gigliotta[1], Paolo Bartolomeo[2], and Davide Marocco[1]

[1] Department of Humanistic Studies, University of Naples "Federico II", Naples, Italy
michela.ponticorvo@unina.it

[2] Sorbonne Université, Institut du Cerveau - Paris Brain Institute - ICM, Inserm, CNRS, AP-HP, Hôpital de la Pitié-Salpêtrière, Paris, France

**Abstract.** Spatial cognition is a function that strongly affects adaptation. This is particularly evident when it is impaired, as often happens after brain injury.

Neglect, or hemispatial visual neglect, is a dramatic consequence of right hemisphere damage that leads patient to ignore the left, controlateral part of the space. It is assessed with tasks and tests that require to direct attention on the whole visual field, both on left and right. Also in healthy people, spatial exploration is not perfectly symmetrical, as witnessed by the phenomenon called pseudo-neglect.

In recent years, these tools have been enhanced by new technological solutions, producing new data.

In this paper, we describe our attempt to use Artificial Intelligence for the assessment of spatial cognition starting from the enhanced version of the Baking Tray Task, the e-BTT.

Results indicate that Artificial Intelligence can be an effective method to analyze these new data thus leading to a more comprehensive assessment.

**Keywords:** Spatial tasks · Neglect · Assessment · Artificial Intelligence · Enhanced tools

## 1 Introduction

Spatial abilities are fundamental for the adaptation of moving organisms [2,25], as they allow to direct attention, to explore and to act in the environment. The importance of these abilities becomes even more evident in the case of impairment deriving from brain injury [29].

Neglect, or hemispatial visual neglect, is a dramatic consequence of right hemisphere damage that leads patient to ignore the left, controlateral part of the space [6,7,37]. Patients with neglect omit almost everything that is controlateral to the injured party - for example following a stroke - and this can emerge using specific tests. We can distinguish 3 forms of neglect which, generally, occur together [15,18,26,30]:


J. M. Ferrández Vicente et al. (Eds.): IWINAC 2022, LNCS 13258, pp. 407–415, 2022.
https://doi.org/10.1007/978-3-031-06242-1_40

(a) Neglect of personal space: when the patient is unaware of body part of opposite to the lesion;
(b) Neglect of the peripersonal space: in which the patient shows a difficulty in perceiving the left/right part of objects very close to him/her, at a distance of about 30 cm;
(c) Neglect of extrapersonal space: which occurs when the patient is unable to recognize objects or people placed in the space far from him/her, more than 30 cm from the patient's body.

Neglect is a complex disorder and it is not always easy to assess it as many different pattern of behaviours [19], different and heterogeneous clinical pictures can be observed.

It is therefore important to use assessment tasks and tests that lead to accurate diagnosis [20,27].

Moreover, also in healthy humans, there is not a perfect symmetry in spatial exploration, as it is evident in the pseudoneglect phenomenon: most people tend to explore space starting from the left side, a bias that can also be found in cancellation and line bisection tasks [33]. It derives from our own brain anatomy and, in particular, from the asymmetries of the ventral attentional networks [10,16,17]. The tools that are used to identify neglect can also be employed to study pseudoneglect in a research context.

## 2 Paper-and-Pencil and Digital Assessment Tools

The most traditional tools to assess neglect in neuropsycological field are the paper-pencil ones, grouped in *ad hoc* batteries, that highlight the presence or absence of a deficit, by comparing the result with the normative value of healthy people and can also of provide information on the overall functioning. For neglect, tests focus on whether the patient shows marked spatial asymmetries, indices of an inability to explore the personal, peripersonal and extrapersonal area.

The most used tests are barrage (cancellation tests), line bisection, copy of drawings, the familiar square description test, ecological tasks as the Comb and Razor test [22,23,28].

Some of these tests and tasks have been translated into their computerized version thanks to the advancement of technology and the use of Machine Learning tools. An example, in this sense, is the BTT (Baking Tray Task) which has been enhanced in its digital version, the E-BTT which will be briefly described in the next paragraph.

## 3 The Baking Tray Task and E-BTT

The BTT, acronym for Baking Tray Test [36] is an ecological test that allows to evaluate deficits in visual-spatial skills by asking participants to simulate a daily activity. Participants are asked to arrange a series of 16 cubes on a 75 × 100 cm support surface as evenly as possible as "if they were biscuits to be placed on a baking sheet". This test is very effective in identifying patients with neglect, by counting how many cubes have been placed on the right and how many on the left. If there is an imbalance, it means that the patient has some problems in spatial processing considering left-right dimension.

It is interesting that there isn't a single correct answer but there are multiple strategies for arranging the cubes [4,10,12,13]. A hybrid digital and physical prototype has been developed by Cerrato and colleagues [10,11,13] as the enhanced version E-BTT. It uses a interface with tangible tools, or concrete objects that the subject can manipulate during the assessment in the peripersonal space. The cubes, provided with ArUco markers [5], can be detected by a camera that transfers the information on cubes positioning to a software.

This version does not alter the simplicity of administration or the ecological value of the original test, but enhances it by giving us with extreme precision some important information such as the test execution time and the disk layout strategies with their spatial coordinates (in terms of X and Y). This allows to recreate the trajectories performed by the various participants thus offering much more information to build the diagnosis on.

Through the use of coordinates it is possible, thanks to the tools provided by machine learning [3], to verify if neural networks [1] are able to reconstruct the spatial arrangement of cubes.

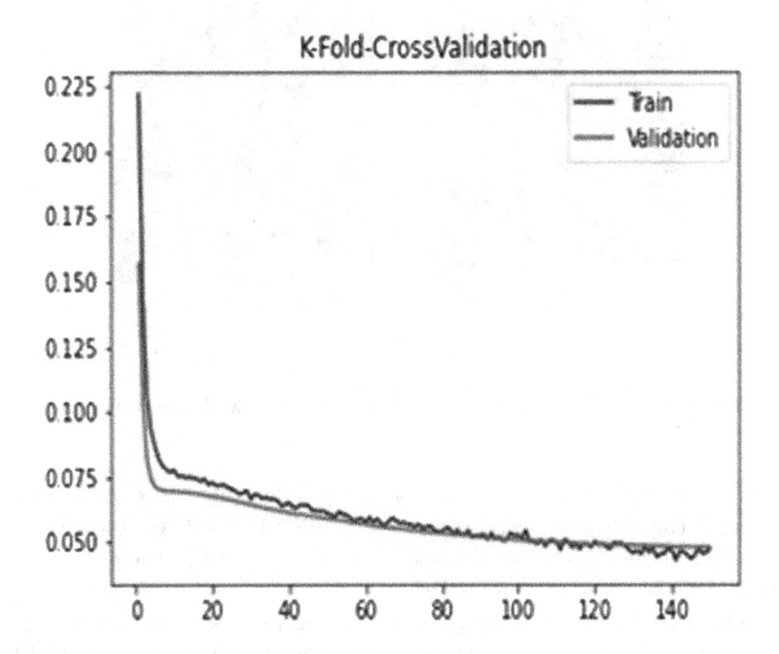


**Fig. 1.** K-Fold-Cross-Validation for the sigmoid network with the comparison between training set and validation set

# 4 Material and Method

## 4.1 Artificial Neural Networks

In order to reconstruct the cubes position, we have run an analysis of E-BTT data using unsupervised learning methods [8].

Two different neural networks were used and compared on their ability to reconstruct the output:

1) Sigmoid: a neural network with a single internal layer with sigmoid activation; it has an input layer, a hidden layer and an output layer with sigmoid activation [24];
2) Autoencoder: an auto-associative neural network model with 3 hidden levels that is associated to a non-linear reduction of the dimensionality [39]; it has an input level with sigmoid activation, a first hidden level with sigmoid activation and a second hidden level with linear activation where the dimensionality reduction occurs, decoding which is mirrored in the encoding, with another hidden level and the output.

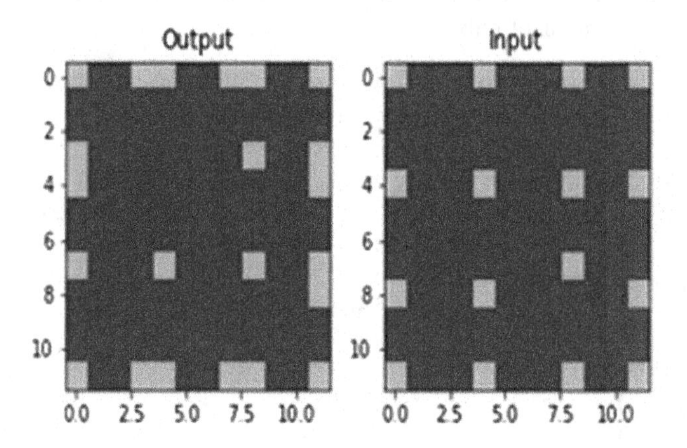


**Fig. 2.** Output of sigmoid network on the left, compared with the input on the right

To obtain better results, dropout [34] has been applied to each layer of the model, one of the most effective and most commonly used regularization techniques for neural networks. The data-set has been divided in the subsets: Train, Validation (70% of observations) e Test set (30%).

The networks then have undergone a learning process to optimize hyperparameters, using a Bayesian optimization (also known as sequential model-based optimization [21]) that uses the results of past evaluations to form a probabilistic model of the objective function and uses this model to choose the next set of hyperparameters values. The probabilistic model is called surrogate model and is represented by p (y — x); y is the performance metric for the model and

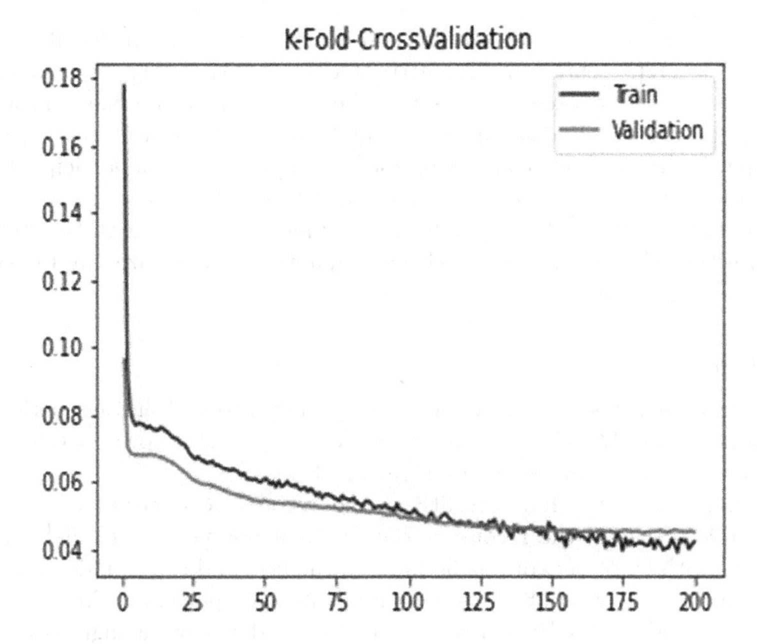


**Fig. 3.** K-Fold-Cross-Validation for the autoencoder with the comparison between training set and validation set

x is the hyperparameters values. Here, the function maps the hyperparameters values to the performance metric chosen by the model (such as the MSE mean standard error) on a validation set. The best performing model is the one that will present the lowest MSE, which is the metric we have selected for this study. To obtain further validation, a 10-Fold Cross Validation was carried out on the best model to avoid any overfitting problem.

For implementation Keras [14] for neural networks and Optuna [35], for the Bayesian approach in Python [38] were used.

### 4.2 The Dataset

The data used for the analyses come from 96 students of the University Federico II of Naples [10,32]. The dataset, therefore, is composed of 96 lines which represent the subjects involved in the test and 32 columns which represent the Cartesian coordinates of the 16 disks; in fact, the first half of the columns represent the ordinates and the remainder the abscissas of the points where the disk has been placed. The dataset only presents subjects who do not have neglect. In the analysis that will be carried out, the order in which the disks were inserted will not be considered. of the test by the subjects, in the overall arrangement of the disks considered.

The data represent the Cartesian coordinates where the disk is present. We have decided to pre-process the data with Histogram2d a Numpy module that allows to transform the coordinate into an image as the ones in Results section. This figure is a representation of a matrix that collects the occurrences in a space divided into as many boxes as are set in the Histogram2d module, and we will have 0 when there are no occurrences and N for the N occurrences present in that point. In our case this transformation was applied for each participant, thus finding a matrix of 0 and 1; it was also decided to set the number of boxes to 48 then the matrices for each subject have 12 times the occurrences and 36 zeros. After creation, the matrices for each participant were vectorized and collected in a single $96 \times 144$ matrix.

### 4.3 Results

The artificial neural networks models were optimized to find the best performing solution (sigmoid: 47 hidden units, 150 epochs of training, MSE = 0.077) The K-Fold-Cross-Validation results are shown in Fig. 1.

The model doesn't show overfitting and reconstructs output as shown in the example of Fig. 2: the reconstruction, even if not perfect, is effective. For autoencoder, the best network ha 90 units in the first and third hidden layer and the innermost layer with 36 units, training lasts 200 epochs and MSE is 0.072. With the K-Fold-Cross-Validation (K = 3), the model shows a slight overfitting in last epochs (Fig. 3). Reconstruction, which is shown in Fig. 4, is almost perfect.

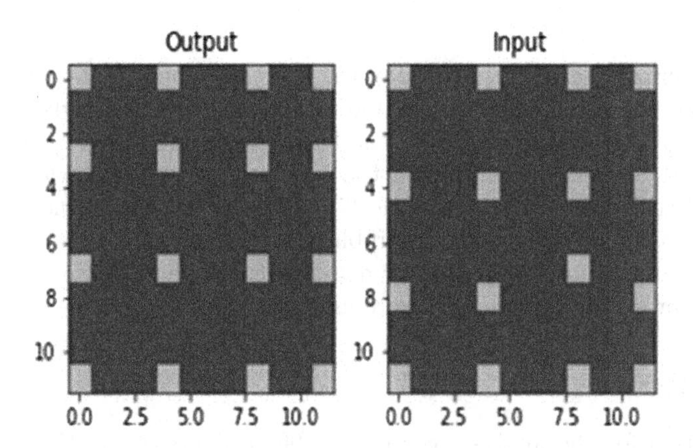


**Fig. 4.** Output of autoencoder on the left, compared with the input on the right

This study shows that the Autoencoder is much more effective in reconstructing the input consisting in the cubes positioning of participants.

## 5 Conclusions and Future Directions

The E-BTT is presented as a tool that enriches the properties of its original paper version. It allows to discriminate, in the assessment, the pathological dispositions from those in the norm. It also allows to investigate the trajectories of exploration in healthy subjects. In this vein, using Artificial Intelligence and Machine Learning is fundamental to reconstruct the possible execution patterns and thus to better understand the alterations in the visual-spatial abilities both in clinical and healthy population.

In future work, our aim is to widen hidden units analysis. The K-Means algorithm [31] will evaluate the existence of common patterns in participants, and the Principal Component Analysis [9] will search for any sub-dimensions that can explain variance in data-set.

**Acknowledgements.** Authors would like to thank Antonietta Argiulo and Federica Somma who were involved in data collection.

**Additional Materials**
More details about the model and the related code can be provided to whom is interested by emailing the authors.

## References

1. Abdi, H., Valentin, D., Edelman, B.: Neural Networks (No. 124). Sage, Thousand Oaks (1999)
2. Allen, G.L.: Spatial abilities, cognitive maps, and wayfinding. In: Wayfinding Behavior: Cognitive Mapping and Other Spatial Processes, p. 4680 (1999)
3. Alpaydin, E.: Machine Learning. MIT Press, Cambridge (2021)
4. Argiuolo, A., Ponticorvo, M.: E-TAN platform and E-baking tray task potentialities: new ways to solve old problems. In: PSYCHOBIT, September 2020
5. Avola, D., Cinque, L., Foresti, G.L., Mercuri, C., Pannone, D.: A practical framework for the development of augmented reality applications by using ArUco markers. In: International Conference on Pattern Recognition Applications and Methods, vol. 2, pp. 645–654. SciTePress, February 2016
6. Bartolomeo, P.: Visual neglect. Curr. Opin. Neurol. **20**(4), 381–386 (2007)
7. Bartolomeo, P., Thiebaut de Schotten, M., Doricchi, F.: Left unilateral neglect as a disconnection syndrome. Cereb. Cortex **17**(11), 2479–2490 (2007)
8. Becker, S.: Unsupervised learning procedures for neural networks. Int. J. Neural Syst. **2**(01n02), 17–33 (1991)
9. Bro, R., Smilde, A.K.: Principal component analysis. Anal. Methods **6**(9), 2812–2831 (2014)
10. Cerrato, A., Ponticorvo, M., Gigliotta, O., Bartolomeo, P., Miglino, O.: BTT-scan: uno strumento per la valutazione della negligenza spaziale unilaterale. Sistemi intelligenti **31**(2), 253–270 (2019)
11. Cerrato, A., Ponticorvo, M.: Enhancing neuropsychological testing with gamification and tangible interfaces: the baking tray task. In: Ferrández Vicente, J.M., Álvarez-Sánchez, J.R., de la Paz López, F., Toledo Moreo, J., Adeli, H. (eds.) IWINAC 2017. LNCS, vol. 10338, pp. 147–156. Springer, Cham (2017). https://doi.org/10.1007/978-3-319-59773-7_16
12. Cerrato, A., Ponticorvo, M., Gigliotta, O., Bartolomeo, P., Miglino, O.: The assessment of visuospatial abilities with tangible interfaces and machine learning. In: Ferrández Vicente, J.M., Álvarez-Sánchez, J.R., de la Paz López, F., Toledo Moreo, J., Adeli, H. (eds.) IWINAC 2019. LNCS, vol. 11486, pp. 78–87. Springer, Cham (2019). https://doi.org/10.1007/978-3-030-19591-5_9

13. Cerrato, A., et al.: E-TAN, a technology-enhanced platform with tangible objects for the assessment of visual neglect: a multiple single-case study. Neuropsychol. Rehabil. **31**(7), 1130–1144 (2021)
14. Chollet, F., et al.: Keras. GitHub (2015). https://github.com/fchollet/keras
15. Committeri, G., et al.: Neural bases of personal and extrapersonal neglect in humans. Brain **130**(2), 431–441 (2007)
16. Facchin, A., Beschin, N., Daini, R.: Rehabilitation of right (personal) neglect by prism adaptation: a case report. Ann. Phys. Rehabil. Med. **60**(3), 220–222 (2016)
17. Gigliotta, O., Malkinson, T.S., Miglino, O., Bartolomeo, P.: Pseudoneglect in visual search: behavioral evidence and connectional constraints in simulated neural circuitry. Eneuro **4**(6) (2017)
18. Guariglia, C., Antonucci, G.: Personal and extrapersonal space: a case of neglect dissociation. Neuropsychologia **30**(11), 1001–1009 (1992)
19. Halligan, P.W., Marshall, J.C.: Spatial compression in visual neglect: a case study. Cortex **27**(4), 623–629 (1991)
20. Halligan, P.W., Robertson, I.: Spatial Neglect: A Clinical Handbook for Diagnosis and Treatment. Psychology Press, Hove (2014)
21. Hutter, F., Hoos, H.H., Leyton-Brown, K.: Sequential model-based optimization for general algorithm configuration. In: Coello, C.A.C. (ed.) LION 2011. LNCS, vol. 6683, pp. 507–523. Springer, Heidelberg (2011). https://doi.org/10.1007/978-3-642-25566-3_40
22. Jewell, G., McCourt, M.E.: Pseudoneglect: a review and meta-analysis of performance factors in line bisection tasks. Neuropsychologia **38**(1), 93–110 (2000)
23. McIntosh, R.D., Brodie, E.E., Beschin, N., Robertson, I.H.: Improving the clinical diagnosis of personal neglect: a reformulated comb and razor test. Cortex **36**(2), 289–292 (2000)
24. Menon, A., Mehrotra, K., Mohan, C.K., Ranka, S.: Characterization of a class of sigmoid functions with applications to neural networks. Neural Netw. **9**(5), 819–835 (1996)
25. Miglino, O., Ponticorvo, M., Bartolomeo, P.: Place cognition and active perception: a study with evolved robots. Connect. Sci. **21**(1), 3–14 (2009)
26. Pizzamiglio, L., et al.: Visual neglect for far and near extra-personal space in humans. Cortex **25**(3), 471–477 (1989)
27. Plummer, P., Morris, M.E., Dunai, J.: Assessment of unilateral neglect. Phys. Ther. **83**(8), 732–740 (2003)
28. Pitzalis, S., Spinelli, D., Zoccolotti, P.: Vertical neglect: behavioral and electrophysiological data. Cortex **33**(4), 679–688 (1997)
29. Purves, D., et al.: Cognitive Neuroscience, vol. 6, no. 4. Sinauer Associates Inc., Sunderland (2008)
30. Shelton, P.A., Bowers, D., Heilman, K.M.: Peripersonal and vertical neglect. Brain **113**(1), 191–205 (1990)
31. Sinaga, K.P., Yang, M.S.: Unsupervised K-means clustering algorithm. IEEE Access **8**, 80716–80727 (2020)
32. Somma, F., et al.: Further to the left: stress-induced increase of spatial pseudoneglect during the COVID-19 lockdown. Front. Psychol. **12** (2021)
33. Somma, F., et al.: Valutazione dello pseudoneglect mediante strumenti tangibili e digitali. Sistemi intelligenti **32**(3), 533–549 (2020)
34. Srivastava, N., Hinton, G., Krizhevsky, A., Sutskever, I., Salakhutdinov, R.: Dropout: a simple way to prevent neural networks from overfitting. J. Mach. Learn. Res. **15**(1), 1929–1958 (2014)

35. Akiba, T., Sano, S., Yanase, T., Ohta, T., Koyama, M.: Optuna: a next-generation hyperparameter optimization framework. In: KDD (2019)
36. Tham, K., Tegner, R.: The baking tray task: a test of spatial neglect. Neuropsychol. Rehabil. **6**(1), 19–26 (1996)
37. Urbanski, M., et al.: Négligence spatiale unilatérale: une conséquence dramatique mais souvent négligée des lésions de l'hémisphère droit. Revue Neurologique, 16 (2007)
38. Van Rossum, G., Drake, F.L., Jr.: Python tutorial. Centrum voor Wiskunde en Informatica Amsterdam, The Netherlands (1995)
39. Wang, W., Huang, Y., Wang, Y., Wang, L.: Generalized autoencoder: a neural network framework for dimensionality reduction. In: Proceedings of the IEEE Conference on Computer Vision and Pattern Recognition Workshops, pp. 490–497 (2014)

# Automatic Diagnosis of Mild Cognitive Impairment Using Siamese Neural Networks

E. Estella-Nonay, M. Bachiller-Mayoral(✉), S. Valladares-Rodriguez, and M. Rincón

Artificial Intelligence Department, UNED, Madrid, Spain
marga@dia.uned.es
http://www.ia.uned.es

**Abstract.** The use of Artificial Intelligence techniques as an aid tool in the medical field is a current and undeniable challenge. In this context, similarity detection methods and Convolutional Neural Networks used in computer vision tasks for feature extraction can greatly contribute to the analysis of medical tests based on freehand drawings. This paper brings together both ideas and proposes the use of Siamese Neural Networks to perform an automatic diagnosis of mild cognitive impairment (MCI) based on the Rey-Osterrieth Complex Figure (ROCF) test. It analyzes the suitability of this type of networks and compares them with an ANN. For this purpose, about 477 drawings collected in a research study in the field of neuropsychology, made by healthy patients or patients with some degree of cognitive impairment, are available. Due to the small number of instances, it is proposed to pre-train the networks with the Transfer Learning technique using a much larger dataset of drawings with similar characteristics.

**Keywords:** The Rey–Osterrieth Complex Figure test · Siamese neural networks · Mild cognitive impairment · Dementia

## 1 Introduction

The increase in life expectancy in recent decades has led to an increase in cases of cognitive impairment mainly associated with age. Early detection of these disorders is essential to minimize their effects. This is a challenge both medically, to find new safe and reliable methods, and technologically, to develop tools to aid in diagnosis. One of the tests capable of aiding in diagnosis is the Rey–Osterrieth Complex Figure (ROCF) test [1] (cf. Fig. 1(a)), in which a patient must perform two tasks. In the first, he/she makes a freehand copy of such a figure. In the second, the subject reproduces the figure from memory. The analysis of these copies involves the use of great personal resources and adds the subjectivity of the expert during the evaluation of criteria such as deformation, proportion or location.


J. M. Ferrández Vicente et al. (Eds.): IWINAC 2022, LNCS 13258, pp. 416–425, 2022.
https://doi.org/10.1007/978-3-031-06242-1_41

Currently, the most promising techniques for the creation of automatic systems are in the field of Deep Learning (DL). Artificial Neural Networks (ANNs) have been successfully implemented in areas such as assistance in magnetic resonance imaging, mammography, disease detection through blood tests or cognitive impairment [2–6]. One of the most successful DL methods is Convolutional Neural Networks (CNNs), which are key to the development of applications in the field of computer vision.

The works that analyze drawings automatically using DL techniques need thousands of instances per class to create an output space based on the observed features. In addition, the probability distribution is difficult to fit by training when there are classes with very similar characteristics. To address these deficiencies, efforts in recent years have focused on the application of similarity detection techniques. These methods are able to calculate differences between instances to determine the membership of the corresponding class. Within these techniques, one-shot learning applications [7] are capable of learning the information of each category simply from a few instances per class and determining the degree of similarity between them. There are different architectures that allow performing classification tasks using one-shot learning, among them we find Siamese Neural Networks (SNN), which are one of those that have obtained good results in classification tasks [8].

SNNs consist of two twin networks, sharing the value of the weights, which are updated simultaneously during the training process. Each of the branches is equivalent to a feature extractor from the input data. Both branches are joined by a function with a metric that calculates the distance between the features. From this value, the degree of similarity between the input data of each branch can be derived using a discriminant function or loss function that is used to update the weights of the networks. The output of the system is a binary value that represents whether or not the inputs belong to the same category. Pairs of images with a binary label indicating their belonging to the same class are used to train the SNNs. In the medical domain, SNNs applications are mainly grouped into three types of tasks: disease diagnosis, assessment of disease severity, and content-based image retrieval. The application of SNNs on drawings or handwritten representations has also been the subject of study for both character recognition and concept recognition from drawings [9–12].

The one-shot learning strategy fits perfectly as a diagnostic solution to the ROCF test since the detection of similarities between a drawing and the ROCF model is a discriminatory criterion of membership in the healthy class. In addition, a small number of available images is a drawback for applying ANNs while it is not for one-shot learning based systems. The objective of this work focuses on the possibility of using SNNs for the automatic generation of a diagnosis based on the differences between the drawing made by the patient (copy) and the ROCF model. The aim is to provide medical personnel with new tools with which to obtain reliable results that optimize the sector's resources and improve diagnosis. It is also taken into account that there are no time restrictions, since it is not a real-time application, nor memory restrictions.

Section 2 describes the datasets used in the study. The architectures and the design phases used for their training are presented in Sect. 3 and the following section shows the results obtained from all the experiments. Finally, Sect. 5 is devoted to conclusions and future work.

## 2 The Datasets

The initial dataset consists of ROCF test drawings collected in different neuropsychological assessment sessions during the research study presented in [13]. Depending on the results obtained on a battery of tests that evaluate different cognitive skills (long-term ans short-term memory, reasoning, attention, planning, etc.), a pacient is assigned the diagnostic profile of Mild Cognitive Impairment (MCI) or healthy. These drawings are scanned and processed for annotation removal. The result is a set that has 887 instances of which only 477 are labeled. So, the ROCF dataset used in this work is just a subset of the initial dataset that collects all labeled instances, which are divided into two classes: healthy and MCI. The instances of both classes are distributed as 58.5% for "healthy" and 41.5% for "MCI". Figure 1(b) and (c) show an example of a drawing made by a healthy person and a person with MCI, respectively.

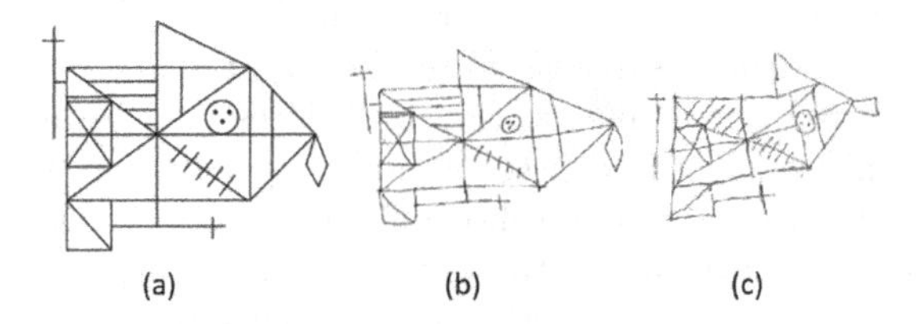


**Fig. 1.** (a) The Rey-Osterrieth Complex Figure (ROCF) test; (b) Drawing of a healthy person and (c) Drawing of an MCI person.

Due to the small size of the dataset, a transfer learning (TL) strategy is used for pre-training the feature extractor stage that is used in all the architectures described in Sect. 3. The dataset used in this phase is the large open source dataset Google Quick Draw [16], a collection of 50 million drawings across 345 categories, contributed by players of the game Quick, Draw!. To restrict the size and training time, a subset of the Google QuickDraw dataset (SQD) comprising 1000 instances of 19 different classes (19,000 instances in total) was selected. The classes were chosen to generalise different line features (straight lines, curved lines, a mixture of both, zig-zag lines, parallel lines and intersections, circles, squares, triangles, complex shapes, etc.). Figure 2 shows examples of drawings included in the SQD dataset. Images were normalized in the range 0 and 1 and scaled to a size of $100 \times 100$ pixels, which greatly reduces the input size to the system while maintaining the detail of the drawings.

**Fig. 2.** Examples of instances of the dataset used in the TL.

## 3 The Architectures

This paper presents two architectures for ROCF drawing classification. The first one is based on an ANN and the second one uses a SNN. Two phases are followed for their training. The first one is dedicated to pre-training a feature extractor using the SQD dataset and, in the second phase, the feature extractor stage obtained in phase 1 is used to build different architectures that are tuned using the ROCF dataset (cf. Fig. 3).

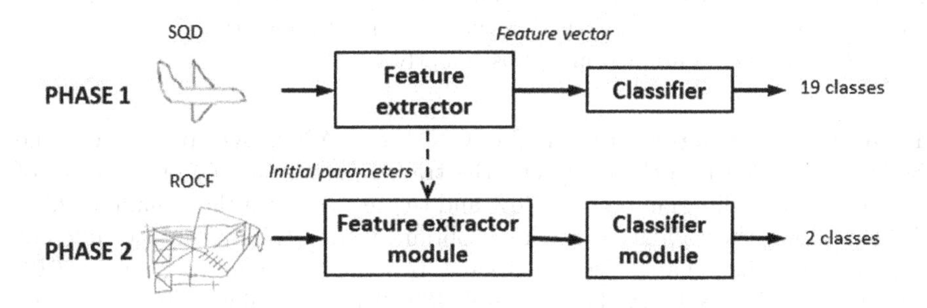


**Fig. 3.** Training Phases: Tranfer learning with SQD dataset and tuning with ROCF dataset.

### 3.1 Phase 1: Pretraining with SQD Dataset

A deep learning architecture for the classification task can be decomposed in two stages sequentially organised: a feature extractor, where the features of interest are obtained, and a classifier, which performs the classification based on the previously obtained features. In this first phase, we use TL to pre-train a CNN (the feature extractor stage) using the SQD dataset in order to find the optimal set of visual features that typically appear in freehand drawings.

Several experiments have been performed with different values of network depth and number of filters. The resulting CNN is composed of 3 layers with filtering (convolution), activation by ReLU and pooling stages. As input it receives an image of size $100 \times 100$ pixels, to which 64 filters are applied in the first layer and 32 in the remaining layers. Regarding the size of the filters, there are works that propose sizes up to $15 \times 15$ that intend to process more context information in the first convolution, and others that decide to use very small $3 \times 3$ filters obviating that information. We have opted for an intermediate solution with $5 \times 5$ filters for the input layer and $3 \times 3$ for the subsequent layers. All of them perform a $2 \times 2$ pooling. The output of the extractor is presented as a vector of 6400 values. The size chosen is also among the largest and the most compact found in previous works for coding drawings [11].

For training the feature extractor, an ANN classifier has been added after the CNN (the classifier stage) with 19 softmax-activated neurons (one output per class). The ANN is composed of 2 FC (fully-connected) intermediate layers, which have a size of 128 neurons with ReLU activation function. We used the Glorot Normal initializer [14] for weight initialization and the Adam optimizer [13] for weight optimization. Finally, due to the mutual exclusivity of the classes, sparse categorical cross-enthropy is used as loss function.

Once the network structure is defined, the experiments focus on the configuration of parameters such as batch size, dropout or learning rate (lr). We first identify the optimal lr of the TL phase, and then, in phase 2, each architecture is trained with the ROCF dataset and an lr chosen from a subset of values lower than the previous one. Thus, the knowledge learned about the general drawing characteristics is maintained and subsequently refined.

**Parametric Configuration of the CNN for ANN Architecture.** The SQD was divided into three subsets: the training set with 60% of instances of each class, the validation set with 20% and the test set with the remaining 20%. Table 1 shows the validation accuracy obtained after training in the different experiments, with a batch size of 8, 16 and 32 and lr values in the range [0.0010, 0.0019]. It is observed how the best result (80.72 %) is obtained after 20 epochs, with a batch size of 16 and an lr of 0.0015.

**Table 1.** Validation accuracy of different combinations of batch size and lr for feature extractor configuration.

| | lr | | | | | | | | | |
|---|---|---|---|---|---|---|---|---|---|---|
| Batch size | 0.0010 | 0.0011 | 0.0012 | 0.0013 | 0.0014 | 0.0015 | 0.0016 | 0.0017 | 0.0018 | 0.0019 |
| 8 | 67.20 | 67.28 | 70.05 | 71.81 | 74.95 | 75.25 | 76.40 | 76.69 | 76.65 | 76.88 |
| 16 | 72.97 | 73.53 | 76.83 | 76.34 | 80.01 | 80.72 | 77.89 | 78.56 | 78.12 | 77.21 |
| 32 | 75.85 | 77.62 | 77.59 | 79.43 | 80.22 | 80.28 | 79.74 | 79.89 | 78.33 | 78.23 |

Keeping the idea of generalization of the feature extractor, experiments have been performed applying the dropout technique. Figure 4 shows the results for training the extractor with the SQD and with the final parameters. The training accuracy without the dropout strategy reaches 99%, decreasing by 5% for a dropout of 20%. However, the difference between the validation accuracy is only 0.1% so that this dropout is maintained in the subsequent training of the next phase.

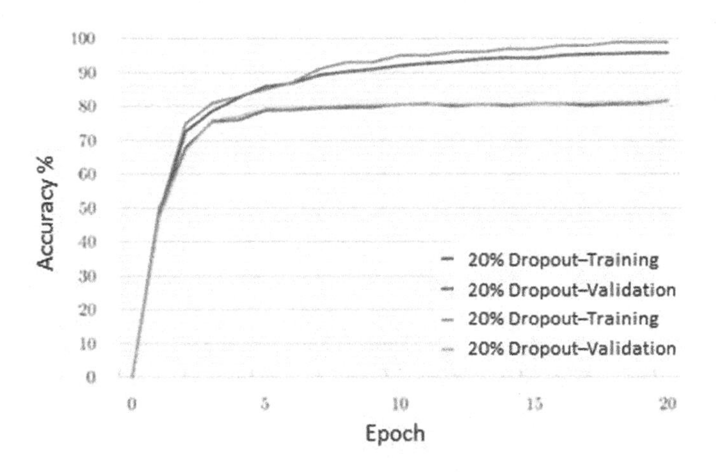


**Fig. 4.** Accuracy from training and validation of the feature extractor for values from 0 to 20% dropout, with optimal values of lr, batch size and epochs.

**Parametric Configuration of the CNN for SNN Architecture.** The SNN is composed of two branches similar to the pretrained CNN (cf. Fig. 6(b)). Taking into account the study [15] which shows the positive impact of FC layers after CNN on image classification tasks, a layer with 4096 neurons activated by the sigmoid function with a L2 regulation of weights and with a regulation factor equal to 0.001, is added to each of the branches. The distance function is calculated in vector form following the formula $|w = x - y|$, where x and y are the vectors defining the features for each of the network branches. The output layer is added just after the vector distance calculation and consists of a neuron activated by the sigmoid function. Binary cross-entropy is taken as the loss function.

The size of the training, validation and test set contains random pairs of all classes of QuickDraw drawings with a size of 5000, 512 and 512, respectively. The content of the pairs varies in each epoch, always guaranteeing exclusivity between sets. Half of the pairs in each set contain instances of the same class and the other half of different ones. In addition, in order to limit learning, the initial lr is modified in each batch following the function:

$$lr_i = lr_{i-1} \frac{1}{1 + 0.00025i} \tag{1}$$

where i is the batch index in an epoch and $\mathrm{lr}_i$ the corresponding learning rate for that batch. Figure 5 shows the differences found during the experiments of the validation accuracy for different batch values as a function of some of the lr values used in this step. Finally, the parameters selected are lr = 0.001, a batch size of 16 and training for 12 epochs.

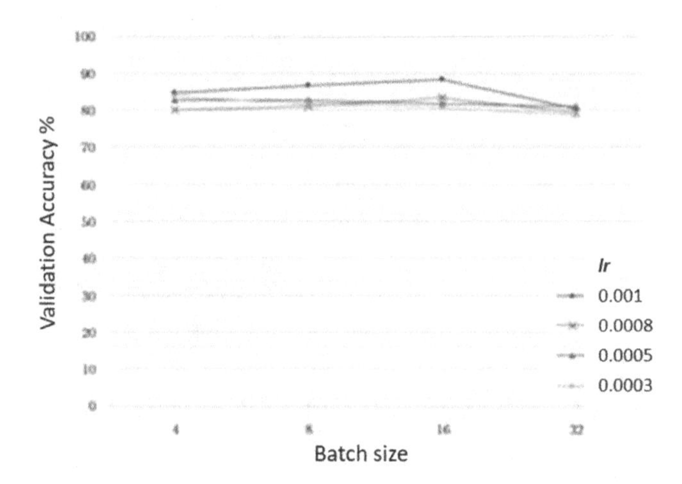


**Fig. 5.** Validation accuracy of the SNN architecture as a function of batch size and with different values of lr.

### 3.2 Phase 2: Tuning with ROCF Dataset

In the second phase, the values of the pre-trained networks are refined with the ROCF dataset and the classifier structure is defined according to the particular DL architecture. To this end, starting from the networks trained for each case in phase 1, several experiments have been carried out to obtain the optimal configuration of the proposed architectures. The training, validation and test sets into which the ROCF dataset has been divided contain the same number of instances for both architectures and maintain a ratio of 70%-15%-15%, respectively.

**ANN Architecture Configuration.** The network consists of a branch similar to the pre-trained network but the softmax output layer is replaced by a single sigmoid-activated neuron in charge of binary classification (cf. Fig. 6(a)). Binary cross-entropy is taken as the loss function. After the different experiments, a batch size of 8 and 30 epochs was chosen. It has been tested with lr values lower than the one chosen in phase 1, selecting 0.001 as the final value.

**SNN Architecture Configuration.** For tunning of the SNN network with the ROCF dataset, we have created pairs with all the instances of the dataset, always using the ROCF model as one of its elements. For the generation of the training, validation and test sets, the proportion between positive and negative instances has been maintained in the number of pairs. The optimal values found correspond to a lr equal to 0.0005 and a batch size of 16 for 50 epochs. The value of lr used in these experiments have also been lower than those set in previous stages.

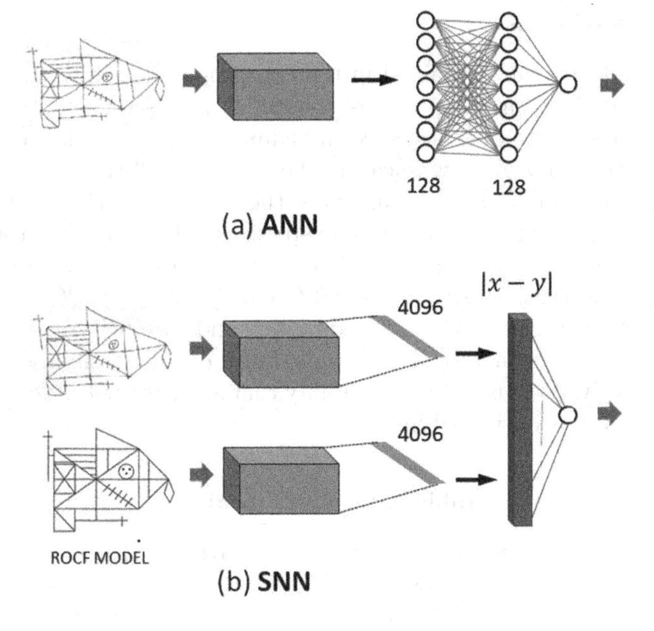


**Fig. 6.** Representation of architectures: (a) ANN; (b) SNN.

## 4 Results

This section presents the results obtained in both phases of the experiments and for each architecture according to the datasets with which it was trained. In order to be able to interpret and compare the results for each architecture, we use, in addition to the accuracy, the measure of sensitivity, which indicates the ability of the system to identify MCI people, and specificity, which indicates the ability to correctly classify "healthy" people.

### 4.1 Results of Phase 1

The results obtained in this first phase of training using the choice of parameters indicated in the previous section, both in the case of the ANN architecture and the SNN architecture, are presented in Table 2. It can be seen that SNN shows better results in training, validation and test than ANN.

**Table 2.** Accuracy obtained in the networks trained in phase 1 and 2.

| Phase 1 | | | | Phase 2 | | | |
|---|---|---|---|---|---|---|---|
| Architecture | Training | Validation | Test | Architecture | Training | Validation | Test |
| ANN | 95.75 | 80.72 | 79.64 | ANN | 60.74 | 58.33 | 56.94 |
| SNN | 98.29 | 88.20 | 85.74 | SNN | 98.29 | 68.05 | 66.66 |

### 4.2 Results of Phase 2

Table 2 shows the results obtained in the two proposed architectures using the parameters indicated in the previous section. It is observed that in the ANN architecture, low accuracy is achieved in training, validation and test, while the SNN architecture improves significantly the results in all three cases.

However, in order to properly interpret the results of each architecture, it is convenient to analyze the sensitivity and specificity. Table 3 shows the confusion matrices for the (a) ANN and (b) SNN architectures. It can be seen that in the ANN architecture, sensitivity is 100% and specificity is 26%, while in the SNN architecture, sensitivity is reduced to 43% and specificity increases to 83%. In both architectures, there is a high probability of misclassifying patients and more importantly, there is a high probability that a patient with signs of cognitive impairment will be considered healthy.

**Table 3.** Confusion matrices.

(a) ANN architectures

| | | Real | |
|---|---|---|---|
| | | Healthy | MCI |
| Prediction | Healthy | 41.7% | 43.0% |
| | MCI | 0.0% | 15.3% |

(a) SNN architectures

| | | Real | |
|---|---|---|---|
| | | Healthy | MCI |
| Prediction | Healthy | 18.1% | 9.7% |
| | MCI | 23.6% | 48.6% |

## 5 Conclusions

In this work, the use of SNN is proposed as a one-shot learning strategy for the diagnosis of cognitive diseases by means of neuropsichological tests based on drawings, since the detection of similarities between a drawing and the ROCF model is a discriminatory criterion of belonging to the class of healthy people. To analyze their adequacy, two architectures have been proposed: ANN architecture and SNN architecture. In addition, to address the problem of dataset size, pre-training of the networks using a subset of the QuickDraw dataset, whose instances share characteristics with the drawings to be analyzed, has been proposed. It is found that the SNN architecture performs better than the ANN architecture in the diagnosis of cognitive impairment in the training, validation and test data sets (98.29%, 68.05% and 66.66%, respectively). However, these results are not satisfactory in terms of sensitivity and specificity and should be improved.

From these experiments it is concluded that SNNs architectures can be considered as a suitable option to help in the automatic diagnosis of the ROCF test, opening a line of research for the use of oneshot recognition in this type of applications. The possibilities offered by these SNNs architectures are enormous and the construction of a more robust system could be achieved with deeper CNNs o another type of loss function, such as triplet loss.

## References


1. Rey, A.: L'examen psychologique dans les cas d'encéphalopathic traumatique. (Les problems.) [The psychological examination in cases of traumatic encepholopathy. Problems] Archives de Psychologie, vol. 28, pp. 215–285 (1941)
2. Graham, S., et al.: AI approaches to predicting and detecting cognitive decline in older adults: a conceptual review. Psychiatry Research, vol. 284 (2020)
3. Bernal, J., Kushibar, K., Asfaw, D., Valverde, S., Arnau, O., Martí, R., Lladó, X.: Deep convolutional neural networks for brain image analysis on magnetic resonance imaging: a review. Artif. Intell. Med. **95**, 64–81 (2019)
4. Pardamcan, D., Cenggoro, T.W., Rahutomo, R., Budiarto, A., Karuppiah, E.: Transfer learning from chest X-Ray pre-trained CNN for learning mammogram data. Procedia Comput. Sci. **135**, 400–407 (2018)
5. Kang, M.J., Kim, S.Y., Na, D.: Prediction of cognitive impairment via deep learning trained with multi-center neuropsychological test data. BMC Medical Informatics Decision Making, vol. 19 (2019)
6. Gorji, H.T., Kaabouch, N.: A deep learning approach for diagnosis of mild cognitive impairment based on MRI images. Brain Sci. **9**, 217 (2019)
7. Li, F.-F., Fergus, R., Perona, P.: A Bayesian approach to unsupervised one-shot learning of object categories. In: Ninth IEEE International Conference on Computer Vision, vol. 2, pp. 1134–1141. Publisher, Location (2010)
8. Vinyals, O., Blundell, C., Lillicrap, T., Wierstra, D.: Matching networks for one-shot learning, In: Neural Information Processing Systems (NII'S) (2016)
9. Amin-Naji, M., Mahdavinataj, H., Aghagolzadeh, A.: Alzheimer's disease diagnosis from structural MRI using Siamese CNN. In: 4th International Conference on Pattern Recognition and Image Analysis (IPRIA), pp. 75–79 (2019)
10. Li, M., Chang, K., Bearce, D.: Siamese neural networks for continuous disease severity evaluation and change detection in medical imaging. npj Digital Med. **3**, 48 (2020)
11. Qi, Y., Song, Y.-Z., Zhang, H., Liu, J.: Sketch-based image retrieval via siamese convolutional neuronal network. In: 2016 IEEE International Conference on Image Processing (ICIP), pp. 2460–2464 (2016)
12. Wang, F., Kang, L., Li, Y.: Sketch-based 3D shape retrieval using convolutional neural networks. In: IEEE Conference on Computer Vision and Pattern Recognition (CVI'R2015), pp. 1875–1883 (2015)
13. García Herranz, S., Díaz Mardomingo, vM.C., Peraita, H.: Neuropsychological predictors of conversion to probable Alzheimer disease in elderly with mild cognitive impairment. J. Neuropsychol. **10**(2), 239–255 (2016)
14. Glorot, X., Bengio, Y.: Understanding the difficulty of training deep feedforward neural networks. In: Proceedings of the Thirteenth International Conference on Artificial Intelligence and Statistics, vol. 9, pp. 249–256 (2014)
15. Shabecr, B., Dubey, S.R., Pulabaigari, V., Mukherjce, S.: Impact of fully connected layers on performance of convolutional neural networks for image classification. Neurocomputing **378**, 112–119 (2020)
16. Jongejan, J., Rowley, H., Kawashima, T., Kim, J., Fox-Gieg, N.: The Quick, Draw! - A.I. Experiment. https://quickdraw.withgoogle.com/ (2016)
17. Kingma, D., Ba, J.: Adam: a method for stochastic optimization. In: International Conference on Learning Representations (2014)

# A Comparison of Feature-based Classifiers and Transfer Learning Approaches for Cognitive Impairment Recognition in Language

González Machorro Monica(✉) and Martínez Tomás Rafael

Universidad Nacional de Educacion a Distancia, Madrid, Spain
mgonzalez8482@alumno.uned.es

**Abstract.** Language provides valuable information in dementia recognition, as language impairment is a common characteristic of early dementia. One current limitation is the lack of data due to the constraints involved in data collection. In this paper, we propose transfer learning methods that address data scarcity and involve the least amount of customization steps. We analyze language in two separate modalities: speech and linguistic information. For the first modality, we employ audio files, and for the second one, transcripts extracted from the audio files. We customize a subset of the Pitt Corpus that contains early Alzheimer's disease (AD) and Mild Cognitive Impairment (MCI) patients. Our proposed methods consist of feature-based classifiers and pre-trained models such as ResNet152, HuBERT, BERT and RoBERTa. Results show that linguistic-based transfer learning methods outperform speech-based transfer learning approaches and conventional classifiers. However, speech-based methods offer a solution that is transcription-free and end-to-end. Our main contribution is to successfully apply Automatic Speech Recognition (ASR) architectures in cognitive impairment recognition.

**Keywords:** Alzheimer's disease · Language impairment · Mild cognitive impairment · Transfer learning

## 1 Introduction

Dementia disease is a general term for degenerative diseases of the brain that interfere in the ability to remember, think, comprehend, and communicate [1]. There are many types of dementia, of which AD accounts for 70% of cases of dementia [2]. Given the severity of the disease, it is imperative to find and provide early dementia diagnosis and treatment to improve patients' living conditions. According to [3], dementia cognitive deficits could be detected up to ten years before its onset. This is possible due to an intermediate state between the changes in cognitive natural aging and dementia disease [4] known as MCI. Researchers


J. M. Ferrández Vicente et al. (Eds.): IWINAC 2022, LNCS 13258, pp. 426–435, 2022.
https://doi.org/10.1007/978-3-031-06242-1_42

consider this period a window of opportunity for early dementia detection [4], since around 38% of MCI patients develop a type of dementia within 5 years [5].

A promising solution for early diagnosis is language analysis, since early types of dementia are characterized by language impairment. For instance, early dementia patients have trouble finding a word, word meanings and hyperfluency. They also present less nouns, repetitions and longer speech pauses [6]. Another advantage of using language as a biomarker of cognitive impairment is the simplicity and non-invasiveness of data collection.

Recent artificial intelligence progress has allowed researchers to find encouraging results investigating speech and linguistic information [7,8]. In the case of speech, research has found a wide number of acoustic features such as articulation rate, speech rate and number of pauses that correlate with AD development [9]. An example is [10], who employed pre-trained models such as Inception v3 and Vision Transformer to classify AD and non-AD patients. The best-performing model obtained an accuracy of 65% [10,11] trained raw audio-based Convolutional Neural Networks (CNNs) and achieved a 74% accuracy [11]. [12] used wav2vec to obtain speech representations and fed them to Support Vector Machine (SVM). This resulted in a 67% accuracy [12].

Linguistic information is explored through human-made and ASR transcripts. For example, [13] investigated text-based CNNs and reported an 88% accuracy. [14] trained a bi-directional Hierarchical Attention Network on transcripts and found an 81% accuracy. [15] adapted a pre-trained BERT model to detect AD from non-AD and achieved an 82% accuracy [15]. Even though this language domain has demonstrated better results than speech-based models, these results depend on how the transcripts are generated. Human-made transcripts have obtained the highest performance there.

Recent work has also looked at linguistic and speech information fusion. [16] normalized scores obtained from SVM models trained on text and audio, and merged the scores by the average, achieving an overall accuracy of 75% [16]. [17] found an 81% accuracy training a SVM model with acoustic and language features [17]. Another example is [18], who applied an early fusion approach by extracting various linguistic and speech features to train a SVM, and reported a 76% accuracy [18].

However, there are still key challenges before one can employ language as biomarker for dementia detection [8]. One of them is data scarcity - a common problem in data collection regarding cognitive disorders. Our main objective is to address this challenge by proposing transfer learning approaches and compare these methods with conventional classifiers. The paper focuses on binary classification of dementia (early AD and MCI) and control samples. Our motivation to jointly study MCI and AD is that accuracy tends to decrease when MCI is studied separately, due to the little data for MCI, an aspect that it is in line with the complexity of distinguishing early stages of dementia for clinicians [7]. We consider this challenge makes the study of MCI an even more pressing matter.

## 2 Proposed Approach

We propose the following methods:

- conventional approaches: SVM, Random Forest (RF) and Gaussian Naive Bayes (GNB),
- transfer learning approaches: for the speech modality, we propose pre-trained ASR HuBERT model and pre-trained ResNet152. Up to our knowledge, these two methods have not yet been applied in our task. In the linguistic modality, we explore pre-trained BERT and RoBERTa models.

### 2.1 Conventional Approaches

We extract speech functionals using the feature set ComParE 2016 [19]. This feature set has been broadly studied in different speech classification tasks [8]. Examples for these features are MFCCs, F0, length of unvoiced segments, loudness, jitter, shimmer, among others. In the case of linguistic features, we extract features using the FLUCALC toolkit from the CLAN program [20], which tracks the frequencies of various fluency indicators such as word repetitions, blockings, retraces, monosyllabic repetitions, percentage of phrase repetitions and typical disfluencies [20].

Given the large feature dimensionality, we apply a filter selection based on a statistical index of feature differences between classes using the Analysis of Variance (ANOVA) ranking scheme. The threshold to select relevant features is to filter those features, of which the p-value is higher than 0.05. Conventional classifiers are developed using scikit-learn version 1.0.1 in Python 3.7.12. Each model is optimized using hyperparameters selected via grid search 5-fold CV on the training set.

For comparison purposes, we also present a baseline model. This model is implemented by training a SVM classifier on two demographic variables - sex and age - without a feature selection technique.

### 2.2 Transfer Learning Approaches

Transfer learning models are developed using Pytorch, version 1.10.0 in Python 3.7.12. Experiments are conducted on Google Colab Pro using a 25 GB NVIDIA Tesla 27 K80 GPU. To account for stochastic initialization, overall performance is averaged across three runs.

**Pre-trained CNNs.** We apply pre-trained CNNs model of the computer vision domain: ResNet152. The reason for choosing this method is that, according to [10], few works have explored models from the computer vision domain in dementia identification [10]. To obtain the spectrograms, we use librosa library, version 0.8.0. Spectrograms are extracted with 128 Mel bands, a hop length equal to 512 and a length of the Fast Fourier Transform (FFT) window of 2,048.

To train the model, we keep the original structure and re-estimate the two last hidden layers to comply with our classification task. After the second to last layer, a dropout layer is adopted with a rate of 0.3. We also set softmax as the activation function in the output layer. We define cross entropy loss as criterion. As optimizer we use Adam with an empirically defined learning rate of 0.001. The batch size to load the images is set to 32. The number of epochs is defined empirically as 35.

**Pre-trained ASR Model.** HuBERT (Hidden Unit Bert) is a self-supervised ASR model. HuBERT uses a "k-means clustering step to obtain noisy labels for Masked Language Model pre-training" [21]. It employs cross-entropy loss as criterion. This architecture is end-to-end, which means that it uses raw audio files and does not require domain-knowledge.

To employ HuBERT, we apply mean as the pooling mode strategy and two linear layers on top of the HuBERT representations (see Fig. 1). We set a 0.3 dropout rate and softmax as activation function. The pre-trained model used is 'hubert-large-ls960-ft' obtained from the HuggingFace's Transformers library. Audio files are loaded using a batch size of 1 and we empirically select the number of epochs as 7. The final optimization steps are 880.

**Pre-trained Transformers Models.** BERT is based on Transformer networks to pre-train bidirectional representations of text [22]. BERT has access to the entire input whereas other architectures such as sequential neural networks usually have access to the input one by one.

We use the model 'bert-based-uncased' to initialize the classifier. This model consists of 12 layers. We add a classification layer at the end to map the BERT output to our task [22]. The maximum sequence length of the input is set to 512 tokens. Empirically, dropout is defined as 0.1. Based on previous literature [22], batch size is set to 16. We use Adam as optimizer and linear scheduling as learning rate. During fine-tuning, we optimize the number of epochs to 5 within a range of 1 to 10.

RoBERTa architecture is built based on BERT's language masking strategy. However, it presents a key difference: it "removes BERT's next-sequence pre-trained objective and trains with much larger mini-batches and learning rates" [23]. The model is initialized with the pre-trained 'roberta-base' model from the HuggingFace's Transformers Library. We add a two-dimensional linear layer with softmax activation function. Dropout is empirically defined as 0.3. We use cross entropy loss as the loss function and Adam as optimizer with a learning rate of 1e–05. Batch size is defined as 16 and the maximum input length is 512, following BERT implementation details. After trying different number of epochs, we set the final value to 7.

## 3 Experiments

### 3.1 Data Set

The data set used is based on the Picture Description task, in which the participant sees an image of a kitchen in black and white and is asked to describe the scene. The data set is the English Pitt Corpus [24] available at DementiaBank. It contains 551 audio files and hand-made transcripts, of which 309 samples belong to the dementia group and 243 to the control group. There are 398 individual participants: 208 dementia, 104 control and 85 patients with unknown diagnosis. The dementia group contains probable AD, MCI and vascular dementia. In this paper, we consider MCI and early AD groups as a broad group ('dementia').

The final training set contains 118 control and 118 dementia samples while the held-out testing set is composed of 30 control and 30 dementia samples. The customized set is not balanced for age, sex and education. For reproducibility purposes, the original audio samples and transcripts are available at DementiaBank upon request, while the exact specification of the subset is available at[1].

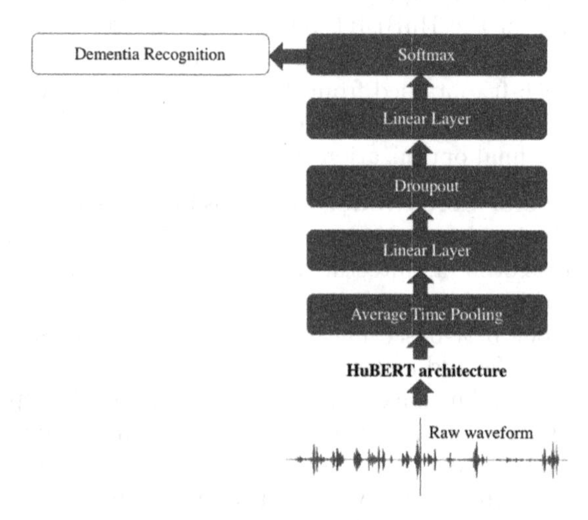


**Fig. 1.** Fine-tuning HuBERT for dementia classification.

## 3.2 Data Processing

The data set is extracted from interviews. Therefore, speaker diarization is carried out using pyAudioAnalysis [25], a Python package that applies a clustering approach to segment based on speaker. The number of clusters is set to 2. For the transcripts, we use the program CLAN to preserve the text information solely from the participant prescinding from the interviewer's transcripts. These transcripts already contain all relevant information to identify speakers.

Furthermore, given the small data set and due to memory constraints, we perform a simple data augmentation method that consists of splitting the audio samples into small segments. We also consider that smaller segments may increase performance since they provide less dimensionality to the models. Accordingly, the audio samples are cropped to a maximum duration of 8 s. Likewise, audio samples are down-sampled to align with the ASR models' sampling rate requirements (16k Hz).

[1] https://github.com/monicagoma/masters_thesis_dementia.

### 3.3 Evaluation Techniques

As we are employing a data set balanced for class, the proposed models are assessed in terms of accuracy, recall, precision and F1. To be consistent with the literature, each feature-based approach is evaluated using 5-fold Cross-Validation (CV). The final 5-fold CV performance is the average result obtained in each fold. Given memory constraints, CV validation is not performed in the transfer learning approaches. The final model's performance is evaluated on the held-out test set that represents 20% of the subset.

## 4 Results

In Table 1, we present our proposed conventional and baseline results in a cross-validation setting and Table 2 shows the results obtained on the held-out test set for the conventional and baseline models. The linguistic feature set outperforms the acoustic modality methods by a large margin(+.11 accuracy). The best-performing methods for the linguistic and acoustic modality are obtained applying SVM. Our overall findings are in line with previous results focusing solely on AD identification [26,27].

**Table 1.** Results of the conventional methods in cross validation for the acoustic and linguistic modality.

| Method | Modality | Accuracy | Class | F1 | Precision | Recall |
|---|---|---|---|---|---|---|
| Baseline | | .63 | Dementia | .70 | .65 | .63 |
| | | | Control | .55 | .51 | .63 |
| SVM | Acoustic | .67 | Dementia | .70 | .66 | .76 |
| | | | Control | .66 | .71 | .61 |
| RF | Acoustic | **.70** | Dementia | .70 | .69 | .58 |
| | | | Control | .70 | .71 | .79 |
| GNB | Acoustic | .64 | Dementia | .59 | .67 | .53 |
| | | | Control | .67 | .61 | .73 |
| SVM | Linguistic | .70 | Dementia | .67 | .65 | .62 |
| | | | Control | .72 | .67 | .78 |
| RF | Linguistic | .63 | Dementia | .59 | .64 | .62 |
| | | | Control | .64 | .64 | .64 |
| GNB | Linguistic | **.74** | Dementia | .74 | .74 | .74 |
| | | | Control | .74 | .74 | .74 |

Table 3 presents the results obtained on the held-out test for the transfer learning approaches. All approaches outperform the baseline model and furthermore the best-performing models (HuBERT and BERT) outperform conventional classifiers. Of special interest is the HuBERT model, which is an end-to-end methodology and has not been previously explored in cognitive impairment classification. Besides, linguistic-based methods outperform acoustic methods by a large margin(+.06 accuracy). Nonetheless, these approaches depend on human-made transcripts, which is a time-consuming technique.

**Table 2.** Results of the conventional methods in held-out test set.

| Method | Modality | Accuracy | Class | F1 | Precision | Recall |
|---|---|---|---|---|---|---|
| Baseline | | .66 | Dementia | .69 | .65 | .73 |
| | | | Control | .64 | .69 | .60 |
| SVM | Acoustic | **.68** | Dementia | .42 | .60 | .92 |
| | | | Control | .42 | .74 | .30 |
| RF | Acoustic | .63 | Dementia | .68 | .68 | .69 |
| | | | Control | .56 | .57 | .55 |
| GNB | Acoustic | .65 | Dementia | .72 | .67 | .77 |
| | | | Control | .54 | .61 | .49 |
| SVM | Linguistic | **.75** | Dementia | .78 | .70 | .87 |
| | | | Control | .72 | .83 | .63 |
| RF | Linguistic | .74 | Dementia | .75 | .62 | .93 |
| | | | Control | .58 | .87 | .43 |
| GNB | Linguistic | .72 | Dementia | .74 | .69 | .80 |
| | | | Control | .69 | .76 | .63 |

**Table 3.** Results of the transfer learning methods in held-out test set.

| Method | Modality | Accuracy | Class | F1 | Precision | Recall |
|---|---|---|---|---|---|---|
| ResNet152 | Acoustic | .68 | Dementia | .76 | .64 | .93 |
| | | | Control | .61 | .88 | .61 |
| HuBERT | Acoustic | **.74** | Dementia | .77 | .77 | .77 |
| | | | Control | .72 | .72 | .71 |
| BERT | Linguistic | **.81** | Dementia | .79 | .85 | .73 |
| | | | Control | .81 | .76 | .87 |
| RoBERTa | Linguistic | .80 | Dementia | .78 | .85 | .73 |
| | | | Control | .82 | .75 | .90 |

## 5 Conclusion

In this paper, we explore speech and linguistic modalities separately. Our findings indicate that AD and MCI obtained similar results as in AD classification. This suggests that AD and MCI can be jointly studied. Our proposed method consists of fine-tuning transfer learning approaches and training conventional classifiers for cognitive impairment classification. We find that transfer learning approaches outperform conventional classifiers and the proposed baseline model.

Even though linguistic approaches outperform acoustic methods, these methods depend on human-made transcripts, which do not provide a potential clinical application. HuBERT model, however, offers a potential solution for cognitive impairment classification since it does not depend on prepossessing steps, domain-knowledge and transcripts. This method also reaches a performance in line with previous work [8]. Although this architecture does not outperform either BERT nor RoBERTa models, we consider it might offer more potential. Further research is needed in order to identify the best audio-length for cognitive impairment identification using the pre-trained HuBERT model. A disadvantage of using this model is the sampling rate alignment since it adds external noise, which can interfere in the pre-trained ASR model implementation.

Future work should aim to study MCI individually. We also intend to explore a multimodal fusion applying our best-performing transfer learning approaches and to obtain transcripts employing an ASR tool. To support the obtained results, future work should also replicate this study with balanced data sets for age, sex and education and investigate other types of data such as conversational speech.

## References

1. What Is Dementia?. Accessed 10 Oct 2021. https://www.cdc.gov/aging/dementia/index.html
2. World Health Organization, Dementia. September 2021. Accessed 10 Oct 2021. https://www.who.int/news-room/fact-sheets/detail/dementia
3. Rajan, K.B., Wilson, R.S., Weuve, J., Barnes, L.L., Evans, D.: Cognitive impairment 18 years before clinical diagnosis of Alzheimer disease dementia. Neurology **85**(10), 898–904 (2015). https://doi.org/10.1212/WNL.0000000000001774, https://www.ncbi.nlm.nih.gov/pubmed/26109713
4. Beltrami, D., et al.: Speech analysis by natural language processing techniques: a possible tool for very early detection of cognitive decline?. Front. Aging Neurosci. **10** (2018).https://www.frontiersin.org/articles/10.3389/fnagi.2018.00369/full
5. Lanzi, A., Bourgeois, M., Wallace, S.: Group external memeory aid treatment for mild cognitive impairment. Alzheimer's I& Dementia **13**(7), 257 (2017). https://dx.doi.org/10.1016/j.jalz.2017.06.121
6. Szatloczki, G., Hoffman, I., Vincze, V., Kalman, J., Pakaski, M.: Speaking in Alzheimer's disease, is that an early sign? importance of changes in language abilities in Alzheimer's disease. Front. Aging Neurosci. **7** (2015). https://www.frontiersin.org/articles/10.3389/fnagi.2015.00195/full
7. Chakraborty, R., Pandharipande, M., Bhat, C., Kopparapu, S.K.: Identification of Dementia Using Audio Biomarkers (2020). https://arxiv.org/abs/2002.12788
8. de la Fuente Garcia, S., Ritchie, C.W., Luz, S.: Artificial intelligence, speech, and language processing approaches to monitoring Alzheimer's disease: a systematic review. J. Alzheimer's Dis. **78**(4), 1547–1574 (2020). https://doi.org/10.3233/JAD-200888, https://www.ncbi.nlm.nih.gov/pubmed/33185605

9. Pastoriza-Dominguez, P., et al: Speech pause distribution as an early marker for Alzheimer's disease. Speech Commun. **136**, 107–117 (2022). https://doi.org/10.1101/2020, https://www.sciencedirect.com/science/article/pii/S0167639321001333.12.28.20248875
10. Ilias, L., Askounis, D., Psarras, J.: Detecting Dementia from Speech and Transcripts using Transformers (2021). https://arxiv.org/abs/2110.14769
11. Cummins, N., et al.: A comparison of acoustic and linguistics methodologies for Alzheimer's dementia recognition. In: Proceedings Interspeech 2020, pp. 2182–86 (2020). https://doi.org/10.21437/interspeech.2020-2635, https://www.isca-speech.org/archive/interspeech_2020/cummins20_interspeech.html
12. Balagopalan, A., Novikova, J.: Comparing acoustic-based approaches for Alzheimer's disease detection. In: Proceedings of Interspeech 2021 (2021). https://doi.org/10.21437/Interspeech.2021-759,https://www.isca-speech.org/archive/interspeech_2021/balagopalan21_interspeech.html
13. Rohanian, M., Hough, J., Purver, M.: Multi-modal fusion with gating using audio, lexical and disfluency features for Alzheimer's dementia recognition from spontaneous speech. In: Interspeech 2020, pp. 2187–2191 (2020). https://doi.org/10.21437/interspeech.2020-2721
14. Di Palo, F., Parde, N.: Enriching neural models with targeted features for dementia detection. In: ACL (2019). https://aclanthology.org/P19-2042/
15. Guo, Y., Li, C., Roan, C., Pakhomov, S., Cohen, T.: Crossing the cookie theft corpus chasm: applying what BERT learns from outside data to the ADReSS challenge dementia detection task. Front. Comput. Sci. (Lausanne), **3** (2021). https://doi.org/10.3389/fcomp.2021.642517, https://doaj.org/article/417d2905f8ed446884c6ff7f860e4453
16. Campbell, E.L. et al.: Alzheimer's Dementia Detection from Audio and Text Modalities (2020). https://arxiv.org/abs/2008.04617
17. Balagopalan, A., Eyre, B., Rudzicz, F., Novikova, J.: To BERT or Not To BERT: comparing speech and language-based approaches for Alzheimer's disease detection. In: Proceedings Interspeech 2020 (2020). https://arxiv.org/abs/2008.01551. https://doi.org/10.21437/Interspeech.2020-2557
18. Clarke, N., Barrick, T.R., Garrard, P.: A comparison of connected speech tasks for detecting early Alzheimer's disease and mild cognitive impairment using natural language processing and machine learning. Front. Comput. Sci. (Lausanne), **3** (2021). https://doi.org/10.3389/fcomp.2021.634360, https://doaj.org/article/ef7ae92f93544eefacafacbab4dfa2cd
19. Schuller, B., et al.: The INTERSPEECH 2013 computational paralinguistics challenge: social signals, conflict, emotion, autism. In: Proceedings of Interspeech 2013 (2013). https://doi.org/10.21437/Interspeech.2013-56, https://www.isca-speech.org/archive/interspeech_2013/schuller13_interspeech.html
20. MacWhinney, B.: The Childes Project. Taylor and Francis (2014)
21. de Lira, J.O., Minnet, T.S., Ferreira P.H., Ortiz, K.Z.: Analysis of word number and content in discourse of patients with mild to moderate Alzheimer's disease. Dementia Neuropsychol. **8**(3), 260–265 (2014). https://doi.org/10.1590/S1980-57642014DN83000010, https://www.ncbi.nlm.nih.gov/pubmed/29213912.57642014DN83000010
22. Balagopalan, A., Eyre, B., Rudzicz, F., Novikova, J.: To BERT or Not To BERT: comparing speech and language-based approaches for Alzheimer's disease detection. In: Proceedings Interspeech 2020 (2020). https://doi.org/10.21437/Interspeech.2020-2557, https://arxiv.org/abs/2008.01551

23. Liu, Y., et al.: RoBERTa: A Robustly Optimized BERT Pretraining Approach (2019). https://arxiv.org/abs/1907.11692
24. Becker, J.T., Boller, F., Lopez, O.L., Saxton, J., McGonigle, K.L.: The natural history of Alzheimer's disease: description of study cohort and accuracy of diagnosis. Arch. Neurol. **51**(6), 585–594 (1994). https://dementia.talkbank.org/access/0docs/Becker1994.pdf
25. Giannakopoulos, T.: pyAudioAnalysis: an open-source python library for audio signal analysis. PloS One **10**(12) (2015). https://doi.org/10.1371/journal.pone.0144610, https://www.ncbi.nlm.nih.gov/pubmed/26656189
26. Luz, S., Haider, F., de la Fuente, S., Fromm, D., MacWhinney, B.: Alzheimer's dementia recognition through spontaneous speech: the ADReSS challenge. In: Proceedings Interspeech, pp. 2172–2176 (2020). https://doi.org/10.21437/Interspeech.2020-2571
27. Luz, S., Haider, F., de la Fuente, S., Fromm, D., MacWhinney, B.: Detecting cognitive decline using speech only: the ADReSSo Challenge (2021). https://doi.org/10.1101/2021.03.24.21254263, https://arxiv.org/abs/2104.09356

# Detection of Alzheimer's Disease Using a Four-Channel EEG Montage

Eduardo Perez-Valero[1,2(✉)], Jesus Minguillon[2,3], Christian Morillas[1,2], Francisco Pelayo[1,2], and Miguel A. Lopez-Gordo[2,3]

[1] Department of Computer Architecture and Technology, University of Granada, 18071 Granada, Spain
{edu,cmg,fpelayo}@ugr.es

[2] Brain-Computer Interfaces Lab (BCI Lab), Research Centre for Information and Communications Technologies (CITIC), University of Granada, 18071 Granada, Spain
{minguillon,malg}@ugr.es

[3] Department of Signal Theory, Telematics and Communications, University of Granada, 18071 Granada, Spain

**Abstract.** Alzheimer's disease (AD) represents the most common form of dementia, hence, its diagnosis and treatment have a heavy socioeconomic impact. Among the medical procedures for AD diagnosis, those based on biochemical markers and medical images stand out. However, data acquisition and processing times associated to these procedures are long, what contributes to growing clinical waiting lists. On the other hand, traditional approaches like neuro-psychological tests, which evaluate cognitive performance, are less powerful since they are influenced by external factors like schooling level or visual health. As an alternative, researchers have proposed the analysis of resting state electroencephalography (EEG), since it is an inexpensive and non-invasive technique to acquire endogenous brain information. Nevertheless, EEG recordings are typically performed using cumbersome electrical setups with many channels (usually 16–64), what hinders the usability of these approaches. The aim of this work is to demonstrate that early-stage AD can be detected using a simple EEG montage, and to elucidate the most relevant EEG channels in terms of classification performance. To this aim, we recorded the resting state EEG of eight patients with early-stage AD and eight healthy controls, and we implemented a binary classifier based on a multilayer perceptron. The cross validation results we obtained suggest that early-stage AD can be detected (F1 score $= 0.89 \pm 0.06$) using an EEG montage consisting of four channels (Fz, C3, Cz, and C4). This opens the possibility for early-stage AD detection in early care services within minutes, using a wearable plug-and-play EEG headset.

**Keywords:** Alzheimer's disease · EEG · Machine learning · Artificial intelligence

This research was supported by projects B-TIC-352-UGR20 (Junta de Andalucia), PGC2018-098813-B-C31 (Spanish Ministry of Science, Innovation and Universities), and the Postdoctoral Fellowship Programme of Junta de Andalucia (PAIDI 2020).


J. M. Ferrández Vicente et al. (Eds.): IWINAC 2022, LNCS 13258, pp. 436–445, 2022.
https://doi.org/10.1007/978-3-031-06242-1_43

## 1 Introduction

Dementia is a neuro-degenerative pathology that encompasses a collection of symptoms regarding cognitive impairment such as loss of memory, language deterioration, and loss of spatial and temporal orientation. This is commonly accompanied by emotional disorders such as personality and mood changes. According to the World Health Organization, more than 55 million people live with dementia, with nearly 10 million new cases every year. Due to the ageing of worldwide population, this number is estimated to rise to 78 million in 2030 and 139 million in 2050. Furthermore, the direct medical and social care costs of dementia are expected to surpass USD 2.8 trillion by 2030. In this context, Alzheimer's disease (AD) is the most common form of dementia since it represents 60–70% of the cases. Although there exists no cure for AD to date, medical treatment can help control the progression of the disease and postpone intellectual decline. This is why early detection is one of the main goals in AD care.

Among the medical procedures for early AD diagnosis, biochemical markers extracted from cerebro-spinal fluid and linked to amyloid plaques and neurofibrillary tangles [19] are the most reliable. However, they have some drawbacks: 1) the risks associated with their invasive nature (e.g., infections, tissue damage, patient reluctance, etc.) that limit their usability, and 2) their complexity in terms of procedures for sample acquisition and processing that extends the diagnosis time and contributes to the growth of clinical waiting lists. Alternatively, techniques based on medical imaging such as single-photon emission computed tomography [8], magnetic resonance imaging [9], and positron emission tomography [13] also stand out. Although the efficacy of image-based techniques is limited for early AD diagnosis, their combination with artificial intelligence algorithms has been proven effective [17]. Nevertheless, like medical procedures based on biochemical markers, these techniques also present limitations, such as long acquisition and processing times, what contributes to increasing diagnosis time and growing waiting lists.

As an alternative to biochemical markers and image-based techniques, and with the aim of performing a non-invasive and fast diagnosis, some authors have proposed approaches based on neuro-psychological tests to evaluate cognitive performance. The most widely used at the clinical level are the Montreal cognitive assessment [16] and the mini-mental state examination [3]. However, these tests have a lower detection capability compared with the techniques mentioned above [12,14]. Additionally, they are influenced by factors such as visual pathologies, educational level and emotional state. In this sense, techniques relying on endogenous information are more advantageous.

To reduce the complexity of data acquisition and processing while keeping the advantages of endogenous information, some authors have presented approaches based on electroencephalography (EEG) [1]. These approaches typically consist in analyzing the resting state brain activity through machine learning methods [2,4,18,20], and have shown classification accuracies in the range of [0.83, 0.97] [6,10,12,21]. However, they have an important drawback regarding usability: EEG recordings are typically performed using cumbersome electrical setups in the electrophysiology facilities available at some hospitals. Furthermore, the use of these facilities is limited due to their scarcity. This makes these approaches impractical for early care services. With respect to the setups, these are cumbersome because of the number of electrodes used in the recordings: the most common EEG montages range from 16 to 64 informative electrodes or channels, plus reference and ground electrodes. Conversely, when EEG data are processed, the information from several channels is removed by means of data dimensionality reduction techniques such as principal component analysis. This step is performed to avoid redundancy of information caused by the short distance between electrodes and the low spatial selectivity of EEG. Data dimensionality reduction is especially noticeable in some applications [7,21].

Therefore, why to record EEG using many channels and then remove the information provided by most of them? A simplified EEG montage consisting of a reduced number of channels could be suitable for some applications. With a view to promoting the use of resting state EEG as a practical and accessible screening tool for early AD diagnosis at the clinical level, we hypothesized that it is possible to detect the disease in early stages using a simplified EEG montage. In this work, we demonstrate that this is feasible by using a four-channel EEG montage along with a classifier based on a multi-layer perceptron (MLP). In addition, we elucidate the most relevant channels in terms of classification performance. This leads us to envision a scenario where early-stage AD can be screened in care services using a wearable plug-and-play headset within minutes, as a prior step to a more comprehensive diagnosis through established biochemical markers.

## 2 Materials and Methods

In this section, we report the main aspects of the methodology followed in this study. We describe the cohorts involved, the experimental procedure, the signal processing steps, and the classification problem we evaluated.

### 2.1 Participants

We conducted the data capture for this study in collaboration with the cognitive and behavioral neurology unit (CBNU) at Hospital Universitario Virgen de las Nieves de Granada (HUVN). Twenty-one participants voluntarily participated in the study. Twelve of the participants were AD patients of the CBNU at HUVN. Following memory complaints, these patients were diagnosed during the year prior to the onset of this study. The diagnosis was based on the presence of positive biomarkers in one of the subsequent two medical evaluations: (a) measurement of Aβ42, p-tau, and total tau in CSF, or (b) β-amyloid PET (florbetaben PET). The remaining nine participants were healthy age-matched controls (HC). We conducted this study in compliance with a protocol accepted by the ethics committee at HUVN. Due to issues during the acquisition and lack of cooperation we discarded the data from five participants, therefore, we considered data from sixteen participants (8 AD and 8 HC) in our analysis.

### 2.2 EEG Acquisition

We recorded the eye-open EEG activity of the participants in resting state. For each participant, we performed two acquisitions: before and after a cognitive task unrelated to this study. Prior to the experiment onset, we asked the participants to read and sign an informed consent, and to remain relaxed and awake during the recordings. Overall, we acquired six minutes of resting state brain activity per participant. For the acquisition, we used the Versatile EEG system by Bitbrain (Zaragoza, Spain), a wireless wearable device with sixteen semi-dry electrodes working at a fixed sampling rate 256 Hz. For the montage, we considered the following electrode positions of the extended 10–20 International System to evenly cover the scalp: Fp1, Fp2, F5, Fz, F6, T7, T8, C3, Cz, C4, P5, Pz, P6, O1, Oz, and O2.

### 2.3 Preprocessing

First, we filtered the raw signals using a bandpass FIR filter with 1–45 Hz passband and zero phase-shift. Then, we split the filtered signals into four-second detrended epochs without overlapping. After this, we performed artifact rejection in two steps. First, we applied the Autoreject algorithm to find an artifact threshold per channel, mark, and reject bad data spans. Autoreject is a data-driven algorithm for artifact processing based on cross-validation and Bayesian optimization that is available as an open-source Python package [11]. Then, we applied independent component analysis to remove blink artifacts. To identify the component related to the blink artifact, we used Fp1 channel as an electro-oculugram proxy, and selected the component that showed the highest correlation with the signal registered from this channel.

### 2.4 Classification

After preprocessing the recordings, we extracted the following features per channel from the resulting epochs: relative power (for the delta, theta, alpha, beta, and gamma bands), spectral entropy, and Hjorth complexity. Consequently, for each participant, we obtained a feature matrix with a row per epoch and a target array with the group label corresponding to each epoch (AD or HC). Then, we averaged the features from every six consecutive epochs in the feature matrix to enhance the signal-to-noise ratio and reduce the size of the matrix. We generated the total dataset through the concatenation of the feature matrices and target arrays of all the participants. For classification, we implemented a two-stage pipeline consisting of: feature scaling and classification. For feature scaling, we used a min-max scaler, and for classification, we implemented a MLP. We found the optimal parameters of the classification pipeline through grid search cross-validation. Throughout this procedure, we followed a leave-one-subject-out strategy to ensure the data from a participant was not included in the training and test sets simultaneously.

### 2.5 Selection of Most Relevant Channels

We wanted to evaluate the feasibility of a reduced montage for the discrimination of AD patients and HC. To this end, we applied a $chi^2$ test to the extracted features. This test evaluates the dependence between each feature and the target, hence, allowing to keep only those features ranked the highest. Thereafter, we summed the scores yielded by the $chi^2$ test for all the features corresponding to each channel. Then, we constructed a topographical map to illustrate the accumulated score per channel. Based on this, we selected the four channels with the highest score, and we repeated the binary classification problem using only the features corresponding to these channels.

## 3 Results and Discussion

In this section we report and discuss the results for the binary classification problem examined in this study and the selection of relevant channels that we performed. Note that the results reported in this section refer to the classification pipeline with the best parameters found via grid-search.

Figure 1 represents the cross-validation performance and the confusion matrix for the sixteen-channel montage. In general, a classification accuracy around 0.85 or higher indicates good classification performance. With this in mind, the classification performance we obtained using this EEG montage, in terms of F1 score ($0.90 \pm 0.05$) and accuracies for both classes (0.86% for AD and 0.84% for HC), is similar to those reported in related studies [10,12,21], and even higher in some cases [6,15]. Moreover, we used a leave-one-subject-out cross validation, which is closer to realistic scenarios than other strategies like K-fold. At the participant level, our classification pipeline reached 100% accuracy for 10 out of 16 participants, with a balance between participants of both classes (5 AD and 5 HC). The model only yielded an accuracy below chance level for two participants (1 AD and 1 HC).

Figure 2 displays the accumulated score yielded from the $chi^2$ test for all the channels. These results indicate that the most relevant channels in terms of classification performance are Cz, Fz, C4, and P6. However, unreported classification results that we obtained indicate that the F1 score and the accuracy for the HC class are unacceptable if only right hemisphere channels are selected. Thus, brain activity from both hemispheres is needed to achieve accurate and balanced classification performance. This requirement is in agreement with studies where channels from both hemispheres were considered [5,6,20]. That is the reason why we selected a hemisphere-balanced configuration including C3. In addition, to minimize the number of channels and after checking that the classification performance was not noticeably affected, we did not include P5 and P6 in the reported four-channel EEG montage. Thereafter, the four channels that we selected and thus we propose for early diagnosis of AD are: Fz, C3, Cz, and C4.

Figure 3 represents the equivalent to Fig. 1 for the four-channel montage. The results we obtained using this EEG montage indicate a classification performance

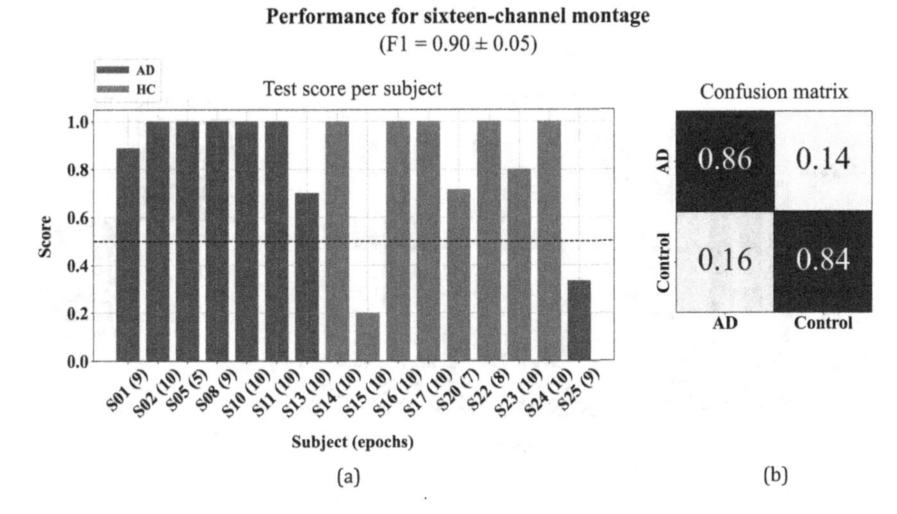


**Fig. 1.** Cross-validation performance for the sixteen-channel montage. (a) Test set accuracy for each cross-validation iteration. Since we followed a leave-one-subject-out strategy, on each iteration, the test set included all the samples from a single participant. (b) Confusion matrix built using the predictions from all the test sets in the cross-validation. The average cross-validation F1 score ± the standard error of the mean are reported in brackets under the figure title.

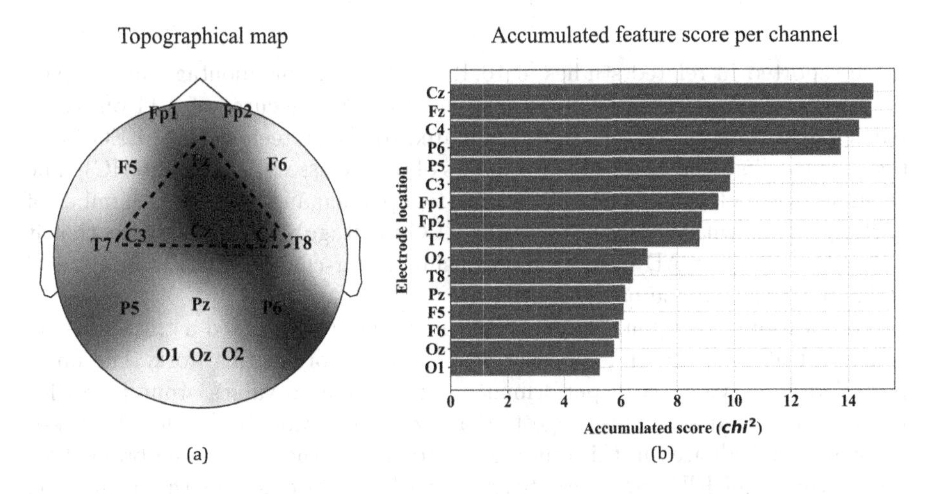


**Fig. 2.** Accumulated feature score per channel. (a) Topographical representation of the accumulated score. The dashed triangle encloses the four channels that we selected for the reduced montage. (b) Bar plot representation of the accumulated score.

at the epoch level that is similar to the one we reported for the sixteen-channel EEG montage in terms of F1 score ($0.89 \pm 0.06$) and accuracy for both classes (0.85% for AD and 0.87% for HC), and similar or even superior compared to

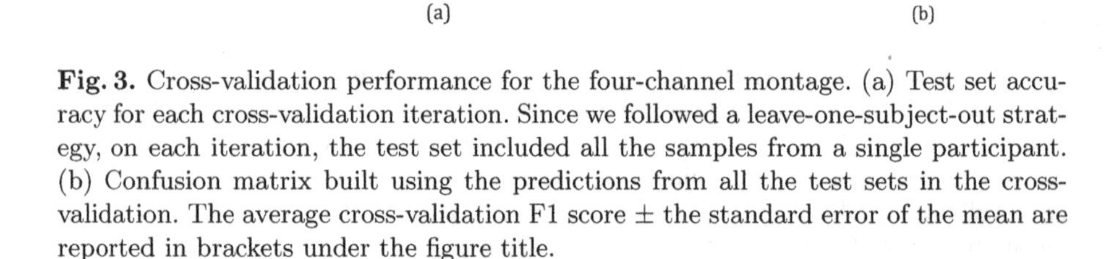


**Fig. 3.** Cross-validation performance for the four-channel montage. (a) Test set accuracy for each cross-validation iteration. Since we followed a leave-one-subject-out strategy, on each iteration, the test set included all the samples from a single participant. (b) Confusion matrix built using the predictions from all the test sets in the cross-validation. The average cross-validation F1 score ± the standard error of the mean are reported in brackets under the figure title.

those reported in related studies [6,10,12,15,21]. For this montage, at the participant level, the classification pipeline yielded 100% accuracy for 11 out of 16 participants (one participant more compared to the sixteen-channel EEG montage), with a balance between participants of both classes (5 AD and 6 HC). The accuracy was below chance level for only two participants (the same number of participants as with the sixteen-channel EEG montage). All this suggests that it is feasible to detect AD in early stages using a four-channel EEG montage and a classification pipeline based on a MLP.

In addition to the four-channel EEG montage, we explored a three-channel configuration (C3, Cz, and C4) that we did not report in this work. The unreported results evidenced a performance degradation to values around 0.80 for F1 score and accuracy. This suggests that Fz is a relevant channel for the classification task evaluated in this study. Despite of this, the results we obtained for the four-channel EEG montage, together with the briefness of the acquisition (a patient could be classified using the average of six 4-second epochs, this is, 24 s of resting state EEG) and the reduced computational cost of the algorithm, lead us to envision the following scenario: a physician can perform a screening of early-stage AD in an early care service within minutes using a wearable headset. Based on the outcome, the physician can then decide if further evaluations based on established biochemical markers are required. Of course, the feasibility of the envisioned scenario has to be validated first through a new study including a higher number of participants.

# 4 Conclusion

In this work we aimed to demonstrate that it is feasible to detect AD in early stages using a simple EEG montage. In addition, we aimed to elucidate the most relevant EEG channels in terms of classification performance. We extracted power, entropy, and complexity features from the resting state EEG of eight healthy participants and eight patients with early-stage AD an we implemented a self-driven binary classification pipeline based on a multi-layer perceptron. We evaluated this pipeline via leave-one-subject-out cross validation, and the results we obtained suggest that early-stage AD can be detected using a four-channel EEG montage consisting of Fz, C3, Cz, and C4. This opens the possibility for early-stage AD to be screened in early care services within minutes, using a wearable plug-and-play headset.

**Acknowledgements.** The authors would like to acknowledge the members of the Cognitive and Behavioral Neurology Unit at Hospital Universitario Virgen de las Nieves de Granada, and the participants who took part in the study for their cooperation in this work.

# References

1. Adeli, H., Ghosh-Dastidar, S., Dadmehr, N.: A spatio-temporal wavelet-chaos methodology for EEG-based diagnosis of Alzheimer's disease. Neurosci. Lett. **444**(2), 190–194 (2008). https://doi.org/10.1016/j.neulet.2008.08.008. https://www.sciencedirect.com/science/article/pii/S0304394008010860
2. Amezquita-Sanchez, J.P., Mammone, N., Morabito, F.C., Marino, S., Adeli, H.: A novel methodology for automated differential diagnosis of mild cognitive impairment and the Alzheimer's disease using EEG signals. J. Neurosci. Methods **322**, 88–95 (2019). https://doi.org/10.1016/j.jneumeth.2019.04.013. http://www.sciencedirect.com/science/article/pii/S0165027019301335, 00000
3. Arevalo-Rodriguez, I., et al.: Mini-Mental State Examination (MMSE) for the early detection of dementia in people with mild cognitive impairment (MCI). Cochrane Database Systematic Rev. **7**, CD010783 (2021). https://doi.org/10.1002/14651858.CD010783.pub3
4. Cassani, R., Estarellas, M., San-Martin, R., Fraga, F.J., Falk, T.H.: Systematic Review on Resting-State EEG for Alzheimer's Disease Diagnosis and Progression Assessment, October 2018. https://doi.org/10.1155/2018/5174815. https://www.hindawi.com/journals/dm/2018/5174815/, iSSN: 0278-0240 Pages: e5174815 Publisher: Hindawi Volume: 2018
5. Cassani, R., Falk, T.H., Fraga, F.J., Cecchi, M., Moore, D.K., Anghinah, R.: Towards automated electroencephalography-based Alzheimer's disease diagnosis using portable low-density devices. Biomed. Signal Process. Control **33**, 261–271 (2017). https://doi.org/10.1016/j.bspc.2016.12.009. http://www.sciencedirect.com/science/article/pii/S174680941630221X, 00000
6. Fiscon, G., et al.: Combining EEG signal processing with supervised methods for Alzheimer' patients classification. BMC Med. Inform. Decision Making **18**(1), 35 (2018). https://doi.org/10.1186/s12911-018-0613-y. https://bmcmedinformdecismak.biomedcentral.com/articles/10.1186/s12911-018-0613-y, 00000

7. Gallego-Jutglà, E., Solé-Casals, J., Vialatte, F.B., Elgendi, M., Cichocki, A., Dauwels, J.: A hybrid feature selection approach for the early diagnosis of Alzheimer's disease. J. Neural Eng. **12**(1), 016018 (2015). https://doi.org/10.1088/1741-2560/12/1/016018, https://iopscience.iop.org/article/10.1088/1741-2560/12/1/016018
8. Górriz, J.M., Segovia, F., Ramírez, J., Lassl, A., Salas-Gonzalez, D.: GMM based SPECT image classification for the diagnosis of Alzheimer's disease. Appl. Soft Comput. **11**(2), 2313–2325 (2011). https://doi.org/10.1016/j.asoc.2010.08.012. url-http://www.sciencedirect.com/science/article/pii/S1568494610002140,00079
9. Hojjati, S.H., Ebrahimzadeh, A., Khazaee, A., Babajani-Feremi, A.: Predicting conversion from MCI to AD using resting-state fMRI, graph theoretical approach and SVM. J. Neurosci. Methods **282**, 69–80 (2017). https://doi.org/10.1016/j.jneumeth.2017.03.006. http://www.sciencedirect.com/science/article/pii/S0165027017300638, 00000
10. Ieracitano, C., Mammone, N., Hussain, A., Morabito, F.C.: A novel multi-modal machine learning based approach for automatic classification of EEG recordings in dementia. Neural Networks **123**, 176–190 (2020). https://doi.org/10.1016/j.neunet.2019.12.006. http://www.sciencedirect.com/science/article/pii/S0893608019303983, 00000
11. Jas, M., Engemann, D.A., Bekhti, Y., Raimondo, F., Gramfort, A.: Autoreject: automated artifact rejection for MEG and EEG data. NeuroImage **159**, 417–429 (2017). https://doi.org/10.1016/j.neuroimage.2017.06.030. https://www.sciencedirect.com/science/article/pii/S1053811917305013
12. Kulkarni, N.N., Bairagi, V.K.: Extracting salient features for EEG-based diagnosis of Alzheimer's disease using support vector machine classifier. IETE J. Res. **63**(1), 11–22 (2017). https://doi.org/10.1080/03772063.2016.1241164. https://doi.org/10.1080/03772063.2016.1241164, 00022 Publisher: Taylor & Francis _eprint
13. Meikle, S.R., Beekman, F.J., Rose, S.E.: Complementary molecular imaging technologies: high resolution SPECT, PET and MRI. Drug Discovery Today Technol. **3**(2), 187–194 (2006). https://doi.org/10.1016/j.ddtec.2006.05.001. https://linkinghub.elsevier.com/retrieve/pii/S1740674906000229, 00044
14. Mendiondo, M.S., Ashford, J.W., Kryscio, R.J., Schmitt, F.A.: Modelling mini mental state examination changes in Alzheimer's disease. Stat. Med. **19**(11-12), 1607–1616 (2000). https://doi.org/10.1002/(SICI)1097-0258(20000615/30)19:11/12⟨1607::AID-SIM449⟩3.0.CO;2-O
15. Morabito, F.C., et al.: Deep convolutional neural networks for classification of mild cognitive impaired and Alzheimer's disease patients from scalp EEG recordings. In: 2016 IEEE 2nd International Forum on Research and Technologies for Society and Industry Leveraging a better tomorrow (RTSI). pp. 1–6, September 2016. https://doi.org/10.1109/RTSI.2016.7740576, 00028 ZSCC: NoCitationData[s2] ISSN: null
16. Nasreddine, Z.S., et al.: The montreal cognitive assessment, MoCA: a brief screening tool for mild cognitive impairment. J. Am. Geriatrics Soc. **53**(4), 695–699 (2005). https://doi.org/10.1111/j.1532-5415.2005.53221.x
17. Pan, D., Zeng, A., Jia, L., Huang, Y., Frizzell, T., Song, X.: Early detection of Alzheimer's disease using magnetic resonance imaging: a novel approach combining convolutional neural networks and ensemble learning. Front. Neurosci. **14**, 259 (2020). https://doi.org/10.3389/fnins.2020.00259, https://www.ncbi.nlm.nih.gov/pmc/articles/PMC7238823/

18. Perez-Valero, E., Lopez-Gordo, M.A., Morillas, C., Pelayo, F., Vaquero-Blasco, M.A.: A review of automated techniques for assisting the early detection of Alzheimer's disease with a focus on EEG. J. Alzheimer's Disease: JAD (2021). https://doi.org/10.3233/JAD-201455
19. Perrin, R.J., Fagan, A.M., Holtzman, D.M.: Multimodal techniques for diagnosis and prognosis of Alzheimer's disease. Nature **461**(7266), 916–922 (2009). https://doi.org/10.1038/nature08538. https://www.nature.com/articles/nature08538, 00590
20. Ruiz-Gómez, S.J., et al.: Automated multiclass classification of spontaneous EEG activity in Alzheimer's disease and mild cognitive impairment. Entropy **20**(1), 35 (2018). https://doi.org/10.3390/e20010035. https://www.mdpi.com/1099-4300/20/1/35, 00019
21. Trambaiolli, L.R., Spolaôr, N., Lorena, A.C., Anghinah, R., Sato, J.R.: Feature selection before EEG classification supports the diagnosis of Alzheimer's disease. Clin. Neurophysiol. **128**(10), 2058–2067 (2017). https://doi.org/10.1016/j.clinph.2017.06.251. http://www.sciencedirect.com/science/article/pii/S1388245717304790

# Evaluating Imputation Methods for Missing Data in a MCI Dataset

Alba Gómez-Valadés Batanero(✉), Mariano Rincón Zamorano, Rafael Martínez Tomás, and Juan Guerrero Martín

Universidad Nacional de Educación a Distancia, 28040 Madrid, Spain
albagvb@dia.uned.es

**Abstract.** Missing data is a recurrent problem in experimental studies, mostly in clinical and sociodemographic longitudinal studies due to the dropout and the negative of some subjects to answer or perform some tests. To address this problem different strategies have been designed to deal with missing values, but incorrect treatment of missing data can result in the database being biased in one or more parameters, compromising the viability of the database and future studies. To solve this problem different imputation techniques have been developed over the last decades. However, there are no regulations or clear guidelines to deal with these situations. In this study, we will analyze and impute a real, incomplete database for the early detection of MCI, where the loss of values on 3 main variables is strongly correlated with the years of studies. The imputation will follow two strategies: assuming that those people would have got a bad scoring if they had taken the test, defining a ceiling score, and a multiple imputation by fully conditional specification. To determine if any kind of bias in mean and variance has been introduced during the imputation, the original database was compared with the imputed databases. Taking a p-value = 0.1 threshold, the database imputed by the multiple imputation method is the one that best preserved the information of the original database, making it the more appropriate imputation method for this MCI database.

**Keywords:** Missing data · Imputation · Multiple imputation

## 1 Introduction

Missing data is a persistent problem mostly in clinical and sociodemographic studies [1,2], mostly in longitudinal studies, where subjects can drop from the study due to illness, ilocation, negative to follow in the study, or death. Between those that keep with the study, there is always the possibility that subjects refuse to do some of the tests [1]. Missing data may seriously compromise inferences from randomized clinical trials, especially if missingness is not at random and if missing data are not handled appropriately, causing potential bias that distorts parameters and relationships between variables which can cause misleading results and conclusions [2–4].


J. M. Ferrández Vicente et al. (Eds.): IWINAC 2022, LNCS 13258, pp. 446–454, 2022.
https://doi.org/10.1007/978-3-031-06242-1_44

To overcome the problems caused by missing data, there has been extensive development of statistical models and software for imputing the data, or by directly analyzing data with missing values [5]. However, in order to select the best option for a dataset, as a first step is necessary to analyze the causes and the distribution of the missing data. Data may be missing due to one of following three reasons: Missing completely at Random (MCAR), Missing at Random (MAR), and Missing Not At Random (MNAR) [6,7]. If the mechanism causing the missing data does not depend on observed or unobserved variables, the data is called MCAR. The deletion of records with missing data does not introduce any bias since the reduced dataset is representative of the original dataset, being the only problem the reduction of the sample size [2,3,8]. If the missing data only depends on the observed variables it is called MAR, where is possible to predict missing values given the existing data [2,3]. However, the deletion of the cases with missing values will generate a biased subdataset. If the distribution of missing data depends on unobserved variables, it is called MNAR. This situation is the most difficult to be detected and to deal with since the relationship is not contained in the database itself and is therefore unknown, and usually causes bias in the subdataset whether the records are imputed or deleted [2,3,9]. Missing data can also be classified as ignorable and non-ignorable. Missing data is ignorable if the probability of observing a data item is independent of the value of that item, and it corresponds with the MCAR and MAR scenarios. If the probability of observing a data item is dependent on the value of that item, the missing data is non-ignorable, and it corresponds with the MNAR scenario [10].

To cope with this missing data problem, especially in scenarios where deletion of missing data is not recommended due to strong dataset reduction and/or introduction of bias, different types of strategies and techniques to estimate the effect on the missing data have been developed [8,9]. The simplest method is the complete case analysis, where all observations with missing data are discarded, but is prone to introduce bias in no-MCAR scenarios [9]. When deletion of records is not considered an adequate option, the data imputation by using the observed data and employing a series of rules is used. The most basic imputation method is to complete the missing values with the mean, median, or mode. Like complete case analysis, this strategy is prone to introduce bias [11], but it can be useful if just a handful of scores are missing. Imputations based on statistical approaches, like logistic regression or k-nearest neighborhoods, have been proposed, but while they may produce better solutions, they may also create a distorted dataset if the assumptions are broken [11]. Finally, multiple imputation is a widely used method because it is a simple and powerful strategy to impute missing data [13]. Currently, two major iterative methods are used for doing multiple imputation: joint modeling (JM) and fully conditional specification (FCS) [11,12]. JM is based on parametric statistical theory and leads to imputation procedures whose statistical properties are known [12]. However, it assumes normality and linearity, and also it is often difficult to realistically specify a joint model, potentially leading to bias [11]. FCS generates imputations by iterating over the conditional densities on a variable-by-variable basis, given an starting point [11,12]. However, depending on the quantity and the distribution of the missing data, sometimes more simple methods can achieve results as good as multiple imputation. In this context, it is preferable to first make a statistical analysis of the missing data distribution, and the possible relationships between variables before selecting a strategy.

In this paper, we will study the missing value distribution and the problem of the imputation of missing data in the database of a longitudinal study of the prevalence and evolution of MCI in a group of monolingual Spanish subjects. Therefore, this database presents the characteristics indicated above and makes it necessary to use these imputation techniques so that the studies carried out with it are valid and informative. In order to select the most appropriate method to avoid losing information or altering the information in future analyses, we will compare two different imputation methodologies.

This study is part of a broader project focused on the comparison of two databases obtained from subjects with different sociodemographic backgrounds, in order to identify the most discriminating variables, as well as those machine learning systems with the best performance in predicting the diagnosis, with the idea of creating an expert system that could help in diagnosis and save time for physicians in the evaluation of certain tests that are currently performed by hand.

## 2 Methodology

### 2.1 Database Description

The database consists of a sample from a large ongoing longitudinal study with the aim of discerning what types of tests, subtests, or sociodemographic variables had the greatest influence when making an early diagnosis of MCI and to determine the prevalence and stability of MCI in the Autonomous Community of Madrid (Spain) [14,15]. As it is shown in Table 1, the database is composed of 947 cases of Spanish monolingual subjects, with 540 cases with missing values. The mean age is nearly 71 years and a standard deviation of 6'31 years. The education, measured in schooling years, has a mean of 11'5 years of schooling, with a standard deviation of 6'45 years. The database is split up to 4 evaluations, with 428 cases on first evaluation, 319 on second evaluation, 171 on third evaluation, 18 on forth evaluation and 11 of unknown evaluation. Each evaluation corresponds to the completion of the entire test battery by the same subject, with an average difference of 1 year between evaluations.

The database used in this study is composed of a total of 32 variables, described in more detail in Table 2, and grouped in identifiers and database metadata, sociodemographic variables, psychologic tests, diagnosis, and independent blocks of screening variables. The screening variables are in turn made up of the groups of MMSE, Verbal Fluency, Graphic tests, Ideomotor Tests, Trail Marking Tests, Rey Figure's tests, and the TAVEC, and integrates the data used to assess if a patient is healthy or MCI, and are the variables in which we will

**Table 1.** Distribution of the Spanish dataset cases and variables.

| | |
|---|---|
| Total cases (without missing values) | 947 (407) |
| Nº assessments | 4 |
| Total of variables | 32 |
| Age (Mean/Standard Deviation) | 70,88/6,31 |
| Years of schooling (Mean/Standard Deviation) | 11,57/6,45 |

focus this study, along with the sociodemographic variables and the relationships between the two groups.

### 2.2 Missing Values Analysis

As a first step, an analysis was performed to determine the distribution of the missing values, identify those variables that concentrate most of them, and identify possible relationships between data and the missing values. A distribution analysis of the missing values across rows was performed in order to detect nearly empty rows that must be discarded. The reason is that more than 40% of missing data makes it impossible to perform a feasible imputation on those rows.

Next, a more in deep analysis of the possible relationship between the variables with the most missing values and the sociodemographic variables of sex, age and schooling was carried out to identify possible links between them, which will be taken into account during the allocation process. The performance of a subject on a test is related to their sociodemographic context, and it is the reason after the fact that the tests are calibrated due to those variables, so it was possible that the missing values were correlated to those variables.

### 2.3 Imputation Strategy

With the data obtained in the analysis, the database was imputed using two different strategies.

The first strategy consists of imputing the dataset using an ad hoc value. The variables with less than 5 missing values will be imputed with the mode. The mode was chosen due to the low number of missing values and the distribution of the data. In the 3 variables in which more values were missing (Trail Marking Test B (TMTB), Rey Figure's score (RFS) and Rey Figure's time (RFT)), a ceiling value was defined. To perform these tests is necessary a certain degree of abstraction. In the case of the TMTB, it is necessary to know both the alphabet and the numbers, and to perform an abstraction in order to be able to alternate between letters and numbers in the correct order. In the case of the Rey Figure, the subject have to copy a completely abstract complex figure in the shortest possible time. Subjects with a low level of education have problems to perform the abstraction necessary to do these tests properly, so their score tends to be worse than subjects with the same mental condition but with a higher education

**Table 2.** Description of the variables used in the database.

| Group | Variables name | General description |
|---|---|---|
| Identifiers | Nº of the column, subject ID evaluation | Allow to identify each case individually |
| Sociodemographic | Sex, years of schooling age | Sociodemographical variables |
| Psychiatric tests | Blessed Yesavage | Evaluation of the mood of the subject, to exclude those unsuitable for the study |
| MMSE | MEC | Spanish version of the Mini-mental test |
| Verbal fluency | Phonetic fluency Animal Clothes Plants Vehicles Semantic Fluency | Maximum number of correct words according to a certain rule said in 1 min |
| Graphic tests | Peak_Loops Constructive Praxias | To make a copy of a series of simple drawings |
| Ideomotor tests | Ideomotor Ideopatic Ideototal | Mimic a verbally described action |
| Trail Marking Tests | Trail Marking Test A Trail Marking Test B | Join with a line a series of numbers or/and letters in ascending order |
| Rey Figure | Rey Figure Score Rey Figure Time | To make a copy the complex Rey Figure |
| TAVEC (Test de Aprendizaje Verbal España-Complutense) | Total List A Total List B Short-term recall Short-term recall with clues Long-term recall Long-term recall with clues Recognition | Evaluation of short-term and long-term memory with and without clues |
| Diagnosis | Profiles | Subject diagnosis: healthy or MCI |

level. Thus, it was assumed that these missing values were due to the inability of those subjects to perform the tests. Therefore, those subjects would have obtained an bad score on those tests. For the TMTB we followed the psychologist criteria, which defines a ceiling score of 300 s for the TMTB when the subject struggles for too long in this test. For the RFS and RFT scores we had no previous ceiling scores, so we obtained it from the data. We analyzed the scores on the range (from 0 to 5 years of schooling) in which the missing values were concentrated, and the worst most common scores were selected. Using the overall worst score was discarded due to the presence of clear outliers for the RFT, and because in the RFS, a discrete variable, the occurrence of very low scores was exceptionally uncommon. Having this into account, we defined a value of 15 for the RFS, and of 210 for the RFT.

The second strategy was to impute the database using multiple imputation with the FCS strategy, in which we impute the dataset 10 times using the FCS algorithm. Then we grouped the results to generate the final imputed database. All variables were used during the multiple imputation.

Finally, mean and variance between original and imputed databases were compared to determine if any type of bias was introduced and, therefore, which imputed database will be more recommendable for future analysis. For the mean comparisons the t-test was used, and for the variance comparison, the Levene test was used. Python programming language was used for this study, using the scikit-learn package for certain parts of the analysis and imputation.

## 3 Results

The date variables of evaluation date and birth-date were removed they are completely redundant with the Age sociodemographic variable.

In the missing value analysis per variable, we found that a big percentage of the variables had a similar quantity of missing values, including sociodemographic variables of sex, age, and schooling, and 3 variables with a number of missing values much higher than the others, TMTB, RFS and RFT. The analysis of the missing values by rows showed the presence of records with more than 50% of missing values. Those rows were discarded since the great percentage made unreliable a correct imputation. In the distribution analysis of missing values that were then carried out after that, the missing values on all variables were greatly reduced, with some variables achieving 0 missing values, with the TMTB, RFS, and RFT variables maintaining a much higher percentage of missing values than the rest of variables.

The distribution analysis of missing values of TMTB, RFS, and RFT showed a great correlation between low education years and missing values. The comparative analysis of the comparison between the global dataset and the subset corresponding to the schooling years between 0 and 5 showed that there are differences in both mean and variance in all variables except one. Since there is a natural bias between subjects with high levels of education and those with low levels of education, the subset focused on years of education from 0 to 5 years will be used in both cases to make the comparison between the original and the imputed database. By using the subtest, we transform the MAR problem into a MCAR, where the distribution and relationship between the original database and the imputed ones can be comparable since the bias caused by years of education is eliminated.

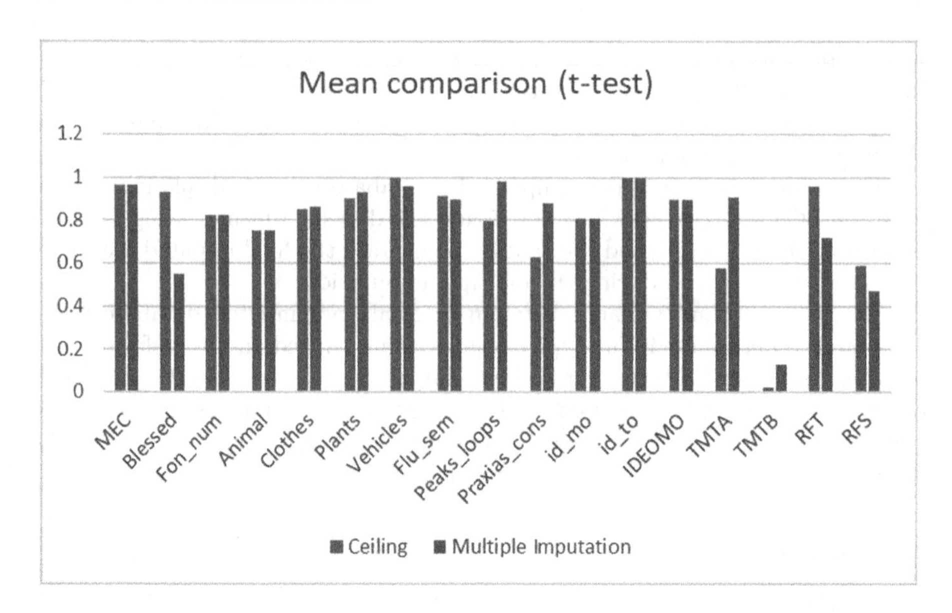


**Fig. 1.** P-values obtained in the comparison of means for ceiling imputation and multiple imputation.

In (Fig. 1) is shown the mean comparison on both imputed databases in all imputed variables. Taken a p-value = 0,1 as the threshold, no significant differences on the mean were founded between the original and the imputed database except on the TMTB, in which p-value = 0,02. It is also noticeable how multiple imputation presents in general better results than ceiling imputation, except in the variables of RFT, RFS and Blessed. On the comparison between variances showed in (Fig. 2), using a p-value = 0,1 as threshold, both methods do not introduce any significant bias in the data. However, it is important to note that multiple imputation introduces less bias than ceiling imputation, except in the Blessed variable. Taking the data obtained from the mean and variance together, we obtain that multiple imputation is a more suitable method to impute the database, making this imputation more suitable for future analysis on the database.

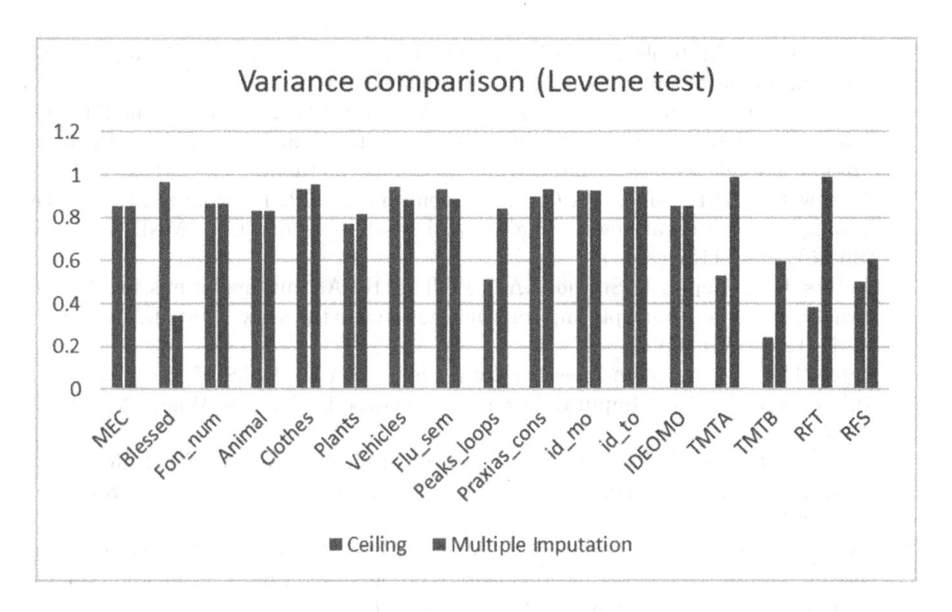


**Fig. 2.** P-values obtained in the comparison of variances for ceiling imputation and multiple imputation.

## 4 Conclusions

In this work we used different imputation method to improve the quality of the MCI dataset, allowing to increase the number of usable cases from 407 to 910 and to work with a no-biased database. The missing values were centered on 3 variables of high interest in the early detection of the MCI, due to its high discrimination and because they are some of the most common tests in MCI screening batteries. This make not recommendable to remove those scores. However, maintaining those variables by eliminating missing records would have generated a bias. The imputation allows these tests to be maintained by losing as little information as possible, allowing the use of this database in future analyses keeping the highest percentage of information. To achieve this, two imputation methods carried out in a real MCI database with missing values. The results showed that, although the ceiling imputation can be useful when the values are lost in a MAR situation and the correlation between values is clear, multiple imputation shows to be completely unbiased in all analyzed aspects. A more in deep analysis of the Blessed variable, the only variable in which multiple imputation obtain a worst score in both mean and variance as ceiling imputation was left as future work.

## References

1. Nguyen, C.D., Carlin, J.B., Lee, K.J.: Model checking in multiple imputation: an overview and case study. Emerg. Themes Epidemiol. **14**(1), 8 (2017)

2. Sterne, J.A.C.: Multiple imputation for missing data in epidemiological and clinical research: potential and pitfalls. BMJ **338**, b2393 (2009)
3. Jakobsen, J.C., Gluud, C., Wetterslev, J., Winkel, P.: When and how should multiple imputation be used for handling missing data in randomised clinical trials - a practical guide with flowcharts. BMC Med. Res. Methodol. **17**(1), 162 (2017)
4. Groenwold, R.H.H., Moons, K.G.M., Vandenbroucke, J.P.: Randomized trials with missing outcome data: how to analyze and what to report. Can. Med. Assoc. J. **186**(15), 1153–1157 (2014)
5. Hughes, R.A., Heron, J., Sterne, J.A.C., Tilling, K.: Accounting for missing data in statistical analyses: multiple imputation is not always the answer. Int. J. Epidemiol. **48**(4), 1294–1304 (2019)
6. Rubin, D.R.: Inference and missing data. Biometrika **63**(3), 581–590 (1976)
7. Rubin, D.B.: Multiple Imputation for Nonresponse in Surveys. Wiley, New York (1987)
8. Dziura, J.D., Post, L.A., Zhao, Q., Fu, Z., Peduzzi, P.: Strategies for dealing with Missing data in clinical trials: from design to analysis. Yale J. Biol. Med. **86**, 343–8358 (2013)
9. Choi, J., Dekkers, O.M., le Cessie, S.: A comparison of different methods to handle missing data in the context of propensity score analysis. Eur. J. Epidemiol. **34**(1), 23–36 (2018). https://doi.org/10.1007/s10654-018-0447-z
10. Marlin, B.M., Roweis, S.T., Zemel, R.S.: Unsupervised Learning with Non-Ignorable Missing. AISTATS (2005)
11. Liu, Y., De, A.: Multiple imputation by fully conditional specification for dealing with missing data in a large epidemiologic study. Int. J. Stat. Med. Res. **4**(3), 287–295 (2019)
12. van Buuren, S.: Multiple imputation of discrete and continuous data by fully conditional specification. Stat. Methods Med. Res. **16**(3), 219–242 (2007)
13. Murray, J.S.: Multiple imputation: a review of practical and theoretical findings. Stat. Sci. **33**(2), 142–159 (2018)
14. Peraita, H., García-Herranz, S., Díaz-Mardomingo, M.C.: Evolution of specific cognitive subprofiles of mild cognitive impairment in a three-year longitudinal study. Curr. Aging Sci. **4**, 171–182 (2011)
15. García-Herranz, S., Díaz-Mardomingo, M.C., Venero, C., Peraita, H.: Accuracy of verbal fluency tests in the discrimination of mild cognitive impairment and probable Alzheimer's disease in older Spanish monolingual individuals. Neuropsychol. Dev. Cogn. Section B, Aging, Neuropsychol. Cogn. **27**(6), 826–840 (2020)

# Automatic Scoring of Rey-Osterrieth Complex Figure Test Using Recursive Cortical Networks

F. J. Pinilla(✉), R. Martínez-Tomás, and M. Rincón

Departamento de Inteligencia Artificial, UNED, 28040 Madrid, Spain
fpinilla20@alumno.uned.es

**Abstract.** Mild Cognitive Impairment (MCI) is a condition which may lead to a more serious neurodegenerative disease called dementia, affecting between 12% and 18% of total Global population aged 60 or older. Neuropsychological tests conducted by professionals allow for early detection of MCI and early treatment of this condition to prevent further development. Several authors have attempted to automate the assessment process of these types of tests, which enables a faster screening of the population and therefore a better prevention of the symptoms of neurodegenerative diseases. However, most of the works published by previous authors rely on classical Machine Learning techniques, which require handcrafted features and their effectiveness depends on the quality of these features. Also, more advanced Deep Learning models used in the automation of these tests require high amounts of training data in order to be accurate, and they are also weak to noise and variability in the data. In this work, we propose a novel approach to automating one of these test called Rey-Osterrieth Complex Figure (ROCF) test, using Recursive Cortical Networks (RCN). The RCN framework provides an improvement over the disadvantages of previously mentioned techniques, presenting resilience to noise and variability, using an automatic hierarchical feature construction instead of hand-crafted features, while using a very small amount of training data. This work describes the properties of RCN and how they can be of use in the development of an automatic scoring algorithm for the ROCF test.

**Keywords:** Mild Cognitive Impairment · MCI · Rey-Osterrieth · ROCF

## 1 Introduction

According to the World Health Organization, at September 2021, over 55 million people live with dementia worldwide, with nearly 10 million new cases every year, with Alzheimer's disease being the most common form of dementia with a contribution of 60%–70% of the cases [1]. Mild Cognitive Impairment (MCI) causes cognitive changes that are serious enough to be noticed by the person


J. M. Ferrández Vicente et al. (Eds.): IWINAC 2022, LNCS 13258, pp. 455–463, 2022.
https://doi.org/10.1007/978-3-031-06242-1_45

affected and by family members and friends but do not affect the individual's ability to carry out everyday activities. MCI may develop into a more serious neurodegenerative disease called "Dementia", which affect between 12% and 18% of total Global population aged 60 or older [2]. Among other elements, a medical workup for MCI includes "Assessment of mental status using brief tests designed to evaluate memory, planning, judgment, ability to understand visual information and other key thinking skills" [2]. In some cases, MCI is an early stage of Alzheimer's or another dementia. In order to prevent further development of this disease, an early detection of symptoms is crucial to start the appropriate treatment as soon as possible [2]. For this reason, reliable and easy screening methods of neurodegenerative symptoms accelerate the early detection of this condition and help with early treatment in those cases where the condition could lead to a more severe neurodegenerative disease.

### 1.1 Neuropsychological Tests

A variety of cognitive tests and assessment tools has been used to diagnose these diseases. The Rey-Osterrieth Complex Figure (ROCF) test is a well-known neuropsychological test for the evaluation of visuospatial constructional ability and visual memory function [3]. The ROCF test can be used to analyse the performance of dementia as well as patients with MCI. The subject has to use a pen or a pencil and paper to copy the complex and abstract geometrical figure, then recall and reproduce the figure from his memory. The ROCF test consists of three trials: copy, 3-min immediate recall, and 30-min delayed recall. However, an evaluation of the ROCF test requires the knowledge and expertise of a qualified psychologist who can assess the results of the test based on both a specific set of guidelines and its own experience at assessing this type of tests.

### 1.2 Automation of Tests

In order to reduce the workload of psychologists and therefore make neuropsychological tests easier and faster to conduct, several authors have provided methods to automatically assess the results of these tests using Machine Learning techniques [4–10]. In Asselborn et al. [4,8], the authors use a Random Forest classifier to diagnose dysgraphia among children, using 53 handwriting features describing various aspects of handwriting with data obtained from a tablet. In Barz et al. [6] authors use Support Vector Machines and gradient boosted decision trees to train of features obtained from a digital pen in the Trail Marking Test and the Snijders-Oomen Non-verbal intelligence test, providing interesting insights on the use of digital pens in neurodegenerative tests. In Dahmen et al. [7] authors also succeed at automating the Trail Marking Test. Their approach uses a tablet to capture the strokes in the drawings of the patients. This data was used in a combination of Linear Regression, Support Vector Machine and Decision Trees in the WEKA framework [16]. Regarding the ROCF test, in Prange et al. [10] authors use a Digital Pen and different Machine Learning algorithms to detect the different components of the figure and automate the scoring process. All of the previous works rely on handcrafted features and very simple Supervised Learning methods.

Only a few works in the literature of neurodegenerative test automation rely on modern Deep Learning techniques. For example, Chen et al. [17] use a dataset of 1315 individuals to train 3 different Deep Neural Network models: VGG16, ResNet-152, and DenseNet-121. Regarding the Rey-Osterrieth Complex Figure test, Cheah et al. [21] use a convolution autoencoder to automatically extract features in the image. However, their low amount of data available makes the task of training this Neural Network more difficult. Deep Learning techniques rely on high volumes of training data and are weak to noise and variability in the data. Given the nature of neuropsychological tests, it is common to find noisy data, with real drawings presenting some variation with respect to the training dataset. The objective of this paper is to provide a theoretical framework for developing an automatic scoring algorithm of the ROCF test based on the use of Recursive Cortical Networks (RCN). This framework, published in 2017 by George et al. [11], is inspired in the structure of the visual cortex of humans. As a hierarchical graphical model, its features are automatically built by learning the best combination of low-level features and combining them into higher-level ones. This solves the problem that classical Machine Learning methods like SVM and Decision Trees present, where the features fed to the models need to be manually crafted. RCN are also resilient to noise and variability in the data thanks to their Loopy Belief Propagation algorithm [18]. Thanks to this type of inference, RCN rely on a smaller training dataset, where only one instance of each possible object needs to be learnt, plus some deformations like searing, small rotations and bigger size changes. This approach helps overcoming the previous disadvantages of both classical Machine Learning methods and the modern Deep Learning models; high volume of data, weakness to noise, deformation and scale variability.

The rest of the paper is structured as follows: Sect. 2 describes the RCN model and its advantages compared to Deep Learning methods. Section 3 describes the ROCF test, its scoring guidelines and the problems that arise when trying to develop an algorithm to automate the scoring process. Section 4 describes the theoretical application of RCN to the automated scoring algorithm. Section 5 summarizes the key points of the paper.

## 2 Methods

Recursive Cortical Networks (RCN) [11] provide a type of structured probabilistic generative model framework based on neuroscientific insights [19,20]. Originally, RCN were specifically designed to break CAPTCHAs. Authors demonstrate how their model outperforms other state-of-the-art Deep Learning methods in the task of breaking CAPTCHAs when only a small amount of training data is available [11]. In general, an RCN model takes as input a fixed size pixel image, identifies all the different parts of the objects present in the image and outputs an image where the contour and the surface of all detected objects are represented. For the application of RCN to the task of breaking CAPTCHAs, a special type of RCN is built where the output of the model is a binary activation matrix containing points that represent the contour of the letter detected by the model.

The strength of RCNs lies on their representation of images using alternating layers containing features or pools of features (Fig. 1a). Learning an object in an RCN model is done by propagating bottom-up activations of the features and pools, and finally capturing the best set of top-level features and pools, along with their position, into a factor graph. Each learnt object is produced by a single example image, so the size of the learning dataset equals the number of learnt objects in the trained RCN model. These objects are represented by the set of learnt top-level factors (pools of features).

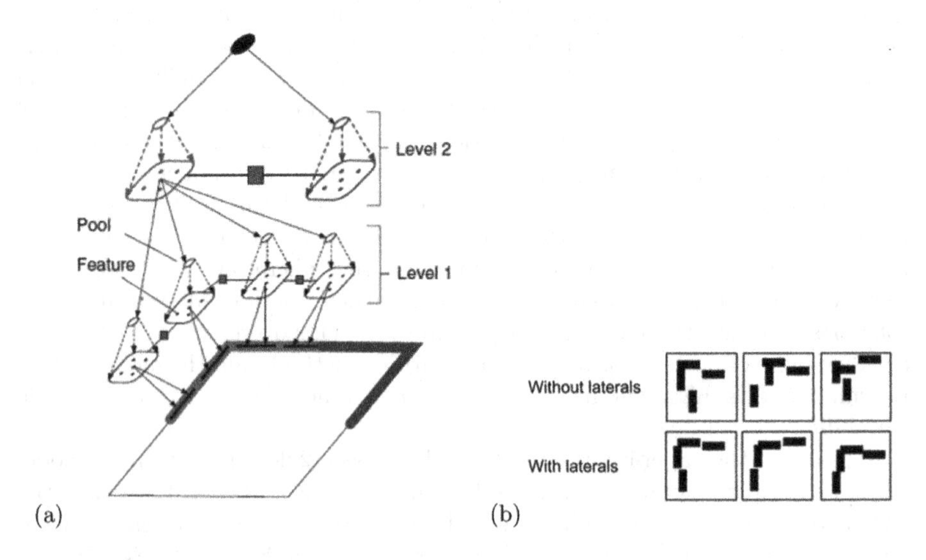


**Fig. 1.** (a) Example of 2 levels with 2 layers each in the RCN representing the corners of the rectangle. Any pool is activated if any of the related features inside the pool is activated. Pools have lateral connections (grey squares) between them to capture spatial relationships. (b) Comparison of the corner of a rectangle with and without laterals. The presence of lateral connection between pools maintains spatial relationship between lower-level features. Images extracted from George D. et al. [11] with permission from the authors.

By capturing the spatial relationship of pools using lateral connections, the RCN model generates a hierarchical probabilistic graph robust to noise and a certain degree of deformation (Fig. 1b). The graph structure of the pooling layers allows inference to be performed using a Loopy Belief Propagation algorithm [18]. This inference process converges to a representation of the most probable object in the image by locating the position of the pools of features in each pooling layer, giving as output an edge map in the form of a top-down binary activation matrix that represents the contour of the detected object (Fig. 2). The edge map returns even parts of the object that are occluded.

**Fig. 2.** Edge map output of an RCN model trained to break Google's reCAPTCHA.

Since RCN is a graphical model that performs Belief Propagation to find the placement of low-level features that best fit a given image, it can converge to a solution that varies slightly from the training data. This makes RCN resilient to noise and variability, and, in contrast to Deep Learning methods, using only a few examples of each element that the model is trained to detect in an image.

## 3 Rey-Osterrieth Complex Figure Test

The Rey-Osterrieth Complex Figure Test (ROCF) is a neuropsychological assessment developed by André Rey in 1941 as a method to evaluate visual-spatial ability and visual memory in brain-injured patients [12]. It was standardized in 1944 by Paul-Alexandre Osterrieth, who introduced a scoring system for the interpretation of a performance [13]. The task involves copying a complex geometric figure: the patients are given a piece of paper and a pencil, and the original figure is placed in front of them. They are asked to reproduce the figure to the best of their ability. Reproducing the ROCF is a complex cognitive task, which involves the ability to organize the figure into units and replicate them in accordance of their topology in the figure and their spatial characteristics. This is why the ROCF provides meaningful data about individual neuropsychological functions and brain dysfunctions, including attention and concentration levels, fine-motor coordination, visual-spatial perception, nonverbal memory and organizational skills [14]. By analysing the patients' performance, it is possible to provide an early evaluation of their neuropsychological dysfunctions. To evaluate a patient's performance, Osterrieth subdivided the original figure in 18 patterns (Fig. 3), each of which is considered separately and scored according to some criteria. All the currently accepted scoring systems for the ROCF test need to be performed by hand in what tends to be a subjective manner, making the reasoning behind the assignment of a certain score open to interpretation [14].

The final score of a test is obtained by adding each of the individual scores assigned to each of the 18 components of the figure. Each component can be assigned up to 2 points, depending on the quality of the lines and their placement in the drawing. In the drawings made by real patients, the professional assigning

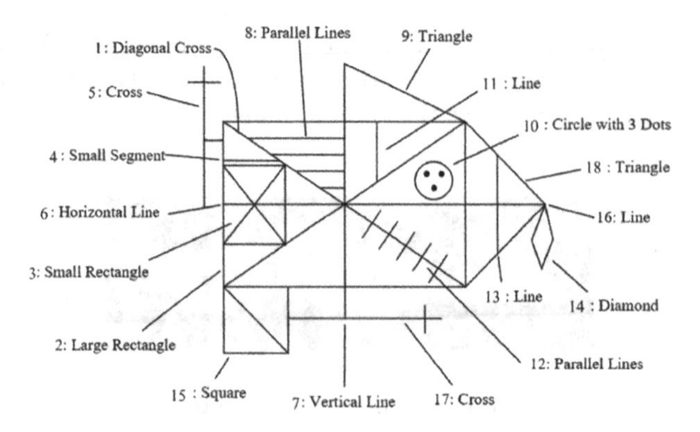


**Fig. 3.** The Osterrieth subdivision of the complex figure.

the score can find uncertainty even in identifying to which component belongs each line. An adequate artificial intelligence algorithm will need to be robust to uncertainty and to the highly variable features present in the real drawings. Recursive Cortical Network may provide the necessary mechanisms to perform a precise segmentation and labelling of the lines in the 18 different components.

## 4 Component Classification Using RCN

Recursive Cortical Networks rely on a structure of layers connected hierarchically from bottom (image) to top (labels). This structure replicates that of the visual cortex in the human brain, allowing for the detection of compositional features in the drawings. For example, different components share the same basic features, corners or straight lines. RCN is able to learn the higher-level features that arise as the composition of several lower-level ones. This property makes RCN a good model for this task, provided that an adequate training dataset is available. Training data for RCN consists in a series of images, each one representing an object that the RCN will detect in the image. During the training, for each training image, the model learns the most efficient way to represent the object contour in the image. The result of the training is an object represented using a graph where the nodes represent the highest-level features of the RCN as a triplet of (feature ID, x position, y position). Figure 4 shows an example of one instance of the capital letter A learnt using the RCN model in the simplest version (2-layer model).

By training the RCN with a wide enough variety of examples of the 18 different components, the RCN can detect them and label the pixels in the image where the lines belonging to each component are. To do this, several steps need to be conducted: First, a selection of all possible candidates found in the image needs to be made. Using a fast message passing algorithm called Forward Pass, for each learnt object in the training dataset, a "Forward-pass score" (fp score) is assigned to that specific label in a given region of the image. In order

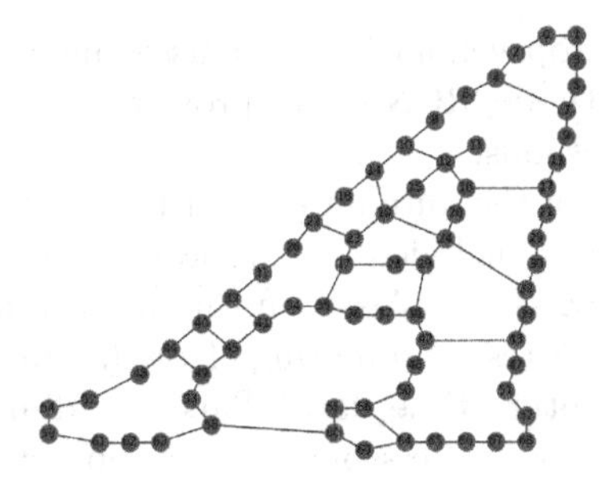

**Fig. 4.** Example of the contour of the capital letter A with 40° searing after training with the RCN. Nodes represent features located at certain positions, and edges represent the lateral connections which provide a constraint in distance.

to find all possible instances of each object over the entire image, this process needs to be repeated on all possible snippets of the image where the object could be found. After all objects of the training dataset have been assigned an fp score all over the image, a selection of the most promising candidates is performed by applying a threshold to the fp score. This threshold is a parameter of the model and needs to be fine-tuned according to each problem where the RCN is applied.

After this candidate selection process has finished, the RCN must select between all the possible candidates detected in an image, and determine which combination of them fits best. This part of the algorithm is called "Scene Scoring" and involves assigning a score to each combination of candidates based on their degree of overlap and the amount of the contour in the image that is explained in total.

Applying the RCN to the ROCF test drawings requires adapting the original code published by the authors [15] in order to detect the lines drawn by the patients. Originally, the authors propose to use a set of 16 different orientations of the Gabor filter. A tuning of the parameters in this set of filters allow to detect lines in different widths. With this improvement, the RCN model will be ready to be applied to the real drawings of the ROCF test, given that the model has already been trained to detect all the 18 components with some variations. The most important part of this application is the Scene Scoring algorithm, where all candidates which where selected in the first part of the algorithm need to be assessed in combination with others. This process allows for discarding all candidates which are redundant and selecting the best combination of candidates that best explain the image.

## 5 Major Conclusion

A novel Artificial Intelligence method has been presented as a tool to help automate Rey-Osterrieth Complex Figure tests scoring. Given the nature of the test, which presents high levels of noise and variability, with low amounts of data to train the model, RCN can overcome these difficulties and might open a new path in the field of neuropsychological test assessments. Together with other Supervised Learning methods, RCN might prove to be useful and achieve better results than previous methods.

Further development of the automated scoring algorithm is still in progress. Among the future work is included the improvement of the current training dataset which is on an early stage. Also, a future improvement in the automation of the scoring of the ROCF test automating algorithm includes the use of tablets or digital pens which capture time data. This new dimension in the data has proven to increase accuracy in the scoring of neuropsychological tests.

## References


1. World Health Organization factsheets - Dementia, September 2021. https://www.who.int/news-room/fact-sheets/detail/dementia. Accessed 3 Mar 2022
2. Alzheimer's Association - Mild Cognitive Impairment. https://www.alz.org/Alzheimers-dementia/what-is-dementia/related_conditions/mild-cognitive-impairment. Accessed 3 Mar 2022
3. Meyers, J., Meyers, K.: Rey complex figure test and recognition trial. San Antonio: The Psychological Corporation
4. Asselborn, T., et al.: Automated human-level diagnosis of dysgraphia using a consumer tablet. NPJ Digital Med. **1**(1), 1–9 (2018)
5. Barz, M., Altmeyer, K., Malone, S., Lauer, L., Sonntag, D.: Digital pen features predict task difficulty and user performance of cognitive tests. In: Proceedings of the 28th ACM Conference on User Modeling, Adaptation and Personalization, pp. 23–32 (2016). https://doi.org/10.1145/3340631.3394839
6. Canham, R.O., Smith, S.L., Tyrrell, A.M.: Automated scoring of a neuropsychological test: the Rey Osterrieth complex figure. In: Proceedings of the 26th Euromicro Conference. EUROMICRO 2000. Informatics: Inventing the Future, vol. 2, pp. 406–413. IEEE. September 2000. https://doi.org/10.1109/EURMIC.2000.874519
7. Dahmen, J., Cook, D., Fellows, R., Schmitter-Edgecombe, M.: An analysis of a digital variant of the Trail Making Test using machine learning techniques. Technol. Health Care **25**(2), 251–264 (2017)
8. Gargot, T., et al.: Acquisition of handwriting in children with and without dysgraphia: a computational approach. PloS One **15**(9), e0237575 (2020)
9. Prange, A., Sonntag, D.: Modeling cognitive status through automatic scoring of a digital version of the clock drawing test. In: Proceedings of the 27th ACM Conference on User Modeling, Adaptation and Personalization, pp. 70–77, June 2019. https://doi.org/10.1145/3320435.3320452
10. Prange, A., Sonntag, D.: Assessing cognitive test performance using automatic digital pen features analysis. In: Proceedings of the 29th ACM Conference on User Modeling, Adaptation and Personalization, pp. 33–43, June 2021. https://doi.org/10.1145/3450613.3456812
11. George, D., et al.: A generative vision model that trains with high data efficiency and breaks text-based CAPTCHAs. Science **358**(6368), eaag2612 (2017)
12. Rey, A.: L'examen psychologique dans les cas d'encéphalopathie traumatique. Archives de psychologie **28**, 112 (1942)
13. Osterrieth, P.A.: Le test de copie d'une figure complexe; contribution a l'etude de la perception et de la memoire. Archives de psychologie **30**, 206–356 (1944)

14. Shin, M.S., Park, S.Y., Park, S.R., Seol, S.H., Kwon, J.S.: Clinical and empirical applications of the Rey-Osterrieth complex figure test. Nat. Protoc. **1**(2), 892–899 (2006)
15. Reference Implementation of Recursive Cortical Networks. https://github.com/vicariousinc/science_rcn. Accessed 3 Mar 2022
16. Hall, M., Frank, E., Holmes, G., Pfahringer, B., Reutemann, P., Witten, I.H.: The WEKA data mining software: an update. ACM SIGKDD Explorations Newsl. **11**(1), 10–18 (2009)
17. Chen, S., Stromer, D., Alabdalrahim, H.A., Schwab, S., Weih, M., Maier, A.: Automatic dementia screening and scoring by applying deep learning on clock-drawing tests. Sci. Rep. **10**(1), 1–11 (2020)
18. Ihler, A.T., Fisher, J.W., III., Willsky, A.S., Chickering, D.M.: Loopy belief propagation: convergence and effects of message errors. J. Mach. Learn. Res. **65**, 905–936 (2005)
19. Lee, T.S., Mumford, D.: Hierarchical Bayesian inference in the visual cortex. JOSA A **20**(7), 1434–1448 (2003)
20. Lamme, V.A., Rodriguez-Rodriguez, V., Spekreijse, H.: Separate processing dynamics for texture elements, boundaries and surfaces in primary visual cortex of the macaque monkey. Cereb. Cortex **9**(4), 406–413 (1999)
21. Cheah, W.T., Chang, W.D., Hwang, J.J., Hong, S.Y., Fu, L.C., Chang, Y.L.: A screening system for mild cognitive impairment based on neuropsychological drawing test and neural network. In: 2019 IEEE International Conference on Systems, Man and Cybernetics (SMC), pp. 3543–3548, October 2019. https://doi.org/10.1109/SMC.2019.8913880

# Influence of the Level of Immersion in Emotion Recognition Using Virtual Humans

Miguel A. Vicente-Querol[1], Antonio Fernández-Caballero[1,2,3], José P. Molina[1,2], Pascual González[1,2,3], Luz M. González-Gualda[4], Patricia Fernández-Sotos[3,4], and Arturo S. García[1,2(✉)]

[1] Instituto de Investigación en Informática, Unidad de Neurocognición y Emoción, 02071 Albacete, Spain

[2] Universidad de Castilla-La Mancha, Departamento de Sistemas Informáticos, 02071 Albacete, Spain

arturosimon.garcia@uclm.es

[3] Biomedical Research Networking Centre in Mental Health (CIBERSAM), 28029 Madrid, Spain

[4] Complejo Hospitalario Universitario de Albacete, Servicio de Salud Mental, 02006 Albacete, Spain

**Abstract.** Facial affect recognition is a skill that involves the ability to perceive and distinguish between different affective facial expressions. This skill is crucial in day-to-day human relationships, and it is negatively affected by various psychiatric disorders. This has led to the development of tests that attempt to measure or train this skill. Virtual Reality (VR) is a technology that has been used for this purpose, as it can immerse the user in a virtual world that can be more similar to the real world than a traditional desktop computer. Therefore, the objective of this paper is twofold. First, a potential improvement in emotion recognition using VR instead of a traditional desktop screen condition is studied. Secondly, the impact that the presentation angle of the virtual faces has on emotion recognition is investigated. To this end, the same application was adapted to be shown in a desktop screen and in a VR environment to 36 mentally healthy participants recruited for each condition. A set of dynamic virtual faces (DVFs) were used as stimuli to show the 6 basic emotions plus the neutral expression. The DVFs were also rotated to be presented to the participants in front and side views. The results show a slight improvement in the VR environment. In addition, the accuracy is better in the front views than in the side views for both conditions, which is consistent with previous studies.

**Keywords:** Emotion recognition · Virtual humans · Immersion level · Emotion recognition


J. M. Ferrández Vicente et al. (Eds.): IWINAC 2022, LNCS 13258, pp. 464–474, 2022.
https://doi.org/10.1007/978-3-031-06242-1_46

## 1 Introduction

Virtual reality (VR) is a technology that intends to overcome the limitations of traditional human-computer interaction. The ultimate goal is to interact in a similar way as human beings normally do in the real world, without the constraints of using devices such as keyboards, mice or flat screens [3]. Therefore, the research community has devoted considerable effort into comparing the performance of a variety of tasks between desktop and VR configurations over the years in an attempt to assess the level of maturity of the VR technology [1,9,14,22]. The improvement in task performance of immersive VR experiences is often associated with an increased sense of presence [25,26]. The sense of presence in VR is related to the feeling of "being there", inside the computer-generated virtual world [2], even if it is a virtual representation of a distant planet [10].

Emotion recognition [5], and more concretely, facial affect recognition [16, 17] is a crucial skill in everyday social interactions. It involves the ability to perceive and distinguish essential types of affective expressions in faces [23]. This ability is hampered by a number of neuropsychiatric disorders such as schizophrenia and autism [19]. Traditional therapies consist of the use of images showing real people displaying different facial emotions which are normally used for training purposes [15]. Virtual humans (VH) have been previously used to generate stimuli for facial affect recognition [11,13]. This technology has several advantages over the traditional approach of using paper pictures, namely the possibility of applying the facial expressions to virtual characters of different race, age and gender, adding animation to the facial expression, include them in different contexts/backgrounds, enriching the data set with images taken from different angles, etc. However, this stimuli are normally used in desktop VR setups and thus not taking advantage of the benefits of the level of immersion that immersive VR provides.

We reckon that the ecological validity of immersive VR is greater than that of desktop VR [21], as the person is sharing the (virtual) space with the virtual human that depicts the emotion, making the experience more real than using a picture or a flat screen. Previous studies have also shown the facial expressions of humans or virtual humans not only on a flat screen but also presented only frontally. As shown in previous work [4,12,27], there are differences emotion recognition when presenting the faces in frontal or side views. We also aim to see whether similar differences can be found using virtual humans in desktop or immersive VR. Therefore, greater benefits can be gained from the use of this technology in future facial affect recognition therapies. We also believe that increasing the level of immersion and therefore the ecological validity of the stimuli used for facial affect recognition therapies will provide better results than traditional approaches. However, there is no previous work up to our knowledge that compares the emotion recognition performance of people using devices that offer different degrees of immersion. Research on emotion recognition in immersive VR has focused mainly on the computer recognising the emotion evoked in the user by the VR simulation in which they are immersed [18]. Therefore, the main aim of our paper is to asses whether there is a benefit in increasing the level of immersion in the identification of facial expression of emotions using virtual humans. An experiment involving 72 healthy participants is described in this paper, comparing the same stimuli presented in desktop and immersive VR.

Once the results of this experiment are analysed, it is our intention to test with patients of neuropsychiatric disorders in order to assess the possible benefits of using immersive VR.

There are two main research questions (RQ) in this work:

- RQ1: Is it possible to recognise emotions represented by virtual humans in an immersive VR setup with a performance similar to that obtained in a desktop VR setup?
- RQ2: Does the viewing angle of the virtual human have a similar effect on emotion recognition in VR as in a desktop setup?

# 2 Materials and Methods

## 2.1 Participants

Two experiments were performed. In the first experiment, participants used a desktop version of the application, whereas in the second experiment participants used the VR version of the application which made use of a head mounted display (HMD). Seventy-two healthy volunteers between 20 and 59 years old were recruited. Thirty-six participated in the desktop study and the other thirty-six participated in the VR study. Having previous diagnosis of mental illness, personal history of medical illness, or first-degree family history of psychosis were causes for exclusion from the study. These participants had no previous experience with the dynamic virtual faces (DVFs) used in this evaluation. The study was conducted in compliance with the guidelines of the Declaration of Helsinki and approved by the Clinical Research Ethics Committee of the Complejo Hospitalario Universitario de Albacete (protocol code 2019/07/073 and date of approval 24 September 2019). Informed consent was obtained from all subjects involved in the study.

## 2.2 Experimental Setup

The computer used was a laptop computer with a 17.3" screen, an Intel Core i7-9750H, 16 GB of RAM and an NVIDIA RTX2070 Super. For the desktop experiment participants used a mouse to select the different choices, whereas for the immersive experiment a gamepad controller was used to select on the alternatives The desktop experiment were performed using the screen of the laptop computer, whereas for the immersive experiment an HMD was needed. FOVE (https://fove-inc.com/), was the selected HMD. It is equipped with a WQHD OLED display of 2560 × 1440 pixels, a field of view of 100°.

### 2.3 Stimuli

The DVFs used for both experiments as stimuli for affect recognition were previously validated by healthy people [7] and people diagnosed with schizophrenia [20]. It includes the six basic emotions (*anger*, *disgust*, *fear*, *joy*, *sadness*, and *surprise*) plus the *neutral* expression. Two race avatars were included (2 Caucasian and 2 African) of about 30 years old and two of old age (male and female for all cases). Further information on how the virtual humans and the emotions depicted were created can be found in a recent paper [8]. The DVFs were presented on a computer screen for the first experiment. For the second experiment the same set of DVFs were utilised, but instead of using a computer screen, the application was adapted to be showed in a HMD [29] for a seated immersive VR experience.

### 2.4 Procedure

Figure 1 shows the flow chart of the experimental procedure. A sociodemographic form (including age, genre, and others) and the Spanish version [24] of the Positive and Negative Affect Schedule (PANAS) [28] form were filled by the participants prior to the start of the experiments in order to exclude participants with an altered mood at the time of performing the task. PANAS is a 20-item questionnaire which measures an individual's positive and negative affect. If the score is less than 25 (PA $< 25$, positive affect score) or more than 35 (NA $> 35$, negative affect score), the participant was excluded from the study.

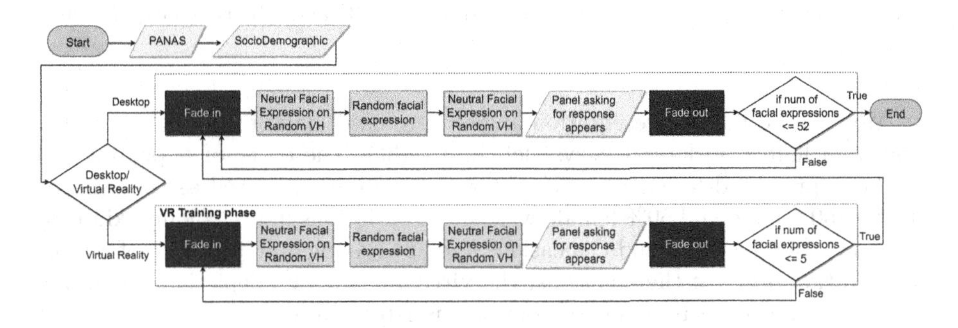


**Fig. 1.** Flowchart describing the experiment procedure.

After filling both forms, each participant had to perform 52 trials (8 for each basic emotion and 4 for the neutral expression). Each trial started with the virtual human showing the neutral expression. Then, this expression was blended to a random emotion among the seven possible alternatives in a process that lasted 0.5 s. The target emotion was displayed for 1.5 s and then it was blended to the neutral expression, again in 0.5 s. After the animation of the emotions ended, a seven-choice panel appeared so that the participants could select one. There was no time limitation for them to choose their answer. Once an emotion was selected, the whole environment faded to a light-grey colour and the process started again with a new target emotion.

The participants performing the VR experiment had to go through a short training phase previous to the beginning of experiment. This way, they could familiarise themselves with the gamepad controller and the VR environment. This training phase consisted on 5 samples of DVFs depicting random emotions. Apart from this initial training phase, the experiments were identical for both the desktop and the VR conditions. The desktop participants selected the correct answer by clicking on the buttons with the name of the emotion that appeared on the screen with the mouse, while the VR participants had to select one alternative with any of the game pad's sticks and then confirm it by pressing any of the buttons. These two different user interfaces are depicted in Fig. 2.

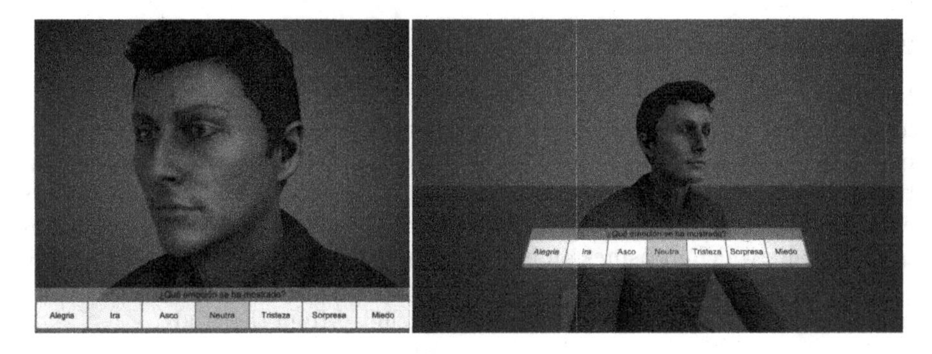


**Fig. 2.** Different user interfaces used in the desktop (left) and VR (right) experiments.

Moreover, different presentation angles were used, half of the times from the front and half from lateral views (left and right). The order of appearance of the faces presented to each participant was randomised, as well as the avatars' gender (50% male and 50% female with slight variations in eye colour, skin tone and hair), race (African and Caucasian) and age (young and old). Therefore, the presentation order of the emotions, the virtual characters depicting them, and the camera angles were different from one user to another.

### 2.5 Data Analysis

Microsoft Excel and IBM SPSS v28 were used for the descriptive analysis of the data (mean, standard deviation and percentages) and the statistical analyses, respectively. Since the emotion recognition data gathered did not follow a normal distribution, non-parametric analysis were used to test differences between the two experimental conditions. The Mann-Whitney test was used to compare data gathered for the desktop condition against the VR condition, while the Wilcoxon Signed Rank test was used to test data gathered within the same condition (front vs side views). A p-value lower than 0.05 was considered statistically significant.

## 3 Results

The correct emotion identification percentages for both testing conditions can be found in Fig. 3, being the average percentages 90.03% for the Desktop and 92.34% for the VR condition. Despite the slight difference, there is no statistically significant difference in the correct identification percentages for the two conditions tested (Mann-Whitney $U = 530.0, p = 0.183$).

Desktop — Participant answer (rows: Correct answer)

| | Neutral | Surprise | Fear | Anger | Disgust | Joy | Sadness |
|---|---|---|---|---|---|---|---|
| Neutral | 97.92 | 0.00 | 0.69 | 0.00 | 0.69 | 0.00 | 0.69 |
| Surprise | 0.69 | 93.06 | 6.25 | 0.00 | 0.00 | 0.00 | 0.00 |
| Fear | 1.39 | 14.93 | 77.78 | 0.00 | 0.00 | 0.00 | 5.90 |
| Anger | 0.69 | 0.35 | 1.04 | 94.44 | 3.12 | 0.00 | 0.35 |
| Disgust | 0.35 | 1.39 | 0.35 | 11.81 | 85.42 | 0.00 | 0.69 |
| Joy | 2.08 | 0.35 | 0.00 | 0.69 | 0.00 | 96.88 | 0.00 |
| Sadness | 4.86 | 5.21 | 2.43 | 0.00 | 2.78 | 0.00 | 84.72 |

Virtual Reality — Participant answer (rows: Correct answer)

| | Neutral | Surprise | Fear | Anger | Disgust | Joy | Sadness |
|---|---|---|---|---|---|---|---|
| Neutral | 97.78 | 0.00 | 0.00 | 0.56 | 0.56 | 0.00 | 1.11 |
| Surprise | 0.00 | 94.72 | 4.44 | 0.00 | 0.00 | 0.00 | 0.83 |
| Fear | 0.56 | 21.39 | 76.67 | 0.00 | 0.83 | 0.00 | 0.56 |
| Anger | 0.00 | 0.28 | 0.28 | 98.89 | 0.00 | 0.28 | 0.28 |
| Disgust | 0.56 | 0.28 | 0.00 | 8.06 | 91.11 | 0.00 | 0.00 |
| Joy | 0.28 | 0.56 | 0.00 | 0.28 | 0.00 | 98.89 | 0.00 |
| Sadness | 2.78 | 1.67 | 1.94 | 3.61 | 1.67 | 0.00 | 88.33 |

**Fig. 3.** Confusion matrices showing the percentages of the correct answers for desktop (left) and VR (right).

Studying the results obtained for each emotion, it can be observed that, in both cases, the emotion with the lowest identification percentage is *fear* (around 77%). This emotion is mostly confused with *surprise* in both cases, but to a greater extent in VR (14.97% in desktop and 21.39% in VR). The rest of the emotions obtained percentages above 84%. Comparing them for both conditions, the differences are generally small, except for *anger* (94.44% vs 98.89% for desktop and VR, respectively) and *disgust* (85.42% vs 91.11%). However, only the difference in *anger* is statistically significant (Mann-Whitney $U = 496.5, p = 0.013$), being non significant for *disgust* ($U = 552.5, p = 0.254$). The test indicates than the percentage for VR is higher.

The differences in the correct emotion identification percentages for the two visualisation angles were also studied. Thus, the desktop front view obtained an average of 90.30% and the side view 89.74%. The differences between the two views were higher for the VR condition, with an average of 94.32% for the front view and 90.42% for the side view. The Wilcoxon Signed Rank test did not find statistically significant differences for the desktop (Wilcoxon $Z = -0.246, p = 0.806$), but the difference was significant for VR ($Z = -2.819, p = 0.005$). The differences between both conditions for the front and side views were also studied. Statistically significant differences were found for the front view (Mann-Whitney $U = 444.5, p = 0.021$), being the percentages higher for VR. No differences were found for the side view ($U = 607.5, p = 0.647$).

Figure 4 shows correct emotion identification percentages for the two visualisation angles and for each emotion for the desktop condition (top row) and the VR condition (bottom row). The percentages are similar to the ones studied in Fig. 3, with *fear* being the emotion than obtains lower percentage of correct emotion identification, being again confused mostly with *surprise*. Similarly to the results obtained previously, *disgust* is confused mostly with *anger*, except for the front view of the VR condition, in which it obtains a percentage of 95.51.

Desktop Front View

| Correct answer \ Participant answer | Neutral | Surprise | Fear | Anger | Disgust | Joy | Sadness |
|---|---|---|---|---|---|---|---|
| Neutral | 100.00 | 0.00 | 0.00 | 0.00 | 0.00 | 0.00 | 0.00 |
| Surprise | 0.70 | 91.55 | 7.75 | 0.00 | 0.00 | 0.00 | 0.00 |
| Fear | 2.05 | 11.64 | 79.45 | 0.00 | 0.00 | 0.00 | 6.85 |
| Anger | 0.00 | 0.69 | 1.38 | 93.79 | 3.45 | 0.00 | 0.69 |
| Disgust | 0.00 | 1.39 | 0.00 | 15.97 | 82.64 | 0.00 | 0.00 |
| Joy | 1.41 | 0.00 | 0.00 | 0.70 | 0.00 | 97.89 | 0.00 |
| Sadness | 3.47 | 3.47 | 3.47 | 0.00 | 2.78 | 0.00 | 86.81 |

Desktop Side View

| Correct answer \ Participant answer | Neutral | Surprise | Fear | Anger | Disgust | Joy | Sadness |
|---|---|---|---|---|---|---|---|
| Neutral | 95.77 | 0.00 | 1.41 | 0.00 | 1.41 | 0.00 | 1.41 |
| Surprise | 0.68 | 94.52 | 4.79 | 0.00 | 0.00 | 0.00 | 0.00 |
| Fear | 0.70 | 18.31 | 76.06 | 0.00 | 0.00 | 0.00 | 4.93 |
| Anger | 1.40 | 0.00 | 0.70 | 95.10 | 2.80 | 0.00 | 0.00 |
| Disgust | 0.69 | 1.39 | 0.69 | 7.64 | 88.19 | 0.00 | 1.39 |
| Joy | 2.74 | 0.68 | 0.00 | 0.68 | 0.00 | 95.89 | 0.00 |
| Sadness | 6.25 | 6.94 | 1.39 | 0.00 | 2.78 | 0.00 | 82.64 |

Virtual Reality Front View

| Correct answer \ Participant answer | Neutral | Surprise | Fear | Anger | Disgust | Joy | Sadness |
|---|---|---|---|---|---|---|---|
| Neutral | 98.86 | 0.00 | 0.00 | 0.00 | 0.00 | 0.00 | 1.14 |
| Surprise | 0.00 | 94.15 | 4.68 | 0.00 | 0.00 | 0.00 | 1.17 |
| Fear | 0.56 | 12.99 | 84.18 | 0.00 | 1.13 | 0.00 | 1.13 |
| Anger | 0.00 | 0.00 | 0.54 | 98.92 | 0.00 | 0.00 | 0.54 |
| Disgust | 1.12 | 0.00 | 0.00 | 3.37 | 95.51 | 0.00 | 0.00 |
| Joy | 0.57 | 1.14 | 0.00 | 0.57 | 0.00 | 97.73 | 0.00 |
| Sadness | 2.27 | 0.57 | 1.70 | 2.27 | 2.27 | 0.00 | 90.91 |

Virtual Reality Side View

| Correct answer \ Participant answer | Neutral | Surprise | Fear | Anger | Disgust | Joy | Sadness |
|---|---|---|---|---|---|---|---|
| Neutral | 96.74 | 0.00 | 0.00 | 1.09 | 1.09 | 0.00 | 1.09 |
| Surprise | 0.00 | 95.24 | 4.23 | 0.00 | 0.00 | 0.00 | 0.53 |
| Fear | 0.55 | 29.51 | 69.40 | 0.00 | 0.55 | 0.00 | 0.00 |
| Anger | 0.00 | 0.57 | 0.00 | 98.85 | 0.00 | 0.57 | 0.00 |
| Disgust | 0.00 | 0.55 | 0.00 | 12.64 | 86.81 | 0.00 | 0.00 |
| Joy | 0.00 | 0.00 | 0.00 | 0.00 | 0.00 | 100.00 | 0.00 |
| Sadness | 3.26 | 2.72 | 2.17 | 4.89 | 1.09 | 0.00 | 85.87 |

**Fig. 4.** Confusion matrices showing the percentages of the correct answers for desktop (top row) and for VR (bottom row) from the front and side views.

We studied the differences in the correct emotion identification percentages for each of the visualisation angles for the emotions in the desktop condition, and the Wilcoxon Signed Rank test could not find any. This was different for the VR condition, as differences were found for *fear* ($Z = -2.972, p = 0.003$) and *disgust* ($Z = -2.782, p = 0.005$). In both cases, the results obtained for the front view are higher than those obtained for the side view. Comparing the two conditions for the front view, we found that *anger* and *disgust* obtained better results for the VR condition (Mann-Whitney $U = 513.0, p = 0.015$ and $U = 502.5, p = 0.044$, respectively). For the side view, the difference was only statistically significant for *joy* ($U = 558.0, p = 0.021$).

# 4 Discussion

This work has highlighted the possible differences in emotion recognition using virtual humans on a computer screen (desktop VR) and using an HMD (immersive VR). In this regards, RQ1 is valid according to the obtained data. It can be observed that the accuracy is slightly higher under the VR condition (92.34% vs 90.03%). Studying the emotions separately, in both conditions *fear* has the lower accuracy and is mainly confused with *surprise*, which is consistent with previous studies [8,12,27,29]. The confusion of *fear* with *surprise* could be due to the fact that these emotions share several action units (AU) according to FACS [6]. The rest of the emotions are better recognised in VR, although the difference is not very high, except for *anger*, *disgust* and *sadness*, where the difference between both conditions is higher in favour of VR. The sense of presence could explain the better accuracy recognition in the VR condition, as the sense of being there increases, objects are perceived in a more realistic way and some features may stand out more and be more noticeable than on a screen (see [25,26]).

With respect to RQ2, previous studies have only used static images shown on a computer screen to the best our knowledge. In general, the accuracy is slightly better in VR for both views (94.32% in front and 90.42% in side view for VR; 90.30% in front and 89.74% in side view for desktop). As in RQ1, this difference could be explained as an increase in the sense of presence. Considering the emotions separately, in Fig. 4 it can be appreciated that *fear* is again the worst recognised emotion, being the percentage of confusion higher in side view than in frontal view. It is confused with *surprise*, which is consistent with previous studies [12,27]. As stated before, *fear* and *surprise* share AUs, and viewing them in profile view could mask the AUs that serve to differentiate between both emotions. The results show that the accuracy obtained for the other emotions are similar in desktop and VR for both views, except for *disgust* and *sadness*. Both of them are mainly confused with *anger*. These results are also partially consistent with previous studies. For instance, *anger* and *disgust* are the emotions with lowest accuracy in both views as identified in [12], and they are confused with *disgust* and *sadness*, respectively. Also, *anger* and *sadness* have lower accuracy than the average and are mainly confused with *disgust* and *neutral* in [27]. Comparing desktop and VR conditions, it is appreciated that the accuracy for *disgust* in the desktop condition is lower in the front view than in the side view. The difference in accuracy in front and side view for both desktop and VR conditions may be explained by the fact that the viewpoints change and it may happen that not all the necessary cues needed to recognise each emotion are visible in the side view.

# 5 Conclusions

The present study has shown a comparison between using virtual humans for emotion recognition in a desktop environment and in VR. It has been corroborated by our results that the sense of presence is higher in VR than in desktop, which is consistent with previous works [25,26]. The sense of immersion and reality have also been found to be enhanced from analysing the data of the different viewpoints. In both conditions the effect is better in the front view than in the side view, but the accuracy for both views is better in the VR condition than in the desktop condition.

**Acknowledgements.** Grants PID2020-115220RB-C21 and EQC2019-006063-P funded by MCIN/AEI/ 10.13039/ 501100011033 and by "ERDF A way to make Europe". This work was also partially supported by CIBERSAM of the Instituto de Salud Carlos III.

## References

1. Baceviciute, S., Terkildsen, T., Makransky, G.: Remediating learning from non-immersive to immersive media: using EEG to investigate the effects of environmental embeddedness on reading in virtual reality. Comput. Educ. **164**, 104122 (2021)
2. Barfield, W., Zeltzer, D., Sheridan, T., Slater, M.: Presence and performance within virtual environments. In: Virtual Environments and Advanced Interface Design, pp. 473–513 (1995)
3. Burdea, G.C., Coiffet, P.: Virtual Reality Technology. Wiley (2003)
4. Busin, Y., Lukasova, K., Asthana, M.K., Macedo, E.C.: Hemiface differences in visual exploration patterns when judging the authenticity of facial expressions. Front. Psychol. **8**, 2332 (2018)
5. Castillo, J.C., Fernández-Caballero, A., Castro-González, Á., Salichs, M.A., López, M.T.: A framework for recognizing and regulating emotions in the elderly. In: Pecchia, L., Chen, L.L., Nugent, C., Bravo, J. (eds.) IWAAL 2014. LNCS, vol. 8868, pp. 320–327. Springer, Cham (2014). https://doi.org/10.1007/978-3-319-13105-4_46
6. Ekman, P., Friesen, W.: Facial Action Coding System: A Technique for the Measurement of Facial Movement. Consulting Psychologists Press, Palo Alto (1978)
7. Fernández-Sotos, P., García, A.S., Vicente-Querol, M.A., Lahera, G., Rodriguez-Jimenez, R., Fernández-Caballero, A.: Validation of dynamic virtual faces for facial affect recognition. PLoS ONE **16**(1 1), 1–15 (2021)
8. García, A.S., Fernández-Sotos, P., Vicente-Querol, M.A., Lahera, G., Rodriguez-Jimenez, R., Fernandez-Caballero, A.: Design of reliable virtual human facial expressions and validation by healthy people. Integr. Comput.-Aided Eng. **27**(3), 287–299, 104122 (2020)
9. García, A.S., Martínez, D., Molina, J.P., González, P.: Collaborative virtual environments: you can't do it alone, can you? In: International Conference on Virtual Reality, pp. 224–233. Springer (2007)
10. García, A.S., et al.: A collaborative workspace architecture for strengthening collaboration among space scientists. In: 2015 IEEE Aerospace Conference, pp. 1–12. IEEE (2015)
11. García, A.S., et al.: Acceptance and use of a multi-modal avatar-based tool for remediation of social cognition deficits. J. Ambient. Intell. Humaniz. Comput. **11**(11), 4513–4524, 104122 (2019). https://doi.org/10.1007/s12652-019-01418-8
12. Guo, K., Shaw, H.: Face in profile view reduces perceived facial expression intensity: an eye-tracking study. Acta Physiol. (Oxf) **155**, 19–28 (2015)

13. Gutiérrez-Maldonado, J., Rus-Calafell, M., González-Conde, J.: Creation of a new set of dynamic virtual reality faces for the assessment and training of facial emotion recognition ability. Virtual Reality **18**(1), 61–71 (2013). https://doi.org/10.1007/s10055-013-0236-7
14. Heldal, I., Steed, A., Schroeder, R.: Evaluating collaboration in distributed virtual environments for a puzzle-solving task. In: HCI International (2005)
15. Kohler, C.G., et al.: Facial emotion recognition in schizophrenia: intensity effects and error pattern. Am. J. Psychiatry **160**(10), 1768–1774, 104122 (2003)
16. Lozano-Monasor, E., López, M.T., Vigo-Bustos, F., Fernández-Caballero, A.: Facial expression recognition in ageing adults: from lab to ambient assisted living. J. Ambient. Intell. Humaniz. Comput. **8**(4), 567–578 (2017). https://doi.org/10.1007/s12652-017-0464-x
17. Lozano-Monasor, E., López, M.T., Fernández-Caballero, A., Vigo-Bustos, F.: Facial expression recognition from webcam based on active shape models and support vector machines. In: Pecchia, L., Chen, L.L., Nugent, C., Bravo, J. (eds.) IWAAL 2014. LNCS, vol. 8868, pp. 147–154. Springer, Cham (2014). https://doi.org/10.1007/978-3-319-13105-4_23
18. Marín-Morales, J., Llinares, C., Guixeres, J., Alcañiz, M.: Emotion recognition in immersive virtual reality: From statistics to affective computing. Sensors **20**(18), 5163 (2020)
19. Marwick, K., Hall, J.: Social cognition in schizophrenia: a review of face processing. Br. Med. Bull. **88**(1), 43–58 (2008)
20. Muros, N.I., et al.: Facial affect recognition by patients with schizophrenia using human avatars. J. Clin. Med. **10**(9), 1904, 104122 (2021)
21. Rapuano, M., Sbordone, F.L., Borrelli, L.O., Ruggiero, G., Iachini, T.: The effect of facial expressions on interpersonal space: a gender study in immersive virtual reality. In: Esposito, A., Faundez-Zanuy, M., Morabito, F.C., Pasero, E. (eds.) Progresses in Artificial Intelligence and Neural Systems. SIST, vol. 184, pp. 477–486. Springer, Singapore (2021). https://doi.org/10.1007/978-981-15-5093-5_40
22. Roberts, D., Wolff, R., Otto, O., Steed, A.: Constructing a gazebo: supporting teamwork in a tightly coupled, distributed task in virtual reality. Presence **12**(6), 644–657 (2003)
23. Russell, J.A.: Is there universal recognition of emotion from facial expression? a review of the cross-cultural studies. Psychol. Bull. **115**(1), 102 (1994)
24. Sandín, B., Chorot, P., Lostao, L., Joiner, T.E., Santed, M.A., Valiente, R.M.: Escalas PANAS de afecto positivo y negativo: Validacion factorial y convergencia transcultural. Psicothema **11**(1), 37–51 (1999)
25. Slater, M., Linakis, V., Usoh, M., Kooper, R.: Immersion, presence and performance in virtual environments: an experiment with tri-dimensional chess. In: Proceedings of the ACM Symposium on Virtual Reality Software and Technology, pp. 163–172 (1996)
26. Stevens, J.A., Kincaid, J.P., et al.: The relationship between presence and performance in virtual simulation training. Open J. Modelling Simul. **3**(02), 41, 104122 (2015)
27. Surcinelli, P., Andrei, F., Montebarocci, O., Grandi, S.: Emotion recognition of facial expressions presented in profile. Psychological Reports (2021)

28. Watson, D., Clark, L.A., Tellegen, A.: Development and validation of brief measures of positive and negative affect: the PANAS scales. J. Pers. Soc. Psychol. **54**(6), 1063–1070, 104122 (1988)
29. del Águila, J., González-Gualda, L.M., Játiva, M.A., Fernández-Sotos, P., Fernández-Caballero, A., García, A.S.: How interpersonal distance between avatar and human influences facial affect recognition in immersive virtual reality. Front. Psychol. **12**, 675515 (2021)

# Influence of Neutral Stimuli on Brain Activity Baseline in Emotional Experiments

Beatriz García-Martínez[1,2](✉) and Antonio Fernández-Caballero[1,2,3]

[1] Universidad de Castilla-La Mancha, Departamento de Sistemas Informáticos, Albacete, Spain
Beatriz.GMartinez@uclm.es
[2] Universidad de Castilla-La Mancha, Instituto de Investigación en Informática de Albacete, Unidad Multidisciplinar en Neurocognición y Emoción, Albacete, Spain
[3] CIBERSAM (Biomedical Research Networking Centre in Mental Health), Madrid, Spain

**Abstract.** In the literature there is a wide number of experiments for induction of emotions using different types of stimuli. Apart from the samples with emotional content, neutral stimuli are typically used to set up a baseline state before or after the presentation of emotional stimuli. In the present study, the effect of these neutral stimuli and the duration necessary for reaching the baseline brain activity was assessed by means of a spectral analysis of electroencephalographic signals. Concretely, the brain activity at the beginning, middle and end of a neutral stimulus was compared with the activity at the end of the previously presented emotional stimulus. The results reported that 30 s of neutral stimulus successfully led to a baseline state after the elicitation of emotions with low arousal or low valence in all brain regions and for all frequency bands, whereas the double of time was necessary for the regulation of emotional states with high arousal or high valence levels. In addition, no statistical differences were found at the end of all neutral stimuli, corroborating the achievement of a common baseline regardless of the emotional stimulus previously shown.

**Keywords:** Electroencephalography · Emotion recognition · Spectral power · Neutral stimulus · Baseline establishment

## 1 Introduction

In the last decades, affective computing has become one of the most relevant research areas in the scientific literature [8]. Its purpose is the development of computational models for emotions recognition that could help to endow automatic systems with emotional intelligence and make them able to identify and interpret human emotional states [17]. The creation of such computational models starts with an experimental procedure in which emotional stimuli are used to


J. M. Ferrández Vicente et al. (Eds.): IWINAC 2022, LNCS 13258, pp. 475–484, 2022.
https://doi.org/10.1007/978-3-031-06242-1_47

elicit different emotional states in the participant. At the same time, the emotional information of the subject is acquired in order to discover the relations among the emotion triggered and the general state of the participant.

Traditional methods for the acquisition of emotional information have been mainly based on physical aspects like speech features and facial gestures [16]. However, recent studies have focused on the assessment of different physiological variables like electrocardiogram (ECG), electro-dermal activity (EDA) or electromyogram (EMG), for the detection of emotions [5]. In the last years, the electroencephalogram (EEG) has been widely analyzed for emotions recognition [1,6]. The main advantage of EEG over all other physiological signals is that EEG measures the electrical activity of the brain, which is the organ responsible for the management of the rest of systems and the source of all bodily responses [5]. Contrarily, other physiological variables are the consequence of the propagation of brain activity to peripheral organs by means of the central nervous system [5].

Many studies for emotions detection have analyzed the EEG recordings from a spectral point of view. Indeed, the EEG frequency spectrum, which ranges between 0.5 45 Hz, is divided into five frequency bands, including *delta* (0.5–4 Hz), *theta* (4–8 Hz), *alpha* (8–13 Hz), *beta* (13–30 Hz) and *gamma* (30–45 Hz) [9]. It is known that changes in the emotional state of a subject are reflected in changes in the prominence of these frequency bands [10]. Therefore, the evaluation of the spectral features of EEG signals allows to detect different emotional states [10].

With respect to experimental procedures, various types of stimuli with emotional content, such as images, sounds, or video clips, among others, are usually used for the elicitation of different emotional states in the subject. The emotion that a certain stimulus evokes is determined according to its associated values of different emotional parameters, such as valence (which represents the pleasantness or unpleasantness produced by a stimulus) and arousal (i.e., the activation or relaxation produced by an emotional stimulus) [19]. In this line, the detection of high and low levels of valence and arousal dimensions is one of the most widely used experimental schemes in the scientific literature [18]. Apart from the stimuli with emotional content, it is also common to show neutral stimuli to the participants in order to set a baseline state before each emotional stimulus. This baseline represents the brain activity of an idle subject, when no emotional processes are being executed [7]. Hence, the process of elicitation of the different emotions starts from the same emotional state of the subject, thus reducing the possible influence of the emotion elicited previously.

In this sense, the present study analyzes the effect of the neutral stimulus for the establishment of a baseline emotional state, with the purpose of verifying whether the neutral stimulus takes the desired effect on the brain activity. Moreover, the visualization time necessary for the achievement of that baseline state will also be evaluated. Concretely, the spectral features of brain activity at the end of an emotional stimulus will be compared with the activity in three different segments (beginning, middle and end) of the posterior neutral stimulus. In order to assess the possible differences in the effect of the neutral sample depending on the emotion elicited by the previous emotional stimulus, samples will be

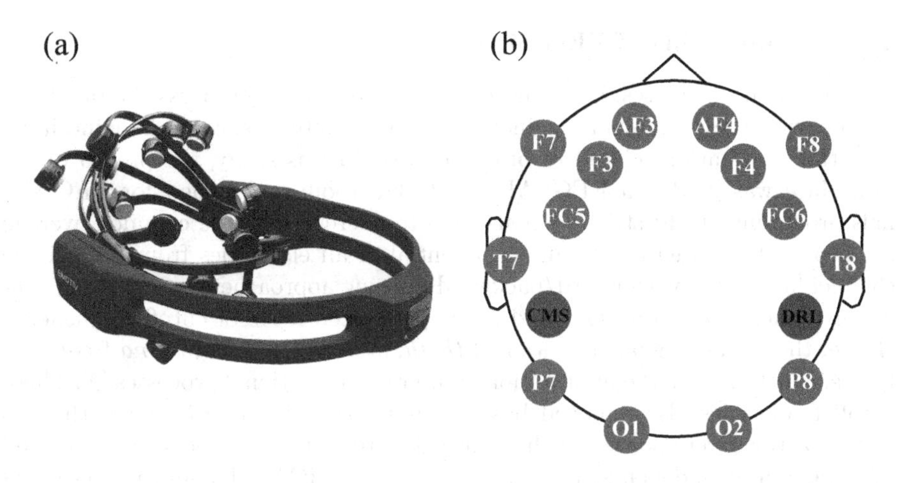


**Fig. 1.** (a) Emotiv EPOC headset. (b) Locations of electrodes (in green) and references (in gray) on the Emotiv EPOC headset. (Color figure online)

analyzed separately according to the associated high or low level of valence and arousal of the emotional clip.

# 2 Materials and Methods

## 2.1 Database

The EEG signals analyzed in this study were extracted from DREAMER, a publicly available database designed for the detection of emotional states with physiological signals recorded with wearable devices [11]. A total of 23 healthy subjects (14 males and 9 females, ages between 22 and 33 years, average 26.6) participated in an experiment consisting of the visualization of 18 movie clips (duration from 65 to 393 s, average 199) with emotional content for the elicitation of happiness, calmness, amusement, excitement, anger, disgust, fear, sadness and surprise. Before each video, a 60-second neutral clip was presented with the purpose of eliciting a neutral state in the participant and establishing a baseline. At the end of each visualization, participants used self-assessment manikins for rating their emotional state in terms of valence and arousal in an intensity scale ranging from 1 to 5.

The EEG signals of this database were recorded with an Emotiv EPOC device. This wearable, wireless and off-the-shelf device includes 16 gold-plated contact electrodes for the measurement of 14 EEG channels plus two references (CMS and DRL), working at a sampling frequency 128 Hz. Concretely, the 14 EEG channels include AF3, F7, F3, FC5, T7, P7, O1, O2, P8, T8, FC6, F4, F8, and AF4, placed according to the International Standard 10–20 system for electrodes location [13]. The Emotiv EPOC device and the distribution of the electrodes over the scalp are shown in Fig. 1.

### 2.2 Preprocessing of EEG Signals

Prior to further analyses, it was necessary to filter and preprocess the raw EEG signals, with the purpose of eliminating noise and interferences and maintaining only the information related to brain activity. In this study, the preprocessing procedure was made on EEGLAB, a Matlab toolbox designed for EEG signals processing [4]. First, the signals were re-referenced to the common average by means of subtracting the mean potential of all electrodes from each single channel [3]. Then, two forward/backward filtering approaches were applied, concretely a high-pass filter 4 Hz and a low-pass filter 45 Hz of cutoff frequencies. Hence, the filtered signals maintained *theta*, *alpha*, *beta* and *gamma* frequency bands, which contain the information of interest in emotional processes [9]. These cutoff frequencies also removed baseline and power line interferences, thus no further actions were needed in this sense. The remaining artifacts were removed by means of an independent component analysis (ICA). These artifacts could be generated by either physical/physiological processes (like eye blinks, facial movements, or heart bumps) or technical aspects (such as electrode pops or bad contact of the electrodes over the scalp). The ICA analysis consisted of firstly decomposing the signal into independent components, and those containing artifactual information were rejected. Therefore, only the information of brain activity remained after this process.

### 2.3 Experimental Procedure

In order to evaluate the effect of the neutral clip, the brain activity during its visualization was evaluated in three different periods. More precisely, three segments of 10 s of length were selected at the beginning (seconds from 1 to 10), middle (from 25 to 35) and end (from 50 to 60 s) of the neutral stimulus. These three periods were compared with the last 10 s of the previous emotional stimulus, with the purpose of verifying whether the neutral clip influenced on brain activity with respect to the clips with emotional content. In addition, in order to evaluate the possible differences in the effect of the neutral stimulus depending on the previous emotion, these analyses were made according to the levels of valence and arousal associated with the prior emotional clip. Precisely, high arousal (HA) stimuli presented an arousal rating $\geq 4$, whereas low arousal (LA) samples obtained an arousal rating $\leq 2$. Samples with a rating of 3 were not considered in this study. The same criteria were used for high valence (HV) and low valence (LV) samples. The total number of samples for each emotional condition was HA $= 166$, LA $= 114$, HV $= 161$ and LV $= 152$.

### 2.4 Feature Extraction

The spectral analysis of the aforementioned segments from both emotional and neutral stimuli was made by means of the power spectral density (PSD), calculated through the Welch's periodogram (Hamming window of 2 s of length, 50% of overlapping, 256 points of resolution). The spectral power of the whole EEG spectrum, namely $SP_{all}$, was obtained for each EEG channel as the area under the PSD curve within 4 45 Hz of frequency:

$$SP_{all} = \sum_{4Hz}^{45Hz} |PSD(f)| \tag{1}$$

The spectral power of each frequency band, $SP_{theta}$, $SP_{alpha}$, $SP_{beta}$, and $SP_{gamma}$, was calculated in a similar manner. Nevertheless, the resulting value in each band was normalized by the power in the whole spectrum, with the purpose of preserving the fluctuations among subjects:

$$SP_B = \frac{1}{SP_{all}} \sum_{f_1}^{f_2} |PSD(f)|, \tag{2}$$

being $B$ a frequency band, and $f_1$ and $f_2$ its lower and higher frequency, respectively.

### 2.5 Statistical Analysis

In order to corroborate the effect of the neutral stimulus, a statistical analysis based on a one-way analysis of variance (ANOVA) was used to evaluate possible statistical differences among the spectral features of the brain activity at the end of the emotional stimulus and the beginning, middle and end of the neutral clip, in the different frequency bands and for the whole spectrum. These tests allowed to verify the effectiveness of the neutral stimulus to modulate the brain activity and establish a baseline. It is important to remark that only statistical significance values of $\rho < 0.05$ were considered as significant.

In addition, an ANOVA analysis was also used to check the statistical differences among the last segment of the neutral stimulus for all samples, regardless of the levels of arousal and valence of the emotional stimulus previously presented. Therefore, it was possible to corroborate if the baseline had been properly established and the brain activity of the subject at the end of the neutral stimulus was the same, independently of the emotional state of the participant before visualizing the neutral clip. As in the previous case, only results of $\rho < 0.05$ were considered as statistically significant.

## 3 Results

Figures 2 and 3 show the results reported by the ANOVA analysis for the assessment of the statistical differences among the end of the stimulus and the beginning, middle and end of the posterior neutral clip. Concretely, Fig. 2 represents those results considering the low or high level of arousal associated to the emotional stimulus, whereas Fig. 3 separates the results for low and high levels of valence of the emotional clip. The statistically significant EEG channels were highlighted in orange circles, while non-statistically significant electrodes were painted in gray.

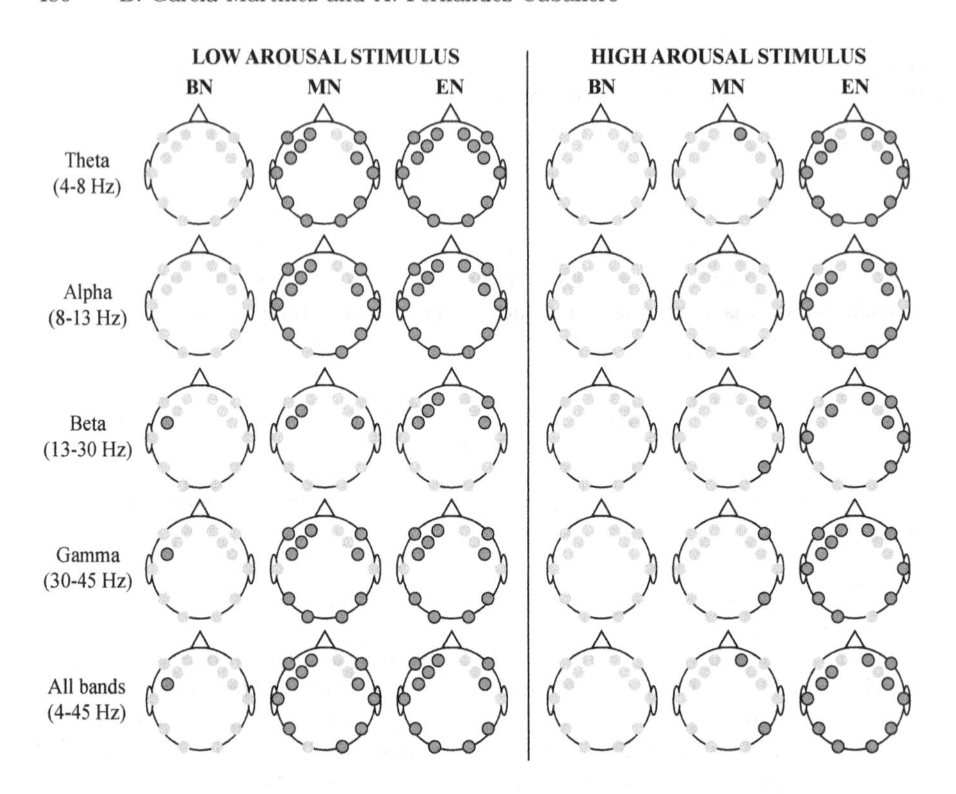


**Fig. 2.** ANOVA results for the comparison among the beginning (BN), middle (MN) and end (EN) of neutral stimulus and the end of the previous emotional clip, depending on its level of arousal. Statistically significant channels are highlighted in orange. (Color figure online)

As can be observed, the outcomes obtained were similar for both emotional dimensions. The comparison between the end of the emotional stimulus and the beginning of the neutral clip (represented as BN in both figures), reported non-statistically significant results for both low and high levels of arousal and valence in almost all EEG channels and all frequency bands. This reveals a similar brain activity at both the end of the emotional clip and the beginning of the neutral video, thus verifying that the short duration of the neutral clip analyzed did not influence on the achievement of a baseline brain activity.

In the case of the comparison between the end of the emotional stimulus and the middle of the neutral clip (namely MN in the figures), many EEG channels from all brain regions were statistically significant for low arousal and low valence stimuli. Hence, the first half of the neutral clip strongly influenced on brain activity and achieved a regulation of the emotional state of the subjects with significant changes statistically measurable. This performance was consistent for all frequency bands, although in the case of *beta* the number of statistically significant channels was slightly lower than for the rest of bands. Nevertheless,

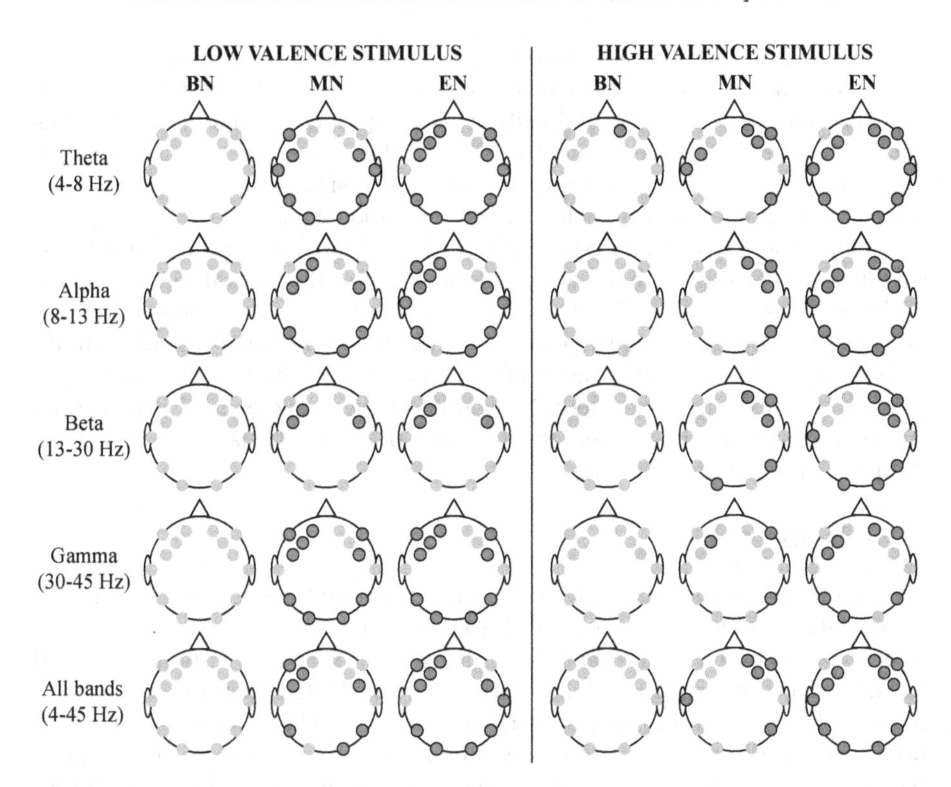


**Fig. 3.** ANOVA results for the comparison among the beginning (BN), middle (MN) and end (EN) of neutral stimulus and the end of the previous emotional clip, depending on its level of valence. Statistically significant channels are highlighted in orange. (Color figure online)

when the previous stimulus presented a high arousal or high valence level, only a few EEG channels reported statistically significant results, thus revealing a lower capability of the neutral stimulus to lead the subject to a baseline state in the middle of its duration, in comparison with the case in which the prior emotional stimulus presented a low arousal or low valence. This difference of the effect of the neutral stimulus in the middle of its presentation among low and high levels of the emotional dimensions is even clearer in the case of arousal, in which only one or two EEG channels reported statistically significant values for the different frequency bands.

On the other hand, when the end of the emotional stimulus was compared with the end of the neutral clip (depicted as EN), the results obtained were similar to the previous analysis in the case of low levels of arousal and valence. In fact, almost all EEG channels reported statistically significant values, except for *beta* band in which the number of relevant channels was slightly lower. Therefore, these results corroborated the changes in the brain activity produced by the visualization of the neutral clip after a low arousal or low valence video.

With respect to the analysis made with stimuli of high level of the emotional dimensions, also a wide number of EEG channels reported statistically significant outcomes for all frequency bands when comparing the end of the stimulus with the end of the posterior neutral clip. In this sense, the visualization of a longer neutral stimulus also achieved statistically significant changes in brain activity after a high arousal or high valence emotional state.

Finally, the ANOVA analysis performed to evaluate the existence of statistical differences among the brain activity achieved after the neutral clip, regardless of the previous emotional stimulus presented, reported no statistically significant outcomes for any of the frequency bands, i.e., no relevant differences among brain activity were obtained after the neutral clip in any of the four scenarios (LA, HA, LV or HV). Therefore, the visualization of the whole neutral stimulus led subjects to a common baseline, independently of the emotional state elicited by the previous emotional stimulus.

## 4 Discussion

The results of the present study have shown that the time of presentation of a neutral stimulus necessary to go back to a baseline state is considerably lower after emotions with low arousal or low valence levels with respect to high arousal or high valence stimuli. Indeed, a duration of around 30 s of neutral clip led subjects to that baseline in brain activity after a LA or LV stimulus, whereas the necessary length of the neutral clip was doubled (around 60 s) for returning to baseline after HA or HV emotional states. These outcomes are in line with previous works stating that strong emotions, typically associated with high levels of arousal and valence, last longer in time and require stronger efforts to reverse the emotional reaction produced [12]. More concretely, intense negative emotions such as fear, stress, or anger have reported the hardest regulation difficulties [12]. As a result, the neutral stimulus for baseline establishment after those emotions should present a longer duration. On the other hand, the reduced recovery time needed after LA or LV emotional stimuli could also be related with possible difficulties for the elicitation of these emotional states [14]. In addition, the tendency of neutral stimuli to present medium-low values of the emotional dimensions, especially in the case of arousal, could also influence on this aspect [2].

The results obtained presented strong similarities for all the frequency bands analyzed. However, the *beta* band presented a slightly lower number of statistically significant channels, thus revealing that the changes in brain activity produced by the neutral stimulus in this frequency band were not as relevant as for the remaining frequencies. In other studies, *beta* band has reported the best results in emotions recognition processes, thus revealing a higher sensitivity of this frequency band to emotion alterations [15]. This could involve a strong elicitation of the emotional states in this frequency band, with the possibility of being associated with more difficulties when trying to reverse the brain activity in *beta* band to a baseline state. In this sense, a longer exposure to a neutral stimulus may be necessary for the establishment of a baseline in *beta* rhythm than for the rest of frequency bands.

Finally, some limitations should be considered in this study. For example, the number of participants and trials from each subject is limited, which hinders the consideration of these results as representative for the whole population. In this sense, new experiments with a higher number of participants should be conducted. Moreover, the reduced number of EEG channels does not allow to thoroughly analyze the activity of the different brain regions, especially in central, parietal and occipital areas. On the other hand, both the emotional and neutral stimuli used in this work were video clips. In future studies, other types of neutral stimuli, such as distractor tasks, will be used in order to evaluate the possible differences in the process of regulation of the emotional states induced.

## 5 Conclusions

This work has analyzed the effect of the neutral stimuli for the establishment of a baseline state in emotional experiments and the influence of the duration of these stimuli depending on the emotional state previously elicited. Results of the spectral analysis of the brain activity revealed that a neutral stimulus with a duration of around 30 s successfully led the participant to a baseline emotional state after the visualization of low arousal or low valence stimuli. On the contrary, in the case of high arousal or high valence clips, the required duration time of the posterior neutral stimulus was around 60 s, thus doubling the time needed for the achievement of the baseline state of the brain activity with respect to the low levels of both emotional dimensions. Furthermore, a statistical analysis revealed that the activity of the brain at the end of the neutral stimulus was similar regardless of the level of arousal and valence of the emotional stimulus previously presented. Therefore, it confirms that the visualization of the neutral clip established a common baseline activity in all the cases.

**Acknowledgements.** Grants PID2020-115220RB-C21 and EQC2019-006063-P, funded by MCIN/AEI/ 10.13039/501100011033/ and "ERDF A way to make Europe". Grant FPU16/03740 funded by MCIN/AEI/10.13039/501100011033/ and "ESF Investing in your future". This work was partially supported by Biomedical Research Networking Centre in Mental Health (CIBERSAM) of the Instituto de Salud Carlos III.

## References

1. Alarcao, S.M., Fonseca, M.J.: Emotions recognition using EEG signals: a survey. IEEE Trans. Affect. Comput. **10**(3), 374–393 (2017)
2. Coan, J.A., Allen, J.J.B.: Handbook of Emotion Elicitation and Assessment. Oxford University Press, Oxford (2007)
3. Cohen, M.X.: Analyzing neural time series data: Theory and practice. MIT press (2014)
4. Delorme, A., Makeig, S.: EEGLAB: an open source toolbox for analysis of single-trial EEG dynamics including independent component analysis. J. Neurosci. Methods **134**(1), 9–21 (2004)

5. Egger, M., Ley, M., Hanke, S.: Emotion recognition from physiological signal analysis: a review. Electron. Notes Theor. Comput. Sci. **343**, 35–55 (2019)
6. García-Martínez, B., Martinez-Rodrigo, A., Alcaraz, R., Fernández-Caballero, A.: A review on nonlinear methods using electroencephalographic recordings for emotion recognition. IEEE Trans. Affect. Comput. **12**(3), 801–820 (2021)
7. Grecucci, A., De Pisapia, N., Kusalagnana Thero, D., Paladino, M.P., Venuti, P., Job, R.: Baseline and strategic effects behind mindful emotion regulation: Behavioral and physiological investigation. PLoS ONE **10**(1), e0116541 (2015)
8. Han, J., Zhang, Z., Schuller, B.: Adversarial training in affective computing and sentiment analysis: recent advances and perspectives. IEEE Comput. Intell. Mag. **14**(2), 68–81 (2019)
9. Ismail, W.W., Hanif, M., Mohamed, S., Hamzah, N., Rizman, Z.I.: Human emotion detection via brain waves study by using electroencephalogram (EEG). Int. J. Adv. Sci. Eng. Inf. Technol. **6**(6), 1005–1011 (2016)
10. Jebelli, H., Hwang, S., Lee, S.: EEG signal-processing framework to obtain high-quality brain waves from an off-the-shelf wearable EEG device. J. Comput. Civ. Eng. **32**(1), 04017070 (2018)
11. Katsigiannis, S., Ramzan, N.: DREAMER: a database for emotion recognition through EEG and ECG signals from wireless low-cost off-the-shelf devices. IEEE J. Biomed. Health Inform. **22**(1), 98–107 (2018)
12. Kim, S.H., Hamann, S.: Neural correlates of positive and negative emotion regulation. J. Cogn. Neurosci. **19**(5), 776–798 (2007)
13. Klem, G.H., Lüders, H.O., Jasper, H.H., Elger, C.: The ten-twenty electrode system of the International Federation. Electroencephalogr. Clin. Neurophysiol. **52**, 3–6 (1999)
14. Koelstra, S., et al.: DEAP: a database for emotion analysis using physiological signals. IEEE Trans. Affect. Comput. **3**(1), 18–31 (2012)
15. Liu, X., Li, T., Tang, C., Xu, T., Chen, P., Bezerianos, A., Wang, H.: Emotion recognition and dynamic functional connectivity analysis based on EEG. IEEE Access **7**, 143293–143302 (2019)
16. Nahid, N., Rahman, A., Ahad, M.A.R.: Contactless human activity analysis across different modalities. In: Contactless Human Activity Analysis. Intelligent Systems Reference Library, vol. 200, pp. 237–270. Springer Nature (2021)
17. Poria, S., Cambria, E., Bajpai, R., Hussain, A.: A review of affective computing: from unimodal analysis to multimodal fusion. Inf. Fusion **37**, 98–125 (2017)
18. Rahman, M.M., et al.: Recognition of human emotions using EEG signals: a review. Comput. Biol. Med. **136**, 104696 (2021)
19. Russell, J.A.: A circumplex model of affect. J. Pers. Soc. Psychol. **39**(6), 1161–1178 (1980)

# Classification of Psychophysiological Patterns During Emotional Processing Using SVM

Andrés Quintero-Zea[1(✉)], Juan Martínez-Vargas[1], Diana Gómez[2], Natalia Trujillo[2], and José D. López[2]

[1] LabMIRP, Parque i, Instituto Tecnológico Metropolitano, Medellín, Colombia
andresquinteroz@itm.edu.co

[2] Universidad de Antioquia UDEA, Calle 70 No. 52-21, Medellín, Colombia

**Abstract.** Emotional processing (EP) of ex-combatants of illegal armed groups plays a significant role in a successful reintegration into society. But determining the links between EP and brain activity in this population is still an open issue due to the subtle physiological differences observed between them and civilians. Consequently, in this work we propose a combined approach with EP psychological assessments and EEG functional connectivity at source level (EEG brain imaging based) that feed a support vector machine. Results show that it is possible to differentiate between psychophysiological patterns of ex-combatants and controls based on their EP, a key component to develop new psychological interventions for this population.

**Keywords:** Coherency · EEG · Emotional processing · Functional connectivity

## 1 Introduction

Emotional processing (EP) is regarded as the way in which an individual processes stimuli or life events with emotional information. EP encompasses a combination of physiological, behavioural, cognitive, and emotional processes to regulate social interaction [16]. Ex-combatants of illegal armed groups involved in disarmament, demobilization, and reintegration (DDR) programs have returned to civil life. However, their successful reintegration have been overshadowed by high prevalence of aggression traits due to their exposition to armed conflict experiences. These traits have been reported to be highly associated with recidivism and have a direct effect on their reintegration into Colombian society [8]. Moreover, previous studies have reported atypical modulation of the EP in ex-combatants during valence recognition using ERP [15,20] and scalp connectivity [14], but the relation between such modulations and the sources of neuronal activity is still an open issue.


J. M. Ferrández Vicente et al. (Eds.): IWINAC 2022, LNCS 13258, pp. 485–493, 2022.
https://doi.org/10.1007/978-3-031-06242-1_48

Here, we investigate cortical Functional Connectivity Network (FCN) in ex-combatants while processing visual stimuli with emotional content. We use electroencephalography (EEG) data to estimate the FCN with the aim of evaluating the brain dynamics on a millisecond time scale inherent to cognitive processes. FCN analysis driven directly over EEG signals recorded at scalp level can lead to misinterpretation of the synchronization measurements. To overcome these limitations, we first determine source level activity solving the EEG inverse problem, and then estimate the FCN between selected regions of interest. Further, we use graph theoretical analysis to investigate two basic topological metrics that have a straightforward neurobiological interpretation: the leaf fraction and the diameter of the graph [19]. These topological metrics are used as inputs for a support vector machine (SVM) classifier, together with psychological evaluations that improve the classification rates of the SVM.

## 2 Materials and Methods

In Fig. 1 we summarize all steps of the proposed methodology. Following sections present further details of each stage.

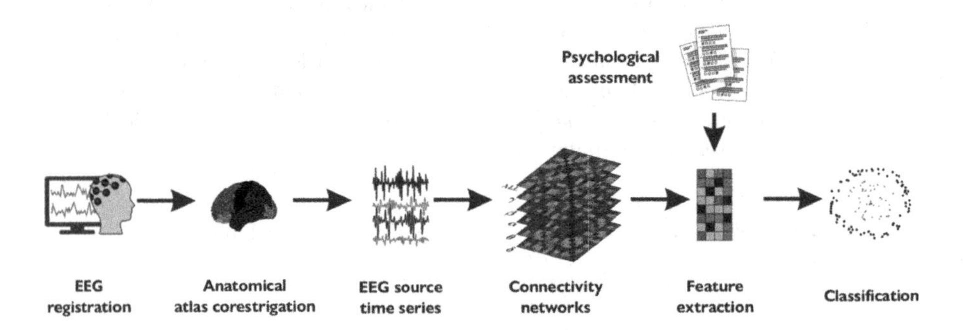


**Fig. 1.** A schematic of the proposed pipeline for the psychophysiological analysis of EP in ex-combatants.

### 2.1 Experimental Data

EEG data for this study were collected from 50 Colombian participants (30 ex-combatants and 20 civilians) during an emotional recognition task (ERT) [20] in which they were asked to classify visual stimuli displayed on a computer screen into three categories according to their valence (positive, negative, or neutral). Table 1 summarizes the demographic information of both groups. The stimuli consisted of 90 pictures taken from the MMI Facial Expression Database [12], and 90 words selected from a Colombian linguistic corpus [13]. Refer to [14,20] for further details regarding the experimental setup. EEG data were collected using a 64-channel Neuroscan SynAmps2 system positioned according to the standard 10–10 system montage at a sampling rate of 1000 Hz.

Additionally, all participants were assessed by the completion of two psychological tests: (i) the social ability scale of Gismero (EHS) [5] to measure the capacity to interact with others in a wide range of social situations. (ii) The reactive- proactive aggression questionnaire (RPQ) [17] to measure aggression traits.

**Table 1.** Demographic information of participants.

| | Ex-Combatants $N = 30$ | Civilians $N = 20$ | $t/Chi2$ $(p)$ | Bayes factor |
|---|---|---|---|---|
| | $M$ $(SD)$ | $M$ $(SD)$ | | |
| Age (Years) | 37.50 (8.22) | 36.15 (9.17) | 0.543 (0.589) | 4.083 |
| Education (Years) | 10.23 (3.02) | 11.05 (2.14) | −1.118 (0.269) | 2.871 |
| Gender (Female: Male) | 2:28 | 2:18 | 0.181 (0.670) | 3.232 |

### 2.2 EEG Data Preprocessing

As a kind of neurophysiological signal, EEG data are high dimensional and contain redundant and noisy information. In this work, the raw data was firstly preprocessed by band-pass filtering between 0.1 to 60 Hz. Next, we extracted epochs from –200 ms to 800 ms around the onset of the stimuli. Such epochs were further baseline corrected, downsampled to 500 Hz, and offline re-referenced to average. Finally, we employed independent component analysis to exclude artifactual components containing heart, eye, or muscle activity. All these steps were performed over 60 electrodes, given that HEO, VEO, CB1 and CB2 electrodes do not record neural activity.

### 2.3 Reconstruction of EEG Sources

EEG brain imaging consists of projecting the scalp recordings into the brain and determine the active regions. This is achieved with the linear model:

$$\boldsymbol{Y} = \boldsymbol{L}\boldsymbol{J} + \boldsymbol{\epsilon} \tag{1}$$

where $\boldsymbol{Y} \in \mathbb{R}^{N_c \times N_t}$ are the measured EEG data from $N_c$ channels at $N_t$ time samples, $\boldsymbol{J} \in \mathbb{R}^{N_d \times N_t}$ is the amplitude of $N_d$ dipolar electrical brain sources, which are distributed across the cortical surface with a fixed orientation perpendicular to it, $\boldsymbol{L} \in \mathbb{R}^{N_c \times N_d}$ is the propagation (lead-field) matrix, and $\boldsymbol{\epsilon} \in \mathbb{R}^{N_c \times N_t}$ is additive noise affecting the measurements. In this work, we used the SPM12 Toolbox [4] to perform the source reconstruction. The EEG forward problem was solved in the New York Head [6]. The cortical surface provided for this anatomy was down-sampled to 2004 vertices. The source reconstruction was conducted using Multiple Sparse Priors following the pipeline proposed in [10].

To define Regions of Interest (ROIs) on this surface, we used a Desikan-Killiany [3] atlas-based segmentation approach, consisting of 68 cortical regions, by co-registering a tesselated high-resolution ICBM152 mesh to the 2004 vertices New York template. Then, we grouped the 68 areas defined by the Desikan–Killiany atlas into 14 broader ROIs. After excluding vertices located outside the 14 ROIs, we retained a total of 1839 vertices for further analysis. Figure 2 depicts the resulting ROIs.

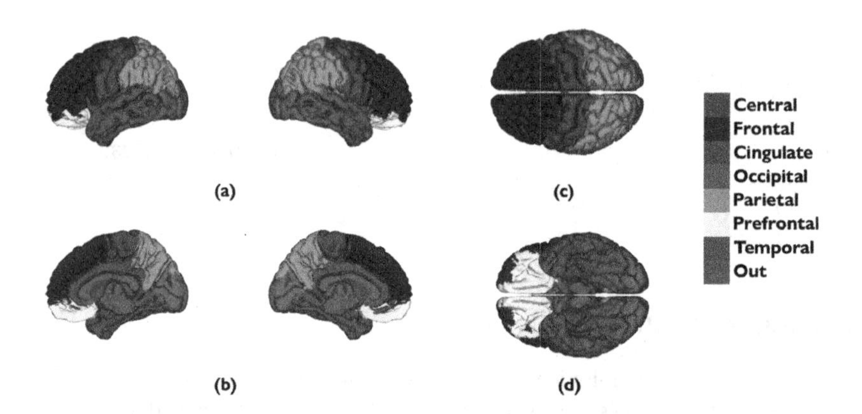


**Fig. 2.** ROIs mapped onto the ICBM152 template anatomy. **(a)** Lateral, **(b)** Sagittal, **(c)** Superior, and **(d)** Inferior views of ROIs defined according to the Desikan–Killiany atlas.

### 2.4 Functional Connectivity Network Construction

The autospectra and cross spectra of each of the averaged ROI time series were computed by means of the Welch's averaged modified periodogram. A sliding Hamming window of 200 ms, with an overlap of 100 ms, was applied to improve the estimation quality. Furthermore, spectra were computed across 60 frequency bins within the frequency range 1 to 60 Hz. This choice covers the standard spectrum of physiologic EEG oscillations: delta (0.1 to 4 Hz), theta (4 to 7 Hz), alpha (7 to 13 Hz), beta (13 to 30 Hz), and gamma (30 to 60 Hz). These spectra were further used to estimate the coherency between all pairs of ROIs for each kind of stimuli.

Coherency is a correlation function and a quantitative measure of the phase consistency between two signals. It is defined as the cross spectral density function normalized by individual auto spectral density functions [11]. Let $X_i(f)$ and $X_j(f)$ be the Fourier transforms of the neural time series $x_i(t)$ and $x_j(t)$ of brain regions (nodes) $i$ and $j$, respectively. Then the cross-spectrum is defined as:

$$S_{ij}(f) = \langle X_i(f) X_j^*(f) \rangle, \tag{2}$$

where $(\cdot)^*$ means complex conjugation and $\langle\cdot\rangle$ means expectation value. Coherency for a given frequency bin $f$ is now defined as:

$$\mathrm{COH}_{ij}(f) = \frac{S_{ij}(f)}{\sqrt{S_{ii}(f)\, S_{jj}(f)}}, \tag{3}$$

where $S_{ii}(f)$ and $S_{jj}(f)$ are the respective autospectra of time series $x_i(t)$ and $x_j(t)$ at frequency bin $f$. To reduce the influence of spurious interactions, we preserved only the imaginary part of the coherency ($\mathrm{iCOH}_{ij}^{f} = \Im(\mathrm{COH}_{ij}(f))$ [11].

We can further estimate the iCOH among each pair of ROIs $p$ and $q$ as

$$\mathrm{iCOH}_{pq}^{f} = \frac{1}{|F|N_p N_q} \sum_{f\in F} \sum_{i\in p} \sum_{j\in q} |\mathrm{iCOH}_{ij}|, \tag{4}$$

where $F$ is the set of frequencies over which iCOH is averaged and $|F|$ its cardinality. $N_p$ and $N_q$ are the numbers of voxels inside the ROIs. For further analysis, resulting pairwise iCOH between $N_r$ ROIs can be conveniently represented by a matrix $\boldsymbol{\Phi} \in \mathbb{R}^{N_r \times N_r}$, where $\phi_{pq}$ is the value of the weighted edges of the graph.

### 2.5 Feature Extraction

FCNs obtained from coherency analysis can be characterized considering all the available information in $\boldsymbol{\Phi}$, possibly yielding misinterpretations in the resulting extracted graph metrics. A common approach to tackle this issue is to filter matrix $\boldsymbol{\Phi}$ maintaining only significant links.

Here, we filter every $\boldsymbol{\Phi}$ matrix by computing their minimum spanning tree (MST) using the Kruskal's algorithm. Then, we computed the leaf fraction and the diameter of each MST matrix per band per task by means of the Brain Connectivity Toolbox [18]. Furthermore, we computed the FCN for all frequency bands. We additionally used three features per subject, extracted from the psychological tests described in Sect. 2.1.

### 2.6 Classification and Evaluation

The classifier adopted here was the support vector machine (SVM) implemented with the libsvm library [2]. The SVM exhibits good classification accuracy based on one of its abilities to select a suitable kernel function [7]. We used a Gaussian kernel to train the SVM, and a grid search method to tune kernel width $\gamma$ and the regularization parameter $C$ of the SVM, which were searched in the range from $2^{-8}$ to $2^{8}$.

For validation purposes of the models, stratified 10-fold cross-validation was used. According to 10-fold cross-validation, the dataset was randomly partitioned in ten equal non-overlapping sets, where nine of them were used alonside as the training set and one as the testing set [9]. The cross-validation process was then repeated ten times, with each of the ten sets used exactly once as the validation data. The ten results from the folds were then averaged to estimate the classification accuracy [1].

## 3 Results

*Psychological Assessments:* We found that ex-combatants exhibited increased values of proactive and reactive aggression scores, which indicates higher levels of aggression in such population. Conversely, ex-combatants have lower GSSS values, suggesting higher social assertion. Table 2 shows statistical results of the social cognition and behavior (SCB) scores from both groups.

**Table 2.** $t$-test results comparing ex-combatants and civilians on SCB scores.

| | Ex-combatants | Civilians | |
|---|---|---|---|
| | $M$ ($SD$) | $M$ ($SD$) | $t$ ($p$) |
| Reactive | 7.33 (3.17) | 5.50 (2.91) | 2.07 (**0.044**) |
| Proactive | 3.63 (3.40) | 1.50 (1.91) | 2.83 (**0.007**) |
| GSSS | 68.10 (15.86) | 80.40 (25.56) | −2.10 (**0.041**) |

Values are given as MEAN (STANDARD DEVIATION).
**Bold values** represent significant $p$-values.

Additionally, we implemented a SVM with the same stratified cross-validation strategy described in Sect. 2.6 to assess the prediction accuracy of these three scores. We obtained an average accuracy of 64%, sensitivity of 56.7%, and specificity of 75%.

*Functional Connectivity Network:* We did not find significant differences for whole brain-averaged level of iCOH, suggesting that the overall level of connectivity did not define a marker. Moreover, a pattern can not be observed since ex-combatants exhibit higher values of iCOH within some frequency bands and lower values within others. Furthermore, there were no differences in topological metrics neither for the word nor face stimuli in none of the frequency bands.

*Classification Results:* Table 3 summarizes the performance of the classifiers with each set of features, it can be seen that the SVM had a poor performance when fed exclusively with the network (Net) features, with the exception of the combination of the five frequency bands (ALL). This poor performance was improved by incorporating the psychological (Psy) assessments. Furthermore, the face stimuli provided greater discriminability compared to the word stimuli. This finding is consistent with those documented from previous research [14,20], and indicates that words might be less important than visual stimuli in the ex-combatants group.

**Table 3.** Results of classification and performance of all SVMs

| Feature set | Frequency band | Accuracy (%) | Sensitivity (%) | Specificity (%) |
|---|---|---|---|---|
| Net-Face | Delta | 54.00 | 50.00 | 60.00 |
| | Theta | 54.00 | 46.67 | 65.00 |
| | Alpha | 52.00 | 46.67 | 60.00 |
| | Beta | 56.00 | 50.00 | 65.00 |
| | Gamma | 64.00 | 63.33 | 65.00 |
| | All | 72.00 | 70.00 | 75.00 |
| Net-Face + Psy | Delta | **72.00** | **73.33** | **70.00** |
| | Theta | **64.00** | **56.67** | **75.00** |
| | Alpha | **76.00** | **70.00** | **85.00** |
| | Beta | **68.00** | **66.67** | **70.00** |
| | Gamma | **80.00** | **73.33** | **90.00** |
| | All | **90.00** | **93.33** | **85.00** |
| Net-Word | Delta | 52.00 | 50.00 | 55.00 |
| | Theta | 50.00 | 50.00 | 50.00 |
| | Alpha | 50.00 | 46.67 | 55.00 |
| | Beta | 60.00 | 50.00 | 75.00 |
| | Gamma | 50.00 | 40.00 | 65.00 |
| | All | 56.00 | 50.00 | 65.00 |
| Net-Word + Psy | Delta | 70.00 | 66.67 | 75.00 |
| | Theta | 62.00 | 53.33 | 75.00 |
| | Alpha | 64.00 | 56.67 | 75.00 |
| | Beta | 62.00 | 56.67 | 70.00 |
| | Gamma | 66.00 | 56.67 | 80.00 |
| | All | 66.00 | 60.00 | 75.00 |

**Bold values** represent best performances per band.

## 4 Conclusions

In this study, we conducted a psychophysiological feature extraction and classification to determine associations between source connectivity and emotional processing in Colombian ex-combatants. We proposed a machine-learning approach to identify a group of features from psychological evaluations and cortical FCNs that provide discriminative information related with emotional processing. The inclusion of such features was in accordance with previous studies and might provide an alternate perspective and new insights into emotional processing in ex-combatants.

Regardless of all its promise, the current work is limited to binary classification and offline analyses of data obtained in controlled environments. Future work might include multiclass classification over data acquired in echological settings to assess EP-related networks in a more reliable way. Moreover, it would be relevant to perform a further multilayer network analysis to evaluate the specific link between different frequency bands and the emotional response to affective stimuli.

**Acknowledgment.** This work was partially funded by MinCiencias project 1222-852-69927, contract 495-2020.

## References

1. Anguita, D., Ghio, A., Ridella, S., Sterpi, D.: K-fold cross validation for error rate estimate in support vector machines. In: Stahlbock, R., Crone, S.F., Lessmann, S. (eds.) Proceedings of The 2009 International Conference on Data Mining, DMIN 2009, 13–16 July, 2009, Las Vegas, USA, pp. 291–297. CSREA Press (2009)
2. Chang, C.C., Lin, C.J.: LIBSVM: A library for support vector machines. ACM Trans. Intell. Syst. Technol. **2**, 27:1–27:27 (2011). Software available at http://www.csie.ntu.edu.tw/~cjlin/libsvm
3. Desikan, R.S., et al.: An automated labeling system for subdividing the human cerebral cortex on MRI scans into gyral based regions of interest. Neuroimage **31**(3), 968–980 (2006). https://doi.org/10.1016/j.neuroimage.2006.01.021
4. Friston, K.J.: Functional and effective connectivity in neuroimaging: a synthesis. Hum. Brain Mapp. **2**(1–2), 56–78 (1994). https://doi.org/10.1002/hbm.460020107
5. Gismero, E.: Escala de Habilidades Sociales (EHS). TEA Publicaciones de Psicología Aplicada, Madrid (2000)
6. Huang, Y., Parra, L.C., Haufe, S.: The New York Head-A precise standardized volume conductor model for EEG source localization and tES targeting. NeuroImage **140**, 150–162 (2016). https://doi.org/10.1016/j.neuroimage.2015.12.019
7. Javaid, H., Kumarnsit, E., Chatpun, S.: Age-related alterations in EEG network connectivity in healthy aging. Brain Sci. **12**(2), 218 (2022). https://doi.org/10.3390/brainsci12020218. http://dx.doi.org/10.3390/brainsci12020218
8. Kaplan, O., Nussio, E.: Explaining Recidivism of Ex-combatants in Colombia. J. Conflict Resolution, p. 002200271664432, May 2016. https://doi.org/10.1177/0022002716644326
9. Klados, M.A., Konstantinidi, P., Dacosta-Aguayo, R., Kostaridou, V.D., Vinciarelli, A., Zervakis, M.: Automatic recognition of personality profiles using eeg functional connectivity during emotional processing. Brain Sci. **10**(5), 278 (2020). https://doi.org/10.3390/brainsci10050278. http://dx.doi.org/10.3390/brainsci10050278
10. López, J., Litvak, V., Espinosa, J., Friston, K., Barnes, G.: Algorithmic procedures for bayesian MEG/EEG source reconstruction in SPM. NeuroImage **84**, 476–487 (2014). https://doi.org/10.1016/j.neuroimage.2013.09.002
11. Nolte, G., Bai, O., Wheaton, L., Mari, Z., Vorbach, S., Hallett, M.: Identifying true brain interaction from EEG data using the imaginary part of coherency. Clin. Neurophysiol. Official J. Int. Federation Clin. Neurophysiol. **115**(10), 2292–307 (2004). https://doi.org/10.1016/j.clinph.2004.04.029
12. Pantic, M., Valstar, M., Rademaker, R., Maat, L.: Web-based database for facial expression analysis. In: 2005 IEEE International Conference on Multimedia and Expo, pp. 317–321. IEEE (2005). https://doi.org/10.1109/ICME.2005.1521424

13. Preseea: Estudios Sociolingüísticos de Medellín. Fase. 1. Corpus Sociolingüístico de Medellín. Technical report, Universidad de Antioquia (2005). http://comunicaciones.udea.edu.co/corpuslinguistico/web/corpus.pdf
14. Quintero-Zea, A., Lopez, J.D., Smith, K., Trujillo, N., Parra, M.A., Escudero, J.: Phenotyping ex-combatants from EEG scalp connectivity. IEEE Access **6**, 55090–55098 (2018). https://doi.org/10.1109/access.2018.2872765
15. Quintero-Zea, A., Sepúlveda-Cano, L.M., Rodríguez Calvache, M., Trujillo Orrego, S., Trujillo Orrego, N., López, J.D.: Characterization Framework for Ex-combatants Based on EEG and Behavioral Features. In: Torres, I., Bustamante, J., Sierra, D.A. (eds.) VII Latin American Congress on Biomedical Engineering CLAIB 2016, Bucaramanga, Santander, Colombia, October 26th-28th, 2016, IFMBE Proceedings, vol. 60, pp. 205–208. Springer Singapore, Singapore (2017). https://doi.org/10.1007/978-981-10-4086-352
16. Rachman, S.: Emotional processing, with special reference to post-traumatic stress disorder. Int. Rev. Psychiatry **13**(3), 164–171 (2001). https://doi.org/10.1080/09540260120074028
17. Raine, A., et al.: The reactive-proactive aggression questionnaire: differential correlates of reactive and proactive aggression in adolescent boys. Aggressive Behav. **32**(2), 159–171 (2006). https://doi.org/10.1002/ab.20115
18. Rubinov, M., Sporns, O.: Complex network measures of brain connectivity: uses and interpretations. NeuroImage **52**(3), 1059–1069 (2010). https://doi.org/10.1016/j.neuroimage.2009.10.003
19. Stam, C.J., Tewarie, P., Van Dellen, E., Van Straaten, E.C.W., Hillebrand, A., Mieghem, P.V.: The trees and the forest: Characterization of complex brain networks with minimum spanning trees. Int. J. Psychophysiol. **92**, 129–138 (2014). https://doi.org/10.1016/j.ijpsycho.2014.04.001
20. Trujillo, S.P., et al.: Atypical modulations of N170 component during emotional processing and their links to social behaviors in ex-combatants. Front. Hum. Neurosci. **11**, 1–12 (2017). https://doi.org/10.3389/fnhum.2017.00244

# Measuring Motion Sickness Through Racing Simulator Based on Virtual Reality

Daniel Palacios-Alonso(✉), Javier Barbas-Cubero, Luis Betancourt-Ortega, and Mario Fernández-Fernández

Escuela Técnica Superior de Ingeniería Informática - Universidad Rey Juan Carlos, Campus de Madrid-Quintana, C/Quintana, 21, 28008 Madrid, Spain
daniel.palacios@urjc.es

**Abstract.** Nowadays, we are experiencing a new boom in virtual reality (VR) technologies led by big video game companies. When these renewed forms of entertainment caught everyone's attention, their most common issues began to be noticed. Motion sickness is a feeling of discomfort triggered by a misalignment between a person's eyes and body when in motion. This condition occurs regularly in VR environments. Although it is a common problem in this context, very few studies have been conducted on this subject. This research work is the first step in a barely explored field that will hopefully help to develop new proposals. The present study aims to find techniques that reduce the negative effects of motion sickness that appear during the use of VR devices and try to identify the main factors that cause them. To this end, a driving simulator with multiple experimental countermeasures has been developed and tested with 13 volunteers to determine the key elements that allow a normative participant to tolerate and overcome the symptoms of this condition. The results will establish a series of recommendations and best practices for further work with virtual reality technologies.

**Keywords:** Motion sickness · Virtual reality · Driving simulator · Best practices

## 1 Introduction

VR simulators render a simulated environment on a small screen built into googles, trying to trick the human brain through sight to immerse it in a simulation. This can easily lead to motion sickness if the inner ear does not perceive the same movements as sight. Sensory conflict theory (SCT) is the dominant theory in this field [1,9,15], proposing that motion sickness experienced during a VR simulation is due to a conflict between the vestibular system and the other senses (mainly sight). Likewise, some articles identify this type of motion sickness as

This research work was partly funded by one intramural project of Rey Juan Carlos University and a contract with the Spanish Defense Ministry (2022/00004/004 and 2021/00168/001, respectively).


J. M. Ferrández Vicente et al. (Eds.): IWINAC 2022, LNCS 13258, pp. 494–504, 2022.
https://doi.org/10.1007/978-3-031-06242-1_49

*pseudo motion sickness* [12]. This is a key point because it implies a difference between diverse types of motion sickness, not only when it comes to avoiding it, but also when measuring its intensity. When we talk about symptoms of dizziness or nausea resulting from VR use, we are probably referring to the aforementioned *pseudo motion sickness.* Therefore, many of the outcomes accomplished from previous studies on motion sickness may not apply in this case. Also, there is evidence linking the occurrence of motion sickness and graphic realism or image refresh rate. However, visual fidelity has not been shown to influence motion sickness [4]. Additionally, motion sickness can also be the result of unfamiliar movements (i.e. to which the organism has not adapted). This theory is known as the *neural mismatch theory* [14]. More experience with VR simulators, for example, could have a positive influence on avoiding motion sickness resulting from their use. The Sang's work proposes the use of music during the simulation session is a significant improvement to reduce the motion sickness [13]. Jaeger and Mourant [5] present some remarkable questions to be taken into account such as the time of exposure to simulation (the longer the time spent in the simulator, the greater the effects of motion sickness suffered by the participants) and different reaction among genders (females were more prone to motion sickness than males in the study). On its turn, Quintana et al.'s work consists of reviewing six research papers dedicated to the negative side effects generated by VR immersion in clinical populations suffering from anxiety. In this way, the authors were able to establish an overview of the prevalence of cyberdizziness in clinical populations; to analyze the validity of the Simulator Sickness Questionnaire (SSQ), used in clinical samples suffering from anxiety and; to better understand the overlap between cyberdizziness and anxiety-induced symptoms [11]. They also obtained some relevant factors to avoid or provoke motion sickness such as the size and weight of the hardware and the field of view offered by the device to the participant.

## 2 Methods and Materials

### 2.1 Objectives

The purpose of this study is based on achieving three main objectives. The first one is to define the relationship between motion sickness and VR technologies, as well as inquiring about correlations between the occurrence of symptoms and the casuistry of the participants. The second aim is to propose and validate the effectiveness of various countermeasures for mitigating the negative effects of motion sickness through the development of a driving simulator. This simulator is addressed to quantify and save the interactions of volunteers going through one or more test sessions. Likewise, a visual and adaptive environment is mandatory to achieve complete immersion in testers/subjects. Finally, the third one is to collect and analyze the data from several sessions to develop a set of best practices for creating safer and more comfortable virtual reality applications.

### 2.2 About VR Driving Simulator

To carry out this study, a VR driving simulator has been developed, which has been equipped with the necessary tools to turn it into the required test scenario. The simulator has hardware, specifically, selected for this project:

1. **VR Goggles**: model *Oculus Quest.* We selected this model for its features and affordable price. The selected model has also two motion controllers, one for each hand, which offer more possibilities when interacting with the VR simulator. It was integrated with the simulator via Oculus*Integration package* [10].
2. **Test station**: a high-end computer with a *NVIDIA 2080*, *intel i7*, and 20 GB *RAM.* The performance required in this aspect was given by the requirements of the VR equipment.
3. **Controllers (steering wheel and pedals)**: model *Logitech G29* with *force-feedback* required for further immersion of the subject whose functionalities were accessible through its *Logitech Gaming SDK package* for Unity [3].

In its turn, the test environment is a software specifically designed for the project objectives. With *Unity3D* as the development platform (starting from the asset package *Standard Assets* [16] whose car physics scripts were modified and adapted to create the core functionalities and systems), a VR simulator has been developed with the following requirements:

1. **Driving on a race track**: The mainstay of our simulator is the driving physics, implemented in a 3D racing circuit; as well as the different challenges and aims to be achieved during the test.
2. **Initial vestibule**: To favor a subtle and progressive adaptation to the VR environment, the simulator has been equipped with a scenario, before the circuit, in which the participant can move around a lobby equipped with mini-games (basketball, target shooting, etc.) and an interactive graphic interface with which to configure the simulation environment (selection of lighting and challenges).
3. **Synchronization of real and virtual environments**: A digital twin model of the *Logitech G29* steering wheel has been created. The simulator offers the possibility of synchronizing the position of the participant in the real environment, using the specifically developed *script* and the model.
4. **Immersion environmental elements**: Different sound effects, feedback forces and vibration effects on the steering wheel, and user-controllable music radio via *hand-tracking.*
5. **Interactive tutorial by *hand-tracking***: A series of infographics and tips were introduced at the beginning of the simulation to inform the participant of the vehicle controls without taking out of the immersion.
6. **Data collection**: The simulator can collect representative data from the trial with each subject and associate it with a time to compare it with recordings, results, and comments from the subject. These data are: (a) Degree of steering wheel rotation. (b) Percentage of acceleration and braking.

7. **Simulation auxiliary display**: To carry out an adequate tracking of the test, an additional screen was added. Thus, the developer team and the participant could see the scene at the same time.

### 2.3 Technical Description of the Simulator

The simulator is divided into two main sections. On the one hand, there is the adaptation room or **initial vestibule**, a scenario where the player can interact with throwable objects. These objects can be used, in several mini-games and activities, to help the participant discover how the simulation world works and tests the hardware used during the driving sessions. On the other hand, there is the **race circuit**, in which the participant using the VR goggles can drive a vehicle around a circular track located on an island. This is the scenario where the tests are carried out within three different lightning (sunrise, sunset and night). Both scenarios include several countermeasures to explore the best ways to mitigate motion sickness symptoms and tools to collect participant data.

The following list enumerates the simulator's features organized by functionality.

- **Interactive interface.** This display provides colliders with the participant's hand. Likewise, it gives the participant the chance of moving through the different simulator scenes.
- **Adaptation room.** (a) Gamified environment using sound and graphic effects. (b) Providing certain objects with interactive physics. (c) Basketball mini-game. (d) Collapsible cube pyramids mini-game. (e) Mobile targets mini-game.
- **Driving system.** (a) A driving game physics system. (b) Modeled and textured racetrack with traffic signals. (c) Driving vehicles in VR, using the simulator peripherals (VR googles, steering wheel and pedals). (d) Synchronization system between the participant's hand and the virtual steering wheel.
- **Immersion.** (a) Sound effects of the environment that reacts to the participant actions. (b) Steering wheel force-feedback that responds to the type of ground the car is driving on. (c) Functional and interactive music radio inside the vehicle. (d) Camera movement and sound effects that react to vehicle collisions.
- **Mechanics.** (a) Time trials system. (b) A scoring system that records the participant's performance during the simulator sessions. (c) Vehicle dashboard feedback, offering the speed, speed limit, and chronometer of the tests. (d) Checkpoint system and reset button of the vehicle's position on the road.
- **Data collection.** (a) Saving system of the volunteer's interactions with the peripherals. (b) Simultaneous rendering of the simulation on the VR googles and on the simulator control computer's screen.

### 2.4 VRSQ Test

The measurement of motion sickness has been complex over the years. Although the MSQ [6] and SSQ [7] questionnaires are frequently used, for this study it was decided to use the VRSQ [8] for its accuracy (there is evidence that the third component of SSQ, nausea, contributes less to the onset of motion sickness in VR simulations [2]). This form narrows the target symptoms to two groups (Table 1).

**Table 1.** Symptoms measured in VRSQ.

| VRSQ symptom | Oculomotor | Disorientation |
|---|---|---|
| 1. General discomfort (GD) | X | – |
| 2. Fatigue (F) | X | – |
| 3. Eyestrain (E) | X | – |
| 4. Difficulty focusing (DF) | X | – |
| 5. Headache (H) | – | X |
| 6. Fullness of head (FH) | – | X |
| 7. Blurred vision (BV) | – | X |
| 8. Dizzy (eyes closed) (DEC) | – | X |
| 9. Vertigo (V) | – | X |
| Total | O | D |

The questionnaire is developed by asking the participant to what extent he/she has felt the symptoms related to motion sickness. The intensity is rated in four ranges (from 0 to 3). Subsequently, the symptom score of the oculomotor subgroup is calculated with the following equation (see Eq. 1):

$$O = \frac{(GD + F + E + DF)}{4} \times \frac{100}{12} \tag{1}$$

Similarly, the disorientation subgroup score is calculated with the following equation (see Eq. 2):

$$D = \frac{(H + FH + BV + DEC + V)}{5} \times \frac{100}{15} \tag{2}$$

Calculating the average of both scores gives the final outcome (see Eq. 3):

$$Total\ Score = \frac{(O + D)}{2} \tag{3}$$

### 2.5 Testing Sessions

To facilitate the coordination and ensure the protection of the participants against the COVID-19 during the testing sessions, they were conducted under an specific and strict protocol. The test sessions are divided into three stages which are explained as follows.

Firstly, participants had to answer a preliminary questionnaire, providing the key information for the fulfillment of their profiles. Some of the questions were: age, previous experience with VR, motion sickness propensity, introversion or extroversion, and usage of glasses or contact lenses. At the end of this questionnaire, participants would be explained the vehicle controls, as well as their objectives for the simulation.

The test consisted of two main objectives: achieve the highest score (based on the quality of their driving) and complete the best lap. Both were measured by comparing the results of all the participants and a historical ranking. The participant would have to run through the circuit twice, one lap for each objective. Nevertheless, the participant would begin the simulation in an environment created to favor an effortless adaptation to the VR headset and its controls (Vestibule). Finally, the participant would answer the last questionnaire to measure the incidence of motion sickness during the simulation.

### 2.6 Corpus

The corpus consists of 13 volunteers divided in two testing sessions. Four of them participated in the experiment twice. Each test session included several subjects of different profiles. Notice that the experience with VR is measured in five grades: 1: no experience, 5: frequent use. Finally, the corpus itself had only two requirements: age range of 18 to 35 years and exclusion of pregnant female.

## 3 Results

The results presented in this study are extracted from the second test session. As it was aforementioned, the participants answered a VRSQ questionnaire after the test. Likewise, the measures of the VRSQ forms, together with the profile of each participant, allowed the comparison of symptom intensity between groups of different anamnesis or simulations with different parameters. The forms yielded the outcomes shown in Table 2.

Looking in detail at the results achieved, it is possible to observe a dispersion in the values accomplished. For example, two participants (see LOR-1992 and DBL-1988) suffered no dizziness at all, obtaining a total score of zero. However, other participants (see SPB-1987, MDJ-1987 and PALF-1990) achieved quite high scores compared to the previous ones; although the first of these two participants surpassed by ten points over the second with the highest degree of motion sickness.

Similarly, the values obtained in each of the sections (oculomotor and disorientation) by the participants were not equal, i.e., there were participants who obtained high scores in one of the two characteristics, but the other had a low or null value. Therefore, the final result of motion sickness (the mean of both features) is not so high. On the other hand, it should be noted that, in the last session, only one out of ten had to abandon the simulator. Consequently, only ten percent of the participants suffered from severe motion sickness.

**Table 2.** VRSQ results of the second test session. *GD: General Discomfort; F: Fatigue; E: Eyestrain; DF: Difficulty focusing; H: Headache; FH: Fullness of head; BV: Blurred vision; DEC: Dizzy with eyes closed; V: Vertigo; O: Oculomotor; D: Disorientation.*

| | *Dropout* | GD | F | E | DF | H | FH | BV | DEC | V | O | D | Total |
|---|---|---|---|---|---|---|---|---|---|---|---|---|---|
| DRC-1993 | No | 0 | 0 | 0 | 1 | 0 | 0 | 0 | 0 | 0 | 8,33 | 0 | 4,17 |
| LOR-1992 | No | 0 | 0 | 0 | 0 | 0 | 0 | 0 | 0 | 0 | 0 | 0 | **0** |
| SPB-1987 | No | 0 | 0 | 2 | 1 | 1 | 0 | 0 | 2 | 2 | 25 | 33,33 | **29,17** |
| PALF-1990 | No | 1 | 1 | 0 | 0 | 0 | 0 | 1 | 0 | 2 | 16,67 | 20 | 18,33 |
| EGF-1992 | **Yes** | 0 | 0 | 0 | 1 | 0 | 0 | 0 | 0 | 2 | 8,33 | 13,33 | 10,83 |
| DBL-1994 | No | 0 | 0 | 0 | 1 | 0 | 0 | 1 | 0 | 0 | 8,33 | 6,67 | 7,5 |
| MDJ-1987 | No | 2 | 1 | 0 | 0 | 0 | 0 | 0 | 0 | 2 | 25 | 13,33 | 19,17 |
| AM-1995 | No | 0 | 0 | 0 | 0 | 0 | 1 | 0 | 0 | 0 | 0 | 6,67 | 3,33 |
| DBL-1988 | No | 0 | 0 | 0 | 0 | 0 | 0 | 0 | 0 | 0 | 0 | 0 | **0** |
| RMZ-1987 | No | 0 | 1 | 0 | 0 | 2 | 1 | 0 | 0 | 1 | 8,33 | 20 | 14,17 |
| Average | 10% | | | | | | | | | | 10 | 11,33 | 10,67 |
| Std. Deviation | 32% | | | | | | | | | | 8,98 | 10,35 | 9,66 |

In order to glimpse possible causes of what happened, a set of graphs comparing oculomotor, disorientation and total score and relating them to the profile of the participants are analysed. Firstly, in Fig. 1, the first introduced concept is based on the experience of participants with VR. Considering four possible values out of five. The group with the most experience using VR goggles is the one that scored lowest on the VRSQ questionnaire (see Fig. 1a). Similarly, those participants who had little or no experience with VR achieved much higher rates than their peers. On the other hand, the information regarding the personal opinion on whether the participant considers him/herself an extrovert or introvert did not offer conclusive results, as can be seen in Fig. 1b.

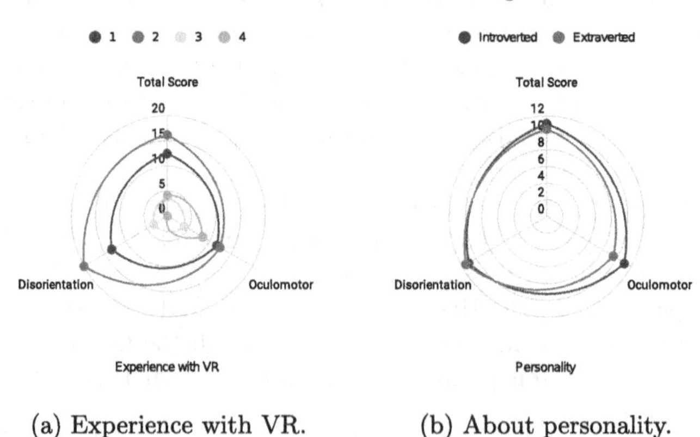


(a) Experience with VR. (b) About personality.

**Fig. 1.** Comparison of VRSQ results depending on the prior experience of the participant and between introverted and extroverted participants.

The next two graphs are based on the relationship with the suggestion of motion sickness and daily driving habit (see Fig. 2). Note that those participants who were not prone to motion sickness scored lower on the VRSQ questionnaire (see Fig. 2a). However, the results obtained related to frequent car use did not yield very conclusive and differential outcomes as shown in Fig. 2b.

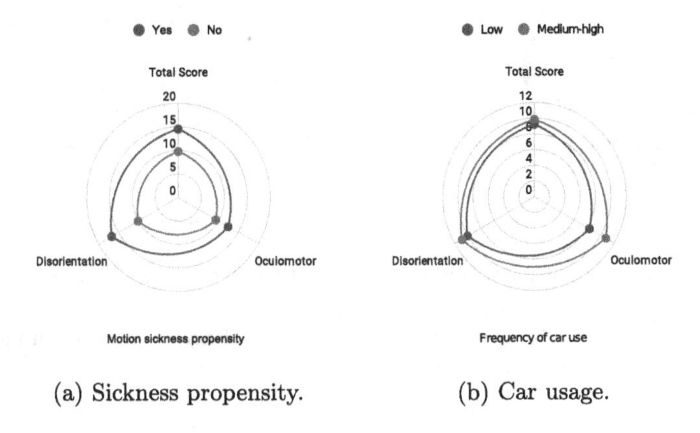


(a) Sickness propensity. (b) Car usage.

**Fig. 2.** Comparison of VRSQ results between participants with or without motion sickness propensity and the car usage.

The third comparative of results presents the relationship between listening to music or not during the test and the type of scenario selected (daylight or sunset) as shown in Fig. 3. As can be seen in the figure on the left (see Fig. 3a, having the music turned on (previously selected by the participant) seems to prevent motion sickness symptoms in a high percentage when compared to those participants who did not have the music turned on. On the other hand, the image on the right shows a curious result, because apparently, the total score obtained is similar with either of the two selected scenarios. Nevertheless, those participants who chose the daytime scenario had more severe symptoms of disorientation than those who chose the nighttime scenario. In the opposite direction occurred when the participants chose the evening illumination, suffering greater oculomotor problems (see Fig. 3b).

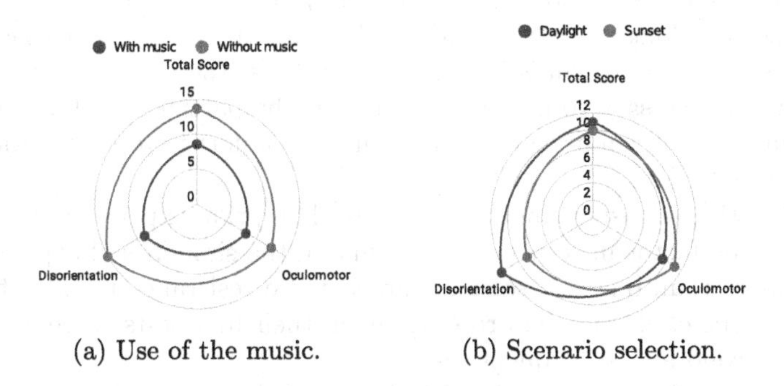


(a) Use of the music. (b) Scenario selection.

**Fig. 3.** Comparison of VRSQ results between participants listening music or not and between simulations with daylight or sunset lighting.

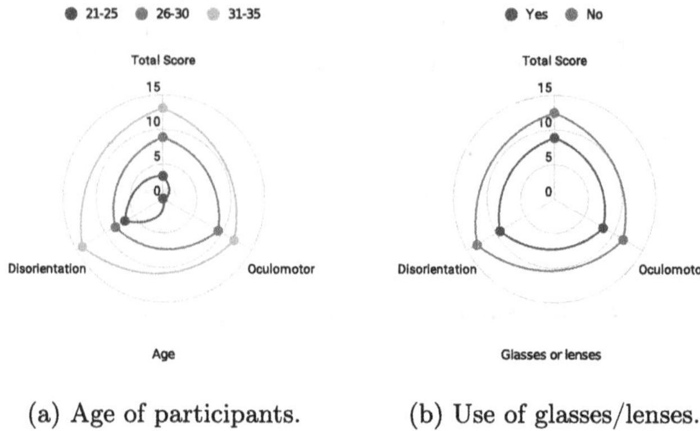


(a) Age of participants. (b) Use of glasses/lenses.

**Fig. 4.** Comparison of VRSQ results between groups of ages and between participants wearing glasses/lenses or not.

The last comparison refers to the use of glasses or contact lenses and the age of the participants as shown in Fig. 4. As mentioned in previous sections, motion sickness tends to occur in older participants than in younger ones. As Fig. 4a shows, the older participants obtained a higher score on the VRSQ questionnaire. On the other hand, in Fig. 4b, participants who daily wear glasses or contact lenses suffered less motion sickness than their peers.

## 4 Conclusions

Analyzing the results, several patterns can be found that may lead to key findings about motion sickness symptoms and VR.

Considering age, elder people are more likely to suffer from feelings of motion sickness and show a low tolerance for long-term sessions with the simulator. The second aspect is music that seems to ameliorate the participants experience during the test.

Thirdly, volunteers who report a tendency to get motion sickness in other contexts such as long road trips, on board a ship or during roller coaster rides, are more susceptible to suffer the motion sickness; therefore the psychological component is necessary to be study further. If the participant thinks too much about symptoms, it can provoke anxiety and discomfort that ends in their actual manifestation.

The fourth promising finding indicates that prior experience with virtual reality is a key factor in terms of usage time of the simulator. People who went through more than one test session showed more resistance to side effects and dizziness. Some of them, who dropped out in their first tests, were even able to finish the second and subsequent sessions.

The fifth conclusion has to do with the use of glasses. Participants who needs to wear them experiences fewer symptoms than people who do not. The main hypothesis suggests that these people are more careful when adjusting the equipment and are more concerned about getting a perfect view of the simulation. In addition, they may be accustomed to the glasses and therefore have fewer problems wearing them. On a psychological and sensory level, the two types of lighting (daytime and sunset) do not show conclusive results.

Sixthly, all the people who went through the simulator test session had a driving license and used their car regularly. The results show that this is not a key factor in the manifestation of symptoms, as there are profiles that suffer some side effects and others that do not show any ailment.

Finally, although the findings look promising, this research hopes to keep exploring this topic in deep, as well as to improve the features and capabilities of the driving simulator in order to be able to conduct new sessions with more volunteers and test new countermeasures in other contexts and conditions.

**Acknowledgements.** The authors would like to thank all participants in tests, especially, CT-Ingenieros workers for their effort and patience.

## References

1. Bos, J.E., Bles, W., Groen, E.L.: A theory on visually induced motion sickness. Displays **29**, 47–57 (2008)
2. Drexler, J.M.: Identification of system design features that affect sickness in virtual environments. University of Central Florida (2006)
3. Gaming, L.: Logitech gaming sdk. https://assetstore.unity.com/packages/tools/integration/logitech-gaming-sdk-6630. Accessed 14 Feb 2022
4. Hettinger, L.J., Riccio, G.E.: Visually induced motion sickness in virtual environments. Presence **1**(3), 306–310 (1992)
5. Jaeger, B.K., Mourant, R.R.: Comparison of simulator sickness using static and dynamic walking simulators. Proc. Hum. Factors Ergon. Soc. Ann. Meeting **45**(27), 1896–1900 (2001)
6. Kennedy, R.S., Graybiel, A.: The dial test: A standardized procedure for the experimental production of canal sickness symptomatology in a rotating environment. US Naval School of Aviation Medicine (1965)
7. Kennedy, R.S., Lane, N.E., Berbaum, K.S., Lilienthal, M.G.: Simulator sickness questionnaire: An enhanced method for quantifying simulator sickness. Int. J. Aviat. Psychol. **3**, 203–220 (1993)
8. Kim, H.K., Park, J., Choi, Y., Choe, M.: Virtual reality sickness questionnaire (VRSQ): motion sickness measurement index in a virtual reality environment. Appl. Ergon. **69**, 66–73 (2018)
9. Nichols, S., Patel, H.: Health and safety implications of virtual reality: a review of empirical evidence. Appl. Ergon. **33**, 251–271 (2002)
10. Oculus: Oculus integration. https://assetstore.unity.com/packages/tools/integration/oculus-integration-82022. Accessed 14 Feb 2022
11. Quintana, P., Bouchard, S., Serrano Zárate, B., Cárdenas-López, G.: Efectos secundarios negativos de la inmersión con realidad virtual en poblaciones clínicas que padecen ansiedad. Revista de psicopatología y psicología clínica **19**(3), 197–207 (2014)

12. Sanchez-Blanco, C., Yañez-Gonzalez, R., Benito Orejas, J.I., Gordon, C.R., Batuecas-Caletrio, A.: Cinetosis. Rev Soc Otorrinolaringol Castilla Leon Cantab La Rioja **28**, 233–251 (2014)
13. Sang, F.D.Y.P., Billar, J.P., Golding, J.F., Gresty, M.A.: Behavioral methods of alleviating motion sickness: effectiveness of controlled breathing and a music audiotape. J. Travel Med. **10**(2), 108–111 (2003)
14. Schmäl, F.: Neuronal mechanisms and the treatment of motion sickness. Pharmacology **91**, 229–241 (2013)
15. Stanney, K.M., Kennedy, R.S.: Human factors in Simulation and Training, pp. 117–127. New York, NY (2009)
16. Technologies, U.: Standard assets. https://assetstore.unity.com/packages/essentials/asset-packs/standard-assets-for-unity-2018-4-32351. Accessed 14 Feb 2022

# Health Applications

# Analysis of the Asymmetry in RNFL Thickness Using Spectralis OCT Measurements in Healthy and Glaucoma Patients

Rafael Berenguer-Vidal[1](✉), Rafael Verdú-Monedero[2], Juan Morales-Sánchez[2], Inmaculada Sellés-Navarro[3], and Oleksandr Kovalyk[2]

[1] Universidad Católica San Antonio, 30107 Murcia, Spain
rberenguer@ucam.edu
[2] Universidad Politécnica de Cartagena, 30202 Cartagena, Spain
[3] Hospital General Universitario Reina Sofía, 30003 Murcia, Spain

**Abstract.** One of the main diseases that affect to the optic nerve is glaucoma, which causes progressive and irreversible damage that reduces the vision field of the patient. The thickness of the retinal nerve fiber layer is an indicator of the status and progression of this illness. A line of research in the early diagnosis of glaucoma is based on the analysis of the asymmetry between the morphological characteristics of both eyes. This article presents preliminary results that start from this hypothesis and use the relative absolute difference between the thickness of the RNFL in both eyes of the same patient as a characteristic of asymmetry. Results indicate that there is a significant difference in the mean value of the asymmetry between healthy patients and those with glaucoma. As future work, the inclusion of automatic methods for the measurement of the RNFL thickness and the use of classification techniques based on these characteristics of asymmetry for the early diagnosis of glaucoma will be developed.

**Keywords:** Glaucoma · OCT · RNFL

## 1 Introduction

Glaucoma is one of the main causes of irreversible blindness in developed countries. It is currently the second cause of blindness in the world and affects one in every two hundred people under the age of fifty and one of every ten over eighty years [9]. Therefore, its early detection, usually with few symptoms or asymptomatic [8], is crucial to delay its evolution and reduce loss of vision.

Glaucoma is a progressive disease of the optic nerve usually caused by high intraocular pressure due to poor drainage of the ocular fluid [6]. Clinically, it results in a progressive and irreversible loss of the visual field that evolves to blindness [4]. The diagnosis of glaucoma is based mainly on the measurement


J. M. Ferrández Vicente et al. (Eds.): IWINAC 2022, LNCS 13258, pp. 507–515, 2022.
https://doi.org/10.1007/978-3-031-06242-1_50

of intraocular pressure by means of tonometry, the examination of the visual field through campimetry, the measurement of the relationship between the area of the cup and the optic disc in fundus images, and the thickness of the retinal nerve fiber layer (RNFL) in optical coherence tomographies (OCT) [4]. Recently, the hypothesis based on the asymmetry between the values of anatomical characteristics of the optic nerve of both eyes as an indicator in the early diagnosis of glaucoma has been resumed [3,5,7]. This work is based on this hypothesis and proposes, as a characteristic to consider, the relative absolute difference in the RNFL thickness of both eyes measured by a Spectralis OCT device.

The rest of the article is structured as follows: Sect. 2 details the characteristics of the OCT dataset as well as its processing. Section 3 describes the results obtained using the proposed characteristic of asymmetry as an indicator to differentiate healthy from glaucoma patients. Finally, Sect. 4 closes the paper with the conclusions and future lines of work.

## 2 Materials and Methods

This Section briefly describes the materials used in this work, i.e., optical coherence tomographies of the papillary circle (2D peripapillary B-scan OCT centered at the optic nerve head), as well as their acquisition procedure. This medical imaging modality is widely used in the diagnosis of glaucoma disease, since it allows to determine the thickness of the RNFL. As described in Sect. 1, the asymmetry between the two eyes, interpreted as the difference in thickness of the RNFL, is proving to be of potential interest in the early diagnosis of glaucoma as well as its evolution over time.

### 2.1 Image Acquisition

In order to validate the hypotheses raised in this paper, a set of spectral domain-OCT (SD-OCT) of the left and right eye of each patient were acquired by the Ophthalmology Service of the Hospital General Universitario Reina Sofía (Murcia, Spain) using a Spectralis OCT S2610-CB (Heidelberg Engineering GmbH, Heidelberg, Germany) from October 2018 to November 2020, being anonymized in accordance with the criteria of the Human Ethics Committee. The dataset includes 159 healthy patients (318 OCTs) and 49 patients with glaucomatous optic neuropathology (98 OCTs). Additionally, there are 33 patients in the dataset who have been excluded from the study, in whom the experts were unable to determine the disease status with sufficient accuracy (classified as suspicious).

This type of imaging uses spectral-domain optical coherence tomography (SD-OCT), also called Fourier-domain OCT (FD-OCT), as a scanning procedure. By means of a super-luminescent diode (SLD) with a wavelength of 870 nm, a cylindrical section of the retinal layers centered on the optic disc, called a B-Scan, is registered. This circular section is divided into different sectors, namely temporal (T), temporal superior (TS), nasal superior (NS), nasal (N), nasal inferior (NI) and temporal inferior (TI), in each of which the thickness of the retinal layers is measured. The range of each sector is detailed in Table 1 and illustrated in Fig. 1. Note that sectors (T) and (N) encompass an angle of 90° while the rest of the sectors comprise an angle of 45°. To simplify the study and processing of these images, the circular B-scan is projected from polar to Cartesian coordinates on the $x$-axis. Table 1 gathers the correspondence between polar values measured in degrees and Cartesian values measured in pixels, being also depicted in the lower part of Fig. 1.

**Table 1.** Range of each sector measured in polar coordinates (degrees) and Cartesian coordinates (pixels).

| Sector | | Polar | | Cartesian | |
|---|---|---|---|---|---|
| | | Min | Max | Min | Max |
| Temporal | (T) | 0° | 45° | 1 | 96 |
| Temporal Superior | (TS) | 45° | 90° | 97 | 192 |
| Nasal Superior | (NS) | 90° | 135° | 193 | 288 |
| Nasal | (N) | 135° | 225° | 289 | 480 |
| Nasal Inferior | (NI) | 225° | 270° | 481 | 576 |
| Temporal Inferior | (TI) | 270° | 315° | 577 | 672 |
| Temporal | (T) | 315° | 360° | 673 | 768 |

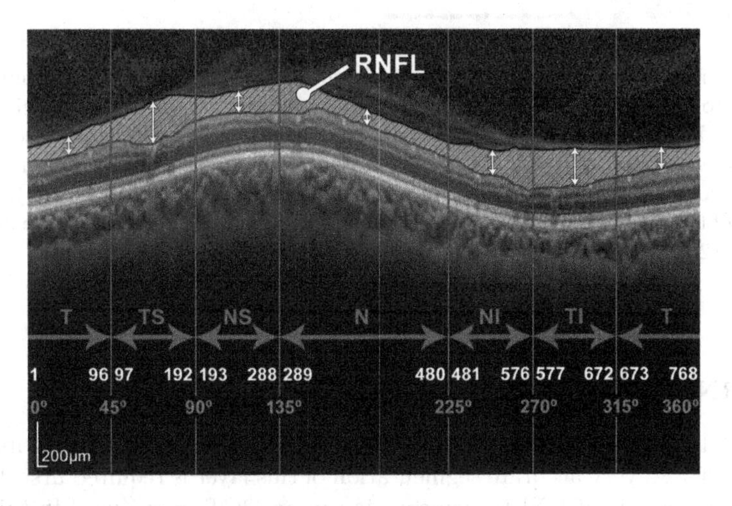


**Fig. 1.** Segmented RNFL in a peripapillary B-scan OCT image. Correspondence of the sectors (T, TS, NS, N, NI and TI) measured in degrees and pixels.

The set of OCTs used in this work have a resolution of $768 \times 496$ pixels, a bit depth of 8 bits/pixel in grayscale and a $z$-scaling of 3.87 μm/pixel. The $x$-scaling depends on the diameter of the peripapillary OCT circumference, being not significant for this study since the main goal is to determine the vertical thickness of the RNFL in the OCT image. The top left of Fig. 2 shows the circular tracing of the SLD beam positioned over the center of the optic nerve, also known as papilla, depicted over the retinal fundus image of the eye. The corresponding Cartesian coordinate projection of the B-scan OCT is shown on the upper right of Fig. 2.

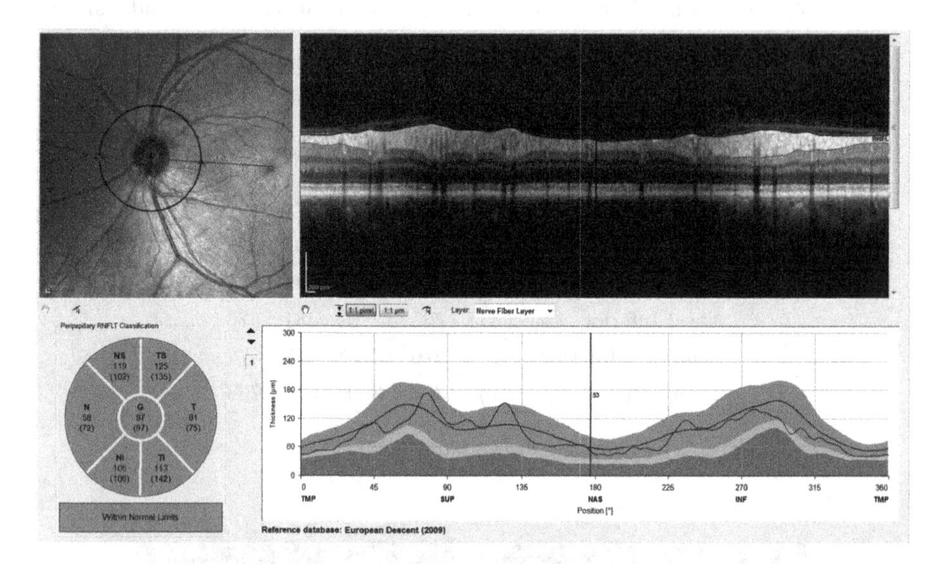


**Fig. 2.** Screenshot provided by the Spectralis software version 6.9.4.0. From left to right and top to bottom: Retinal fundus photography centered on the optic disc with B-scan indicated by the green circle. 2D peripapillary B-scan OCT on Cartesian coordinates with the segmentation of the RNFL. Estimated mean values for RNFL layer thickness for the sectors (T, TS, NS, N, NI and TI), as well as the overall mean (G); rectified outline of the RNFL thickness and reference values according to the database European Descent (2009). (Color figure online)

## 2.2 RNFL Segmentation and Thickness Calculation

In order to determine the thickness of the RNFL layer on the 2D peripapillary B-scan OCT, a very accurate segmentation of this layer is required first. Figure 1 shows an example of RNFL segmentation in an OCT image used in this work. This segmentation process can be very challenging due to image characteristics such as speckle noise and low contrast. A comprehensive review of the methods commonly used in this process is detailed in [1]. Specifically, the method chosen for this work is the algorithm provided by the Spectralis software version 6.9.4.0, as illustrated in top left of Fig. 2.

From the results of this segmentation algorithm, the thickness between the upper and lower boundary of the RNFL is calculated and scaled from pixels to μm according to the vertical resolution of the OCT. The circumpapillary RNFL thickness is then averaged by sectors (T, TS, NS, N, NI, and TI) and the mean value of thickness in the global circumpapillary contour (G) is also calculated.

### 2.3 Calculation of Thickness Asymmetry

The RNFL thickness asymmetry $\triangle_{S,i}$ is defined as the absolute difference of the layer thickness in sector $S$ between the eyes of the $i$-th patient,

$$\triangle_{S,i} = |w^{\mathtt{l}}_{S,i} - w^{\mathtt{r}}_{S,i}|, \tag{1}$$

where $S$ denotes the specific sector (T, TS, NS, N, NI, TI or G) over which the asymmetry is calculated, $i$ indicates the patient number, and $w^{\mathtt{l}}_{S}$ and $w^{\mathtt{r}}_{S}$ indicate the mean thickness of the RNFL layer in the sector $S$ calculated for the left and right eye, respectively. This paper also proposes as a characteristic of asymmetry, the relative thickness asymmetry $\delta_{S,i}$, calculated as,

$$\delta_{S,i} = |w^{\mathtt{l}}_{S,i} - w^{\mathtt{r}}_{S,i}|/w^{\mathtt{l}}_{S,i}, \tag{2}$$

which allows this asymmetry to be measured proportionally to the thickness of the RNFL.

## 3 Results

The thickness asymmetry $\triangle_{S,i}$ and the relative thickness asymmetry $\delta_{S,i}$ have been calculated for each sector of the B-Scan OCT as well as for the overall RNFL of all patients. These patients were initially classified in two groups, healthy and unhealthy individuals, following expert criteria. For the statistical characterization of the results, the mean value and standard deviation of $\triangle_{S,i}$ and $\delta_{S,i}$ have been computed, in each of the two groups of patients. Furthermore, the $p$-value of this dataset is also calculated in order to determine the statistical significance of the hypothesis indicating that the RNFL thickness asymmetry is statistically related to the occurrence of glaucoma disease. Note that the $p$-value is estimated for all sectors, for both the asymmetry $\triangle_{S,i}$ and relative asymmetry $\delta_{S,i}$ values. All these values are gathered in Table 2 and depicted in Fig. 3.

As can be easily noticed, both thickness asymmetry $\triangle_S$ and relative thickness asymmetry $\delta_S$ are notably lower in the group of healthy individuals than in the group of glaucoma patients. This can be observed in the mean value and the standard deviation of the two asymmetry characteristics. Moreover, the $p$-values of these dataset are substantially small, which leads us to state that there is a sufficient statistical significance between this characteristic of asymmetry and the occurrence of glaucoma. In particular, comparing both characteristics of

**Table 2.** RNFL thickness asymmetry $\triangle_S$ (mean and standard deviation in μm) and relative thickness asymmetry $\delta_S$ (mean and standard deviation). Values calculated for each sector of the eyes (TS, T, TI, NS, N and NI) as well as for the whole layer (G) for the group of healthy and unhealthy individuals. The *p*-value columns give the *p*-values of $\triangle_S$ and $\delta_S$ for healthy and unhealthy patients in each sector.

| Sector | Thickness asymmetry $\triangle_S$ | | | Relative asymmetry $\delta_S$ | | |
|---|---|---|---|---|---|---|
| | Healthy | Unhealthy | *p*-value | Healthy | Unhealthy | *p*-value |
| TS | $12.7 \pm 11.3$ | $35.5 \pm 29.4$ | $3.1 \cdot 10^{-6}$ | $0.09 \pm 0.09$ | $0.43 \pm 0.46$ | $4.9 \cdot 10^{-6}$ |
| T | $7.0 \pm 7.0$ | $15.3 \pm 13.1$ | $1.6 \cdot 10^{-4}$ | $0.10 \pm 0.10$ | $0.35 \pm 0.50$ | $9.1 \cdot 10^{-4}$ |
| TI | $11.6 \pm 9.3$ | $41.5 \pm 32.5$ | $1.8 \cdot 10^{-7}$ | $0.08 \pm 0.07$ | $0.62 \pm 0.80$ | $3.1 \cdot 10^{-5}$ |
| NS | $13.9 \pm 9.6$ | $30.0 \pm 25.0$ | $2.2 \cdot 10^{-4}$ | $0.14 \pm 0.12$ | $0.42 \pm 0.34$ | $4.8 \cdot 10^{-6}$ |
| N | $9.8 \pm 8.3$ | $18.3 \pm 20.1$ | $9.1 \cdot 10^{-3}$ | $0.13 \pm 0.11$ | $0.30 \pm 0.33$ | $5.3 \cdot 10^{-4}$ |
| NI | $12.8 \pm 10.5$ | $28.1 \pm 22.2$ | $2.4 \cdot 10^{-5}$ | $0.12 \pm 0.12$ | $0.38 \pm 0.31$ | $1.2 \cdot 10^{-6}$ |
| G | $4.0 \pm 4.6$ | $22.2 \pm 16.7$ | $9.7 \cdot 10^{-10}$ | $0.04 \pm 0.05$ | $0.32 \pm 0.26$ | $1.2 \cdot 10^{-9}$ |

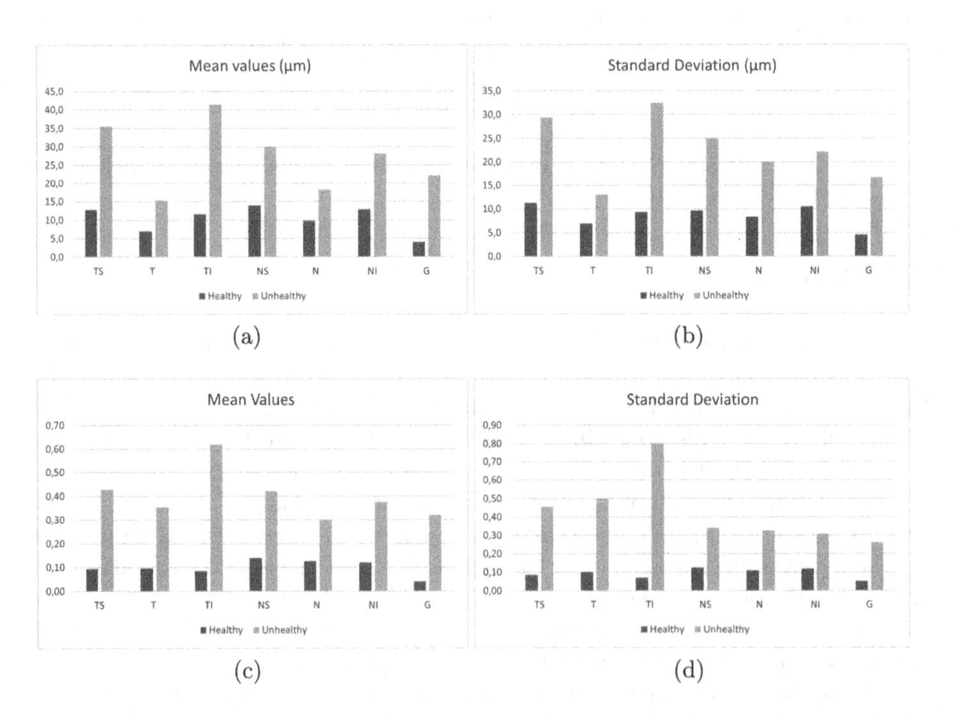


**Fig. 3.** Comparison of asymmetry between eyes of healthy (blue) and unhealthy (red) patients. Results shown for each eye sector (TS, T, TI, NS, N and NI), as well as the overall average of the RNFL (G). Top row: (a) mean and (b) standard deviation of absolute asymmetry $\triangle_S$. Bottom row: (c) mean and (d) standard deviation of relative asymmetry $\delta_S$. (Color figure online)

asymmetry, the $p$-value of $\delta_S$ is substantially smaller for all sectors than the $p$-value calculated for $\triangle_S$, which lead us to use $\delta_S$ as a discriminant variable for the classification of patients.

Although the statistical significance in all sectors is high according to its small $p$-value, the layer asymmetry of the whole layer $\delta_G$ gives a better result with a $p$-value of $1.2 \cdot 10^{-9}$. This can also be seen in Fig. 4, which depicts a box plot of the relative thickness asymmetry $\delta_S$ for all sectors. As can be seen in

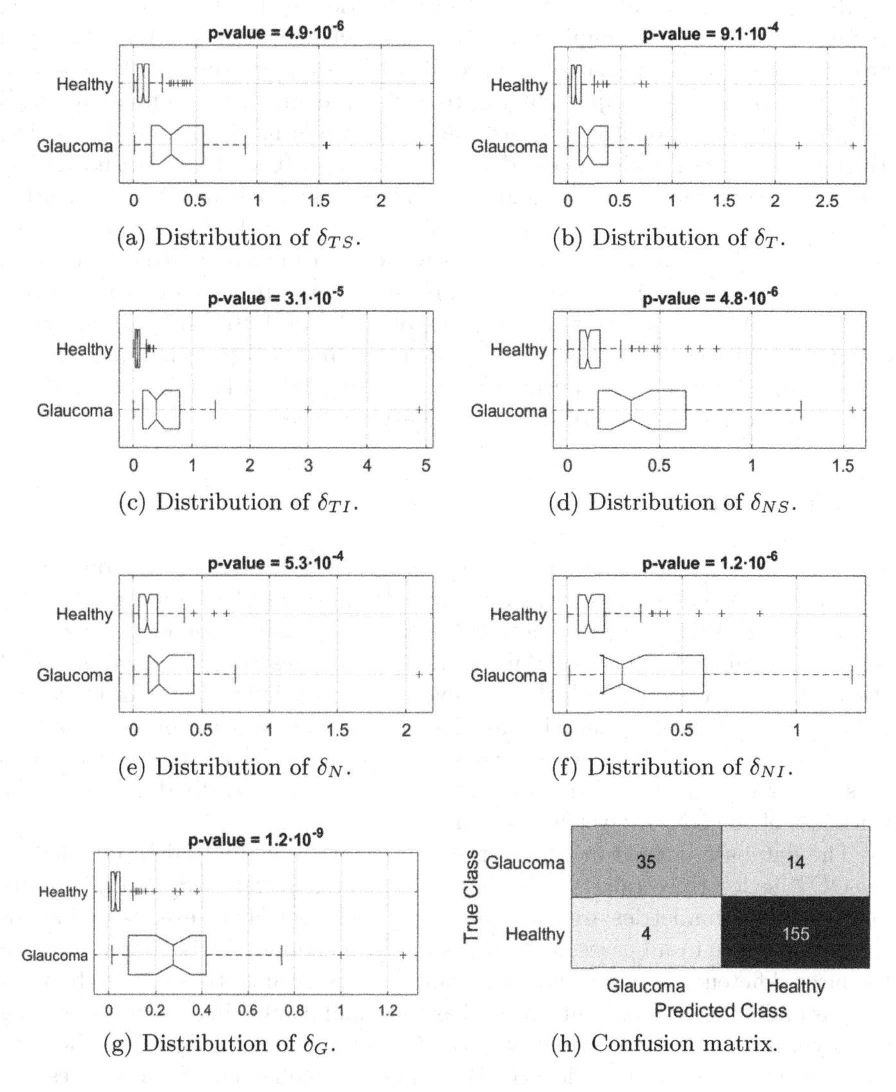


(a) Distribution of $\delta_{TS}$. (b) Distribution of $\delta_T$. (c) Distribution of $\delta_{TI}$. (d) Distribution of $\delta_{NS}$. (e) Distribution of $\delta_N$. (f) Distribution of $\delta_{NI}$. (g) Distribution of $\delta_G$. (h) Confusion matrix.

**Fig. 4.** Box plots of the relative thickness asymmetry $\delta_S$: (a–f) for individual sectors (TS, T, TI, NS, N and NI) and (g) for the whole layer (G). (h) Confusion matrix resulting from the application of the threshold $\gamma_G = 0.163538$.

this figure, the distributions of healthy and unhealthy individuals are sufficiently disjunct, especially in the TS, NS and NI sectors, which means that the asymmetry in any of these sectors could be sufficiently significant for a classification of patients. Nevertheless, it is the distribution for the whole layer G which provides a significantly greater separation, providing a lower $p$-value as previously seen. This leads us to use $\delta_G$ as a discriminative parameter in the diagnosis of glaucoma, since it provides a greater distinction between healthy and unhealthy individuals.

Regardless of the more advanced classification methods to be considered in future work, a first simple threshold-based classification method has been proven in this work, which can be seen like the simplest version of a decision tree. The aim of this approach was to reach preliminary results, that could be interpretable by ophthalmologists and achievable in their clinical practice. To this end, the Classification and Regression Trees (CART) algorithm [2] has been applied to maximize the highest number of matches of the classification algorithm considering $\delta_G$, giving a threshold of $\gamma_G = 0.163538$. Applying this threshold to the dataset collected in this work, the confusion matrix depicted in Fig. 4(h) is obtained. As can be seen, this matrix indicates a false negative rate of 28.57% (14/49) and a false positive rate of 2.52% (4/159). Both false negative and false positive rate can be optimized in future work by combining the $\delta_G$ with asymmetry values in specific sectors or using other classification methods that consider not only a threshold for the asymmetries.

## 4 Conclusions

The thickness of the retinal nerve fiber layer (RNFL) has been proposed as a parameter to be considered in the early diagnosis of glaucoma disease. The thickness of the RNFL can be determined by the segmentation of peripapillary spectral-domain OCT images. This work hypothesizes that the asymmetry of the RNFL thickness between the two eyes is substantially greater in glaucoma patients than in healthy individuals. This study uses as a parameter of asymmetry the absolute difference in the thickness of the RNFL between the two eyes, $\triangle_S$, although the relative asymmetry, $\delta_S$, obtained by dividing $\triangle_S$ by the thickness of the RNFL layer is also analyzed.

The database defined in [1] and the segmentation of the RNFL provided by the OCT device, Spectralis (version 6.9.4.0), are used in this study. Both absolute and relative asymmetries are calculated for each sector of the eye, as well as for the whole layer. In all cases, sufficient statistical significance is observed, given the large difference between the mean values of the asymmetries of healthy individuals and unhealthy patients, as well as the small $p$-value in all sectors. Among the asymmetry parameters analyzed, the relative asymmetry $\delta_G$ is the one with the highest statistical significance. By using a classifier based on a threshold over the parameter $\delta_G$, a false positive rate of 2.52% and a false negative rate of 28.57% are achieved. These preliminary results, although not sufficiently accurate to be used as the sole diagnostic method, give reason to expect that in combination with other asymmetry parameters or other classification methods, a simple and sufficiently accurate method of glaucoma diagnosis based on the RNFL thickness asymmetry can be derived.

**Acknowledgments.** This work has been partially supported by Spanish National projects AES2017-PI17/00771, AES2017-PI17/00821 (Instituto de Salud Carlos III), and Regional project 20901/PI/18 (Fundación Séneca).

## References

1. Berenguer-Vidal, R., et al.: Automatic segmentation of the retinal nerve fiber layer by means of mathematical morphology and deformable models in 2d optical coherence tomography imaging. Sensors **21**(23) (2021). https://doi.org/10.3390/s21238027
2. Breiman, L., Friedman, J., Stone, C.J., Olshen, R.A.: Classification and Regression Trees. CRC Press (1984)
3. Budenz, D.L.: Symmetry between the right and left eyes of the normal retinal nerve fiber layer measured with optical coherence tomography (an aos thesis). Trans. Am. Ophthalmol. Soc. **106**, 252–275 (2008)
4. Kwon, Y.H., Fingert, J.H., Kuehn, M.H., Alward, W.L.M.: Primary open-angle glaucoma. N. Engl. J. Med. **360**(11), 1113–1124 (2009)
5. Mahmudi, T., Kafieh, R., Rabbani, H., Mehri, A., Akhlaghi, M.R.: Evaluation of asymmetry in right and left eyes of normal individuals using extracted features from optical coherence tomography and fundus images. J. Med. Signals Sens. **11**(1), 12–23 (2021)
6. Mantravadi, A.V., Vadhar, N.: Glaucoma. Primary Care: Clinics in Office Practice **42**(3), 437–449 (2015). Primary Care Ophthalmology
7. Ong, L.S., Mitchell, P., Healey, P.R., Cumming, R.G.: Asymmetry in optic disc parameters: the Blue Mountains Eye Study. Invest. Ophthalmol. Visual Sci. **40**(5), 849–857 (1999)
8. Sommer, A., et al.: Clinically detectable nerve fiber atrophy precedes the onset of glaucomatous field loss. Arch. Ophthalmol. **109**(1), 77–83 (1991)
9. Vos, T., et al.: Global, regional, and national incidence, prevalence, and years lived with disability for 310 diseases and injuries, 1990–2015: a systematic analysis for the Global Burden of Disease Study 2015. The Lancet **388**(10053), 1545–1602 (2016)

# Performance Evaluation of a Real-Time Phase Estimation Algorithm Applied to Intracortical Signals from Human Visual Cortex

Fabrizio Grani[1(✉)], Cristina Soto-Sanchez[1], Alfonso Rodil Doblado[1], Maria Dolores Grima[1], Fernando Farfan[3], Mikel Val Calvo[1], Leili Soo[1], Dorota Waclawczyk[1], Jose Manuel Ferrandez[4], Pablo Gonzalez[1], María Dolores Coves[1], Arantxa Alfaro[1], and Eduardo Fernández[1,2]

[1] Bioengineering Institute, Universidad Miguel Hernandez, Elche, Spain
fgrani@umh.es
[2] CIBER Research Center on Bioengineering, Biomaterials and Nanomedicine (CIBER BBN), Madrid, Spain
[3] Universidad Nacional de Tucuman, Departmento de Bioingenieria Fac de Ciencias Exactas y Technologia, Tucumán, Argentina
[4] Universidad Politecnica de Cartagena, Madrid, Spain

**Abstract.** Cortical visual prostheses are a subgroup of visual prostheses which use electrical stimulation of the occipital cortex to evoke visual percepts in profoundly blind people. The stimulation approaches are usually open-loop, meaning that the stimulation is not controlled by any other factor. However, closed-loop approaches have shown advantages in many neural prosthesis. In the case of cortical visual prosthesis, the closed-loop approach can be based on the phase of local field potentials recorded by the electrodes. Indeed, previous studies have shown that it is easier to induce perception through stimulation at certain phases of brain oscillations.

Here, we evaluated the performance of a real-time phase estimator algorithm applied to local field potentials recorded with intracortical microelectrodes inserted in the occipital cortex of a blind human volunteer. Phase estimation was more accurate at certain phases than others. The error of the estimated phase was in the range $\pm 20°$.

These results should be taken into account when implementing phase-locked stimulation approaches in cortical visual prosthesis. Indeed, the phase estimation accuracy represents the limitation of the closed-loop stimulation approach.

**Keywords:** Visual prosthesis · Multielectrode recordings · LFPs · Closed-loop stimulation

## 1 Introduction

Loss of vision affects millions of persons worldwide. Electrical stimulation of the visual cortex induces the perception of points of light, called phosphenes, and


J. M. Ferrández Vicente et al. (Eds.): IWINAC 2022, LNCS 13258, pp. 516–525, 2022.
https://doi.org/10.1007/978-3-031-06242-1_51

this finding established the physiological basis for the present efforts to develop a visual prosthesis for the blind [2,5,10,14,15]. The advantage of the cortical approach with respect to other vision restoration techniques such as retinal stimulation, is that the visual pathway does not need to be intact. The visual cortex can be stimulated via superficial electrodes [2], or with intracortical electrodes such as the Utah Electrode Array (UEA) [7]. UEAs showed the possibility to elicit complex visual perception in monkeys [9] and humans [12] with lower levels of stimulation current than using superficial electrodes.

UEAs can be used at the same time to stimulate the cortex and to collect brain signals, opening the possibility to adjust the stimulation based on the brain state under the electrodes (closed-loop stimulation approach). Chen *et al.* [9] showed that it is possible to determine the minimum current to elicit visual perception in monkeys stimulating in the primary visual cortex (V1) from the signals recorded in V4. This approach simplifies the tuning of a visual prosthesis, since thresholds can be determined without the user's feedback.

Studies showed that there are phases in which brain stimulation is more effective than others, thus a phase locked stimulation would increase the chances to get a neural response [3]. Cagnan *et al.* [6] improved the outcomes of deep brain stimulation to control Parkinson disease by locking the stimulation to the hands tremor phase. Furthermore, the energy of electromyographic signals generated in muscles followed by transcranial magnetic stimulation (TMS) of the motor cortex depends on the phase of the brain oscillations at the moment of stimulation [17].

In the visual cortex, the phase of brain oscillations indicates a state in which the brain is more responsive to stimulation. Intracortical stimulation in the visual cortex of rats increases the firing rates of neurons close to the stimulation point only at specific local field potential (LFP) phases [1]. In humans, the probability to elicit visual perception through TMS of the visual cortex was higher at specific phases of the EEG signals recorded from the occipital cortex [11].

In this context, the application of phase locked stimulation in a cortical visual prosthesis based on intracortical electrodes, would decrease the current needed to elicit perception, thus improving the safety of the device. In open-loop stimulation, the perception threshold is defined as the current intensity at which the participant reports perception at least 50% of the times. In order to achieve higher perception percentages, it is necessary to increase the current. For the previous reported literature, if stimulation is provided at the right phase of the LFP, 100% of perception could be achieved without increasing the injected current. Further studies are needed to understand which is the LFP phase that could ensure 100% perception. Many repetitions of the stimulus at threshold current should be repeated in order to get a statistical distribution of the phase when perception happens and when it does not happen.

In order to translate the phase specific response to stimulation into a practical application in visual prosthesis, a real-time algorithm for phase estimation is needed. Many methods have been proposed in literature [4,8,13,16]. All these algorithms share a common working principle: after filtering the brain signals, they estimate the future of the signals to determine the phase at the current time through Hilbert transform.

All the methods proposed are evaluated on specific datasets, but it is of fundamental importance to understand the performance of real-time phase detection on data as similar as possible to the one acquired in a real case scenario. In this manuscript, the performance of the method proposed by Blackwood *et al.* in [4] applied to real data acquired from the visual cortex of a blind participant will be evaluated. The obtained results are relevant in order to understand to which extent phase-locked stimulation could improve cortical visual prosthesis.

## 2 Methods

### 2.1 Experiment, Data Acquisition and Preprocessing

An intracortical microelectrode array with 96 channels has been implanted in the visual cortex of a blind volunteer. The participant was 57 years old with complete blindness for the last 16 years. The microelectrode array stayed implanted during a period of six months. The participant gave written consent before any experiment. Procedures were approved by the Hospital General Universitario de Elche Clinical Research Committee and registered at ClinicalTrials.gov (NCT02983370).

The main objective of the implant was to study the feasibility of a cortical visual prosthesis to restore visual perceptions in blind subjects. More details about the study can be found in [12]. The experimental setup consisted of a current stimulator (Cerestim 96 multichannel microstimulation system, Blackrock Microsystems Inc., Salt Lake City, UT) and a neural recording system (NeuroPort data acquisition system, Blackrock Microsystems Inc., Salt Lake City, UT). Data were sampled at 30 kHz.

During the six months of implantation, experimental sessions were performed daily and electrical stimulation was provided to the participant through the UEA. Neural signals were acquired simultaneously. Each experimental session was preceded by 1- to 2-min of spontaneous neural recordings.

The results reported in this manuscript are based on the spontaneous activity data. In total we have 43 spontaneous activity recordings on different days. Each of them with 96 recording channels.

As reported in previous literature [11], the 5–15 Hz bandwidth is the frequency band where phase dependent response to stimulation happens. According to Shannon Sampling theorem, in order to get reliable data at those frequencies we need at 30 Hz as sampling frequency, therefore we down-sampled the collected data from 30 kHz to 100 Hz. We then divided the signals in 0.5 s overlapping windows (0.49 s overlap) from which we estimated the phase at the last time point of the window. The estimated phase was then compared to the phase extracted using all the signals' time course.

### 2.2 Real Time Phase Estimation Algorithm

In this section we briefly explain the algorithm proposed by Blackwood *et al.* [4], that we used to evaluate the performance of real-time phase estimation on our data.

The data in the current observation window are filtered using a causal second order Butterworth filter (bandpass filter between 4 and 15 Hz). An autoregressive (AR) model of order 10 is then fitted to the data using the Yule-Walker method. The AR model is then used to forecast the future of the data. In our case, we estimated 6 time points (0.06 s in the future 100 Hz).

The phase is calculated through Hilbert transform of the data in the current observation window concatenated to the forecasted data. The resulting angle at the current time is the phase estimation.

### 2.3 Performance Indexes

All the estimated phases have been compared to the real phase of the signal. In order to compute the real phase we bandpass filtered the signal (not divided in time windows in this case) with a second order Butterworth filter both in the forward and backward directions between 4 and 15 Hz. The real phase was then extracted at each time with the Hilbert transform.

The performance was then evaluated by comparing the difference between estimated and real phase (the mean difference is the circular mean), and by the circular variance of the estimated phases. The closer the two measures are to 0, the better the performance of the algorithm. Among the two measures, the circular variance is the more important since it describes how stable is the estimation. Indeed, with a stable estimation we can correct the circular mean deviations by adding a constant to the estimate. Each measure was evaluated on the difference between estimated and real phase, and on the estimated phase at specific real phase angles (0°, 90°, 180°, 270°). For each of these angles, the phase estimated when the real phase was equal to that angle ±5° was considered.

As previously shown [17], the performance of the phase estimation is proportional to the SNR of the oscillation in the frequency of interest. The observation was also confirmed as the data showed a linear relation between performance of the algorithm and 4–15 Hz SNR. The Wald test was used in order to test the significance of the proposed linear fit.

## 3 Results

The circular histograms in Fig. 1 show the performance of the algorithm on the signals acquired from channel 0 in one recording day. Figure 1A shows the distribution of the difference between estimated and real phase, while Fig. 1B,C,D and E show the distribution of the estimated phase when the real phase was respectively 0°, 90°, 180° and 270°. Circular mean and variance for every case are reported in the figure.

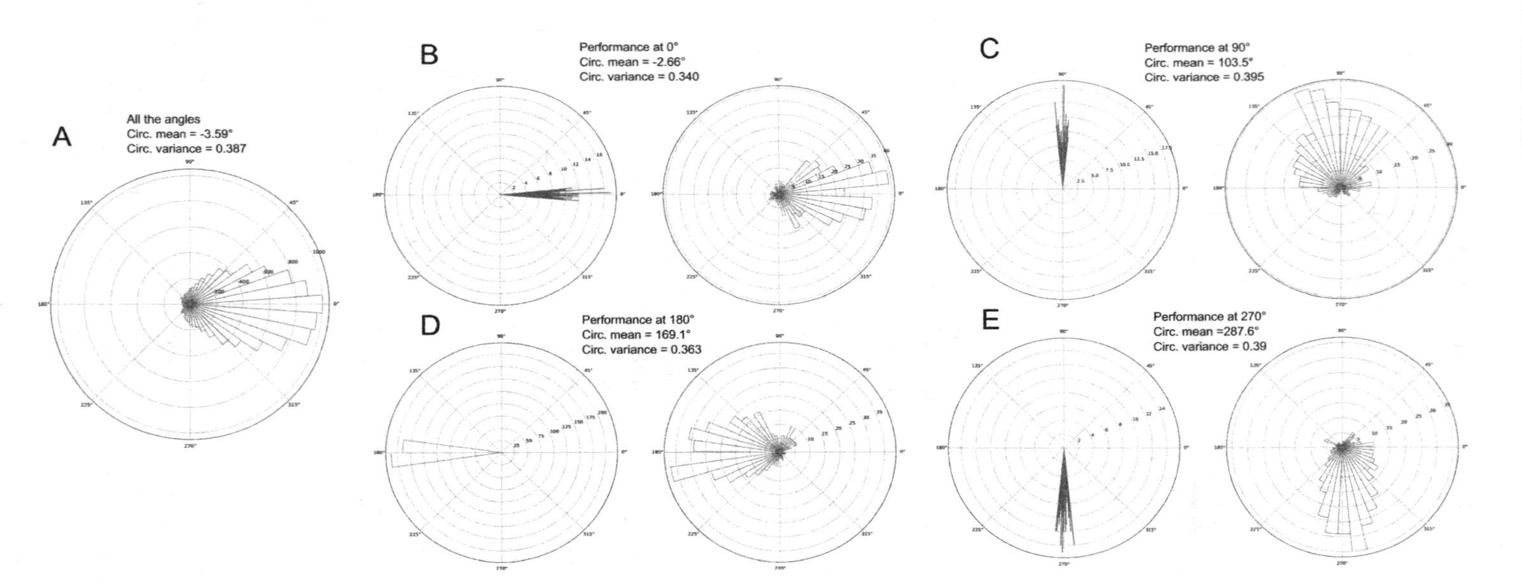


**Fig. 1.** Performance of the phase estimation algorithm on the data collected on one single day from channel 0. A) Circular distribution of the difference between real and estimated phase (all angles are included here). B-E) Circular distribution of the real (left) and estimated (right) phase at specific angles (0°, 90°, 180°, 270°) .

Fig. 2 reports the performance of the algorithm on different channels of one single day (top part) and different days (bottom part). The estimated phase (circular mean) was consistently smaller than the real phase at 0° and 180°, while it was consistently bigger than the real phase at 90° and 270°. This is true for different channels (Fig. 2 top left) and for different recording days (Fig. 2 bottom left). A consistent error between predicted and real phase allows to apply a constant correction factor to the algorithm, reducing the estimation error.

**Table 1.** Average performances of the algorithm. Circular mean values are in radians, circular variance values are without measurement unit.

| | Desired | Day 0 | All days |
|---|---|---|---|
| All angles circ. mean | 0 | –0.057 | –0.005 |
| All angles circ. var | 0 | 0.34 | 0.33 |
| 0° circ. mean | 0 | –0.101 | –0.090 |
| 0° circ. var | 0 | 0.31 | 0.273 |
| 90° circ. mean | 1.57 | 1.81 | 1.69 |
| 90° circ. var | 0 | 0.34 | 0.33 |
| 180°circ. mean | 3.14 | 3.01 | 2.81 |
| 180° circ. var | 0 | 0.33 | 0.28 |
| 270° circ. mean | –1.57 | –1.32 | –1.38 |
| 270° circ. var | 0 | 0.36 | 0.34 |

The right part of Fig. 2 shows the mean and standard deviation of circular variance across channels (top) and days (bottom) for the phase estimated at specific angles (0°, 90°, 180°, 270°) and for all the angles. The circular variance measures the spread of the estimation error around its mean and it can't be compensated as the circular mean, therefore it is the most important performance index. Our data show a smaller circular variance at 0° and 180°, meaning that the algorithm performs better at these phases. Table 1 summarizes the average values of the data shown in Fig. 2.

Figure 3 shows that the circular variance decreases (better algorithm performances) with the increase of SNR, confirming the results reported in [17].

The linear fit (continuous lines in the figures) between circular variance and SNR is statistically significant for every angle considered (Pearson correlation coefficients r and Wald test p values reported in Table 2) .

**Table 2.** Significance of the decrease of circular variance with respect to SNR.

| | r | p |
|---|---|---|
| All angles | –0.75 | 1.65e–18 |
| 0° | –0.63 | 7.06e–12 |
| 90° | –0.71 | 7.37e–16 |
| 180° | –0.75 | 1.04e–18 |
| 270° | –0.62 | 1.49e–11 |

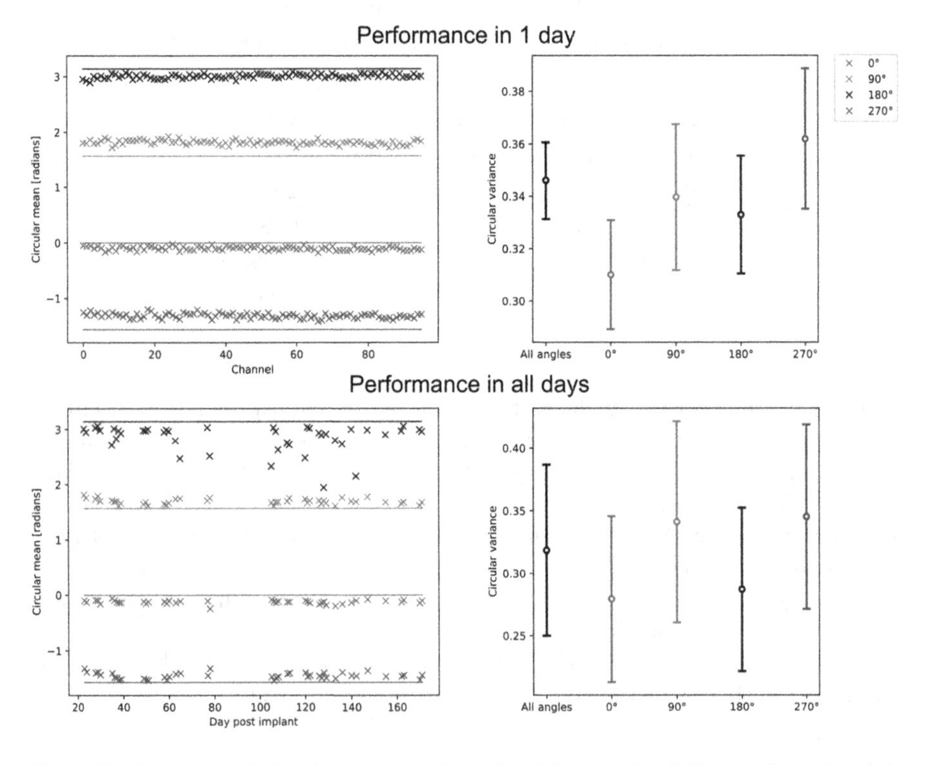


**Fig. 2.** Performance of the phase estimation algorithm on the different channels of the day shown in Fig. 1 (top), and on all the acquisition days (bottom). Circular mean values are depicted on the left (x), with the continuous line representing the expected value. Circular variance value are depicted on the right. The error bars show mean and standard deviation of the circular variance for all the channels in 1 single day (top right) and for all the channels and days (bottom right).

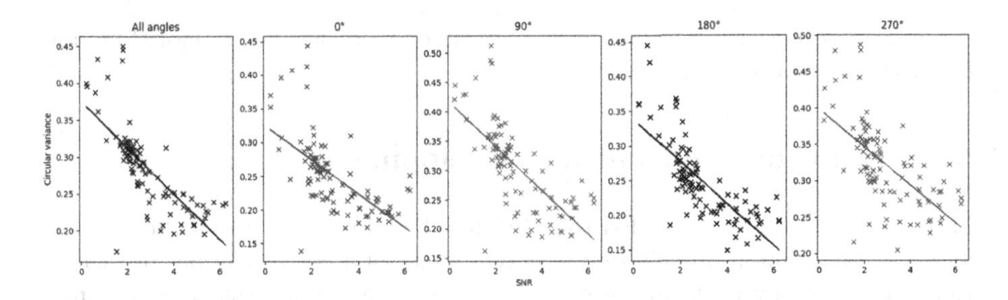


**Fig. 3.** The circular variance decreases with the increase of the SNR in the frequency band of the oscillation of interest. Data shown for 1 single recording day. Each x represents one channel, the lines represent the linear fit of the data.

## 4 Discussion

In this work we showed the performance of the real-time phase estimator algorithm proposed by Blackwood *et al.* [4] on neural data acquired with an intracortical electrode array inserted in the visual cortex of a blind participant. In the context of cortical visual prosthesis, stimulating at the correct phase of the LFP could increase the probability to induce phosphene perception even at low current intensities, increasing the performance and reliability of this technology, but more studies are still needed.

Dugue' *et al.* [11] showed that brain oscillation frequencies, whose phase influence the probability of achieving phosphene perception with TMS, are between 5–15 Hz. A 10 Hz oscillation has a period equal to 0.1 s. This means that the maximum time we would have to wait before stimulating if we want to target a certain phase is 0.1 s, thus limiting 10 Hz the maximum frame rate at which the prosthesis could induce different phosphenes. The use of phase-locked stimulation approach could decrease the current needed to elicit perception, but at the same time it limits the phosphenes frame rate. Anyway, 10 Hz frame rate could be reasonable for a cortical visual prosthesis.

The results show that the algorithm tends to underestimate the phase at 0° and 180° while it overestimates the phase at 90° and 270° (Fig. 2 left). Once the desired stimulation phase is known, the algorithm could be easily compensated for this error by the addition of a constant to the estimated phase.

The circular variance in all the dataset ranges from 0.22 to 0.43, with a mean value of 0.33. A circular variance of 0.33 corresponds to a variation around the circular mean of ±20°, which in 10 Hz wave corresponds to a possible stimulation timing error of ±5 ms. The delay of the stimulator to send the pulse must be added to this 5 ms delay. The stimulator delay changes depending on the hardware used and should be taken into account when applying the algorithm in a real-life scenario.

Algorithm stimulation timing error plus the stimulator delay represent the limit of real-time phase-locked stimulation in our present approach. This limit can surely be overcome by the development of more accurate algorithms and faster stimulation devices.

## 5 Conclusions and Future Development

Further studies are still needed to determine how beneficial would be a phase-locked stimulation approach for the development of a cortical visual prosthesis. Here, the accuracy that can be obtained in real-time phase estimation using a simple autoregressive forecasting model was shown.

Future works include:

1. Determination of the best LFP phase to induce visual perception through intracortical electrical stimulation;
2. Study of new algorithms to improve LFP forecasting and therefore the phase estimation accuracy using artificial intelligence;
3. Implementation of the algorithm in real-time using an electrical stimulator to calculate the real time delay between the stimulation and the desired phase.

**Acknowledgements.** We would like to thank B.G. and her husband for their extraordinary commitment to this study. This project has received funding by grant RTI2018-098969-B-100 from the Spanish Ministerio de Ciencia Innovación y Universidades, by grant PROMETEO/2019/119 from the Generalitat Valenciana (Spain), by the European Union's Horizon 2020 research and innovation programme under the Marie Skłodowska-Curie grant agreement No 861423 (enTRAIN Vision) and by grant agreement No. 899287 (project NeuraViPer).

## References

1. Allison-Walker, T.J., Ann Hagan, M., Chiang Price, N.S., Tat Wong, Y.: Local field potential phase modulates neural responses to intracortical electrical stimulation. In: Proceedings of the Annual International Conference of the IEEE Engineering in Medicine and Biology Society, EMBS, pp. 3521–3524 (2020). https://doi.org/10.1109/EMBC44109.2020.9176186
2. Beauchamp, M.S., et al.: Dynamic stimulation of visual cortex produces form vision in sighted and blind humans. Cell **181**(4), 774–783.e5 (2020). https://doi.org/10.1016/j.cell.2020.04.033
3. Bergmann, T.O.: Brain state-dependent brain stimulation. Front. Psychol. **9**, 2108 (2018). https://doi.org/10.3389/FPSYG.2018.02108
4. Blackwood, E., Lo, M.C., Alik Widge, S.: Continuous phase estimation for phase-locked neural stimulation using an autoregressive model for signal prediction. In: Proceedings of the Annual International Conference of the IEEE Engineering in Medicine and Biology Society, EMBS 2018-July, pp. 4736–4739 (2018). https://doi.org/10.1109/EMBC.2018.8513232
5. Brindley, G.S., Lewin, W.S.: The sensations produced by electrical stimulation of the visual cortex. J. Physiol. **196**(2), 479–493 (1968). https://doi.org/10.1113/jphysiol.1968.sp008519

6. Cagnan, H., et al.: Stimulating at the right time: phase-specific deep brain stimulation. Brain **140**(1), 132–145 (2017). https://doi.org/10.1093/BRAIN/AWW286, https://academic.oup.com/brain/article/140/1/132/2732724
7. Campbell, P.K., Jones, K.E., Huber, R.J., Horch, K.W., Normann, R.A.: A silicon-based, three-dimensional neural interface: manufacturing processes for an intracortical electrode array. IEEE Trans. Biomed. Eng. **38**(8), 758–768 (1991). https://doi.org/10.1109/10.83588
8. Chen, L.L., Madhavan, R., Rapoport, B.I., Anderson, W.S.: Real-time brain oscillation detection and phase-locked stimulation using autoregressive spectral estimation and time-series forward prediction. IEEE Trans. Biomed. Eng. **60**(3), 753–762 (2013). https://doi.org/10.1109/TBME.2011.2109715
9. Chen, X., Wang, F., Fernandez, E., Roelfsema, P.R.: Shape perception via a high-channel-count neuroprosthesis in monkey visual cortex. Science **370**(6521), 1191–1196 (2020). http://science.sciencemag.org/
10. Dobelle, W.H., Mladejovsky, M.G., Evans, J.R., Roberts, T.S., Girvin, J.P.: Braille reading by a blind volunteer by visual cortex stimulation. Nature **259**(5539), 111–112 (1976). https://doi.org/10.1038/259111a0, https://pubmed.ncbi.nlm.nih.gov/1246346/
11. Dugué, L., Marque, P., VanRullen, R.: The phase of ongoing oscillations mediates the causal relation between brain excitation and visual perception. J. Neurosci. **31**(33), 11889–11893 (2011). https://doi.org/10.1523/JNEUROSCI.1161-11.2011
12. Fernández, E., et al.: Visual percepts evoked with an Intracortical 96-channel microelectrode array inserted in human occipital cortex. J. Clin. Invest. **131**(23)(2021). https://doi.org/10.1172/JCI151331, http://www.jci.org/articles/view/151331
13. Mansouri, F., Dunlop, K., Giacobbe, P., Downar, J., Zariffa, J.: A fast EEG forecasting algorithm for phase-locked transcranial electrical stimulation of the human brain. Front. Neurosci. **11**, 1–14 (2017). https://doi.org/10.3389/fnins.2017.00401
14. Roelfsema, P.R.: Writing to the mind's eye of the blind. Cell **181**(4), 758–759 (2020). https://doi.org/10.1016/j.cell.2020.03.014
15. Schmidt, E.M., Bak, M.J., Hambrecht, F.T., Kufta, C.V., O'Rourke, D.K., Vallabhanath, P.: Feasibility of a visual prosthesis for the blind based on intracortical microstimulation of the visual cortex. Brain **119**(2), 507–522 (1996). https://doi.org/10.1093/brain/119.2.507, https://pubmed.ncbi.nlm.nih.gov/8800945/
16. Wodeyar, A., Schatza, M., Widge, A.S., Eden, U.T., Kramer, M.A.: A state space modeling approach to real-time phase estimation. eLife **10**, 1–28 (2021). https://doi.org/10.7554/eLife.68803
17. Zrenner, C., Galevska, D., Nieminen, J.O., Baur, D., Stefanou, M.I., Ziemann, U.: The shaky ground truth of real-time phase estimation. NeuroImage **214**(December 2019), 116761 (2020). https://doi.org/10.1016/j.neuroimage.2020.116761

# Electrical Stimulation Induced Current Distribution in Peripheral Nerves Varies Significantly with the Extent of Nerve Damage: A Computational Study Utilizing Convolutional Neural Network and Realistic Nerve Models

Jinze Du[1,3](✉), Andres Morales[2,3], Pragya Kosta[3], Jean-Marie C. Bouteiller[2,3], Gema Martinez[4], David Warren[5], Eduardo Fernandez[4], and Gianluca Lazzi[1,2,3]

[1] Department of Electrical Engineering, University of Southern California, Los Angeles, CA 90089, USA
jinzedu@usc.edu

[2] Department of Biomedical Engineering, University of Southern California, Los Angeles, CA 90089, USA

[3] Institute for Technology and Medical Systems Innovation (ITEMS), Keck School of Medicine, University of Southern California, Los Angeles, CA 90089, USA

[4] Institute of Bioengineering, University Miguel Hernandez, Elche and CIBER-BBN, Madrid, Spain

[5] Department of Biomedical Engineering, University of Utah, Salt Lake City, UT 84112, USA

**Abstract.** Although electrical stimulation is an established treatment option for multiple central nervous and peripheral nervous system diseases, its effects on the tissue and subsequent safety of the stimulation are not well understood. Therefore, it is crucial to design stimulation protocols that maximize therapeutic efficacy while avoiding any potential tissue damage. Further, the stimulation levels need to be adjusted regularly to ensure that they are safe even with the changes to the nerve due to long-term stimulation. Using the latest advances in computing capabilities and machine learning approaches, we developed computational models of peripheral nerve stimulation based on very high-resolution cross-sectional images of the nerves. We generated nerve models constructed from non-stimulated (healthy) and over-stimulated (damaged) rat sciatic nerves to examine how the current density distribution is affected by nerve damage. Using our in-house numerical solver, the Admittance Method (AM), we computed the induced current distribution inside the nerves and compared the current penetration for healthy and damaged

This work was supported by the NIBIB of the National Institute of Health Grant No. 5R01EB029271, and an unrestricted grant to the Department of Ophthalmology from Research to Prevent Blindness, New York, NY.


J. M. Ferrández Vicente et al. (Eds.): IWINAC 2022, LNCS 13258, pp. 526–535, 2022.
https://doi.org/10.1007/978-3-031-06242-1_52

nerves. Our computational results indicate that when the nerve is damaged, primarily evidenced by the decreased nerve fiber packing, the current penetrates deeper inside the nerve than in the healthy case. As safety limits for electrical stimulation of biological tissue are still debated, we ultimately aim to utilize our computational models to determine refined safety criteria and help design safer and more efficacious electrical stimulation protocols.

**Keywords:** Computational modeling · Electrical stimulation · Peripheral nerve · Tissue safety

## 1 Introduction

Electrical neural stimulation has been shown to be effective in the treatment of various medical conditions, such as retinal stimulation for vision restoration and deep brain stimulation for essential tremors, epilepsy, and Parkinson's disease [2,26]. Electrical stimulation of peripheral nerves has been used to treat chronic pain and aid treatment of many diseases [8,10,24,25].

Various studies have been performed to investigate the tissue damage caused by the electrical neural stimulation [1]. However, conventional safety standards only account for stimulation current waveform and tissue-electrode interface area and disregard actual induced current distribution across different tissue locations [14,22]. Electrical stimulation induces current density inside the tissue based on the material properties of the various tissue types. Therefore, details of heterogeneous anatomy can play a vital role in determining the possibility of tissue damage. Accurate representation of neural stimulation through computational modeling could provide insights into how various stimulation parameters impact the safety and efficiency of specific tissues. Though there are multiple computational models for peripheral nerves, most of the models are oversimplified and represent nerve fascicles by effective media properties [11,19,20]. Therefore, better computational models are needed that incorporate the heterogeneous fascicles with fine details like nerve fibers and myelination of varying diameters for more accurate predictions.

In our previous work, we built a multi-scale computational model of a realistic healthy rat sciatic nerve with the help of convolutional neural network (CNN) segmentation of cross-sectional nerve images [9]. Utilizing this model, we simulated the electrical stimulation of the nerve by cuff electrodes and found that the distance between the cuff electrodes significantly influences the current penetration depth inside the nerve. In this work, we build two separate computational models to represent healthy and damaged nerves using images from non-stimulated and stimulated nerve slices. First, we use CNN segmentation to label various tissue types of the nerve and thus, build anatomically accurate, heterogeneous, high-resolution nerve models that include fine details such as axons and myelin. Then, we quantify the morphological properties of the nerves, such as fiber density and fiber packing, to highlight the changes that occurred in the damaged nerve. Finally, we use multi-scale computational models based on Admittance Method (AM) [5] to predict the current distribution inside the healthy and damaged nerves due to stimulation via cuff electrodes. As chronic peripheral nerve stimulation often involves nerve morphological changes during the process, it is essential to consider morphological details when predicting stimulation safety. Therefore, we analyze how the induced current density distribution and the possibility of further tissue damage are impacted when the stimulation levels are left unchanged in the case of early nerve damage.

## 2 Methods

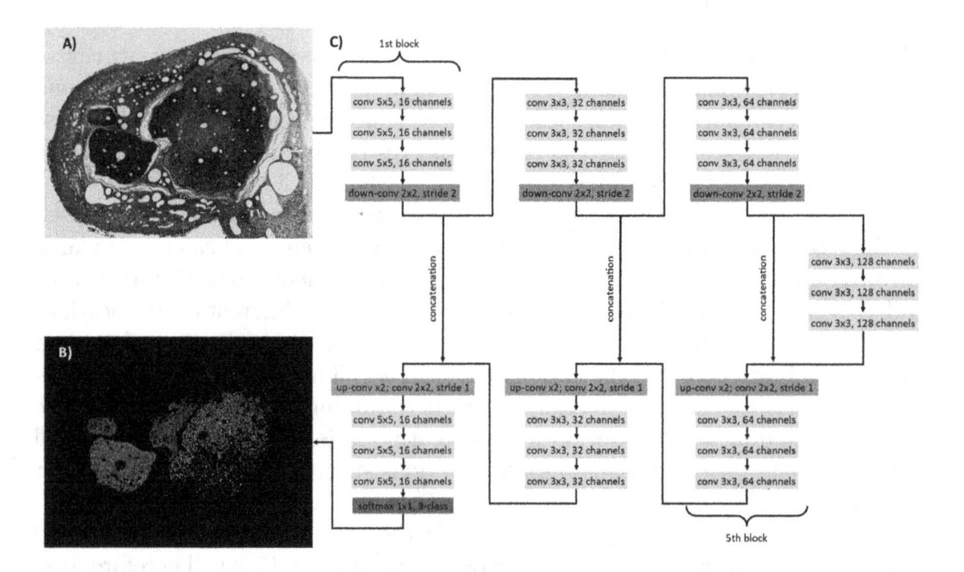


**Fig. 1.** CNN segmentation of cross-sectional image of peripheral nerve. (A) a cross-sectional image of rat sciatic nerve, (B) segmented image of nerve cross-section containing multiple fascicles populated by axons and myelin of various morphologies (grey represents myelin and white represents axon), (C) CNN U-Net architecture used to generate the segmented image.

### 2.1 CNN Segmentation of Peripheral Nerve Cross-sectional Images

Current computational models of peripheral nerve stimulation are based on heterogeneous fascicles populated by artificially generated axons or very simplified homogeneous fascicles [11,12,19,20]. To build more realistic nerve stimulation models from actual peripheral nerves, we used high-resolution confocal imaging to acquire cross-sectional images of rat sciatic nerve. To replicate the details of axon location and morphology from these tissue samples, the histological images were segmented using a convolutional neural network (CNN) known as AxonDeepSeg [27]. As shown in Fig. 1, the architecture consists of 7 "blocks" each containing 3 convolutional layers. The blocks on the top path, feeding to the right, are each followed by a down-convolutional layer to further expand the field of view of the subsequent block. Symmetrically, the blocks on the bottom path, feeding back to the left, are preceded by up-convolutional layers that also take in output from the corresponding block in the top path. This U-Net architecture allows the network to make structural analysis over larger fields of view but also localise the tissue segmentations to pixel-wise resolution. Once the network was trained on a relatively small sample of manually segmented images, we were able to obtain fast pixel-wise labeling of myelin and axon. The segmented cross-sectional slice of nerve was then ready to be discretized and extruded to build a pseudo-3D nerve model for electrical stimulation simulation.

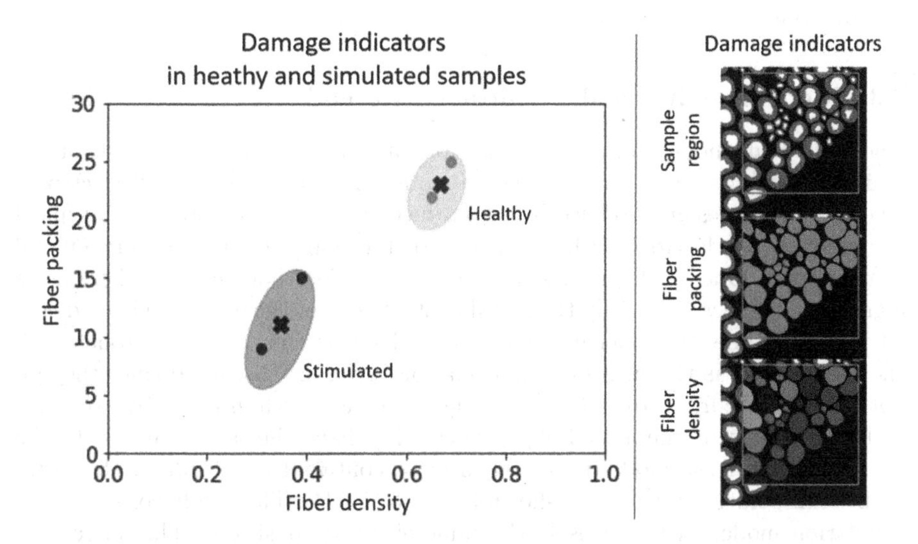


**Fig. 2.** Fiber packing and fiber density are two well recognized parameters for determination of the extent of nerve damage. The calculation of these metrics are shown on the right. To ensure the samples used for modeling are good general representations, we use these two metrics to select one healthy nerve sample and one damaged nerve sample from total of six samples, all of the similar type of cells fall within the range of twice the standard deviation.

## 2.2 Nerve Image Selection

In order to compare the current distribution inside healthy nerve and damaged nerve, two different nerve models are constructed from two nerve microscopy images from actual rat sciatic nerve stimulation experiments: one healthy and one with a certain extent of the damage. To further rule out the individual differences between different samples, we developed two assessing metrics to help ensure these two images are a good representation of their own classes: Fiber packing and fiber density. As shown in Fig. 2, using a 200 μm × 200 μm (1600 px × 1600 px) window, fiber density is the number of cells inside (cells cut by the window are counted fractionally) divided by the window area. Fiber packing uses the same window but divides the total cell area inside by the window area. The value for window area was adjusted to account for when windows included regions outside fascicles or occupied by veins where neurons would not be expected. These two metrics are essential in nerve degeneration and changes in nerve conduction [6,7,21]. Making use of these metrics, we selected one healthy and one slightly damaged nerve sample from a total of six samples within twice of standard deviation, as shown in Fig. 2. The selected healthy nerve image is from control samples with cuff electrodes only implanted but not stimulated; the selected damaged nerve image is taken from a sample stimulated by a 1.2 mA current source.

### 2.3 Model Building and Admittance Method

The multi-scale model considered for peripheral nerve stimulation consists of two main components: the segmented nerve model and the model of cuff electrode. The cuff electrodes are modeled based on the cuff electrode design from a typical commercial cuff electrode with an inner diameter of about 2 mm, with two metal contact wires. One of which works as a source cuff electrode and the other one as a ground cuff electrode. Only the metal contact part of the electrode is modeled. At the same time, the other non-conductive elements that are not in touch with the tissue, such as the surrounding insulation layer, are discarded since they do not impact the current distribution in the tissue, as shown in Fig. 3 (E). In order to better represent the actual experiments that have the nerve sutured to the electrode, the nerve model is placed in close contact to one side of the nerve with saline solution filled in, as shown in Fig. 3 (C)(H). This study considers two simulation models with precisely the same electrode positions. The model size is $800 \times 800 \times 1000$ voxels in x, y, z dimensions, and the resolution is 3.1 μm in all three dimensions. The model is discretized in cubic voxels, and a unique material index represents each voxel. The material properties of the nerve model are taken from [4,15] and described in Table 1.

The multi-resolution admittance method (AM) [3,5,13] is used to compute electric field values at each node of the computation grid [16–18,23]. AM defines a matrix describing the admittance (G), or resistance, throughout the model. The resistance of each node is defined by the diagonal components of the matrix, while the surrounding values define the resistances between nodes, producing a sparse, diagonal matrix. The admittance values are computed using the conductivity and the distance between nodes in the x, y, and z directions, as described in Eq. 1.

$$g_x^{i,j,k} = \sigma_x^{i,j,k} \frac{\Delta y \Delta z}{\Delta x} \tag{1}$$

**Table 1.** Tissue properties

| Tissue type | Conductivity $(\sigma_x, \sigma_y, \sigma_z)$S/m |
|---|---|
| Perineurium | (0.01, 0.01, 0.01) |
| Myelination | $(2 \times 10^{-4}, 5 \times 10^{-9}, 5 \times 10^{-9})$ |
| Intracellular space | (0.33, 0.33, 0.33) |
| Axoplasm | (0.91, 0.91, 0.91) |
| Epineurium | (0.1, 0.1, 0.1) |
| Nerve membrane | (0.02, 0.02, 0.02) |
| Intracellular space | (0.91, 0.91, 0.91) |
| Saline solution | (1.45, 1.45, 1.45) |
| Extracellular space | (0.33, 0.33, 0.33) |

A current vector (I) is defined with current values applied to whichever nodes contain a source. A voltage vector (V) can then be solved for using G and I in Eq. 2. A multi-threaded Python program using a biconjugate gradient algorithm is developed to construct the matrices and solve the matrix equation with accelerated speed.

$$GV = I \tag{2}$$

A 3D multi-resolution meshing algorithm is executed prior to the field simulations in order to reduce the complexity of the problem without impacting the accuracy of the solution. In this, a high level of details and fine resolution is maintained near the nerve periphery, in locations proximal to the contact electrode, and the voxel size is increased further away from nerve periphery where fine resolution is unnecessary. Therefore, along the center of the nerve bundle, the resolution would be coarser, whereas along the periphery of the nerve (i.e. closer to the electrodes and fascicle edges), the resolution would remain fine. In this way, the number of nodes and edges are decreased and the computational complexity of the system is reduced. AM simulations provide voltage values at every node of the model. In our multi-resolution scheme, network nodes are located at the vertices of voxels. Because conductivity value is considered constant inside each voxel, trilinear interpolation is used to calculate the voltage at arbitrary points inside a voxel from the values at its vertices. Once voltage values have been interpolated back to unit voxel, electric field, charge density, and current density could be calculated at any point in the model.

# 3 Results

## 3.1 Current Distribution Inside Two Nerve Models

Our in-house Python AM platform was used to perform the interpolation calculation in healthy and damaged nerve models. Starting from the two selected images, which are good representations of their own kind shown in Fig. 3 (A) and

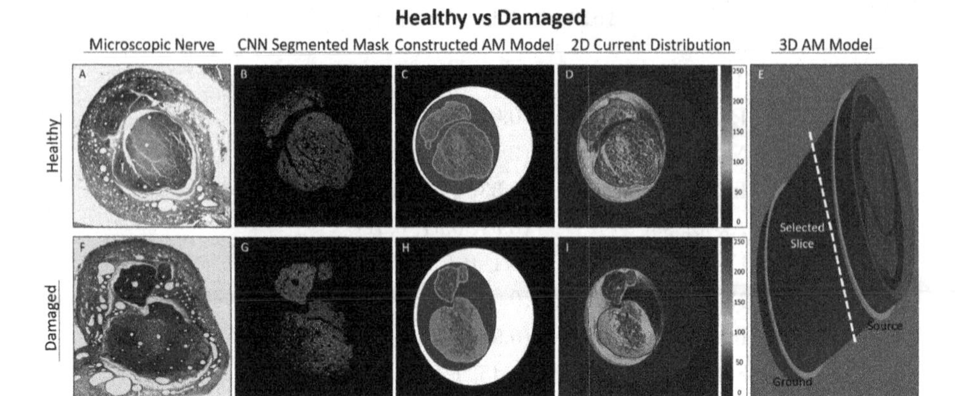


**Fig. 3.** One healthy and one damaged nerve model are constructed from representative nerve microscopic images (A)(F). The damaged nerve image has a lower cell density. In order to construct the 3D model with source and ground electrodes (E), a convolutional neural network was applied to segment the myelin and axons (B)(G). Current density distributions near the source electrode are plotted for both models. It shows that damaged nerve models have a higher current density at the nerve periphery than the healthy nerve model due to the decrease of the cell population (D)(I).

(F), first CNN segmentation is performed to identify the axon and myelin in the nerve. The healthy nerve model has a higher cell density, whereas the damaged nerve model has a lower density closer to its nerve boundary due to potential damage. With all the considerations, electric current density distribution inside the two nerve models is calculated and plotted as in Fig. 3 (D)(I). The 2D slices are selected close to the source electrode. It is shown that inside both nerves, the current density is highest closer to the nerve edge, which is closer to the electrode, and current density decreases as it goes further away. Damaged nerve models have a higher current density at the nerve periphery than the healthy nerve model due to the decrease of the cell population.

### 3.2 Current Density in Different Nerve Components

With the constructed very high-resolution model with realistic nerve morphology, in addition to the general current distribution pattern illustrated above, we are able to extract current density values on different nerve components. Figure 4 (A) illustrates the current density distribution inside the healthy nerve model, and on the right Fig. 4 (B) shows the current density distribution inside the damaged nerve model. Several major nerve components are selected and illustrated here: myelin, nerve intracellular space, axoplasm, and epineurium. From the comparison, it is clear that the current density values on axoplasm in the damaged model are much higher than in the healthy nerve model, which is because of the decrease of the highly resistive myelinated fiber population. Also, due to the less densely populated myelinated axons, it is easier for the current to penetrate deeper to the nerve's center and thus enter the more conductive axoplasm in the damaged model. Thus, the current density on axoplasm is much higher in the damaged nerve model than the healthy one. Furthermore, the current density values generally decrease as they get further away from the electrode due to the electrode being the source of stimulation. Also, we can barely see much current on the myelin, despite the large population number, which is due to the high resistivity of myelin tissue that is hard for the current to flow through.

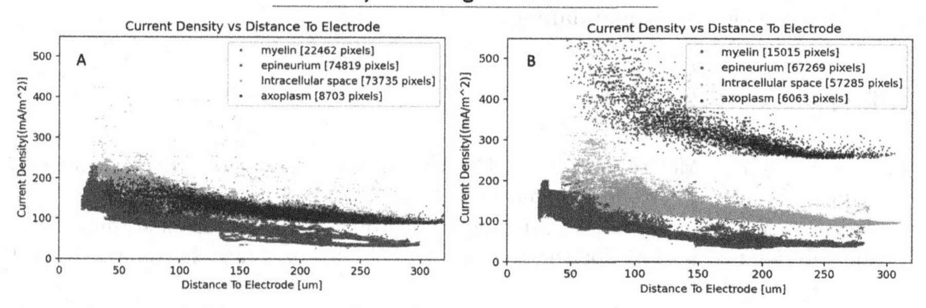


**Fig. 4.** Current density distributions on different nerve components are illustrated: healthy (A) and damaged (B). The current density values on axoplasm in the damaged model are much higher than in the healthy nerve model.

## 4 Discussion

Current density has always been considered as one of the most crucial factors related to nerve damage induced during electrical stimulation. However, our knowledge of the exact current density distribution inside the nerve has always been elusive. A computational study using the Admittance Method multi-scale computational platform was conducted to help provide additional insights. After segmenting cross-sectional images of healthy and damaged sciatic nerves using a convolutional neural network and building realistic peripheral nerve models from the segmentation results, we developed realistic cuff electrode models with different damage extent nerve models.

Due to the complexity of the constructed computational models, it is unrealistic to construct multiple healthy and damaged nerve samples for comparison purposes. Thus, to rule out the singularity cases and make this study more generally applicable, we assessed the damage extent using different metrics in three different healthy nerve samples and three damaged nerve samples. All the samples are within two times of standard deviation of their own types. Furthermore, each sample we selected is right in the middle of their own types to help better represent the general case.

The simulation results found that the current distribution values on axoplasm in the damaged model are higher than the healthy model. Our results suggest that the peripheral nerve's accurate, high-resolution anatomical features significantly affect the current distribution inside the nerve. A very high resolution and accurate model of the peripheral nerve may play a critical role in assessing neurostimulation devices' efficacy and safety considerations. Besides implications on the recruitment of target fibers, the significant variations due to the damage extent of models have implications on the level of stimulation that are considered safe and do not result in axonal damage. Thus, with the proposed method and models, criteria for safe and effective peripheral neurostimulations can be established, especially for long-term chronic stimulation where nerve morphology might change due to chronic damage.

## References

1. Agnew, W.F., McCreery, D.B.: Considerations for safety with chronically implanted nerve electrodes. Epilepsia **31**, S27–S32 (1990)
2. Benabid, A.L., Chabardes, S., Mitrofanis, J., Pollak, P.: Deep brain stimulation of the subthalamic nucleus for the treatment of Parkinson's disease. Lancet Neurol. **8**(1), 67–81 (2009)
3. Bingham, C.S., et al.: Admittance method for estimating local field potentials generated in a multi-scale neuron model of the hippocampus. Front. Comput. Neurosci. **14**, 72 (2020)
4. Butson, C.R., Miller, I.O., Normann, R.A., Clark, G.A.: Selective neural activation in a histologically derived model of peripheral nerve. J. Neural Eng. **8**(3), 036009 (2011)
5. Cela, C., et al.: A multiresolution admittance method for large-scale bioelectromagnetic interactions. Ph.D. thesis, North Carolina State University (2010)
6. Christensen, M.B., Tresco, P.A.: Differences exist in the left and right sciatic nerves of naïve rats and cats. Anat. Rec. **298**(8), 1492–1501 (2015)
7. Comin, C.H., et al.: Statistical physics approach to quantifying differences in myelinated nerve fibers. Sci. Rep. **4**(1), 1–11 (2014)
8. Doucet, B.M., Lam, A., Griffin, L.: Neuromuscular electrical stimulation for skeletal muscle function. Yale J. Biol. Med. **85**(2), 201 (2012)
9. Du, J., Morales, A., Paknahad, J., Kosta, P., Bouteiller, J.M.C., Fernandez, E., Lazzi, G.: Electrode spacing and current distribution in electrical stimulation of peripheral nerve: a computational modeling study using realistic nerve models. In: 2021 43rd Annual International Conference of the IEEE Engineering in Medicine & Biology Society (EMBC), pp. 4416–4419. IEEE (2021)
10. Grill, W.M., Kirsch, R.F.: Neuroprosthetic applications of electrical stimulation. Assist. Technol. **12**(1), 6–20 (2000)
11. Kosta, P., Mize, J., Warren, D.J., Lazzi, G.: Simulation-based optimization of figure-of-eight coil designs and orientations for magnetic stimulation of peripheral nerve. IEEE Trans. Neural Syst. Rehabil. Eng. **28**(12), 2901–2913 (2020)
12. Kosta, P., Warren, D.J., Lazzi, G.: Selective stimulation of rat sciatic nerve using an array of mm-size magnetic coils: a simulation study. Healthc. Technol. Lett. **6**(3), 70–75 (2019)

13. Loizos, K., Lazzi, G., Lauritzen, J.S., Anderson, J., Jones, B.W., Marc, R.: A multi-scale computational model for the study of retinal prosthetic stimulation. In: 2014 36th Annual International Conference of the IEEE Engineering in Medicine and Biology Society, pp. 6100–6103 (2014). https://doi.org/10.1109/EMBC.2014.6945021
14. McCreery, D.B., Agnew, W.F., Yuen, T.G., Bullara, L.: Charge density and charge per phase as cofactors in neural injury induced by electrical stimulation. IEEE Trans. Biomed. Eng. **37**(10), 996–1001 (1990)
15. McNeal, D.R.: Analysis of a model for excitation of myelinated nerve. IEEE Trans. Biomed. Eng. BME- **23**(4), 329–337 (1976)
16. Paknahad, J., Loizos, K., Humayun, M., Lazzi, G.: Responsiveness of retinal ganglion cells through frequency modulation of electrical stimulation: a computational modeling study*. In: 2020 42nd Annual International Conference of the IEEE Engineering in Medicine Biology Society (EMBC), pp. 3393–3398 (2020)
17. Paknahad, J., Loizos, K., Humayun, M., Lazzi, G.: Targeted stimulation of retinal ganglion cells in epiretinal prostheses: a multiscale computational study. IEEE Trans. Neural Syst. Rehabil. Eng. **28**(11), 2548–2556 (2020)
18. Paknahad, J., Loizos, K., Yue, L., Humayun, M.S., Lazzi, G.: Color and cellular selectivity of retinal ganglion cell subtypes through frequency modulation of electrical stimulation. Sci. Rep. **11**(1), 1–13 (2021)
19. RamRakhyani, A.K., Kagan, Z.B., Warren, D.J., Normann, R.A., Lazzi, G.: A μm-scale computational model of magnetic neural stimulation in multifascicular peripheral nerves. IEEE Trans. Biomed. Eng. **62**(12), 2837–2849 (2015)
20. Raspopovic, S., Capogrosso, M., Micera, S.: A computational model for the stimulation of rat sciatic nerve using a transverse intrafascicular multichannel electrode. IEEE Trans. Neural Syst. Rehabil. Eng. **19**(4), 333–344 (2011)
21. Sandell, J.H., Peters, A.: Effects of age on nerve fibers in the rhesus monkey optic nerve. J. Comp. Neurol. **429**(4), 541–553 (2001)
22. Shannon, R.V.: A model of safe levels for electrical stimulation. IEEE Trans. Biomed. Eng. **39**(4), 424–426 (1992)
23. Stang, J., et al.: Recent advances in computational and experimental bioelectromagnetics for neuroprosthetics. In: 2019 International Conference on Electromagnetics in Advanced Applications (ICEAA), pp. 1382–1382 (2019)
24. Stein, R.B., Peckham, P.H., Popović, D.: Neural prostheses: replacing motor function after disease or disability. Oxford University Press (1992)
25. Weiner, R.L.: The future of peripheral nerve neurostimulation. Neurol. Res. **22**(3), 299–304 (2000)
26. Yue, L., Weiland, J.D., Roska, B., Humayun, M.S.: Retinal stimulation strategies to restore vision: fundamentals and systems. Prog. Retinal Eye Res. **53**, 21–47 (2016)
27. Zaimi, A., Wabartha, M., Herman, V., Antonsanti, P.L., Perone, C.S., Cohen-Adad, J.: Axondeepseg: automatic axon and myelin segmentation from microscopy data using convolutional neural networks. Sci. Rep. **8**(1), 1–11 (2018)

# Statistical and Symbolic Neuroaesthetics Rules Extraction from EEG Signals

M. Coccagna[1], F. Manzella[2], S. Mazzacane[1], G. Pagliarini[2], and G. Sciavicco[2](✉)

[1] Department of CIAS Interdepartmental Research Center, University of Ferrara, Ferrara, Italy
[2] Department of Mathematics and Computer Science, University of Ferrara, Ferrara, Italy
guido.sciavicco@unife.it

**Abstract.** Neuroaesthetics, as defined by Zeki in 1999, is the scientific approach to the study of aesthetic perceptions of art, music, or any other experience that can give rise to aesthetic judgments. One way to understand the processes of neuroaesthetics is studying the electroencephalogram (EEG) signals that are recorded from subjects while they are exposed to some expression of art, and study how the differences among such signals correlate to the differences in their subjective judgments; typically, such studies are conducted on limited data with a purely statistical signal analysis. In this paper we consider a larger data set which was previously used in an experiment on beauty perception; we apply a novel machine learning-based data analysis methodology that allows us to extract symbolic *like/dislike* rules on the voltage at the most relevant frequencies from the most relevant electrodes. Our approach is not only novel in this particular area, but it is also reproducible and allows us to treat large quantities of data.

**Keywords:** Interpretable machine learning · Neural activity classification · Modal decision trees · Neural correlates

## 1 Introduction

*Neuroaesthetics* was defined by Zeki [21] as the scientific approach to the study of the perception of beauty in a broad sense, and later formalized as a field of study by Nalbantian [14]. The approaches to neuroaesthetics vary very much in the recent scientific literature, but they can be essentially divided into *top-down* processes, in which the essence of beauty undergoes an axiomatic treatment in which the subjective feeling is broken down in its constituting elements, and *bottom-up* ones, in which some kind of objective data is analyzed and related to the subjective expression of beauty. While the former, e.g. as in [20], is certainly fascinating from an epistemological point of view (i.e., it tries to answer the question of whether *beauty can be defined*), the latter has the advantage of being based on objective data, being systematic, and taking advantage from modern analysis techniques. For these reasons, bottom-up approaches are more common.


J. M. Ferrández Vicente et al. (Eds.): IWINAC 2022, LNCS 13258, pp. 536–546, 2022.
https://doi.org/10.1007/978-3-031-06242-1_53

Bottom-up strategies aimed to understand how the brain interprets the sense of beauty can be classified into (functional) MRI (*magnetic resonance*) image processing and interpretation, and EEG (*electroencephalogram*) signal processing. In the first case, the specific aim is to understand the which are the involved areas of the brain, how they are activated, and by what are they influenced. In [4], Di Dio, Macaluso, and Rizzolatti designed and carried on an experiment using functional MRI; by analyzing the data produced by 14 subjects, they concluded that the sense of beauty is mediated by two non-mutually exclusive processes: one based on a joint activation of sets of cortical neurons (objective beauty), the other based on the activation of the amygdala, driven by one's own emotional experiences (subjective beauty). Functional MRI was also used by Chatterjee et al. [1], in which the authors used a AI-based system to create artificial face stimuli in order to measure the response to beauty in the brains of 13 subjects. Ishizu and Zeki, in [8], considered functional MRI data from 21 subjects, and established that the experience of beauty is transversal to being stimulated by, for example, vision or hearing, at least in terms of functional brain area. Huang et al., on the other hand, asked the question of whether during the vision of art the experience of beauty is influenced by the apriori knowledge of an artwork being an original masterpiece or rather a copy [7], and their result seem to indicate that such a knowledge has zero or very low influence. Jacobs, Renken, and Cornelissen asked the question of whether the subjective sense of beauty is influenced by the internal state of the subject [9], and the concluded that, instead, there seems to be areas specifically devoted to aesthetic assessment, irrespective of the stimulus type. Kühn and Gallinat researched and investigated concurrence across 40 studies reporting brain regions which seem correlated with self-reported judgements of subjective pleasantness [10]. A few, more recent contributions focus specifically on the use of statistical/machine learning related techniques used to assess the problem of establishing the brain mechanisms that regulate the sense of beauty. Among these, we mention Reng and Geng's work [15], based on a learning process founded on labels' distributions among attractiveness rating, and Seresinhe, Prais, and Moat's contribution [18], based on using a deep learning strategy to understand and predict the sense of scenic of a subject looking at an outdoor image. When it comes to EEG signal processing, on the other hand, most work is quite recent. In [6] Hadjidimitriou and Hadjileontiadis used different feature extraction approaches and classifiers and focused on the discrimination between subjects' EEG responses to self-assessed liked or disliked music. In [2], Chew, Teo, and Mountstephens used neural networks to classify EEG signals recorded from subjects while viewing 3D shapes and expressing their level of appreciation. Finally, in [5], Guo et al. designed and executed an experiment to measure and classify EEG signals recorded while the subjects were exposed to 3D prototype led lamps, with the aim of predicting the subjects' sense of like/dislike. Li and Zhang's review [22] on computational neuroaesthetics lists and discusses most of this work.

At least two considerations can be drawn from reviewing the literature. First, most of the existing work on computational neuroaesthetics focuses on artificial, 2D images or 3D shapes, designed with experimental purposes, and not submerged into a realistic context. Second, EEG signals have been analyzed with

**Fig. 1.** Selected electrodes disposition

functional machine learning techniques that do not offer interpretability of the results, but only their statistical value, and do not allow, in general, the extraction of rules. The data used in this paper have been collected during a real-world experiment, analysing the explicit and implicit reactions of participants, using different kinds of sensors, during their visit to the exhibition *Painting affections: sacred painting in Ferrara between the '500 and the '700*, set up at the Estense Castle in Ferrara from the 26th of January to the 26th of December 2019 [12,16] (ISRCTN: 70216542). The EEG data are part of a larger bio-signals data set recorded to evaluate the participants' reactions during the observation of paintings, including ECG (*electrocardiogram*), EDA (*electrodermal activity*), two different tools for EEG recording, the *gaze pattern* during the observations, the participants' main features (age, gender, education, familiarity with art, etc.) and their explicit judgments about paintings. Here, we focus on a total of 248 trials (16 subjects, exposed up to 18 paintings each), and we analyze their EEG recording by means of a new machine learning technique, called *temporal random forests* [13,17], with the purpose of, first, establishing which electrodes the EEG signal carries the most information about the subjects' personal beauty experience, and, second, if symbolic rules can be extracted from the classification models (Fig. 1).

## 2 Data Origin, Description, and Preparation

EEG data have been recorded through two main tools, to analyse their affordability and easiness of use: a dry electrode EEG cap *Waveguard*$^{TM}$*touch* by

*ANT Neuro* in the 64-channel variant, as well as a *Myndplay* headsets, managed through *OpenVibe* software. In this paper, we focus on the recording from the cap only, and our data set consists of 16 people (10 males, 6 females, average age 32.19). For each subject a single long recording was made from the beginning to the end of the exhibition, therefore a slicing operation was necessary based on the timing the subject was standing in front of each painting to collect only the EEG data relative to a specific painting (about 60 s each). Being recorded at a sampling rate 512 Hz, each resulting data slice consisted of a time series of 30720 points. Each subject was asked to express a integer in the range [0, 50] as an expression of his/her *liking score* of the painting. Since not all pairs subject/painting were recorded, the resulting data set is composed by 248 instances, each instance being described by 64 time series (one for each electrode) of 30720 points and labeled by the related liking score.

The EEG data was processed by applying the Short Time Fourier Transform (STFT), using different parametrizations for the two phases: *best electrode selection* and *knowledge extraction*. For the former, the range of frequencies 0–256 Hz was explored, corresponding to the entire available spectrum for a sampling rate 512 Hz (i.e., the Nyquist frequency 256 Hz [19]). This spectrum was divided into 60 equally wide bands of frequencies ($F_1, \ldots, F_{60}$), with a resulting width of 4.267 Hz. The STFT time window size was set to 50 ms with a step time of 25 ms, and the resulting time series were 2665 points long. For the knowledge extraction phase a narrower frequency range was fixed (0–51.2 Hz), composed by five relevant wave patterns as suggested by the literature: $\delta$, ranging in 0–4 Hz (usually associated with slow-wave sleep), $\theta$, ranging in 4–7 Hz (usually associated with phase 1 and 2 of non-REM sleep and with REM sleep), $\alpha$, ranging in 7–13 Hz (usually associated with waking state with closed eyes and instants prior to fall asleep), $\beta$, ranging in 14–30 Hz (usually associated with intense mental activity), and $\gamma$, ranging in 30–49 Hz (usually associated with states of particular stress). After trying multiple configurations for the width of the STFT frequency window, the one providing the best results was ultimately found to be 3.01 Hz, which divides the whole frequency range into 17 bands ($\overline{F}_1, \ldots, \overline{F}_{17}$). Similarly, multiple settings for the time window parameter were explored, and we finally selected a size of 300 ms and a step time of 100 ms. The resulting time series were 638 points long. In the end, we found that the first quarter of each series, corresponding to the first 1 s, actually contains all relevant information, and therefore we limited the data set to the first 159 points, further reduced to 15 points by means of a moving average filter.

The scores among the data set were non-normally distributed ($p = 10^{-56}$, Kolmogorov-Smirnov test). For the purpose of learning, we treated the resulting problem as a classification problem, by discretizing the liking scores into three categories (*dislike*, *neutral*, *like*), with thresholds, respectively, of 16 and 33, both included, and, finally, by excluding the middle category (neutral). The resulting data set was composed by 81 *positive* instances (*like*) and 86 negative instances (*dislike*).

**Table 1.** Twenty-five statistical measures for time series (including 22 measures from [11]).

| Measure | Symbol |
|---|---|
| Mean | $M$ |
| Max | $MAX$ |
| Min | $MIN$ |
| Mode of $z$-scored distribution (5-bin histogram) | $Z5$ |
| Mode of $z$-scored distribution (10-bin histogram) | $Z10$ |
| Longest period of consecutive values above the mean | $C$ |
| Time intervals between successive extreme events above the mean | $A$ |
| Time intervals between successive extreme events below the mean | $B$ |
| First $1/e$ crossing of autocorrelation function | $FC$ |
| First minimum of autocorrelation function | $FM$ |
| Tot. power in lowest 1/5 of frequencies in the Fourier power spectrum | $TP$ |
| Centroid of the Fourier power spectrum | $CE$ |
| Mean error from rolling 3-sample mean forecasting | $ME$ |
| Time-reversibility statistic $\langle(x_{t+1-x_t})^3\rangle t$ | $TR$ |
| Automutual information ($m = 2$, $\tau = 5$) | $AI$ |
| First minimum of the automutual information function | $FMAI$ |
| Proportion of successive differences exceeding $0.04\sigma$ (Mietus 2002) | $PD$ |
| Longest period of successive incremental decreases | $LP$ |
| Entropy of two successive letters in equiprobable 3-letter symbolization | $EN$ |
| Change in correlation length after iterative differencing | $CC$ |
| Exponential fit to successive distances in 2-d embedding space | $EF$ |
| Ratio of slower timescale fluctuations that scale with DFA (50% sampling) | $FDFA$ |
| Ratio of slower timescale fluctuations that scale with linearly rescaled range fits | $FLF$ |
| Trace of covariance of transition matrix between symbols in 3-letter alphabet | $TC$ |
| Periodicity measure of (Wang et al. 2007) | $PM$ |

## 3 Statistical Analysis Phase

We modeled this problem as a multivariate temporal series classification problem. Naïve symbolic treatment of time series consists of simple feature extraction for further analysis; typical examples of interesting features are minimum, maximum, or average of a series. As suggested [3,11], elementary features can be combined with more elaborate ones, in order to derive a systematic, statistical treatment of pure temporal data. Prior to the knowledge extraction phase, we aimed to highlight which electrodes and which measures are more prone to have

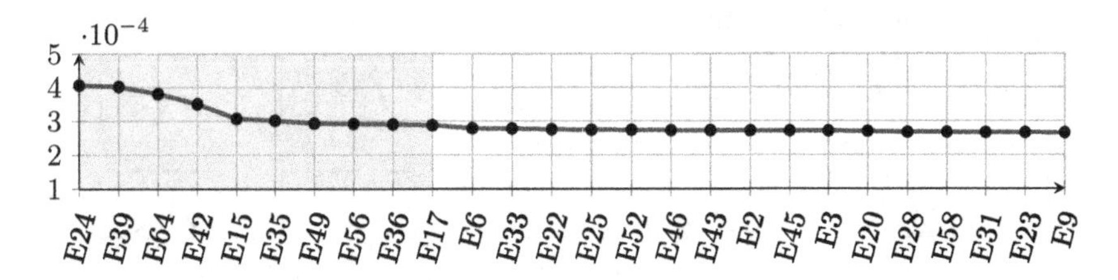


**Fig. 2.** Normalized mean variance of each electrode shown in descending order.

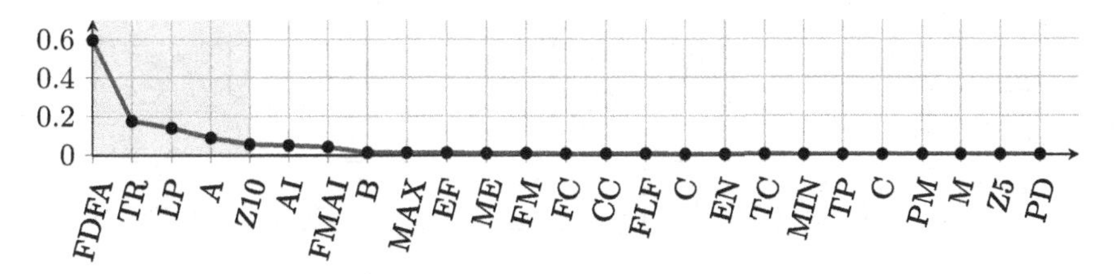


**Fig. 3.** Normalized mean variance of each measure shown in descending order.

a role in the problem. To this end, we performed a statistical analysis based on the assumption that, when coupled with a given measure, a frequency is informative if it displays a high variance across the data set. We proceeded as follows:

- We first considered the electrodes listed as $E_1, \ldots, E_{64}$ in Fig. 2, and we computed the variance of the value of each one of the measures in Table 1 for each one of the electrodes, on each one of the frequencies $F_1, \ldots, F_{60}$; then we aggregated the result by electrode by averaging the variances, and sorted the electrodes themselves. As a result, we obtained a list of electrodes that displayed, on average, the highest variance across the whole spectrum of frequencies. We proceeded by selecting the first 10 electrodes, which are the ones highlighted in Fig. 2.
- Then, we averaged the variance of each of the measures in Table 1 across all frequencies $F_1, \ldots, F_{60}$ and the 10 selected electrodes, to identify the 5 most informative ones, as in Fig. 3.

Each pair electrode/measure gives can be compared among the two classes (like,dislike) across each of the 60 frequencies band. An graphical example of such a comparison can be seen in Fig. 4, namely for $F_3$, and for the five most informative electrodes. As it can be noticed, the differences are subtle, which is an indication of the difficulty of the problem if approached with purely, univariate statistical methods.

## 4 Knowledge Extraction

Time series classification can be approached in several ways. Methods for classifying time series can be roughly divided into *symbolic* and *functional*. Symbolic

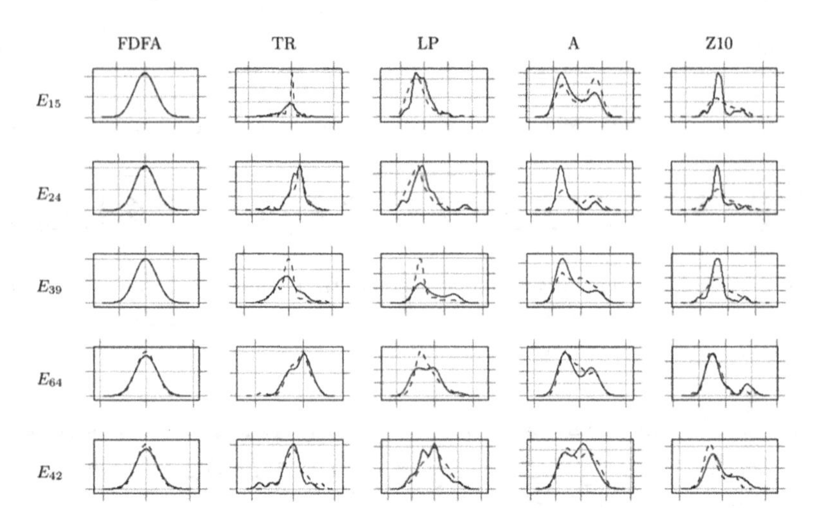


**Fig. 4.** Distribution of each of the selected variables by measure.

methods aim to extract a logical characterization of the classes in terms of the behaviour of the series, while functional ones approach the classification problem by extracting a mathematical function of the series. Time series classification methods can also be divided into *native* or *feature-based*. Native methods consider time series as they are, without performing any modification of the signals. Feature-based method, on the other hand, focus on extracting statistically interesting measures of the signals and use those for the classification phase. Feature-based methods are far more common, and they are both symbolic and functional; their major drawback is the lack of interpretability of the results in the functional case, and the low predictive capabilities in the symbolic one. Native methods are more scarce, and the most common ones among them, that is, distance-based methods, do not offer, in general, a real grasp of the underlying problem, despite their general good behaviour in terms of performances.

In [13,17], a new class of symbolic, native time series classification methods was proposed. Despite their short history, *temporal decision trees* showed a good compromise between interpretability and performances. The key points that define temporal decision trees are:

- They follow the general pattern and schema of conventional decision trees. Decisions are taken on a data set in order to maximize the amount of *information gain* in a greedy fashion, starting from the original training data set and obtaining, at each step, smaller, and more informative subsets. When the data set associated with a node is too small, or too pure in terms of class, it is converted into a leaf, and labeled with the majority class (generating, as in the classical case, a certain amount of misclassifications). Classical techniques, up to and including *pre- and post-pruning*, can be applied, at least in a limited form, to propositional and temporal decision trees alike.

| symbol | Allen's relation | graphical representation |
|---|---|---|
| $\langle A \rangle$ | $[x,y]R_A[z,t] \Leftrightarrow y = z$ | |
| $\langle L \rangle$ | $[x,y]R_L[z,t] \Leftrightarrow y < z$ | |
| $\langle B \rangle$ | $[x,y]R_B[z,t] \Leftrightarrow x = z, t < y$ | |
| $\langle E \rangle$ | $[x,y]R_E[z,t] \Leftrightarrow y = t, x < z$ | |
| $\langle D \rangle$ | $[x,y]R_D[z,t] \Leftrightarrow x < z, t < y$ | |
| $\langle O \rangle$ | $[x,y]R_O[z,t] \Leftrightarrow x < z < y < t$ | |

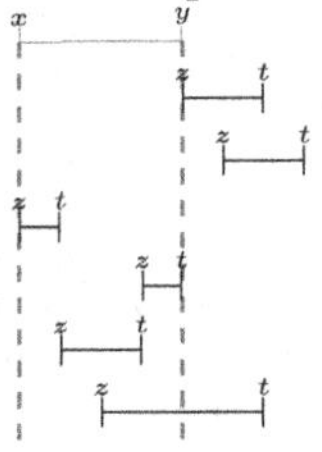

**Fig. 5.** Allen's interval relations and their notation in temporal decision trees.

- Unlike conventional decision trees, decisions are relativized to intervals on time series. So, while conventional decision trees treat times series by extracting features from them, and then taking decisions on such features, temporal decision trees take decisions directly on time series, in a native way. Consider, for example, the mean; while a conventional decision tree may separate the data set using the fact that the mean of a specific variable on the whole time period exceeds some value (e.g. *if the mean value of a variable is more than that value, then...*), a temporal decision tree may do so using the existence of an interval in which the mean of a specific variable exceeds some value (e.g. *if the mean value of a variable is more than that value between the instants $x$ and $y$, then...*).
- Like conventional decision trees, a temporal decision tree has a clear logical interpretation, but makes use of a more complex logic than propositional logic, which allows one to express properties over intervals, and their relations. There are thirteen relations between two intervals, known as *Allen's* relations (see Fig. 5, in which we show only the six *direct* relations of the type $\langle X \rangle$; their inverses, denoted with $\langle \overline{X} \rangle$, can be obtained by switching the roles of each interval, and the thirteenth, *equals*, is denoted $\langle = \rangle$), and a temporal decision tree is able to learn interval patterns which we can formalize using suitable symbols to denote Allen's relations (see Fig. 5, first column).

In [13,17] it was shown that temporal decision trees perform better than their propositional counterparts, and, while retaining a very high level of interpretability, are able to extract classification models that are comparable with those extracted by non-interpretable approaches.

We used temporal decision trees, in their *random forest* generalization, to build models based with the selected electrodes and the selected measures, using the frequency bands $\bar{F}_1, \ldots, \bar{F}_{17}$ (from $\delta$ to $\gamma$). We used forests with 100 trees, run in 10-fold cross validation mode, with four different configurations in terms of selected frequencies. The results of our experiment can be seen in Table 6. As it can be noticed, the best results are obtained with $\beta$ frequencies ($\bar{F}_6$ to $\bar{F}_{10}$), followed by $\gamma$ ($\bar{F}_{11}$ to $\bar{F}_{17}$), and then by the combination with all frequencies. The standard deviations across the 10 repetitions are not too high, indicating that the experiment is quite solid. Some of the seeds (e.g., seed 2) show very

| | accuracy | | | | | κ index | | | |
|---|---|---|---|---|---|---|---|---|---|
| seed | *all* | $\beta$ | $\gamma$ | $\beta+\gamma$ | seed | *all* | $\beta$ | $\gamma$ | $\beta+\gamma$ |
| 1 | 68.75 | 70.62 | 68.75 | 69.38 | 1 | 37.50 | 41.25 | 37.50 | 38.75 |
| 2 | 73.12 | 77.50 | 70.00 | 70.62 | 2 | 46.25 | 55.00 | 40.00 | 41.25 |
| 3 | 66.88 | 68.75 | 61.88 | 66.88 | 3 | 33.75 | 37.50 | 23.75 | 33.75 |
| 4 | 57.50 | 60.00 | 58.13 | 58.13 | 4 | 15.00 | 20.00 | 16.25 | 16.25 |
| 5 | 65.00 | 64.38 | 66.88 | 62.50 | 5 | 30.00 | 28.75 | 33.75 | 25.00 |
| 6 | 66.25 | 68.12 | 68.12 | 67.50 | 6 | 32.50 | 36.25 | 36.25 | 35.00 |
| 7 | 61.88 | 63.75 | 60.00 | 66.25 | 7 | 23.75 | 27.50 | 20.00 | 32.50 |
| 8 | 70.00 | 66.88 | 68.12 | 68.75 | 8 | 40.00 | 33.75 | 36.25 | 37.50 |
| 9 | 72.50 | 71.88 | 68.12 | 72.50 | 9 | 45.00 | 43.75 | 36.25 | 45.00 |
| 10 | 64.38 | 68.12 | 70.00 | 70.62 | 10 | 28.75 | 36.25 | 40.00 | 41.25 |
| avg | 66.63 | **68.00** | 67.31 | 66.00 | avg | 33.25 | 36.00 | 32.00 | 34.63 |
| std | 4.79 | 4.82 | 4.33 | 4.27 | std | 9.58 | 9.64 | 8.66 | 8.54 |

**Fig. 6.** Results.

high values (e.g., accuracy 77.5%). More important than the statistical value of our models are the rules that can be extracted from them. As we have recalled, these have the form of formulas in the logic of Allen's relations. Among the many possible rules that a model such as our encompasses, we have selected two examples for dislike ($d$) and like ($l$):

$$\varphi_d = [L](A(A_{20}) \geq -0.7 \rightarrow ([B](TR(A_{51} > -3.04 \rightarrow \langle\overline{A}\rangle(Z10(A_{80} < 0.66)))))$$
$$\varphi_l = [L]A(A_{122} > -0.71) \wedge [L]TR(A_6 < 3.65) \wedge [L](TR(A_{84} < 4.30))$$

where, in the first rule, which has an estimated validation confidence of 0.75, $A_{20}$ (resp., $A_{51}, A_{80}$) is the pair $E_{64}/\bar{F}_7$ (resp., $E_{35}/\bar{F}_{11}$, $E_{36}/\bar{F}_{13}$), and in the second rule, which has an estimated confidence of 0.8, $A_{122}$ (resp., $A_6, A_{84}$) is the pair $E_{56}/\bar{F}_3$ (resp., $E_2/\bar{F}_6$, $E_{15}/\bar{F}_{16}$).

## 5 Conclusions

In this paper we have applied a new methodology to extract knowledge from EEG signals, in order to study a neuroaesthetic problem. Our data consisted of the EEG recordings of subjects exposed to artistic paintings, and we analyzed them to extract rules that allows one to establish if the subject liked, or disliked, the painting. In a way, this work contributes to the more general neuroaesthetic problem of synthesizing a *theory of beauty*. While our results must be considered only preliminary, they constitute a first step towards using a new generation of knowledge extraction methods, by means of which one trades, in a limited way, some degree of performance of a prediction model (i.e., reliability of the prediction) to obtain, again in a limited way, explicit knowledge.

## References

1. Chatterjee, A., Thomas, A., Smith, S., Aguirre, G.: The neural response to facial attractiveness. Neuropsychology **23**(2), 135–143 (2009)

2. Chew, L., Teo, J., Mountstephens, J.: Aesthetic preference recognition of 3D shapes using EEG. Cogn. Neurodynam. **10**(2), 165–173 (2016)
3. Christ, M., Braun, N., Neuffer, J., Kempa-Liehr, A.: Time series feature extraction on basis of scalable hypothesis tests. Neurocomputing **307**, 72–77 (2018)
4. Dio, C.D., Macaluso, E., Rizzolatti, G.: The golden beauty: brain response to classical and renaissance sculptures. Plos One **11**, 1–9 (2007)
5. Guo, F., Li, M., Hu, M., Li, F., Lin, B.: Distinguishing and quantifying the visual aesthetics of a product: an integrated approach of eye-tracking and EEG. Int. J. Ind. Ergon. **71**, 47–56 (2019)
6. Hadjidimitriou, S., Hadjileontiadis, L.: Toward an EEG-based recognition of music liking using time-frequency analysis. IEEE Trans. Biomed. Eng. **59**(12), 3498–3510 (2012)
7. Huang, M., Bridge, H., Kemp, M., Parker, A.: Human cortical activity evoked by the assignment of authenticity when viewing works of art. Front. Hum. Neurosci. **5**, 1–20 (2011)
8. Ishizu, T., Zeki, S.: Toward a brain-based theory of beauty. Plos One **6**(7), 1–10 (2011)
9. Jacobs, R., Renken, R., Cornelissen, F.: Neural correlates of visual aesthetics - beauty as the coalescence of stimulus and internal state. Plos One **7**(2), 1–8 (2012)
10. Kühn, S., Gallinat, J.: The neural correlates of subjective pleasantness. Neuroimage **61**, 289–294 (2012)
11. Lubba, C., Sethi, S., Knaute, P., Schultz, S., Fulcher, B., Jones, N.: Catch22: canonical time-series characteristics - selected through highly comparative time-series analysis. Data Min. Knowl. Discov. **33**(6), 1821–1852 (2019)
12. Coccagna, M.: Environment and people perceptions: the experience of nevart, neuroestethics of the art vision. In: Proceeding of Global Challenges in Assistive Technology: Research, Policy & Practice, pp. 204–205 (2019)
13. Manzella, F., Pagliarini, G., Sciavicco, G., Stan, I.: Interval temporal random forests with an application to COVID-19 diagnosis. In: Proceedings of the 28th International Symposium on Temporal Representation and Reasoning. LIPIcs, vol. 206, pp. 7:1–7:18. Schloss Dagstuhl - Leibniz-Zentrum für Informatik (2021)
14. Nalbantian, S.: Neuroaesthetics: neuroscientific theory and illustration from the arts. Interdisc. Sci. Rev. **33**(4), 357–368 (2008)
15. Ren, Y., Geng, X.: Sense beauty by label distribution learning. In: Proceedings of the 26th International Joint Conference on Artificial Intelligence, pp. 2648–2654 (2017)
16. Mazzacane, S. .: Neuroaesthetics of art vision: an experimental approach to the sense of beauty. J. Clin. Trials **10**, 1–8 (2020)
17. Sciavicco, G., Stan, I.: Knowledge extraction with interval temporal logic decision trees. In: Proceedings of the 27th International Symposium on Temporal Representation and Reasoning. LIPIcs, vol. 178, pp. 9:1–9:16. Schloss Dagstuhl - Leibniz-Zentrum für Informatik (2020)
18. Seresinhe, C., Preis, T., Moat, H.: Using deep learning to quantify the beauty of outdoor places. R. Soc. Open Sci. **4**(7), 1–14 (2017)
19. Shannon, C.: Communication in the presence of noise. Proc. IRE **37**(1), 10–21 (1949)
20. Sidhu, D., McDougall, K., Jalava, S., Bodner, G.: Prediction of beauty and liking ratings for abstract and representational paintings using subjective and objective measures. Plos One **13**(7), 1–15 (2018)

21. Zeki, S.: Inner Vision: An Exploration of Art and the Brain. Oxford University Press (1999)
22. Li, R., Zhang, J.: Review of computational neuroaesthetics: bridging the gap between neuroaesthetics and computer science. Brain Inform. **7**(1), 1–17 (2020). https://doi.org/10.1186/s40708-020-00118-w

# Brain Shape Correspondence Analysis Using Variational Mixtures for Gaussian Process Latent Variable Models

Juan P. V. Minoli[1,2](✉), Álvaro A. Orozco[1], Gloria L. Porras-Hurtado[2], and Hernán F. García[2]

[1] Automatics Research Group, Universidad Tecnológica de Pereira, Pereira, Colombia
juanpablo.velasquez1@utp.edu.co
[2] Salud Comfamiliar Research Group, Comfamiliar Risaralda, Pereira, Colombia

**Abstract.** Analyzing brain structures in the medical imaging field poses challenging problems due to neurological diseases' heterogeneity. Besides, measuring brain changes quantitatively in neurodevelopmental is crucial to evaluate clinical outcomes correctly. From a computer-vision perspective, establishing correspondences between shapes often requires computing similarity measures that, in most cases, are unavailable. This paper proposes an unsupervised probabilistic framework for shape correspondence analysis on brain structures by using variational unsupervised learning. The probabilistic framework comprehensively captures the form of brain shapes from surface descriptors. Then, we learned clustered latent space representations of surface descriptors by using mixtures distributions for Gaussian process latent variable models to avoid computing similarity measures, which classify the resulting latent vectors to establish group-wise correspondences. The experimental results show how the proposed model captures non-linearities in non-rigid 3D shapes even when they present occlusion or partialities. These results demonstrated that the proposed model is suitable for shape correspondence analysis.

**Keywords:** Brain modelling · Shape correspondence · Variational inference · Gaussian process latent variable models

## 1 Introduction

Magnetic resonance imaging (MRI) is a prevalent technique frequently used to investigate possible abnormalities in different human tissues, such as the brain. Lately, this medical study has become one of the preferred tools for specialists as it allows to produce high-resolution images with different perspectives in both 2D and 3D, MR devices use potent magnets and radiofrequency pulses instead of ionizing radiation [13].

Magnetic resonance problems lead to several approaches based on signal processing and data analysis. Some of the early studies generally examined a small set of brain images and, as a result, did not yield generalized solutions that could


J. M. Ferrández Vicente et al. (Eds.): IWINAC 2022, LNCS 13258, pp. 547–556, 2022.
https://doi.org/10.1007/978-3-031-06242-1_54

work with different groups of subjects. Also, the principal component analysis (PCA) method was employed on a small set of images to achieve reasonably high accuracy rates and improve recognition systems' performance, thus increasing the number of samples with which it was possible to work. For the classification part, support vector machines (SVM) commonly achieve high hit rates in different tasks, but the increased changes in brain structures do not allow a reasonable hit rate. MRI analysis brings with it a massive amount of vertices due to the different slices and perspectives it can capture; therefore, analyzing certain types of complex structures tends to be a computationally challenging task [17].

Different medical treatments need the analysis of complex volumetric structures challenging to see for the naked eye. Then, this leads to a lousy performance in treatments for other diseases such as Alzheimer's or Parkinson's and infectious diseases such as meningitis and multiple sclerosis. Although this task is of vital importance, it is still performed manually due to the complexity of automatically determining specific areas. Because of this, generally, the specialists in the area only have a previously labeled atlas of the brain to compare the magnetic resonance images (MRI) [9].

Shape correspondence methodologies help to quantify additional information about shapes, such as volume changes, shape comparison, and texture transfer, to name a few. Most matching methods for medical imaging problems focus on computing multiple similarity metrics with prior information to build an approximation. Such as geodesic contours and bag-of-words [14]. Point-wise is one of them, which find closer neighbors by solving a linear assignment problem to obtain correspondences [12]. Also, pair-wise methods, which usually become computationally expensive where their more classic formulation takes the form of a QAP (quadratic assignment problem) [10]. However, these approaches only work for objects of the same size and with prior information, which leads to poor accuracy in shape correspondences processes in non-rigid and variants brains.

Among the critical aspects of the research problem is the lack of a complete and robust methodology capable of capturing correspondences on 3D shapes related to non-rigid brain structures. Computing correspondences between shapes using methods on state-of-the-art still presents various problems, such as computational intractability, lack of robustness to non-linearities, and the need for larger datasets. This paper introduces an unsupervised methodology based on Warped Mixture Models using variational inference. That captures group-wise correspondences from latent representations of 3D shape descriptors. In particular, our contribution is to provide a new model for shape correspondence analysis of non-rigid shapes associated with brain structures in the context of neurodegenerative diseases.

## 2 Materials and Methods

### 2.1 Normalized Geodesic Error for Evaluation Measure

Evaluation of the correspondence quality, we refer to the Princeton benchmark protocol [11]. To evaluate the accuracy of a predicted map, $f : \mathcal{M}_1 \rightarrow \mathcal{M}_2$

with respect to a ground-truth[1] map, $f_{true} : \mathcal{M}_1 \rightarrow \mathcal{M}_2$. Where, $\mathcal{M}_1$ and $\mathcal{M}_2$ correspond with shapes that we want to match. We compute for every point $\mathbf{y}$, the geodesic distance, $d_{\mathcal{M}_2}(f(\mathbf{y}), f_{\text{true}}(\mathbf{y}))$ between its predicted map, $f(\mathbf{y})$, and its true correspondence, $f_{true}(\mathbf{y})$. We calculate these geodesic distances into an error measure in Eq. 1.

$$\operatorname{Err}(f, f_{\text{true}}) = \sum_{\mathbf{y} \in \mathcal{M}_1} d_{\mathcal{M}_2}(f(\mathbf{y}), f_{\text{true}}(\mathbf{y})), \tag{1}$$

where $d_{\mathcal{M}_2}(f(\mathbf{y}), f_{\text{true}}(\mathbf{y}))$ is normalized by $\sqrt{\operatorname{Area}(\mathcal{M}_2)}$. We also generate plots to examine the distributions of errors, where the $x$-axis represents a varying geodesic distance threshold[2], $\varepsilon$, and the $y$-axis shows the average percentage of points for which $d_{\mathcal{M}_2}(f(\mathbf{y}), f_{\text{true}}(\mathbf{y})) < \varepsilon$.

### 2.2 Scale-invariant Heat Kernel Signature (SI-HKS)

To model complex shapes, we need to capture some surface descriptors. Here, we model the brain structures as Riemannian manifolds with heat properties [4]. Heat shape descriptor maintains invariance under a different class of transformations. That is helpful for shape analysis because non-rigid shapes have many different sizes but may share similar forms. The SI-HKS is a scale-invariant version of the heat kernel descriptor based on diffusion scale-space analysis to describe a given shape. Let us define $\mathbf{Y} = \{\mathbf{y}_n\}_{n=1}^{N}$, where $\mathbf{y}_n$ is a 3D point. The solution of the heat differential equation, describes the amount of heat on a given shape at point $\mathbf{y}$ in time $t$. This solution is called the heat kernel and is denoted by $K_{\mathbf{Y},t}(\mathbf{y}, \mathbf{y}')$. Thus, the heat kernel signature can be described as Eq. 2.

$$h(\mathbf{y}, t) = K_{\mathbf{Y},t}(\mathbf{y}, \mathbf{y}') = \sum_{i=0}^{\infty} e^{-\lambda_i t} \varphi_i^2(\mathbf{y}), \tag{2}$$

where $\lambda_i$ are eigenvalues, and $\varphi_i$ are the corresponding eigenfunctions of the Laplace-Beltrami operator, satisfying $\Delta_{\mathbf{Y}} \cdot \varphi_i = \lambda_i \varphi_i$. Then, the solution is transformed to a logarithmic scale and taken to a Fourier space as shown in the equations to permit a scale-invariant version.

### 2.3 Mixtures for Gaussian Process Latent Variable Models

The Fig. 1 shows a graphical representation for the model we proposed. This model is a variational extension from Warped Mixture Models (WMM). Also, some priors were eliminated to facilitate the inference. Our model can be more scalable than WMM because it uses a variational approximation of the posterior densities instead of sample techniques. This is a clear vantage in shape correspondence analysis such that the shapes usually have a huge number of vertices.

[1] Real correspondence label in both shapes.

[2] Different levels of geodesic error in which it is evaluated what percentage of matches are located on it.

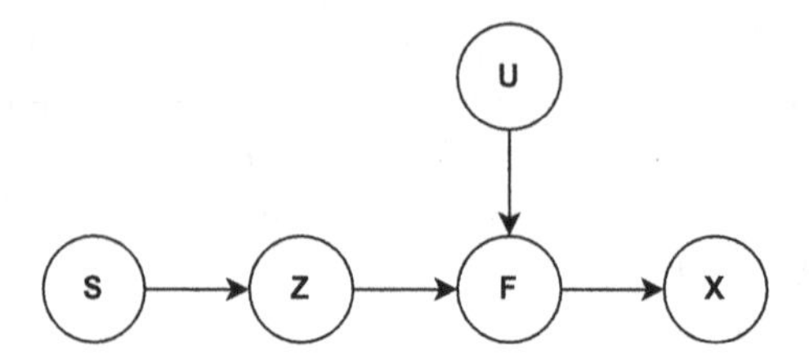


**Fig. 1.** Graphical representation in proposed model.

This model is a mixture representation between a Gaussian Process Latent Variable Model (GPLVM) and Gaussian Mixture Models (GMM). The Eq. (3) shows the join distribution from the proposed model.

$$p(\mathbf{X}, \mathbf{F}, \mathbf{U}, \mathbf{Z}, \mathbf{S} \mid \mathbf{Z}_u) = p(\mathbf{X} \mid \mathbf{F}) p(\mathbf{F} \mid \mathbf{U}, \mathbf{Z}, \mathbf{Z}_u) p(\mathbf{U} \mid \mathbf{Z}_u) p(\mathbf{Z} \mid \mathbf{S}) p(\mathbf{S}), \quad (3)$$

where $\mathbf{X} \in \mathbb{R}^{N \times p}$ is the observed space (with columns $\{\mathbf{x}_{:,j}\}_{j=1}^{p}$ ) (SI-HKS), $\mathbf{F} \in \mathbb{R}^{N \times p}$ GP non-linear functions (with columns $\{\mathbf{f}_{:,j}\}_{j=1}^{p}$ ), $\mathbf{U} \in \mathbb{R}^{M \times p}$ GP non-linear functions over inducing points (with columns $\{\mathbf{u}_{:,j}\}_{j=1}^{p}$ ), $\mathbf{Z} \in \mathbb{R}^{N \times q}$ latent space, $\mathbf{S} \in \mathbb{R}^{N \times K}$ is the cluster responsability using a $1 - of - K$ representation and $\mathbf{Z_u} \in \mathbb{R}^{M \times q}$ are the inducing points. We define then the specific priors over the GP like [7].

$$p\left(\mathbf{X} \mid \mathbf{F}, \beta^{-1}\right) = \prod_{j=1}^{p} \mathcal{N}\left(\mathbf{x}_{:,j} \mid \mathbf{f}_{:,j}, \beta^{-1}\mathbf{I}_n\right), \quad (4)$$

$$p\left(\mathbf{f}_{:,j} \mid \mathbf{u}_{:,j}, \mathbf{Z}, \mathbf{Z}_u\right) = \mathcal{N}\left(\mathbf{f}_{:,j} \mid \mathbf{a}_j, \mathbf{\Sigma}_f\right), \quad (5)$$

is the conditional GP prior, with $\mathbf{a}_j = \mathbf{K}_{fu}\mathbf{K}_{uu}^{-1}\mathbf{u}_{:,j}$ and $\mathbf{\Sigma}_f = \mathbf{K}_{ff} - \mathbf{K}_{fu}\mathbf{K}_{uu}^{-1}\mathbf{K}_{uf}$. We used subindex $f$ to denote covariance function over the input parameters. Further,

$$p\left(\mathbf{u}_{:,j} \mid \mathbf{Z}_u\right) = \mathcal{N}\left(\mathbf{u}_{:,j} \mid \mathbf{0}, \mathbf{K}_{uu}\right), \quad (6)$$

is the marginal GP prior over the inducing variables. In the above expressions, $\mathbf{K}_{uu}$ denotes the covariance matrix constructed by evaluating the covariance function on the inducing points, $\mathbf{K}_{uf}$ is the cross-covariance between the inducing and the latent points and $\mathbf{K}_{fu} = \mathbf{K}_{uf}^{T}$. Finally, the priors over the latent space and responsibilities are defined in Eq. (7).

$$p(\mathbf{Z} \mid \mathbf{S}) = \prod_{n=1}^{N} \prod_{k=1}^{K} \mathcal{N}\left(\mathbf{z}_n \mid \boldsymbol{\mu}_k, \mathbf{R}_k\right)^{s_{nk}}, p(\mathbf{S}) = \prod_{n=1}^{N} \prod_{k=1}^{K} \pi_k^{s_{nk}}, \quad (7)$$

where, $\boldsymbol{\mu}_k$, $\mathbf{R}_k$ and $\pi_k$ are the mean, covariance matrix and mixing coefficient of each Gaussian distribution.

### 2.4 Variational Inference

In this section we will show the variational approximations of posterior distribution and the Lower Bound like optimization objective function. This model not permits the use of common techniques like mean-field because this leads to intractable calculations [7]. For that reason we present variational posterior distributions which behavior is similar to the proposed priors. We can now apply variational inference to approximate the true posterior, $p(\mathbf{F}, \mathbf{U}, \mathbf{Z}, \mathbf{S} \mid \mathbf{X})$ with a variational distribution of the form,

$$q(\mathbf{F}, \mathbf{U}, \mathbf{Z}, \mathbf{S}) = p(\mathbf{F} \mid \mathbf{U}, \mathbf{Z})q(\mathbf{U})q(\mathbf{Z} \mid \mathbf{S})q(\mathbf{S}). \tag{8}$$

The distribution $p(\mathbf{F} \mid \mathbf{U}, \mathbf{Z})$ does not have a variational representation $q$ because it makes some mathematical derivations easier.

$$q(\mathbf{Z} \mid \mathbf{S}) = \prod_{n=1}^{N}\prod_{k=1}^{K} \mathcal{N}\left(\mathbf{z}_n \mid \boldsymbol{\nu}_{nk}, \boldsymbol{\Lambda}_{nk}\right)^{s_{nk}}, q(\mathbf{S}) = \prod_{n=1}^{N}\prod_{k=1}^{K} \gamma_{nk}^{s_{nk}}, \tag{9}$$

where $\boldsymbol{\Lambda}$ is a diagonal covariance matrix and $\sum_{k=1}^{K} \gamma_{nk} = 1$. The marginal variational distribution over $\mathbf{z}$ is.

$$q(\mathbf{Z}) = \sum_{\mathbf{S}} q(\mathbf{Z} \mid \mathbf{S})q(\mathbf{S}) = \prod_{n=1}^{N}\sum_{k=1}^{K} \gamma_{nk}\mathcal{N}\left(\mathbf{z}_n \mid \boldsymbol{\nu}_{nk}, \boldsymbol{\Lambda}_{nk}\right). \tag{10}$$

### 2.5 The Evidence Lower Bound (ELBO)

In this section we show the closed-form solution of the ELBO that we found based on the proposed priors and variational posteriors. The ELBO is defined as.

$$\begin{aligned} \mathcal{F}(q(\mathbf{Z} \mid \mathbf{S}), q(\mathbf{U}), q(\mathbf{S})) &= \int q(\mathbf{F}, \mathbf{U}, \mathbf{Z}, \mathbf{S}) \log \frac{p(\mathbf{X}, \mathbf{F}, \mathbf{U}, \mathbf{Z}, \mathbf{S})}{q(\mathbf{F}, \mathbf{U}, \mathbf{Z}, \mathbf{S})} \mathrm{d}\mathbf{S}\mathrm{d}\mathbf{Z}\mathrm{d}\mathbf{F}\mathrm{d}\mathbf{U} \\ &= \hat{\mathcal{F}}(q(\mathbf{Z}), q(\mathbf{U})) - \langle \mathrm{KL}(q(\mathbf{Z} \mid \mathbf{S}) \| p(\mathbf{Z} \mid \mathbf{S})) \rangle_{q(\mathbf{S})} - \mathrm{KL}(q(\mathbf{S}) \| p(\mathbf{S})). \end{aligned} \tag{11}$$

where $\langle \cdot \rangle$ is a shorthand for expectation. Next, we will show the solution for the final three terms in Eq. (11).

**First Term.** The Eq. (12) shows the analitical approximation for the first term, in this case we optimally calculate the value for the distribution $q(\mathbf{U})$; eliminating its presence in the result Eq. like [7].

$$\begin{aligned} \hat{\mathcal{F}}_j(q(\mathbf{Z})) &= \frac{N}{2}\log\beta + \frac{1}{2}\log|\mathbf{K}_{uu}| - \frac{1}{2}\mathbf{y}_{:,j}^{\top}\mathbf{W}\mathbf{y}_{:,j} - \frac{N}{2}\log 2\pi \\ &- \frac{1}{2}\log\left[|\beta\boldsymbol{\Psi}_2 + \mathbf{K}_{uu}|\right] - \frac{\beta\psi_0}{2} + \frac{\beta}{2}\operatorname{tr}\left(\mathbf{K}_{uu}^{-1}\boldsymbol{\Psi}_2\right), \end{aligned} \tag{12}$$

where $\mathbf{W} = \beta \mathbf{I}_n - \beta^2 \boldsymbol{\Psi}_1 \left(\beta \boldsymbol{\Psi}_2 + \mathbf{K}_{uu}\right)^{-1} \boldsymbol{\Psi}_1^\top$ and the additional quantities are defined as

$$\psi_0 = \operatorname{tr}\left(\langle \mathbf{K}_{ff} \rangle_{q(\mathbf{Z})}\right), \quad \Psi_1 = \langle \mathbf{K}_{fu} \rangle_{q(\mathbf{Z})}, \quad \Psi_2 = \langle \mathbf{K}_{uf} \mathbf{K}_{fu} \rangle_{q(\mathbf{Z})}. \tag{13}$$

The above quantities are kernel convolutions with a closed-form RBF solution.

**Second Term.** The second term is an expectation over a kullback-leibler divergence $\langle \mathrm{KL}(q(\mathbf{Z} \mid \mathbf{S}) \| p(\mathbf{Z} \mid \mathbf{S})) \rangle_{q(\mathbf{S})}$ and could be express like Eq. (14)

$$\sum_{n=1}^{N} \sum_{k=1}^{K} \frac{\gamma_{nk}}{2} \left( \log \frac{|\mathbf{R}_k|}{|\boldsymbol{\Lambda}_{nk}|} - q + \operatorname{tr}(\mathbf{R}_k^{-1} \boldsymbol{\Lambda}_{nk}) + (\boldsymbol{\nu}_{nk} - \boldsymbol{\mu}_k) \mathbf{R}_k^{-1} (\boldsymbol{\nu}_{nk} - \boldsymbol{\mu}_k)^\intercal \right). \tag{14}$$

**Third Term.** The third term is a Kullback-Leibler divergence over multinomial distributions, which solution is defined in (15).

$$\mathrm{KL}(q(\mathbf{S}) \| p(\mathbf{S})) = \sum_{n=1}^{N} \sum_{k=1}^{K} \gamma_{nk} \left( \log \gamma_{nk} - \log \pi_k \right). \tag{15}$$

# 3 Results

We test our approach in three different datasets like Tosca non-rigid world [3], SHREC'16 [6] and Neonatal Brain Atlas [8]. We evaluated the overall performance of our technique regarding the Princeton benchmark protocol to quantify the percentage of correspondences in different shapes.

## 3.1 Tosca Non-rigid World Dataset

We train our technique on the TOSCA dataset. This dataset incorporates a set of three-dimensional non-rigid shapes in many different poses for correspondence experiments. We need to use the triangular faces and a listing of vertex $XYZ$ coordinates as our enter data to calculate the SI-HKS. We show the results of the training model in different examples in the shape correspondence task in Fig. 2 with the latent space generated for the model. The graphical results exhibit that our system can find correspondences in non-rigid shapes.

We evaluate the performance of the training model in shape correspondences tasks using the Princeton benchmark protocol as shown in Sect. 2.1. The Fig. 3 shows a comparison between state-of-art methods like, [19] with symmetric maps, [11] with symmetric maps and, functional maps [15] with symmetric maps against ours. Finally, we also report the geodesic error when the 90% of correspondences are reached in Table 1. And, we compare the error with another methods in state-of-art such as product manifold filter (PMF) [18], Blended intrinsic maps (BIM) [11], spectral generalized multidimensional scaling (SGMDS) [1], functional maps (FM) [15], and Random Forest (RF)[16].

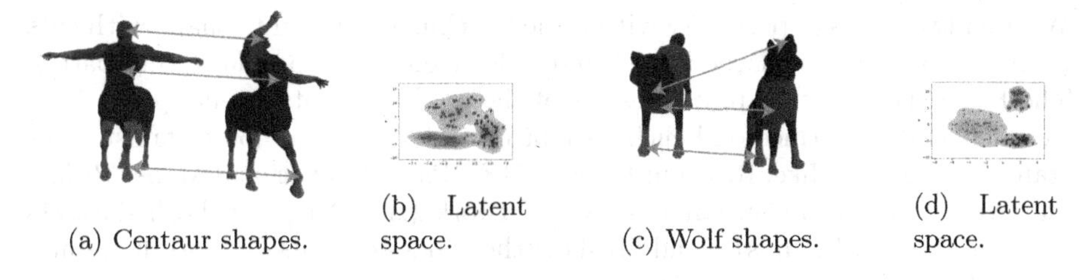

(a) Centaur shapes. (b) Latent space. (c) Wolf shapes. (d) Latent space.

**Fig. 2.** Tosca non-rigid world dataset correspondence

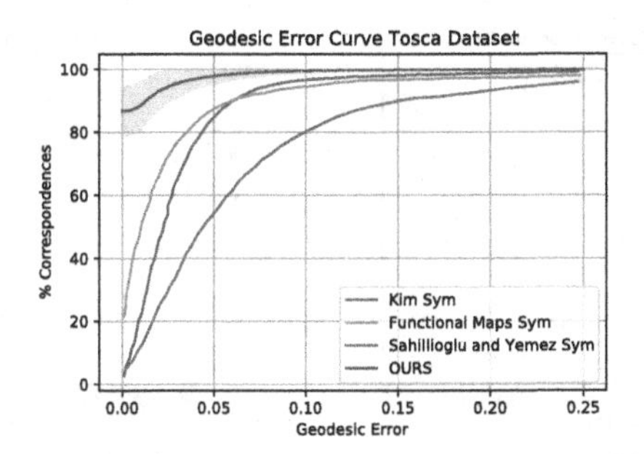


**Fig. 3.** Princeton bechmark protocol for shape correspondence analysis

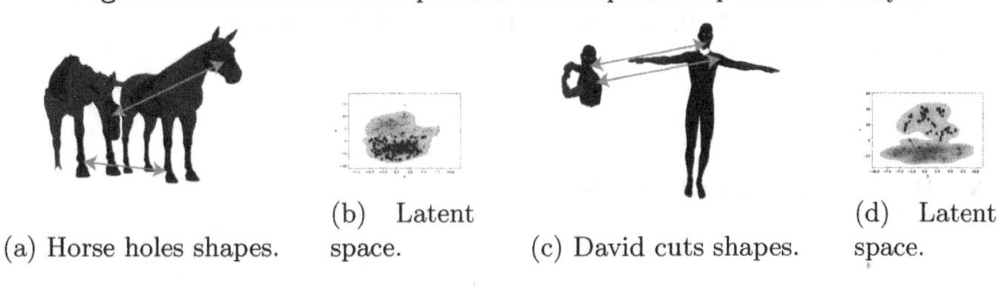

(a) Horse holes shapes. (b) Latent space. (c) David cuts shapes. (d) Latent space.

**Fig. 4.** SHREC'16 dataset correspondence

### 3.2 SHREC'16 Dataset

Since different brain diseases can destroy the brain matter [5], it is necessary to test the performance of our model against non-rigid shapes with a parcellation.

**Table 1.** Geodesic error for which 90% of correspondences are reached.

| Approach | Geodesic error |
|---|---|
| PMF | 0.030 |
| BIM | 0.060 |
| SGMDS | 0.040 |
| FM | 0.065 |
| RF | 0.045 |
| **Ours** | **0.0132** |

We used two forms of partiality within the benchmark related to shape with cuts (removing a couple of large parts) and holes (removing the many tiny parts). The Fig. 4 shows the graphical results of the model with its latent space. Also, we calculated the Princeton benchmark protocol for cuts and holes against other state-of-art models like, Random Forests (RF), Partial Functional Maps (PFM), and Anisotropic Convolutional Neural Network (ACNN) [2] in both datasets (holes and cuts). Figures 5a and 5b show the geodesic errors for both holes and cuts datasets.

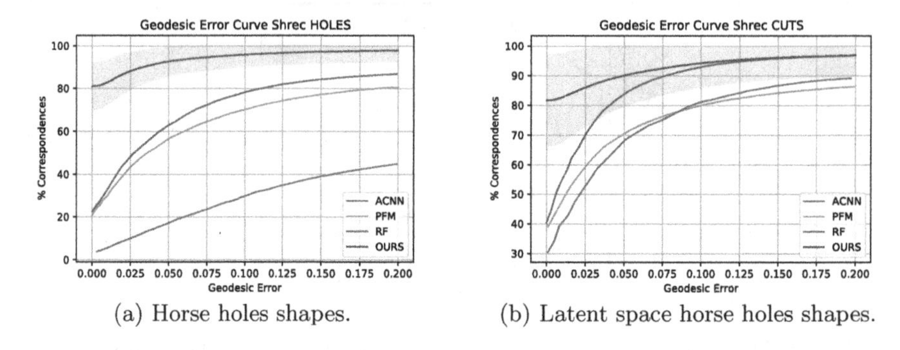


(a) Horse holes shapes. (b) Latent space horse holes shapes.

**Fig. 5.** Princeton benchmark for holes and cuts.

### 3.3 Brain Structure Dataset

(a) Brain in shape correspondence. (b) Latent space brains.

**Fig. 6.** SHREC'16 dataset correspondence

We use a neurodevelopmental dataset called BrainAsphixyaCOMF developed within the funded project by MINCIENCIAS with code 497984467090, and the Neonatal Brain Atlas [8], to compare the differences between a healthy and a sick brain. The Fig. 6 shows the correspondence analysis between a healthy and an unhealthy brain. This image also exhibits the latent space constructed by the model.

## 4 Conclusions

We present a Gaussian Process Unsupervised Model for shape correspondence analysis between non-rigid shapes. Here, we used Scale-Invariant Heat Kernel Signatures to describe the curvature in the non-rigid surface. We then propose a new model based mainly on Gaussian Process strategies to construct clustering assignments in a latent space. The variational inference was the strategy we chose to calculate the parameters in the model. Because it makes the model more scalable and reliable to establish meaningful relations between shapes, that is especially needed in shape correspondence problems. Also, the experimental results with different datasets showed that our approach establishes shape correspondences in different 3D datasets. The model matches non-rigid and partial surfaces more accurately than other correspondence methods. In addition, our methodology provides a shape correspondence strategy that generates a latent space that could present additional information about input data. As future works, we plan to extend this version to a fully-Bayesian approximation that allows a better inference of each parameter.

**Acknowledgements.** This research was developed under the project: "DESARROLLO DE UN SISTEMA AUTOMÁTICO DE ANÁLISIS DE VOLUMETRÍA CEREBRAL COMO APOYO EN LA EVALUACIÓN CLÍNICA DE RECIÉN NACIDOS CON ASFIXIA PERINATAL" financed by MINCIENCIAS with code COL497984467090.

## References

1. Aflalo, Y., Dubrovina, A., Kimmel., R.: Spectral generalized multidimensional scaling. Int. J. Comput. Vis. **118**(3), 380–392 (2016)
2. Boscaini, D., Masci, J., Rodolá, E., Bronstein, M.M.: Learning shape correspondence with anisotropic convolutional neural networks. arXiv, Cornell University (2016)
3. Bronstein, A.M., Bronstein, M.M., Kimmel, R.: Calculus of nonrigid surfaces for geometry and texture manipulation. IEEE Trans. Vis. Comput. Graph **13**(5), 902–913 (2007)
4. Bronstein, M.M., Kokkinos, I.: Scale-invariant heat kernel signatures for non-rigid shape recognition. In: Proceedings of CVPR, pp. 1704–1711 (2010)
5. Chouvatut, V., Boonchieng, E.: Brain tumor's approximate correspondence and area with interior holes filled. In: 2017 14th International Joint Conference on Computer Science and Software Engineering (JCSSE), pp. 1–5 (2017)
6. Cosmo, L., Rodolá, E., Bronstein, M.M., Torsello, A., Cremers, D., Sahillioglu, Y.: Shrec'16: Partial matching of deformable shapes. In: Eurographics Workshop on 3D Object Retrieval (2016)
7. Damianou, A.C., Titsias, M.K., Lawrence, N.D.: Variational inference for latent variables and uncertain inputs in gaussian processes. J. Mach. Learn. Res. (2016)
8. Gousias, I.S., et al.: Magnetic resonance imaging of the newborn brain: manual segmentation of labelled atlases in term-born and preterm infants. NeuroImage **62**(3), 1499–1509 (2012)

9. Katayama, Y., Oshima, H., Kano, T., Kobayashi, K., Fukaya, C., Yamamoto, T.: Direct effect of subthalamic nucleus stimulation on levodopa-induced peak-dose dyskinesia in patients with parkinson's disease. Stereotact. Funct. Neurosurg. **84**, 176–179 (2006)
10. Kezurer, I., Kovalsky, S.Z., Basri, R., Lipman, Y.: Tight relaxation of quadratic matching. In: Computer Graphics Forum, vol. 34(5), 115–128 (2015)
11. Kim, V.G., Lipman, Y., Funkhouser, T.A.: Blended intrinsic maps. TOG **30**(4), 1–12 (2011)
12. Lahner, Z., et al.: Efficient deformable shape correspondence via kernel matching. arXiv, Cornell University (2017)
13. Legaz-Aparicio, A.-G., et al.: Efficient variational approach to multimodal registration of anatomical and functional intra-patient tumorous brain data. Int. J. Neural Syst. **27**(6), 1750014 (2017)
14. Liang, L., Szymczak, A., Mingqiang, W.: Geodesic spin contour for partial near-isometric matching. Comput. Graph. **146**, 156–171 (2015)
15. Ovsjanikov, M., Ben-Chen, M., Solomon, J., Butscher, A., Guibas., L.: Functional maps: a flexible representation of maps between shapes. ACM Trans. Graph. **31**(4), 1–11 (2012)
16. Rodolá, E., Bulò, S.R., Windheuser, T., Vestner, M., Cremers., D.: Dense non-rigid shape correspondence using random forests. In: Proceedings of the 2014 IEEE Conference on Computer Vision and Pattern Recognition, CVPR 2014, pp. 4177–4184 (2014)
17. Talo, M., Yildirim, O., Baloglu, U.B., Aydin, G., Acharya, U.R.: Convolutional neural networks for multi-class brain disease detection using MRI images. Comput. Med. Imaging Graph. **78**, 101673 (2019)
18. Vestner, M., et al.: Efficient deformable shape correspondence via kernel matching. In: 2017 International Conference on 3D Vision (3DV), pp. 517–526 (2017)
19. Sahillioglu, Y., Yemez, Y.: Coarse-to-fine combinatorial matching for dense isometric shape correspondence. Comput. Graph. Forum **30**(5), 1461–1470 (2011)

# Explainable Artificial Intelligence to Detect Breast Cancer: A Qualitative Case-Based Visual Interpretability Approach

M. Rodriguez-Sampaio(✉), M. Rincón, S. Valladares-Rodriguez, and M. Bachiller-Mayoral

Artificial Intelligence Department, UNED, Madrid, Spain
soniavr@dia.uned.es
http://www.ia.uned.es/

**Abstract.** Nowadays, research in the field of artificial intelligence is focusing on the explainability of the developed algorithms, mainly neural networks. This trend is known as XAI and brings certain advantages such as increased confidence in the decision-making process, improved capacity for error analysis, verification of results and possibility of model refinement, among others. In this work we have focused on interpreting the predictions of recently developed deep learning models through different visualization techniques. The use case we introduce is the detection of breast cancer through the classification of mammographies, since the medical field is widely benefited by the contributions of XAI methods. Furthermore, the target neural networks are based on recent and poorly explored architectures: EfficientNet, designed to improve the performance of convolutional networks.

**Keywords:** Explainable Artificial Intelligence · Interpretability · Deep learning · Mammography · Breast cancer detection

## 1 Introduction

The popularity of deep learning continues to rise: new architectures are being designed, novel uses of existing ones are revealed or they are being modified with new computational mechanisms. The literature describes [1] advances and possible lines of future work. However, for many people, these models remain black boxes, making it difficult to use them in real scenarios beyond experimental assumptions. This situation creates the need to develop techniques for explaining the reasoning that allow the behavior of these systems to be described, so that they can be fully understood even by people with no expertise in this or related fields. The aim is to improve transparency, confidence, fairness and understanding of model bias, objectives pursued by XAI (i.e., eXplainable Artificial Intelligence).


J. M. Ferrández Vicente et al. (Eds.): IWINAC 2022, LNCS 13258, pp. 557–566, 2022.
https://doi.org/10.1007/978-3-031-06242-1_55

The use of this type of explanation techniques to improve model development is a considerable advantage that also broadens the perspective on the scope of application, providing complementary information beyond model prediction.The ability to explain AI models in a concise and easy-to-understand manner by non-experts in the field is vital to be able to employ them safely, ethically and reliably. In the field of computer vision, it is common to find studies on the development of methods to determine the regions of the input image used by the model when making predictions, thus making it possible to identify the errors made by the system. However, their use in real problems is limited to providing an evidence on whether the model has correctly arrived at the result, similar to the usual accuracy metrics. Thus, their potential is not fully exploited.

Among the existing possibilities within the field of XAI, we have focused on visualization techniques that facilitate the interpretability of model predictions. In other words, it is a matter of explaining in a visual way the relationship between the input images and the result of the classification performed by the network. Colloquially it could be understood as giving an answer to the question: *Why does this input produce this particular output?*. This type of post-development explanation has become popular and has led to the creation of a wide variety of methods. There are many taxonomies proposed in the literature that classify them as [2] explain: 1) Purpose of the explanation; 2) Explanation control; 3) Explanation families; 4) and Explanation estimation methods.

On the other hand, medical imaging has become indispensable in the early detection or diagnosis of diseases. Images based on X-rays, tomography, ultrasound and magnetic resonance imaging (MRI) are mainly employed. This is why many researches in artificial vision focus their efforts on the analysis of this type of image datasets, providing different approaches to automatic diagnosis and avoiding human errors committed by the misinterpretation of results due to inexperience or fatigue, among others. However, there are many difficulties in developing diagnostic support systems. Accuracy is particularly important and a rigorous process of validation of results must be carried out, usually by a person who is a specialist in the pathology or imaging method, but not necessarily trained in AI. It is at this point where interpretability methods play a crucial role, allowing to reduce the ambiguity of the learning process and strengthening the credibility of the machine learning models used.

There are several cases of studies of different applications of XAI in medical imaging tasks [3], summarizing the benefits that these approaches bring to both clinical specialists and AI practitioners. One of the cases mentioned in this study is breast cancer. This is one of the most common types of cancer among women whose early detection makes a clear difference in the effectiveness of treatment and, therefore, in mortality reduction. The most widespread type of imaging for preventive check-ups is mammography, although in some cases it is difficult for radiologists to identify lesions and they must refer the patient for a biopsy, a more invasive and costly procedure. Therefore, the development of machine learning models that facilitate the mammography analysis process could be considered almost indispensable. Both [4] and [5] show clear examples of the benefits that this area of research can bring, allowing to reduce the ambiguity of the learning process and strengthening the credibility of the classifiers used to detect breast cancer on mammographies. Thus, the main goal of this work is to perform a qualitative analysis of the predictions of several neural networks based on different learning mechanisms, through visual techniques of interpretability, differentiating the categories of "malignant" or "benign".

In Sect. 2 the proposed system and experimental setup are described, where the different datasets of images which have been used are also defined. The experiments conducted, as well as their results and analysis are included in Sects. 3 and 4 respectively. Finally, the conclusions drawn are presented in Sect. 5.

## 2 Materials and Methods

**Datasets.** Since the objective of the present study is the use of explanation of reasoning techniques, we have reduced the problem to the most basic task: a binary classification without using the segmentation or other data that may be present. Thus, we decided to use two datasets of mammograms and to focus on the task of classifying relevant findings into the categories 'malignant' and 'benign':

- INbreast[6]: dataset of full-field digital mammographies subsequently digitized through different methods. There are a total of 115 cases, where we randomly choose 90 patients for training and 25 for testing.
- MNIST[1]: dataset of images of handwritten digits. It has been chosen since it shares with mammographies the property of being composed of grayscale images

**Architecture.** In order to provide a deep learning system that brings more confidence to the user and allows a better understanding of the results to target users, the authors have proposed the structure represented in Fig. 1. The first flow, in green, is the usual one in the development of neural networks through supervised learning; a dataset of images is used to train the model, which will then be responsible for making predictions about new images. The second flow, in purple, represents the explanation of the predictions. This explanation is achieved by applying a series of scoring techniques to the model.

In this way, the end user of the system is presented with two types of responses: the class assigned to the input image and a visual explanation of which areas of this image have influenced the result. This second visualization allows the professionals being to verify in a simple way whether the network correctly identifies the areas of interest. In the case of breast cancer detection, the specialist would simply have to check for abnormalities in the indicated region to be sure of the correctness of the diagnosis. This explanation provides transparency and confidence, increasing the analytical skills of medical professionals.

[1] http://yann.lecun.com/exdb/mnist/.

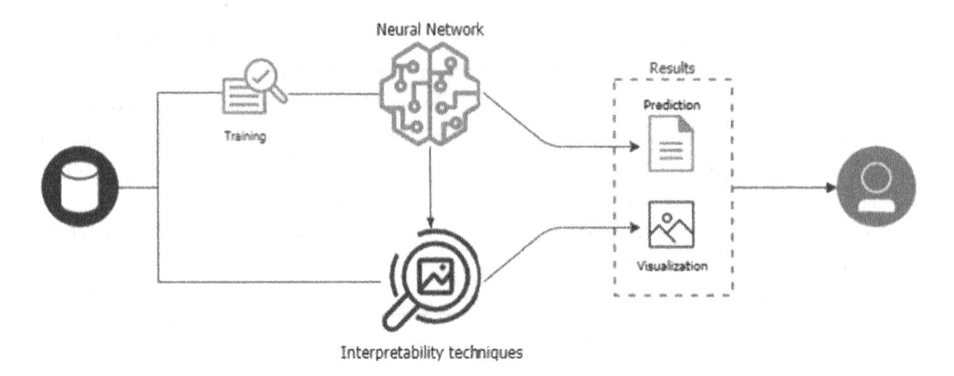


**Fig. 1.** System architecture.

**Deep Learning Model: EfficientNet.** This model, based on Convolutional Neural Networks (i.e., CNNs onwards), has been chosen beause is usually applied to solve classification problems. The CNN design procedure is based on developing neural models with a fixed resource cost, then arbitrarily expanding the size of the network when more resources become available or increasing the resolution of the input images during the training or evaluation phases, requiring tedious manual adjustment and not always achieving optimal performance. In contrast, the EfficientNet model employs a dynamic scalability method, allowing for improved accuracy without sacrificing efficiency. This technique allows unified scaling of network width, depth and resolution through a fixed dataset of coefficients, the values of which are experimentally determined for each dimension (cf. Fig. 2, that shows traditional scaling methods (b)-(d) compared to this compound method (e)). Furthermore, the focus in terms of improving convolutional neural networks is not so much on the novelty of the operations performed, but rather on increasing their efficiency and reducting execution costs.

**Implementation and Experimental Setup.** The system on which the experiments have been executed has 4 T V100 GPUs, two with 16 GB of RAM and two with 32 GB and an Intel(R) Xenon(R) Silver 4210 CPU @ 2.20 GHz. In order to facilitate the reproduction of this research, the development has been based on open source software. The programming language used is Python and the API Keras, included in the Tensorflow library, commonly used to build deep learning models.

The EfficientNet implementation is the official one, available in the tensorflow.keras.applications package, using Tensorflow instead of Pytorch. Furthermore, Transfer Learning strategy consists of using pre-trained models, in this case on the Imagenet dataset. Once loaded, these network layers are frozen and a classifier is added at the end, we have chosen to vary this classifier for each of the cases.

Tests have been performed with the following different optimizers: SGDW [8] and RAdam [9], with the parameters described in the previous section. Although

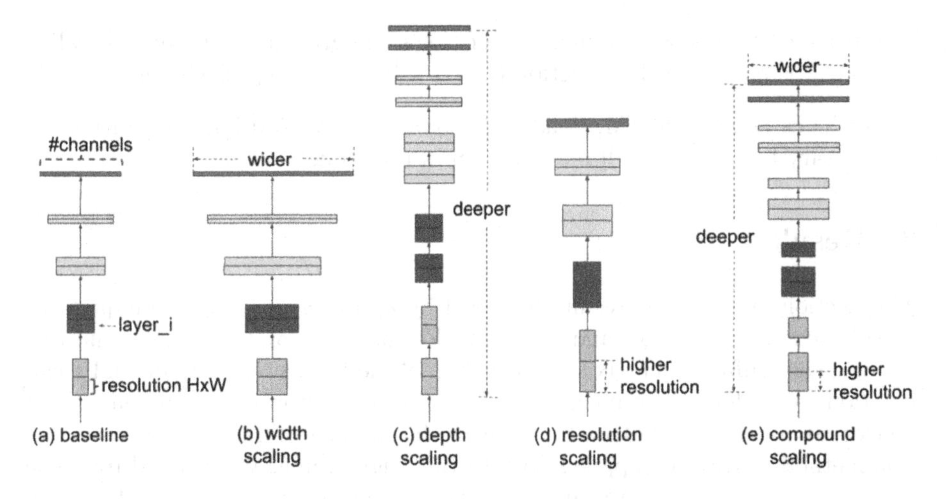


**Fig. 2.** Different neural network scaling methods [7].

the variant of Adam reached high values of accuracy in the first stages of training, in some cases it was not possible to converge, obtaining a model unable to generalize properly, therefore the results obtained with SGDW, more stable thanks to the use of momentum, are presented. The main metric used during training is categorical accuracy, while losses are calculated using categorical cross-entropy, as usual. The batch-size varies depending on the size of the dataset used being 16, 32 or 64. However, all images are resized to $224 \times 224$ in RGB format in order to use the same input tensor in the neural networks and to avoid possible alterations in the results. Finally, the total duration of the training is set at 300 epochs.

**Visualization Techniques.** Once the network has been trained, the weights with which the best accuracy was obtained on the validation subset and applied together with the interpretability methods on an image extracted from the test subset[2]. In addition to the original image, the Python OpenCV[3] library is used to apply certain transformations to it and study the results. It is required some custom configuration for their proper functioning:

- Occlusion Sensitivity: the size of the patch applied must be specified, in our case this value is 15.
- SmoothGrad: the parameters used refer to the number of samples with noise to be generated from the original image and the standard deviation of the normal distribution to generate such noise. The values 5 and 0.5 are set, respectively.

[2] Instead of a division into three subsets, only two subsets are used, one of which is used both for validation during training and for testing.

[3] https://opencv.org.

- Integrated Gradients: the number of times the gradients are applied, called "steps", is passed to this function (i.e., a value of 20 steps is chosen).

In the case of GradCAM, GradCAM++ and ScoreCAM implementations, it is necessary to make a small modification in the model.

## 3 Results

This section presents the results obtained by applying the XAI techniques. To complement them, the loss and accuracy plots obtained on the training and test subsets, the confusion matrix and the values of the F1-Score, precision and recall functions are added. The images will be organized according to the dataset of images to which they belong and the model applied. The different visualization methods have been applied both to the original images extracted from the test dataset and to some variations of them obtained performing the following transformations:

- Color transformations: log transformation; Gaussian noise addition; Speckle noise addition; and addition of a black rectangle.
- Spatial transformations: Warp affine; and 180º rotation of the original image.

### 3.1 INbreast

We propose to use INbreast dataset. These images have a much higher resolution as they were created digitally, yet we reduced their size to reduce computational cost. With the image divided into network input patches, it was easier for the untrained eye to identify the region with the anomaly. Applying interpretability techniques on this occasion also serves to test whether the models encounter a difficulty analogous to that of the human eye in recognizing these areas of interest.

The obtained accuracy using EfficientNet is 68.18%, also the results of the gradient saliency maps in the natural and integrated gradients stand out, due to the low intensity of the attributions on the image. In this case it seems that the results of the last columns of the Fig. 3, corresponding to the CAM-based methods are closer to the area where the mass is located than in the case of the previous dataset of images.

### 3.2 MNIST

The best classifier published on the MNIST image dataset[4] achieves an accuracy of 99.87%. Taking this value as a reference, the results obtained by the models used are mediocre, but the objective in this case is to check which areas of the input image affect the decision taken in a classification problem with multiple categories through the proposed visualization methods.

[4] https://paperswithcode.com/sota/image-classification-on-mnist.

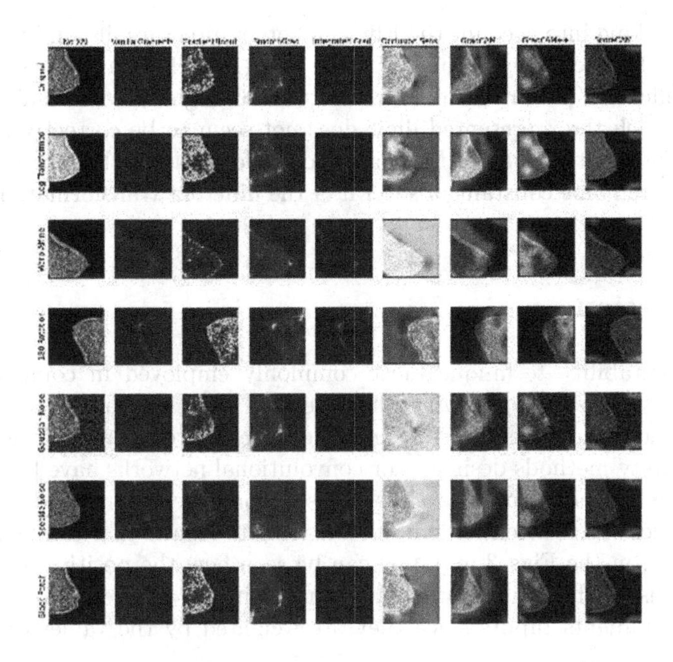

**Fig. 3.** Visualization techniques applied to EfficientNet (INbreast).

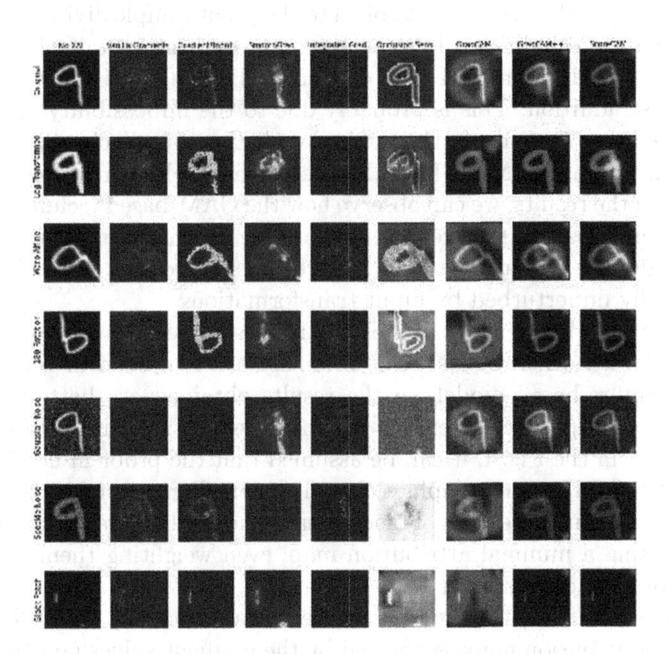

**Fig. 4.** Visualization techniques applied to EfficientNet (MNIST). X-axis both figures: No XAI; vanilla gradients; Gradient*Input; Smooth Grad; Integrated Grad.; Occlusion Sens.; GradCAM; GradCAM++; & ScoreCAM

Again, taking into account the EfficientNet architecture, it can be considered that the 96.05% accuracy obtained by the ENet model in this dataset implies a good classification performance. Moreover, we can highlight the results of Score-CAM, in which the represented digit does not seem to be correctly located. In contrast, the result of GradCAM, in addition to focusing on the representation of the number remains constant in several of the different transformations applied (cf. Fig. 4).

## 4 Discussion

The interpretability technique most commonly employed in computer vision tasks consists of simply showing the attention scores obtained by the layers that implement these mechanisms, ignoring the effect of the rest of the layers, so interpretability methods designed for convolutional networks have been applied. Nevertheless, it should be noted that these approaches are not adapted to the particularities of self-attention modules, leading to results that can be confusing. For instance, in the Figs. 3 and 4, it can be seen how the positive and negative contributions of the weights are interchanged in the gradient-based methods, except in Gradients Inputs since they are weighted by the values of the input image.

As for the occlusion sensitivity method, there is no agreement on how the perturbation should be correctly applied to the input sample division. In this case it has been applied individually to each patch, resulting in a display that is not very accurate and is clearly affected by variations in the input image, especially in cases of noise addition. This is probably due to the impossibility of associating a small occlusion in one of the patches with the overall classification result. However, in the case of MNIST (cf. Figure 4) where a higher degree of accuracy is achieved in the results, we can observe how the CAM-based techniques focus on the characteristic zones that represent the "nine", highlighting the circular zone and the angle that it forms with the straight line. Moreover, these visualizations are practically unperturbed by input transformations.

Regarding the EfficientNet model, the results are similar to those that could be obtained with other CNNs. Not as much data is needed in the training as for the attention-based model, so the results obtained are better in terms of accuracy in all experiments, except for the classification with INbreast. Based on the images in the Fig. 4, it can be assumed that the problem lies in the small difference between mammographies classified as malignant and those considered benign, as well as in the excess of black areas. The gradient values are practically null, generating a minimal attribution map, even weighting them by the input values there is no clear answer.

On the other hand, the analysis of the visualizations in the MNIST case is similar, the attribution maps generated by the gradient values and the occlusion technique highlight the edges of the digit, while GradCAM draws a circular map starting from the center of the digit and encompassing it almost completely. It is easy to interpret the reasoning followed by the network in both cases, visualizing the areas it identifies as relevant in the representations, identifying both the shape of the 9 and the outline of the dog and classifying them correctly.

Finally, from the point of view of the transformations applied to the input image, it can be observed that in some cases the results of the visualizations vary considerably. This effect may have the same origin as that of adversarial attacks against neural networks. This type of attack consists in the imperceptible alteration to the human eye of an input image to the network that manages to completely alter the result of the prediction made. Applying XAI techniques can help to build more robust models against this type of attacks.

## 5 Conclusions

The post-hoc interpretability of neural networks performed in this work provides a potential application to computer-aided detection of tumor signs in mammograms. For this purpose, one of the latest neural network models has been selected: EfficientNet, a novel architecture and, therefore, scarcely studied. In addition, the results of interpretability techniques that perform relevance attribution following different patterns and constructing complementary visualizations that help to understand the reason for the decision taken by the model are presented.

After experimental evaluation, it can be concluded that the results in the mammogram classification task, with low accuracy values, are adequately complemented by the visualizations produced, providing a possible explanation and allowing to change the direction of development in an informed manner. In particular, it may be beneficial to employ annotations of the dataset to crop the images and reduce them to the region of interest, as well as to employ thresholding techniques that convert the image to binary values.

Moreover, due to the analysis performed is qualitative, based on visual perception, it could be possible to provide quantitative information by the XAI algorithms, considering the visualizations generated as segmentations and comparing them with the ground-truth. Thus, it seems clearly necessary to adapt most of the methods developed for convolutional neural networks to other types of architectures such as the Vision Transformer, allowing to extract information from the different attention blocks and combine them in a final attribution map, instead of simply accumulating and extracting them from the final layer. This would help to obtain clearer relevance visualizations, such as those extracted from the EfficientNet model.

To conclude, interpretability is a feature of deep learning models that is increasingly desirable and sometimes mandatory in legal terms. The visualization of the areas of interest considered by the model in the input images allows the development of more robust models against adversarial attacks, allowing decisions to be made based on how certain input affects the learning process. Combining this approach with the attention mechanism used in the transformer would enable a more widespread use of artificial intelligence in healthcare applications, taking into consideration the individual contribution of the details present in the analyzed image and allowing professionals to verify the diagnosis made quickly and directly. Nevertheless, more research is needed in order to implement interpretability models reliable, valid, and ready to be used in clinical practice.

**Acknowledgements.** The authors gratefully acknowledge research project PID2019-110686RB-I00 of the State Research Program Oriented to the Challenges of Society.

## References

1. Dong, S., Wang, P., Abbas, K.: A survey on deep learning and its applications. Comput. Sci. Rev. **40**, 100–379 (2021)
2. Gilpin, L.H., Bau, D., Yuan, B.Z., Bajwa, A., Specter, M.A., Kagal, L.: Explaining explanations: An overview of interpretability of machine learning. In: 2018 IEEE 5th International Conference on Data Science and Advanced Analytics (DSAA), pp. 80–89 (2018)
3. Singh, A., Sengupta, S., Lakshminarayanan, V.: Explainable deep learning models in medical image analysis. J. Imaging **6**(6), 52 (2020)
4. Lenis, D., Major, D., Wimmer, M., Berg, A., Sluiter, G., Bühler, K.: Domain aware medical image classifier interpretation by counterfactual impact analysis. In: Martel, A.L., et al. (eds.) MICCAI 2020. LNCS, vol. 12261, pp. 315–325. Springer, Cham (2020). https://doi.org/10.1007/978-3-030-59710-8_31
5. Lamy, J.-B., Sekar, B., Guezennec, G., Bouaud, J., Séroussi, B.: Explainable artificial intelligence for breast cancer: a visual case-based reasoning approach. Artif. Intell. Med. **94**, 42–5 (2019)
6. Moreira, I.C., Amaral, I., Domingues, I., Cardoso, A., Cardoso, M.J., Cardoso, J.S.: Inbreast: toward a full-field digital mammographic database. Acad. Radiol. **19**(2), 236–248 (2012)
7. Tan, M., Le, Q.V.: Efficientnet: rethinking model scaling for convolutional neural networks. ArXiv, abs/1905.11946 (2019)
8. Loshchilov, I., Hutter, F.: Decoupled weight decay regularization. In: 7th International Conference on Learning Representations, ICLR 2019, New Orleans, LA, USA, 6–9 May 2019 (2019)
9. Liu, L., et al.: On the variance of the adaptive learning rate and beyond. In: 8th International Conference on Learning Representations, ICLR 2020, Addis Ababa, Ethiopia, 26–30 April 2020 (2020)
10. AL-Antari, M.A., Han, S.M., Kim, T.-S.: Evaluation of deep learning detection and classification towards computer-aided diagnosis of breast lesions in digital X-ray mammograms. Comput. Meth. Prog. Biomed. **196**, 105584 (2020)

# Evaluation of a Gaussian Mixture Model for Generating Synthetic ECG Signals During an Angioplasty Procedure

Anderson I. Rincon Soler[1,2], Pedro D. Arini[1,2], Santiago F. Caracciolo[1,2], Fernando Ingallina[3], and María Paula Bonomini[1(✉)]

[1] Instituto Argentino de Matemática, "Alberto P. Calderón", CONICET, Buenos Aires, Argentina
paula.bonomini@conicet.gov.ar

[2] Instituto de Ingeniería Biomédica, Facultad de Ingeniería, Universidad de Buenos Aires, Buenos Aires, Argentina

[3] Instituto de Investigaciones Médicas Dr. Alfredo Lanari, Universidad de Buenos Aires, Buenos Aires, Argentina

**Abstract.** Mathematical models have intensively been used in the generation of synthetic electrocardiogram (ECG) signals. They are used to test and optimize different algorithms for patient monitoring and heart disease detection. In this work, we present a Gaussian mixture model that allows the generation of heartbeats with ischemic alterations, induced by an occlusion procedure performed in the right coronary artery. Realizations obtained using the Gaussian mixture model were compared against real signals using cross-correlation, obtaining average values equal to 94.2%. Confidence intervals, at 99% level, calculated using ST values from the synthesized heartbeats agreed with the real ST values in 100% cases for each group (ECG Lead and ST-level). In this sense, the model proposed allows the synthesis of heartbeats with complex morphologies that cannot be obtained with more traditional models.

**Keywords:** Heartbeat synthesis · Coronary artery occlusion · Reversible myocardial ischemia · Non-supervised classification

## 1 Introduction

Myocardial ischemia is a cardiovascular disease that appears in consequence to deprivation of perfusion in cardiac tissue. Depending on the extent of the affected tissue, it can be stratified into levels, each of these being reflected in alterations of the electrocardiographic pattern, especially in the ST segment and the T wave [1,2,22]. The utilization of algorithms to detect or classify myocardial ischemia is of great interest for the clinical practice. However, there are very little data from controlled myocardial ischemia to train and validate these algorithms, including a few public datasets [5,13].


J. M. Ferrández Vicente et al. (Eds.): IWINAC 2022, LNCS 13258, pp. 567–575, 2022.
https://doi.org/10.1007/978-3-031-06242-1_56

Electrocardiographic signal modelling turns out to be a valuable resource for an automated and controlled recreation of an endless number of clinical scenarios, including myocardial ischemia. However, to the best of our knowledge, there are no currently ECG models adapted well enough to simulate ischemia dynamics. McSharry *et al.* presented the ECGSyn model based on a set of coupled nonlinear differential equations that properly describe normal and arrhythmogenic conditions [3,15,16]. However, the model suited well to simulate temporal dynamics, but fails at recreating pathological ECG morphologies. Recently, we showed a well-detailed methodology to synthesize heartbeats with ST-T segment variability in patients undergoing a percutaneous transluminal coronary angioplasty (PTCA) in the right coronary artery (RCA) by modifying the ECGSyn model [18]. This approach performed well at simulating mild ischemia, but became limited in the recreation of morphologies from severe ischemia, which led to the need of a new approach that came out with improved computational models. In this order or ideas, Louis *et al.* proposed a methodology based on Gaussian Mixture Models (GMMs) to generate artificial heartbeats, from patients under normal conditions, modeled using a fixed number of Gaussian densities [10,11]. These artificial signals were used to train and test security systems based on biometric parameters.

In this paper, we propose a methodology to train and optimize GMMs in order to generate ECG signals with significantly morphological variability induced by an angioplasty procedure. Data from the Right Coronary Artery (RCA) was used. In almost all cases, four well defined levels of ischemia were obtained with characteristic morphologies that take place during an ischemic scenario. Because of this, these synthetic ischemic ECG signals can be used in algorithms for ischemia detection, heartbeat segmentation and ECG delineation.

## 2 Materials and Methods

### 2.1 Database

The STAFFIII database was used in this study [5,13]. This data set is freely available at the PhysioNet web site and contains ECG recordings acquired in patients undergoing elective PTCA procedure in one of their major coronary arteries. All signals were recorded with a sampling rate and amplitude resolution of 1000 Hz and 0.6 μV, respectively, employing custom-made equipment by Siemens-Elena AB [8,23]. In this work, we use ECG signals belonging to occlusions in the RCA artery.

### 2.2 Pre-processing

A low pass (LP - Butterworth, 6th order, 80 Hz) and notch (Butterworth, 2nd order, 60 Hz) filters were applied to minimize artifacts and power-line interference. Also, a cubic spline filter was used to attenuate ECG baseline drift. Thereafter, R-peak locations were detected in each ECG lead using a wavelet-based ECG delineator [12]. Relative to the R peak occurrences, the J-points were located for each heartbeat and ST elevation was measured. Additionally, we used these R locations to implement a heartbeat segmentation procedure through a fixed temporal window starting at this point and reaching 350 ms, on each side.

### 2.3 Inclusion and Exclusion Criteria

ECG signals from patients with occlusion in the RCA were carefully evaluated by an expert cardiologist, in order to eliminate confounding ECG records, such as those with myocardial infarction, left or right bundle branch block, ventricular pre-excitation and secondary repolarization abnormality. In summary, we just kept those ECG records without concomitant pathology for the analysis. As a consequence of this selection process, the number of patients was drastically reduced from 59 to 12 subjects. These signals are a representative sample of the ST-T complex alterations that take place during myocardial ischemia induced by an angioplasty procedure. To analyze the occlusion in the RCA, electrocardiographic leads II, III and aVF were used. These combinations were selected because they are the most useful leads to map the physiological behavior of the RCA occlusion according to the cardiological criteria [2,7,14,20].

### 2.4 ST Classification Criteria

In each of the aforementioned ECG recordings, ST-segment elevation was measured using the mean value of a 5 ms window around the point J+80 $ms$ for an ECG register with a cardiac frequency less than or equal to 120 $bpm$ ($J$+60 $ms$ in the case of cardiac frequency being greater than this value) [21]. In addition, each heartbeat was segmented and classified according to the ST level measurement obtained in the previous step. According to this information, four ST-segment elevation levels were defined as:

- **L1**: $50\ \mu V \leq ST\text{-}level \leq 200\ \mu V$
- **L2**: $200\ \mu V < ST\text{-}level \leq 400\ \mu V$
- **L3**: $400\ \mu V < ST\text{-}level \leq 700\ \mu V$
- **L4**: $700\ \mu V < ST\text{-}level$

On each group of data, a QRS template was constructed using the median value among 20 non-consecutive QRS complexes with a cross-correlation coefficient ($\rho$) greater than 98%. In this process, an 80 $ms$ window centered in the R peak was analyzed at every heartbeat. After this, the new incoming QRS complexes were compared against the QRS-template using the same correlation technique. Heartbeats with less than 98% cross-correlation value were discarded. This process helps to eliminate signals that have no physiological validity and at the same time align the heartbeats respect to their R-peaks.

### 2.5 Gaussian Mixture Model

Gausian mixture models are probabilistic models used to group data with similar characteristics, without using a-priori information of the data. Thus, GMMs constitute an unsupervised classification method; however, under our approach, they will be used for data generation, in the sense that we are seeking aleatory variables that represent ECG samples statistically. GMMs can be defined as the weighted sum of K Gaussian densities [4], as shown in Eq. 1.

$$p(\mathbf{x}) = \sum_{i=1}^{K} \Pi_i \mathcal{N}(\mathbf{x}|\mu_i, \Sigma_i) \tag{1}$$

where $\Pi_i$ are the weights of the Gaussian density functions, subject to the condition $\Sigma_{i=1}^{K} \Pi_i = 1$, $\mathbf{x}$ is the feature vector with dimension $N$ and $\mathcal{N}(\mathbf{x}|\mu_i, \Sigma_i)$ corresponds to a probability density function, defined by Eq. 2, with mean vector $\mu_i$ and covariance matrix $\Sigma_i$.

$$\mathcal{N}(\mathbf{x}|\mu_i, \Sigma_i) = \frac{1}{2\pi^{N/2}|\Sigma_i|^{1/2}} exp\left\{-\frac{1}{2}(\mathbf{x} - \mu_i)^T \Sigma_i^{-1}(\mathbf{x} - \mu_i)\right\} \tag{2}$$

In order to reduce the complexity of the GMMs, each heartbeat was down-sampled to 250 $Hz$. This helped obtain a better performance in the modeling stage. Afterwards, We have trained three different GMMs, one per lead and their corresponding four ST levels. Thus, every GMM was designed with $K = 8$ Gaussian components and $N = 175$ dimensions in each component. Notice that $N$ has reduced from 700 to 175 due to the down-sampling process. In this way, we can obtain synthesized heartbeats with specific alterations in their ST segment by generating random realizations from each one this mixture components.

### 2.6 Model Training

Heartbeats were grouped according to ECG lead and ST-elevation level. The ST levels were described in Sect. 2.4. Next step, heartbeats of each group were aligned using the R-peak as a reference. Each of the aforementioned groups of data were split using a 70%-30% ratio. These were called training and testing groups, respectively. The number of clusters for each GMM was set to 8. This value was obtained as a result of a complexity analysis performed through calculation of the Bayesian Information Criteria (BIC) [4,19]. Surprinsingly, BIC provided a number of clusters with physiological sense, since 8 accounts for 4 ST-levels, each allowing for positive and negative T-waves. Finally, a synthesized heartbeat is generated by selecting a specific trained GMM and generating random realizations from one of their eight clusters.

### 2.7 Validation

Based on our previous work [18], a number of 600 realizations were generated for each combination of ECG lead and ST level. This number of signals was determined using a sample size test configured to discriminate real and synthesized heartbeats with at least 10% of difference in the ST-level. So, t-test was used with 95% of power and significance level of 0.005 [6]. Subsequently, cross-correlation was used as a measure of similarity between the synthesized data and the respective test groups, defined in the previous section. Here, each artificial beat was compared against each real beat and finally an average correlation value was calculated. On the other hand, the ST values for the artificial signals were measured and confidence intervals, at a 99% level, were calculated for this information. In this way we can corroborate that Gaussian mixture models can generate sets of signals that meet the required morphological characteristics.

# 3 Results

Using the methodology described in Sect. 2.4 and Sect. 2.6 it was possible to obtain synthesized heartbeats for all cases. Figure 1 shows an example of different real and synthesized ECG signals with alterations in the ST-segment due to occlusions in the RCA for the four ST-level analyzed.

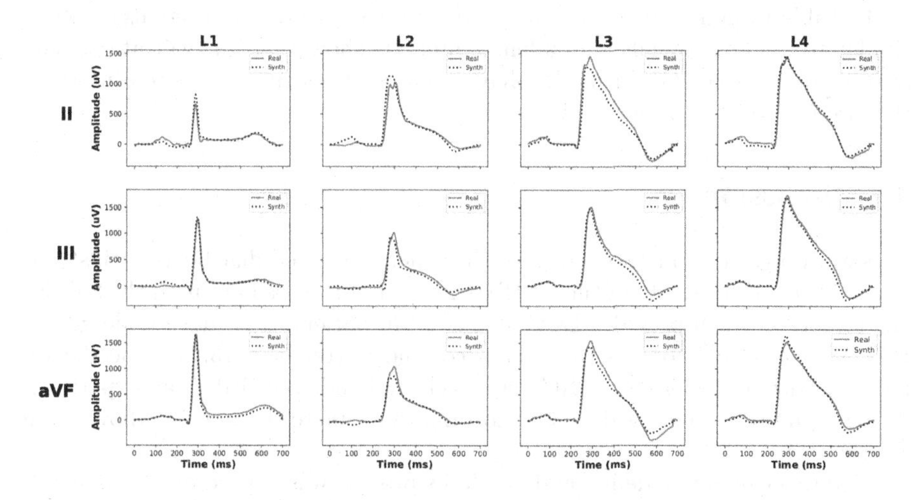


**Fig. 1.** A comparison of the ECG signal morphology obtained using the GMM trained models and some real signals used for testing. These representations belong to ST changes by occlusion of the right coronary artery (RCA).

In Fig. 1 it is possible to see how the dynamic model can recreate well-defined ischemic ECG morphologies in subjects that receive a PTCA procedure due to RCA occlusion. In some cases, differences in the amplitude of real versus synthesized signals arise, but this is an expected behavior if we consider the variability among the signals used in this study.

On the other hand, results for mean correlation coefficient ($\overline{\rho}_{R,Syn}$) and ST-level confidence intervals ($CI_{ST_l}$), calculated for the ST level values in the synthesized beats, are presented in Table 1.

**Table 1.** Mean value of the correlation coefficient between real and synthesized beats ($\overline{\rho}_{R,S}$) in (%), ST-level confidence intervals ($CI_{ST_l}$) in $\mu V$ for the synthesized heartbeats and at each combination of lead and ST-level. The statistical tests associated with the correlation analysis were highly significant as they rejected the null hypothesis of no relationship between the data in at least 95% of the comparisons made between real and synthesized heartbeats.

| Lead | Parameter | ST classification level | | | |
|---|---|---|---|---|---|
| | | L1 | L2 | L3 | L4 |
| II | $\overline{\rho}_{R,S}$ | 81.6 | 92.9 | 99.3 | 99.8 |
| | $CI_{ST_l}$ | $91 \leq 98 \leq 103$ | $299 \leq 304 \leq 309$ | $553 \leq 560 \leq 567$ | $755 \leq 762 \leq 770$ |
| III | $\overline{\rho}_{R,S}$ | 84.8 | 95.9 | 98.8 | 99.3 |
| | $CI_{ST_l}$ | $102 \leq 106 \leq 110$ | $280 \leq 287 \leq 295$ | $506 \leq 516 \leq 525$ | $785 \leq 796 \leq 807$ |
| aVF | $\overline{\rho}_{R,S}$ | 85.2 | 93.7 | 99.4 | 99.7 |
| | $CI_{ST_l}$ | $92 \leq 95 \leq 98$ | $254 \leq 261 \leq 268$ | $553 \leq 567 \leq 582$ | $773 \leq 784 \leq 794$ |

In Table 1, high correlation values indicates that synthesized signals are quite similar to the real signals. Confidence intervals show that all realizations that were made with the GMM models have their ST-levels in the desired intervals, that were presented in Sect. 2.4.

## 4 Discussion

This work presents a methodology for the generation of cardiac beats with strong morphological variability of the ST-T segment. For this we have used the STAFF III database, which provides ECG signals from patients who have undergone a PTCA procedure in one of their main coronary arteries [8,13,17]. Specifically, patients with occlusion in the RCA were selected and classified by an expert cardiologist in order to discard ECG recordings with more than one cardiovascular disease.

The ECG signal modeling and synthesis process was carried out using Gaussian mixture models (GMMs). This type of models are widely used in unsupervised cluster classification tasks; however, there are few works where the main application is to generate new data. It is important to highlight previous works by Luis *et al.*, where GMMs allow the replication of heartbeats from healthy patients [9–11]. In this way, the validation of several ECG-based biometric security techniques is performed. In addition, the synthesis of ECG signals with morphological variability due to ventricular ischemia is a subject currently under development and there are not many references on the subject.

In this work, ECG signals with presence of ventricular ischemia due to RCA occlusion were classified using different levels of ST-segment elevation. Herein, only ECG leads mapping occluded regions were analyzed [20]. Each of the ECG leads were processed and cardiac beats with different ST-segment levels were extracted. The data sets used showed correlation values above 80%, which allows us to establish that the GMM models effectively learn the morphological characteristics of the signals with ventricular ischemia. Percentages above 90% are not obtained in all situations, and this is a direct consequence of the high inter-patient variability. Nevertheless, these correlation values indicate that there is a good representation of the different types of ischemic morphologies by the GMM models, as can be seen in Fig. 1. Additionally, in the results of Table 1 we see that the confidence intervals for the ST level of the synthesized beats are within the working ranges defined in Sect. 2.4. In this way, we know that all simulations meet the objectives of the work. Compared to previous studies that aboard this modelling scenarios [18], this work has shown improved performance of GMMs for learning and representing complex ECG signal morphologies.

## 5 Conclusions

We obtained realistic ECG signals in different electrocardiographic leads for several morphologies according to an ischemic process in some of the main coronary arteries. In particular, GMMs allowed to synthesize ischemic ECG signals from patients undergoing a PTCA procedure in the RCA. In this way, it was possible to generate, in a controlled manner, a set of ECG signals with significant ST-segment variability. These variations follow physiological ischemic criteria. Then, synthesized signals can be used to test different algorithms for automatic ST detection and measurement.

**Acknowledgments.** This work was supported by Universidad de Buenos Aires (UBACyT 20020130100485BA), Agencia (MINCyT PICT 2145 2016) and by Consejo Nacional de Investigaciones Científicas y Técnicas (PIP 2014-2016 112-20130100552CO), Argentina.

**Conflicts of Interest.** The authors declare that there are no conflicts of interest

## References

1. Bernardo, D.D., Murray, A.: Explaining the T-wave shape in the ECG. Nature **403**(6765), 40–40 (2000)
2. Cinca, J., Noriega, F.J., Jorge, E., Alvarez-Garcia, J., Amoros, G., Arzamendi, D.: ST-segment deviation behavior during acute myocardial ischemia in opposite ventricular regions: observations in the intact and perfused heart. Heart Rhythm Official J. Heart Rhythm Soc. **11**(11), 2084–2091 (2014)
3. Clifford, G.D., McSharry, P.E.: A realistic coupled nonlinear artificial ECG, BP, and respiratory signal generator for assessing noise performance of biomedical signal processing algorithms. In: Fluctuations and Noise in Biological, Biophysical, and Biomedical Systems II. SPIE (2004)
4. Géron, A.: Hands-On Machine Learning with Scikit-Learn, Keras, and TensorFlow, 2nd Edition. O'Reilly Media, Inc. (2019)

5. Goldberger, A.L., et al.: PhysioBank, PhysioToolkit, and PhysioNet: components of a new research resource for complex physiologic signals. Circulation **101**(23), e215–e220 (2000)
6. Hickey, G.L., Grant, S.W., Dunning, J., Siepe, M.: Statistical primer: sample size and power calculations—why, when and how?†. Eur. J. Cardio-Thoracic Surg. **54**(1), 4–9 (2018)
7. Horacek, B., Wagner, G.: Dynamic Electrocardiography, chap. Spatial patterns of ST segment shift during myocardial ischaemia, pp. 250–259. Blackwell Publishing (2004)
8. Laguna, P., Sörnmo, L.: The STAFF III ECG database and its significance for methodological development and evaluation. J. Electrocardiol. **47**(4), 408–417 (2014)
9. Louis, W., Komeili, M., Hatzinakos, D.: Continuous authentication using one-dimensional multi-resolution local binary patterns (1DMRLBP) in ECG biometrics. IEEE Trans. Inf. Forensics Secur. **11**(12), 2818–2832 (2016)
10. Louis, W., Komeili, M., Hatzinakos, D.: Real-time heartbeat outlier removal in electrocardiogram (ECG) biometrie system. In: 2016 IEEE Canadian Conference on Electrical and Computer Engineering (CCECE), pp. 1–4 (2016)
11. Louis, W., Abdulnour, S., Haghighi, S.J., Hatzinakos, D.: On biometric systems: electrocardiogram gaussianity and data synthesis. EURASIP J. Bioinform. Syst. Biol. 2017 (2017)
12. Martinez, J.P., Almeida, R., Olmos, S., P, R.A., Laguna P.: A wavelet-based ECG delineator: evaluation on standard databases. IEEE Trans. Biomed. Eng. **51**(4), 570–581 (2004)
13. Martínez, J.P., Pahlm, O., Ringborn, M., Warren, S., Laguna, P., Sörnmo, L.: The STAFF III database: ECGs recorded during acutely induced myocardial ischemia. Comput. Cardiol. **44**, 133–266 (2017)
14. Mason, J.W., Hancock, E.W., Gettes, L.S.: Recommendations for the standardization and interpretation of the electrocardiogram. Circulation **115**(10), 1325–1332 (2007)
15. McSharry, P.E., Clifford, G.D., Tarassenko, L., Smith, L.A.: A dynamical model for generating synthetic electrocardiogram signals. IEEE Trans. Biomed. Eng. **50**(3), 289–294 (2003)
16. McSharry, P.E., Clifford, G.D.: A comparison of nonlinear noise reduction and independent component analysis using a realistic dynamical model of the electrocardiogram. In: Second International Symposium on Fluctuations and Noise, vol. 5467 (2004)
17. Perry, R.A., et al.: Balloon occlusion during coronary angioplasty as a model of myocardial ischaemia: reproducibility of sequential inflations. Eur. Heart J. **10**(9), 791–800 (1989)
18. Rincon Soler, A.I., Bonomini, M.P., Fernández Biscay, C., Ingallina, F., Arini, P.D.: Modelling of the electrocardiographic signal during an angioplasty procedure in the right coronary artery. J. Electrocardiol. **62**, 65–72 (2020)
19. Schwarz, G.: Estimating the dimension of a model. Ann. Stat. **6**(2), 461–464 (1978)
20. Sgarbossa, E.B., Birnbaum, Y., Parrillo, J.E.: Electrocardiographic diagnosis of acute myocardial infarction: current concepts for the clinician. Am. Heart J. **141**(4), 507–517 (2001)
21. Taddei, A., Costantino, G., Silipo, R., Emdin, M., Marchesi, C.: A system for the detection of ischemic episodes in ambulatory ECG. In: Computers in Cardiology 1995. IEEE (1995)

22. Wang, J.J., Pahlm, O., Warren, J.W., Sapp, J.L., Horáček, B.M.: Criteria for ECG detection of acute myocardial ischemia: sensitivity versus specificity. J. Electrocardiol. **51**(6), S12–S17 (2018)
23. Warren, S.G., Wagner, G.S.: The STAFF studies of the first 5 minutes of percutaneous coronary angioplasty balloon occlusion in man. J. Electrocardiol. **47**(4), 402–407 (2014)

# Automatic Left Bundle Branch Block Diagnose Using a 2-D Convolutional Network

Axel Wood[1], Marcos Cerrato[1], and María Paula Bonomini[2(✉)]

[1] Instituto Tecnológico de Buenos Aires (ITBA), Buenos Aires, Argentina
{awood,jcerrato}@itba.edu.ar
[2] Instituto Argentino de Matemática Alberto P. Calderón (IAM), CONICET, Buenos Aires, Argentina
paula.bonomini@conicet.gov.ar

**Abstract.** Left bundle branch block (LBBB) patients are the population that benefits most from cardiac resynchronization therapy (CRT), a therapy applied in heart failure. However, CRT presents about 40% non-responders rates. A plausible explanation to this fact, is a precarious LBBB diagnosis. QRS duration is currently one of three pillars in LBBB diagnosis. However, ECG morphology is severely altered in the presence of LBBB, affecting seriously the process of ECG delineation. Thus, QRS duration becomes a highly unreliable measure in LBBB diagnosis. Herein, we propose a LBBB classification framework complettely independent of temporal measures. In this line, a 2-D convolutional network (CNN) was utilized to separate strict LBBB patients from (not strict/not) LBBB patients, obtained from a subset of the Multi-center Autonomic Defibrillator Implantation (MADIT) trial. In order to fit the 2-D architecture, we fed the CNN with 10 s- spectrograms, constructing and validating 6 separated unilead models, one per precordial lead. From all analyzed models, the one using lead $V_1$ turned out to be the most informative. The latter, produced an 89% accuracy and 90% positive predictive value. These results encourage the use of such statistical models to provide a more reliable and automated LBBB diagnosis.

**Keywords:** LBBB diagnosis · CRT · Non-responders rate

## 1 Introduction

Electrical stimulation has become a powerful tool in many biomedical areas such as sensory restoration, cardiac dysfunction, pain relief or movement aid [1,6,7]. In particular, within the cardiac stimulation field, the advent of cardiac resynchronization therapy (CRT) has imposed a change of paradigm. This therapy includes a third lead in the left ventricle in order to induce more synchronized, and thus, more efficent interventricular contraction. Even though beneficial for heart failure patients, CRT remained with a 30–40% non-responders rate [5]. It has been postulated then, that CRT indication should be better tuned so


J. M. Ferrández Vicente et al. (Eds.): IWINAC 2022, LNCS 13258, pp. 576–585, 2022.
https://doi.org/10.1007/978-3-031-06242-1_57

that a higher portion of CRT recipients can benefit from it. Under this light, different views have arisen. On the one hand, a growing body of evidence has reported a high correlation between intraventricular dyssynchrony and CRT clinical outcome. On the other hand, since most LBBB (whether primary LBBB or secondary to heart failure) are referred to CRT, many researchers have simply dedicated themselves to improve LBBB diagnosis. By seeking a finer tune in LBBB diagnosis, some reports have utilized wavelets transform while others have relied on vectorcardiographic analysis [8,12].

From the study of the relationship between cardiac dyssynchrony and CRT clinical output, attempts to quantify the activation dynamics of the posterior wall of the left ventricle have been made [2], specially when septal stimulation emerged as an alternative to right ventricular pacing [1,10,11]. In addition, many non invasive models were developed to explain the delays in the free wall of the left ventricle. Vereckei et al. has measured time differences between intrinsecoid deflections of electrocardiographic leads [14], our group accomplished a mixed effects model purely based on electrocardiographic features to explain left ventricular activation measured by corronary sinus mapping [3] and implemented an algorithm to quantify the spatial variance in a set of electrocardiographic leads [4].

A particular caveat in LBBB diagnosis, whether a strict or a conventional criteria is applied, is the measure of QRS duration, since QRS morphologies are severely altered in LBBB patients, hampering the delineation process. Therefore, unreliable QRS offsets and onsets, cannot guarantee reliable QRS durations. In order to overcome this flaw, and become independent of ECG delineation, we tested a method for LBBB diagnosis entirely based on the ECG's spectrograms. In this work, we have addressed the usefulness of a 2-D convolutional network to differentiate healthy from LBBB patients on the frequency domain.

## 2 Materials and Methods

**Study Population.** The LBBB database contains a subset (n = 602) of heart failure patients included in the MADIT-CRT clinical trial, publicly available at the THEW project, from University of Rochester (Rochester, NY) [9]. From this subset, 330 patients have strict LBBB, 193 present not strict LBBB and 79 show no LBBB at all. The 12-lead high-resolution ECGs were recorded before CRT implantation using 24-hours Holter recorders (H12+, Mortara Instruments, Milwaukee, WI, USA) with Mason-Likar lead configuration (the Mortara system provides 10 electrodes and records 8-lead signals [$I$, $II$, $V_1$-$V_6$] the other 4 leads are calculated). The first 20-minute ECG signals were recorded while the patients were in a supine position. However, only 10 s excerpts were made publicly available, which is the recording length we have worked on. The sampling frequency was 1 kHz and amplitude resolution 3.75 microVolts.

**Signal Processing.** In this work, a 0.5–45 Hz Butterworth order 4 filter was applied to the ECG signals, in order to remove line interference and high frequency noise. In addition, signals, originally sampled at 1 KHz, were downsampled 100 Hz in order to reduce computational costs.

After preconditioning of the signals, we performed the spectrogram of the recordings. Spectrograms were performed with a 120-sample (1.2 s) Tukey window with shape parameter of 0.25 and 60 samples overlap, resulting in the spectrogram represented in Fig. 1, for the particular case of $V_3$ lead.

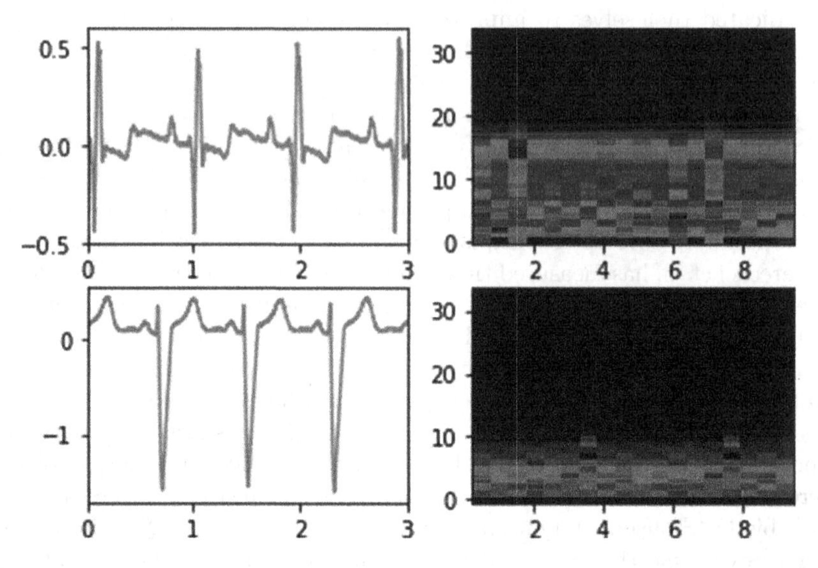


**Fig. 1.** Spectrograms for two representative non LBBB (top) and strict LBBB (bottom) patients. ECG traces are displayed on the left column while spectrograms are presented on the right column. Notice the increase of energy in higher frequencies for the patients with intact conduction while LBBB patients have diminished energy in the same frequency range.

**CNN Architecture.** As for the CNN, it consists of an input layer with the size of our training set followed by 3 convolutional layers of 20 filters with a kernel size of (3,3) and a Relu activation function. Each of these layers is followed by a 2D Max-pooling layer. The output of our third convolutional layer then passes through a flattening layer whose function is to convert the data into a one-dimensional matrix for input to the next layer. The output of the convolutional layers is flattened in order to create a single long feature vector. Then, we have 2 sets of dense layers with dropout layers the first with 64 neurons and a dropout of 0.5 and the second with 32 neurons and a dropout of 0.5 as well. Finally, with a Dense Layer with 2 neurons and a sigmoid activation function that works as the output layer for our model. Figure 2 shows the CNN architecture used in this work.

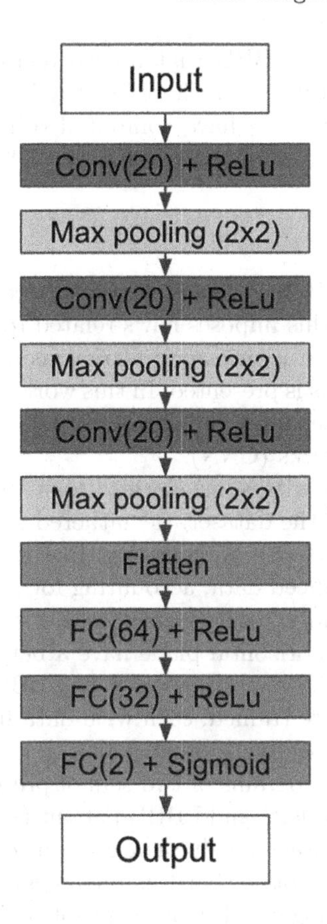


**Fig. 2.** The architecture of the proposed 2D - CNN model

**Data Splitting.** The dataset contains 330 patients with "strict LBBB", 193 with "not strict LBBB" and 79 with no LBBB at all ("not LBBB"). Due to the unbalanced nature of these data, we gathered together into a single "not LBBB" class those patients with "not LBBB" and "not strict LBBB". This way, we ended up with a much more balanced dataset, accounting for 330 LBBB patients and 272 not LBBB patients.
Afterwards, data split was performed, consisting of 70% (n = 421) training records, 18% (n = 108) validation records and 12% (n = 73) test records. The training and validation sets were used to train the network and the test set will be used to validate the network.

**Model Performance.** In order to assess the performance of each of the six models analyzed, the confusion matrix from each model was calculated. True Positive (TP), True Negative (TN), False Positive (FP) and False Negative (FN) cases were extracted. On these figures, different metrics have been computed. Sensitivity (or recall), also known as the true positive rate, measuring the proportion of individuals with LBBB that are correlated with the true positive rate, and specificity, providing the ratio of healthy individual correctly classified as non-LBBB patients. We also have computed other indicators, useful when evaluating models such as accuracy, precision and F1 value.

## 3 Results

As mentioned above, LBBB diagnosis is currently accomplished by analysing the patient's QRS duration. This imposes flaws related to unreliable fidutial points, on which the QRS duration is measured. To overcome this bug, an alternative method for LBBB diagnosis is presented. In this work, we developed an algorithm independent on ECG delineation that classifies LBBB using QRS spectra and convolutional neural networks (CNN).

Prior to describe the results, some clarifications should be made. Firstly, and due to unbalanced data in the dataset, we gathered together those patients with "not LBBB" and "not strict LBBB" in a "Not LBBB" class. This way, we ended up with a much more balanced data, accounting for 330 LBBB patients and 272 not LBBB patients. Secondly, we evaluated the network in the precordial leads, since the left leads in the horizontal plane have a better view of the free wall of the left ventricle, which shows a large delay in LBBB patients. In order to do this, 6 different models were trained, each with data from every precordial lead, and evaluated with its respective test set.

Figure 3 shows the specrograms of two sets of precordial leads ($V_1$ to $V_6$) for a paradigmatic not LBBB (left) and LBBB patient (LBBB). Notice the increase of energy in V1 for the intact conduction case, in contrast to a lower energy in the LBBB counterpart. Notice that this phenomenon holds for all 6 leads. In order to maintain a physiological view of the problem, Fig. 4 shows the respective ECG traces from Fig. 3.

After training and validation of each of the six models, performance was evaluated on the test sets. Such metrics for every model are shown in Table 1. Notice that the best result was obtained in the first model, with lead $V_1$. It is important to note that the results shown above evaluate the network already trained with the previously separated test data, which were unknown for the CNN.

From Table 1, it can be seen that V1 was the lead that provided the higher performance. Thus, Fig. 5 shows the accuracy of the training and validation data sets after training the network for 300 epochs, for the specific case of the first model (lead $V_1$). On the other hand, Fig. 5 shows the V1-lead confusion matrix for the test set. As the test set represents 12% of our total data (602), it is composed of a total of 73 items of data. The hyper-parameters of the model were as follows:

- Batch size = 20
- Epochs = 300
- Optimizer = Adam
- Learning rate = 0.0001
- Dropout rate= 0.5

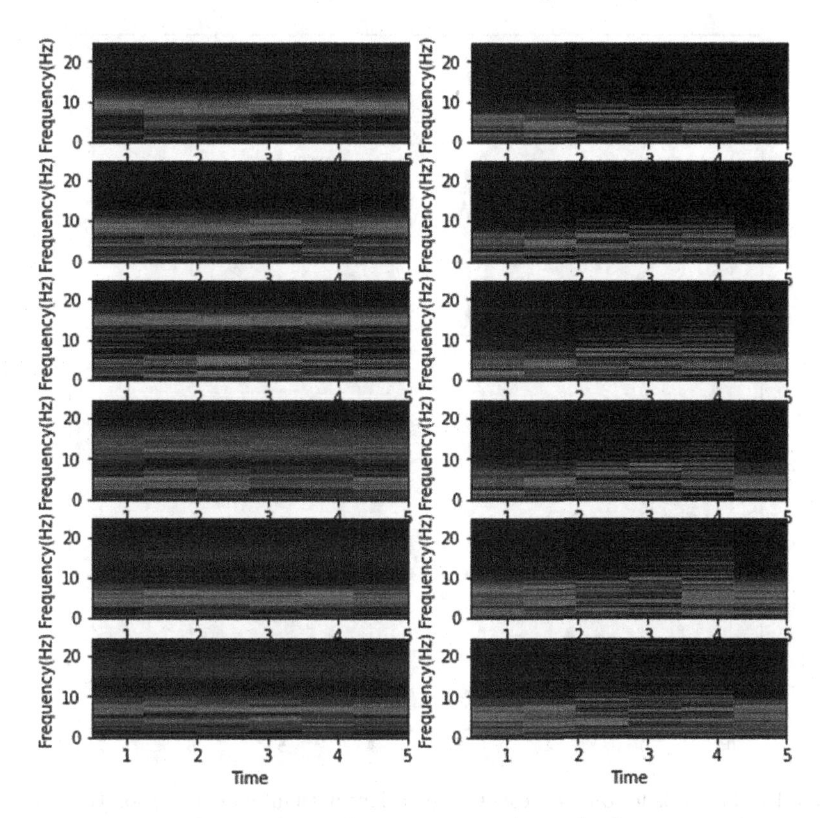


**Fig. 3.** Spectrograms for V1, V2, V3, V4, V5 and V6 leads, from top to bottom. For the spectrograms, a 120 ms Tukey window was utilized with a 60 samples overlap on the 10-s recordings bandpass-filtered and downsampled 100 Hz. Left: Intact conduction patient. Right: LBBB patient.

**Table 1.** Accuracy, Sensitivity, Specificity, Precision and F1-score for the test sets of every model representing each precordial lead.

| Derivation | Accuracy | Sensitivity | Specificity | Precision | F1 |
|---|---|---|---|---|---|
| $V_1$ | 0.89 | 0.9 | 0.88 | 0.9 | 0.9 |
| $V_2$ | 0.79 | 0.76 | 0.81 | 0.84 | 0.79 |
| $V_3$ | 0.73 | 0.59 | 0.91 | 0.89 | 0.71 |
| $V_4$ | 0.63 | 0.59 | 0.69 | 0.71 | 0.64 |
| $V_5$ | 0.7 | 0.85 | 0.5 | 0.69 | 0.76 |
| $V_6$ | 0.78 | 0.83 | 0.72 | 0.79 | 0.81 |

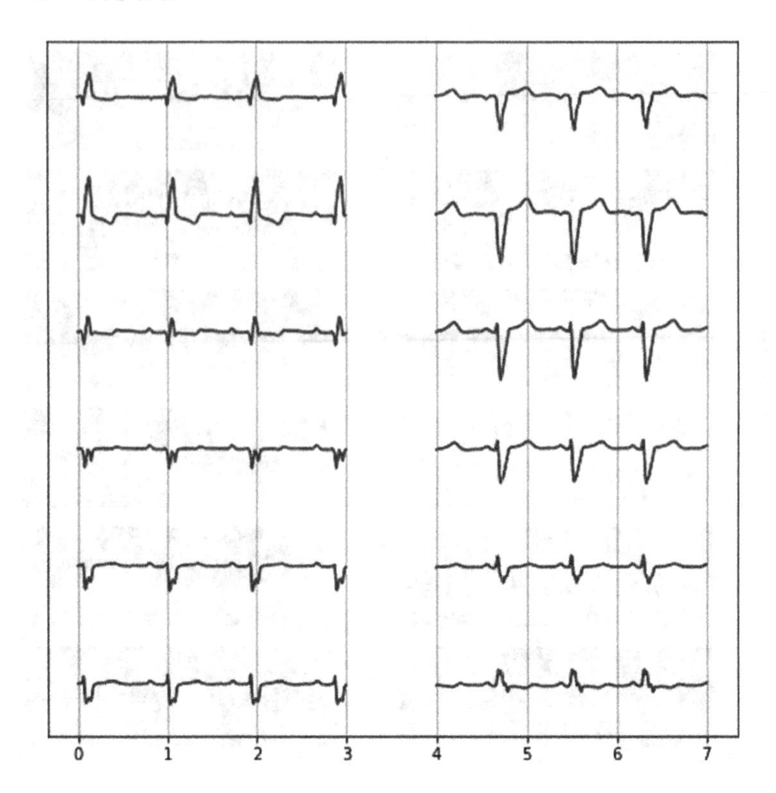


**Fig. 4.** ECG records for one representative patient with intact conduction (left) and one representative LBBB patient (right). Six precordial leads used for model construction were displayed, $V_1$ to $V_6$, from top to bottom.

From Fig. 5, it is noticeable that 100 epochs is a sensible cutoff for 100 epochs, since from 100 epochs on, the training curve starts separating from the validation curve, suggesting an over-fitting behaviour. Notice also that the confusion matrix for the same V1 lead confirms a higher sensitivity than specificity, observation that remained true for V1, V5 and V6 leads. This fact sheds light on the selection of precordial lead to use when trying to assess an automatic diagnosis, since in this particular case of CRT indication, sensitivity takes over specificity.

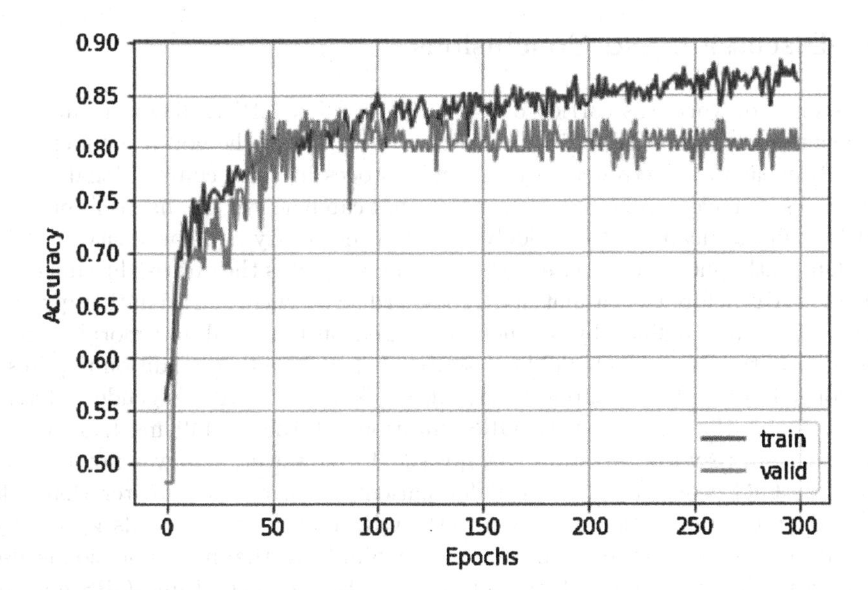


**Fig. 5.** CNN training and validation curves for the $V_1$-lead model. The model hyper-parameters were the following: Batch size = 20, epochs = 300, learning rate = 0.0001 and Dropout rate = 0.5.

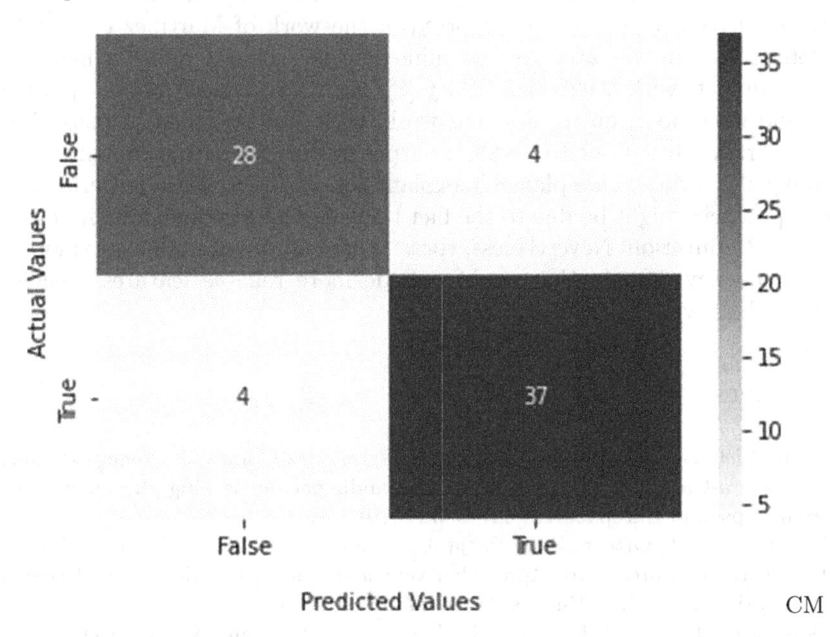


**Fig. 6.** Confusion matrix for model evaluation with test datasets (n = 73) ($V_1$)

## 4 Discussion and Conclusions

This piece of work was carried out in order to refine LBBB diagnosis and consequently, improve CRT indication, with a reduction in the non-responders rate [5]. By means of electrocardiogram signal processing and convolutional neural networks we have developed an algorithm that can perform the diagnosis of strict LBBB automatically, with a widely accepted sensitivity and specificity.

Currently, the criteria established by Strauss et al.is the one used to diagnose strict LBBB before the patient undergoes cardiac resynchronization therapy [13]. This criterion is defined by 3 conditions based on temporal and morphological features of the ECG which will be described below. The first condition requires a prolonged QRS whose temporal range depends on the patient's gender. Strauss et al. found that women with QRS durations of 130 to 149 ms had greater benefit from resynchronization therapy (CRT). However, this was not true for men, as CRT was more beneficial for patients with a QRS greater than 140 ms. The second condition is related to the QRS morphology of leads $V_1$ and $V_2$, specifically related to the size of the R wave. Similarly, the third condition is also related to the QRS morphology and imposes the presence of mid-QRS notches or slurs in some leads. All of the above conditions have improved the rate of correct diagnosis of strict LBBB.

Throughout the work, the best accuracy was achieved for the V1 lead, with a 90% sensitivity, close to that reported in the work of Martinez et al. where wavelets transform was used to determine whether the 3 conditions mentioned above were met, with a 92% sensitivity [8]. However, a much better specificity was found in comparison to Martinez work, with 88% specificity versus 65.1%. Also, accuracy turned out to be 89%, outperforming that from Martinez work, which equaled 79.5%. One plausible explanation to the increase in performance in our approach, might be due to the fact that we avoided time-domain features such as QRS duration. Nevertheless, these results encourage further exploration in the frequency domain that could provide more reliable features specific of LBBB pathology.

## References

1. Barba-Pichardo, R., Moriña-Vázquez, P., Fernández-Gómez, J., Venegas-Gamero, J., Herrera-Carranza, M.: Permanent his-bundle pacing: seeking physiological ventricular pacing. Europace **12**, 527–533 (2010)
2. Bonomini, M.P., Ortega, D.F., Barja, L.D., Mangani, N.A., Paolucci, A., Logarzo, E.: Electrical approach to improve left ventricular activation during right ventricle stimulation. Medicina (Buenos Aires) **77**, 7–12 (2017)
3. Bonomini, M., Ortega, D., Barja, L., Logarzo, E., Mangani, N., Paolucci, A.: ECG parameters to predict left ventricular electrical delay. J. Electrocardiol. **51**, 844–850 (2018)
4. Bonomini, M., Ortega, D., Barja, L., Mangani, N., Arini, P.: Depolarization spatial variance as a cardiac dyssynchrony descriptor. Biomed. Signal Process. Control **49**, 540–545 (2019)

5. Chung, E.S., et al.: Results of the predictors of response to CRT (*PROSPECT*) trial. Circulation **117**(20), 2608–2616 (2008)
6. Ferrandez, J.M., Liano, E., Bonomini, P., Martinez, J., Toledo, J., Fernandez, E.: A customizable multi-channel stimulator for cortical neuroprosthesis. In: 2007 29th Annual International Conference of the IEEE Engineering in Medicine and Biology Society, pp. 4707–4710 (2007)
7. Ferrandez, J., Alfaro, A., Bonomini, P., Tormos, J., Concepcion, L., Pelayo, F., Fernandez, E.: Brain plasticity: feasibility of a cortical visual prosthesis for the blind. In: Proceedings of the 25th Annual International Conference of the IEEE Engineering in Medicine and Biology Society (IEEE Cat. No.03CH37439), vol. 3, pp. 2027–2030 (2003)
8. Martín-Yebra, A., Martínez, J.: Automatic diagnosis of strict left bundle branch block using a wavelet-based approach. PLoS ONE **14**, e0212971 (2019)
9. Moss, A.J., et al.: Cardiac-resynchronization therapy for the prevention of heart-failure events. New England J. Med. **361**(14), 1329–1338 (2009)
10. Ortega, D., et al.: Novel implant technique for septal pacing. a noninvasive approach to nonselective his bundle pacing. J. Electrocardiol. **63**, 35–40 (2020)
11. Ortega, D.F., Barja, L.D., Logarzo, E., Mangani, N., Paolucci, A., Bonomini, M.P.: Nonselective his bundle pacing with a biphasic waveform. enhancing septal resynchronization. Europace **20**, 816–822 (2018)
12. Pérez-Riera, A., et al.: Re-evaluating the electro-vectorcardiographic criteria for left bundle branch block. Ann. Noninvasive Electrocardiol. **24**, e12644 (2019)
13. Strauss, D.G.., Selvester, R., Wagner, G.: Defining left bundle branch block in the era of cardiac resynchronization therapy. Am. J. Cardiol. **107**(6), 927–934 (2011)
14. Vereckei, A., et al.: Novel electrocardiographic dyssynchrony criteria improve patient selection for cardiac resynchronization therapy. Europace **20**, 97–103 (2018)

# QRS-T Angle as a Biomarker for LBBB Strict Diagnose

Beatriz del Cisne Macas Ordónez[1,2], José Manuel Ferrández-Vicente[3], and María Paula Bonomini[1,2](✉)

[1] Instituto Argentino de Matemática "Alberto P. Calderón" (IAM), CONICET, Buenos Aires, Argentina
paula.bonomini@conicet.gov.ar
[2] Instituto de Ingeniería Biomédica (IIBM), Fac. de Ingeniería, Universidad de Buenos Aires, Buenos Aires, Argentina
bmacas.ext@fi.uba.ar
[3] Dpto. Electrónica, Tecnología de Computadoras y Proyectos, Univ. Politécnica de Cartagena, Cartagena, Spain
jm.ferrandez@upct.es

**Abstract.** Currently, the diagnosis of left bundle branch block (LBBB) is based on the duration of the QRS complexes. However, aberrant QRS complexes, characteristic of LBBB, obscure the delineation process, and therefore, unreliable QRS durations are produced, due to erratic fiducial points. To overcome this flaw, we propose the QRS-T angle as an alternative LBBB biomarker, which is independent of QRS duration. A subset of heart failure patients with and without LBBB was compared to a set of healthy patients, all obtained from the Telemetric and ECG Holter Warehouse Project. The methodology included signal filtering, construction of QRS and T loops and obtention of depolarization and repolarization dominant vectors. Finally, the QRS-T angle in the different 2D and 3D spaces were calculated. Results show that patients with LBBB have much larger QRS-T angles compared to Controls. Moreover, QRS-T angles in the $XY$ plane and the $XYZ$ volume were the markers that best separated LBBB from intact conduction patients, with accuracies of 100% and 96.9%, respectively. These results set the foundations to further investigate the QRS-T angle as a biomarker for the diagnosis of left bundle branch block.

**Keywords:** QRS-T angle · Left Bundle Branch Block (LBBB) · Vectorcardiogram

## 1 Introduction

Electrical stimulation has provided several solutions in the biomedical field, including sensory restoration and cardiac dysfunction [11,12,21]. Particularly in cardiology, the first pacemaker was implanted in Sweden in 1957, and since then, the career of implantable pacemakers has not stopped. Normally, pacing catheters are placed at the tip of the right ventricle (apex) to ensure the mechanical stability that chronic pacing requires. However, evidence on the deleterious


J. M. Ferrández Vicente et al. (Eds.): IWINAC 2022, LNCS 13258, pp. 586–594, 2022.
https://doi.org/10.1007/978-3-031-06242-1_58

effects of pacing in the apex has recently been generated [2]. For this reason, the search for alternative cardiac pacing sites has intensified in recent years, from which His bundle pacing (septal pacing) has emerged as the most physiological and convenient alternative. Over time, mechanically stable sheaths and catheters were achieved that allowed chronic septal stimulation schemes [1]. In this context, the study of intraventricular dyssynchrony for the selection of the optimal pacing site became vitally important, since the interventricular septum is relatively extensive [20]. In addition, the evaluation of cardiac dyssynchrony today determines the selection of those patients who would potentially benefit from biventricular pacing therapies, such as cardiac resynchronization therapy (CRT) [7]. Currently, QRS duration is the only electrocardiographic parameter of cardiac dyssynchrony, so only those patients with a wide QRS (>120 ms) are recommended for CRT (Dickstein et al., 2008). Paradoxically, 43% of patients with heart failure and narrow QRS (<120 ms) have mechanical dyssynchrony on echocardiography. Furthermore, when dyssynchrony is well documented by all echocardiographic methods, response to therapy tends to be favorable regardless of QRS duration [8].

Recently, a new approach has emerged, aiming to evaluate CRT clinical outcome based on LBBB diagnosis. Many reports focus on reducing CRT non-responders rate by improving its indication. Since most LBBB patients are likely to receive CRT, then, refining LBBB diagnosis would have a positive impact on the amount of patients that would effectively benefit from CRT [17]. In this piece of work, we computed the QRS-T angles on LBBB patients and compared them with those from native conduction subjects in different spaces.

## 2 Materials and Methods

**Study Population.** The data were provided by the ECG Telemetry and Holter ECG Warehouse (THEW), two databases are the subject of the present study, one containing the records of healthy subjects (Health group) and a second database containing the records of patients with left bundle-branch block (LBBB group). The database E-OTH-12–0602-024 [18], contains 602 recordings of heart failure patients included in the Multi-center Autonomic Defibrillator Implantation (MADIT-CRT) clinical trial, publicly available at the THEW project, from University of Rochester (Rochester, NY). The 12-lead high-resolution ECGs were recorded before CRT implantation using 24-h Holter recorders (H12+, Mortara Instruments, Milwaukee, WI, USA) with Mason-Likar lead configuration (the Mortara system provides 10 electrodes and records 8-lead signals [$I$, $II$, $V_1$-$V_6$] the other 4 leads are calculated). The first 20-min ECG signals were recorded while the patients were in a supine position. The sampling frequency was 1 kHz and amplitude resolution 3.75 $\mu$V. The database E-HOL-03–0202-003 (Intercity Digital Electrocardiogram Alliance - IDEAL) [22],contains 202 recordings of healthy individuals captured from Holter recordings (SpaceLab-Burdick Inc.) of around 24 h. The ECG signals have a sampling frequency 200 Hz, an amplitude resolution of 10 $\mu$V, and have been captured using a three pseudo-orthogonal leads configuration ($x$, $y$, $z$).

**Preprocessing.** In the preprocessing stage, a fourth-order Butterworth band-pass filter was used in the range of 0.5–40 Hz, in order to remove baseline drift and high-frequency noise. For the LBBB group, VCG records were reconstructed by means of the inverse Dower matrix as shown in Table 1. This transformation poses the orthogonal leads, $x$,$y$,$z$, as linear combinations of the eight linearly independent ECG leads. It is worth mentioning that this transform only applied to the LBBB group, since recordings from the Control group were already obtained from the Frank orthogonal lead system.

**Table 1.** Transformation matrix for Inverse Dower transformation (IDT).

| Lead | V1 | V2 | V3 | V4 | V5 | V6 | I | II |
|---|---|---|---|---|---|---|---|---|
| X | −0.172 | −0.074 | 0.122 | 0.231 | 0.239 | 0.194 | 0.156 | −0.010 |
| Y | 0.057 | −0.019 | −0.106 | −0.022 | 0.041 | 0.048 | −0.227 | 0.887 |
| Z | −0.229 | −0.310 | −0.246 | −0.063 | 0.055 | 0.108 | 0.022 | 0.102 |

The control and LBBB signals were delineated by means of a wavelet-based algorithm using the WT-delineator library in Python [15]. The delineation algorithm detects the onset and offset of the QRS and T-wave, from which the QRS and T loops were constructed.

It is important to note that patients from the LBBB group come from the MADIT trial, including all kind of heart failure patients. With the help of independent physicians, all 602 heart failure patients were classified into three classes: "strict LBBB", "not strict LBBB" and "not LBBB". Therefore, from the E-OTH-12-0602-024 database, only the first 200 records were utilized in order to balance the data with the Control group (n = 200). Afterwards, from the first 200 delineated records, only those with "strict LBBB" and aceptable fidutial points were preserved, keeping 47 electrocardiographic records. Finally, 60 VCG records from the Control group were obtained.

**QRS-T Angles.** QRS segments were obtained by applying a window that spanned between the median of the QRS onsets and offsets of all beats in the recording. This window was centered on the QRS peak of every QRS complex. Afterwards, an ensemble was constructed with the above QRS segments and averaged in order to obtain a QRS template on every vectorcardiographic lead $x$,$y$,$z$. A similar procedure was accomplished in order to derive a T-wave template.

Following obtention of the QRS and T templates, the vectorcardiographic loops were constructed and aligned to the same origin, and dominant vectors defined as the maxima euclidean norm progressing between the onset and offset of the QRS and T loops in each of the following coordinate systems $xy$, $xz$, $yz$ and $xyz$, as shown in Eq. 1 for the $XY$ plane case:

$$opt = \arg\max_{i} \left\{ d[i] = \sqrt{(x[i] - x[0])^2 + (y[i] - y[0])^2} \right\}, 0 <= i <= t_{off} \quad (1)$$

where $x[0]$ and $y[0]$ are the QRS and T loop origins and $t_{off}$ denotes the duration of the QRS or the T-wave templates, respectively. Dominant depolarization and repolarization vectors were then defined as $\mathbf{QRS}_{opt}/\mathbf{T}_{opt} = \{x[opt], y[opt]\}$. Once the QRS and T dominant vectors were determined, the angle between both was computed for every coordinate system as follows:

$$\theta_{QRS-T} = \left\{\frac{180}{pi}\right\} \arccos \frac{< \mathbf{QRS}_{opt}, \mathbf{T}_{opt} >}{|\mathbf{QRS}_{opt}| * |\mathbf{T}_{opt}|} \quad (2)$$

where $< \mathbf{QRS}_{opt}, \mathbf{T}_{opt} >$ denotes the inner product of the dominant vectors and $|\mathbf{QRS}_{opt}|$, $|\mathbf{T}_{opt}|$ their $L_2$ norms.

## 3 Statistical Analysis

Descriptive data are presented as mean ±SD. Normality was assessed by means of the Shapiro-Wilks test. For comparison between two independent groups, the Mann-Whitney U Test was utilized.

To test the informative power of angles in the different planes, we constructed a simple Tree decision algorithm with 13 nodes and tested it on every coordinate space $XY$, $XZ$, $YZ$ and $XYZ$. Data splitting was carried out as follows: 70% (n = 75) of the data was used for the training and remaining 30% (n = 32) for testing. The classification scheme was binary "LBBB" vs "not LBBB", where "not LBBB" gathered together "not strict LBBB" and "not LBBB" instances. In order to reduce the effect of overfitting, a cross-validation scheme was set to 5 folds.

## 4 Results

### 4.1 QRS-T Angles

Figure 1 represents the electrical currents within the heart during a cardiac cycle in terms of its vectorcardiogram, showing the ventricular contraction (QRS complex forming the large loop) and the relaxation phase (T waves including the small loop) for representative Control and LBBB patients. Notice that in the LBBB case, a widening of the QRS complex as well as an inversion in the T-wave can be confirmed, opposite to the Control case, where narrow and consistent QRS and T-waves appear.

In analogy, Fig. 2 shows the loops generated from VCG for a healthy and LBBB patient. Notice in the LBBB patient, that T-wave inversion is reflected in a repolarization loop pointing to the opposite side than the depolarization loop. This produces enlarged QRS-T angles, in average greater than 90°. Note also the widened QRS complex for the pathological case, which in turn produces QRS loops with higher energy, as can be appreciated in Fig. 2. Finally, note the twisted morphology of the depolarization loop in the LBBB case, evidencing a much more intrincated electrical path, in contrast with those with native conduction (Controls), the cardiac electrical vector shows a clearly defined trajectory.

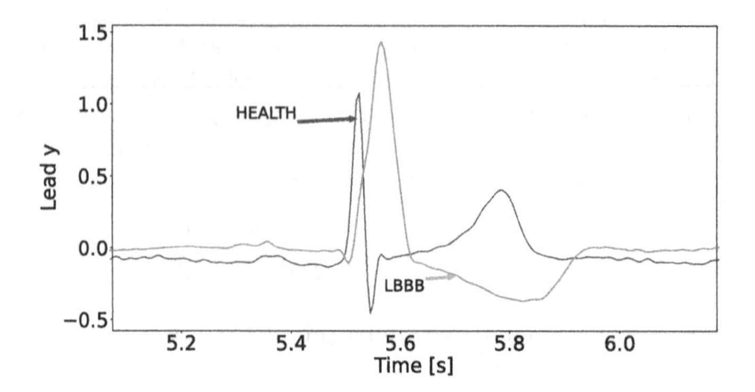


**Fig. 1.** VCG Control Group/ LBBB Group

Figure 3 illustrates the average QRS-T angles in the different planes, as well as in the XYZ volume, resulting in angles under 70° for the Control group. This result is confirmed by Cortez y Schlegel in his study [9] where the Dower-related reconstruction results in a spatial QRS-T angle of 66° to 81°.

On the contrary, in the case of subjects with left bundle branch block, the median QRS-T angles were greater than 146° in every 2D and 3D representations. Note that QRS-T angles resulted statistically significant in all four analyzed spaces according to the Mann-Whitney U test, with p-values lesser than 0.05.

### 4.2 Classification Results

Figure 4 shows the confusion matrices obtained for the test sets on every constructed model, each related to a different space. Notice that each model represents a Tree algorithm with 13 nodes, fed by the angles in one out of four spaces analyzed in this piece of work: $XY$, $ZY$ and $XZ$ planes and the $XYZ$ volume. Surrounding the confusion matrices there are the performance metrics. Bottom corner: accuracy, Top right: precision for Controls, Top middle: precision for LBBB, Bottom left: specificity for "LBBB" class, Bottom middle: sensitivity for "LBBB" class. Notice that the $XY$ plane turned out to be the most informative when separating LBBB from Controls (accuracy = 100%), followed by the volume $XYZ$ (accuracy = 96.9%).

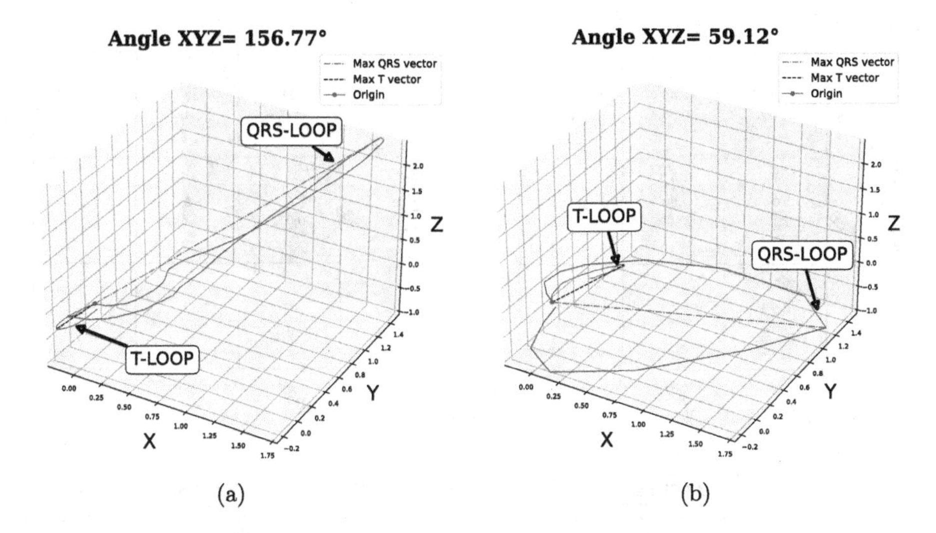


**Fig. 2.** VCG angles: (**a**) LBBB Group. (**b**) Health Group

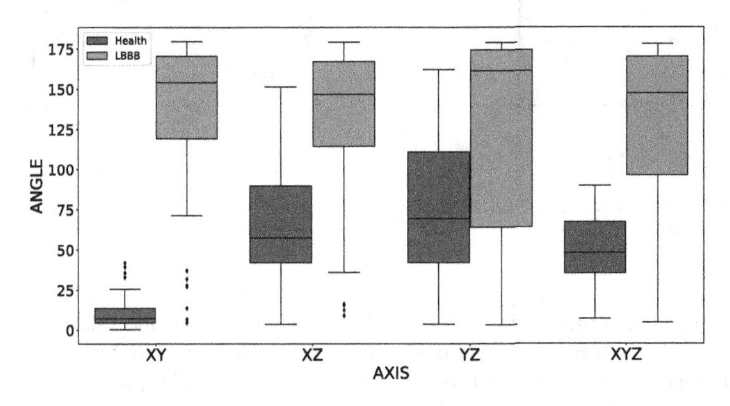


**Fig. 3.** Intercuartile distribution of QRS-T angles in different coordinate systems for the LBBB and Control group. QRS-T in the LBBB group was significantly increased with respect to the Control group (Mann-Whitney U test, $p < 0.05$) in all analyzed spaces.

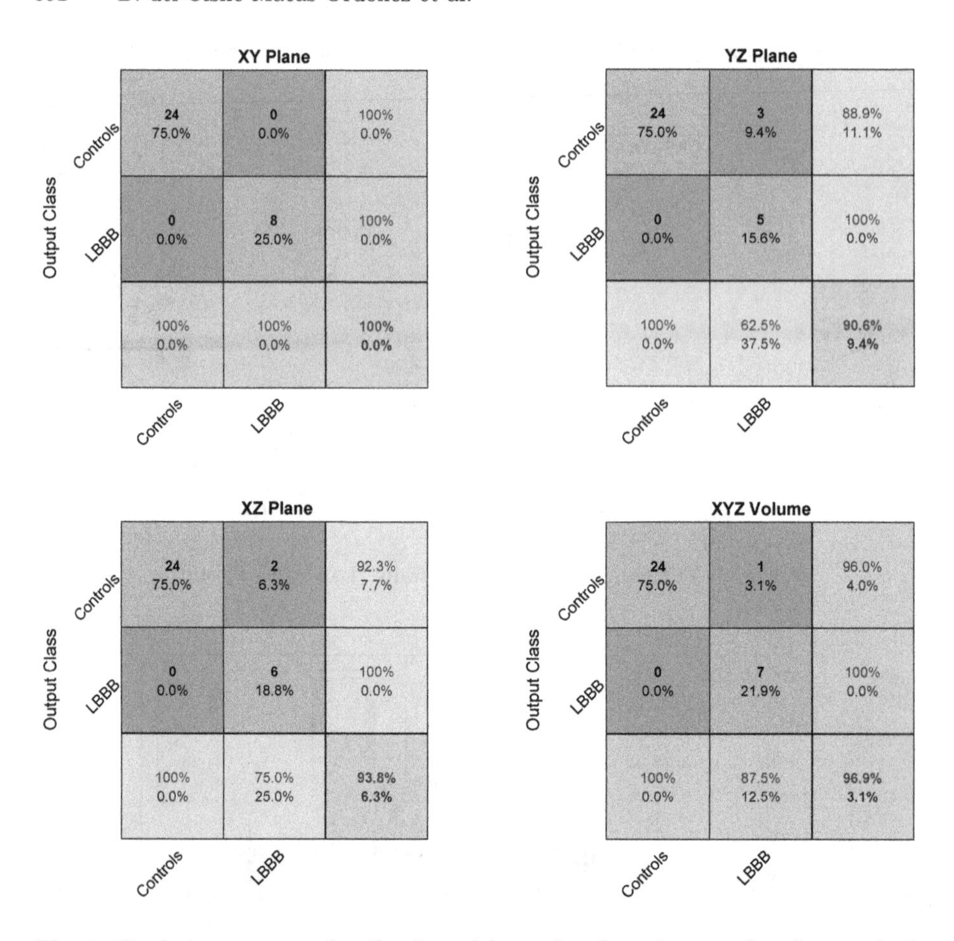


**Fig. 4.** Confusion matrices for the 4 models analyzed, each one related to a single feature or angles in a plane (or volume). Surrounding the $2 \times 2$ confusion matrix there are the performance metrics. Bottom corner: accuracy, Top right: precision for Controls, Top middle: precision for LBBB, Bottom left: specificity for "LBBB" class, Bottom middle: sensitivity for "LBBB" class.

# 5 Discussion and Conclusions

In this study, an algorithm using the WT-delineator library in Python was used. The wavelet-based model introduced by martinez et all [16], has the highest performance in both sensitivity and accuracy (Se = 99.8% and P+ = 99.86%), which is why it was chosen for the present study. It is important to note that despite the good sensitivity of the delineator used, a manual validation of the detected fidutial points was necessary due to the nature of the ECG recordings of patients with LBBB, where the detection of fidutial points is severely hampered by aberrant QRS morphologies.

The concept of the spatial QRS-T angle has been known for a long time [10] and it has recently regained interest as an independent predictor of cardiac death [14]. It has also been shown that the enlarged QRS-T angle predicts ventricular arrhythmias, since narrow spatial angle were associated with a lower risk of ventricular arrhythmias [13]. The main objective of the present study is to set the basis for the measurement of QRS-T angles with an LBBB-diagnosis purpose.

In a normal myocardium, the QRS and T-vector loops point in the same spatial direction, but with changes in ventricular depolarization and repolarization, the vectors point in opposite spatial directions [19]. This is specially true in LBBB, and one of the foundations in postulating the QRS-T angle as a biomarker for LBBB. In this case, we set the distribution values for both populations in different spaces, analyzing either planar and spatial QRS-T angles. We postulate that the VCG is a more compact space that will describe with minimum dimensions the trajectory of the cardiac electrical activity. However, more posibilities should be analyzed, such as preferential planes obtained by PCA, for instance, in order to infer which representation suits best the problem, as in [3,4].

Many efforts have been done to quantify cardiac dyssynchrony in the last decade. Our group pioneered on septal stimulation [21] and proposed a no invasive models to predict electrical activation of the left ventricle measured invasively [2,5,6]. However, these models, since they were based on electrophysiologic studies, accounted for few data, generally under 50 patients. This opportunity, the THEW project allowed us to analyze a higher amount of data in order to have more realistic results. Even though the study presented in this work is a foundational analysis, further studies should improve the delineation process of LBBB patients in order to complete the analysis for the entire dataset, and be able to separate not strict from strict from not LBBB. Results herein encourage such further efforts.

## References


1. Barba-Pichardo, R., Moriña-Vázquez, P., Fernández-Gómez, J., Venegas-Gamero, J., Herrera-Carranza, M.: Permanent his-bundle pacing: seeking physiological ventricular pacing. Europace **12**, 527–533 (2010)
2. Bonomini, M.P., Ortega, D.F., Barja, L.D., Mangani, N.A., Paolucci, A., Logarzo, E.: Electrical approach to improve left ventricular activation during right ventricle stimulation. Medicina (Buenos Aires) **77**, 7–12 (2017)
3. Bonomini, M.P., Corizzo, S.J., Laguna, P., Arini, P.D.: 2D ECG differences in frontal vs preferential planes inpatients referred for percutaneous transluminal coronary angioplasty. Biomed. Signal Process. Control **11**, 97–106 (2014)
4. Bonomini, M., Arini, P., Gonzalez, G., et al.: The allometric model in chronic myocardial infarction. Theor. Biol. Med. Model. **9**, 15–22 (2012)
5. Bonomini, M., Ortega, D., Barja, L., Logarzo, E., Mangani, N., Paolucci, A.: ECG parameters to predict left ventricular electrical delay. J. Electrocardiol. **51**, 844–850 (2018)

6. Bonomini, M., Ortega, D., Barja, L., Mangani, N., Arini, P.: Depolarization spatial variance as a cardiac dyssynchrony descriptor. Biomed. Signal Process. Control **49**, 540–545 (2019)
7. Bonomini, M., Villarroel-Abrego, H., Garillo, R.: Spatial variance in the 12-lead ECG and mechanical dyssynchrony. J. Interv. Card. Electrophysiol. **62**, 479–485 (2021)
8. Breithardt, O., Kühl, H., Stellbrink, C.: Acute effects of resynchronisation treatment on functional mitral regurgitation in dilated cardiomyopathy. Heart **88**, 440 (2002)
9. Cortez, D.L., Schlegel, T.T.: When deriving the spatial QRS-T angle from the 12-lead electrocardiogram, which transform is more frank: regression or inverse dower? (2010), https://pubmed.ncbi.nlm.nih.gov/20466388/
10. Draper, B.W.H., Peffer, C.J., Stallmann, F.W., Littmann, D., Pipberger, H.V.: The corrected orthogonal electrocardiogram and vectorcardiogram in 510 normal men (frank lead system), http://ahajournals.org
11. Ferrandez, J.M., Liano, E., Bonomini, P., Martinez, J., Toledo, J., Fernandez, E.: A customizable multi-channel stimulator for cortical neuroprosthesis. In: 2007 29th Annual International Conference of the IEEE Engineering in Medicine and Biology Society, pp. 4707–4710 (2007). https://doi.org/10.1109/IEMBS.2007.4353390
12. Ferrandez, J., et al.: Brain plasticity: feasibility of a cortical visual prosthesis for the blind. In: Proceedings of the 25th Annual International Conference of the IEEE Engineering in Medicine and Biology Society (IEEE Cat. No.03CH37439), vol. 3, pp. 2027–2030 (2003). https://doi.org/10.1109/IEMBS.2003.1280133
13. Jan, C., et al.: Predicting ventricular arrhythmias in patients with ischemic heart disease clinical application of the ECG-derived QRS-T angle. Circulation: Arrhythmia Electrophysiol. **2**(5), 548–554 (2009). https://doi.org/10.1161/CIRCEP.109.859108, http://ahajournals.org
14. Kardys, I., Kors, J.A., Meer, I.M.V.D., Hofman, A., Kuip, D.A.M.V.D., Witteman, J.C.M.: Spatial QRS-T angle predicts cardiac death in a general population. Eur. Heart J. **24**(14), 1357–1364 (2003). https://doi.org/10.1016/S0195-668X(03)00203-3, https://academic.oup.com/eurheartj/article/24/14/1357/501772
15. Ledezma, C.A.: WTdelineator (2021), https://github.com/caledezma/WTdelineator
16. Martínez, J.P., Almeida, R., Olmos, S., Rocha, A.P., Laguna, P.: A wavelet-based *ECG* delineatior: *E*valuation on standard databases. IEEE Trans. Biomed. Eng. **51**(4), 570–581 (2004)
17. Martín-Yebra, A., Martínez, J.: Automatic diagnosis of strict left bundle branch block using a wavelet-based approach. PLoS ONE **14**, e0212971 (2019)
18. Moss, A.J., et al.: Cardiac-resynchronization therapy for the prevention of heart-failure events. New England J. Med. **361**(14), 1329–1338 (2009)
19. Oehler, A., Feldman, T., Henrikson, C.A., Tereshchenko, L.G.: QRS-T angle: a review. Ann. Noninvasive Electrocardiol. **19**, 534–542 (2014). https://doi.org/10.1111/anec.12206
20. Ortega, D., et al.: Novel implant technique for septal pacing. a noninvasive approach to nonselective his bundle pacing. J. Electrocardiol. **63**, 35–40 (2020)
21. Ortega, D.F., Barja, L.D., Logarzo, E., Mangani, N., Paolucci, A., Bonomini, M.P.: Nonselective his bundle pacing with a biphasic waveform. enhancing septal resynchronization. Europace **20**, 816–822 (2018)
22. Telemetric, of Rocher Medical Center, H.E.W.U.: E-hol-03-0202-003, http://thew-project.org/Database/E-HOL-03-0202-003.html

# Variable Embedding Based on L–statistic for Electrocardiographic Signal Analysis

Lucas Escobar-Correa[1(✉)], Juan Murillo-Escobar[2], Edilson Delgado-Trejos[1], and David Cuesta-Frau[3]

[1] AMYSOD Lab –Parque i, CM&P Research Group, Instituto Tecnológico Metropolitano ITM, 050034 Medellín, Colombia
lucasescobar213859@correo.itm.edu.co, edilsondelgado@itm.edu.co

[2] Department of Exact and Applied Sciences, GI2B Research Group, Instituto Tecnológico Metropolitano ITM, 050034 Medellín, Colombia
juanmurillo@itm.edu.co

[3] Technological Institute of Informatics, Universitat Politecnica de Valencia, Alcoi Campus, 03801 Alcoi, Spain
dcuesta@disca.upv.es

**Abstract.** In this paper, a variable embedding approach for reconstructing attractors of dynamical systems is proposed, using the L–statistic based on noise amplification. Particularly, the variable manifold is obtained from a time-series using delay coordinates and an embedding vector, the last one, is constructed based on a L–statistic matrix which contains the local reconstruction quality of whole attractor. The embedding vector contains the optimal embedding dimension for each point in the manifold. This approach were performed on electrocardiography databases, we obtain the first four statistical moments for the embedding dimension vectors and apply statistical tests to distinguish between normal and pathological signals. Results shown significant differences that lead to new classification strategies, infer about functional states, and establish a new path for processing signals with high embedding dimensions, i.e., high computer complexity.

**Keywords:** Variable embedding · Embedding dimension · Signal analysis

## 1 Introduction

Signals derived from deterministic nonlinear systems, e.g. physiological signals, represent a high computational cost for nonlinear processing routines, which usually leads to non-feasible algorithm implementation, due to low development

---

This work is presented in partial fulfillment of the requirements for the "Call for the strengthening of vocations and training in ST&I for economic reactivation in the framework of the 2020 post-pandemic" No. 891 of MinCiencias- Colombia.


J. M. Ferrández Vicente et al. (Eds.): IWINAC 2022, LNCS 13258, pp. 595–604, 2022.
https://doi.org/10.1007/978-3-031-06242-1_59

of effective optimization routines that should simplify the computational complexity involved in nonlinear analysis [2,8,10].

The increase of the embedding dimension also increase the computational effort required to compute nonlinear features from the response of the dynamic systems [5,7,16,16,33]. In fact, nonlinear analysis is based on attractors obtained by time delay reconstructions [26], where conventional procedures are based on fixed embedding dimensions. Likewise, the use of a variable embedding dimension (i.e. the embedding dimension changes over time) was proposed since 1998 by Judd and Mess [12]. However, variable embedding has not been widely applied in literature such the practical use has not been proved. Several approaches to find variable embedding has been proposed, such as the isometric feature mapping [27], local false nearest neighbors [1], and others methodologies based on unsupervised learning [24,34].

In this paper, an experimental setup is proposed in order to show that the variable embedding could decrease the computational requirements involved in reconstructed attractor processing in high dimensional embedding spaces. In this regard, a variable embedding approach is proposed which is based on L–statistic [30] and is tested in electrocardiographic signal analysis.

## 2 Materials and Methods

### 2.1 Electrocardiography Databases

Electrocardiographic signals (ECG) are of interest for studying and testing the proposed approach of this work due to their impact in the identification of functional states in human health. Table 1 shows the 4 employed databases: MIT-BIT Normal Sinus Rhythm (NSR), MIT-BIT Arrhythmia (ARH), MIT-BIT Atrial Fibrillation (AF) and Congestive Hearth Failure (CHF). Altogether, 99 signals were chosen for this study, specifically those acquired in the second modified lead (MLII). Finally, because of these databases were sampled at different frequencies, all signals had to be re-sampled 128 Hz.

**Table 1.** Electrocardiography databases

| Database | # of signals | ADC resolution | Sampling frequency | link |
|---|---|---|---|---|
| MIT-BIT normal sinus rhythm | 18 | 12 | 128 | Normal |
| MIT-BIT Arrhythmia | 46 | 11 | 360 | Arrhythmia |
| MIT-BIT Atrial Fibrillation | 20 | 12 | 250 | Atrial |
| Congestive Heart Failure | 15 | 12 | 250 | Congestive |

ADC: Analog to digital conversor

## 2.2 Phase Space Reconstruction

The phase space is a multidimensional space that contains all the possible system states and the trajectory that the system variables follow within the space is called attractor [9]. It is important to study the phase space (or state space) in order to perform accurate predictions, to know the intrinsic dynamics of a system and to infer about their functional state. However, such an attractor is not available for practical applications, as equations that model the system dynamics are difficult to achieve or are not able to measure (observe) all the variables in the system or both.

The time delay reconstruction is the most commonly used method to reconstruct the state space from available observations [20], where the state trajectory is obtained by means of a ordered set of $m-$dimensional vectors extracted from the observations corresponding to the system response $x$, by delaying $\tau$ sampling unities, thus: $\overrightarrow{s}(t) = \{x(t), x(t-\tau), x(t-2\tau), \ldots, x(t-(m-1)\tau)\}$, where $x(t)$ is an observation of the system at the $t$ instant and $\overrightarrow{s}(t)$ is a vector in the state space. Therefore, an attractor is the result of embedding a signal in a state space, with $m-$dimensional points temporally sorted, as $\mathbf{S} = \{\overrightarrow{s}(1), \overrightarrow{s}(2), \ldots, \overrightarrow{s}(n)\}$.

The parameters involved in the phase space reconstruction, $m$ and $\tau$, must be rigorously selected to guarantee that the obtained phase space will be a diffeomorphic copy to the real embedding of the analysis phenomenon if the model equations would be available. In this aim, several approaches have been proposed [8,10,13,17,21,22,26], but these developments do not involve relevant intrinsic features that can contain the signals to embed, causing unreliable results.

## 2.3 L–statistic

L–statistic is a reconstruction quality measure of an attractor, based on minimizing the complexity in the reconstruction and a concept of noise amplification, in order to make the reconstruction method independent of the reconstruction quality [30]. Thus, the quantification is performed by a combination of measures that penalize relevance $\sigma$ (inspired on noise amplification [4]) and irrelevance $\alpha$ (normalization factor required to adjust $\sigma$). Additionally, this approach has three free parameters, the prediction horizon ($T_{Max}$), the Theiler window ($Th_w$, used to avoid correlated points) and a number of nearest neighbors ($k$).

This method starts taking a point in the reconstructed attractor at the time $t$ ($\overrightarrow{s}(t)$) and considering $s'$ as the first component of the vector $\overrightarrow{s}(t)$. The estimation of the noise amplification is performed by recurring to the $k$ nearest neighbors, considering a $Th_w$, as the minimum temporal separation between the attractor point and the $k$ neighbors, where a neighborhood of $k+1$ points, $B_k(\overrightarrow{s})$, is defined [28]. In order to approximate the conditional variance of the prediction function, the average value of the $k+1$ points at the time $t+T$, in the neighborhood, is computed as:

$$u_k(T, \overrightarrow{s}) = \frac{1}{k+1} \sum_{\overrightarrow{g} \in B_k(\overrightarrow{s})} g'(T) \tag{1}$$

where $g'(T)$ is the first component of the vector $\overrightarrow{g}$ at the time $t+T$. Therefore, the conditional variance is approximated by

$$E_k^2(T,\overrightarrow{s}) = \frac{1}{k+1} \sum_{\overrightarrow{g} \in B_k(\overrightarrow{s})} [g'(T) - u_k(T,\overrightarrow{s})]^2 \tag{2}$$

And the $k-$neighbourhood size is estimated as

$$\epsilon_k^2(\overrightarrow{s}) = \frac{2}{k(k+1)} \sum_{\substack{\overrightarrow{g},\overrightarrow{h} \in B_k(\overrightarrow{s}) \\ \overrightarrow{g} \neq \overrightarrow{h}}} \left|\overrightarrow{g} - \overrightarrow{h}\right|^2 \tag{3}$$

Thus, $E_k^2$ is averaged over a range of $T$ in $[0, T_{Max}]$ and the noise amplification for a point in the attractor is:

$$\sigma_k^2(\overrightarrow{s}) = \frac{E_k^2(\overrightarrow{s})}{\epsilon_k^2(\overrightarrow{s})} \tag{4}$$

In this approach, a normalization factor is estimated to penalize the irrelevance, which can be understood as

$$\alpha_k^2 = \left[\sum_{i=1}^{n} \epsilon_k^{-2}(\overrightarrow{s}_i)\right]^{-1} \tag{5}$$

where $n$ is the number of points in the attractor. In this sense, combining $\sigma_k(\overrightarrow{s})$ and $\alpha_k$, the local L–statistic at time $i$, is:

$$L_i = \log_{10}\left(\alpha_k \sigma_k(\overrightarrow{s}_i)\right) \tag{6}$$

### 2.4 Proposed Variable Embedding Approach

Several approaches of variable embedding have emerged due to conditions that usually appear on time-series analysis, such as multiple timescales, noisy data or other important considerations, since these inconveniences have not been completely solved by methods implemented for non-uniform embedding [8,10]. Alternatively, variable embedding theory assumes that is not necessary to use the same embedding dimension on all the attractor, this means, attractors made up by segments with less reconstruction complexity (represented with a smaller dimension) and portions with high reconstruction complexity (represented with a higher dimension) [12]. As a result, a different embedding dimension on each point or section of the attractor can be estimated.

In this paper, a variable embedding approach that uses the L–statistic presented in [30] is proposed. For this, an L–statistic matrix is computed, which contain the quality reconstruction metric for every attractor point at different embedding dimension. Consequently, a local embedding dimension vector $\overrightarrow{m}$ based on the the L–statistic matrix is achieved, as detailed below.

Let $\mathbf{S}_{m,\tau}$ be a reconstructed attractor with an embedding dimension $m$ and a time delay $\tau$. Given values for $k$, $T_{Max}$ and $Th_w$, the local L-statistic $\overrightarrow{L}_m = \{l_1, l_2, \cdots, l_n\}$ is obtained, for every point in $\mathbf{S}_{m,\tau}$. In this way, the L–statistic matrix is composed by local L-statistic vectors at different embedding dimensions, $\mathbf{L} = \{\overrightarrow{L}_{m_1}, \overrightarrow{L}_{m_2}, \cdots \overrightarrow{L}_{m_q}\}$, and a vector of tested $m$ values $\overrightarrow{m}^{test} = \{m_1, m_2, \cdots m_q\}$ is used. Summarizing, L–statistic matrix has dimensions $n \times q$ and can be structured as:

$$\mathbf{L} = \begin{bmatrix} l_{1,m_1} & l_{1,m_2} & \cdots & l_{1,m_q} \\ l_{2,m_1} & l_{2,m_2} & \cdots & l_{2,m_q} \\ \vdots & \vdots & \vdots & \\ l_{n,m_1} & l_{n,m_2} & \cdots & l_{n,m_q} \end{bmatrix} \tag{7}$$

Once, $\mathbf{L}$ is obtained, vector $\overrightarrow{m}$ can be performed. With this purpose, the average L–statistic at time $i$ is taken over a square window $TW$ and for every tested embedding dimension (i.e. the average of the columns in a sub-matrix of $\mathbf{L}$, composed by rows in the time interval $[i - TW, i + TW]$), achieving $\overrightarrow{L}_i^{mean}$ as follow

$$\overrightarrow{L}_i^{mean} = \left\{ \sum_{j=i-TW}^{i+TW} \mathbf{L}_{j,p} \quad \forall p \in [1, q] \right\} \tag{8}$$

With a low value of L–statistic, a better reconstruction quality is achieved, thus, the position of the minimal value of $\overrightarrow{L}_i^{mean}$ represents the optimal embedding dimension at the $i$ point, resulting the $\overrightarrow{m}$ vector, as the following optimization product

$$\overrightarrow{m}_i = \overrightarrow{m}_p^{test}; \quad \min_p \overrightarrow{L}_{i,p}^{mean} \tag{9}$$

## 2.5 Experimental Framework

The longest signal found in the databases was made up by 231000 points, so each signal was cropped taking a random segment of the 10% from the total of points. Likewise, a random noise was added with an amplitude relation 1/1000 to NSR signals, in order to avoid quantification errors related to the analog to digital conversion.

Then, the vector $\overrightarrow{m}$ was obtained using the proposed approach. In this regard, the parameters employed to compute the L–statistic were: $k = 3$, $Th_w = 3$ and $T_{Max} = 50$, following the recommendation given by [15,30]. Moreover, $m$ was tested in the interval $[2, 9]$ for a window $TW = 3$. It is important to clarify that in this paper we do not intend to select $\tau$, but as the employed databases are widely used in the literature, we take $\tau = 1$, since this is the most commonly used and allows signals differentiation [6]. At this point, each signal per database has their respective manifold and embedding dimension vector.

Finally, to verify if the $\overrightarrow{m}$ provide information about the functional states in human health, we compute the first four statistical moments (mean, variance, skewness and kurtosis) of $\overrightarrow{m}$. Next, we use the Mann-Whitney U test to check if there is significant difference between healthy (NSR signals) and pathological subjects (ARH, AF and CHF signals). Also, Tukey's range test were used to verify difference between the four groups.

# 3 Results and Discussion

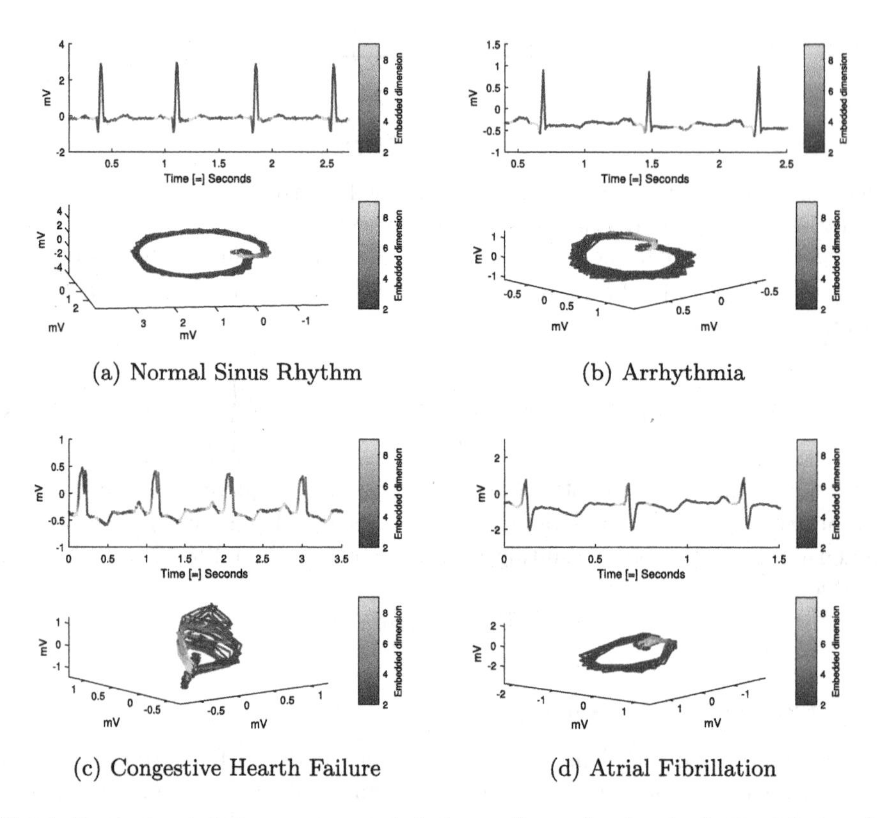


(a) Normal Sinus Rhythm

(b) Arrhythmia

(c) Congestive Hearth Failure

(d) Atrial Fibrillation

**Fig. 1.** Reconstructed phase spaces and electrocardiography signals obtained from subjects with different health states. The line color indicates the local embedding dimension obtained with the proposed approach

Figure 1 shows a 3-dimensional attractor, the original signal for an example signal in each database, and the local embedding dimension projected to the temporal signal. The obtained attractor shows a well defined circular trajectory with a knot, except for CHF signals, where the orbits diverge. The embedding dimension along of circular trajectory is 2, this finding is coherent, because with 2 dimension is enough to unfold the manifold without crosses in the orbits [12,26]. On the other hand, when the trajectories converge in the knot, the local embedding dimension increase, because, it is necessary more dimension to unfold the manifold. Moreover, when the local embedding dimension is projected to the original temporal signal, it is possible to observe that the QRS complex represents a low embedding dimension, as well as the P and T waves, however, the previous moments to the P and T waves show an increment in the embedding dimension. This observation could be further study to develop automatic segmentation algorithm to detect P or T waves and QRS Complex, due to this methods are highly interest in many biomedical applications [3,11,25,32].

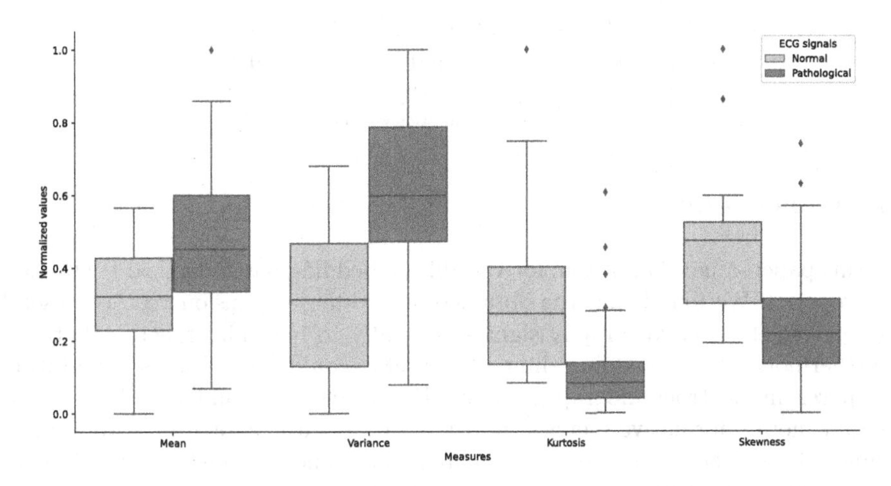


**Fig. 2.** Boxplot of different features obtained from embedding dimension vectors for Normal and Pathological electrocardiography signals

Figure 2 shows the boxplots for the normalized values of the four statistical moments, in healthy (NSR signals) and pathological subjects (ARH, AF and CHF signals). Visually, a difference between these kind of signals can be found, specially for the last three moments (variance, skewness and kurtosis). Moreover, the Mann-Whitney U [18] found that this difference are statistical significant (Mean p-value = 0.0086 , Variance p-value = 2.78e−6, Skewness p-value = 5.14e−6 and Kurtosis p-value = 1.41e−5).

Table 2 shows the obtained p-values for the multiple comparisons of the Tukey's range test [14]. The mean of embedding dimension vector fail to identify most of the comparisons, only found statistical difference between NSR and ARH. In another way, variance, skewness and kurtosis, show statistical significance to identify NSR from each of the studied pathologies. However, the obtained features from embedding dimension vector can not distinguish between pathologies, although ARH vs AF could be distinguished by means of variance and skewness. The discriminant information contained in the embedding dimension vector could be used in classification tasks and the detection of abnormalities in ECG signals continues like an active research field [19,23,29,31].

**Table 2.** Tukey's range test p-values for the multiple comparison of features derived from the embedding vector for different classes of electrocardiography signals

| Comparison | Mean | Variance | Skewness | Kurtosis |
|---|---|---|---|---|
| NSR vs ARH | 0.005* | 5.75e-8* | 1.9e-7* | 1.01e-7* |
| NSR vs AF | 0.73 | 0.016 * | 0.025* | 0.011* |
| NSR vs CHF | 0.08 | 1.57e-4* | 2.43e-4* | 1.29e-4* |
| ARH vs AF | 0.09 | 0.027* | 0.033* | 0.054 |
| ARH vs CHF | 0.97 | 0.898 | 0.93 | 0.95 |
| AF vs CHF | 0.47 | 0.3.691 | 0.34 | 0.39 |

NSR: Normal sinus rhythm, ARH: Arrhythmia,
AF: Atrial fibrillation, CHF: Congestive heart failure.
Values with *, are those where a signification were found

## 4 Conclusions

In this paper, a novel approach for variable embedding based on Uzal L–statistic is proposed. We found that the obtained embedding dimension vector is useful to analyse electrocardiography signals, specially to recognize functional states. Furthermore, the embedding dimension vector could be use for segmentation purposes in electrocardiography signal. Future work, should be addressed to obtain more informative metrics from the variable attractor like entropy measures. Also it is necessary a deeper research on parameter's tuning and its impact in the reconstruction process of the variable attractor.

**Conflicts of Interest.** The authors declare no conflict of interest.

## References

1. Abarbanel, H.D.I., Kennel, M.B.: Local false nearest neighbors and dynamical dimensions from observed chaotic data. Phys. Rev. E **47**(5), 3057–3068 (1993). https://doi.org/10.1103/PhysRevE.47.3057
2. Bandt, C., Pompe, B.: Permutation entropy: a natural complexity measure for time series. Phys. Rev. Lett. **88**(17), 174102 (2002). https://doi.org/10.1103/PhysRevLett.88.174102
3. Buś, S., Jędrzejewski, K.: Two stage SVD-based method for QRST waves cancellation in atrial fibrillation detection. In: 2019 Signal Processing Symposium (SPSympo), pp. 24–28 (2019). https://doi.org/10.1109/SPS.2019.8882032
4. Casdagli, M., Eubank, S., Farmer, J., Gibson, J.: State space reconstruction in the presence of noise. Physica D **51**(1), 52–98 (1991). https://doi.org/10.1016/0167-2789(91)90222-U
5. Cuesta-Frau, D., Murillo-Escobar, J.P., Orrego, D.A., Delgado-Trejos, E.: Embedded dimension and time series length. practical influence on permutation entropy and its applications. Entropy **21**(4), 385 (2019). https://doi.org/10.3390/e21040385

6. Espinosa, R.A., Calderón, R.: Choice of tau in the estimation of ApEn and SampEn entropy of EMG, ECG and EEG signals. In: 2019 XXII Symposium on Image, Signal Processing and Artificial Vision (STSIVA), pp. 1–5 (2019). https://doi.org/10.1109/STSIVA.2019.8730216
7. Han, M., Ren, W., Xu, M., Qiu, T.: Characteristic lyapunov exponents and smooth ergodic theory. Russian Math. Surv. **32**(4), 55–114 (1977). https://doi.org/10.1070/rm1977v032n04abeh001639
8. Han, M., Ren, W., Xu, M., Qiu, T.: Nonuniform state space reconstruction for multivariate chaotic time series. IEEE Trans. Cybern. Ser. 1–14 (2018). https://doi.org/10.1109/tcyb.2018.2816657
9. Henry, B., Lovell, N., Camacho, F.: Nonlinear dynamics time series analysis. Nonlinear Biomed. Signal Process. Dyn. Anal. Model. **2**, 1–39 (2000)
10. Hirata, Y., Aihara, K.: Dimensionless embedding for nonlinear time series analysis. Phys. Rev. E **96**(3), 032219 (2017). https://doi.org/10.1103/PhysRevE.96.032219
11. Johnson, L.S., Persson, A.P., Wollmer, P., Juul-Möller, S., Juhlin, T., Engström, G.: Irregularity and lack of p waves in short tachycardia episodes predict atrial fibrillation and ischemic stroke. Heart Rhythm **15**(6), 805–811 (2018). https://doi.org/10.1016/j.hrthm.2018.02.011
12. Judd, K., Mees, A.: Embedding as a modeling problem. Physica D **120**, 273–286 (1998). https://doi.org/10.1016/S0167-2789(98)00089-X
13. Kennel, M.B., Brown, R., Abarbanel, H.D.I.: Determining embedding dimension for phase-space reconstruction using a geometrical construction. Phys. Rev. A **45**(6), 3403–3411 (1991). https://doi.org/10.1103/PhysRevA.45.3403
14. Keselman, H.J., Rogan, J.C.: The Tukey multiple comparison test: 1953–1976. Psychol. Bull. **84**(5), 1050–1056 (1977). https://doi.org/10.1037/0033-2909.84.5.1050
15. Kraemer, K.H., Datseris, G., Kurths, J., Kiss, I.Z., Ocampo-Espindola, J.L., Marwan, N.: A unified and automated approach to attractor reconstruction. New J. Phys. **23**, 033017 (2021). https://doi.org/10.1088/1367-2630/abe336
16. Little, D.J., Kane, D.M.: Permutation entropy with vector embedding delays. Phys. Rev. E **96**(6), 062205 (2017). https://doi.org/10.1103/PhysRevE.96.062205
17. Ma, H., Han, C.: Selection of embedding dimension and delay time in phase space reconstruction. Front. Electr. Electron. Eng. China **1**(1), 111–114 (2006). https://doi.org/10.1007/s11460-005-0023-7
18. McKnight, P.E., Najab, J.: Innovation and intellectual property rights. In: Weiner, I.B., Craighead, W.E. (eds.) The Corsini Encyclopedia of Psychology. John Wiley & Sons, Inc. (2010). https://doi.org/10.1002/9780470479216.corpsy0524
19. Moridani, M., Abdi Zadeh, M., Shahiazar Mazraeh, Z.: An efficient automated algorithm for distinguishing normal and abnormal ECG signal. IRBM **40**(6), 332–340 (2019). https://doi.org/10.1016/j.irbm.2019.09.002
20. Packard, N.H., Crutchfield, J.P., Farmer, J.D., Shaw, R.S.: Geometry from a time series. Phys. Rev. Lett. **45**(9), 712–716 (1980). https://doi.org/10.1103/PhysRevLett.45.712
21. Pecora, L.M., Moniz, L., Nichols, J., Carroll, T.L.: A unified approach to attractor reconstruction. Chaos Interdisciplinary J. Nonlinear Sci. **17**(1), 013110 (2007). https://doi.org/10.1063/1.2430294
22. Sauer, T., Yorke, J.A., Casdagli, M.: Embedology. J. Stat. Phys. **65**(3–4), 579–616 (1991). https://doi.org/10.1007/bf01053745

23. Sepulveda, J., Murillo-Escobar, J., Urda-Benitez, R., Orrego-Metaute, D., Orozco-Duque, A.: Atrial fibrillation detection through heart rate variability using a machine learning approach and poincare plot features. In: VII Latin American Congress on Biomedical Engineering CLAIB 2016, vol. 60, pp. 24–28 (2017). https://doi.org/10.1007/978-981-10-4086-3_142
24. Song, Y., Nie, F., Zhang, C.: Semi-supervised sub-manifold discriminant analysis. Elsevier **29**(13), 1806–1813 (2008). https://doi.org/10.1016/j.patrec.2008.05.024
25. Stone, J., Mor-Avi, V., Ardelt, A., Lang, R.M.: Frequency of inverted electrocardiographic t waves (cerebral t waves) in patients with acute strokes and their relation to left ventricular wall motion abnormalities. Am. J. Cardiol. **121**(1), 120–124 (2018). https://doi.org/10.1016/j.amjcard.2017.09.025
26. Takens, F.: Dynamical Systems and Turbulence (Lecture Notes in Mathematics), vol. 898, chap. Detecting strange attractors in turbulence, pp. 366–381. Springer-Verlag (1981). https://doi.org/10.1007/bfb0091903
27. Tenenbaum, J.B., de Silva, V., Langford, J.C.: A global geometric framework for nonlinear dimensionality reduction. Sciencemag **290**, 2319–2323 (2000). https://doi.org/10.1126/science.290.5500.2319
28. Theiler, J.: Spurious dimension from correlation algorithms applied to limited time-series data. Phys. Rev. A **34**(3), 2427–2432 (1986). https://doi.org/10.1103/PhysRevA.34.2427
29. Ullah, A., Rehman, S.u., Tu, S., Mehmood, R.M., Ehatisham-ul haq, M.: A hybrid deep CNN model for abnormal arrhythmia detection based on cardiac ECG signal. Sensors **21**(3) (2021). https://doi.org/10.3390/s21030951
30. Uzal, L.C., Grinblat, G.L., Verdes, P.F.: Optimal reconstruction of dynamical systems: a noise amplification approach. Phys. Rev. A **84**(3), 016223 (2011). https://doi.org/10.1103/PhysRevE.84.016223
31. Venkataramanaiah, B., Kamala, J.: ECG signal processing and KNN classifier-based abnormality detection by VH-doctor for remote cardiac healthcare monitoring. Soft. Comput. **24**(22), 17457–17466 (2020). https://doi.org/10.1007/s00500-020-05191-1
32. Wijaya, C., Andrian, H.M., Christnatalis, T.M., Turnip, A.: Abnormalities state detection from p-wave, QRS complex, and t-wave in noisy ECG. J. Phys. Conf. Ser. **1230**(1), 012015 (2019). https://doi.org/10.1088/1742-6596/1230/1/012015
33. Wolf, A., Swift, J.B., Swinney, H.L., Vastano, J.A.: Determining lyapunov exponents from a time series. Physica D **16**(3), 285–317 (1985). https://doi.org/10.1016/0167-2789(85)90011-9
34. Yan, S., Wang, H., Fu, Y., Yan, J., Tang, X., Huang, T.S.: Synchronized submanifold embedding for person-independent pose estimation and beyond. IEEE Trans. Image Process. **18**(1), 202–210 (2009). https://doi.org/10.1109/TIP.2008.2006400

# Uniform and Non-uniform Embedding Quality Using Electrocardiographic Signals

Juan P. Restrepo-Uribe[1], Diana A. Orrego-Metaute[2](✉), Edilson Delgado-Trejos[1], and David Cuesta-Frau[3]

[1] AMYSOD Lab –Parque I, CM&P Research Group, Instituto Tecnológico Metropolitano ITM, 050034 Medellín, Colombia
juanrestrepo196291@correo.itm.edu.co, edilsondelgado@itm.edu.co

[2] Department of Exact and Applied Sciences, GI2B Research Group, Instituto Tecnológico Metropolitano ITM, 050034 Medellín, Colombia
dianaorrego@itm.edu.co

[3] Technological Institute of Informatics, Universitat Politecnica de Valencia, Alcoi Campus, 03801 Alcoi, Spain
dcuesta@disca.upv.es

**Abstract.** Uniform embedding techniques have limitations for the reconstruction of the phase space of nonlinear time series whose dynamics is not completely known, so new embedding techniques have been developed based on non-uniform methodologies. This work compares the reconstruction quality supported by the Uzal cost function, of three electrocardiography databases. For the uniform reconstruction, Average Mutual Information was implemented to find the time delay ($\tau$) and False Nearest Neighbor and Average False Neighbor were tested to find the attractor dimension ($m$). For the non-uniform reconstruction, the algorithm of Hankel Singular Value Decomposition was implemented. The results showed that the non-uniform embedding, based on Hankel Singular Value Decomposition, provides a better quality in the reconstruction of the phase space.

**Keywords:** Time delay reconstruction · Non-uniform embedding · Nonlinear time series analysis · Uzal cost function

## 1 Introduction

With nonlinear time series analysis it is possible to study the behavior of chaotic systems [13,17]. Phase space (or state space) reconstruction is usually the first step in the study of dynamical systems, for inferring information about a dynamic model from observations. In this way, The Taken's seminal embedding theorem establishes that under certain conditions, using the time delay

This work is presented in partial fulfillment of the requirements for the "Call for the strengthening of vocations and training in ST&I for economic reactivation in the framework of the 2020 post-pandemic" No. 891 of MinCiencias- Colombia.


J. M. Ferrández Vicente et al. (Eds.): IWINAC 2022, LNCS 13258, pp. 605–614, 2022.
https://doi.org/10.1007/978-3-031-06242-1_60

reconstruction from a one-dimensional signal, it can be obtained a diffeomorphic copy of the original system attractor [25]. This reconstruction preserves relevant geometrical and dynamical invariants, although the optimal selection of the embedding dimension $m$ and the time delay $\tau$ is a critical issue for a proper reconstruction.

The literature reports two perspectives for the estimation of these key parameters: In first place, $m$ and $\tau$ can be estimated independently, where, the Average Mutual Information (AMI) and False Nearest Neighbors (FNN) methods are the most known [9]. The subjective choice of several parameters in FNN leads to different estimations of the embedding dimension. In this regard, Cao introduces a modified algorithm denominated as Averaged False Neighbors (AFN) to avoid this subjectivity. This method has shown that for a sufficiently long noise free data set, $\tau$ and the minimum $m$ are almost independent, and $\tau$ can be set arbitrarily [5]. Secondly, the estimation of $m$ and $\tau$ can be simultaneously performed using an embedding window $\tau w = (m - 1)\tau$, where both parameters are correlated, because of that real data sets are finite and noisy [6]. Cao-Cao (C-C) is the most popular method to estimate $\tau w$ by applying the correlation integral based on the BDS statistic [1]. Although this method lacks theoretical support, in practice shows some advantages, such as simple operation, lower algorithm complexity, reliability for less data and better robustness [7]. The uniform and non-uniform reconstruction problem has been referred by different authors [7], receiving more and more attention the non-uniform embedding techniques [16,17,26]. Although uniform embedding is simple and effective for classic chaotic dynamical systems, it could lead to poor performance for complex time series with multiple periodicities, and often fail in real data [12], for dealing these issues, non-uniform embedding uses different time delays through a vector of positive integers that lies in a embedding space [27]. Some methods have been developed to take advantage the structure of the Hankel matrix, like as, Eigensystem Realization Algorithm (ERA) [18], Singular Spectrum Analysis (SSA) [2] and nonlinear Laplacian spectrum analysis [14]. Likewise, Staniszewski and Polanski in [24] used Hankel Singular Value Decomposition (HSVD) as a method of preprocessing the Magnetic Resonance Spectroscopy.

In this paper, a comparison between uniform and non-uniform embedding is presented, where a non-uniform methodology based on Singular Value Decomposition (SVD) of the Hankel matrix is implemented, in order to identify dominant modal content of data represented in delay coordinates [22]. This comparison is carried out on electrocardiographic signal with evidence of normal or abnormal functional states.

## 2 Materials and Methods

### 2.1 Databases

The evaluation of the embedding quality from nonlinear time series was performed using three electrocardiographic signals (ECG) databases. It is important to note that the derivation used in the experimental setup was the Modified

Lead II (MLII), which corresponds to the sinus rhythm. The sampling frequency was different for every signal, and for this reason, a resampling was implemented, taking the minimum frequency (i.e. 128 Hz).

- *MIT-BIT normal sinus rhythm:* Database formed by signals from healthy subjects. 18 signals were acquired (5 men and 13 women between 20 and 50 years old) at a sampling frequency 128 Hz [15].
- *MIT-BIT arrhythmia Database:* The database contains 48 signals of some types of arrhythmia. The sampling frequency 360 Hz, re-sampled 128 Hz. Two signals that did not have the MLII were eliminated [21].
- *MIT-BIT atrial Fibrillation Database:* Database composed of 25 ECG signal records of subjects with atrial fibrillation, mostly paroxysmal. The signals were acquired at a sampling frequency 250 Hz, re-sampled 128 Hz, 5 subjects were also eliminated from the database because they did not have the MLII [20].

### 2.2 Preprocessing Data

An approach based on pseudoperiodic surrogate data was applied in order to determine the nonlinearity existence in the ECG signals selected for this study [8,19]. The discriminating statistics for null hypothesis testing was the Lempel-Ziv complexity, that in a binary sequence determines the pattern repeatability. On the other hand, signal segments contain 1% of the total signal data, which were embedded using uniform and non-uniform approaches.

### 2.3 Uniform Embedding

**Average Mutual Information (AMI):** This measure seeks to explore the independence level among the reconstructed orbits, which should be as independent as possible [11]. In this sense, the mutual information $I(x(t), x(t+\tau))$ is estimated, where for two-time series can be expressed as

$$I(x(t), x(t+\tau)) = \sum_{i,j} p_{ij}(\tau) \log\left(\frac{p_{ij}(\tau)}{p_i p_j}\right) \tag{1}$$

where, $p_i$ is the probability that $x(t)$ is contained in the histogram of $x$, while $p_{ij}(\tau)$ is the probability that $x(t)$ is contained in bin $i$ and $x(t,\tau)$ is in bin $j$ [28]. Here, $p_{ij}(\tau)$ just depends on $\tau$ and the AMI function also depends on the histogram specifications (width and position of the bins).

**False Nearest Neighbors (FNN):** It is possible to reconstruct the original multidimensional space by embedding a one-dimensional time series as follows [28]:

$$y(t) = (x(t), x(t+\tau), \cdots, x(t,(m-1)\tau)) \tag{2}$$

where $m$ represents the maximum embedding dimension. This method calculates the distance of the neighbors. If the distance is considerable, it is a false neighbor, and the reconstruction in a higher dimension must be evaluated. If the distance is not considerable, is a true neighbor, and it is considered that the reconstruction reached the maximum dimension [28]. A parameter defined in [5] for examining the false neighbors without the threshold is expressed as:

$$a(i,m) = \frac{\left\| y_i(m+1) - y_{n(i,m)}(m+1) \right\|}{\left\| y_i(m) - y_{n(i,m)}(m) \right\|} \tag{3}$$

where $\|.\|$ is the maximum norm, $n(i,m)$ is an integer such that the $m$-dimensional time-delay vector $y_{n(i,m)}(m)$ is the nearest neighbor of $y_i(m)$.

**Average False Neighbors (AFN):** Several studies have recommended the use of the mean of the distances of the trajectories, considering that methods based on distance must determine the threshold to consider the false neighbors, which has a high computational cost [5]. Although FNN is a standard method to estimate these parameters for one-dimensional time series, this approach is also subjective to some extent, given the sensitivity to the threshold parameters when determining the false nearest neighbors [4]. The Averaged False Nearest Neighbor method solves the threshold problem, estimating the average change of distances between any point and its nearest neighbor and increasing the embedding dimension to identify if the nearest neighbor of each individual point is false. AFN is defined by the follow averaged quantity [23]:

$$E(m) = \frac{1}{N - m\tau} \sum_{i=1}^{N-m\tau} a(i,m) \tag{4}$$

### 2.4 Non-uniform Embedding: Hankel Singular Value Decomposition (HSVD)

The HSVD method is based on applying singular value decomposition (SVD) to the Hankel matrix ($H$), derived from the original time series. The $U$ and $V$ matrices obtained from the application of SVD are ordered hierarchically by their ability to model the columns and rows of $H$. Likewise, the vector $\sum$ contains the number of singular vectors to be used from the $U$ and $V$ matrix. The following is the mathematical expression of the idea described for the application of HSVD [3,10]:

$$x(t) = [a_{t1}, a_{t2}, a_{t3}, a_{t4}, \cdots, a_{tq}] \rightarrow H = \begin{bmatrix} a_{t1} & a_{t2} & \cdots & a_{tp} \\ a_{t2} & a_{t3} & \cdots & a_{tp+1} \\ a_{t3} & a_{t4} & \cdots & a_{tp+1} \\ \vdots & \vdots & \vdots & \vdots \\ a_{tq} & a_{tq+1} & \cdots & a_{tp+m} \end{bmatrix} \rightarrow U \times \sum \times V \tag{5}$$

### 2.5 Reconstruction Quality

Uzal et al. [26] proposed a cost function $L_k$ to guide the search of the optimal reconstruction. $L_k$ measures conditional probability using nearest neighbors. This measure has the advantage that one does not need the original state space for its computation. The measure is made up of two components:

$$L_k = R_k + I_k \quad (6)$$

where $r_k$ in the measures the redundancy and $I_k$ is the measures the irrelevancy. The reconstruction with lower cost function is smoother and, therefore, has better predictability.

## 3 Results and Discussions

Data preprocessing showed statistical difference with a *p-value* $< 0.05$ Thus indicating that electrophysiological signals are in principle not noise.

**Table 1.** Selected parameters for the EMG signals embedding

| Data base | Uniform | | | | Non-uniform |
|---|---|---|---|---|---|
| | AMI-FNN | | AMI-AFN | | HSVD |
| | $m$ | $\tau$ | $m$ | $\tau$ | $m$ |
| MIT-BIT atrial Fibrillation Database | 3 to 8 | 1 to 9 | 2 to 9 | 1 to 8 | 4 |
| MIT-BIT normal sinus rhythm | 6 to 14 | 1 to 5 | 2 t 6 | 1 to 4 | 4 |
| MIT-BIT Arrhythmia Database | 2 to 15 | 1 to 9 | 2 to 6 | 1 to 10 | 4 |

Table 1 shows the results of the AMI- FNN, AMI-AFN uniform embedding and HSVD non-uniform embedding methods. In the case of the uniform embedding methods, it was determined that the embedding parameters $m$ and $\tau$ corresponded to a range. This considering that each signal was analyzed individually, and each signal segment analyzed showed different values for $m$ and $\tau$. However, for the non-uniform embedding, the dimension of the attractors was determined to be the same in all the signals. This is an important aspect, since the HSDV method shows better stability when representing the dynamics of the ECG signals. It should be noted that the literature has reported that uniform embedding techniques are useful for the study of nonlinear dynamics in non-real systems. In contrast, these techniques are not considered to be able to adequately represent the dynamics of real systems such as ECG signals [28].

Table 2 shows the average Uzal Cost Function obtained in each analyzed ECG database. The results showed that uniform embedding techniques provide

**Table 2.** Uzal Cost Function results

| Data base | Uniform | | Non-uniform |
|---|---|---|---|
| | AMI-FNN | AMI-AFN | HSVD |
| MIT-BIT atrial Fibrillation Database | $-0.51 \pm 0.32$ | $-0.53 \pm 0.31$ | $\mathbf{-1.85 \pm 0.06}$ |
| MIT-BIT normal sinus rhythm | $-0.41 \pm 0.40$ | $-0.45 \pm 0.41$ | $\mathbf{-1.86 \pm 0.03}$ |
| MIT-BIT Arrhythmia Database | $-0.41 \pm 0.21$ | $-0.45 \pm 0.21$ | $\mathbf{-1.91 \pm 0.23}$ |

lower reconstruction quality than that provided by the HSVD method. Likewise, when analyzing each of the reconstruction methods independently, it was observed that for AMI-FNN and AMI-AFN the reconstruction quality reached the maximum values with the MIT-BIT Atrial Fibrillation Database, while for HSVD the best reconstructions were obtained with the MIT-BIT Arrhythmia Database. However, the average values of the Uzal cost function do not differ notably between each of the databases.

Figure 1 shows the individual results of the Uzal Cost Function and the box plot for each of the databases considering the different methods applied. It can be observed that the behavior of the Uzal Cost Function presents similar values when considering the non-uniform embedding methods. On the contrary, the non-uniform embedding method presents notably lower values in the cost function. This translates into a better quality in the reconstruction. Figure 1(a) shows a peculiarity in the reconstruction quality. The presence of a minimum value in the Uzal Cost Function is evident in the uniform embedding methods related to the signal 5 of the MIT-BIT normal sinus rhythm database. Analyzing the uniform embedding results for the signal, it is evidenced that the $m$ and $\tau$ are at the upper limit of the ranges shown in Table 1. Considering the above, it is feasible to consider that the reconstruction presented a certain degree of redundancy, which led to the improvement of the embedding quality.

Figures 1(b), 1(d) and 1(f) show the box plot obtained from the comparison of the Uzal cost function values. It can be seen that there is a clear statistical difference between the uniform and non-uniform reconstruction methods used in this work. Moreover, as previously commented, the quality of the reconstruction obtained by the HSVD method reaches lower values than those obtained by the application of AMI-FNN and AMI-AFN. The presence of outliers is also observed, which are related to the signals where the cost function values were significantly reduced.

Figure 2 shows an example of a phase state reconstruction of a signal obtained from the MIT-BIT normal sinus rhythm database.

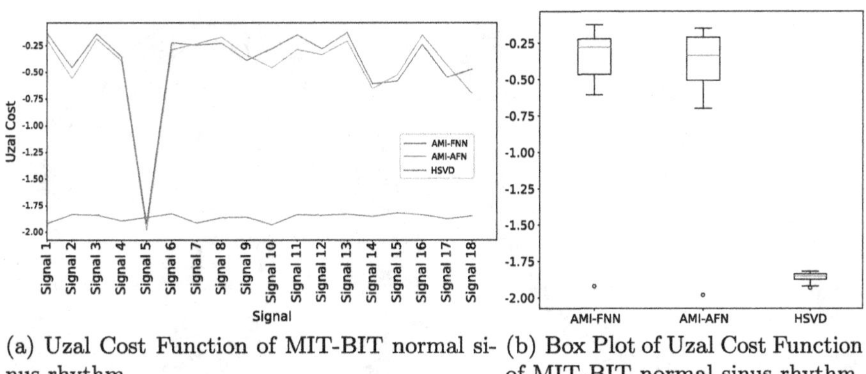


(a) Uzal Cost Function of MIT-BIT normal sinus rhythm (b) Box Plot of Uzal Cost Function of MIT-BIT normal sinus rhythm

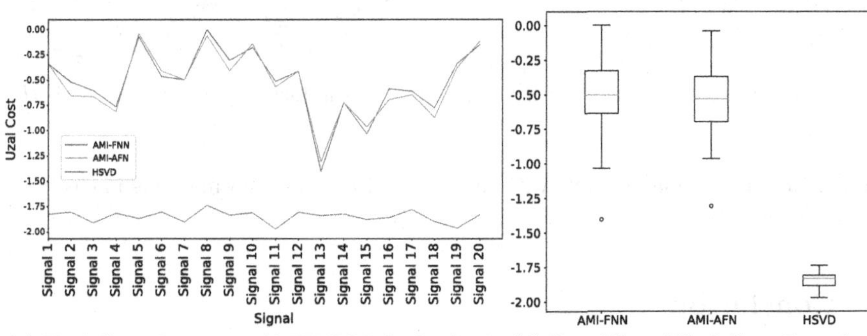


(c) Uzal Cost Function of MIT-BIT Arrhythmia Database (d) Box Plot ofUzal Cost Function of MIT-BIT Arrhythmia Database

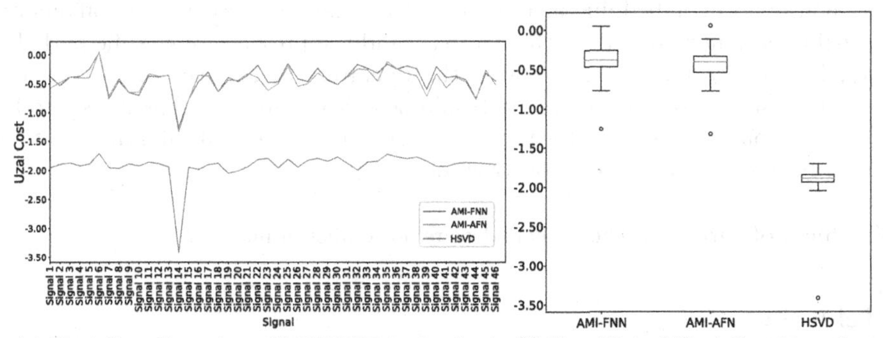


(e) Uzal Cost Function of MIT-BIT Arrhythmia Database (f) Box Plot of Uzal Cost Function of MIT-BIT Arrhythmia Database

**Fig. 1.** Results obtained for each signal of the Uzal cost function.

(a) Uniform reconstruction with $\tau = 1$ (b) Non-uniform reconstruction with $H = 10$

**Fig. 2.** Reconstruction example with uniform and non-uniform methods in electrocardiography signals

# 4 Conclusion

Uniform embedding methods provide a weak reconstruction quality than HSVD as a non-uniform embedding method on ECG signals. The reconstruction quality probably does not depend on the database and the presence of specific pathologies, but is directly associated with the dynamics of the electrophysiological system as such. However, more electrophysiological signal databases need to be explored, considering a greater number of embedding methods and metrics that quantify the quality of the reconstruction.

**Conflicts of Interest.** The authors declare no conflict of interest.

# References

1. Broock, W.A., Scheinkman, J.A., Dechert, W.D., LeBaron, B.: A test for independence based on the correlation dimension. Econ. Rev. **15**(3), 197–235 (1996). https://doi.org/10.1080/07474939608800353
2. Broomhead, D.S., Jones, R.: Time-series analysis. Proc. Royal Soc. London. A. Math. Phys. Sci. **423**(1864), 103–121 (1989). https://doi.org/10.1098/rspa.1989.0044
3. Brunton, S.L., Brunton, B.W., Proctor, J.L., Kaiser, E., Kutz, J.N.: Chaos as an intermittently forced linear system. Nat. Commun. **8**(1), 1–9 (2017). https://doi.org/10.1038/s41467-017-00030-8
4. Cao, H.: Development of techniques for general purpose simulators. Stanford University (2002)

5. Cao, L.: Practical method for determining the minimum embedding dimension of a scalar time series. Physica D **110**(1–2), 43–50 (1997). https://doi.org/10.1016/S0167-2789(97)00118-8
6. Chelidze, D.: Reliable estimation of minimum embedding dimension through statistical analysis of nearest neighbors. J. Comput. Nonlinear Dyn. **12**(5) (2017). https://doi.org/10.1115/1.4036814
7. Delage, O., Bourdier, A., et al.: Selection of optimal embedding parameters applied to short and noisy time series from rössler system. J. Mod. Phys. **8**(09), 1607 (2017). https://doi.org/10.4236/jmp.2017.89096
8. Delgado-Trejos, E., Hurtado Jaramillo, J.S., Guarín, D.L., Orozco, Á.Á.: Datos sustitutos pseudoperiódicos en señales de voz para determinar dinámicas subyacentes. Ingeniería y Desarrollo **31**(2), 185–201 (2013)
9. Dhanya, C., Kumar, D.N.: Nonlinear ensemble prediction of chaotic daily rainfall. Adv. Water Resources **33**(3), 327–347 (2010). https://doi.org/10.1016/j.advwatres.2010.01.001
10. Donoho, D.L., Gavish, M.: The optimal hard threshold for singular values is $4/\sqrt{3}$ (2013)
11. Fraser, A.M., Swinney, H.L.: Independent coordinates for strange attractors from mutual information. Phys. Rev. A **33**(2), 1134 (1986). https://doi.org/10.1103/PhysRevA.33.1134
12. Gao, C., Lin, Q., Ni, J., Guo, W., Li, Q.: A nonuniform delay-coordinate embedding-based multiscale predictor for blast furnace systems. IEEE Trans. Control Syst. Technol. (2020). https://doi.org/10.1109/TCST.2020.3023072
13. Gao, Z.K., Small, M., Kurths, J.: Complex network analysis of time series. EPL (Europhysics Letters) **116**(5), 50001 (2017). https://doi.org/10.1209/0295-5075/116/50001
14. Giannakis, D., Majda, A.J.: Nonlinear laplacian spectral analysis for time series with intermittency and low-frequency variability. Proc. Natl. Acad. Sci. **109**(7), 2222–2227 (2012). https://doi.org/10.1073/pnas.1118984109
15. Goldberger, A., et al.: The MIT-BIH normal sinus rhythm database. Circulation **101**(23), e215–e220 (2000)
16. Gómez-García, J.A., Godino-Llorente, J.I., Castellanos-Dominguez, G.: Non uniform embedding based on relevance analysis with reduced computational complexity: application to the detection of pathologies from biosignal recordings. Neurocomputing **132**, 148–158 (2014). https://doi.org/10.1016/j.neucom.2013.01.059
17. Han, M., Ren, W., Xu, M., Qiu, T.: Nonuniform state space reconstruction for multivariate chaotic time series. IEEE Trans. Cybern. **49**(5), 1885–1895 (2018). https://doi.org/10.1109/TCYB.2018.2816657
18. Juang, J.N., Pappa, R.S.: An eigensystem realization algorithm for modal parameter identification and model reduction. J. Guidance, Control, Dyn. **8**(5), 620–627 (1985). https://doi.org/10.2514/3.20031
19. Lancaster, G., Iatsenko, D., Pidde, A., Ticcinelli, V., Stefanovska, A.: Surrogate data for hypothesis testing of physical systems. Phys. Rep. **748**, 1–60 (2018). https://doi.org/10.1016/j.physrep.2018.06.001
20. Moody, G.: A new method for detecting atrial fibrillation using RR intervals. Comput. Cardiol. 227–230 (1983)
21. Moody, G.B., Mark, R.G.: The impact of the MIT-BIH arrhythmia database. IEEE Eng. Med. Biol. Mag. **20**(3), 45–50 (2001). https://doi.org/10.1109/51.932724
22. Pan, S., Duraisamy, K.: On the structure of time-delay embedding in linear models of non-linear dynamical systems. Chaos: An Interdisciplinary J. Nonlinear Sci. **30**(7), 073135 (2020). https://doi.org/10.1063/5.0010886

23. Ramdani, S., Casties, J.F., Bouchara, F., Mottet, D.: Influence of noise on the averaged false neighbors method for analyzing time series. Physica D **223**(2), 229–241 (2006). https://doi.org/10.1016/j.physd.2006.09.019
24. Staniszewski, M., Polański, A.: Hankel singular value decomposition as a method of preprocessing the magnetic resonance spectroscopy. arXiv preprint arXiv:2103.15754 (2021). https://doi.org/10.48550/arXiv.2103.15754
25. Takens, F.: Detecting strange attractors in turbulence. In: Rand, D., Young, L.-S. (eds.) Dynamical Systems and Turbulence, Warwick 1980. LNM, vol. 898, pp. 366–381. Springer, Heidelberg (1981). https://doi.org/10.1007/BFb0091924
26. Uzal, L.C., Grinblat, G.L., Verdes, P.F.: Optimal reconstruction of dynamical systems: a noise amplification approach. Phys. Rev. E **84**(1), 016223 (2011). https://doi.org/10.1103/PhysRevE.84.016223
27. Vlachos, I., Kugiumtzis, D.: Nonuniform state-space reconstruction and coupling detection. Phys. Rev. E **82**(1), 016207 (2010). https://doi.org/10.1103/PhysRevE.82.016207
28. Wallot, S., Mønster, D.: Calculation of average mutual information (AMI) and false-nearest neighbors (FNN) for the estimation of embedding parameters of multidimensional time series in matlab. Front. Psychol. 1679 (2018). https://doi.org/10.3389/fpsyg.2018.01679

# Decoding Lower-Limbs Kinematics from EEG Signals While Walking with an Exoskeleton

Javier V. Juan[1], Luis de la Ossa[1], Eduardo Iáñez[1,2], Mario Ortiz[1,2](✉), Laura Ferrero[1], and José M. Azorín[1,2]

[1] Brain-Machine Interface Systems Lab, Miguel Hernández University of Elche, Elche, Spain
{javier.juanp,lossa,eianez,mortiz,lferrero,jm.azorin}@umh.es
[2] Centro de Investigación en Ingeniería de Elche-I3E, Miguel Hernández University of Elche, Elche, Spain
http://bmi.edu.umh.es/

**Abstract.** Neurorehabilitation has gradually become one of the most hopeful tools in some kind of injuries and diseases during the last decade. Several studies have shown that conscious movement effected by patients with mobility difficulties, assisted by a clinical device such as an exoskeleton, contributes positively to their mobility recovery, shortening the rehabilitation times and improving its results. Besides, other studies have hypothesized that the motor cortex is particularly active during specific phases of gait cycle. In this study, a multilinear regression model has been applied to eight users in order to decode lower limb kinematics from EEG signals, reaching an average similitude between real and decoded trajectories (Pearson Correlation Coefficient) of 0.35 and up to 0.42 after optimizing the model parameters. These results encourage us to undertake a deeper analysis of the multilinear regression model as well as consider other processing approaches to perform the decoding in the future.

**Keywords:** Decoding · EEG · Lower limb kinematics · Multidimensional linear regression · Neurorehabilitation · Exoskeleton · Gait

## 1 Introduction

Electroencephalography allows specialists to record neural activity from user's brain [1]. This technique can register electrical pulses generated by user's brain, which are known as electroencephalographic signals (EEG signals),

This study is part of a project that has received funding from the European Union's Horizon 2020 research and innovation programme, via an Open Call issued and executed under Project EUROBENCH (grant agreement No 779963).


J. M. Ferrández Vicente et al. (Eds.): IWINAC 2022, LNCS 13258, pp. 615–624, 2022.
https://doi.org/10.1007/978-3-031-06242-1_61

whereas through invasive methods (intracortical) or through non-invasive methods (directly from the scalp). Thus, BCIs are a specially useful support in neural rehabilitation procedures [1,2], as far as they provide the chance to analyze and interpret patients' EEG signals. For patients who require neural motor neurorehabilitation (mostly Spinal Cord Injury -SCI- and stroke patients), this support is even greater given that a correct interpretation of neural activity suppose the detection of the patient's movement intention. These detections are essential for assuring a conscious control of the clinical devices employed in the motor rehabilitation, for instance a lower-limb exoskeleton. According to several studies [3,4], a deeper neural implication from the patients during their motor rehabilitation entails a more realistic immersion in their motor tasks, which carries a more satisfactory rehabilitation performance and results. This is because these patients have lost some of their neural connections between their muscles and their brain, so it is logic that the biggest neural implication from the patients, the most complete will be the recovery of their movement control.

Following these lines, many researches have been focused in assuring a safe and precise exoskeleton control from EEG signals [5,6]. First of all, recent studies assert that the motor cortex is particularly active during concrete phases of the human gait cycle [7]. Thus, it has been showed several times that it is possible to control an exoskeleton by analyzing user's EEG signals [5,6,8]. However, most of these models can only recognize user's motor imagery, for example the walking intention, so in the best case the exoskeleton can only be activated or deactivated by analyzing user's EEG signals and perform a predetermined task, such as walking or climbing stairs [9–11]. Regarding neurorehabilitation therapies, this sort of models are often enough for accomplishing all the tasks, but sometimes their limitations may result conflictive in the rehabilitation process.

Trying to overcome this issue, during the last years some studies have intended to reach one step further by directly decoding limb trajectories [12,13]. Some of these approaches have generated very promising results. In this regard, regression models seems to be a wise way of facing the problem. There are many recent examples of this, where the application of a multilinear regression model to the EEG signals achieves a reliable lower-limb trajectories decodification [12,14,15]. However, the great parameters variability that offers this type of regression models makes far way complicated to obtain perfect decoding algorithms.

For all these reasons, in this paper is presented a lower-limb kinematics decoding model from EEG signals, based on the application of multilinear regression. Besides, it contains an analysis of which parameters generate better results, as well as the algorithm performance and its variability between eight subjects.

## 2 Materials and Methods

In this section, an experimental procedure employed for the data acquisition is presented, as well as the sensorization deployed focused on lower-limbs kinematics. Furthermore, the multidimensional linear regression model used for kinematics reconstruction is presented.

### 2.1 Subjects

Eight able-bodied subjects have participated in the study, formed by six males and two females, with ages between 20 and 26 years (23'3±2'3). All of them received information about the study, as they signed a consent, the Helsinki declaration and an ethical committee approved by the Universidad Miguel Hernández of Elche.

### 2.2 Procedure

Subjects were asked to walk during 20 s, stop for a few of seconds, walk another 20 s, stop again for a couple of seconds, and walk another 20 s more (three stretches of twenty seconds walking); all that while wearing the H3-Exoskeleton (Technaid S.L.) (see Fig. 1). That procedure forms a complete run. Three of the subjects conducted seven runs (U5, U7, U8), one of them completed thirteen runs (U3) and the remaining four made sixteen complete runs (U1, U2, U4, U6). During the runs, EEG signals and lower-limbs joints information were simultaneously recorded.

**Register:** For the experiments, a non-invasive EEG recording system was used, consisted of BrainProducts' actiChamp amplifier and BrainProducts' actiCap 32 (see Fig. 1). The acquisition of EEG signals was done with a 32 active electrodes configuration, according to 10/10 international system, as illustrates Fig. 2, with a sampling frequency 500 Hz. The ground electrode was placed in the right earlobe; moreover, a reference electrode was positioned in the left earlobe too, in order to improve the register quality. Additionally, a high-pass hardware filter over 0'1Hz was used in order to reject gait artifacts, as well as a notch hardware filter 50 Hz for erasing power supply component.

**Inertial Measurement Units:** Regarding the lower-limbs trajectories, seven Inertial Measurement Units (IMUs) from Techanid S.L. (Tech-MCS) have been used. The seven IMUs are placed on: one in lumbar the area, one per thigh, one per shin, and one per ankle (see Fig. 3); all of them are connected to the Tech-HUB (Technaid S.L.). This setup allows us to get trajectories from both right and left hips, knees and ankles. The acquisition sample frequency 30 Hz.

**Exoskeleton:** Exoskeleton Exo-H3 (Technaid S.L., see Fig. 1) was employed during the experimental sessions. Users wore it while walking in the tests. This device counts with six engines (both hips, both knees and both ankles), what allows users to walk without any effort, just like a lower-limb handicapped person would do.

### 2.3 Data Preprocessing

First of all, lower-limbs trajectories are obtained. For that purpose, Direction Cosine Matrix (DCMs) are extracted from the IMUs' registered data. These matrix are then multiplied by another predetermined DCMs with the aim of rotate the original DCMs computed by the IMUs and make them match with Technaid reference systems. Afterwards, these resulting matrix are transformed into quaternions, so that Technaid kinematic architecture can be applied, calculating the lower-limbs trajectories. Once computed, lower-limbs kinematics are resampled from 30 Hz to 500 Hz to match EEG sampling frequency.

According to G. Garipelli et al. [16], information related to Slow Cortical Potentials is located above 0'1Hz. Then, both EEG signals and lower-limbs trajectories are band-pass filtered by a zero-phase 4th-order Butterworth filter between 0'1–2 Hz. In this regard, A. Úbeda ensures in [12] that employing only the electrodes from the central parietal cortex barely change the decodification performance, but it does considerably decreases the computational times. As a consequence, in this study only fifteen electrodes have been employed for these analysis, which are FC5, FC1, FC2, FC6, C3, Cz, C4, CP5, CP1, CP2, CP6, P3, Pz and P4, as shown in Fig. 2. Moreover, after filtering, both EEG signals and lower-limbs trajectories are standardized according to the next expression:

$$EV[t] = \frac{V[t] - \bar{V}}{SD_V} \tag{1}$$

In this standardization, for every signal $EV$ is subtracted, for each time sample $V[t]$, the mean of the whole signal $\bar{V}$ and then the result is divided by the standard deviation $SD_V$. This preprocessing method has been applied to all the trials from all the subjects.

### 2.4 Decoding Method

With the aim of decoding lower-limbs kinematics from EEG signals (see Fig. 4), a multidimensional linear regression is applied to those signals, according to next equation:

$$x[t] = a + \sum_{n=1}^{N} \sum_{k=0}^{L} b_{nk} \cdot S_n[t - G \cdot k] \tag{2}$$

In this regression, $x[t]$ is the kinematic state (the lower-limbs angles) in the time sample $t$, $a$ and $b$ are the weights of the linear regression, $N$ is the number of channels (16 as explained in 2.3), $L$ is the number of lags, $Sn$ is the standardized voltage measured in electrode $n$, and $G$ is the gap between lags. For further details, please refer to [17].

The way of employing this algorithm has been generating a regression model for each subject with EEG and lower-limbs kinematics data from all the trials except the last one. Thereby, these last trials have served to validate the regression performance.

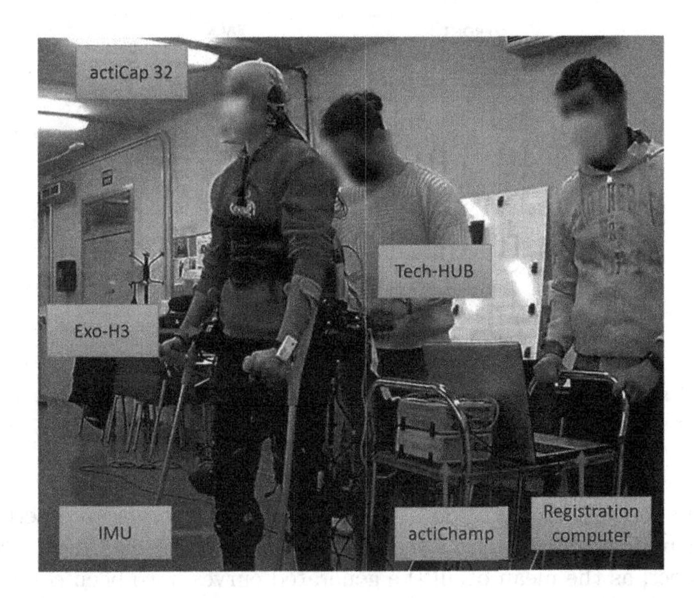


**Fig. 1.** Experimental setup employed during the tests. Subjects walked wearing the Exo-H3 while their EEG signals and lower-limbs trajectories were registered.

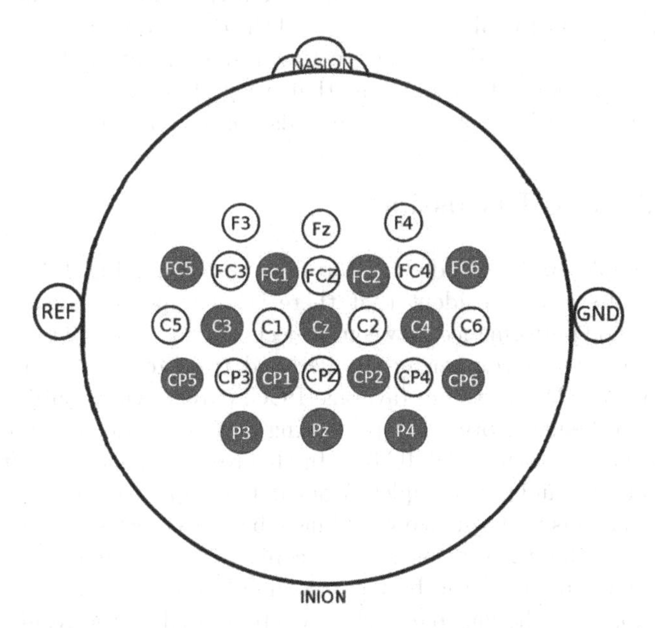


**Fig. 2.** Electrodes distribution employed during the EEG registers, following standard 10/10. Ground (GND) electrode was placed in the right earlobe, as well as a reference (REF) electrode was added to the left earlobe. Four more active electrodes where placed around users' eyes, contemplating the chance of rejecting blink artifacts. The electrodes whose names appear in white are the ones used in the analysis.

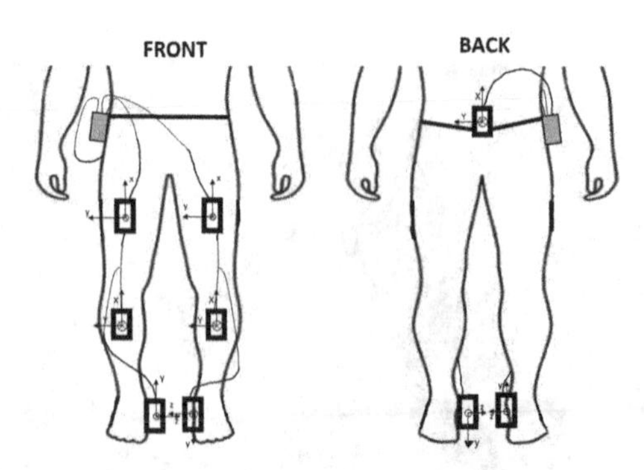


**Fig. 3.** IMUs distribution employed for acquiring lower-limbs kinematics. Trajectories from both right and left hips, knees and ankles were measured.

For the analysis, $L$ has been defined as 10, and in order to assert the best algorithm performance, a gap sweep between 1 and 150 has been applied for every subject, as the mean of all the generated curves have been computed (see Fig. 5), determining which gap value offers better results for each lower-limb of each subject. To that end, it has been compared the Pearson Correlation Coefficients (PCCs) between de real lower-limb trajectories (obtained with the IMUs setup in the last trial of each user) and the decoded trajectories (generated when applying the regression model to that last run, out of the model), of all the cases. It is important to highlight that a lag of 10 combined with a gap of 150 generates time windows of three seconds, for a sampling rate 500 Hz.

## 3 Results and Discussion

A pair of examples of the decoded trajectories are shown in Fig. 4. In this concrete graph, it results clearly evident that there is an important dispersion in the decoding method performance between subjects. In this case, U2 presents slightly deficient results when comparing the decoded trajectories with the real ones obtained with the IMUs system (average PCC: 0'371); whilst subject U8 seems to have an excellent performance, presenting decoded trajectories quite similar to the real ones (average PCC: 0'705). In this respect, it is very surprising the fact that user U8 could only complete 7 of the 16 complete trials, fact that makes us suppose that this performance would have been even better if U8 had finished the whole test. This theory seems to be confirmed when analyzing Table 1. In that table, it is presented the best gap value obtained for every lower-limb of every user applying the gap parameter sweep, as well as the resulting PCC of comparing the decoded lower-limb trajectory computed with that gap value and the real one, for the last trials of each user.

In Fig. 5, the plots in the right half shows the average gap tendency from all the 8 subjects, for each lower-limb. There, it can be checked how, for every lower-limb, the optimal gap value tends to 150, the maximum contemplated in the gap

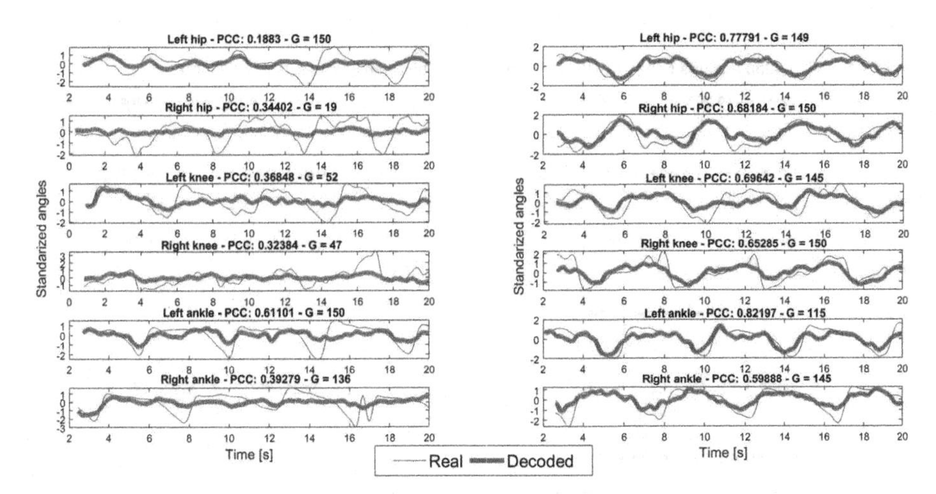


**Fig. 4.** Comparative of the decoded lower-limbs trajectories between subjects U2 (left) and U8 (right). Decoded trajectories are represented in thick lines, whereas real trajectories are plotted in thin lines. The gap value used in each decodification has been selected executing a previous gap sweep.

sweep. Even though, this do not mean that the maximum gap value will always be the best for every user, as shows Table 1. Besides, after analyzing Table 1, it can be concluded that, generally, users who did not complete all the trials (U3, U5, U7, and U8) have lower accuracy in their lower-limbs decodification. This is probably due to the lack of data in their models for the regression, as there are less walking stretches recorded to be compared with the EEG signals and locate patterns for the regression models. Another important aspect in the table is the enormous dispersion existent between gap values for every user. Only by examining the table is hardly possible to determine a tendency in the optimal gap values, despite that in Fig. 5 it does can be found a clear tendency.

Regarding the dispersion, in Table 2 it is represented the performance of each lower-limb for every user when using the average optimal gap value obtained from the whole group of subjects. In this case, for instance, the decoded trajectories of left knee do not reach a PCC with the real ones of 0.5 ($0.49\pm0.25$), whilst that average similitude is reached when applying personalized regression methods for each user ($0.52\pm0.19$). The rest of the lower-limbs do not improve their results neither, as in Table 1 has been selected the optimal results by executing a gap sweep. Thus, the same gap value may fit for some regression models, but not for others. Therefore, generating custom regression applications for every subject improves general results.

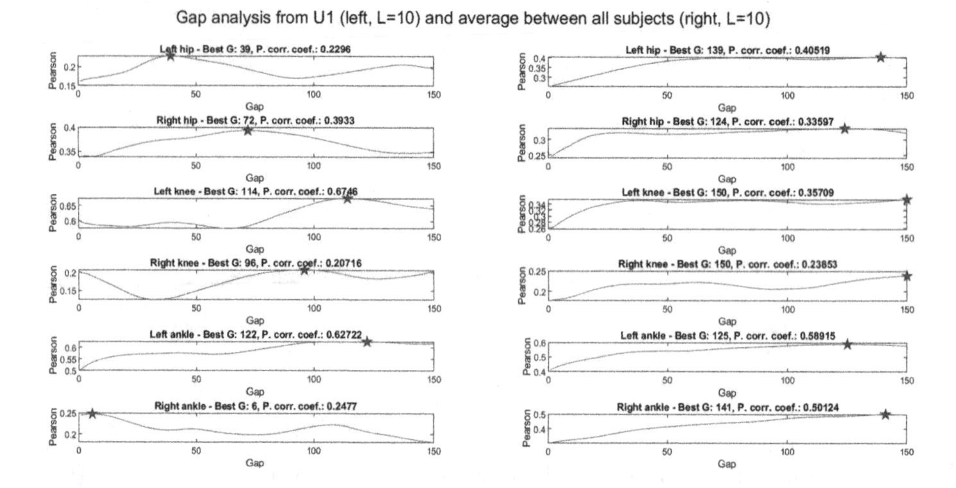


**Fig. 5.** Results from a gap parameter sweep, from user U1 (left) and the mean of all the subjects (right). For every lower-limb from both cases, the peak (gap value which presents better decoding performance) has been marked with an star.

**Table 1.** Lower-limbs decodification results. For every lower-limb from each user, it contains the gap value that has presented a higher PCC during the decodification, as well as this coefficient, and the processing time window associated with that gap value. The mean and standard deviation (STD) between subjects have also been added to the table.

| Limb | Item | U1 | U2 | U3 | U4 | U5 | U6 | U7 | U8 | Mean | STD |
|---|---|---|---|---|---|---|---|---|---|---|---|
| Right ankle | PCC | 0.248 | 0.393 | 0.312 | 0.384 | 0.423 | 0.483 | 0.306 | 0.599 | 0.393 | 0.111 |
| | GAP | 6 | 136 | 143 | 147 | 67 | 150 | 150 | 145 | 118 | 53 |
| | Time [s] | 0.110 | 2.450 | 2.576 | 2.648 | 1.208 | 2.702 | 2.702 | 2.612 | 2.126 | 0.955 |
| Left ankle | PCC | 0.627 | 0.611 | 0.269 | 0.506 | 0.475 | 0.596 | 0.241 | 0.822 | 0.519 | 0.193 |
| | GAP | 122 | 150 | 32 | 129 | 95 | 150 | 48 | 115 | 105 | 44 |
| | Time [s] | 2.198 | 2.702 | 0.578 | 2.324 | 1.712 | 2.702 | 0.866 | 2.072 | 1.8943 | 0.7961 |
| Right knee | PCC | 0.207 | 0.324 | 0.362 | -0.143 | 0.053 | 0.399 | -0.132 | 0.653 | 0.215 | 0.276 |
| | GAP | 6 | 47 | 57 | 136 | 13 | 1 | 150 | 150 | 70 | 65 |
| | Time [s] | 0.110 | 0.848 | 1.028 | 2.450 | 0.236 | 0.020 | 2.702 | 2.702 | 1.2620 | 1.1776 |
| Left knee | PCC | 0.675 | 0.369 | 0.265 | 0.433 | 0.255 | 0.224 | 0.017 | 0.696 | 0.367 | 0.231 |
| | GAP | 6 | 52 | 147 | 150 | 41 | 101 | 1 | 145 | 80 | 63 |
| | Time [s] | 0.110 | 0.938 | 2.648 | 2.702 | 0.740 | 1.820 | 0.020 | 2.612 | 1.449 | 1.140 |
| Right hip | PCC | 0.393 | 0.344 | 0.305 | 0.290 | 0.322 | 0.331 | 0.047 | 0.682 | 0.339 | 0.173 |
| | GAP | 6 | 19 | 17 | 99 | 117 | 7 | 1 | 150 | 52 | 60 |
| | Time [s] | 0.110 | 0.344 | 0.308 | 1.784 | 2.108 | 0.128 | 0.020 | 2.702 | 0.9380 | 1.078 |
| Left hip | PCC | 0.230 | 0.188 | 0.315 | 0.121 | 0.416 | 0.148 | 0.636 | 0.778 | 0.354 | 0.240 |
| | GAP | 6 | 150 | 7 | 57 | 136 | 95 | 94 | 149 | 87 | 59 |
| | Time [s] | 0.110 | 2.702 | 0.128 | 1.028 | 2.450 | 1.712 | 1.694 | 2.684 | 1.564 | 1.058 |

**Table 2.** Decoding performance (PCC between real and decoded trajectories) for each limb of every subject, using the average optimal gap value for the analysis. It is also included the processing time windows in seconds as well as the mean and the standard deviation values.

| Limb | Gap | Time | U1 | U2 | U3 | U4 | U5 | U6 | U7 | U8 | Mean | STD |
|---|---|---|---|---|---|---|---|---|---|---|---|---|
| Right ankle | 118 | 2.126 | 0.212 | 0.378 | 0.299 | 0.357 | 0.384 | 0.476 | 0.276 | 0.558 | 0.368 | 0.111 |
| Left ankle | 105 | 1.892 | 0.626 | 0.561 | 0.230 | 0.485 | 0.466 | 0.576 | 0.190 | 0.813 | 0.493 | 0.205 |
| Right knee | 81 | 1.460 | 0.201 | 0.283 | 0.317 | −0.191 | −0.112 | 0.321 | −0.152 | 0.630 | 0.162 | 0.288 |
| Left knee | 94 | 1.694 | 0.643 | 0.331 | 0.247 | 0.397 | 0.214 | 0.217 | −0.131 | 0.662 | 0.323 | 0.256 |
| Right hip | 60 | 1.082 | 0.389 | 0.278 | 0.164 | 0.188 | 0.265 | 0.233 | −0.057 | 0.631 | 0.261 | 0.197 |
| Left hip | 91 | 1.640 | 0.171 | 0.148 | 0.149 | 0.056 | 0.380 | 0.148 | 0.635 | 0.712 | 0.300 | 0.250 |

## 4 Conclusions

In this paper, a multidimensional linear regression algorithm has been developed for lower-limb trajectories reconstruction through EEG signals. This regression model has also been tested and some analysis have been applied in order to optimize the algorithm and its parameters. Nevertheless, notwithstanding this progress it is evident that there is a big dispersion between subjects performance. This problem seems to be eased when executing gap parameter sweeps for each user and lower-limb, in order to determine which are the optimal gap values in every case. However, trying to minimize computational times by deleting this sweeps and applying always fixed gap values dos not seem to be accurately feasible for now, although an optimal gap rising tendency has been found.

As a consequence of this, further investigation research should be undertaken with the intention of improve the algorithm and the parameter values selection, like executing lag parameter sweeps, or different kinds of preprocessing. Additionally, an electrode study could help in improving the algorithm performance, because it is possible that other parts of the human brain contain valuable neural information about lower-limb kinematics, or another type of patterns that enhance regression accuracy.

Anyhow, results like the obtained with subject U8 (up to 0'82 Pearson Correlation Coefficient) prove that kinematics decodification from EEG could be possible, despite the fact that it still remain many challenges to solve.

## References


1. Slutzky, M.W.: Brain-machine interfaces: powerful tools for clinical treatment and neuroscientific investigations. Neuroscientist **25**(2), 139–154 (2019). https://doi.org/10.1177/1073858418775355
2. Mak, J., Wolpaw, J.R.: Clinical applications of brain-computer interfaces: current state and future prospects. IEEE Rev. Biomed. Eng. **2**, 187–199 (2010)
3. Cramer, S.C.: Repairing the human brain after stroke. II. Restorative therapies. Ann. Neurol. **63**(5), 549–560 (2008). https://doi.org/10.1002/ANA.21412
4. Gharabaghi, A.: What turns assistive into restorative brain-machine interfaces? Front. Neurosci. (OCT), 456 (2016). https://doi.org/10.3389/FNINS.2016.00456

5. Millan, J.D.R., et al.: Combining brain-computer interfaces and assistive technologies: state-of-the-art and challenges. Front. Neurosci. **4**, 161 (2010)
6. Li, Y.D., Hsiao-Wecksler, E.T.: Gait mode recognition and control for a portable-powered ankle-foot orthosis. In: 2013 IEEE 13th International Conference on Rehabilitation Robotics (ICORR), vol. 2013 (2013). https://doi.org/10.1109/ICORR.2013.6650373
7. Castermans, T., Duvinage, M.: Corticomuscular coherence revealed during treadmill walking: further evidence of supraspinal control in human locomotion. J. Physiol. **591**, 1407–1408 (2013)
8. Contreras-Vidal, J.L., Grossman, R.G.: NeuroRex: a clinical neural interface roadmap for EEG-based brain machine interfaces to a lower body robotic exoskeleton. In: Proceedings of the 2013 35th Annual International Conference of the IEEE Engineering in Medicine and Biology Society (EMBC), pp. 1579–1582 (2013). https://doi.org/10.1109/EMBC.2013.6609816
9. Ortiz, M., Ferrero, L., Iáñez, E., Azorín, J.M., Contreras-Vidal, J.L.: Sensory integration in human movement: a new brain-machine interface based on gamma band and attention level for controlling a lower-limb exoskeleton. Front. Bioeng. Biotechnol. **8**(September) (2020). https://doi.org/10.3389/fbioe.2020.00735
10. Ferrero, L., Quiles, V., Ortiz, M., Iáñez, E., Azorín, J.M.: A BMI based on motor imagery and attention for commanding a lower-limb robotic exoskeleton: a case study. Appl. Sci. **11**(9) 4106 (2021). https://doi.org/10.3390/APP11094106
11. Kaplan, Zhigulskaya, D., Kiriyanov, D.A.: Studying the ability to control human phantom fingers in P300 brain-computer interface. Bull. Russ. State Med. Univ. (2016)
12. Úbeda, A., et al.: Decoding knee angles from EEG signals for different walking speeds. In: Proceedings of the 2014 IEEE International Conference on Systems, Man, and Cybernetics (SMC), vol. 2014-Janua, no. January, pp. 1475–1478 (2014). https://doi.org/10.1109/smc.2014.6974123
13. Gui, K., Liu, H., Zhang, D.: Toward multimodal human-robot interaction to enhance active participation of users in gait rehabilitation. IEEE Trans. Neural Syst. Rehabil. Eng. **25**(11), 2054–2066 (2017). https://doi.org/10.1109/TNSRE.2017.2703586
14. Úbeda, A., et al.: Single joint movement decoding from EEG in healthy and incomplete spinal cord injured subjects. In: Proceedings of the 2015 IEEE/RSJ International Conference on Intelligent Robots and Systems (IROS), vol. 2015-Decem, pp. 6179–6183 (2015). https://doi.org/10.1109/IROS.2015.7354258
15. Mercado, L., Azorín, J.M., Platas, M., Úbeda, A., Quiroz, G.: Offline lower-limb kinematic decodification by segments of EEG signals. In: Proceedings of the 40th International Conf. of the IEEE Eng. in Medicine and Biology Society, pp. 2398–2401, Honolulu, HI, 17–21 July 2018. ISBN:978-1-5386-3646-6
16. Garipelli, G., Chavarriaga, R., Millán, J.D.R.: Single trial analysis of slow cortical potentials: a study on anticipation related potentials. J. Neural Eng. **10**(3) (2013). https://doi.org/10.1088/1741-2560/10/3/036014
17. Úbeda, A., Hortal, E., Iáñez, E., Planelles, D., Azorín, J.M.: Passive robot assistance in arm movement decoding from EEG signals. In: Proceedings of the 6th Annual International IEEE EMBS Conference on Neural Engineering, pp. 895–898 (2013)

# EEG Signals in Mental Fatigue Detection: A Comparing Study of Machine Learning Technics VS Deep Learning

Halima Ettahiri[1,2](✉), José Manuel Ferrández Vicente[1], and Taoufiq Fechtali[2]

[1] Departamento de Electrónica, Tecnología de Computadores y Proyectos - Campus la Muralla, Edif. Antigones, Universidad Politécnica de Cartagena, 30202 Cartagena, Murciaa, Spain
ettahirihalima@gmail.com

[2] Department de Biosciences, Exploration Fonctionnelle Intégrée., Université Hassan II casablanca, Faculté de sciences et techniques de Mohammedia, 30003 Mohamedia, Casablanca, Morocco

**Abstract.** Mental fatigue is a complex disorganization that affects the human being efficiency in work and daily activities as (driving, exercising etc.). To discern that fatigue, the encephalography is routinely used, several automatic procedures have been deploying conventional approaches to support neurologists in mental fatigue detection episodes e.g. (sleepy vs normal). The aim of this work is the use of the EEG's data to understand how the mental fatigue can affect the conductor's behavior, lot of methods are involved to understand these data as machine learning approach and deep learning method. The data is organized as follow: EEG data of 10 normal people and other 10 people who are deprived from sleep, the recording time is 7 min in each session, and the experiment includes three sessions for each person, none of the volunteers n have a mental history and none of them are on medication. The main of this study is to compare the different methods for the analysis of EEG signals for the detection of fatigue, using machine learning and deep learning.

**Keywords:** Mental fatigue · EEG signals · Dataset

## 1 Introduction

The proposed framework for automatic sleep detection using EEG signals based on deep learning is presented in Fig. 1. It consists of three major parts (i) cutting the input signal into a sub-signal using an interval, (ii) a band of DNN model where each distinguished signal is arranged by DNN model, and (iii) making the final decision about the final classification (Matrix).

Generally, DL framework required a big amount of data to train the system, but in this study, a small amount of data is used. To deal with this obstacle, data augmentation is introduced in the next subsection.


J. M. Ferrández Vicente et al. (Eds.): IWINAC 2022, LNCS 13258, pp. 625–633, 2022.
https://doi.org/10.1007/978-3-031-06242-1_62

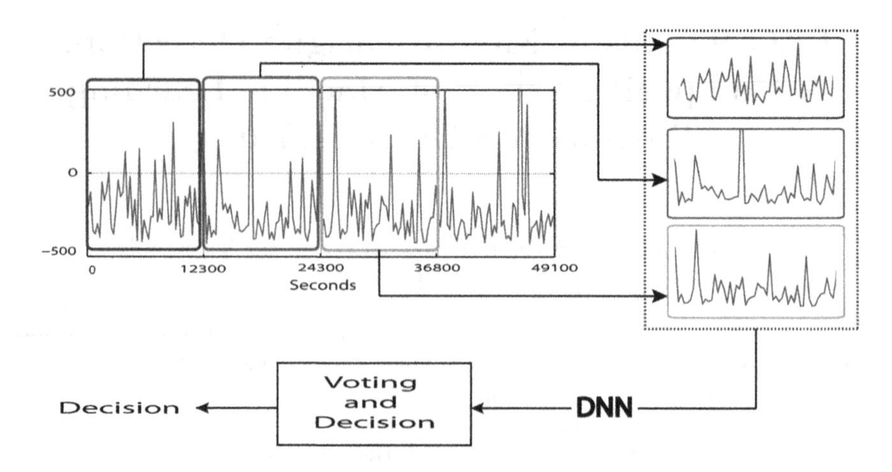


**Fig. 1.** Division of signal architecture

## Dataset Augmentation

The dataset trained in this project is used for mental fatigue detection. The EEG signals are recorded using 8 electrode placement system (F3, F4, Fz, C3, C4, Cz, Pz, T8) Fig. 2, the acquisitions are performed based on the OpenVibe acquisition server.

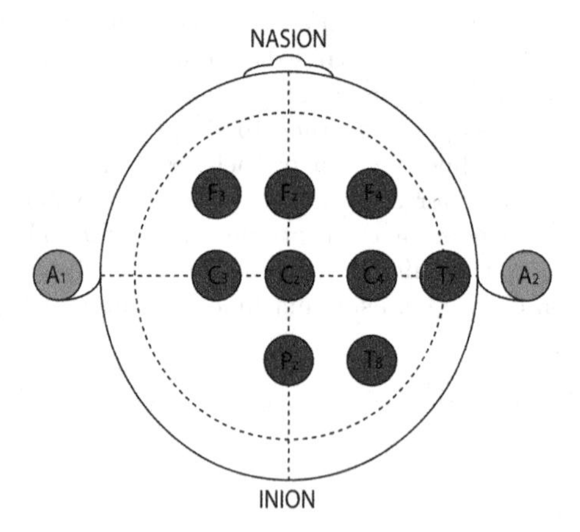


**Fig. 2.** Placement of used electrodes. Blue colour: electrodes used in the experience. (Color figure online)

The amount of data collected in this experiment is insufficient to train a deep model (CNN in our case). Because collecting EEG signals needs a long duration and many volunteers; so is decided to augment the schemes to increase the data size for training CNN model in order to generate a better classification. Each record in this dataset has 7 min of recording EEG signal (420 s), the record is a General Data Format (GDF) that contain the data recorded from 8 electrodes, a transformation via Open Vibe from GDF files to CSV files is adapted.

The obtained data is divided into sub-data disjoint training and testing sets, each signal (420 s for each signal) is divided into 28 sub-signals (15 s for each sub-signal), in total 560 sub-signals are obtained for the whole the dataset.

**Architecture Model CNN**

CNN is an ensemble of superimposed covers (layers) that involve linear and non-linear processes [7]. The head structure parts of that CNN model are pooling layer, non-linear Rectified Linear Units (ReLU), and convolutional layer. Thus, the layer is connected to a regular multilayer neural network called a fully connected layer, then, a loss layer is placed as the back-end. CNN is known for its prominent results for many applications as the image classification and natural language processing [14].

The suggested network for automatic mental fatigue recognition that uses EEG brain signals consists of first by the data augmentation, and the ensemble of the classified sub-signals by 1D CNN model. To validate the obtained results, the fusion matrix is applied.

The data augmentation schemes are introduced in Sect. 4, each EEG signal is divided into 28 sub-signals, in total 560 sub-signals are generated.

The P-1D-CNN structure of the CNN model absorbs structures of EEG signals by an automatically from data, the classification is processed from the beginning until the end, unlike traditional classifications that are hand-engineered. Firstly, the selected features are extracted, then the features subdivision is applied. Finally, they are delivered to a classifier in order to perform classification. A convolutional layer composed of many channels (feature maps) is the major element of a CNN model [6]. The yield of every neuron in a channel is the result of a convolution action with a kernel (those things that are common to all the neurons of every channel) of the static receptive field on input signal or feature maps (1D signals) of the preceding convolutional layer. In this manner, the convolutional neural network evaluates how a signal can learn the discriminative information from the hierarchy. In CNN, instead of the traditional classification, the kernels are well predefined e.g. (wavelet transformation). kernels are learned from data. However, the original idea of CNN with shared kernels has the benefit of a substantial reduction of the parameters number compared to entirely connected models [1]. The latest apparition of making CNN deeper can increase the use of an incredibly huge parameters number to its complexity which caused the over-fitting for the small dataset [13]. Therefore, in this study, a small amount of data of EEG signals for mental fatigue detection is used. In order to handle this obstacle, two plans are used: (i) a new data augmentation scheme is proposed, (ii) the use of a small dataset for deep CNN model is exploited the EEG signal is 1-Dimension time series. Thus, to analyse it, a pyramidal model 1D-CNN is analysed, called P-1D-CNN as shown in the generic architecture presented by Fig. 3, which is a bit different from the CNN traditional models, it doesn't comprise of any reducing layer. In addition, bigger strides in convolution layers help to reduce redundant or unnecessary features. In this case, the convolutional layers acquire a hierarchy from the low features to high-level ones in the input signal. Semantic representation is with the high-level features supplied as input to the softmax classifier in the last layer to predict the relevant class of the input EEG signal.

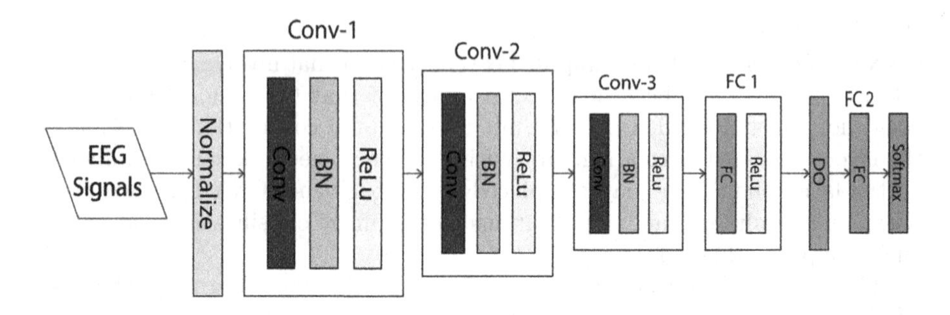


**Fig. 3.** Architecture 1 P CNN

In most of the cases, the CNN model is organized by finding ways to define a good method where there is a small number of kernels in the low-level layers, and where a high-level layer had a large number of kernels. Therefore, this model implies a vast number of learnable parameters, which means that the complexity is too high. Instead, the adoption of a pyramidal architecture for deep 2-D CNN [12] is proposed to avoid the risk of over-fitting. A huge number of kernels are taken in a Conv1 layer after they are reduced by a constant number in Conv2 and Conv3 layers. To test the efficiency of the pyramidal CNN structure, eight models are considered, four of them are from pyramid architecture. The last fully connected layer has two or three nodes, depending on the EEG brain signal classification (e.g. tired or normal) [3]. With the help of those models, the manner in which an accurately designed model can lead to high performance even when there are fewer parameters, which is riskier in terms of over-fitting. The input signals are normalized with zero mean and unit variance. This normalization helps to get fast convergence and to avoid local minima. The normalized input is processed by three convolutional blocks, where each block consists of three layers: Convolutional layer (Conv), Batch normalization layer (BN), and non-linear activation layer (ReLU) [4]. The output of ReLU layer in the third block is passed to a fully connected layer (FC1) that is followed by a ReLU layer and another fully connected layer (FC2). In order to avoid over-fitting, a dropout is used before FC2. The output of FC2 is supplied to a softmax layer, which has the purpose of a being classifier and predicts the class of the input signal. The number of neurons in the final layer will change in accordance with the number of classes to classify (e.g. normal vs sleepy)(Binary Class). In the following subsections, a brief explanation of two of the main layers is provided. The 1D-convolution operation is generally used to filter 1D signals (e.g. time series) for deducting discriminative features.

Neural networks are inspired by the concept of neurons, they correspond to a number of input values that are performed by the synapses, and those values are analysed in the neurons. Moreover, the neuron doesn't just appear as an output that weighted sum, since the computation associated with a cascade of neurons would then be a simple linear algebra operation [11]. Figure 1 shows a diagrammatic of CNN used in this classification.

The input layers receive the values from the electrodes F3, Cz, Tz and spread them for the neurons in hidden layers (layers in the middle) of the network Fig. 4. The amount are from the layers in the middle spread to network outputs layers which are in this case (sleepy or normal).

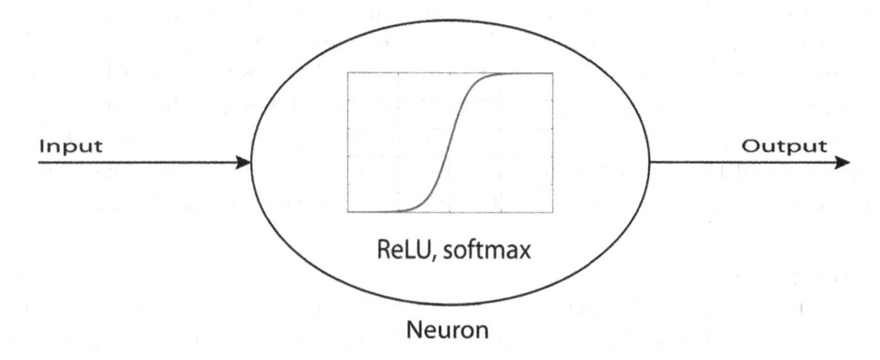


**Fig. 4.** Model of the CNN

**Evaluation of the Model Proposed with k-Fold Cross-validation**

The neural network model has to be evaluated by the scikit-learn for training data, which has a great ability to estimate the models employing multiple procedures. To evaluate machine learning models, the great standard is k-fold cross-validation. Firstly, the model evaluation procedure architecture is explained. In this case, the number of folds is fixed at 10 (an excellent by default), and then, data is mixed before partitioning them (shuffle the data). Thus, the evaluation of the model (estimator) on the dataset is deploying 10 fold procedure (k fold). The Evaluation of the model takes approximately 10 s and returns an object that helps to describe the evaluation of the 10 constructed models for each dataset split.

$$k_{fold} = K_{Fold}\ \ (n\_splits = k, shuffle = True, random\_state = seed)\ \ k = 10; \tag{1}$$

The results obtained and standard deviation are regrouped for model accuracy in the dataset. This problem has a reasonable estimation of the performance of the model on unseen data. also within the realm of known top results.

## Experimental Design, Materials and Methods

### Laboratory Experience

Before applying any application of the model, the main idea of this experiment with the used dataset needs to be clear. In this experiment, the sleep deprivation impact on mental fatigue is assessed. In fact, 20 volunteers are selected to participate in this study, none of them had a medical history, neurological or psychiatric disorders or drug addiction. Thus, the participants are randomly assigned to one of the two groups: (i) the normal sleep group and (ii) the deprivation sleep group.EEG signals are collected using the Open VIBE interface, this interface was developed by INRIA, and the used material is the famous cap ENOBIO 8.

After collecting EEG signals from 20 people using the ENOBIO 8 which allows us to perform tests in three sessions, each session duration is 7 min (total 21 min per volunteer), with 40 trials each session, in each trial the volunteer must imagine the movement of their right and left hand. Therefore, the EEG signal analysis is important in order to get conclusions for different volunteers groups. In fact, the EEG signals are able to define mental fatigue and its impact. The application of DL (CNN model) is adapted to the problem of EEG classification and compared to those obtained using SVM and linear discriminant analysis.

### Participants

In the first part of this experiment, five healthy men and five healthy women aged between 26 and 36 years old participate voluntarily. In the second part, ten volunteers from different countries take part in the study aged between 25 and 30 years old. In the two experiment groups, the volunteers were informed of the experiment. Also, no ethical review or approval was required for this study and all the subjects did not suffer from any mental illnesses or disorders that can disturb the collected results.

The volunteers selection has consisted of recruiting from different genders. In order to get better EEG signals, only measured signals without noises are considered. Sometimes, noises appear because of cosmetic products, hair gel and curly hair. Also, the age range is important to suit case drivers.

### Data Acquisitions

EEG data is recorded in SICOMO laboratory (Polytechnic University of Cartagena, Spain)and the OpenVibe acquisition protocol is respected, the device (ENOBIO 8) with 8 electrodes placed is presented in (Fig. 2). The sampling frequency is set 500 Hz. Thus, EEG signals recorded are filtered using a pass-band filter [1–30 Hz], it's called filtering in the frequency domain. The purpose of a filter is to eliminate noise components from signals. The distinction between noise and signal would be simple if the signal was well known. In this work, the original signals arrive simultaneously to the eight electrodes. The signal, in this case, can be expressed as follow:

$$X = AS + N \tag{2}$$

where $X = [X1, \ldots, XN]$ is signals matrix and $S = [S1, \ldots, SN]$ is the original data.

The most famous tested brain wave rhythm is the alpha wave, this rhythm can be observed clearly on the occipital and posterior zones. The frequency can variate 8 Hz and 12 Hz. Due to the results with ten volunteers (selection from the normal group), the alpha rhythm is important in the central and posterior regions.

Alpha rhythm characterizes a state of appeased conscience and is mainly emitted when the subject has their eyes closed, the mean powers present the apparition of Alpha waves for the normal group during the experiment, those rhythms characterize light relaxation and calm awakening [5].

The second selection from the sleepy group (deprivation group) is characterized by low frequencies (less 4 Hz), those waves reflect deep relaxation and can especially be reached by experienced meditators, they are called theta waves. The brain activities (measured by the electroencephalogram) show the functioning similarity between sleep deprivation and meditating [7]. Often, meditators find that during a retreat, where they practice intensively, they have less need to sleep. When a volunteer does not sleep for one night, the next night will be enriched by dreams and the paradoxical sleep becomes less recuperative. In fact, when one volunteer meditates regularly, he can generate normal sleep rhythm more quickly. Therefore, after studying this data with different methods, The results are shown in table Table 1. As a result, the CNN proposed approach gives better accuracy (97,3%) campared with other methods.

**Table 1.** Accuracies using LDA, SVM, KNN, and TREE classifiers

| | Tree | LDA | SVM | KNN |
|---|---|---|---|---|
| BP | 0.78 | 0.87 | 0.88 | 0.78 |
| CSP | 0.91 | 0.93 | 0.92 | 0.83 |
| BP+CSP | 0.95 | 0.93 | 0.92 | 0.86 |

## Discussion

Detecting fatigue in drivers, mainly multi-level drowsiness, is a hard problem that is often handled by using neurophysiological signals as the better approach for building a well-founded system [9]. In this setting, electroencephalogram (EEG) signals are the most important source of data to achieve successful detection.

In this paper, we first review EEG signal features used in the literature for a fatigue detection, then we focus on reviewing the applications of EEG features and machine learning approaches in driver fatigue detection [8], also a section about systems of deep learning used to detect mental fatigue and finally we discuss the open challenges and opportunities in improving driver fatigue detection based on EEG.

This paper aims to review the features extracted from the EEG signal and the applications [2] of these features to the problem of driver drowsiness detection. We compare the features since the large number of features described in the literature makes it difficult to understand their interrelationships, and makes it difficult to choose the right ones for the given problem. as known, there is no similar work that covers all the features discussed in this paper [10]. After we detailed the experience, we continue with the review of driver fatigue detection systems based on machine learning's system. The main goal is to compare insight into the most used EEG features and recent deep learning approaches for fat igue detection, which would allow us to identify possibilities for further improvements of fatigue detection systems. Finally, the main contribution of our work is like the following, the application of deep learning approaches give us a better accuracy than the other machine learning methods.

**Acknowledgements.** Grant PID2020-115220RB-C22 funded by MCIN/AEI/ 10. 13039/501100011033 and, as appropriate, by "ERDF A way of making Europe", by the "European Union" or by the "European Union NextGenerationEU/PRTR".

## References

1. Biswal, S., Sun, H., Goparaju, B., Westover, M.B., Sun, J., Bianchi, M.T.: Expert-level sleep scoring with deep neural networks. J. Am. Med. Inform. Assoc. **25**(12), 1643–1650 (2018)
2. Chen, J., Wang, H., Hua, C.: Assessment of driver drowsiness using electroencephalogram signals based on multiple functional brain networks. Int. J. Psychophysiol. **133**, 120–130 (2018)
3. Craik, A., He, Y., Contreras-Vidal, J.L.: Deep learning for electroencephalogram (EEG) classification tasks: a review. J. Neural Eng. **16**(3), 031001 (2019)
4. Daniela, T., Alessandro, C., Giuseppe, C., Fabio, M., Cristina, M., Michele, F., et al.: Lack of sleep affects the evaluation of emotional stimuli. Brain Res. Bull. **82**(1–2), 104–108 (2010)
5. Ferrara, M., et al.: The role of sleep in the consolidation of route learning in humans: a behavioural study. Brain Res. Bull. **71**(1–3), 4–9 (2006)
6. Haidar, R., McCloskey, S., Koprinska, I., Jeffries, B.: Convolutional neural networks on multiple respiratory channels to detect hypopnea and obstructive apnea events. In: 2018 International Joint Conference on Neural Networks (IJCNN), pp. 1–7. IEEE (2018)
7. Hussain, M., Aboalsamh, H., Abdul, W., Bamatraf, S., Ullah, I., et al.: An intelligent system to classify epileptic and non-epileptic EEG signals. In: 2016 12th International Conference on Signal-Image Technology & Internet-Based Systems (SITIS), pp. 230–235. IEEE (2016)
8. Jiao, Y., Deng, Y., Luo, Y., Lu, B.L.: Driver sleepiness detection from EEG and EOG signals using GAN and LSTM networks. Neurocomputing **408**, 100–111 (2020)
9. Lee, B.G., Lee, B.L., Chung, W.Y.: Mobile healthcare for automatic driving sleep-onset detection using wavelet-based EEG and respiration signals. Sensors **14**(10), 17915–17936 (2014)

10. Min, J., Wang, P., Hu, J.: Driver fatigue detection through multiple entropy fusion analysis in an EEG-based system. PLoS one **12**(12), e0188756 (2017)
11. Sze, V., Chen, Y.H., Yang, T.J., Emer, J.S.: Efficient processing of deep neural networks: a tutorial and survey. Proc. IEEE **105**(12), 2295–2329 (2017)
12. Ullah, I., Hussain, M., Aboalsamh, H., et al.: An automated system for epilepsy detection using EEG brain signals based on deep learning approach. Expert Syst. Appl. **107**, 61–71 (2018)
13. Urtnasan, E., Park, J.U., Lee, K.J.: Multiclass classification of obstructive sleep apnea/hypopnea based on a convolutional neural network from a single-lead electrocardiogram. Physiol. Measure. **39**(6), 065003 (2018)
14. Zammouri, A., Moussa, A.A., Mebrouk, Y.: Brain-computer interface for workload estimation: assessment of mental efforts in learning processes. Expert Syst. Appl. **112**, 138–147 (2018)

# Application of RESNET and Combined RESNET+LSTM Network for Retina Inspired Emotional Face Recognition System

Mahmudul Huq[2], Javier Garrigos[2], Jose Javier Martinez[2], Jose Manuel Ferrandez[2(✉)], and Eduardo Fernández[1,2,3]

[1] Department of Electronics, Computer Architecture and Projects, Technical University of Cartagena, Cartagena, Spain
[2] Bioengineering Institute, Universidad Miguel Hernandez, Elche, Spain
jm.ferrandez@upct.es
[3] CIBER Research Center on Bioengineering, Biomaterials and Nanomedicine (CIBER BBN), Madrid, Spain

**Abstract.** Various Facial Expression Recognition (FER) systems have been studied in the field of computer vision and machine learning to encode expression information from facial representations. In this research paper, a facial emotion recognition system is proposed, addressing automatic face detection and facial expression recognition using 1) Residual Neural Network (RESNET) and 2) Combined Residual Neural Network + Long Short-Term Memory (Combined RESNET+LSTM). The architectures of RESNET and Combined RESNET+LSTM are inspired by the human retina structure and human primary visual cortex structure. The proposed architectures are compared with each other. They are tested using a challenging public database.

**Keywords:** Face recognition · Machine learning · Residual neural network · Recurrent neural network

## 1 Introduction

Facial emotion recognition can be used todays for many different applications that covers from neuromarketing to visual neuroprosthesis [10,11]. The following two strategies can be used to design facial emotion recognition (FER) systems: 1) image-based FER and 2) video-based FER. Even though video-based FER systems convey a lot of emotional data due to their dynamic features, they require a more comprehensive analysis that is still difficult to apply in real-time systems. As a result, the FER methodology is first tested on static images using a simpler approach. However, because to many environmental variables such as brightness, occlusions, and body posture, which might impact the information available in the processed image, FER remains a difficult task in these situations.


J. M. Ferrández Vicente et al. (Eds.): IWINAC 2022, LNCS 13258, pp. 634–646, 2022.
https://doi.org/10.1007/978-3-031-06242-1_63

Generally, FER image-based systems consist of following three major stages: 1) Face detection and pre-processing, 2) Feature extraction, and 3) Classification. The face detection and pre-processing stage seeks to detect the face with the goal of removing superfluous information such as the background to recognize the facial expression. The feature extraction procedure results in a characteristic vector containing the relevant characteristics being used to represent the pre-processed image. The appearance-based method and the geometric-based method are the two basic kinds of feature extraction strategies in FER. Using well-known feature extractors like Gabor filters and local binary patterns (LBP), the appearance-based technique seeks to detect textural information from face photos. Geometric properties-based strategies use all the observed facial landmarks to represent geometric information such as angles and distances. The collected features are then classified to identify face emotions.

Compared to traditional methods based on the feature extraction techniques dis-cussed before, Residual Neural Networks (RESNETs) based techniques have achieved the FER state of the art recognition rates [14]. RESNETs are based on the shared-weight architecture and translation invariance properties of convolution kernels, which scan the hidden layers [17].

RESNETs were inspired by the organic structure of a visual cortex, which com-prises configurations of simple and complicated cells. Recurrent Neural Network (RNN) is a class of neural network that allows previous outputs to be used as inputs while having hidden status [2]. A RNN is a class of artificial neural net-works where connections between nodes form a directed graph along a temporal sequence. This allows it to exhibit temporal dynamic behavior. Long Short-Term Memory (LSTM) is a type of an artificial recurrent neural network (RNN) architecture used in the field of deep learning.

In this research paper, we will be proposing a facial emotion recognition system addressing automatic face detection and facial expression recognition using 1) Residual Neural Network (RESNET) and 2) Combined Residual Neural Network + Long Short-Term Memory (Combined RESNET+LSTM). The architectures of RESNET and Combined RESNET+LSTM will be based on the human retina + primary visual cortex structure. It would also be interesting to compare the proposed networks with other networks trained for the FER2013 challenge. [1]

The rest of the research paper is organized as follows: In the Sect. 2 we will dis-cuss on the related works done in the fields of facial emotion recognition, RESNET and RNN. We will be discussing on the proposed solution in detail in the Sect. 3. Section 4 will contain the discussion on the experiments done on the proposed solution and the results of the experiments. In the Sect. 5 we will have discussion. The paper will end with conclusion in Sect. 6.

## 2 Related Works

Several studies have been conducted in the FER sector to increase emotion estimation accuracy. Deep learning has outperformed traditional feature extraction techniques in comparison to traditional methodologies. Deep learning training methodologies can be classified into two categories: single network learning and ensemble network learning.

For feature extraction and recognition, the single network learning category uses only one architecture. Mollahosseini et al. [3] constructed a Convolution Neural Network (CNN) architecture for FER that included two convolutional layers and four inception layers. Guo et al. [16] developed a Deep Neural Network with Relativity Learning (DNNRL) that consisted of three convolutional layers, three inception layers, and was trained using the triple loss. Devries et al. [15] presented a multi-task neural network for simultaneous facial land-mark identification and facial expression recognition. Yang [18] proposed the DLSVM for FER application by replacing the softmax layer with a linear Support Vector Machine (SVM). Using the Class Activation Mapping Technique (CAM) and a visualization model based on the Densenet-BC architecture. Using the Class Activation Mapping Technique (CAM) and a visualization model based on the Densenet-BC architecture, Lian et al. [19] investigated the role of each face region in facial expression identification.

Levi and Hassner [8] proposed a revolutionary deep ensemble network-based technique. LBP codes are extracted and mapped to a 3D space via Multidimensional Scaling (MDS) after being applied varying radius settings. Following that, the original RGB images and mapped codes are used to train a collection of 20 CNN models, with the projected emotion determined by a weighted average of the output of each model. For FER Kim et al. [4] proposed combining information from non-aligned and aligned faces. They introduced an Alignment Mapping Network (AMN) to estimate aligned states of non-alignable faces using a set of 9 Deep Convolutional Neural Networks (DCNNs) The emotion prediction is computed using the average or majority voting rules of DCNNs. Similarly, Pramerdorfer et al. [7] trained an ensemble of 8 CNNs using the FER 2013 database and achieved the best state of the art performance.

Benamara et al. [6] proposed a facial emotion recognition system that addressed both automatic face detection and facial expression recognition individually. In comparison to an ensembling strategy, facial expression recognition is performed by a set of only four deep convolutional neural networks, while miss-labeled training data is dealt with using a label smoothing algorithm [5].

## 3 Methods

The Architecture of Human Retina Inspired RESNET and Combined RESNET+LSTM based Facial Emotion Recognition (FER) system is consisted of the following 3 stages: (1) Face Detection and Pre-Processing. (2) Deep Feature Extraction and Classification using Residual Neural Network (RESNET). (3) Deep Feature Extraction and Classification using Combined RESNET+LSTM (Fig. 1).

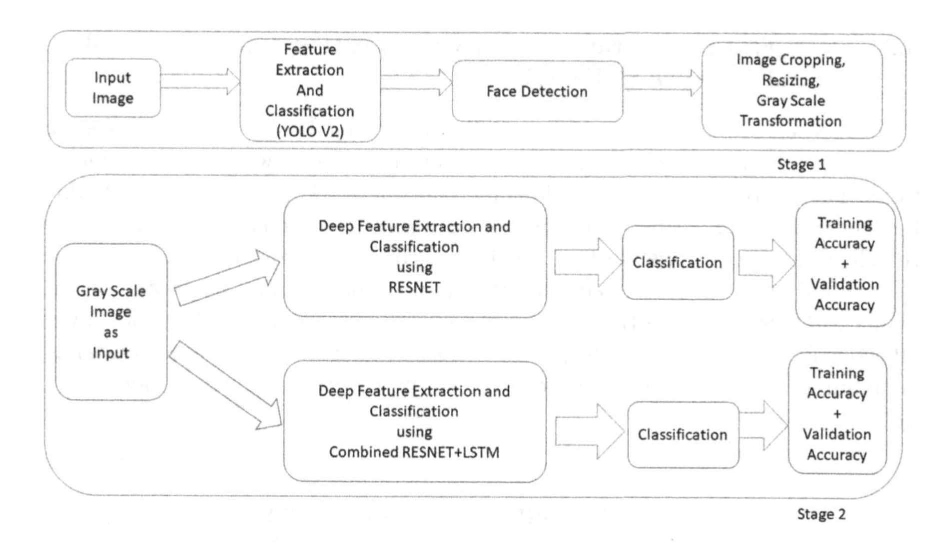


**Fig. 1.** Complete architecture of human retina inspired RESNET and combined RESNET+LSTM based FER system.

### 3.1 Imaging Pre-processing

The aim of the face detection stage is to reduce the amount of data fed into Convolution Neural Networks (CNNs) that identify facial emotions. Several object detection techniques have been proposed in the literature for the deep learning stage, which can be divided into two key approaches: 1) Region Proposal Based Network. 2) One Shot Based Network.

The two stages of the region proposal-based network approach are as follows: 1) First, the possible regions in the input picture that contain objects are chosen. 2) After that, the chosen input image is fed into a CNN for feature extraction and classification to describe the object and its bounding box coordinates. Examples of the region proposal-based network approach include RCNN and its improved versions Fast RCNN and Faster RCNN . In contrast to the first approach, the one shot-based network approach only requires a single stage for detection and classification. Overall, the one shot-based methods are faster than the computationally costly area proposal-based methods, which use two stages for object detection.

We chose to use two face detection methodologies for our FER system: 1) Viola Jones Method [13]. 2) YOLO V2 Method [9].The Viola Jones method is a non-deep learning-based method. It is a fast and accurate method achieving a performance of 76.8 per cent [5]. The YOLO V2 method is a deep learning-based method to deal with face detection by training the network on the fully annotated face database WIDER DB [12]. Cropped faces are converted from RGB to grayscale space and resized to a $48 \times 48 \times 1$ pixels resolution for RESNET architecture and resized to a $224 \times 224 \times 3$ pixels for combined RESNET+LSTM architecture.

### 3.2 Deep Feature Extraction and Classification Using Residual Neural Network (RESNET)

The deep feature extraction and classification using Residual Neural Network (RESNET) implements a deep learning neural network with residual connections. The use of residual connections increases network gradient flow and allows for the training of deeper networks. A residual network is a kind of Directed Acyclic Graph (DAG) network in which the primary network layers are bypassed by residual or shortcut connections. Residual connections allow parameter gradients to travel more easily from the output layer to the network's prior layers, allowing for the training of deeper networks. Higher accuracies on increasingly demanding tasks may be achieved because of the greater network depth (Fig. 2).

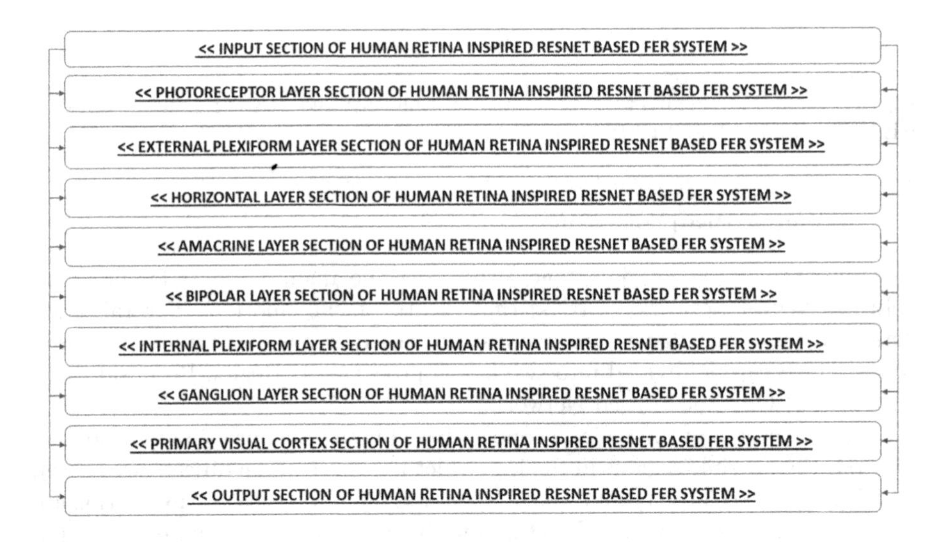


**Fig. 2.** Architecture of deep feature extraction and classification using RESNET.

The residual network architecture of the deep feature extraction and classification using RESNET used in our research paper is based on the structure of human retina. A typical human retina is consisted of the following ten layers: 1) The Input Layer. 2) The Layer of Photoreceptor Cell. 3) The External Plexiform Layer (EPL). 4) The Layer of Horizontal Cell. 5) The Layer of Amacrine Layer. 6) The Layer of Bipolar Cell. 7) The Inner Plexiform Layer (IPL). 8) The Layer of Ganglion Cell. 9) The Primary Visual Cortex. And 10) The Output Layer.

The residual network architecture of the deep feature extraction and classification using RESNET is consisted of following two components: 1) The Main Branch. And 2) The Residual Connections. The main branch of the deep feature extraction and classification using RESNET is consisted of ten sections. An initial section of the main branch is consisted of the image input layer, convolutional layer, batch normalization layer, and ReLU layer. These layers are connected sequentially. A middle section is consisted of eight stages of convolutional layers with different feature sizes. The stages of the middle section are the following: 1) The Layer of Pho-toreceptor Cell consisting of feature size of NetWidth-by-NetWidth. 2) The External Plexiform Layer (EPL) consisting of feature size of 2*NetWidth-by-2*NetWidth. 3) The Layer of Horizontal Cell consisting of feature size of 3*NetWidth-by-3*NetWidth. 4) The Layer of Amacrine Layer consisting of feature size of 4*NetWidth-by-4*NetWidth. 5) The Layer of Bipolar Cell consisting of feature size of 5*NetWidth-by-5*NetWidth. 6) The Inner Plexiform Layer (IPL) consisting of feature size of 6*NetWidth-by-6*NetWidth. 7) The Layer of Ganglion Cell consisting of feature size of 7*NetWidth-by-7*NetWidth. 8) The Primary Visual Cortex consist-ing of feature size of 8*NetWidth-by-8*NetWidth. Each stage contains 1 convolu-tional unit which contains two 2-by-2 convolutional layers with activations, two ReLU layers and a batch normalization layer.

The network width NetWidth of the RESNET architecture is a 1-by-10 vector of values [16 18 20 22 24 28 30 32 34 36]. The vector of NetWidth of the RESNET architecture is used for 10-fold cross validation. A final section is consisted of the global average pooling layer, fully connected layer, softmax layer, and classification layer. The eight convolutional units of the main branch are bypassed by residual connections. The remaining connections and outputs of the convolutional units are added one by one. When the sizes of the activations change, the residual connections must also contain 1-by-1 convolutional layers. Parameter gradients can flow more freely from the output layer to the network's earlier layers with residual connections, enabling for the training of deeper networks.

### 3.3 Deep Feature Extraction and Classification Using Combined Residual Neural Network+Long Short-Term Memory (RESNET+LSTM)

The deep feature extraction and classification using combined RESNET+LSTM implements a deep learning neural network with 1) The pretrained RESNET, such as GoogleNet, 2) LSTM, and 3) Assembly network consisting of pretrained RESNET and LSTM (Fig. 3).

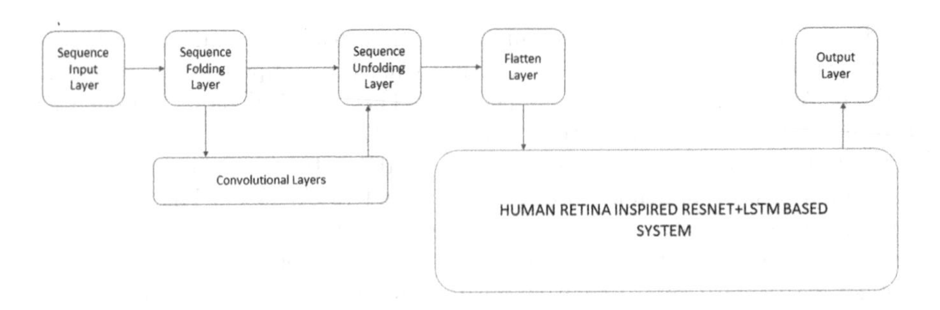


**Fig. 3.** Architecture of deep feature extraction and classification using RESNET.

The combined RESNET+LSTM is consisted of the following sections: 1) A sequence input layer for inputting input image sequences. 2) Existing convolutional layers of GoogleNet to extract features. Convolutional operations are applied to each image independently, this means a sequence folding layer is followed by the convolution layers. 3) A sequence unfolding layer and a flatten layer are used to restore the sequence structure and to reshape the output to image vector sequences. 4) LSTM layers followed by an output layer are included to classify the resulting image vector sequences.

The deep feature extraction and classification using combined RESNET+LSTM architecture used our research paper is based on the structure of human retina. A typical human retina is consisted of the following eight layers: 1) The Layer of Photoreceptor Cell. 2) The External Plexiform Layer (EPL). 3) The Layer of Horizontal Cell. 4) The Layer of Amacrine Cell. 5) The Layer of Bipolar Cell. 6) The Inner Plexiform Layer (IPL). 7) The Layer of Ganglion Cell. And 8) The Primary Visual Cortex.Each of the eight layers of the human visual system based combined RESNET+LSTM architecture are consisted of the following sections: 1) The Layers of Photoreceptor Cell, EPL, Horizontal Cell, Amacrine Cell, Bipolar Cell, and IPL are each represented by the BiLSTM layers consisting of 1000 hidden layers. 2) The Primary Visual Cortex is represented by a BiLSTM layer consisting of $2 \times 1000$ hidden layers.In the combined RESNET+LSTM architecture, 90 per cent of the images were used for training and 10 per cent of the images were used for validation.

## 4 Results

### 4.1 Deep Feature Extraction and Classification Using Residual Neural Network (RESNET)

The deep feature extraction and classification using RESNET is consisted of following two components: 1) The Main Branch. And 2) The Residual Connections.We have used 24.688 images from the FER2013 database to test the deep feature extraction and classification using RESNET architecture. The images are classified as angry, disgust, happy, neutral and surprise. A computer with

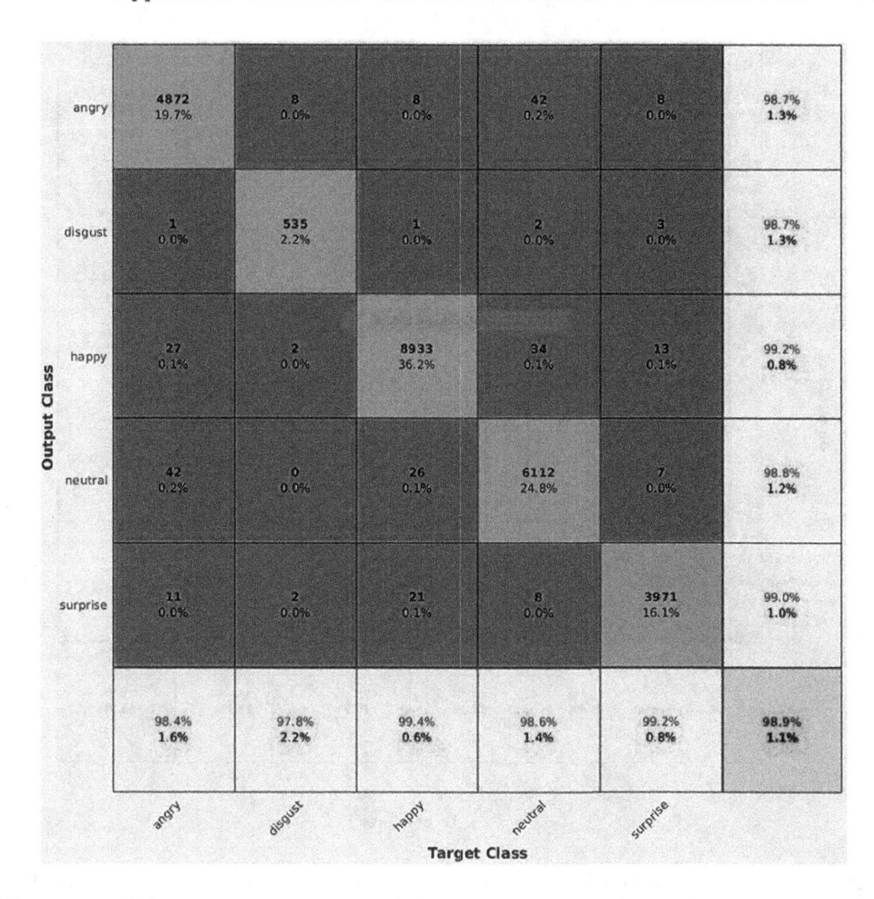


**Fig. 4.** 10-fold cross training matrix of deep feature extraction and classification using RESNET.

Linux operating system is used to run the algorithm. We have applied 10-fold cross validation technique. It took 5760 min to train and to validate the 24688 images. Our mini batch size was 1024. The learning rate of the RESNET architecture was 1e–1. The training accuracy of the RESNET architecture was 99%. The validation accuracy of the RESNET architecture was 75.3% (Figs. 4 and 5).

### 4.2 4.2 Deep Feature Extraction and Classification Using Combined Residual Neural Network+Long Short-Term Memory (RESNET+LSTM)

The deep feature extraction and classification using combined RESNET+LSTM implements a deep learning neural network with 1) The pretrained RESNET, such as GoogleNet, 2) LSTM, and 3) Assembly network consisting of pretrained RESNET and LSTM. We have used 24688 images from the FER2013 database to validate the combined RESNET+LSTM architecture. A computer with Linux

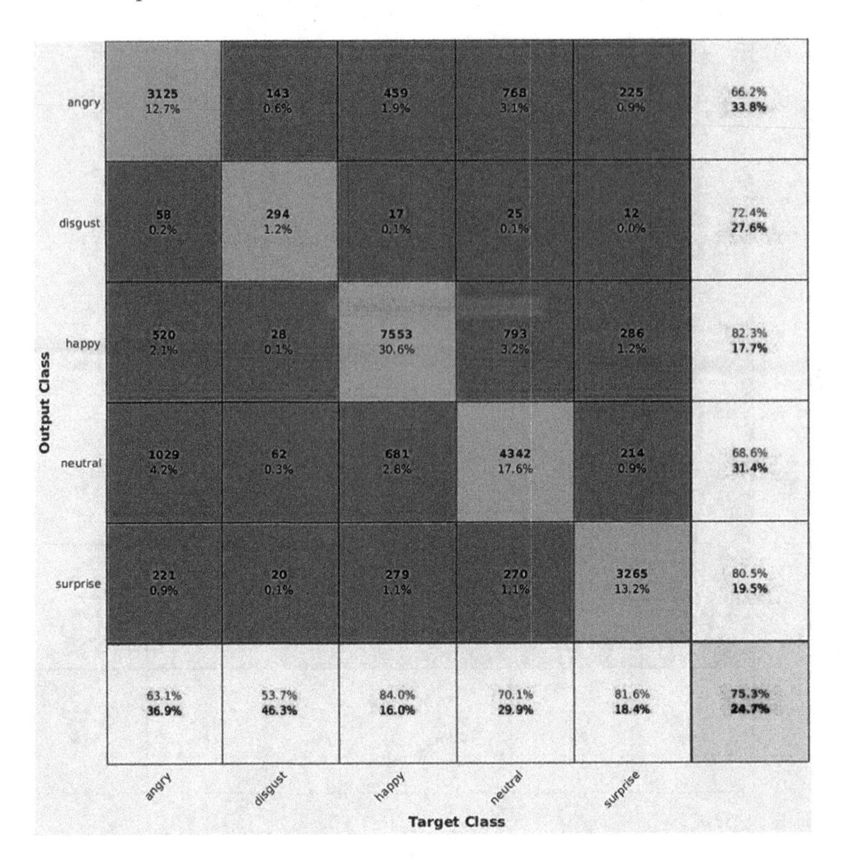


**Fig. 5.** 10-fold cross validation matrix of deep feature extraction and classification using RESNET.

operat-ing system is used to run the algorithm. It took 3912 min to train and to validate the 24688 images. Here, 90% of the images were used for training and 10 % of the images were used for validation. Our mini batch size was 16. The learning rate the combined RESNET+LSTM architecture was 1e–4. The training accuracy of the combined RESNET+LSTM architecture was 99.5%. The validation accuracy of the combined RESNET+LSTM architecture was 72% (Figs. 6, 7 and 8).

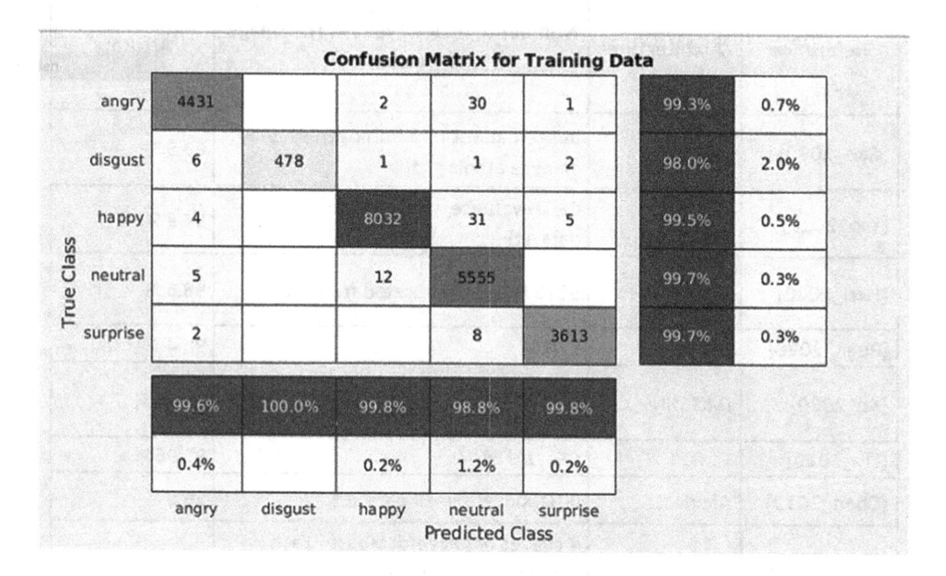


**Fig. 6.** Training matrix of deep feature extraction and classification using combined RESNET+LSTM

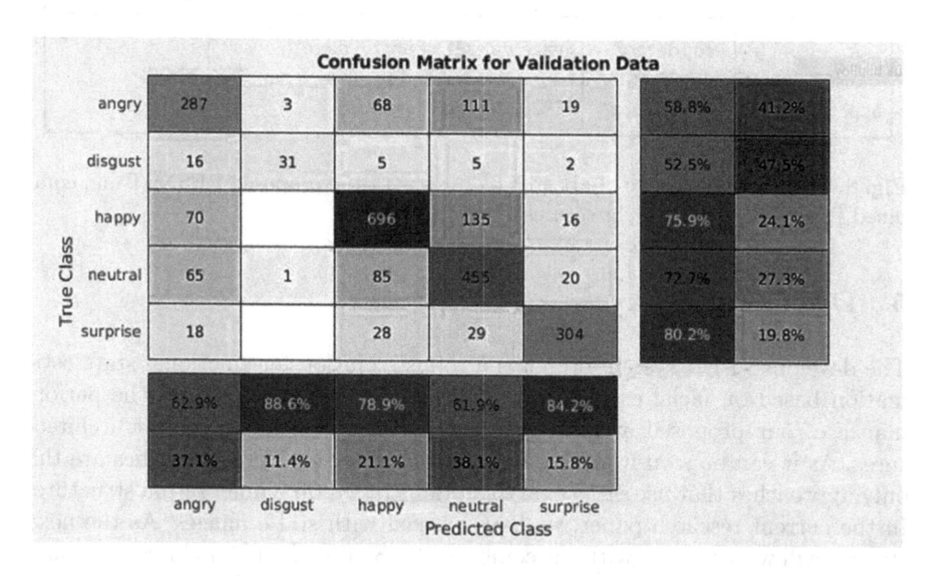


**Fig. 7.** Validation matrix of deep feature extraction and classification using combined RESNET+LSTM.

| Reference | Architecture | Number of Image Types in the Image Dataset | Accuracy |
|---|---|---|---|
| [Gao_2020] | VGG16 | Image dataset of 5 homogeneous chinese ethnic group | 80.5 % |
| [Yao_2020] | VGGNet | Casia-webface and LWF [Small image dataset] | 94.8 % |
| [Jafri_2020] | LivenessNet | 3248 images generated from video | 98.5 % |
| [Ruan_2020] | Lenet-5 | SWUN | 98.4 % |
| [Xu_2020] | IMISCNN | Casia-webface and LWF [Small image dataset] | 99.36 % |
| [Liu_2020] | VGG11 | CK+, JAFFE | 95.96 % |
| [Chen_2019] | AlexNet | IIITD Kincet [Small data set) | 86.0 % |
| [Khan_2019] | AlexNet | 4 classes of image database. Each containing 1000 images] | 97.95 % |
| [Ke_2018] | VGGNet | CMU-PIE | 97.55 % |
| Our RESNET approach | Proprietory RESNET architecture | Image dataset of angry, disgust, happy, neutral and surprise, from FER2013 image database | 75.3 % |
| Our RESNET+LSTM approach | Proprietory RESNET+LSTM architecture | Image dataset of angry, disgust, happy, neutral and surprise, from FER2013 image database | 72.0 % |

**Fig. 8.** Comparison of state-of-art architectures with the proposed RESNET and combined RESNET+LSTM architectures.

## 5 Discussion

The developed FER system proposes a robust solution for emotional state estimation based on facial expressions. Following Figure demonstrates the performance of our proposed architectures compared with state-of-the-art architectures. As it can be seen from the comparison table that our approaches are the only approaches that use proprietary solutions based on human retina structure. In the current research paper, we have worked with static images. As the next step we will work further with the combined RESNET+LSTM architecture using realistic color videos. Such arrangement will enable us the human retina more realistically. Since the proposed the RESNET and combined RESNET+LSTM architectures are based on the human retina structure, in the future we will work on the alteration of the architectures due to the retinal diseases. We will incorporate spiking neural network in the existing RESNET and combined RESNET+LSTM architectures for understanding the cause-effect relationship between retinal disease and human retinal structure.

As a future we will concentrate ourselves on the following improvements: (1) We will work with the realistic color video database. (2) We will improve the validation accuracy of the deep feature extraction and classification using RESNET architecture. (3) We will improve the validation accuracy of the deep feature extraction and classification using combined RESNET+LSTM architecture.

## 6 Conclusions

In this research paper, a facial emotion recognition system was proposed, addressing automatic face detection and facial expression recognition using 1) RESNET and 2) combined RESNET+LSTM. The architectures of RESNET and combined RESNET+LSTM were based on the human retina and primary visual cortex structure. The proposed architectures were compared with each other. They are tested using the challenging database FER2013.

The suggested FER system offers an advanced solution for expanding numerous computer vision applications, and it has the potential to improve human-robot interactions (HRI), particularly affective robotics, by making the robot aware of the users' emotional state. This could be utilized to help persons with autism and Alzheimer's disease, as well as a companion for the elderly.

**Acknowledgements.** This project has received funding by grant RTI2018-098969-B-100 from the Spanish Ministerio de Ciencia Innovación y Universidades and by grant PROMETEO/2019/119 from the Generalitat Valenciana (Spain), and by Grant PID2020-115220RB-C22 funded by MCIN/AEI/ 10.13039/501100011033 and, as appropriate, by "ERDF A way of making Europe", by the "European Union" or by the "European Union NextGenerationEU/PRTR".

## References

1. Fer dataset (2013). https://www.kaggle.com/msambare/fer2013/
2. Amidi, A.S.A.: Vip cheatsheet: Reccurent neural network (2019). https://stanford.edu/ shervine
3. Mollahosseini, A.D., Chan, M.M.: Going deeper in facial expression recognition using deep neural networks. In: IEEE Winter Conference on Applications of Computer Vision WACV, pp. 1–10 (2016)
4. Kim, B.-K., Dong, S.-Y., Roh, J., Kim, G., Lee, S.-Y.: Fusing aligned and non-aligned face information for automatic affect recognition in the wild: a deep learning approach. In: IEEE Conference on Computer Vision and Pattern Recognition Workshops CVPRW (2016)
5. Benamara, N.K., et al.: Real-time emotional recognition for sociable robotics based on deep neural networks ensemble. In: Ferrández Vicente, J.M., Álvarez-Sánchez, J.R., de la Paz López, F., Toledo Moreo, J., Adeli, H. (eds.) IWINAC 2019. LNCS, vol. 11486, pp. 171–180. Springer, Cham (2019). https://doi.org/10.1007/978-3-030-19591-5_18

6. Benamara, N.: Real-time facial expression recognition using smoothed deep neural network ensemble. Integrat. Comput.-Aid. Eng. **28**(1), 97–111 (2021)
7. Pramerdorfer, C., Kampel, M.: Facial expression recognition using convolutional neural networks: State of the art. ArXiv:1612.02903 (2016)
8. Levi, G., Hassner, T.: Emotion recognition in the wild via convolutional neural networks and mapped binary patterns. In: International Conference on Multimodal Interaction ICMI, pp. 503–510 (2016)
9. Redmon, J., Farhadi, A.: Yolo9000: Better, faster, stronger. In: IEEE Conference on Computer Vision and Pattern Recognition CVPR 11486 (2017)
10. Ferrandez, J.M., et al.: Brain plasticity: feasibility of a cortical visual prosthesis for the blind. In: Proceedings of the 25th Annual International Conference of the IEEE Engineering in Medicine and Biology Society vol. 3, pp. 2027–2030 (2003)
11. Ferrandez, J.M., Liaño, E., Bonomini, P., Martinez, J.J., Toledo, J., Fernandez, E.: A customizable multi-channel stimulator for cortical neuroprosthesis. In: 29th Annual International Conference of the IEEE Engineering in Medicine and Biology Society, pp. 4607–4710 (2007)
12. Multimedia Laboratory, D.o.I.E.T.C.U.o.H.K.: Wider face dataset. http://shuoyang1213.me/WIDERFACE/ (2019)
13. Viola, P., Jones, M.: Rapid object detection using a boosted cascade of simple features. In: IEEE Computer Society Conference on Computer Vision and Pattern Recognition CVPR 11486 (2001)
14. Li, S.W.D.: Deep facial expression recognition: a survey. IEEE Trans. Affect. Comput. **1**(1) (2020)
15. Devries, T., Biswaranjan, K.T., Graham W.: Multi-task learning of facial landmarks and expression. In: Canadian Conference on Computer and Robot Vision (2014)
16. Guo, Y., Tao, D., Yu, J., Xiong, H., Li, Y., Tao, D.: Deep neural networks with relativity learning for facial expression recognition. In: IEEE International Conference on Multimedia Expo Workshops (ICMEW)) (2016)
17. LeCun, Y., Bottou, L., Bengio, Y., Haffner, P.: Gradient-based learning applied to document recognition. Proc. IEEE **86**(22) (1998)
18. Yang, Y.: Deep learning using linear support vector machines. ArXiv:1306.0239 (2013)
19. Lian, Z., Li, Y., Tao, J.-H., Huang, J., Niu, M.-Y.: Expression analysis based on face regions in real-world conditions. Int. J. Autom. Comput. **17**(1), 96–107 (2019). https://doi.org/10.1007/s11633-019-1176-9

# Author Index